OXFORD HANDBOOK

T0177450

CLIMATE CHANGE AND SOCIETY

OXFORD HANDBOOK OF

CLIMATE CHANGE AND SOCIETY

Edited by

JOHN S. DRYZEK,
RICHARD B. NORGAARD,
AND DAVID SCHLOSBERG

OXFORD
UNIVERSITY PRESS

OXFORD

UNIVERSITY PRESS

Great Clarendon Street, Oxford OX2 6DP
United Kingdom

Oxford University Press is a department of the University of Oxford.
It furthers the University's objective of excellence in research, scholarship,
and education by publishing worldwide. Oxford is a registered trade mark of
Oxford University Press in the UK and in certain other countries

Published in the United States of America by Oxford University Press
198 Madison Avenue, New York, NY 10016, United States of America

British Library Cataloguing in Publication Data
Data available

Library of Congress Cataloging in Publication Data
Data available

ISBN 978-0-19-968342-0

ACKNOWLEDGEMENTS

The editors wish to thank all of the authors of this Handbook not only for their efforts in producing their chapters, but also for their attempts to make their approaches and arguments accessible to those in other fields, and to a broader audience. More generally, we want to express our appreciation for all of those who have taken on the issue of climate change in their disciplines, and brought a full array of approaches, frames, discourses, and tools to bear in helping us understand—and begin to respond to—the many crucial relationships between climate change and society. We received helpful suggestions and comments from an audience at a symposium at the University of Copenhagen in December 2009, thanks to Kasper Møller Hansen for making that possible. Thanks also to Richard Howarth, Daniel Bromley, Hayley Stevenson, and several anonymous reviewers for advice; and to Jeff Joyce for help with the cover image.

Our editor at Oxford University Press, Dominic Byatt, supported the idea of this Handbook from the very beginning, and helped with every step of the way thereafter. Also at OUP, Lizzy Suffling and Sarah Parker were exemplary in their attention to detail.

Finally, we must acknowledge the recent loss of one of the best exemplars of interdisciplinary understanding, public communication, and broad passion on the issue of climate change, Professor Stephen Schneider. His wide array of talents are needed now more than ever, and while he is already sorely missed, the best tribute to his work will be the expanded efforts of others dedicated to seeing his mission through.

CONTENTS

PART IV: SOCIAL IMPACTS

PART V: SECURITY

PART VI: JUSTICE

PART VII: PUBLICS AND MOVEMENTS

PART VIII: GOVERNMENT RESPONSES

PART IX: POLICY INSTRUMENTS

PART X: PRODUCERS AND CONSUMERS

PART XI: GLOBAL GOVERNANCE

PART XII: RECONSTRUCTION

List of Contributors

W. Neil Adger, Professor, Tyndall Centre for Climate Change Research, School of Environmental Sciences, University of East Anglia, Norwich, UK.

Walter F. Baber, Director, Graduate Center for Public Policy and Administration, California State University, Long Beach.

Karin Bäckstrand, Associate Professor of Political Science, Lund University.

Paul Baer, Assistant Professor, School of Public Policy, Georgia Institute of Technology.

Jon Barnett, Professor, Department of Resource Management and Geography, The University of Melbourne.

Robert V. Bartlett, Gund Professor of Liberal Arts, Political Science Department, University of Vermont.

David Benson, Lecturer Centre for Social and Economic Research on the Global Environment (CSERGE), School of Environmental Sciences, University of East Anglia, Norwich, UK.

Frank Biermann, Professor and Head, Department of Environmental Policy Analysis, Institute for Environmental Studies, VU University Amsterdam, The Netherlands, and Chair, Earth System Governance Project.

Katrina Brown, Professor of Development Studies, School of International Development, University of East Anglia, Norwich UK.

Harriet Bulkeley, Professor of Geography, Durham University.

Armin Bunde, Professor, Institut für Theoretische Physik, Justus-Liebig Universität, Giessen, Germany.

Sanjay Chaturvedi, Professor of Political Science at the Centre for the Study of Geopolitics, Department of Political Science and Honorary Director, Centre for the Study of Mid-West and Central Asia, Punjab University, India.

Peter Christoff, Associate Professor, Department of Resource Management and Geography, University of Melbourne.

Mark Diesendorf, Associate Professor and Deputy Director, Institute of Environmental Studies, University of New South Wales.

Simon Dietz, Deputy Director, Grantham Research Institute on Climate Change and the Environment, and Lecturer in Environmental Policy, Department of Geography and Environment, London School of Economics and Political Science.

Lisa Dilling Ph.D., Assistant Professor, University of Colorado-Boulder, Environmental Studies & Center for Science and Technology Policy Research.

Timothy Doyle, Chair and Professor of Politics and International Relations in the Research Centre for Politics, International Relations and Environment, Keele University, UK; and Chair and Professor in the Indo-Pacific Governance Research Centre, School of History and Politics, University of Adelaide, Australia.

John S. Dryzek, Australian Research Council Federation Fellow and Professor of Political Science, Australian National University.

Riley E. Dunlap, Regents Professor of Sociology, Oklahoma State University.

Hallie Eakin, Assistant Professor, School of Sustainability, Arizona State University.

Robyn Eckersley, Professor, School of Social and Political Sciences, University of Melbourne.

Daniel A. Farber, Sho Sato Professor of Law and Chair, Energy and Resources Group University of California, Berkeley.

Robert Melchior Figueroa, Associate Professor, Department of Philosophy and Religion Studies, University of North Texas.

Andrew Foss, Consultant, NERA Economic Consulting.

Stephen M. Gardiner, Associate Professor, Department of Philosophy and Program on Values in Society, University of Washington, Seattle.

Nils Gilman, Senior Consultant, Monitor Group.

Ian Gough, Professorial Research Fellow, London School of Economics.

Maarten Hajer, Professor of Public Policy at the University of Amsterdam and Director of PBL Netherlands Environmental Assessment Agency.

Elizabeth G. Hanna, Fellow, National Centre for Epidemiology & Population Health, The Australian National University; and Senior Fellow, Centre for Risk & Community Safety, Royal Melbourne Institute of Technology.

Paul G. Harris, Professor of Global and Environmental Studies, Hong Kong Institute of Education.

David Harrison, Jr., Ph.D., Senior Vice President, NERA Economic Consulting.

Richard B. Howarth, Pat and John Rosenwald Professor, Dartmouth College.

Dale Jamieson, Director of Environmental Studies, Professor of Environmental Studies and Philosophy, Affiliated Professor of Law, Environmental Studies Program, New York University.

Sheila Jasanoff, Pforzheimer Professor of Science and Technology Studies, John F. Kennedy School of Government, Harvard University.

Andrew Jordan, Professor of Environmental Politics, Tyndall Centre for Climate Change Research, University of East Anglia, Norwich, UK.

Sivan Kartha, Senior Scientist, Stockholm Environment Institute.

Laurel Kearns, Associate Professor, Sociology of Religion and Environmental Studies, The Theological School and Graduate Division of Religion, Drew University.

Per Klevnas, Senior Consultant, NERA Economic Consulting.

Ronnie D. Lipschutz, Professor of Politics, University of California, Santa Cruz.

Timothy W. Luke, University Distinguished Professor, Department of Political Science, Virginia Polytechnic Institute & State University.

Aaron M. McCright, Associate Professor of Sociology, Lyman Briggs College, Department of Sociology, and Environmental Science and Policy Program, Michigan State University.

Corina McKendry, Ph.D. candidate in Politics, University of California Santa Cruz.

James Meadowcroft, Professor in the School of Public Policy and in the Department of Political Science, Carleton University, Ottawa, and Canada Research Chair in Governance for Sustainable Development.

Robert Mendelsohn, Edwin Weyerhaeuser Davis Professor of Forest Policy, Yale University.

Susanne C. Moser, Susanne Moser Research & Consulting; University of California-Santa Cruz (Institute for Marine Sciences); and Stanford University (Woods Institute).

Matthew C. Nisbet, Associate Professor of Communication and Affiliate Associate Professor of Environmental Science, American University, Washington, DC.

Kari Marie Norgaard, Assistant Professor of Sociology and Environmental Studies, University of Oregon.

Richard B. Norgaard, Professor of Energy and Resources, University of California, Berkeley.

Matthew Paterson, Professor of Political Science, University of Ottawa.

Colin Polsky, Associate Dean for Undergraduate Research & Active Pedagogy, Associate Professor of Geography, Director, HERO NSF REU Site Program, Clark University.

Simone Pulver, Assistant Professor of Environmental Studies, University of California, Santa Barbara.

Daniel Radov, Associate Director, NERA Economic Consulting.

Doug Randall, Managing Partner, Monitor 360.

Paul Routledge, Reader in Human Geography, School of Geographical and Earth Sciences, University of Glasgow.

Mark Sagoff, Director, Institute for Philosophy and Public Policy at George Mason University in Fairfax Virginia.

David Schlosberg, Professor of Government and International Relations, University of Sydney.

Miranda A. Schreurs, Professor and Director, Environmental Policy Research Centre (FFU), Department of Political and Social Sciences, Otto Suhr Institute for Political Science, Freie Universität Berlin.

Peter Schwartz, Chairman, Global Business Network.

Will Steffen, Executive Director, Climate Change Institute, The Australian National University.

Nico Stehr, Karl Mannheim Professor for Cultural Studies, Zeppelin University, Friedrich-shafen, Germany.

Clive L. Spash, Professor of Public Policy & Governance in the Department of Socio-Economics at WU Vienna University of Economics and Business, Austria, and Professor II in Department of International Environment and Development Studies (Noragric), Norwegian University of Life Sciences, Norway.

Andrew Szasz, Professor of Sociology, University of California at Santa Cruz.

Wytske Versteeg, Researcher, Department of Political Science, University of Amsterdam.

Hans von Storch, Director, Institute of Coastal Research, Helmholtz Zentrum Geesthacht, Geesthacht, Germany.

James Waters, Ph.D. researcher, Tyndall Centre for Climate Change Research, School of Environmental Sciences, University of East Anglia, Norwich, UK.

Spencer Weart, Emeritus Historian, Center for History of Physics, American Institute of Physics.

Rüdiger Wurzel, Department of Politics, University of Hull, Hull, UK.

Oran R. Young, Professor of Environmental Governance and Institutions, Bren School of Environmental Science and Management, University of California, Santa Barbara.

Anthony Zito, Reader in Politics and Co-Chair of the Jean Monnet Centre of Excellence at Newcastle University, Newcastle, UK.

PART I

INTRODUCTION

....

CHAPTER 1

CLIMATE CHANGE AND SOCIETY: APPROACHES AND RESPONSES

....

JOHN S. DRYZEK, RICHARD B. NORGAARD,
AND DAVID SCHLOSBERG

CLIMATE change presents perhaps the most profound challenge ever to have confronted human social, political, and economic systems. The stakes are massive, the risks and uncertainties severe, the economics controversial, the science besieged, the politics bitter and complicated, the psychology puzzling, the impacts devastating, the interactions with other environmental and non-environmental issues running in many directions. The social problem-solving mechanisms we currently possess were not designed, and have not evolved, to cope with anything like an interlinked set of problems of this severity, scale, and complexity. There are no precedents. So far, we have failed to address the challenge adequately. Problems will continue to manifest themselves—both as we try to prevent and as we try to adapt to the consequences of climate change—so human systems will have to learn how better to respond. One of the central social, political, and economic questions of the century is: how then do we act?

In this Handbook we have brought together a representation of the best scholars on climate change and society. We identified the key approaches and selected authors to represent and engage with their literatures in a manner that would be informative and interesting to scholars in other areas and to newcomers as well. We have encouraged authors to make linkages between approaches and to other chapters. We hope the Handbook will contribute to the integration of understanding needed to tackle so systemic and complex a problem as the relationship between climate change and society. At the same time, the Handbook is by no means a synthesis, nor does it provide a unified diagnosis of what is wrong (and right) with contemporary human systems, an integrated and coherent program for research, or a singular blueprint for collective action. While we have views of our own on such questions, some of which will come through in this introductory chapter, there is no unified line followed by our authors as they address the complex relationship between people, societies, and the natural world. Most (not all) agree on the magnitude and severity of the problems. But there are substantial differences when it comes to identifying what matters, what is wrong, what is right, how it got to be that way, who is responsible, and, not least, what should be done.

Commissioning, reading, and editing these contributions has left us acutely aware of the limitations of human knowledge—and the major constraints on intelligent human action—when it comes to complex social-ecological systems. Climate change is, as Steffen explains in his opening chapter, a truly diabolical problem. It is additionally devilish in the mismatch between human capacities to act and the scale, scope, and immediacy of collective action seemingly demanded. Nevertheless we have to start somewhere, and we have aspired in this Handbook to commission and compile the best available set of intellectual resources for the multiple tasks ahead. Given the complexity of what we face, no single volume can offer commentary on absolutely everything that is needed. Yet we have aspired to a measure of comprehensiveness in addressing the range of ways climate change plays out in the social realm.

Our main task is, then, to lay out the various ways that climate change affects society, and what society might do in response. The authors represent a variety of disciplinary understandings and intellectual frameworks that can be brought to bear. In this chapter we introduce the key topics, themes, layers, and issues, before concluding with a discussion of our chosen structure. We begin with the science that first identified climate change as a problem, and how it is received by and in society and government.

1 SCIENCE AND SOCIETY

While the effects of climate change—floods, drought, heat stress, species loss, and ecological change—can be experienced very directly, their conceptualization as connected phenomena with common causes is due to climate science, which therefore plays a very basic part when it comes to climate change and society. Natural scientists (such as Steffen in his chapter) tell us that there is now consensus in the climate science community about the reality of climate change, and near consensus on its severity and the broad range of attendant harms and risks. But that consensus does not of course mean the science is then accepted as the basis for policy. Climate science does not provide certain future projections of risks and damages. The projections are entangled in assumptions about how human systems respond over time—as well as natural ones. Climate is an outcome of a complex geo-atmospheric-ecological system, and complex systems always have a capacity to surprise by behaving in unanticipated ways. Climate change, furthermore, is only one of a range of interacting phenomena of global environmental change caused or affected by human activity. We may indeed be entering the unknown territory of an 'anthropocene' era where people drive truly major changes in global systems. Thus while the broad sweep of history shows climate change being taken ever more seriously as an issue within the scientific community and eventually far beyond (see Weart's chapter), we are dealing with complex processes with uncertain outcomes rather than simple facts, and the public and politicians have difficulty seeing the drivers to collective action in any simple way. The agendas of climate science are now affected by larger social and political processes (see the chapter by von Storch et al.). Thus scientific findings and their action implications must seek validation not just within the scientific community itself, but also within the larger society, and different political systems have different means for validation (see Jasanoff's chapter).

But even getting to the point of taking science seriously can be difficult. The Intergovernmental Panel on Climate Change (IPCC) famously uses language seeped in uncertainty to qualify its predictions (likely, very likely, virtually certain, etc.), and there are a few dissenting scientists who claim there is little evidence of major and imminent climate change. As Dunlap and McCright discuss in their chapter, a thoroughly organized campaign has successfully used such scientific uncertainty to create political uncertainty, with those who fund the case against the reality of climate change having a massive stake in the fossil fuel economy. Skepticism is in some countries joined to a right-wing ideology such that, because climate change requires coordinated collective action of the kind that is anathema in this ideology, climate change should not exist. More insidiously, skepticism may also give the impression that it is empowering ordinary people to be able to question the assertions of a scientific elite. Any lapses in the practice and content of this science (of the sort alleged but unproven in the stolen e-mails from the University of East Anglia in 2009, and the admission of a mistaken claim about the rate of melting of Himalayan glaciers in an IPCC report) are seized upon by these ideologists to discredit climate science in its totality. The media, looking for 'balance' amid controversy, gives as much airtime to skeptics as it does to climate scientists and others who point to the reality and scale of change (Boykoff and Boykoff 2004). Science moves to the center of political controversy, and scientists respond in varied ways (Schneider 2009).

Unsurprisingly, scientists feel harassed by the attacks of organized skeptics and denialists. To the extent scientists respond with further insistence on the consensus within the scientific community about the veracity of their claims, the more they play into their critics' hands. The net result is that science enters a spiral of politicization. Scientists themselves in many cases cannot avoid becoming political actors, as they fight for the credibility of what they do in the larger public arena. Not surprisingly, they can and do make many false steps in this arena, and much can be done to improve the communication of science to the public (see Moser and Dilling in this volume). They are also faced with the quandary over whether to admit to uncertainties in the range of their own findings—and so leave themselves open to critics who discredit the scientists' lack of confidence—or to claim certainty greater than that actually warranted by these findings. Admission of a degree of uncertainty is the norm among colleagues, but fodder for skeptics. One thing we do know is that simply insisting on the rightful authority of science as the guide to action has failed. But the natural sciences are not the only politicized disciplines.

2 FROM SCIENCE TO ECONOMICS

What do scientific findings mean in human terms? An answer is given by economics, which can attach cost estimates to the current impacts and projections of future impacts of climate change. One such set of estimates is provided in the chapter by Mendelsohn, who comes up with relatively low estimates, with costs concentrated among the rural poor in developing countries. Economists such as Nicholas Stern in his famous 2006 report to the government of the United Kingdom come up with much higher estimates. A lot turns on seemingly technical factors such as the rate of discount used to calculate a present value for future costs. Depending on the discount rate chosen, we can end up with massive differences in the size of the present value of future costs, and so radically different implications for climate policy. The choice of discount rate turns out to be a major ethical issue, not just a technical economic matter (see the chapters by Howarth and R. Norgaard). Further contestation arises once we move beyond the confines of standard economic analysis to contemplate other ethical issues (Dietz's chapter), pertaining (for example) to basic human needs, and the distribution of burdens and benefits of action and inaction across rich and poor, within and across national boundaries, as well as between generations. Sagoff argues in his chapter that the asymmetry of burdens and benefits across generations means that economic thinking should not be at the core of climate policy analysis.

Once we get past controversies over cost estimates and distributions, economics also provides a powerful set of analytics for thinking about the choice of policy instruments to achieve the desired level of mitigation (expressed in terms of targets and timetables for total greenhouse gas emissions). The consensus among economists—at least those steeped in the neoclassical paradigm that dominates the discipline—is that market-based instruments are the most efficient, and in particular emissions trading or cap-and-trade (see the chapter by Harrison et al.). Governments have begun to experiment with such schemes, established for some time in connection with non-greenhouse pollutants such as sulfur dioxide in the United States, more recently extended to greenhouse gases and CO_2 in particular, especially in the European Union (see the chapter by Jordan et al.). Emissions trading requires that some authority sets a cap on total emissions, then issues permits for quantities that add up to that cap. These permits can then be traded, such that companies for which reducing pollution is expensive can buy permits from those for which reductions are cheaper. The economic theory is very clear, but the politics and policy making is much murkier. Even before we get to monitoring and compliance, polluters with sufficient political power will demand exemptions and/or free permits for themselves. So when emissions trading schemes are proposed or introduced, it is common to find whole economic sectors exempted (for example, agriculture in Australia), or established dirty industries (for example, coal-burning electricity generators) favored at the expense of more efficient but less established competitors (see Spash's chapter).

These real-world politics notwithstanding, market discourse is increasingly pervasive and powerful. It informs many discussions of national policy instruments, and extends to global policy and emissions trading across national boundaries. The discourse affects the content of global governance arrangements, which can even be privatized as carbon traders seek to escape international governmental authority (see Paterson's chapter). Market logic extends too to offsets, whereby polluters can compensate for their greenhouse gas emissions by paying somebody else, for example, to plant trees that will absorb an equal quantity of emissions. What actually happens at ground level in countries where there is weak monitoring capacity is another matter entirely. Unlike conventional markets where one party of the transaction can complain, or at least never transact with the other party again, both parties in offset transactions have every incentive to give misleading information to the public on the real number of trees planted and their actual effectiveness in offsetting climate impacts. Again, complexity rules. But whatever their consequences for mitigation, new kinds of climate markets present many opportunities for traders to become wealthy, becoming a constituency pushing for further marketization (see Spash's chapter).

National governments are embedded in market economies that constrain what they can do, and the social realm is often limited by economistic frames and discourse. However, markets are not necessarily just a source of constraint. Markets are made up of producers and consumers who might themselves change their behavior in ways that reduce emissions. The most important producers here are large corporations. Why might they change their ways? In part, if they thought the world was moving in a low-carbon direction (whether by choice or necessity), positioning themselves to take advantage of this shift might be profitable. Of course this positioning would need to be more than the kind of rhetoric that enabled (for example) BP to market itself as 'Beyond Petroleum'—at least until an oil spill in the Gulf of Mexico in 2010 exposed a range of problems in its public relations approach (in addition to its safety practices). While there may be money to be made in producing goods for a low-greenhouse gas economy, the problem is that currently there is much more money to be made in climate-unfriendly activities. Corporate responses to the challenge of climate change have been highly variable (see Pulver's chapter), and there is little reason to suppose a significant number of corporations will play a leadership role if governments do not. The only corporations that do have a clear financial incentive to take the risks of climate change very seriously are insurance companies. This is especially true of the big reinsurance companies with potentially high exposure to damages caused by extreme weather events. The high hopes once vested in insurance companies by some analysts (Tucker 1997) on this score seem so far to have produced little in the way of comprehensive action.

A decarbonizing economy would of course have to involve changes in patterns of consumption, whether induced by government policy and price increases, or chosen by consumers through changing mores. Such basic individual and broad cultural changes that affect consumption have been promoted by a variety of social movements, religious actors, and celebrities. Many environmental organizations focus on consumer behavior—from the individual level up to the decarbonization and transition of towns and regions—both as a source of direct change and as a clear economic and political statement. The 'green governmentality' identified by Lipschutz and McKendry in their chapter would help mold citizens of a new ecological order, whose consumption demands could look quite different from those characteristic of industrial society. However, as Szasz points out in his chapter, consumption choices are limited by the social-economic structure, which conditions the range of easy options that individual consumers have. Luke also insists we understand the dangers of such forms of such behavioral control, even if it does look green. At any rate, changing consumer habits are no substitute for coordinated collective action.

3 THE PUBLIC REALM, AND ITS PROBLEMS

In a world where the legitimacy of public policies and other collective actions rests in large measure on the democratic credentials of the processes of their production, it matters a great deal what publics think, and what actions they consequently support, or are willing to undertake themselves. Initially, many climate scientists, policy makers, and activists thought that the key here was simply getting publics to understand the facts by providing information (the point behind Al Gore's 2006 documentary film *An Inconvenient Truth*, for example). Yet as Moser and Dilling point out in their chapter, just providing information normally has little impact on behavior. And trying to instill fear in publics about possible impacts often turns out to be counter-productive, as people switch off. Most people get their information via the media, but as already noted there are structural features of mainstream media (the reporting only of controversy, which requires two opposing sides) that are problematic when it comes to communicating climate change. Face-to-face dialogue might work much better in terms of prompting people to think through the issues seriously; but that is extraordinarily hard to organize on any scale involving more than a handful of people. Thus there remain many failures in public cognition of the complex phenomena attending climate change (see Jamieson's chapter). Public opinion polls often show that people do care, and do want something to be done (see Nisbet's chapter); but there is no necessary urgency. In practice, many issues of more immediate concern (and which impose far fewer burdens of cognition) trump climate change when it comes to (for example) voting behavior.

Interlinked psychological, social, cultural, and political-economic processes can lead even those who in the abstract accept the need for action to in practice come to believe that they personally—or even the society in which they live—have no need to do anything that will impose any major disruptions on their own lives (see the chapter by K. Norgaard). Information, scientific or otherwise, is often processed through the lens of existing beliefs formulated in areas of life remote from climate science. Those beliefs can be very powerful, for better or for worse. Religious beliefs are particularly important in this respect (see Kearns's chapter). Sometimes religious beliefs line up on the side of ideological skepticism of the kind we have already noted; but sometimes these beliefs can join with the need for action (as in the 'creation care' movement among US evangelical Christians).

Publics should not however be understood as simply *mass* publics, which are problematic when it comes to mastering complex issues simply by virtue of their mass nature. A public can also be a *concerned* public organized around an issue; Nisbet in his chapter estimates the concerned 'issue public' on climate change to constitute around 15 percent of Americans—quite high in comparison to other political issues. Publics of this sort can be found at many levels: local, national, transnational, and global. They are organized in many different ways, ranging from community groups to the translocal solidarity identified by Routledge in his chapter to global networks of activists depicted by Lipschutz and McKendry in their chapter. They also demand a range of behavioral and policy changes, from a radical transition to a post-carbon lifestyle to basic democratic demands for more public participation in decision making. Concerned publics almost by definition are geared for action in the way mass publics most of the time are not. But the extent of their influence in the face of structural political forces and powerful recalcitrant actors remains highly uncertain. Publics are often vocal and visible—for example, at meetings of the Conference of the Parties (COPs) of the United Nations Framework Convention on Climate Change (UNFCCC), or at local city council meetings, but that does not mean they are decisive. And yet, in the face of the intransigence of many governments, such non-governmental publics continue to provide ideas, energy, and pressure necessary to respond to climate change.

4 JUSTICE AND VULNERABILITY

Increasingly, concerned publics advance a discourse of climate justice. The political philosopher John Rawls (1971) once famously proclaimed that justice should be the first virtue of social institutions. Itself disputable, that ideal remains a distant aspiration when it comes to climate change. Considerations of justice have often been marginalized in favor of economic efficiency and aggregate welfare in public policies and intergovernmental negotiations. Yet climate justice does inform policy debates and positions taken in negotiations, as well as political activism.

The debate around climate justice has revived an argument within justice theory about the adequacy of proposing principles for ideal situations of the kind Rawls himself proposed. The alternative task for theory involves addressing major pressing and concrete social and political problems, concerning human rights, poverty, and now the changing climate. Increasingly, justice frameworks are being used in the development of climate policy strategies.

The fact is that existing vulnerabilities will be exacerbated by climate change. The costs of climate change and the unintended effects of some policy responses to it will not be evenly distributed, and we need, at the outset, some way to measure the vulnerabilities to be experienced in such an unequal way (see Polsky and Eakin's chapter). Many of the direct costs of climate change itself will, as Mendelsohn points out in his chapter, be felt by the poor in developing countries. Those costs are sufficiently severe to undermine human security in terms of rights and basic needs (see Barnett's chapter). Climate change can have many substantial direct impacts on human health, and many secondary impacts if health problems undermine the adaptive capacity of social systems (see Hanna's chapter). Many indigenous communities, already living on the margins, are particularly vulnerable (see Figueroa's chapter). Many initiatives done for the sake of global mitigation—such as biofuels and offsets—have negative impacts on the well-being of the rural poor in developing countries by taking land away from food production. These people are of course those with the least political power in global politics in general, and when it comes to climate change in particular. They may have justice on their side, but that alone will not give them an effective voice.

Environmental ethicists and climate justice theorists have examined the moral challenges that attend climate change, and what ought to be done in response. Beyond the science, the economic arguments, the policy differences, and the actions and frames of the various actors in the climate change drama, lies a normative dimension of the crisis. Emerging norms of justice may play a number of roles in regulating the relationships of the whole range of human actors as they confront climate change. As Gardiner in his chapter summarizes, questions of justice concern the procedures around which decisions are made, the unfairness of the distribution of existing vulnerabilities to climate change and the fair distribution of benefits and burdens in the present and near future (see also Baer's chapter), the extent and nature of our obligations to both those within and outside our own country (international or cosmopolitan justice), responsibility to future generations (or intergenerational justice—see Howarth's chapter), and even the potential injustice done to nature itself.

The discourse of climate justice increasingly pervades questions of global governance of climate change. For example, the concept of international justice takes nations as its basic unit of ethical considerability—and as such, national governments can deploy this discourse when it suits their interests to do so. So developing countries can point to the history of fossil fuel use on which developed countries built their economies, such that fairness demands that it is the developing countries that should shoulder the burden of mitigation. The response on the part of the wealthy countries is that for most of this history, their governments had no awareness that what they were doing could change the climate, and so ought not to be held uniquely responsible for future mitigation. This kind of response can be augmented by reference to the huge numbers of rich consumers in China, India, and Brazil who can have their own profligate lifestyles protected so long as justice is conceptualized in inter*national* terms—'hiding behind the poor'. Effective global action on mitigation could benefit from taking a more cosmopolitan approach to justice, one in which people rather than nations are the subjects of moral considerability and responsibility (see discussions in chapters by Harris, Baer, and Gardiner). Here, obligations of justice surpass those owed only to those in our own country. Given global climate change, such nationalist limits begin to look irrelevant—as our individual actions affect people outside our own nations, our obligations exceed those borders as well. In this light, rich consumers in China have a global climate responsibility equal to that of rich consumers in the United States. Pragmatically, as Harris points out, if it introduced measures to restrict the emissions of its own rich, China would then have more credibility in international negotiations when it asked the US cut its emissions. This is just one example of how ethical considerations could have real practical importance. The larger point is that while the discourse of climate justice can be put in the service of those most vulnerable to the effects of climate change, it can also facilitate resolution of collective problems.

5 GOVERNMENTS

Negotiating a context defined by concerned publics, experts, lobbyists, and structural limits on what they can do, governments can choose to act on climate issues. Some of them already do. Dealing with major climate change issues has however never been a part of the core priorities of any government. Of course environmental policy has been a staple of government activity (especially in developed countries) since the 1960s. But it remains the case that the environment is not core business in the same way that the economy is. Governments acted swiftly and with the expenditure of vast sums of money in response to global financial crisis in 2008-9. They have never shown anything like this urgency or willingness to spend on any environmental issue. The difference is easily explained: the first concern of any government in a market economy is always to maintain the conditions for economic growth, which normally also means maintaining the confidence of markets in the government's own operations (Lindblom 1982). The second concern of most governments in developed countries has been to operate and finance a welfare state (see Gough and Meadowcroft's chapter), which itself is predicated upon continued economic growth. The core security imperative of government—protection against external threats—has receded with the increasing rarity of war between states, but remains important. Failure on one of these core priorities has the potential for swift catastrophe for any government, be it in terms of fiscal crisis and punishment by voters at the polls, or (in the case of security) erosion or even loss of sovereignty. Failure when it comes to climate change, where the risks, burdens, and benefits are distributed in complex fashion across space and time, does not yet mean anything at all comparable in the immediacy of its consequences for government.

While none of them performs adequately, some national governments do perform better when it comes to climate policy than others, though this variation is not easily explained (see Christoff and Eckersley's chapter). Historically the 'coordinated market economies' of northern Europe, accompanied by political systems that work on the basis of consensus rather than majority rule, have on most indicators done better when it comes to environmental performance in general than their more liberal counterparts in the Anglo-American countries, and that is reflected in climate policies. The surprising development here is that the UK has shown signs of trying to break the mold. In stark contrast to its counterparts in the United States and Australia, the leadership of the Conservative Party in the UK has decided to try to appeal to green voters. In the face of the failure—or in the US in the 2000s the blatant refusal—of national governments to substantively address the issue, subnational governments (US states such as California, regions, cities, and localities) have in many cases adopted policies to reduce emissions of greenhouse gases (see the chapter by Bulkeley). However, while insisting on the importance of subnational action, even its most ardent enthusiasts would not see it as a substitute for effective national (and international) policy action. The multi-leveled generation of the problem, and the sting of its impacts, demand multi-level governance (see Farber's chapter).

To date, very few national governments look at all like decarbonizing their economy, or redesigning energy systems to reverse growth in energy consumption (see the chapters by Diesendorf and Christoff/Eckersley). While countries like the UK, Iceland, Denmark, Spain, and Portugal have taken significant steps to increase conservation and the generation of carbon-free energy, they are still below 30 percent clean energy generation, and economic downturn may impede future progress. China deserves watching closely in these terms, because of the size and growth of its proportion of global emissions, its vulnerability to the effects of climate change, and uncertainty about the kind of political-economic development trajectory that it could take in future. Despite its seeming refusal to countenance any infringement on its sovereignty of the sort that agreeing as part of a global process to cut its emissions would connote, China could decide to make substantial unilateral cuts (see Schreurs's chapter). Chinese policy for the moment remains dominated by the economic growth imperative, but some of those exasperated by the kind of stalemate so common in liberal democratic states think that Chinese style authoritarianism might be capable of more decisive action. However, actually implementing such decisions amid complex circumstances may prove beyond the capacity of authoritarianism,

In the context of the UNFCCC, the G77 group of countries claimed a voice for the developing world in general (see Kartha's chapter). However, when it came to the Copenhagen Accord, China dropped the G77 for which it had been a spokesman in favor of a G2 deal with the United States. The governments that compose the G77 generally stress their right to very conventional forms of economic growth that may themselves do little for their rural poor. So state-based action does not exhaust the possibilities for the most vulnerable, which might also include (for example) building translocal solidarities as described in Routledge's chapter, or mobilizing collectively to resist damaging outside initiatives.

What could induce national governments to do better? Aside from international agreements (of which more shortly), there is some scope for reframing climate issues in ways that would make effective national government action more likely. That reframing might involve recognition of the security dimension of climate change. Climate change can, as Gilman et al. point out in their chapter, threaten the security of populations and vital systems, even in some cases threaten the sovereign integrity of states (if for example there are catastrophes on their borders). Conceptualizing energy security as energy independence may also be helpful, as it would mean freedom from reliance on unstable and/or authoritarian foreign countries. Security could also refer to the basic security of human needs, as argued by Barnett in his chapter. The 'securitization' of climate issues also has its critics, such as Doyle and Chaturvedi, who in their chapter criticize the concept of 'climate refugees' for its construction of vulnerable people as security threats. A security framing does mean emphasizing threat and so fear, in a way that Moser and Dilling in their chapter have identified as problematic in moving public opinion. And as a comprehensive frame for climate issues, it probably makes most sense for the United States—a global superpower with security interests in all parts of the world that could therefore be affected by impacts of climate change that are only locally catastrophic. Yet such a frame failed to help the US develop a climate policy, despite being invoked (if weakly) by the Obama administration both before COP-15 in 2009, and after the BP oil spill in the Gulf of Mexico in 2010.

Another possible reframing might involve more widespread adoption of a discourse of ecological modernization, which puts economic growth and environmental protection in a mutually reinforcing, positive-sum relationship—rather than their traditional zero-sum conflict. In this light, mitigation might actually be an economically profitable option. This particular reframing has been adopted most extensively in the coordinated market economies of northern Europe (and Japan), and as Hajer and Versteeg point out in their chapter, can now also be found very prominently in international negotiations on climate change. But as they also note, there can be a large gap between discourse structuration and discourse institutionalization, where the discourse adopted actually conditions the content of public policies. A more radical reframing would see national governments adopting resilience rather than economic growth as their core priority (see the chapter by Adger et al.); but that is a more distant prospect, as it would involve a wholly new imperative, rather than modification of existing imperatives.

Despite the reframings that have occurred, they have not yet led to the broad type of action necessary to avoid large-scale climate change and deal with its growing impacts.

6 GLOBAL ACTION (AND INACTION)

Neither coordinated collective action nor discursive reframings can stop at the national level. Climate change involves a complex global set of both causal practices and felt impacts, and as such requires coherent global action—or, at a minimum, coordination across some critical mass of global players. Without such coordination, there is substantial incentive for every player to seek to impose the burdens of mitigation on others, while seeking to take as free a ride as possible on their efforts. Enough players doing this will of course result in little in the way of effective action. Such is the status quo.

The United Nations Framework Convention on Climate Change was established in 1992 to organize negotiations that eventually involved just about all the world's states. In 1997 the Kyoto protocol seemed to commit many of the World's developed countries—the 'Annex One' states—to reductions in the absolute level of greenhouse gases that they emitted by 5.2 percent overall by 2012, in relation to a baseline of 1990. But Kyoto failed to deliver much in the way of actual reductions. The world's largest emitter, the United States, did not ratify the agreement, which imposed no obligations at all on developing countries. So at the time of writing, the world's two largest economies and largest emitters, the USA and China, are not covered by Kyoto. These are also two of the states that cling most tightly to a notion of sovereignty that cannot be diminished by global governance. Even those states that did ratify the Protocol generally fell far short of the commitments they had registered. After Kyoto the UNFCCC process made its torturous way forward, with expectations centered on the 15th Conference of the Parties (COP-15) in Copenhagen in 2009, when representatives of 190 states gathered. What happened at the eleventh hour in Copenhagen was that G190 was supplanted by G2. China and the United States, two of the most problematic participants in the prior negotiations and when it comes to the very idea of global governance in general, produced a Copenhagen Accord with no binding targets for anyone and no enforcement mechanism for the weak targets that were proclaimed. While most countries agreed to take note of the Accord, few did so with any enthusiasm, or with any intention to do anything much in consequence.

This Handbook goes to press in the shadow of the disappointing outcome of COP-15. Our authors disagree about the best response to this kind of disappointment, and the very weak international climate regime that it leaves in place. Biermann suggests a number of ways to strengthen the regime, including the establishment of a World Environment Organization on a par with the World Trade Organization, a strengthening (rather than abandonment) of the UNFCCC itself, and a stronger institutionalized role for civil society organizations (many of which push for stronger action on the international stage). Young suggests institutionalization of fairness principles of a sort that would induce more serious participation from China and key developing countries. Harris suggests that a cosmopolitan interpretation of fairness might be a circuit-breaker in international negotiations because it would enable China to demonstrate that it sought to impose burdens of mitigation on its own wealthy citizens. China would then have more credibility when it demanded that developed nations commit to more effective emissions reductions. Young also suggests more attention to intersections with other regimes (such as that for international trade) in a way that would induce more mitigation, and perhaps an enhanced role for effective minilateralism—negotiation among a small number of key parties. While at first glance this looks exclusive, that could be ameliorated to the degree representatives of those likely to suffer most from climate change are also at the table. Baber and Bartlett suggest that a common law approach to the establishment of international environmental norms may be just as productive as negotiation of international treaties—though the time scale on which any such bottom-up approach could work makes that insufficient in and of itself.

While these and other ideas for its improvement are being canvassed, Paterson in his chapter points out that what is happening in practice is that the international climate regime is being marketized. Whether in the context of internationally agreed targets and timetables or outside such agreements, emissions trading and offsets grow in prominence, to the point they are poised to dominate global climate governance. This may well continue whether or not such use of markets is ultimately effective in containing climate change.

Analysis of the global climate regime might focus on particular deficiencies and proposals for reform, but it is also worth taking a step back to consider the whole idea of a comprehensive, inclusive, negotiated, global approach to climate change mitigation. Perhaps that is asking more than the international system is capable of delivering. Comprehensive self-transformation of the basic parameters of the international system has only ever been negotiated in the wake of total war: the Treaty of Westphalia in 1648, the Congress of Vienna in 1815, the Versailles settlement in 1919, Bretton Woods in 1945. The first three of these concerned only security; the fourth added economics. While comparisons are sometimes made between climate change and war (e.g. Lovelock 2010), there is no total war-like catastrophe to spur global action; and even if there were, there is no obvious mechanism to ensure that mitigation would be at the top of the agenda.

Perhaps we need to think in very different terms about the coordination of a global response. Such terms would recognize the inherent complexity of multi-level governance in the global system, and the multiple points of leverage. It would involve attending to the roles that stakeholder communities, shared norms, evolving discourses, local practices, and regional agreements, could play—while not necessarily renouncing global negotiation in its entirety. This sort of thinking has barely begun (but see Bäckstrand's chapter). The problem is that the pace at which the mechanisms it identifies could change and take effect in positive fashion may be too slow to match the pace at which climate change is arriving. In addition, governance mechanisms need to be anticipatory rather than reactive when it comes to future change. Governments are not used to acting in this kind of way; nor do more diffuse governance mechanisms necessarily compensate.

7 ORGANIZATION OF THIS HANDBOOK

The complexity of the issues of climate change and society means that an element of arbitrariness is inescapable when breaking down the whole into component areas of scholarship, and then ordering those areas. The interconnections are many and strong. There are few independent subsystems of scholarship with significant findings that stand on their own. Responding effectively to the challenges of climate change will require coordination of efforts across different ways of looking at the problems. Understanding all the social dimensions of climate change requires us to embrace these complexities and interrelationships. Nevertheless, publishing the contributions between covers requires putting them in a linear order. We have chosen to do as follows.

Part II, 'The Challenge and its History,' lays out the key challenges climate change presents, and how matters got to be that way. Complexity means that a range of perspectives and discourses can be brought to bear (in both the history of climate change and the rest of this Handbook). The climb up the scientific agenda took place over a century. The climb up the political agenda was slow, but eventually reached a point where climate change became the archetypical environmental problem. These ascents have been accompanied by changing conceptualizations of climate and the way it plays into social, political, and economic discourses that condition the responses of actors and institutions. The impact of those discourses now itself merits critical scrutiny.

Natural science is obviously central when it comes to understanding climate change and responding to it, but the relationship of the science to society and public opinion as addressed in Part III, 'Science, Society, and Public Opinion,' proves problematic. The natural sciences themselves need to understand the complex relationship between 'pure' science and the way that scientific agendas interact with society at large. Knowledge claims are processed in politically variable ways. In the face of organized skepticism, conventional ways of communicating science to the public have come unstuck. We know what does not work when it comes to communicating climate change; we know much less about what does work.

Part IV turns to 'Social Impacts.' Economists have devoted a great deal of effort to estimating the present and likely future costs of climate change. Some economists (represented here by Mendelsohn) reach modest estimates. Much turns not just on technical matters such as choice of a discount rate, but also on what kind of economic paradigm ought to be applied. Even economists who reach relatively small estimates of total costs recognize that particular vulnerable populations such as the rural poor in developing countries and indigenous peoples living in ecosystems at the margins of industrial society may be hit hardest, be it in terms of health, livelihood, or culture. So costs need to be understood not only in economic terms, but also in broader social and cultural terms.

Many of the negative social impacts of climate change (and of adaptive responses to it) will be felt in the form of an undermining of the 'Security' of nations and peoples, and these issues are addressed in Part V. In one sense, it is a matter of the security of collectivities such as nations, populations, and the social and economic systems that support them. Security concerns therefore range from national security to basic human needs. The securitization of climate change and the very use of categories such as 'climate refugees' also have their critics.

Threats to human security are just one kind of social justice issue that arises in connection with climate change; a range of issues is covered in Part VI, 'Justice.' These issues include the distribution of benefits and burdens across nations and, perhaps more fundamentally, across people, but climate justice also entails issues of basic needs, procedures, corrective justice, and the nature of the obligation of those living in the present to future generations. Justice is in part a matter for philosophical analysis, but can also be used to challenge utilitarian economic analysis, influence international policy discourse, and rally social movements.

Such movements are just one kind of relevant public. The range of 'Publics and Movements' is addressed in Part VII. At an aggregate level public opinion exists in terms of percentages of people concerned about or willing to respond to climate change. Only the most engaged participate in movements, which can be organized locally, nationally, and globally, and in networks transcending these levels. The impact of movements in promoting cultural change may however be blunted by psychological and sociological denial mechanisms. Opinion and activism on climate change do not exist in isolation, but are also affected by factors such as economic interests and religious beliefs.

Responding more or—more often—less effectively to concerns raised by publics and movements, the actions of governments do of course matter a great deal, and are the subject of Part VIII, 'Government Responses.' Performance currently varies substantially across different countries. The case of China gets special treatment, because of the size and growth of its economy, its authoritarian response to climate issues, and its potentially massive international impact. In an era of multi-level governance, responsibility for action is going to be shared across different levels, subnational, regional, local, national, and international. The way states are currently organized to facilitate economic growth and, at least in most developed countries, provide social welfare constrains the possibilities for effective action on climate change, and the positions governments can adopt and targets to which they can commit in international negotiations. From the perspective of the governments of the Global South, without developed welfare states and without the history of growth that made them possible, matters look very different indeed.

The 'Policy Instruments' governments can deploy to meet their obligations are analyzed in Part IX. Market-based instruments, especially cap-and-trade, offsets, and carbon taxes, are especially prominent in the recommendations of economists, and in some cases the actions of governments. The most extensive experience with such instruments when it comes to climate change is in Europe, so that experience receives special attention. The redesign of energy systems is high on the list of possible policy initiatives.

'Producers and Consumers,' the subjects of Part X, can both respond to the policy instruments of government and take actions on their own initiative in the context of climate change. Our authors examine the role of corporations and consumers in both impeding and facilitating action against climate change.

Public and movements, national and subnational governments, producers and consumers all have roles to play in climate change mitigation and adaptation, but much still turns on what happens at the global level. Especially after the frustrations and failures evident in UN-based negotiations, rethinking 'Global Governance,' is central, and the topic of Part XI. Our authors look at the problematic history and performance of such governance, the lessons we might draw from existing global regimes, the moral foundations of alternative institutional arrangements, and the role of international law.

Finally, Part XII, 'Reconstruction,' contemplates the reworking of political, economic, and social arrangements as we adapt to the reality of coming climate change. The emphasis is on new forms of governance (especially at the global level), and more resilient social-ecological systems. After all of the challenges, opinions, impacts, actors, and responses, the task, of course, is to look forward to adaptation, transition, and rebuilding a society immersed in climate change.

8 CONCLUSION

The broad scope of this Handbook encompasses a range of issues and approaches beyond the basic science of climate change, from the philosophical to the political, from the psychological to the sociological, from the historical to the geographical, from the economic to the legal. On how science is disseminated, on how we assign economic value, on how states negotiate and govern, on the meaning of justice, and on the experience of those affected by climate change, we see contested concepts, frames, meanings, and responses.

As we said at the outset, climate change presents perhaps the most profound and complex challenge to have confronted human social, political, and economic systems. It also presents one of the most profound challenges to the way we understand human responses. In this collection, we have tried to lay out the variety and complexity of the issues at the intersection of climate change and human society. Our goal has been to be as comprehensive as possible within the limits of space. We offer the reader a broad-ranging collection of ways to think about one of the most difficult issues we human beings have brought upon ourselves in our short life on the planet.

REFERENCES

BOYKOFF, M. T., and BOYKOFF, J. M. 2004. Balance as bias: Global warming and the US prestige press. *Global Environmental Change* 14: 125–36.

LINDBLOM, C. E. 1982. The market as prison. *Journal of Politics* 44: 324–36.

LOVELOCK, J. 2010. Interview with James Lovelock. *The Guardian*. Online at <http://www.guardian.co.uk/science/2010/Mar/29/james-lovelock-climate-change> (accessed 29 March 2010).

RAWLS, J. 1971. *A Theory of Justice*. Cambridge, MA: Harvard University Press.

SCHNEIDER, S. H. 2009. *Science as a Contact Sport: Inside the Battle to Save Earth's Climate*. Washington, DC: National Geographic.

TUCKER, M. 1997. Climate change and the insurance industry: The cost of increased risk and the impetus for action. *Ecological Economics* 22(2): 85–96.

PART II

THE CHALLENGE
AND ITS HISTORY

CHAPTER 2

A TRULY COMPLEX AND DIABOLICAL POLICY PROBLEM

WILL STEFFEN

1 INTRODUCTION

CLIMATE change is like no other environmental problem that humanity has ever faced. Ross Garnaut, in his exhaustive review of the climate change problem for the Australian Government, called it a 'diabolical policy problem' (Garnaut 2008: xviii) and concluded his report with the statement: 'On a balance of probabilities, the failure of our generation would lead to consequences that would haunt humanity until the end of time' (Garnaut 2008: 597). Nicholas Stern, who carried out the first comprehensive economic analysis of the climate change problem, said that 'this (climate change) is an externality like none other. The risks, scales and uncertainties are enormous. . . . There is a big probability of a devastating outcome' (Stern 2009).

Perhaps no other problem—environmental or otherwise—facing society requires such a strong interdisciplinary knowledge base to tackle; research to support effective policy-making and other actions must cut across the full range of natural sciences, social sciences (including economics), and humanities. The research remit is so large because (i) a shift in global climate represents a fundamental change in the life support system for humans—the basic physical, chemical, and biological conditions necessary for life; and (ii) climate change cuts to the core of contemporary society—energy systems, lifestyles, institutions and governance, forms of economic organization, and basic values.

2 NATURE OF CLIMATE CHANGE

Climate change is a complex problem by its very nature, and will confound any attempts by policy makers or environmental managers to simplify it. Complexity here refers to a system that is characterized by multiple driving forces, strong feedback loops, long time lags, and abrupt change behavior. As an example of multiple driving forces, the current observed warming trend is driven primarily by a suite of greenhouse gases, including methane, nitrous oxide, and tropospheric ozone in addition to carbon dioxide (the most important of the gases). In contrast, simple systems are dominated by linear cause-effect phenomena. Six features of climate change are particularly important in terms of its complexity.

First, climate change is truly global in that it is centered around the two great fluids—the atmosphere and the ocean—that transport material and energy around the planet. Greenhouse gases, the emission of which represent the primary human influence on the climate system (IPCC 2007a), are well mixed in the atmosphere; emissions from any particular location are transported around the Earth in a matter of weeks. Ocean circulation connects far distant parts of the Earth; a slowdown or shutdown of the thermohaline circulation in the north Atlantic Ocean (often popularly known as the Gulf Stream) would cause regional cooling in northern Europe but would increase the rate of warming in much of the southern hemisphere (IPCC 2007a). Changes associated with the atmosphere and the ocean are often called 'systemic global changes' because of the mixing power of these fluids.

By contrast, changes in land cover, for example conversion of forest to cropland, invariably change the functioning of the terrestrial biosphere in the context of the climate system. While these effects are primarily local and regional in scale, when aggregated they are sometimes referred to as 'cumulative global changes.' Changes in land cover occur at specific locations, and have long been the province of local and regional decision making within the context of national sovereignty. They can, however, have global impacts via teleconnections in the atmosphere-ocean system. For example, large-scale deforestation of the Amazon rainforest would affect temperature and precipitation over Tibet (Snyder et al. 2004).

Human-driven climate change operates on a time scale that is beyond the experience of decision makers today. Many of the projections of changes in climate are carried out on a century timescale out to 2100, which is so far in the future that it is meaningless for nearly all political or economic analyses, based on discount rates that are normally used. Yet 2100 is, in fact, a rather early waypoint in the trajectory of contemporary climate change. The human-driven changes to the climate system that have occurred since the industrial revolution up to the present will still be discernible at least 1,000 years into the future, regardless of the future trajectory of emissions (Solomon et al. 2009).

Below are several characteristics of the climate system that operate on timescales that are significantly longer than those typical of human affairs.

Carbon dioxide. In addition to being well mixed in the atmosphere, carbon dioxide has a very long residence time—on average, about 100 years. This means, however, that a significant fraction of the carbon dioxide emitted at any one time will still be present 500 years into the future (IPCC 2007a). The policy implication of this characteristic of carbon dioxide is that delays in reducing emissions lead to an accumulation of the gas in the atmosphere that will continue to influence climate for a very long time.

Temperature increase. Even if greenhouse gas emissions could be reduced to zero tomorrow, the global average temperature would continue to rise for several decades into the future. This inertial effect also implies that the rate of temperature increase over the next two or three decades is largely insensitive of the level of emissions over that period (IPCC 2007a). Thus, policy decisions taken now will not have demonstrable effects on the trajectory of climate change until mid-century. For example, deep emission cuts now, with their likely economic and social costs, will not yield benefits for two or three decades.

Sea-level rise. Until recently, the primary factor driving sea-level rise has been thermal expansion of the oceans (Domingues et al. 2008). However, as climate change continues, melting and dynamic changes in the large polar ice sheets on Greenland and Antarctica will become increasingly important. A 2°C rise in temperature above pre-industrial would likely lead to an eventual sea-level rise of about 25 meters above current levels (Dowsett and Cronin 1990; Shackleton et al. 1995). However, although the 2 °C temperature rise would be realized this century, it would take many more centuries or even a millennium or two for the full sea-level rise to be realized (IPCC 2007a).

Extinction of biological species. Climate change is projected to lead to an enhanced rate of extinctions, probably increasing the current, high rate of extinctions by a factor of 10 (MA 2005). Extinctions are irreversible; once a species is lost, it cannot be retrieved. This represents the ultimate human impact on the global environment.

Arguably the most demanding of the challenges facing climate change negotiators is the suite of equity issues that separate the perspectives of various countries and regions (see Gardiner's chapter in this volume). Climate change is inherently unfair.

The long residence time of carbon dioxide leads to one of the most profound of the inequities. The cumulative emissions from about 1750 to present drive the currently experienced level of climate change and will continue to dominate the trajectory until about mid-century. About 75 percent of these cumulative emissions come from the OECD countries and the former Soviet Union (Figure 2.1; Raupach et al. 2007). In effect, the

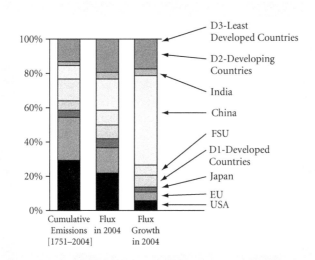

FIGURE 2.1 Various aspects of human carbon emissions by country/region (Raupach et al. 2007); FSU is the Former Soviet Union. The first column shows the cumulative emissions from the beginning of the industrial revolution to 2004. It is these stocks of carbon in the atmosphere that are largely driving observed climate change. The second column shows the flow rate of human carbon emissions into the atmosphere in 2004. The third column shows the annual rate in 2004 by which the flows of carbon into the atmosphere are growing.

wealthy countries have largely consumed the capacity of the atmosphere to absorb the wastes of industrial metabolism, leaving very little for the developing world to use in their quest to bring their populations out of poverty. This presents a dilemma of profound moral and ethical dimensions (see Baer in this volume).

The equity issue can be also be cast in terms of per capita emissions rather than national aggregates, thus focusing on the principle of equal rights to the atmosphere for each individual human regardless of where he or she lives. Although China has recently over-taken the USA as the largest emitter of carbon dioxide in a nationally aggregated sense, the per capita figures tell a different story. The average American emits over 20 tons of carbon dioxide per annum compared to less than 4 tons for the average Chinese citizen (UNDP 2007). One approach to bridging this gap is the 'contract and converge' strategy, aiming for a per capita emission entitlement in 2050 of about 2 tons of carbon dioxide for each human on Earth (Richardson et al. 2009).

Other aspects of the equity issue, which are temporal and inter-species in nature, raise important questions. What are the obligations of our generation to those to come in the future? (See Howarth's chapter in this volume). Do they have a fundamental right to an environment they can live in? Do humans have the ethical right to allow their actions to eradicate other living organisms?

The definition of what constitutes 'dangerous climate change' highlights another aspect of the equity issue. The distribution of the consequences of climate change is highly uneven around the world (IPCC 2007b). Developing countries are suffering the impacts of climate change much more than the industrialized world. There are several reasons for this. First, many of the industrialized countries lie in the northern mid- and high latitudes, where low temperatures limit important aspects of the economy, such as agriculture. Thus, modest levels of climate change are beneficial. Second, although rainfall is increasing overall with a warmer climate, regions in the sub-tropics that are prone to drought are experiencing intensifying and prolonged droughts with climate change (IPCC 2007a). With the excep-tion of Australia, these regions primarily consist of developing countries. Third, because of their higher wealth and levels of education, industrialized countries have higher adaptive capacity than developing countries. A synthesis of these types of argument lies behind the '2 °C guardrail' (limiting temperature rise to no more than 2 °C above pre-industrial levels), first proposed by the European Union and then adopted by the Copenhagen Accord in 2009.

Small island states represent a special case. Because many of them are low lying, they are exceptionally vulnerable to sea-level rise. For the most vulnerable of them, such as Kiribati, the climate system has already moved into what they consider to be the dangerous zone. The small island states, for example, argued vigorously at the COP-15 for a 1.5 °C guardrail because of their extreme vulnerability to sea-level rise.

One of the most striking scientific advances over the past decade is the analysis of the climate as a complex system. A prominent feature of a complex system is threshold/abrupt change behavior, in which an apparently small, insignificant change in a forcing variable can trigger an unexpectedly large and rapid or irreversible change in a major feature of the climate system—a so-called 'tipping element' (Lenton et al. 2008). This type of behavior is especially dangerous in the context of anthropogenic climate change because an apparently safe change in a variable, such as carbon dioxide concentration in the atmosphere, can suddenly trigger a massive

FIGURE **2.2** Map of potential climatic 'tipping elements' (after Lenton et al. 2008). Tipping elements are regional-scale features of the climate that could exhibit threshold-type behavior in response to human-driven climate change—that is, a small amount of climate change at a critical point could trigger an abrupt and/or irreversible shift in the tipping element. The consequences of such shifts in the tipping element for societies and ecosystems are likely to be severe. Question marks indicate systems whose status as tipping elements is particularly uncertain.

impact in some part of the climate system with very serious or even catastrophic consequences for humanity (Figure 2.2). Three examples of potential tipping elements are:

Large polar ice sheets. The Greenland and West Antarctic ice sheets, which together carry enough land-based ice to raise global sea levels by 13 metres if they totally disappeared, are vulnerable to modest increases in global average temperature above pre-industrial. Temperature rises of around 2 or 3 °C may be close to the tipping point for these ice sheets, although it would take many centuries or perhaps a couple of millennia for them to disappear completely (Gregory and Huybrechts 2006; Lenton et al. 2008).

Amazon rainforest. The world's largest contiguous rainforest is subject to rapid conversion to a grassland or savanna ecosystem if the climate warms and dries sufficiently, or if human-driven deforestation reaches a critical level, currently thought to be around 20 percent (Oyama and Nobre 2003; Foley et al. 2007). The combination of a warming climate and continuing deforestation may further hasten the tipping of the Amazon rainforest into a savanna.

South Asian monsoon system. Past evidence shows that this monsoon system can oscillate between wet and dry states. This behavior is crucial for the well-being of over a billion people, as the population in the region has risen sharply over the past century when the monsoon system has consistently been in a wet state. Modeling suggests that the monsoon is vulnerable to changes in aerosol concentration (pollution) over the subcontinent coupled to a warming global climate; the flip to a dry state could occur, without warning, as rapidly as one year (Zickfield et al. 2005).

Such behavior challenges the 2 °C definition of dangerous climate change. Furthermore, it presents profound challenges for institutional frameworks and legal systems, which have experience in dealing with simple cause-effect aspects of science but are ill-equipped to cope with complex systems. These challenges will be described in detail in the next two sections.

3 SPECIFIC CHALLENGES FOR GOVERNANCE

The characteristics of the climate system and the nature of the human influence on climate lead to profound challenges for governance. A recent review (Young and Steffen 2009) has analyzed the most important of these challenges, some of which are highlighted below.

The core science of climate change—the reality of the greenhouse effect, the observed warming of the Earth's surface over the past century, and the dominant role of human emissions of greenhouse gases in driving the observed changes—is beyond doubt in the credible climate science community, although challenged in the popular media and the blogosphere by non-experts. However, many uncertainties remain in more detailed aspects of the science, uncertainties that can have important implications for governance. Here are two examples: (i) The land and the ocean 'carbon sinks' currently absorb over half of the human emissions of carbon dioxide, thus acting as a powerful brake on the rate of temperature increase (Canadell et al. 2007). However, the future behavior of these sinks is highly uncertain, with the ocean sink already showing signs of weakening and the land sink projected to weaken later this century (Le Quéré et al. 2009). In addition, additional new sources of carbon may be activated later this century, but 'if and when' are highly uncertain. One of the most important of these potential new sources is the large amount of methane stored in the permafrost of the northern high latitudes; large-scale release of this methane could cause a sharp acceleration of warming, perhaps adding over a degree to the global mean temperature rise (Tarnocai et al. 2009). (ii) A critical uncertainty that bedevils the international negotiations on emission reductions surrounds the sensitivity of the climate system to a given concentration of greenhouse gases in the atmosphere. That is, even if we could stabilize greenhouse gas concentrations at a precise, desired level—say, 450 ppm CO_2-equivalent—we could not be sure of the precise level to which the temperature will eventually rise. For example, for 450 ppm CO_2-equivalent, there is only a 50 : 50 chance that the ultimate temperature increase will be below 2 °C above pre-industrial level (Hare and Meinshausen 2006).

The MRV (Measurement, Reporting, Verification) issues were highly contested at the COP-15 in Copenhagen, and when applied to biological gain or loss of carbon (in contrast to industrial emission reductions), climate science has a significant role to play. Measuring the amount of carbon lost from or stored in terrestrial ecosystems has always been a technically challenging aspect of emissions reporting, but has become even more important as carbon becomes a financial commodity. Reliable and precise measurements of the change in carbon storage in soils, where two-thirds of all terrestrial carbon resides, is particularly challenging.

The multiple scales at which climate change is manifest pose serious challenges for governance. The governance challenges associated with mitigation (emissions reductions) are well known. As CO_2 is a well-mixed gas globally, only an international solution, involving at least the most important emitters in a coordinated way, will begin to address the nature of the challenge. Although many of the individual actions required to reduce emissions will need to occur at local and regional (subnational) scales, most of the policy settings and financial instruments will need to be nationally determined and administered.

The adaptation imperative presents a different type of scalar challenge. At fine scale, climate change is highly differentiated, with much heterogeneity in the way in which climate is actually experienced—rainfall patterns, cyclones, bushfires, and other extreme events. Adapting to climate change is thus strongly a local and regional issue, with most of the policy and management burden falling on local and state/provincial jurisdictions. However, some of the broad-scale policy settings as well as much of the funding for adaptation may well need to come from national governments. Multi-level interactions in governance are thus required, especially when the inevitable interactions and trade-offs between mitigation and adaptation activities arise. A well-known example is the simultaneous need to adapt terrestrial ecosystems to continue to provide food and conserve biodiversity under a changing climate while increasing their capacity to store carbon and produce biofuels.

The nature of nonlinearities in the climate system, described above, strongly suggests that an early warning system would be very useful from a governance perspective. However, science is still a long way from being able to provide the knowledge base for early warning systems. Some preliminary analyses suggest that complex, dynamical systems slow down in terms of their natural fluctuations (measured mathematically as an increase in autocorrelation) when they are approaching a tipping point (Dakos et al. 2008). Such behavior has been observed in past abrupt changes in the climate system and offers hope for being able to anticipate tipping points in future. But even if science could provide the basis for a reliable early warning system, how quickly and decisively could contemporary society respond to avert disaster? In terms of governance, a further complicating factor is the long timescale and thus irreversibility (in a human timeframe) of many of the tipping elements in the climate system. Once a tipping point has been crossed, there is no way for humanity to reverse the change, no matter how deleterious or even catastrophic the new behavior of the climate system might be. This feature of the climate system argues for a careful application of the precautionary principle.

The scientific community is accelerating its efforts to understand the climate system, and even the IPCC reports—the 'gold standard' of scientific information—need to be updated soon after they are published (e.g. Richardson et al. 2009; Steffen 2009). For example, just as the international policy community is gradually coalescing around the 2 °C guardrail and a consequent need for stabilization of greenhouse gas concentrations at no more than 450 ppm CO_2-equivalent, the scientific community is moving towards a target of 350 ppm CO_2-equivalent to avoid serious or even catastrophic climate impacts (e.g. Rockström et al. 2009; Smith et al. 2009; Figure 2.3). Much of this rapidly accumulating knowledge is interdisciplinary, aiming to understand complex system behavior through integration of natural and

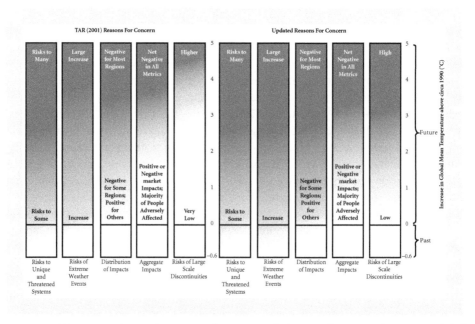

FIGURE 2.3 Diagram relating the potential impacts of climate change to the rise in global average temperature. Zero on the temperature scale corresponds approximately to 1990 average temperature, and the bottom of the temperature scale to pre-industrial average temperature. The level of risk or severity of potential impacts increases with the intensity of grey colour. The panel on the left is from Smith et al. in the IPCC Third Assessment Report (2001). The panel on the right is an updated version from Smith et al. (2009), using the same methodology as the TAR, based on expert judgement.

social science research. Much is directly policy relevant. For example, in terms of adaptation, just as local councils and the insurance industry are planning for a global mean sea-level rise of around 0.5 meter by 2100, the more recent analyses are suggesting that rises of upwards of 1 meter or more are likely (Rahmstorf 2007; Rahmstorf et al. 2007). In summary, the climate system is moving faster than science can understand, and new scientific knowledge is being generated at a rate with which governance is struggling to keep up.

4 NOVEL APPROACHES TO MEETING THE CLIMATE CHANGE CHALLENGE

The targets-and-timetable approach to reducing greenhouse gas emissions has dominated the policy dialogue for the last two decades. However, very recently the scientific community has proposed a somewhat different approach, based on aggregate emissions, that may facilitate the interaction between the scientific and the policy communities (e.g. Meinshausen et al. 2009). But what if attempts to reduce greenhouse gas emissions fall well short of the targets needed to avoid what most people consider to be dangerous climate change? Given this distinct possibility, there is a growing global dialogue on the possibility of geoengineering approaches to meet the climate change challenge (Royal Society 2009).

Conceptualizing the mitigation challenge as a limit on total, aggregate global emissions rather than on percentage reductions at certain time intervals has, in some ways, simplified the challenge and connected it more directly to the scientific underpinning (Meinshausen et al. 2009; Allen et al. 2009; England et al. 2009; Rogelj et al. 2009). The concept is simple. Starting from a normative judgement about the level of climate change humanity is willing to accept—say, a 2 °C increase in global mean temperature above pre-industrial—the concentration of greenhouse gases in the atmosphere to limit the temperature to this level can be calculated, and from that the allowable cumulative emissions of additional greenhouse gases can be obtained. There are two important caveats. First, there is still significant uncertainty surrounding the temperature increase associated with any given concentration of greenhouse gases in the atmosphere, so the relationship between cumulative emissions and temperature increase is a probabilistic one. Second, it is assumed that the fraction of human emissions of carbon dioxide currently absorbed by the ocean and land sinks and thus removed from the atmosphere will continue into the future as cumulative emissions rise (but see section 3 above).

A probabilistic analysis based on this approach (Meinshausen et al. 2009) and assuming the 2 °C guardrail shows that to keep the probability of exceeding the 2 °C limit to 25 percent, humanity should emit no more than a total of 1,000 Gt CO_2 (Gt = Gigaton = 1 billion tons) in the 2000–50 period. Given that we have already emitted about 350 Gt in the 2000–9 period, there are only 650 Gt of permitted emissions remaining over the next 41 years to stay within the 1,000 Gt limit. This is an exceptionally challenging target! If we accept only a 50 : 50 chance of limiting the temperature rise to 2 °C or less, then the permitted emissions through the 2000–50 period become 1,440 Gt CO_2.

How are we tracking towards this cumulative emission target as we approach the end of the first decade of the twenty-first century? An analysis of stated national positions on emissions reductions going into the COP-15 meeting in Copenhagen showed that intentions at that time fell well short of what is required (Rogelj et al. 2009). For example, aggregating the stated commitments of Annex I countries as a group give an emission reduction by 2020 of 8–14 percent below 1990 levels. This figure would have to be 25–40 percent to be on track to stay within the 2 °C guardrail. When the intentions of developing countries are included in the analysis, the conclusion is even clearer: the current pathway gives virtually no chance of limiting warming to 2 °C or less.

With the likelihood that climate change may well move into the 'dangerous zone' later this century, increasing attention is being given to geoengineering approaches (Royal Society 2009). The term 'geoengineering' is applied to a range of possible technologies or methodologies that can be divided into two groups: (i) techniques that remove carbon dioxide from the atmosphere, and (ii) techniques that modify the radiation balance at the Earth's surface by changing the amount of incoming solar radiation that the Earth absorbs. The scientific bases for the two approaches are fundamentally different from the perspective of Earth as a complex system. The first approach attempts to address the source of the problem by removing CO_2 from the atmosphere, while the second attempts to manipulate the functioning of the Earth System itself. From a risk perspective, the second approach is

Table 2.1 The nine proposed planetary boundaries, showing the Earth System process or subsystem, the control variable (parameter), the proposed boundary value, the current status, and the pre-industrial value (Rockström et al. 2009). The rows shaded dark grey indicate boundaries that humanity has already transgressed.

PLANETARY BOUNDARIES

Earth-system process	Parameters	Proposed boundary	Current status	Pre-industrial value
Climate change	(i) Atmospheric carbon dioxide concentration (parts per million by volume)	350	387	280
	(ii) Change in radiative forcing (watts per metre squared)	1	1.5	0
Rate of biodiversity loss	Extinction rate (number of species per million species per year)	10	<100	0.1-1
Nitrogen cycle (part of a boundary with the phosphorus cycle)	Amount of N_2 removed from the atmosphere for human use (millions of tonnes per year)	35	121	0
Phosphorus cycle (part of a boundary with the nitrogen cycle)	Quantity of P flowing into the oceans (millions of tonnes per year)	11	8.5-9.5	-1
Stratospheric ozone depletion	Concentration of ozone (Dobson unit)	276	283	290
Ocean acidification	Global mean saturation state of aragonite in surface sea water	2.75	2.90	3.44
Global freshwater use	Consumption of freshwater by humans (km^3 per year)	4,000	2,600	415
Change in land use	Percentage of global land cover converted to cropland	15	11.7	Low
Atmospheric aerosol loading	Overall particulate concentration in the atmosphere, on a regional basis		To be determined	
Chemical pollution	For example, amount emitted to, or concentration of persistent organic pollutants, plastics, endocrine disrupters, heavy metals and nuclear waste in, the global environment, or the effects on ecosystem and functioning of Earth system thereof		To be determined	

far more dangerous as it entails potentially deleterious or even catastrophic side effects that are very difficult or even impossible to anticipate a priori.

Examples of carbon dioxide removal techniques include direct engineered capture of CO_2 from the free atmosphere, enhancement of natural carbon sinks in terrestrial ecosystems, and the enhancement of oceanic uptake of CO_2 by increasing the amount of micronutrients like iron. The last approach, however, carries a high risk of significant impacts on the structure and functioning of marine ecosystems. The most prominent example of changing the radiation balance is by injecting sulphate aerosols into the lower stratosphere. This approach, however, has some severe side effects that are already well known—it would do nothing to counteract the increasing acidity of the ocean, and it could reduce precipitation in the Asian and African summer monsoon systems and impact the food supplies of billions of people (Robock et al. 2008).

The governance implications of geoengineering, dealt with at some length in the Royal Society analysis (2009), are enormous. For example, many of the geoengineering approaches are transboundary in nature, especially those that modify the Earth's radiation balance, and will require new international institutions and/or mechanisms that are not yet in place. Others, such as the proposed iron fertilization of the ocean, could be handled by existing instruments (the International Law of the Sea in this case) but they would likely require significant modification.

An opposite approach to geoengineering is the planetary boundaries framework (Rockström et al. 2009), which focuses on the complex-system nature of the Earth and, in particular, on the risk for abrupt and/or irreversible changes in important features of the Earth System. Rather than trying to engineer solutions to global environmental change after it has occurred, the planetary boundaries approach attempts to define the 'safe operating space' for humanity by defining 'no-go zones' in a global environmental context. The initial analysis has identified nine planetary boundaries, one of which relates directly to climate change (Table 2.1). This boundary has been proposed as 350 ppm CO_2 concentration and, concurrently, a + 1 watt per meter squared increase in radiative forcing (the aggregate of all of the factors—natural and anthropogenic—that influence the energy balance at the Earth's surface). Thus, we are currently in overshoot.

5 Reconceptualizing the Climate Change Problem

The climate change problem has been cast largely as one of changing the energy systems of contemporary society away from fossil fuel-based systems towards low- or no-carbon systems. There is no doubt scientifically that the emission of greenhouse gases, predominantly carbon dioxide, from the combustion of fossil fuels lies at the heart of the climate change problem. Meeting the challenge of reducing these emissions is undoubtedly the highest priority climate mitigation action facing humanity. However, there is a significant—and growing—body of scholarship that focuses on global change rather than only on climate change, and views climate change as a symptom of a much deeper problem that centers on the fundamental relationship of humanity with the rest of nature (Steffen et al. 2004).

One of the most striking conceptual frameworks now used to describe the changing relationship between humanity and our environment is that of the Anthropocene, a new geological era proposed by Nobel Laureate Paul Crutzen (2002). The concept of the Anthropocene encompasses climate change, but goes on to consider the many other changes to the global environment that have occurred since the industrial revolution—changes in global element cycles such as nitrogen and phosphorus, the rapid loss of biodiversity, the changes in the water cycle, the vast changes to the Earth's land cover, the depletion of many of the world's fisheries, and so on—all of which are driven ultimately by human numbers and human activities. Taken together, these changes demonstrate unequivocally that the human enterprise has now become so powerful in terms of its impact at the global scale that it rivals some of the great forces of nature (Figures 2.4 and 2.5; Steffen et al. 2004).

Global change has clearly moved the planetary environment out of its 10,000-year-old Holocene state and into the new state of the Anthropocene (Steffen et al. 2007). This has enormous implications for the future of humanity, as our societies and civilizations, and the ecosystems on which we depend for essential services, are tuned to the environmental envelope of the Holocene. For example, agricultural systems are now finely tuned (optimized) for the temperature ranges and rainfall patterns of the last century or two; our emergency management services have been resourced and trained to deal with the natural disasters of the recent past; and even our own fundamental physiology is not equipped to deal with the temperature extremes associated with a 2 or 3 °C rise in average temperature above the long-term Holocene average. As we move out of the Holocene envelope, we are sailing into planetary *terra incognita*, with an uncertain outcome in terms of the viability of contemporary civilization beyond this century, or even the next few decades.

As humanity moves more deeply into the unknown world of the Anthropocene, the question arises as to whether our society will even survive this transition, or will collapse as many other civilizations have done in the past. This existential question is driving a new area of scholarship that is reconceptualizing history by integrating palaeo-environmental research with anthropology, archaeology, and history (Costanza et al. 2007). The aim is to explore the reasons that some earlier civilizations collapsed in the face of environmental stresses of various kinds, while other civilizations, facing similar stresses and constraints, engineered relatively smooth transformations into different societies that were much better equipped to deal with the stresses. Such knowledge can inform humanity's present situation, and suggest pathways that may help guide us towards a more sustainable future.

Although we can learn much from the past, the present situation is fundamentally different in that contemporary society is much more interconnected at the global scale than ever before. If contemporary, globalized society collapses, there is no alternative waiting in the wings to rescue humanity. Our society is strongly driven by a core value of continuing economic growth and ever-increasing material wealth for larger numbers of humans. The implicit assumptions behind this core value are that the Earth's resources are essentially infinite (or can be made so through substitution) and its capacity to absorb societies metabolic wastes is also limitless. Climate change and other environmental changes are challenging this assumption.

At its most fundamental level, then, climate change may represent the canary in the coal mine for our own species. Is it another environmental problem to be solved at the margins of society in its continual march of progress, or does it signal the end of the era of ever-expanding population, continuous economic growth, and increasing material wealth?

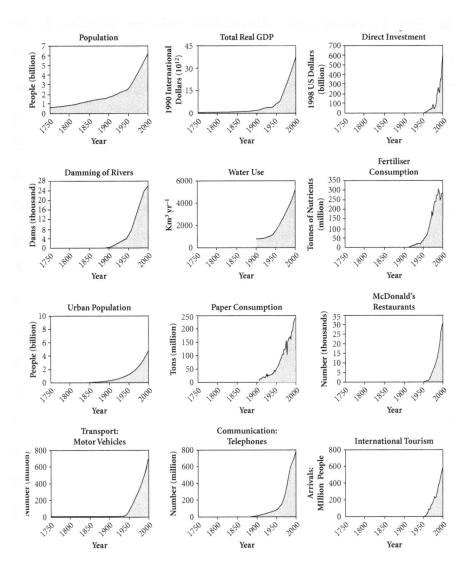

FIGURE 2.4 The increasing rates of change in human activity since the beginning of the industrial revolution. Significant increases in rates of change occur around the 1950s in each case and illustrate how the past fifty years have been a period of dramatic and unprecedented change in human history. From Steffen et al. (2004), which includes references to the individual databases from which the panels are derived.

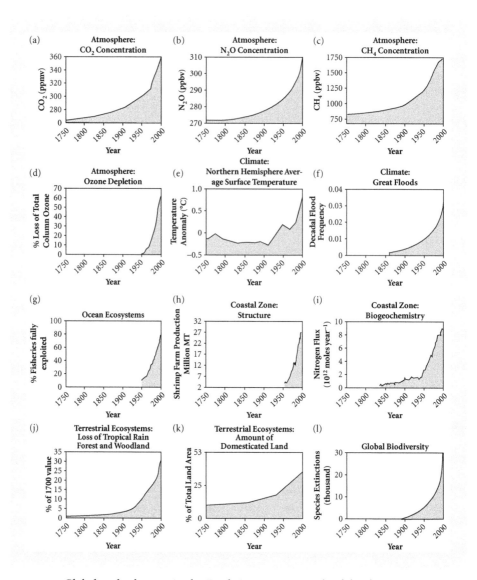

FIGURE 2.5 Global-scale changes in the Earth System as a result of the dramatic increase in human activity: (a) atmospheric CO_2 concentration; (b) atmospheric N_2O concentration; (c) atmospheric CH_4 concentration; (d) percentage total column ozone loss over Antarctica, using the average annual total column ozone, 330, as a base; (e) northern hemisphere average surface temperature anomalies; (f) natural disasters after 1900 resulting in more than ten people killed or more than 100 people affected; (g) percentage of global fisheries either fully exploited, overfished, or collapsed; (h) annual shrimp production as a proxy for coastal zone alteration; (i) model-calculated partitioning of the human-induced nitrogen perturbation fluxes in the global coastal margin for the period since 1850; (j) loss of tropical rainforest and woodland, as estimated for tropical Africa, Latin America, and South and Southeast Asia; (k) amount of land converted to pasture and cropland; and (l) mathematically calculated rate of extinction. From Steffen et al. (2004), which includes references to the individual databases from which the panels are derived.

REFERENCES

ALLEN, M. R., FRAME, D. J., HUNTINGFORD, C., JONES, C. D., LOWE, J. A., MEINSHAUSEN, M., and MEINSHAUSEN, N. 2009. Warming caused by cumulative carbon emissions towards the trillionth tonne. *Nature* 458: 1163–6.

CANADELL, J. G., LE QUÉRÉ, C., RAUPACH, M. R., FIELD, C. R., BUITENHUIS, E., CIAIS, P., CONWAY, T. J., GILLETT, N. P., HOUGHTON, R. A. and MARLAND, G. 2007. Contributions to accelerating atmospheric CO_2 growth from economic activity, carbon intensity, and efficiency of natural sinks. *Proceedings of the National Academy of Sciences (USA)* 104: 18866–70.

COSTANZA, R., GRAUMLICH, L., STEFFEN, W., CRUMLEY, C., DEARING, J., HIBBARD, K., LEEMANS, R., REDMAN, C., and SCHIMEL, D. 2007. Sustainability or collapse: What can we learn from integrating the history of humans and the rest of nature? *Ambio* 36: 522–7.

CRUTZEN, P. J. 2002. Geology of mankind: the anthropocene. *Nature* 415: 23.

DAKOS, V., SCHEFFER, M., VAN NES, E. H., BROVKIN, V., PETOUKHOV, V., and HELD, H. 2008. Slowing down as an early warning signal for abrupt climate change. *Proceedings of the National Academy of Sciences (USA)* 105: 14308–12.

DOMINGUES, C. M., CHURCH, J. A., WHITE, N. J., GLECKLER, P. J., WIJFFELS, S. E., BARKER, P. M. and DUNN, J. R. 2008. Improved estimates of upper-ocean warming and multi-decadal sea-level rise. *Nature* 453: 1090–3.

DOWSETT, H. J. and CRONIN, T. M. 1990. High eustatic sea level during the middle Pliocene: Evidence from southeastern U.S. Atlantic coastal plain. *Geology* 18: 435–8.

ENGLAND, M. H., SEN GUPTA, A., and PITMAN, A. J. 2009. Constraining future greenhouse gas emissions by a cumulative target. *Proceedings of the National Academy of Sciences (USA)* 106: 16539–40.

FOLEY, J. A., ASNER, G. P., COSTA, M. H., COE, M. T., DEFRIES, R., GIBBS, H. K., HOWARD, E. A., OLSON, S., PATZ, J., RAMANKUTTY, N., and SNYDER, P. 2007. Amazonian revealed: forest degradation and loss of ecosystem goods and services in the Amazon Basin. *Frontiers in Ecology and Environment* 5: 25–32.

GARNAUT, R. 2008. *The Garnaut Climate Change Review: Final Report.* Cambridge: Cambridge University Press.

GREGORY, J. M., and HUYBRECHTS, P. 2006. Ice-sheet contributions to future sea-level change. *Philosophical Transactions of the Royal Society of London, Series A* 364: 1709–31.

HARE, W., and MEINSHAUSEN, M. 2006. How much warming are we committed to and how much can we avoid? *Climatic Change* 75: 111–49.

IPCC (Intergovernmental Panel on Climate Change) 2001. *Climate Change 2001: Impacts, Adaptation and Vulnerability.* Contribution of Working Group II to the Third Assessment Report of the Intergovernmental Panel on Climate Change, ed. J. J. MCCARTHY, O. F CANZIANI, N. A. LEARY, D. J. DOKKEN, and K. S. WHITE. Cambridge: Cambridge University Press.

——2007a. *Climate Change 2007: The Physical Science Basis.* Contribution of Working Group I to the Fourth Assessment Report of the Intergovernmental Panel on Climate Change, ed. S. SOLOMON, D. QIN, M. MANNING, Z. CHEN, M. MARQUIS, K. AVERYT, M. M. B. TIGNOR, H. L. MILLER, JR, and Z. CHEN. Cambridge: Cambridge University Press.

——2007b. *Climate Change 2007: Impacts, Adaptation and Vulnerability.* Contribution of Working Group II to the Fourth Assessment Report of the Intergovernmental Panel on Climate Change, ed. M. L. PARRY, O. F. CANZIANI, J. P. PALUTIKOF, P. J. VAN DER LINDEN, and C. E. HANSON. Cambridge: Cambridge University Press.

LENTON, T. M., HELD, H., KRIEGLER, E., HALL, J. W., LUCHT, W., RAHMSTORF, S., SCHELLN-HUBER, H. J. 2008. Tipping elements in the Earth's climate system. *Proceedings of the National Academy of Sciences (USA)* 105: 1786–93

LE QUÉRÉ, C., RAUPACH, M. R., CANADELL, J. G., MARLAND, G., BOPP, L., CIAIS, P., CONWAY, T. J., DONEY, S. C., FEELY, R. A., FOSTER, P., FRIEDLINGSTEIN, P., GURNEY, K., HOUGHTON, R. A., HOUSE, J. I., HUNTINGFORD, C., LEVY, P. E., LOMAS, M. R., MAJKUT, J., METZL, N., OMETTO, J. P., PETERS, G. P., PRENTICE, I. C., RANDERSON, J. T., RUNNING, S. W., SARMIENTO, J. L., SCHUSTER, U., SITCH, S., TAKAHASHI, T., VIOVY, N., van der WERF, G. R. and WOODWARD, F. I. 2009. Trends in the sources and sinks of carbon dioxide. *Nature Geoscience* 2: doi: 10.1038/NGEO689.

MA (Millennium Ecosystem Assessment) 2005. Ecosystems and Human Well-being: Biodiversity Synthesis. Island Press, Washington DC.

MEINSHAUSEN, M., MEINSHAUSEN, N., HARE, W., RAPER, S. C. B., FRIELER, K., KNUTTI, R., FRAME, D. J., and ALLEN, M. R. 2009. Greenhouse-gas emission targets for limiting global warming to 2 °C. *Nature* 458: 1158–62.

OYAMA, M. D. and NOBRE, C. A. 2003. A new climate-vegetation equilibrium state for tropical South America. *Geophysical Research Letters* 30: 2199, doi: 10.1029/2003GL018600.

RAHMSTORF, S. 2007. A semi-empirical approach to projecting future sea-level rise. *Science* 315: 368–70.

——CAZENAVE, A., CHURCH, J. A., HANSEN, J. E., KEELING, R. F., PARKER, D. E., SOMERVILLE, R. C. J. et al. 2007. Recent climate observations compared to projections. *Science* 316: 709.

RAUPACH, M. R., MARLAND, G., CIAIS, P., LE QUÉRÉ, C., CANADELL, J. G., KLEPPER, G., and FIELD, C. B. 2007. Global and regional drivers of accelerating CO_2 emissions. *Proceedings of the National Academy of Sciences (USA)* 104: 10288–93.

RICHARDSON, K., STEFFEN, W., SCHELLNHUBER, H. J., ALCAMO, J., BARKER, T., KAMMEN, D. M., LEEMANS, R., LIVERMAN, D., MUNASINGHE, M., OSMAN-ELASHA, B., STERN, N., and WAEVER, O. 2009. *Synthesis Report. Climate Change: Global Risks, Challenges & Decisions*. Summary of the Copenhagen Climate Change Congress, 10–12 March 2009. University of Copenhagen.

ROBOCK, A., OMAN, L., and STENCHIKOV, G. L. 2008. Regional climate responses to geoengineering with tropical and Arctic SO_2 injections. *Journal of Geophysical Research* 113: D16101. doi: 10:1029/2008JD010050.

ROCKSTRÖM, J., STEFFEN, W., NOONE, K., PERSSON, Å., CHAPIN, III, F. S., LAMBIN, E. F., LENTON, T. M., SCHEFFER, M., FOLKE, C., SCHELLNHUBER, H. J., NYKVIST, B., de WIT, C. A., HUGHES, T., van der LEEUW, S., RODHE, H., SÖRLIN, S., SNYDER, P. K., COSTANZA, R., SVEDIN, U., FALKENMARK, M., KARLBERG, L., CORELL, R. W., FABRY, V. J., HANSEN, J., WALKER, B., LIVERMAN, D., RICHARDSON, K., CRUTZEN, P., and FOLEY, J. A. 2009. A safe operating space for humanity. *Nature* 461: 472–5.

ROGELJ, J., HARE, B., NABEL, J., MACEY, K., SCHAEFFER, M., MARKMANN, K., and MEINSHAUSEN, M. 2009. Halfway to Copenhagen, no way to 2 °C. *Nature Reports Climate Change* 3: 81–3.

Royal Society 2009. Geoengineering the climate: science, governance and uncertainty. RS Policy Document 10/09 (issued September 2009).

SHACKLETON, N. J., HALL, J. C. and PATE, D. 1995. Pliocene stable isotope stratigraphy of ODP Site 846. Pp. 337–56 in *Proceedings of the Ocean Drilling Program, Scientific Results*, vol. 138. Ocean Drilling Program, College Station, TX.

SMITH, J. B., SCHNEIDER, S. H., OPPENHEIMER, M., YOHE, G. W., HARE, W., MASTRANDREA, M. D., PATWARDHAN, A., BURTON, I., CORFEE-MORLOT, J., MAGADZA, C. H. D., FUSSEL, H.-M., PITTOCK, A. B., RAHMAN, A., SUAREZ, A., and VAN YPERSELE, J.-P. 2009. Assessing dangerous climate change through an update of the Intergovernmental Panel on Climate Change (IPCC) 'reasons for concern'. *Proceedings of the National Academy of Sciences (USA)* doi/10.1073/pnas.0812355106

SNYDER, P. K., DELIRE, C., and FOLEY, J. A. 2004. Evaluating the influence of different vegetation biomes on the global climate. *Climate Dynamics* 23: 279–302.

SOLOMON, S., PLATTNER, G.-K., KNUTTI, R., and FRIEDLINGSTEIN, P. 2009. Irreversible climate change due to carbon dioxide emissions. *Proceedings of the National Academy of Sciences (USA)* 106: 1704–9.

STEFFEN, W. 2009. *Climate Change 2009: Faster Change & More Serious Risks.* Department of Climate Change, Australian Government.

——SANDERSON, A., TYSON, P. D., JÄGER, J., MATSON, P., MOORE III, B., OLDFIELD, F., RICHARDSON, K., SCHELLNHUBER, H. J., TURNER II, B. L. and WASSON, R. J. 2004. *Global Change and the Earth System: A Planet Under Pressure.* The IGBP Book Series, Berlin: Springer-Verlag.

——CRUTZEN, P. J., and MCNEILL, J. R. 2007. The Anthropocene: Are humans now overwhelming the great forces of Nature? *Ambio* 36: 614–21.

STERN, N. 2009. Plenary presentation. Climate change: global risks, challenges & decisions, Copenhagen 10–12 March 2009.

TARNOCAI, C., CANADELL, J. G., SCHUUR, E. A. G., KUHRY, P., MAZHITOVA, G., and ZIMOV, S. 2009. Soil organic carbon pools in the northern circumpolar permafrost region. *Global Biogeochemical Cycles*, 23: GB2023.

UNDP (United Nations Development Program) 2007. Human Development Report 2007/2008. *Fighting Climate Change: Human solidarity in a divided world.* United Nations, New York.

YOUNG, O., and STEFFEN, W. 2009. The Earth System: Sustaining planetary life support systems. Pp. 295–315 in F. S. Chapin III, G. P. Kofinas, and C. Folke (eds.), *Principles of Ecosystem Stewardship: Resilience-Based Natural Resource Management in a Changing World.* New York: Springer-Verlag.

ZICKFELD, K., KNOPF, B., PETOUKHOV, V., and SCHELLNHUBER, H. J. 2005. Is the Indian summer monsoon stable against global change? *Geophysical Research Letters*, 32: doi: 10.1029/2005GL022771.

CHAPTER 3

....

THE NATURE OF THE PROBLEM

....

DALE JAMIESON*

1 INTRODUCTION

....

ALMOST everything about climate change is contested: whether it is occurring, whether it is anthropogenic, whether it is a problem, whether it is soluble, what would be the solutions, and even what would even count as a solution. While not every view is as good as every other, it is important to understand why the juxtaposition of climate and humanity provides such fertile soil for a diversity of interpretations and perspectives. In this chapter I characterize some of these interpretations and perspectives, and explain what it is about climate and humanity that supports them. I provide my own view of the nature of the problem, but my most robust conclusion is that it is extremely difficult for creatures like us to arrive at common understandings about climate change, much less to respond in ways that involve acting in concert.

2 CHANGE AND STABILITY

....

By definition, climate change involves change, but it is not easy to identify either climate or change, either empirically or conceptually. Climate is an abstraction from the weather that people experience. In a highly variable system, it is difficult to distinguish climate change from variability. During the twentieth century there was a 148 °C difference between the warmest and coldest recorded temperatures. In Rapid City, South Dakota, the temperature once dropped 26 °C in 15 minutes; in Spearfish, South Dakota the temperature rose 27 °C in

* An earlier version of this chapter was delivered as the Wayne Morse Lecture at the University of Oregon. I am grateful to Margaret Halleck for making this lecture possible, and to everyone who participated in the discussion. I also thank John S. Dryzek, Richard B. Norgaard, and David Schlosberg for their comments on an earlier draft.

2 minutes. In New York City there is on average about a 25 °C difference between summer and winter temperatures; in Berlin, the day/night gradient is usually about 10 °C. As I write these words, there is a 48 °C difference between Barrow, Alaska, and Singapore. Against this background, a 1.4–6.4 °C warming over this century, which the Intergovernmental Panel on Climate Change (IPCC) predicts, is difficult to detect and does not seem very dramatic. If you want climate change, get on an airplane. Or just wait a few hours, not a century.

Of course this response embodies confusions, but it also expresses a deep truth. The very idea of climate change involves a particular paradigm—call it the 'stability/change' paradigm. From an array of data points that could fairly be described as 'all over the place,' certain values are identified as anomalous, and then questions are asked about whether these anomalies are harbingers of change. Were we to drop the stability/change paradigm, we might see the record as displaying spatial and temporal variation rather than anomalies. Indeed, even from within the stability/change paradigm, we need only to shift the temporal dimension to see change as variability. For example, what appears to be climate change from a nineteenth-century baseline may appear to be variability from the perspective of millennia. This is why examining the paleo-climate record sometimes induces people to become climate change skeptics.[1]

There are considerations concerning the basic science of the climate system that support viewing the data in terms of the stability/change paradigm, but it is important to recognize that even raising the question of climate change involves interpreting the climate record in a particular way.[2] Climate data do not come marked 'change' or 'variability,' anymore than the works of fourteenth-century Florentine artists are stamped with the words, 'Renaissance artwork.'

3 PROBLEMS

Even if we accept that climate change is occurring, as we should, we do not yet have a problem. Dramatic changes occur all the time that we do not consider to be problems (e.g. summer changes to autumn, black holes devour stars). What is minimally required for a change to be a problem is that it adversely affects what we care about. But this may not be enough. Some people would say that if the climate change that is now under way were due to purely natural (i.e. non-anthropogenic) factors, then it would not be a problem. There are two distinct grounds one might have for this view. Some people may think that it is necessary for a change to be a problem that it is caused by human agency. On this view, natural occurrences that affect us adversely are unfortunate, but they are not problems. A second, more plausible ground is that if we cannot in some way remedy a change that adversely affects us, then the change is not a problem. On this view, problems imply solutions. For example, dying prematurely of a curable disease is a problem; being mortal is not. There are people who think that nothing can be done about climate change. If they also have the view that problems require solutions, then they do not think that climate change is a problem, however regrettable they may think it is that it is occurring. Of course they are wrong in thinking that we cannot respond to climate change in ways that make a difference. But what counts as making a difference depends enormously on what exactly we think the problem is.

4 FRAMING

Viewing climate as changing and seeing change as a problem involves framing our knowledge and experience of climate in a particular way. The meaning of climate is extremely dense, and so there are many ways in which people frame climate and climate change (Hulme 2009).

At the deepest level, for some people, climate change is a version of the biblical story of Adam's Fall, though the connection is not usually acknowledged or even perhaps consciously in mind. In that story Adam and Eve were banished from the Garden of Eden because Eve, having been seduced by a serpent, disobeyed God and ate the forbidden fruit from the tree of the knowledge of good and evil. Their descendents—all of us—bear the mark of this original sin, though there is hope of redemption through a messiah or savior.[3] The Garden of Eden is the stable climate regime, untouched by humanity. The serpent is industrial civilization, which has given us the forbidden fruit of disposable consumer goods which satisfy our immediate desires, and the greenhouse emissions that they entail. Original sin is expressed in the fact that once we are introduced to this shallow consumer culture, there is no turning back for ourselves or our children. Redemption requires an apocalypse on some views, or merely a 'cap and trade' system on others. Some think that Al Gore is the messiah, while others put their faith in the IPCC (though in the wake of Climategate, many have found their faith shaken). Rather than the messiah, other people see Al Gore as the Antichrist and they see climate change as a hoax rather than as a crisis having 'the potential to end human civilization as we know it.'[4] For them climate change is a conspiracy on the part of scientists, bankers, and politicians seeking to line their pockets, and a cynical pretext for those who want to cede American sovereignty to the United Nations.[5] Interestingly, while both of these framings can be seen as theological in inspiration, they are typically expressed in the language of science.[6]

Scientists were extremely influential in the initial framing of the climate change issue. While some concerns were expressed earlier, until the 1980s climate change was mainly portrayed as a matter of scientific curiosity and research (Wearth, this volume). In an extremely influential 1957 article, Roger Revelle and Han Suess wrote, without a hint of alarm or serious concern:

> Human beings are now carrying out a large-scale geophysical experiment of a kind that could not have happened in the past nor be repeated in the future . . . Within a few centuries we are returning to the atmosphere and oceans the concentrated organic carbon stored in sedimentary rocks over hundreds of millions of years. (Revelle and Suess 1957: 27)

The 1988 Toronto Conference on 'The Changing Atmosphere' marked a turning point. There, an international group of scientists and government officials called for a 20 percent reduction of carbon dioxide emissions by 2005 from the 1988 baseline (Environment Canada, WMO, and UNEP 1988). That same year the quasi-scientific, quasi-political IPCC was created, in part to bring science to bear on climate policy, but also to blunt the activism that was beginning to emerge in the scientific community (Agrawala 1998).

In the run-up to the 1992 Rio Earth Summit large oil, coal, and auto companies felt threatened by the possibility of an international agreement that would lead to phasing out fossil fuels. They mobilized scientists who were already on their payroll, and enlisted the support of a handful of academic and government scientists including Fred Singer, Patrick Michaels, and Richard Lindzen (Leggett 1999; Hoggan 2009). In the background was a small group of prestigious physicists who had been active in cold war politics, and saw the emerging concern with climate as a threat to American national sovereignty, their own anti-regulatory values, and the institutional power of their own discipline and fields of research. One of these scientists, William Nierenberg, chaired an early National Academy of Sciences study of climate change (National Academy of Sciences 1983), and went on to co-found the George Marshall Institute with Frederick Seitz and Robert Jastrow. The Marshall Institute was originally established to support Ronald Reagan's Strategic Defense Initiative, but soon became an institutional home for scientists with anti-environmental agendas. Several of those involved in climate change denial also worked to cast doubt on the science behind acid precipitation and ozone depletion, and also worked for the tobacco industry in its campaign to refute the claim that second-hand smoke causes cancer (Oreskes and Conway 2010).[7]

Framing climate change as a scientific issue invites these kinds of responses. Skepticism about scientific claims is generally appropriate and often contributes to scientific progress. However, it is not always easy to distinguish a constructive skeptic from a dogmatic denier, especially when a claim is said to be uncertain. Since science is fallible and probabilistic, uncertainty is always lurking in the background, though it is often ignored ('blackboxed'). It is essential to scientific practice that we take some claims as fixed, since progress would not be possible if every proposition were problematized in every investigation. Nevertheless, when the stakes are high and society is fractious, deviant scientists, especially those who are backed by interest groups, will pry open the black box. Scientific uncertainty, rather than being a cause of controversy, is often a consequence of controversy.[8] For this reason as well as others, science, which often has a privileged role in identifying issues of public concern, cannot by itself bring such issues to policy closure (Jamieson 1996; Sarewitz 2004).

In addition, scientists have not always been effective advocates for their causes, nor generally very adept in dealing with the broad social landscape in which the climate policy drama is enacted (Moser and Dilling in this volume).[9] The media, because of the way in which they frame the issue, have often made things extremely difficult for mainstream climate scientists. For many years in the United States the science of climate change was covered as if it were a political story: every assertion had to be balanced by a counter-assertion. Allegations of fraud or misconduct particularly catch media attention because they fit the 'he said/she said' model even better than competing scientific claims.[10] After the IPCC won the Nobel Peace Prize in 2007 the political frame on climate science began to recede, but not for long. Controversies over hacked e-mails from the Climate Research Unit at East Anglia University, allegations about IPCC Chair R. K. Pachauri's role as a corporate consultant, and the discovery of errors in the 2007 IPCC report have brought the political frame roaring back.[11] In any case, the scientific framing of the climate change issue has led to many people feeling confused, unmoved, or even alienated from the discussion.

From the beginning, many American environmental groups had difficulty with climate change. The American environmental movement has deep roots in the anti-nuclear movement, and concerns about climate change seem to strengthen the case for nuclear power. Moreover, some environmental groups sensed, correctly, that climate change would be a difficult issue to champion, compared to such vivid and immediate issues as clean air and water, and even the protection of charismatic species.[12]

5 PRUDENCE AND POLICY

Those who want to take action on climate change often frame it as an issue about self-interest, where 'our' self-interest is thought of as the aggregate of the interests of each of us. One way of trying to understand our interests, from this point of view, is through an economic assessment of the aggregate expected damages of climate change and the costs of avoiding them. Nordhaus (2008), working in this tradition, tells us that optimal climate policy would involve a carbon tax of about $17 per ton in 2005, ramping up to $270 per ton in 2100.

There are obvious problems with this approach. It assumes that all preferences are commensurable and can be monetized, yet diverse values are at risk from climate change, not only income and economic assets, but also biodiversity and social solidarity (for example). It is not clear that all such values can be monetized, or even meaningfully placed on the same scale. Indeed some would say that even if in some sense this could be done, it would be wrong to do so, just as it would be wrong to monetize the value of a friend or lover (Sagoff, this volume).[13] Even if we leave these concerns aside, the idea that one can know enough to reliably calculate the benefits and costs of climate change and climate stabilization policies into the distant future is patently absurd. We can barely predict the state of the economy from one quarter to another.[14]

The Stern Review provides a different perspective on the economics of climate change, claiming that the optimal carbon tax now is $311 per ton (Stern 2006).[15] The core of the difference between Stern and Nordhaus concerns how to value costs and benefits that occur in the further future. Nordhaus discounts them at 3 percent for pure time preference, declining to 1 percent in 300 years; he derives these rates from what he takes to be people's actual discounting behavior. Stern rejects pure time preference altogether on ethical grounds. His point is that those in the further future who will bear the costs and benefits of present policies are different people from those who bear them at present. When time horizons are so long, discounting for pure time preference does not express an attitude about how consumption should be scheduled over time, but rather expresses an attitude about how different people should be valued, and this involves questions of ethics. The fact that present people may value future people less than they value themselves should not for that reason alone be built into an economic analysis without further scrutiny, any more than the fact that people have racial or gender biases should be built into an economic analysis.

There are other, more subtle differences between Stern and Nordhaus. While both aggregate damages and work to identify the marginal social cost of carbon, Nordhaus does this in order to identify economically efficient mitigation strategies, while Stern is more interested in evaluating pathways that avoid unacceptable atmospheric concentrations of greenhouse gases while identifying trade-offs. Nordhaus is working towards a global benefit–cost analysis. He takes this to be an empirical exercise and seems confident about what a competent study can hope to achieve. Stern, on the other hand, views climate change as a risk management problem involving great uncertainties and diverse values, not all of which can be quantified. He thinks that the ethical dimensions of the problem are so central that we should not be very confident about what even the best economic study can hope to achieve.

One way of thinking about risk management is in terms of insurance. As Steve Schneider has said, 'we buy fire insurance for our house and health insurance for our bodies. We need planetary sustainability insurance.'[16]

However there are important disanalogies between investing in climate protection and purchasing insurance. First, insurance compensates for losses that are suffered; it does not directly mitigate or prevent losses. Fire insurance, for example, does not reduce the probability of fire occurring or diminish the damage that a fire would cause if it were to occur, but reducing greenhouse gas emissions is supposed to reduce the probability and severity of climate change damages. Second, we have no actuarial tables for the climate protection market in the way that we have for accidents and fires. We have very little idea about the specific impacts of climate change on societies like ours, living on planets like this, much less data about how specific changes in the composition of the atmosphere are likely to bring about specific impacts. Finally, insurance is typically purchased by an agent to benefit herself or, in some cases, those whom she loves or to whom she feels responsible. But in this case, we would be asking people who are now living very well, who under many scenarios have adequate resources for adaptation, to buy insurance that will mainly benefit poor people who will live in the future in some other country; and to do this primarily on the basis of predictions about the future based on climate models, expert reports, and so on. Rich people for the most part do not love or feel responsible for their poor contemporaries, especially those who live across national boundaries, much less those who will live in the future.

There are other reasons for doubting that the case for responding aggressively to climate change can be made simply on prudential grounds. This approach views the human community as a single agent, and compares the aggregate costs and benefits of various policies. Human communities are diverse, involving individuals with different interests, and are not (in the economists' sense) perfectly rational, or even in many cases aspiring to be so. Any climate change will have distributional effects, and the model of humanity as a single agent, presupposed by the prudential perspective, cannot adequately reflect such distributional conflicts.[17]

6 MORALITY AND JUSTICE

Distributional concerns are the terrain of ethical concepts, and the idea that climate change is fundamentally a matter of ethics has been gaining traction in recent years. Nobel Peace Prize co-recipients Al Gore and R. K. Pachauri have both endorsed it.[18] One version of this view is that climate change is a matter of individual moral responsibility; another version is that it presents questions of justice between states.

The idea that the problem of climate change is fundamentally a matter of individual moral responsibility is inspired by the insight that at its core the problem is that some people are appropriating more than their share of a global public good and harming other people by causally contributing to extreme climatic events such as droughts, hurricanes, and heat waves. Moreover, much of this behavior is unnecessary, even for maintaining the profligate lifestyles of the global rich. Though this view is plausible, once we begin to model climate change on more familiar cases of individual moral responsibility, significant differences begin to emerge.

A paradigm case of individual moral responsibility is one in which an individual acting intentionally harms another individual; both the individuals and the harm are identifiable; and the individuals and the harm are closely related in time and space.[19] Consider Example 1, the case of Jack intentionally stealing Jill's bicycle.[20] The individual acting intentionally has harmed another individual, the individuals and the harm are clearly identifiable, and they are closely related in time and space. If we vary the case on any of these dimensions, we may still see the case as posing a moral problem, but its claim to be a paradigm moral problem weakens. Consider some further examples. In Example 2, Jack is part of an unacquainted group of strangers, each of which, acting independently, takes one part of Jill's bike, resulting in the bike's disappearance. In Example 3, Jack takes one part from each of a large number of bikes, one of which belongs to Jill. In Example 4, Jack and Jill live on different continents, and the loss of Jill's bike is the consequence of a causal chain that begins with Jack ordering a used bike at a shop. In Example 5, Jack lives many centuries before Jill, and consumes materials that are essential to bike manufacturing; as a result, it will not be possible for Jill to have a bicycle. While it may still seem that moral considerations are at stake in each of these cases, this is less clear than in Example 1, the paradigm case with which we began. The view that morality is involved is weaker still, perhaps disappearing altogether for some people, if we vary the case on all these dimensions at once. Consider Example 6: acting independently, Jack and a large number of unacquainted people set in motion a chain of events that causes a large number of future people who will live in another part of the world from ever having bikes. For some people the perception persists that this case poses a moral problem. This is because it may be thought that the core of what constitutes a moral problem remains: Some people have acted in a way that harms other people. However, most of what typically accompanies this core has disappeared. In this case it is difficult to identify the agents and the victims or the causal nexus that obtains between them; thus, it is difficult for the network of moral concepts (for example, responsibility, blame, and so forth) to gain traction.

These 'thought experiments' help to explain why many people do not see climate change as a moral problem. For climate change is not a matter of a clearly identifiable individual acting intentionally so as to inflict an identifiable harm on another identifiable individual, closely related in time and space. Structurally, climate change is most analogous to Example 6: A diffuse group of people is now setting in motion forces that will harm a diffuse group of future people.

There is a deeper problem about whether contributing to climate change is a matter of individual moral responsibility. The paradigm that I have been discussing views the causation of harm as being at the center of what makes an act a matter of moral concern. Even if harm causation is neither necessary nor sufficient for an act or omission to be of moral concern, the view that some such connection exists has been very influential in modern moral philosophy.[21] However, recent work in social psychology suggests that when it comes to construing an act or omission as within the domain of morality, other considerations are just as important to people as harm causation. Jonathan Haidt and his colleagues have claimed that considerations involving fairness and reciprocity, in-group and loyalty, authority and respect, and purity and sanctity are, in addition to considerations about the causation of harm, at the foundation of morality as conceived by most people.[22] Since these considerations can come apart, often people will deny that harm-causing activity is within the moral domain, while at the same time considering behavior that does not cause harm to be of moral import. Daniel Gilbert brings these considerations to bear on the question of climate change when he writes that

> global warming doesn't . . . violate our moral sensibilities. It doesn't cause our blood to boil (at least not figuratively) because it doesn't force us to entertain thoughts that we find indecent, impious or repulsive. When people feel insulted or disgusted, they generally do something about it, such as whacking each other over the head, or voting. Moral emotions are the brain's call to action. Although all human societies have moral rules about food and sex, none has a moral rule about atmospheric chemistry. And so we are outraged about every breach of protocol except Kyoto. Yes, global warming is bad, but it doesn't make us feel nauseated or angry or disgraced, and thus we don't feel compelled to rail against it as we do against other momentous threats to our species, such as flag burning. The fact is that if climate change were caused by gay sex, or by the practice of eating kittens, millions of protesters would be massing in the streets.[23]

Rather than being a matter of individual moral responsibility, climate change can be seen as presenting a problem of global justice. Ugandan President Yoweri Museveni has been quoted as saying that climate change is 'an act of aggression by the rich against the poor.'[24] The data seem to bear him out. The rich countries of the North do most of the emitting, but the poor countries of the South do most of the dying (Patz et al. 2005).

When we look at some countries in particular the case becomes more vivid. A recent paper suggests that climate change will lead to a 1 meter change in sea level by the end of the century (Grinsted et al. 2010). Such a sea level rise will flood one-third of Bangladesh's coastline and create an additional 20 million environmental refugees. Saline water will intrude even further inland, fouling water supplies and crops, and harming livestock. This will occur as cyclones and other natural disasters become more frequent and perhaps more intense. In order to begin to adapt to climate change by building embankments, cyclone shelters, roads, and other infrastructure, it is estimated that four billion dollars would be required. Yet Bangladesh's total national budget in 2007 was less than $10 billion. Bangladesh suffers in all these ways, yet its carbon dioxide emissions per capita are one-twentieth of the global average. Such facts seem to lead to the conclusion that climate change poses questions of global justice.

However, there are complications. Since the atmosphere does not attend to national boundaries and a molecule of carbon has the same effect on climate wherever it is emitted, climate change is largely caused by rich people, wherever they live, and suffered by poor people, wherever they live. A recent study suggests that global carbon emissions can be reduced 50 percent by 2030, simply by reducing the emissions of the richest one-sixth of the people in the world (Chakravarty et al. 2009). These high emitters are roughly distributed equally in four regions: the US, the OECD minus the US, China, and the non-OECD minus China. On this view, there is as much emissions reduction to be done among high-emitting Chinese as there is among high-emitting Americans, and more emissions reduction to be done in both of these countries than in the European Union.[25]

Moreover, since poor people will suffer most from climate change, wherever they live, it is plausible to suppose that they too are distributed around the globe. The societal factors that caused Hurricane Katrina to be so devastating in New Orleans—high levels of inequality, large populations living in poverty, poor public services, and so on—will lead to similar consequences in the future. Indeed, there is reason to suppose that poor people in the United States will suffer more from climate change than similarly situated people in a country such as Cuba, which has less inequality and a more effective public sector in responding to climate and weather-related disasters (Mas Bermejo 2006).

Just as the problem of climate change has some of the dimensions of problems of morality but strays from the paradigm, so too with justice. In several important respects, causing climate change is not like one country unjustly invading another country. The nation-state is one level of social organization that is relevant to addressing climate change because it is casually efficacious, but the nation-state is not the primary bearer or beneficiary of ethical responsibilities in this regard.

7 POLITICS AND GOVERNANCE

Another perspective on climate change emphasizes the ways and the extent in which it challenges our systems of governance. Since the end of the Second World War, humans have attained a kind of power that is unprecedented in history. While in the past entire peoples could be destroyed, now all people are vulnerable. While once particular human societies had the power to upset the natural processes that made their lives and cultures possible, now people have the power to alter the fundamental global conditions that permitted human life to evolve and that continue to sustain it. There is little reason to suppose that our systems of governance are up to the task of managing such threats (Speth and Hass 2006; Adger and Jordan 2009; Hulme 2009: ch. 9).

Thus far, the most systematic attempts at climate governance have been through the international system, taking nation-states as primary agents.[26] The crowning achievement in climate governance is the Framework Convention on Climate Change (FCCC), opened for signature at the Rio Earth Summit in 1992, and now ratified by 192 countries. The parties to the FCCC committed themselves to stabilizing 'greenhouse gas concentrations in the atmosphere at a level that would prevent dangerous anthropogenic interference with the climate system.'[27] The Kyoto Protocol provided mechanisms for beginning to implement this commitment. The Kyoto Protocol was opened for signature in December 1997, and has been ratified by 187 countries, but not by the United States, the second largest emitter among nations. While its modest targets are likely to be met, it is unclear to what extent this will be due to the Protocol.[28] The December 2009 Copenhagen Climate Conference was supposed to result in a successor agreement to the Kyoto Protocol, but it ended in disarray and confusion. While some remain confident that this approach to climate governance will continue and bear fruit, many others are skeptical.

There are many specific problems with the existing governance structure (e.g. the requirement for consensus, the crude division between developed (Annex I) and developing (non-Annex I) countries, etc.). However, the heart of the problem is that climate change has many of the properties of being the world's largest collective action problem, and it is difficult for any country that is responsive to its citizens to do its part in securing the global public good of climate stability. In part, this is because of self-interest. People as individuals want climate to be stabilized, but they also want to benefit from their own greenhouse gas emissions while others reduce their emissions. High-emitting rich countries do not want developing countries to follow in their footsteps, but developing countries want rich countries to take the first steps in reducing emissions. Even among the rich countries there is a 'you first, then me' attitude. To a great extent, this behavior simply follows from the logic of a collective action problem: for each of us, defection dominates cooperation, however others act.

Climate change also poses an intergenerational collective action problem (Gardiner 2003). Since every generation benefits from its own emissions but the costs are deferred to future generations, they have an incentive not to control their emissions. Moreover, since each generation (except the first) suffers from the emissions of previous generations, benefiting from their own present emissions may even appear to be just compensation for what they have suffered. But of course, this reasoning leads to the continuous build-up of greenhouse gases in the atmosphere over time.

Indeed, these problems are even worse than they seem, for climate change does not involve just single, intra- and intergenerational collective action problems. Jurisdictional boundaries and competing scales cause multiple, overlapping, and hierarchically embedded collective action problems. A vast variety of behaviors by individuals, nations, and other entities affect climate, but they are governed by an equally vast array of different regimes with different mandates and even in many cases different parties. For example, decisions about trade and intellectual property affect greenhouse gas emissions, but each of these areas is governed by its own legal regimes. While this may seem abstract, we witness policy failures and dysfunctions driven by the same dynamics on a daily basis with respect to simpler problems. When a city provides services for residents who live in outlying areas and do not pay city taxes, this is an example of the sort of problem that occurs with respect to climate change.

Well-functioning democracies act in the interests of the governed rather than on behalf of all those whose interests are affected. The benefits from the activities that cause climate change primarily accrue to those who are members of particular political communities, while the costs are primarily borne by those who are not. In the case of climate change, costs are borne by those who live beyond the borders of the major emitters, future generations, animals, and nature. Perhaps surprisingly, this seems relatively well understood by the American public, as the Figure 3.1 indicates.[29]

One reason it is difficult to reform and restructure governance is because people and institutions have strong status quo biases (Samuelson and Zeckhauser 1988). The United States, which has been particular unresponsive with respect to climate change, has a political system designed to strengthen this bias. Some of these elements are constitutional: for example, the division between the three branches of government, and the two, independent, legislative branches. Practices have also developed that strengthen status quo biases, such as the Senate filibuster, and the system of campaign finance.

One way of solving or softening these collective action problems, even when institutional mechanisms are not available, is through love, sympathy, and empathy. However, these seem in short supply in diverse, fragmented, modern societies; and in their more systematized forms as ethical systems and principles of justice, they are not fully adequate responses to the problem of climate change, as we discussed in the previous section.

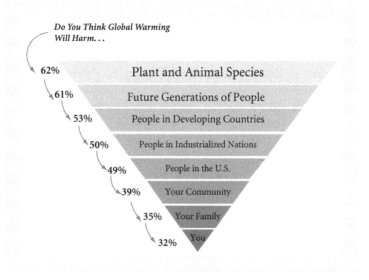

Do You Think Global Warming Will Harm...

62% Plant and Animal Species

61% Future Generations of People

53% People in Developing Countries

50% People in Industrialized Nations

49% People in the U.S.

39% Your Community

35% Your Family

32% You

FIGURE 3.1 Climate change in the American mind

8 COGNITION AND COMPLEXITY

In the background of our attempts to address climate change is the fact that evolution did not design us to solve or even to recognize this kind of problem. We have a strong bias towards dramatic movements of middle-sized objects that can be visually perceived, and climate change does not typically present itself in this way.

The onset of climate change is gradual and uncertain rather than immediate and obvious. Increments of climate change are usually barely noticeable, and even less so because we re-norm our expectations to recent experiences. Some have suggested that the strong reaction to the severe winter of 2009–10 in Eurasia and the United States can be partly explained by the fact that as the world warms, people lose their memory of cold winters. Bizarrely enough, against the background of a warming world, a winter that would not have been seen as anomalous in the past is viewed as unusually cold, thus as evidence that a warming is not occurring. In fact, regional data from a single season is not the sort of evidence that could overturn a climatological theory like global warming.[30] Global warming does not mean that every region will become warmer, nor does it mean that every day in a warmer world will be warmer than a comparable day at present. Schneider (1989) explains this with a gambling metaphor: global warming loads the dice in favor of increased temperatures, changes in precipitation, and extreme climatic events, but it doesn't determine the outcome. A global warming increases the probability for particular regions to be affected by these changes, but does not necessarily bring about such changes in every season of every year in every region. The basic problem here is that climate change is a technical, complex issue that is best represented probabilistically. Many people, probably most people, are not scientifically equipped to understand more than the rudiments of the problem, and all of us are bad at probabilistic thinking, at least when we are thinking intuitively.

Another feature of climate change that makes it difficult for us to respond is that its causes and effects are geographically and temporally unbounded. Earth system scientists study the earth holistically and think on millennial timescales and beyond, but this perspective is foreign to most people. Most of us pay little attention to events that occur beyond national boundaries, unless they are 'one-off' disasters. The idea that turning up my thermostat in New York will affect people living in Malaysia in a thousand years is virtually beyond comprehension to most of us.[31]

Climate change will have multiple, sometimes paradoxical, indirect effects, and many of its impacts on human welfare will be relatively invisible. Effects of climate change will include sea level rises and increased frequency of droughts, storms, and extreme temperatures. In some regions these effects may also include an increased frequency of cold days. In addition to these first-order impacts, climate change will have indirect, second-order impacts such as species extinctions and changes in agricultural patterns, as well as third-order impacts affecting social and political relationships, and human and national security. Many people will be killed or harmed by first-order effects but many more will be affected by the second- and third-order effects that are mediated by economic status, food availability, disease burdens, and so on. However, many of these effects will be relatively invisible since they will involve 'statistical' rather than 'identifiable' lives (Schelling 1968). Climate change will cause the deaths of many people, but there will be no obituary that will say that Dale Jamieson (for example) died yesterday, cause of death: climate change. While we can be very responsive to individual victims we have difficulty empathizing with statistical victims. We mobilize huge resources around highly publicized cases of little girls falling into wells, while we do comparatively little to save children when they are the invisible victims of policy choices.

9 CONCLUDING REMARKS

The juxtaposition of climate and humanity provides fertile ground for a diversity of interpretations and perspectives. Seeing climate as changing and construing it as a problem requires assumptions that are not shared by everyone. Even among those who agree that climate change is a problem there are serious differences about solutions, reflecting in part disagreements about causes. If the problem is fundamentally one of global governance, then new agreements and institutions are what are needed. If the problem is market failure, then carbon taxes or a cap and trade system is what is required. If the problem is primarily a technological failure, then we need an Apollo program for clean energy or perhaps geoengineering. If climate change is just the latest way for the global rich to exploit the global poor, then the time has come for a global struggle for justice. If there was ever a problem to which Kaplan's 'law of the instrument' applied, this is it.[32]

My view, is that climate change is an unprecedented problem that exhibits some dimensions of familiar problems, but in novel combination, and some new features as well. Thus climate change poses fundamental challenges to our existing systems of value, ways of knowing, and institutions of governance. What we can hope for is not that 'the' problem of climate change will be solved, but that we will learn from our failures to prevent or seriously mitigate climate change thus far, and go on to manage the change that is under way, and rebuild our institutions and ways of life so that they are appropriate to the high-population, high-consumption, high-technology world in which we now live.

NOTES

1. This can be seen on various climate contrarian websites; for a somewhat more sophisticated version, see Richard Lindzen, 'Resisting Climate Hysteria,' available at <http://www.quadrant.org.au/blogs/doomed-planet/2009/07/resisting-climate-hysteria>.
2. Indeed, among the interpretations is the fact that the IPCC understands climate change as change in the climate system rather than as change in climate (thought of as an abstraction from weather); see Hajer and Versteeg in this volume for discussion. For discussion of how expectations of change and stability affect experience see Weber (2010).
3. This is a Christian flourish on a Jewish story, but of course there are many different interpretations of this story throughout the Judeo-Christian-Islamic tradition.
4. <http://ourchoicethebook.com/chapter1>
5. These views are all over the web; as examples, visit the following sites: <http://www.globalwarminghoax.com/news.php>, <http://climatedepot.com/>. Some may think that the language of 'messiah' and 'Antichrist' is excessive, even in the service of a metaphor, but these concepts are very active at least in the political imagination of many people in the United States. Three US presidents and two vice-presidents have won the Nobel Peace Prize, and they have all been messianic figures in the eyes of some while reviled by others. According to a recent Harris survey 14 percent of Americans say that President Obama may be the Antichrist (<http://www.livescience.com/culture/obama-anti-christ-100325.html>); this is about half the number of those who strongly approve of his presidency (<http://www.rasmussenreports.com/public_content/politics/obama_administration/obama_approval_index_history>).
6. The validity of Michael Mann's 'hockey stick' figure of the temperature record is one of the central battlegrounds between those who believe in anthropogenic climate change and those who deny it. For an introduction to the controversy, visit <http://www.realclimate.org/index.php/archives/2005/02/dummies-guide-to-the-latest-hockey-stick-controversy/>.
7. At times the contrarian influence has been felt not just in government policy but also in the management of scientific information and the treatment of government scientists. James Hansen (2009) tells his own story, but see also Bowen 2008.
8. The growth of scientific knowledge can also give rise to additional uncertainty. In unpublished work Jessica O'Reilly discusses an example of this regarding estimates of the vulnerability of West Antarctic ice sheets to disintegration.
9. For first-person accounts of the experience of two influential scientist-advocates, see Schneider 2009 and Hansen 2009.
10. Many dedicated scientists have been damaged by their treatment in the media. A particularly egregious and well-documented case concerns the treatment of Ben Santer in the so-called 'Chapter 8' controversy, in which he was accused of secretly altering text in an IPCC report in order to exaggerate the case for anthropogenic climate change. For his account, see <http://www.realclimate.org/index.php/archives/2010/02/close-encounters-of-the-absurd-kind/>. For a scholarly treatments, see Lahsen 2005 and Oreskes and Conway 2010: ch. 6.
11. For a more nuanced discussion see Boykoff 2008. See also <http://www.realclimate.org/index.php/archives/2010/02/whatevergate/>

12. Exceptions are the Environmental Defense Fund and the Natural Resources Defense Council, which began working on this issue in the late 1980s, largely because of their scientific advisors (Michael Oppenheimer and Dan Lashof). Eventually, for reasons that cannot be explored here, virtually all American environmental groups began advocating for climate stabilization policies, but they have been under vigorous attack for how they have framed the issue (see Shellenberger and Nordhaus 2007).

13. For what may be a contrary view, see Becker 1991.

14. Of course, sometimes long-term predictions are more reliable than short-term predictions, at least at a very general level. For example, my prediction that every reader of this chapter will be dead by 2100 is much more reliable than any prediction I could make about the deaths of particular readers in the next twelve months. However, what is needed for a benefit–cost analysis that is helpful with respect to climate policy is detailed predictions that are reliable both in the short term, and throughout the century or even longer.

15. For a third view of the economics of climate change, see <http://realclimateeconomics. org/briefs.html>.

16. <http://www.pnas.org/content/102/44/15725.full>. Weitzman 2007 discusses the insurance analogy in his review of Stern. While Diamond 2005 does not explicitly employ this analogy, this book has become a popular *locus classicus* for the view that we have prudential reasons to be concerned about environmental degradation.

17. Appeals to environmental security are another version of a prudential approach that I cannot discuss here. For discussion, see Part V of this volume.

18. Gore claims this in his Academy Award winning film, *An Inconvenient Truth*; R. K. Pachauri indirectly makes this point in his Nobel Lecture (given on behalf of the IPCC, which actually won the prize), available at <http://nobelprize.org/nobel_prizes/peace/ laureates/2007/ipcc-lecture_en.html>. For websites devoted to climate justice, see <http://climateethics.org/>, <http://www.ecoequity.org/>. Schneider 2009 and Hansen 2009 also see climate change as posting questions of ethics and values but their treatments of such claims are not very detailed or sophisticated. For a collection of academic papers of this topic see Gardiner et al. 2010. See also Part VI of this volume.

19. I briefly discuss other paradigms of moral responsibility in Jamieson, 2010.

20. I first introduced this series of examples in Jamieson 2007.

21. The most thorough treatment of the normative significance of harm causation is Joel Feinberg's magisterial four-volume work (1984–8). Though criminal law is Feinberg's main concern, much of what he says applies to morality as well.

22. For an introduction to this work visit <http://faculty.virginia.edu/haidtlab/mft/index. php?t=home> 123.

23. <http://www.randomhouse.com/kvpa/gilbert/blog/200607its_the_end_of_the_world_as_we. html>

24. See *The Economist*, 10 May 2007, p. 123.

25. For somewhat different calculations that sustain the basic point that those responsible for changing the climate are well represented in countries throughout the world, see Jamieson 2010 and Grubler and Pachauri 2009.

26. For criticism of this approach, see Harris in this volume.

27. Framework Convention on Climate Change, Article 2, available at <http://unfccc.int/ essential_background/convention/background/items/1353.php>.

28. The former claim is due to the Netherlands Environmental Assessment Agency; visit <http://www.pbl.nl/en/dossiers/Climatechange/FAQs/index.html?vraag=6&title=Will%

20countries%20with%20an%20emission%20target%20meet%20their%20Kyoto%20target
%3F>. For a skeptical view, see the 'Hartwell Paper,' available at <http://www.lse.ac.uk/
collections/mackinderProgramme/theHartwellPaper/Default.htm>.

29. Figure 3.1 is based on figure 22 in a report from the Yale Project on Climate Change
and the George Mason University Center for Climate Change Communication,
available at <http://www.climatechangecommunication.org/images/files/Climate_Change_
in_the_American_Mind.pdf>. I regret that I have been unable to identify the person who
constructed it.

30. See James Hansen, Reto Ruedy, Makiko Sato, and Ken Lo 'If It's That Warm, How Come
It's So Darned Cold?,' available at <http://www.columbia.edu/~jeh1/Mailings/2010/
20100115_Temperature2009.pdf>. It is also noteworthy that the occurrence of an El
Niño is at least part of the explanation for the winter weather of 2009–10.

31. For evidence that the warming that has already occurred will affect the future of the planet
for more than a millennium, see Solomon et al. 2009.

32. 'Give a small boy a hammer, and he will find that everything he encounters needs
pounding' (Kaplan 1968: 28).

References

ADGER, W. N., and JORDAN, A. J., eds. 2009. *Governing Sustainability*. Cambridge: Cambridge
University Press.

AGRAWALA, S. 1998. Context and early origins of the Intergovernmental Panel on Climate
Change. *Climatic Change* 39: 605–20.

BECKER, G. 1991. *A Treatise on the Family*. Cambridge, MA: Harvard University Press.

BOWEN, M. 2008. *Censoring Science: Dr. James Hansen and the Truth of Global Warming*.
New York: Plume.

BOYKOFF, M. 2008. Lost in translation? United States television news coverage of anthropogenic climate change, 1995–2004. *Climatic Change* 86: 1–11.

CHAKRAVARTY, S., CHIKKATUR, A., DE CONINCK, H., PACALA, S., SOCOLOW, R., and TAVONI,
M. 2009. Sharing global CO_2 emission reductions among one billion high emitters.
Proceedings of the National Academy of Sciences of the United States of America 106:
11885–8.

DIAMOND, J. 2005. *Collapse*. New York: Penguin Books.

Environment Canada, WMO, and UNEP. 1988. *The Changing Atmosphere Conference Proceedings*, No. 710. Toronto: Environment Canada, WMO, and UNEP.

FEINBERG, J. 1984–8. *The Moral Limits of the Criminal Law*. New York: Oxford University
Press.

GARDINER, S. 2003. The pure intergenerational problem. *The Monist: Special Issue on Moral
Distance* 86: 481–500.

—— CANEY, S., JAMIESON, D., and SHUE, H. 2010. *Climate Ethics: Essential Readings*. New
York: Oxford University Press.

GRINSTED, A., MOORE, J. C., and JEVREJEVA, S. 2010. Reconstructing sea level from paleo and
projected temperatures 200 to 2100 AD. *Climate Dynamics* 34: 461–72.

GRUBLER, A., and PACHAURI, S. 2009. Problems with burden-sharing proposal among one billion high emitters. *Proceedings of the National Academy of Sciences of the United States of America* 106: E124.

HANSEN, J. 2009. *Storms of my Grandchildren: The Truth about the Coming Climate Catastrophe and our Last Chance to Save Humanity*. New York: Bloomsbury.

HOGGAN, J. 2009. *Climate Cover-up: The Crusade to Deny Global Warming*. Vancouver: Greystone Books.

HULME, M. 2009. *Why we Disagree about Climate Change: Understanding Controversy, Inaction and Opportunity*. Cambridge: Cambridge University Press.

JAMIESON, D. 1996. Scientific uncertainty and the political process. *Annals of the American Academy of Political and Social Science* 545: 35–43.

—— 2007. The moral and political challenges of climate change. Pp. 475–82 in L. DILLING and S. MOSER (eds.), *Creating a Climate for Change: Communicating Climate Change and Facilitating Social Change*. New York: Cambridge University Press.

—— 2010. Climate change, responsibility, and justice. *Science and Engineering Ethics*.

KAPLAN, A. 1964. *The Conduct of Inquiry: Methodology for Behavioral Science*. San Francisco: Chandler.

LAHSEN, M. 2005. Technocracy, democracy, and U.S. climate politics: The need for demarcations. *Science, Technology & Human Values* 30: 137–69.

LEGGETT, J. 1999. *The Carbon War: Global Warming and the End of the Oil Era*. New York: Routledge.

MAS BERMEJO, P. 2006. Preparation and response in case of natural disasters: Cuban programs and experience. *Journal of Public Health Policy* 27: 13–21.

National Academy of Sciences, Carbon Dioxide Assessment Committee. 1983. *Changing Climate*. Washington, DC: National Academy of Sciences.

NORDHAUS, W. 2008. *A Question of Balance*. New Haven: Yale University Press.

ORESKES, N., and CONWAY, E. 2010. *Merchants of Doubt: How a Handful of Scientists Obscured the Truth on Issues from Tobacco Smoke to Global Warming*. New York: Bloomsbury.

PATZ, J. A., CAMPBELL-LENDRUM, D., HOLLOWAY, T., and FOLEY, J. 2005. Impact of Regional Climate Change on Human Health. *Nature* 438: 310–17.

REVELLE, R., and SUESS, H. E. 1957. Carbon dioxide exchange between atmosphere and ocean and the question of an increase of atmospheric CO_2 during the past decades. *Tellus* 9: 18–27.

SAMUELSON, W., and ZECKHAUSER, R. J. 1988. Status quo bias in decision making. *Journal of Risk and Uncertainty* 1: 7–59.

SAREWITZ, D. 2004. How science makes environmental controversies worse. *Environmental Science & Policy* 7: 385–403.

SCHELLING, T. C. 1968. The life you save may be your own. Pp. 127–62 in S. B. Chase (ed.), *Problems in Public Expenditure Analysis*. Washington, DC: The Brookings Institution.

SCHNEIDER, S. H. 1989. *Global Warming: Are we Entering the Greenhouse Century?* San Francisco: Sierra Club Books.

—— 2009. *Science as a Contact Sport: Inside the Battle to save Earth's Climate*. Washington, DC: National Geographic Society.

SHELLENBERGER, M., and NORDHAUS, T. 2007. *Break through: From the Death of Environmentalism to the Politics of Possibility*. Boston: Houghton Mifflin.

SOLOMON S., PLATTNER G. K., KNUTTI R., FRIEDLINGSTEIN P. 2009. Irreversible climate change due to carbon dioxide emissions. *Proceedings of the National Academy of Sciences of the United States of America* 106: 1704–9.

SPETH, G., and HASS, P. 2006. *Global Environmental Governance*. Washington, DC: Island Press.

STERN, N. 2006. *The Economics of Climate Change: The Stern Review*. Cambridge: Cambridge University Press.

WEBER, E. U. 2010. What shapes perceptions of climate change? *Wiley Interdisciplinary Reviews: Climate Change* 1(3): 332–42.

WEITZMAN, M. 2007. A review of the Stern review on the economics of climate change. *Journal of Economic Literature* 45: 703–24.

CHAPTER 4

....

THE POVERTY OF CLIMATE ECONOMICS

....

MARK SAGOFF*

A *New Yorker* cartoon illustrates the intergenerational aspect of climate change. It shows an Eskimo mother, father, and young child as they wave a tearful goodbye to an old man, presumably a grandparent, whom they have placed on an ice floe. The family itself stands on a floating piece of ice. Which generation is responsible for the plight of which?

I want to argue that the intergeneration aspect of climate change makes economic reasoning about it more problematic than one might think. Economic reasoning looks for ways more efficiently to allocate or exchange property rights and—sometimes by determining those rights—to resolve collective action dilemmas. I shall argue that this kind of analysis cannot apply to our relations to people in the further future because we have no reason to trust them and, even if we did, they can do nothing for us. If ability to pay is a prerequisite of willingness to pay (WTP), moreover, then future generations cannot be willing—because they are not able—to pay us anything. Since they do not yet exist, they cannot have property rights. Even if they did exist, if they are made destitute by climate change, they will not be in a position to pay anyway. There can be no 'benefit of trade' or 'reciprocity of advantage' with them. I shall argue that the passivity of future generations undermines economic instruments to 'cap' and 'trade' greenhouse gas (GHG) allowances. The economic value of these allowances is more likely to reflect bets on the likelihood of enforcement than the marginal costs of 'clean' energy technologies.

This chapter will argue that an efficient allocation of resources, since it depends on exhausting the benefits of trade among the people who can trade (the living), cannot in any direct way respond to the needs or interests of future generations. An efficient policy therefore cannot be a sustainable policy. That economic theory is limited in this way suggests we must rely on other reasons and rationales to justify a response to climate change.

* The author gratefully acknowledges support from the National Science Foundation Grant No. SES 0924827. The views expressed are those of the author alone and not necessarily of any foundation or agency.

1 CLIMATE CHANGE IS NOT A COLLECTIVE ACTION PROBLEM

According to one commentator, 'Climate change is a collective action problem *par excellence*' (Harris 2007: 196). One can see the appeal of this analysis. In 1965, Mancur Olsen in *The Logic of Collective Action* showed that when each individual acts on self-interest, for example, to 'free ride' on the more socially motivated action of others, public goods will not be produced. Olson wrote, 'Unless the number of individuals in a group is quite small, or unless there is coercion or some other special device to make individuals act in their common interest, rational, self-interested individuals will not act to achieve their common or group interests' (1965: 2). In a much-cited article popularizing this analysis, Garrett Hardin argued that the rational proclivity of each individual to except him or herself from cooperation (to 'free ride' on the rest) made the destruction of public goods the likely result of liberty. 'The only kind of coercion I recommend is mutual coercion, mutually agreed upon by the majority of the people affected' to preserve or provide a public good (Hardin 1968: 1243–8).

A little reflection, however, suggests that the 'tragedy of the commons' analysis does not fit the problem of climate change. In the typical collective-action problem, such as managing a commons or preventing defections in a 'prisoner's dilemma' game, each person will gain if all cooperate and all will lose if each acts in his or her own individual self-interest. In the case of climate change, however, people alive today and through the next generation will sacrifice, for example, by forgoing the consumption of inexpensive fossil fuels. A different collection of people, whom one might call 'posterity,' will benefit. Olsen defines a 'group' as 'a number of individuals with a common interest.' It is not clear, however, that people alive today share a common interest with posterity. People alive today have a conflicting interest—not a common interest—with those who will inhabit the earth after we are all dead.

The coercion necessary to solve a collective action problem is justified by the mutual reciprocity of advantage, that is, the idea that each person gains more by the restriction of the freedom of others than he or she loses by accepting that same restriction. In the context of climate change, however, the winners and the losers are different—so different, in fact, that those who make sacrifices (or accept restrictions) may be long gone before posterity appears to enjoy the fruits of the sacrifices earlier generations had made for them. If one assumes—as I shall for the sake of argument—that those who make sacrifices to mitigate climate change will all be dead before those who benefit from their sacrifices appear, no relevant 'common interest' exists to establish a collective action dilemma. This hardly implies that we have no responsibility for the further future. The justification for sacrifice, however, would seem to lie in extraordinary altruism rather than in enlightened self-interest.

I am hardly the first to argue that climate change cannot be analyzed in terms of the logic of collective action. Stephen Gardiner, for example, has noted that climate change represents a lagging phenomenon 'because some of the basic mechanisms set in motion by the greenhouse effect—such as sea level rise—take a very long time to be fully realised.' Because of this, the mechanism of mutual coercion mutually agreed upon is unavailable. The obstacle to bargaining 'arises because the parties do not coexist, and so seem unable to even influence each other's behaviour through the creation of appropriate coercive institutions' (Gardiner 2006).

The irrelevance of the logic of collective action to climate change becomes even more apparent when we reflect that we have no way to control the behavior of future generations. Even if we conceive of them as partners in arranging a system of controls that greatly reduces GHGs, there is no way to bind them to whatever discipline we manage to exert over ourselves. We have to 'go first' in this game. They could undo what we had done.

2 An Objection

To this argument a reader might object that a collective action problem arises because we can act individually to provide our great-grandchildren some things, like trust funds, but not other things, like a stable climate. A collective action problem, however, arises only if two conditions are met: (1) individuals can achieve some common goal together that they cannot achieve alone; and (2) they gain more from the sacrifice of others than they lose from making the same sacrifice. To see the importance of the second condition, consider the example of light pollution. We would all like to be able to see the stars at night—but in cities we cannot because there is too much diffused light pollution from household illumination, streetlights, and headlamps. If I turn out my lights and wander about in the dark, I do little to restore the splendor of the stars. Since everyone is in the same situation, we might seem to have a collective action problem. To get to see the stars in all their glory, we all have to turn off our lights—car lights, street lights, and houselights. Is the collective gain worth the individual cost? People may not think it worth the candle—in other words, they might rationally believe that the collective benefit of a magnificent starscape is not worth the individual cost of stumbling around in the dark.

We may think about climate change the same way. We may share a fine sentiment that it would be nice, if all else were equal, that our great-great-grandchildren had the same 'option' to screw up the climate as we have today. Yet few of us may be willing even in common with others to make the necessary sacrifices to act on that shared sentiment. To show otherwise, one may need to provide evidence that political conditions have changed since 22 February 1977, when President Jimmy Carter, voted out of office after one term, in a cardigan sweater urged the nation to conserve energy. As of this writing, John Boehner (Republican, Ohio), who has publicly scoffed at the idea that carbon dioxide is a pollutant, is likely to become the Speaker of the House.

3 Climate Change is Not a Market Failure

In a much discussed report issued in 2006, British economist Nicholas Stern described climate change as 'a unique challenge for economics: it is the greatest and widest-ranging market failure ever seen.' William Nordhaus (2009) stated, 'Emissions of carbon dioxide are externalities, i.e, social consequences that are not accounted for in the market place.'

Jonathan Weiner (2003) described climate change as 'a classic market failure, an "externality" that ordinary market operations ... will not correct.' Law professor Jedediah Purdy (2008) has written, 'Climate change threatens to be, fairly literally, the externality that ate the world.'

Climate change represents an 'externality' if we mean any cost to any party not directly involved in a decision or activity that affects him or her. Since we cannot bargain with people in generations past, however, their effects on us—which are pervasive and ubiquitous—cannot be considered market failures although these effects are 'externalities' in an extended sense. Is an 'unpriced' harm to unborn or hypothetical people a market failure? A market fails when it does not implement gains that can be achieved through trade. A market that implements the gains that can be achieved through trade may ignore the interests of those who cannot trade. Individuals not yet born cannot trade. An efficient market, then, may ignore the interests of later generations.

The idea of market exchange or market activity—and therefore the model of market failure—depends on the existence of WTP to acquire goods and willingness to accept (WTA) compensation to relinquish them. An economist who asked respondents how much they would accept as payment from future generations to mitigate GHG emissions could not expect an answer. The only payments that can be made to current people must come from current sources—thus people would have to pay themselves. If WTP is the way we measure welfare effects, moreover, it is completely meaningless in the case of future generations.

In an influential analysis, Ronald Coase (1960) argued that the fundamental reason that externalities arise lies in the costs third parties would have to bear to enter or influence the relevant activities or decisions. Richard Zerbe and Howard McCurdy (2000) have written, 'The externalities on which market failure analysts tend to focus are defined by transactions costs. In essence, externalities exist because the transactions costs of resolving them are too high. In this sense, every story about externalities and market failures is also a story about transactions costs.'

What kind of story about transaction costs explains climate change as a market failure? In the usual case of pollution, one may refer to the costs of bargaining with or bringing a legal action against a polluter. With climate change the problem is that the victims do not yet exist and thus that the concept of a transaction cannot apply. If there cannot be transactions there cannot be associated costs, however large or even infinite. Because no one can tell a plausible story about market exchange or transaction costs, no one may be able to show that climate change is a market failure at least as Coase understood that concept. This may have more to do with metaphysics than with markets.

The essential problem is that future generations play a passive role in our decisions. We can affect them but they cannot affect us. They are epiphenomenal. Our relation to them is not a market relationship but an ethical, political, or spiritual one. A market analogy—the idea of market failure—provides a poor model for understanding climate change and might have the untoward effect of misleading us about the motives and reasons that justify GHG limits. Since future generations do not exist, they cannot pay for anything. Nor can they accept payment. Nor are they likely to be able and thus willing to pay us anything when they come into being because they will not be able to find us (whatever they pay in 'travel costs') or by that payment alter what we had already done. To invoke a market-based justification may be to defeat regulation.

4 PROPERTY RIGHTS

Consider the hoary Pigouvian example of the uncompensated damage done to a woods by sparks from railway engines. The operator of the railroad owns the right to use the woods in this way—as a catchment for engine sparks—as long as there is no legal rule against it and the wind blows the sparks this way or that. One may regret that things work this way; if one owns the forest one might sue the railroad to enjoin this right—but the right belongs to the railroad until it is taken (by a legal judgment, for example) and awarded to the owner of the forest.

Both the railroad owner and the forest owner have equal rights to uses of the forest but, because the land cannot in fact be used simultaneously to grow trees and catch sparks (let us assume), these rights conflict. The forest owner sues. Whether the courts impose an injunction, award damages, or just let the chips (or sparks) fall where they may is a question to be answered by judges and juries through the incremental wisdom of common law. One cannot show within economic theory that one party (the forester) really or originally or fundamentally 'owns' the right to its use of the forest, while the other creates an illegitimate 'externality.' As Coase argued, it takes two to create an 'externality'—the forest owner contributes by insisting on growing trees, for example, just where the railroad engine throws sparks (Coase 1960). Each side has an equal property right—the legal system or the government does not create the right asserted by the railroad owner or by the forester but decides which of these uses prevails when they conflict.

Now consider the analogy between the Pigouvian railroad and climate change. The GHGs we emit will harm future generations just as the sparks the railroad emits harm the forester. This tells us nothing, however, about property rights. These have to be adjudicated; they are not determined by God or found in nature. The relevant property rights belong to those who use them—emitters of GHGs included—until a judgment, rule, or settled expectation decides which right should prevail. The legal, social, and political institutions that typically adjudicate conflicts, however, are not good at recognizing future generations.

In liberal political theory, property rights to previously unowned aspects of nature have been associated with priority of acquisition and with the labor needed to develop wild lands and raw materials. The famous Lockean proviso that no one could acquire resources out of the commons except if there were as much and as good for others—or 'no prejudice to any others'—has often been repeated. For example, Michael Otsuka (2003: 24) has written, 'You may acquire previously unowned worldly resources if and only if you leave enough so that everyone else can acquire an equally advantageous share of unowned worldly resources.' Who is it that 'everyone else' includes? If 'everyone else' includes all future generations then there is a paradox. At some point at least in the distant future virtually any currently unowned resource may become scarce. If one had to assure an 'equally advantageous share' to future as well as contemporary potential claimants, no one might ever be able to take anything from the commons. But if one had to assure only the share of contemporary claimants, then possibly catastrophic acquisitions or alterations of the commons with long lag times or delayed effects may be permitted.

Consider an analogy. My neighbor at the top of the road maintains a magnificent garden which I enjoy each time I drive by it. I believe I have the right to gaze at her garden—a property right, if you will, in that use of it. What is wrong with that? Is it that I fail to compensate her for her efforts? That I have not reached an agreement with her beforehand? It is clear that she owns her garden from the point of view of the land and I own it from the point of view of the spectacle, at least when I view it. Since there is no conflict between her using the garden to putter and my using it to admire, we keep our rights and nothing is done to characterize or codify them.

Now suppose that scientists found out that when passers-by admire a garden they deplete it in some way so that gardeners a hundred years from now will have to plant different kinds of flowers to withstand further admiration or build high walls to keep admirers from seeing them. At that point we face a puzzle. How should we deal with the conflict between my admiring the garden now and the ability of my neighbor's great-grandchildren (if they still live in the house) to raise the same sort of flowers or to do so without building high walls? Do homeowners in the further future have a right to grow the kinds of gardens we grow now in the places we grow them? How do these future people claim or exercise those rights? How do these rights cancel my own—turn my enjoyment of the garden from a right into an uncompensated harm or an 'externality'? To be sure I have no right to harm my neighbor, but my enjoyment of her garden is harmless to her. I do not share the environment with future generations as I do with my neighbor. It is unclear how I share the environment with them.

Future generations may not have claims against those who acquire or privatize unowned resources even when this may be prejudicial to them. On this view, people could continue to emit GHGs right up to the time that the ill consequences weigh on them and their contemporaries. This seems to be the view of Hugo Grotius (1583–1645): '[H]e who is not yet born, can have no right, as that Substance which is not yet in Being has no Accidents. Wherefore if the People (from whose Will the Right of Government is derived) should think fit to alter that will, they cannot be conceived to injure those that are unborn, because they have not as yet obtained any Right' (Grotius 1625). According to Matthias Risse (2010), Grotius held that 'the domain of what is commonly owned simply is whatever is left to any given generation. It is up to each generation how much it leaves behind.' This seems as far as market-based notions take us. We must look to concepts of equity not efficiency to find a basis for dealing with climate change.

5 CAP AND TRADE

I have argued so far that climate change cannot usefully be diagnosed in terms of a collective action problem, a market failure problem, or a problem in defining and exchanging property rights. One can concede this and propose nevertheless that a market model should inform our response to the threat of climate change. Let us say that society decides to limit GHGs to an 'acceptable' level. One could contend that a market should be constructed to allocate efficiently the emissions that are allowed in order to minimize the pain to consumers.

So much has been written on the 'cap and trade' strategy of reducing GHG emissions that one hesitates to add another word. It may be worth noting nevertheless that the same deficiency—namely the passivity of future generations—that defeats the market failure model as a means of justifying a 'cap' also defeats the attempt to trade 'allowances' under that cap. Consumers today will see the cost of a 'cap' in their energy bills—but future generations will experience the benefits. The problem is that future generations, who are the beneficiaries of the 'cap,' can do nothing to defend it. In the absence of their political support, prices of 'allowances' may reflect not the marginal costs of GHG abatement but the bets speculators place on whether a regulatory regime will be sustained.

One could argue that companies that own 'grandfathered' rights and other credits or allowances would defend their value by resisting political efforts to inflate their number. On the contrary, the history of SO_2 markets in the US and carbon trading in Europe suggests that firms regard allowances as hedges against future regulation, so they tend to 'bank' the allowances they do not need rather than try to generate them for sale. Firms might also hedge allowances against price swings, 'short' them, or trade them on secondary and derivative markets in a cat's cradle of obscure transactions in which their original purpose may be lost.

'The experience with SO2 trading,' as Michael Hanemann (2010) has written, 'is regularly cited by proponents of emission trading as an important reason for applying this [method] to GHGs.' Advocates of cap-and-trade strategies often appeal to 'the SO_2 allowance trading system to control acid rain . . . and the subsequent success of that program: it reduced emissions faster than expected, at far lower costs' (Weiner 2008). The reasons that SO_2 emissions fell faster than expected and at lower costs, however, had little to do with the '"trading" mechanism inaugurated under Title IV of the 1990 Clean Air Act Amendments' (Hanemann 2010). The 'market' mechanism accomplished very little with respect to the decline in SO_2 emissions (Kreutzer 2006; Ellerman and Dubroeucq 2004).

Emissions fell over the last two decades chiefly because power plants, previously mandated to install scrubbers needed to protect the market for high-sulfur or 'dirty' coal, were allowed flexibility in the means they used to achieve regulatory limits (Ackerman and Hassler 1981). The cost of transporting low-sulfur from Wyoming's Powder River Basin fell by half as a result of railroad deregulation (Ellerman 2003). According to one study, 'flexibility to use low-sulfur coal was responsible for about 80 percent of the decline in marginal abatement costs, while technical change was responsible for about 20 percent' (Burtraw and Szambelan 2009). There is no evidence that R&D led to any patents or other signs of invention as a result of the SO_2 trading program (Taylor et al. 2005). Power plant operators relied on familiar and proven technology—not on innovation—to make mandated reductions. 'What emerges from the experience with Title IV is that [abatement] costs are lower for reasons beyond the ability to trade emission reductions' (Ellerman 2003).

The vast and profound differences between the control of SO_2 and the challenge of GHGs are often noted. The preponderance of SO_2 emissions in the US have come from just a few hundred companies that generate electricity and which were already used to and subjected to regulation. In the US only about 34 percent of GHG emissions come from electric utilities, however, with transportation and agriculture accounting for much of the rest (US EPA 2009). While sulfur dioxide is a single pollutant, GHGs comprise about twelve different gases, with many interelasticities among them. Acid rain is a relatively local or regional problem; climate change represents a global problem that requires a broadly enforceable international treaty. Reductions in SO_2 were achieved with existing capital stock and technology and without changes in consumer behavior; the health and environmental benefits were quickly realized. The control of GHGs will require revolutionary changes in capital stock, technology, and consumer behavior; the benefits involve long lag times and consist in catastrophes averted, not visible improvements from the status quo. The reductions of SO_2 achieved under Title IV did not result from the 'trading' mechanism and, even if they did, this would tell us nothing about its relevance to mitigating climate change.

What makes the SO_2 experience relevant, however, is the extent to which the prices of sulfur allowances blew like straws in winds of politics and litigation. If SO_2 markets teach any lesson, it is that the price of an allowance is likely to vary less with the marginal cost of abatement than with the uncertainty of enforcement. 'There is strong evidence of the close connection between regulatory uncertainty and structural shifts in allowance market prices,' Dallas Burtraw has written of market-based approaches to both SO_2 and nitrous oxide (NOx). 'Recently the SO_2 and NOx markets have been volatile and prices have fallen precipitously. These programs have been undermined by substantial regulatory uncertainty' (Burtraw and Szambelan 2009).

The short history of the European Union Emissions Trading Scheme (ETS), inaugurated in 2005, teaches the same lesson. For trading to affect behavior, firms that emit GHGs must not simply hold them to comply with regulation but generate them for sale by reducing emissions. They will do this only if they believe that allowances will have economic value. This economic value, in turn, will depend on how many allowances political authorities may create.

During the first phase of ETS trading (2005–7), the gross over-allocation of allowances caused prices to plunge; in 2007, they dropped effectively to zero (MacKenzie 2009). Early in the second phase (2008–12) of trading, prices have remained depressed owing to 'heavy selling and a realisation that the market was oversupplied with allowances' (Point Carbon 2010). The beneficiaries of regulation, generations in the further future, cannot enter the market to retire allowances no matter how much they would be worth to them. After Copenhagen and in view of the impasse in the US Congress, the price of an allowance cannot overcome skepticism (Point Carbon 2010).

The basic problem is that GHG allowances are created out of political whole cloth and allocated accordingly. There is an unlimited supply of this fabric. What would lead us to think that when prices rise politicians will stand up to populist majorities who want cheap energy? To imagine that political actors will hold the line is to assume that they are not economic actors; it is to assume that they will put the interests of future generations before their own reelections. It would be heartening to find examples of this phenomenon.

6 COMMAND AND CONTROL

In comparison to market-based or market-structured strategies, non-market or 'command-and-control' approaches for reducing GHG emissions look good. Making automobiles more fuel efficient, however, is not the answer. More efficient cars have the initial effect of decreasing demand for—and thus lowering the price of—petroleum. As a result more people will buy cars and drive them farther to take advantage of lower fuel prices. If you pay for a hybrid SUV, for example, you just enable me to drive another five miles on your nickel. The Obama administration has proposed that beginning in model year 2016, manufacturers of new cars will have to achieve a fleet average limit of 250 grams of carbon emitted per mile driven. If climate is the problem, reducing grams-per-mile could be better than increasing miles-per-gallon.

Similarly, the government could require that major producers of electricity provide power from sources—whatever they are—that on average produce half or less of the carbon emissions per kilowatt hour associated with burning coal. Electric utilities might achieve this average, for example, by switching to gas or investing in nuclear plants. One could expect companies themselves to discover a trading mechanism. Those firms that need to burn coal will—to achieve a better 'average'—buy interests in firms that produce electricity in other ways, thus subsidizing low-carbon production. There is much to be said for mandates for more efficient consumer technologies, but this could lead to more consumption. As people switch, for example, from incandescent to compact fluorescent bulbs, they may not bother to turn off their lights, since it is so cheap to leave them on. It seems churlish, however, to suppose that people will not care. Mandates for more efficient appliances overall make sense.

The American Recovery and Reinvestment Act of 2009 tried another promising strategy. It threw an awful lot of public money—largely borrowed from China—at the wall of technology in the hope that some of it would stick. Maybe it will. A carbon (or GHG) tax is widely discussed as an intervention into the market that would raise revenue for the government and therefore which politicians—who could then give their constituents more pork-barrel projects—may support over the long run. Governments can increase GHG allowances to infinity, thus giving emitters rights they can use to resist any meaningful regulation that might emerge. A tax, in contrast, usually goes up and cannot be reduced below zero. It is commonplace to say, moreover, that a carbon tax makes more sense than a payroll tax since it penalizes emissions, which we do not like, rather than work, which we want to encourage. A tax on carbon, like a payroll tax, represents a government intervention in the market and not a market-based mechanism or strategy. It can be classified, then, as command-and-control regulation.

7 CONCLUSION

In this chapter, I have argued that economic theory cannot provide a useful way—either a model, method, or metaphor—to think about climate change. Climate change is not a collective action problem because individuals today have no common interest that would lead them to believe they have more to gain overall as individuals from the restrictions placed on others than they have to lose from accepting those restrictions themselves.

I have also argued that climate change cannot be understood as a market failure since no market relationship could possibly engage present generations with those in the further future. Nor can the problem plausibly be understood in terms of the allocation or distribution of property rights since there is no theory on which to ascribe that kind of right to hypothetical individuals who lack existence, agency, and identity. Future generations could no more assert property rights against us than we might assert them against those who preceded us. Future generations are not agents in a relevant sense; it is unclear how they can exercise rights (Steiner 1983). The property rights they may possess will govern their relations only with each other not with us.

This chapter has also argued that cap-and-trade approaches must fail because governments can write allowances like scrip. Politicians might show some constraint if future generations made it worthwhile for them to do so, for example, by hosting fundraising events or helping to get out the vote. This is unlikely to happen. In order to get 'buy-in' to a mitigation market, political authorities must buy out interest groups. The buy-out might include everything from giving 'hot air' allowances to the Russians, to creating 'offset' opportunities for 'green' projects, to 'grandfathering' rights for BP and EXXON. This is more like a feeding frenzy than like a crash diet.

An analogy is often drawn with indulgences issued by papal and other authorities in medieval times (Goodin 1994). In the Middle Ages, indulgences forfeited their credibility in part because the pope lost his monopoly control of them. All kinds of authorities issued them. How will governments who contest everything else with one another convene on a single agency with power to create allowances that could be worth billions? Proponents of cap-and-trade approaches believe that political actors will agree on a way to limit the number of allowances they create and on a way to allocate them acceptable to all parties. This belief may represent the most heroic 'assumption of the can-opener' in the history of economic thought.

Climate change cannot be seen as a collective action problem, because people today seem to have more to lose individually than to gain collectively by instituting meaningful GHG reductions. I have argued that climate change cannot be understood as a market failure because no market is intergenerational in the relevant way. Economic theory is based in the concept of market exchange—but markets cannot enlist generations in the further future. No one analyzes the follies or favors of generations past as 'external' costs or benefits to us. Likewise, generations in the future will not regard our actions in that way. They may regard these actions as a moral failure, a cultural failure, a spiritual failure, or as a political failure but not as a market failure. There is no possible market that could have failed. One may question my assumption that the catastrophe will not kick in for at least eighty years—but it would be interesting if this empirical assumption were the principal shortcoming of my argument.

Future generations represent the constituency that is most affected by climate change and whose interests are the most served by policies intended to mitigate it. They do not vote. They do not trade. They do not exist. Any economic model that is consistent with its own normative presupposition—the idea of exhausting the benefits of trade—must conclude that they do not count. Insofar as a strategy to control climate change is based on ideas such as property, exchange, markets, WTP, efficiency, and the like, it may conceptually lead nowhere—except perhaps to despair—if applied to the problem of regulating GHGs in response to the threat of climate change. What one generation owes another—if we analogize future generations to separate populations that replace each other—is more reasonably analyzed in terms of moral attitudes ranging between indifference and altruism than in terms of economic concepts such as efficiency and exchange.

Economists believe that exchange defines the good—so that the concept of a perfectly competitive market provides a guide to social and resource policy. A perfectly competitive market, however, cannot include future generations because they do not yet exist. There are plenty of good reasons to care about future generations and to reduce GHGs but these reasons elude economic thought. An ethical view of the matter—one that invokes responsibility, decency, compassion, and justice—will go further than price theory to take the interests of future generations seriously. This is just the difference between ethics and economics.

References

ACKERMAN, B. A., and HASSLER, W. T. 1981. *Clean Coal/Dirty Air: Or How the Clean Air Act Became a Multibillion-Dollar Bail-out for High-Sulfur Coal Producers and What Should Be Done about it.* New Haven: Yale University Press.

BURTRAW, D., and SZAMBELAN, S. F. 2009. US emissions trading markets for SO_2 and NOx. *Resources for the Future*, 9–40.

COASE, R. 1960. The problem of social cost. *Journal of Law and Economics* 3: 1–44.

ELLERMAN, A. D. 2003. Ex post evaluation of tradable permits: the US SO_2 cap-and-trade program. Working Paper CEEPR 03–003, Massachusetts Institute of Technology.

—— and DUBROEUCQ, F. 2004. The sources of emission reductions: Evidence from US SO_2 emissions from 1985 through 2002. Working Paper CEEPR 04–001, Massachusetts Institute of Technology.

GARDINER, S. 2006. A perfect moral storm: climate change, intergenerational ethics and the problem of moral corruption. *Environmental Values* 15: 397–413.

GOODIN, R. 1994. Selling environmental indulgences. *Kyklos* 47(4): 573–96.

GROTIUS, H. 1625. *De jure belli ac pacis* (On the law of war and peace). P. II.4.X.2. in R. Tuck (ed.), *The Rights of War and Peace*. Liberty Fund (2005).

HANEMANN, M. 2010. Cap-and-trade: a sufficient or necessary condition for emission reduction? *Oxford Review of Economic Policy* 26: 225–52.

HARDIN, G. 1968. The tragedy of the commons. *Science* 162: 1243–8.

HARRIS, P. G. 2007. Collective action on climate change: The logic of regime failure. *Natural Resources Journal* 47: 195–224.

HILLEL, S. 1983. The rights of future generations. Pp. 151–65 in D. Maclean and P. Brown (eds.), *Energy and the Future*. Totowa: Rowman & Littlefield.

KREUTZER, J. 2006. Cap and trade: a behavioral analysis of the sulfur dioxide emissions market. *New York University Annual Survey of American Law* 62: 125–62.

MACKENZIE, D. 2009. Making things the same: Gases, emission rights and the politics of carbon markets. *Accounting, Organizations and Society* 34(3–4): 440–55.

NORDHAUS, W. D. 2009. Economic issues in designing a global agreement on global warming. Keynote Address. Prepared for: *Climate Change: Global Risks, Challenges, and Decisions*, 10–12 March 2009, Copenhagen, Denmark. Available at: <http://nordhaus.econ.yale.edu/documents/Copenhagen_052909.pdf> [accessed 6 August 2010].

OLSEN, M. 1965. *Logic of Collective Action*. Cambridge, MA: Harvard University Press.

OTSUKA, M. 2003. *Libertarianism without Inequality*. Oxford: Oxford University Press.

Point Carbon. 2010. Carbon 2010: Return of the sovereign. In: *Carbon Market Insights 2010*, 2–4 March, 2010, Amsterdam, Netherlands. Available at: <http://www.pointcarbon.com/polo-poly_fs/1.1420234!Carbon%202010.pdf> [accessed 6 August 2010].

PURDY, J. 2008. Climate change and the limits of the possible. *Duke Environmental Law & Policy Forum* 18 (Spring).

RISSE, M. In press. The grounds of justice: An inquiry about the state in global perspective. Available at: <http://habermas-rawls.blogspot.com/2010/04/forthcoming-book-by-mathias-risse.html> [accessed 6 August 2010].

STEINER, H. 1983. The Rights of future generations. In D. MacLean and P. G. Brown (eds.), *Energy and the Future*. Maryland Studies in Public Philosophy. Totowa, NJ: Rowman & Littlefield.

STERN, N. 2006. *The Stern Review: The Economics of Climate Change, Executive Summary* [Online]. Available at: <http://www.hm-treasury.gov.uk/media/4/3/executive_summary.pdf> [accessed 5 August 2010].

TAYLOR, M. R., RUBIN, E. L., and HOUNSHELL, D. A. 2005. Regulation as the mother of innovation: The case of SO_2 control. *Law & Policy* 27(2): 348–78.

US Environmental Protection Agency. 2009. *Inventory of US Greenhouse Gas Emissions and Sinks: 1990–2007*, EPA 430-R-09-004, Washington, DC.

WEINER, J. B., and STEWART, R. B. 2003. Practical climate change policy. *Issues in Science and Technology* 20 (Winter): 71–8.

WIENER, J. 2008. Radiative forcing: Climate policy to break the logjam in environmental law. *NYU Environmental Law Journal* 17.

ZERBE, R., and McCURDY, H. 2000. The end of market failure. *Regulation* 23(2): 10–14.

CHAPTER 5

THE DEVELOPMENT OF THE CONCEPT OF DANGEROUS ANTHROPOGENIC CLIMATE CHANGE

SPENCER WEART

FROM ancient times people suspected that over the course of centuries human activity could change the climate of a territory. For example, Theophrastus, a pupil of Aristotle, told how the draining of marshes had made a particular locality more susceptible to freezing, and he speculated that lands became warmer when the clearing of forests exposed them to sunlight. Millennia later Count C.-F. Volney, traveling in the United States around 1800, was told by settlers everywhere from Kentucky to upstate New York that the local climate had grown warmer and milder promptly after the forests were cleared. In the late nineteenth century Americans and Europeans, farmers and scientists alike, debated whether cutting down forests brought more rainfall—or perhaps less.[1]

Meanwhile the discovery of ice ages in the distant past proved that climate could change all by itself, perhaps even globally, with momentous consequences for human life. It seemed obvious that such massive changes could only be caused by forces vastly beyond anything that humanity could command. The first to contradict this was the Swedish scientist Svante Arrhenius. In 1896 he published a calculation that humanity was burning coal in such prodigious quantities that the carbon dioxide gas (CO_2) added to the Earth's atmosphere could raise the planet's average temperature significantly over the next thousand years or so.[2] This was only one of many speculations about future climate change, and to most experts far from the most likely. They found plausible arguments to deny that human emissions could perturb the global climate system. Most people took it for granted that a 'balance of nature' would prevent any rapid and severe change in the planet's physical systems.

Through the first half of the twentieth century, the handful of scientists who thought about 'greenhouse effect' warming (as it came to be called) supposed that if it ever did happen it would be all for the good. In chilly Sweden, Arrhenius (1908: 63) claimed that the world 'may hope to enjoy ages with more equable and better climates.' He and others also noted that CO_2 might act as a fertilizer to improve crop yields. Whatever changes might result from human industry would be 'progress,' a beneficent advance. In any case nobody worried about the impacts of a climate change that, if it happened at all, would not become apparent for centuries.

1 A NEW CONCERN EMERGES
(1930–1969)

The idea of climate change began to worry some people in the 1930s. The 'Dust Bowl' drought in the United States resonated with the claims of some historians that dry spells lasting for centuries had laid low ancient Near Eastern civilizations. Such spells were seen as phases of a natural cycle with unknown causes. Only one lone engineer, G. S. Callendar, offered arguments that greenhouse warming was on the way. He expected it would be beneficial; looking centuries ahead, he predicted that crops would grow more abundantly, and 'the return of the deadly glaciers [of the ice ages] should be delayed indefinitely' (Callendar 1938: 236; Fleming 2007).

The advent of nuclear weapons in 1945 brought a sudden realization that humanity could be mastering forces capable of planetary effects. In the 1950s newspapers speculated about controlling climate with atom bombs, or with other means such as rain-making by spraying substances into the atmosphere. The new attitude was reflected by two scientists, Roger Revelle (Revelle and Seuss 1957) and Gilbert Plass (1956), who separately undertook studies that overturned the old arguments against Arrhenius's hypothesis. It was theoretically possible, after all, that CO_2 could build up in the atmosphere and bring warming. Painstaking measurements by Charles D. Keeling (1960, 1970) drove home the point in 1960 by showing that the level of the gas was in fact rising, year by year.

People had also started to realize that human population and industry were increasing exponentially: the rate of emissions to be expected by the end of the century would be an order of magnitude above the rate in Arrhenius's day. Revelle publicly speculated that in the twenty-first century the greenhouse effect might exert 'a violent effect on the earth's climate' (Time 1956). He thought the temperature rise might eventually melt the Greenland and Antarctic icecaps, raising sea level enough to flood coastlines. In 1957 he told a Congressional committee that the greenhouse effect might someday make the Arctic Ocean ice-free and turn Southern California and Texas into 'real deserts' (Revelle 1957: 104–6). This was obviously more science fiction speculation than reasoned scientific prediction. In any case the twenty-first century was so far away!

In 1963 a path-breaking meeting on 'Implications of Rising Carbon Dioxide Content of the Atmosphere' was convened by a private foundation (Conservation Foundation 1963). Participants in the meeting began to see greenhouse warming as something 'potentially dangerous' to the planetary environment. The group noted the risk of sea-level rise, but otherwise could scarcely say what dangers might appear.

Enough scientists became interested in global warming to catch the attention of the US President's Science Advisory Committee. In 1965 the group reported that 'By the year 2000 the increase in atmospheric CO_2 ... may be sufficient to produce measurable and perhaps marked changes in climate ... ' Without attempting to say anything specific, they remarked dryly that the resulting changes 'could be deleterious from the point of view of human beings' (President's Science Advisory Committee 1965: 126–7). Although such studies were scarcely reported by the press, the minority of citizens who followed science news could read about the prospects for blessings or harms from greenhouse warming, along with more traditional speculations about the next ice age. In 1966 a panel of the US National Academy of Sciences warned against 'dire predictions of drastic climatic changes.' The panel remarked that the geological record showed many swings of temperature, which 'had locally catastrophic effects, [but] they did not stop the steady evolution of civilization.' Like other scientists, the Academy panel saw no need for policy actions. What all experts did call for was a greater research effort to determine how real the problem might be.[3]

Meanwhile a few scientists were devising simple mathematical models of the climate. They turned up feedbacks that could make the system surprisingly vulnerable to severe change under slight perturbations. Others (e.g. Broecker 1966), studying ancient pollens and fossil shells, found that severe climate change had happened in the past within as little as a century or two. The sense of instability was reinforced by attempts to model the general circulation of the atmosphere with computers. A 1967 calculation (Manabe and Wetherald 1967) suggested that the average global temperature might rise a few degrees within the next century. The calculations were plainly speculative, however, and the next century still seemed far away.

2 Scientific Evidence Builds up (1970–1987)

In the early 1970s, the rise of environmentalism accompanied public doubts about the benefits of human activity for the planet. Smoke in city air and pesticides on farms were no longer tokens of 'progress' but threatened regional or even global harm. A feeling spread that modern technology brought not only practical but moral problems, polluting and mistreating the natural order. This harked back to mythical modes of thinking. Tribal peoples often attributed natural disasters, such as an unusually severe winter, to human misdeeds: someone had violated an incest taboo or the like. Just so was the Flood of Noah called down upon humanity by our sins. Chinese dynasties were shaken when people blamed devastating floods on the corruption of mandarins; European communities declared days of public penance as an answer to droughts. Climate change caused by human emissions could seem like a scientifically validated retribution for abusing Mother Nature.

One reflection of the growing misgivings was a landmark project to study 'Man's Impact on the Global Environment,' conducted at the Massachusetts Institute of Technology in 1970. Among other matters the scientists looked into greenhouse warming. They concluded it might bring 'widespread droughts, changes of the ocean level, and so forth,' but could not get beyond such vague worries (SCEP 1970: 18). The problem, to this point discussed mainly in the United States, was internationalized through a meeting of scientists in Stockholm the following year. Their conclusions were similar to those of earlier studies, but they added a warning about climate instability: if the Arctic Ocean's ice cover disappeared we might pass a point of no return, changing the world's weather in ways impossible to predict (Wilson and Matthews 1971).

Climate change became a matter of broad public concern in the early 1970s when savage droughts afflicted the American Midwest, devastated the Russian wheat crop, and brought starvation upon millions in Africa. Were these events entirely natural? Alongside the greenhouse effect, some scientists pointed to deforestation and overgrazing. A few warned that human activity was putting so much dust and smog into the atmosphere that it could block sunlight and cool the world (e.g. Bryson 1973, 1974). The mass media were confused, sometimes predicting a balmy globe with coastal areas flooded as the ice caps melted, sometimes warning of the prospect of a catastrophic new Ice Age.

Experts scoffed at the media exaggerations, but they were coming to believe that human activities could indeed affect the global climate. While insisting that the state of climate science was too primitive to make any predictions, the majority increasingly found global warming plausible (Peterson et al. 2008). An important advance came from a team led by Syukuro Manabe (Manabe et al. 1975), a sophisticated computer calculation that plausibly showed a temperature rise of several degrees for doubled CO_2, a level that would be reached sometime in the twenty-first century. Many began to call for serious study of the likely impacts. One example was a 1977 report on 'Energy and Climate' from a panel of geophysicists convened by the US National Academy of Sciences. The panel got fairly specific. On the positive side, the Arctic Ocean might eventually be opened to shipping. On the negative side, there would be 'significant effects in the geographic extent and location of important commercial fisheries . . . marine ecosystems might be seriously disrupted.' Stresses on the polar ice caps might lead to a surge of ice into the sea, bringing a 'rise in sea level of about 4 meters within 300 years.' As for agriculture, there would be 'far-reaching consequences,' possibly 'human disasters' like the recent African droughts. However, the panel made clear they could not foresee what would actually happen. They concluded vaguely that 'world society could probably adjust itself, given sufficient time and a sufficient degree of international cooperation. But over shorter times, the effects might be adverse, perhaps even catastrophic' (National Academy of Sciences 1977: 8–14). Two years later another Academy panel said much the same, and took brief note of an additional threat— the rise of CO_2 in the atmosphere would make the oceans more acidic. Here too they found the consequences beyond guessing (National Academy of Sciences 1979: 24–7).

Up to this point the United States had dominated climate science, as it dominated most research in the first decades following the Second World War. But now that other nations had recovered from the war's devastation, international discussions began to dominate the discourse. The driving force, as one observer remarked, was 'a small group of "entrepreneurs," who promoted what they viewed as global rather than national interests.' They organized a series of quasi-official international meetings that were increasingly influential (Bodansky 1997: 4.1.6). The first landmark was the 1979 World Climate Conference, which assembled more than 300 experts from more than 50 countries in Geneva. Their views were diverse, and they managed to reach a consensus only that there was a 'serious concern that the continued expansion of man's activities'—including in particular emissions of CO_2— 'may cause significant extended regional and even global changes of climate.' Effects might become visible by the end of the century, so governments should start preparing for 'significant social and technological readjustments' (WMO 1979: 1–2).

As research on the impacts of warming advanced, more details emerged. For example, an elaborate 1983 study of sea-level rise by the US Environmental Protection Agency concluded that by the end of the twenty-first century the world 'could confidently expect major coastal impacts, including shoreline retreat . . . flooding, saltwater intrusion, and various economic effects' (Hoffman et al. 1983). A 1983 US Academy report covered now-familiar categories like effects on agriculture, but went on to point out, for example, that global warming would worsen the 'excess human death and illness' that came with summer heat waves. Also, climate shifts 'may change the habitats of disease vectors.' Finally and most important, 'In our calm assessments we may be overlooking things that should alarm us.' For there might be effects that nobody had imagined, effects all the more dangerous because they would take the world by surprise. As usual the Academy did not recommend any policy initiatives, aside from the customary plea for more research (National Academy of Sciences 1983: 45, 50, 53).

In 1985 experts from twenty-nine countries both rich and poor gathered for intense discussions in Villach, Austria. They reviewed the research that had accumulated in recent years. There was a discovery from ice cores drilled in Greenland that CO_2 and temperature levels had risen together within mere decades at the end of the last ice age. There was an impressive recent rise of the actual global temperature, to a level above anything in the historical record. Moreover, a group led by Veerabhadran Ramanathan (Ramanathan et al. 1985) had shown that methane and other trace greenhouse gases were rising rapidly and could have as great an effect as CO_2 itself. The climate changes that had been predicted for a century in the future would in fact come on twice as fast—within the Villach experts' own lifetimes. Pointing out that 'the rate and degree of future warming could be profoundly affected by governmental policies,' the conference's concluding statement called on governments to consider taking action—such as a 'global convention' to prevent too much global warming (Bolin et al. 1986: x–xii; Long and Iles 1997). Climate science was no longer just a matter for scientists.

3 PRESSURE FOR POLICY ACTION (1988–2001)

A few scientists tried to reach policy makers and the public not just by issuing reports, but directly. The press began to pay attention in the summer of 1988, as the United States suffered under severe heat and drought. Television news reported how the NASA scientist James Hansen told a Congressional committee he felt quite sure that 'the greenhouse effect is here.' Other scientists were more cautious, continuing to predict that unambiguous evidence would emerge only around the end of the century. Nevertheless reporters unexpectedly swarmed to the 'World Conference on the Changing Atmosphere' that convened in Toronto. This was yet another meeting by invitation of scientist experts—not official government representatives, who would have had a much harder time reaching a consensus. The Toronto Conference's report concluded that changes in the atmosphere 'represent a major threat to international security.' For the first time, a group of prestigious scientists called on the world's governments to set strict, specific targets for reducing greenhouse gas emissions. By 2005, said the experts, the world should push its emissions some 20 percent below the 1988 level.[4]

For the first time, substantial majorities of adults in developed countries told poll-takers that they had heard about the greenhouse effect, and many were at least somewhat worried about it. Politicians took note. In the US Congress and other bodies, hearings were held and new policies were proposed. The first major world leader to take a determined position was the United Kingdom's Prime Minister Margaret Thatcher, who in a September 1988 speech described global warming as a key issue. Attention from the politically powerful 'Greens' in Germany and elsewhere in continental Europe added to the issue's legitimacy. One immediate consequence was a 1989 meeting in Hanover, Germany, where environmentalists from Europe and the United States discussed ways to work together. The result was the Climate Action Network, a loose coalition of non-governmental organizations; within two decades the network was coordinating strategy among more than 360 NGOs around the world. What had begun as a research puzzle had become a serious question of policy choices (Nolin 1999; Beuermann and Jäger 1996).

Effective steps to restrain global warming would have major economic and political consequences. As a problem of the 'commons,' where actions that seemed profitable to each individual separately could bring harm to all jointly, greenhouse gas emissions could only be restricted by governmental regulation such as mandates or taxation. That prospect aroused the implacable enmity of the fossil fuel industries and their industrial allies. They were joined by conservatives who abhorred government regulation and despised anything smacking of environmentalism. In the United States and to a much lesser extent elsewhere these groups backed up their opinions with large sums of money for advertising, lobbying, and reports that mimicked scientific publications. Much of the funding was channeled through the Global Climate Coalition, formed in 1989 to tell politicians and the public that climate science was far too uncertain to justify action (Gelbspan 2004; Mooney 2005; Goodell 2006: ch. 9). The publicists made much of the normal uncertainties of frontier research, trumpeting the occasional scientific paper that challenged one or another element of the emerging majority position. Media reporters satisfied their traditional demand for 'balance' (or at any rate for dramatic conflict) by giving nearly as much attention to a few insistent skeptics as they gave to the far more numerous and productive scientists who were growing increasingly worried (Boykoff and Boykoff 2004).

The Reagan administration in the United States and others did not welcome the policy pressures arising from the reports of international conferences, academy committees, and other self-elected bodies of scientists. Policymakers preferred to get advice from a more conservative and politically controllable body. Lengthy negotiations led to the creation of the Intergovernmental Panel on Climate Change (IPCC). The IPCC's final statements, unlike earlier reports of scientific panels, would be crafted by people who participated not only as science experts but as official representatives of their governments. All conclusions would have to be unanimous.

It looked like a plan for paralysis. But the scientists on the IPCC had long experience in working together to reach conclusions based on reasoning and evidence, and they found it easier than they had expected to reach a consensus. They debated research findings in correspondence and workshops, crafted draft reports, solicited reviews by virtually every climate expert in the world, and hammered out conclusions. However, the official report had to say not only what essentially all the experts agreed was scrupulously correct, but also what the representatives of all the governments found politically acceptable. The first IPCC Report, issued in 1990, managed to conclude only that the world had indeed been warming.

The scientists were not confident that the change was caused by the greenhouse effect, but they warned that if so, the world could well grow a few degrees hotter by the middle of the next century. The report only hinted at economically sound steps that the world might start taking to reduce the threat.[5]

The governments had established a cyclic international process in which the IPCC would compile research and, roughly twice a decade, issue a consensus statement. That would lay a foundation for international negotiations, which would guide individual national policies. The IPCC's first report thus led to the Second World Climate Conference, held in 1990. The conference issued a strong call for policy action, which induced the United Nations General Assembly to call for negotiations toward an international agreement that might restrain global warming. The result was a 1992 gathering of world leaders in Rio de Janeiro—the United Nations Conference on Environment and Development, dubbed the 'First Earth Summit.' The great majority of countries, led by the western Europeans, wanted mandatory limits on greenhouse gas emissions. But no agreement could work without the United States, where the administration of George H. W. Bush rejected any binding targets or timetables. Negotiators papered over disagreements to produce a compromise, formalized as the United Nations Framework Convention on Climate Change. The convention, signed by more than 150 states, solemnly promised to work toward 'stabilisation of greenhouse gas concentrations in the atmosphere at a level that would prevent dangerous anthropogenic interference with the climate system' (Mintzer and Leonard 1994; Hecht and Tirpak 1995).

Few governments followed up with anything more than inexpensive plans to improve energy efficiency. No politician wanted to offend economic interests for something that, after all, might never happen. Skepticism was not unreasonable. In particular, a few reputable scientists published claims (loudly amplified by industrial public relations efforts and right-wing institutions, e.g. Seitz et al. 1989) that the observed rise of global temperature was caused by a rise in the Sun's activity that had been observed in the same decades. The only evidence that greenhouse gases were the cause was buried in computer models with many arbitrary and uncertain features.

During the 1990s the problems raised by skeptics were answered. Solar activity ceased to rise, but global temperature soared faster than ever. A gigantic volcanic eruption of smoke and gas in the Philippines in 1991 gave Hansen et al. (1992) an opportunity to apply their computer model to predicting the temporary effects on world climate, and by 1995 they were proved correct. Meanwhile other computer teams passed a variety of other tests of their ability to reproduce real-world data. In particular, the specific pattern of temperature rise in different regions of the world that the models calculated was a good match for the actual historical record—if and only if the rise of greenhouse gas levels was put into the models. Scientists hailed this as a clear 'signature' of the greenhouse effect (Santer et al. 1996). Still more striking news came from Antarctic ice cores, which showed that the level of CO_2 and methane in the atmosphere had gone up and down in lockstep with temperature through past ice ages, pointing to powerful biological and geochemical feedbacks. Indeed a variety of studies of ancient climates revealed the same sensitivity to a doubled CO_2 level—roughly 3 °C of warming, give or take 50 percent—that the computer models had calculated (Lorius et al. 1990; Hoffert and Covey 1992).

These studies had an impact on the IPCC's next report, issued in 1995. After grueling negotiations the delegates unanimously agreed to tell the world that 'The balance of evidence suggests that there is a discernible human influence on global climate.' The weaselly wording showed the strain of political compromises that had watered down the scientists' original draft, but the message was unmistakable. More likely than not, global warming was upon us. The news media reported this widely, reinvigorating public debate (IPCC 1996).

In 1997 the IPCC issued a report on what climate change might mean for each region of the globe. For example, the experts concluded that Africa was 'the continent most vulnerable to the impacts of projected changes.' That was not just because so many parts of Africa were already water stressed, subject to tropical diseases, and so forth, but still more because Africa's 'widespread poverty limits adaptation capabilities.' By contrast, the carefully managed agricultural systems of Europe and North America might even contrive to benefit from a modest warming and rise in the level of CO_2, although the developed nations would certainly suffer some harmful impacts as well (Watson et al. 1997: 6). A few experts, notably in Russia, predicted significant benefits from global warming, while others insisted that nothing would happen that humanity could not easily handle. These assessments, and the publics they addressed, could see the impacts as manageable because they were looking no more than half a century or so ahead. The twenty-second century was so far away! Surely by then, humanity would have taken control of its emissions so that CO_2 would not rise to three or four times the pre-industrial level . . . wouldn't we?

The IPCC's conclusions cast a long shadow over the next major conclave, the 1997 UN Conference on Climate Change held in Kyoto, Japan. This was a policy and media extravaganza attended by nearly 6,000 official delegates and thousands more representatives of environmental groups and industry, plus a swarm of reporters. The negotiations almost broke down in frustration and exhaustion. Yet the IPCC's conclusions could not be brushed aside, and dedicated efforts by many leaders were capped by a dramatic intervention when US Vice President Al Gore flew to Kyoto on the last day and pushed through a compromise—the Kyoto Protocol. The agreement exempted poor countries from any commitment, and pledged wealthy countries to cut their emissions significantly (Oberthür and Ott 1999). Much of the world public thought the arrangement was fair. But the Global Climate Coalition organized a lobbying and public relations campaign in the United States, and Congress refused to take any action. That gave other governments an excuse to avoid tough measures, casting blame on the United States for any failure to get started.

4 CONSENSUS ON SCIENCE BUT NOT POLICY
(2001 AND AFTER)

The next round of the IPCC process led to a Third Report, issued in 2001. Arduous negotiations ended in unanimous and blunt conclusions. The world was rapidly getting warmer, at a rate probably 'without precedent during at least the last 10,000 years,' most likely because of the increase in greenhouse gases. With global emissions of CO_2 rising faster than previous reports had considered, the range of warming that the IPCC predicted for the late twenty-first century ran from 1.4 °C up to a shocking 5.8 °C (10 °F). This range was not for the traditional doubled CO_2 level, which was now expected to arrive around mid-century, but for the still higher levels that would surely come after 2070 unless the world took action (IPCC 2001).

The new American President, George W. Bush, publicly renounced the Kyoto Protocol, but most of the world's governments remained publicly committed to act. At an international meeting held in Bonn in July 2001, 178 governments (but not the United States) negotiated an agreement for implementing the Kyoto Protocol. The stated goal was to return greenhouse gas emissions to roughly the 1990 rate within a decade. Scarcely anyone believed the world would really achieve that. And if somehow it did happen, at the 1990 rate of emissions the greenhouse gases in the atmosphere would still continue to rise.

Reality now descended upon the abstract world of impact studies. Instead of future possibilities, some experts began to estimate the role that global warming might have played in one or another actual disaster. It turned out that because of unexpected complexities, the rich nations were not as safe as some had thought. One example: in 2003 a heatwave of unprecedented scope killed tens of thousands in Europe. Nobody had foreseen that old people could not save themselves when the traditional August vacation emptied the cities. Another example: bark beetles, no longer controlled by winter freezes, devastated millions of acres of forests from Alaska to Arizona, leaving the weakened timber prey to an unprecedented outbreak of forest fires. Nobody had prepared for these particular impacts of global warming. Another spectacle was the break-up of ice sheets near Antarctica, raising a threat that sea levels would rise faster than the IPCC had predicted, perhaps driving millions of refugees from their homes. Still more impressively, the ice cover of the Arctic Ocean also dwindled faster than models had predicted.

Public opinion in most countries increasingly accepted that global warming had arrived, and feared the consequences, although substantial numbers continued to deny there was a problem. Denial was especially strong in the United States, where corporate lobbyists, right-wing institutions, and their allies in the media continued to call global warming a myth, if not a deliberate hoax. Even in the United States, however, key elites were becoming convinced of the need for action. Pressed by environmentalist groups as well as by general public opinion and their own employees, prominent corporations pulled out of the Global Climate Coalition. Some European firms, notably oil giant BP under the far-sighted John Browne, had already decided (as he put it in 1997), that 'it falls to us to take precautionary action now.' A growing number of leading corporations pledged to limit their own emissions. Most important, legal restrictions on emissions seemed inevitable. A wise corporation would take the lead in discussing just which business operations should be taxed or regulated. In the fore was the insurance industry, whose traditional job was to make sure that catastrophic contingencies were included in corporate business plans.[6]

There was an even more sobering way to frame climate change—as a security threat. For example, a 2003 report commissioned by intellectuals in the Pentagon warned of a risk that 'mega-droughts, famine and widespread rioting will erupt across the world.... abrupt climate change could bring the planet to the edge of anarchy as countries develop a nuclear threat to defend and secure dwindling food, water and energy supplies.' If you thought like a military officer (or an insurance company executive), the IPCC's approach—concentrating on what everyone could agree was likely, while setting aside less likely but still possible scenarios—was not 'conservative' at all, but irresponsible (see Gilman et al., this volume).[7]

That was what some scientists declared. A debate that had simmered since the 1970s was coming into full public view. Should scientists tell the public only what was totally certain, or should they warn people to prepare, on the 'insurance' model, for conceivable catastrophes (Schneider 2009; Hansen 2009)? What was the worst that could happen? A widely seen movie (*The Day after Tomorrow*, Roland Emmerich 2004) pictured a collapse of the ocean circulation that brought an instant ice age. The well-known environmentalist James Lovelock warned of 'the possibility that global heating will all but eliminate people from the Earth' (Lovelock 2009: 6). No climate expert believed such things were at all possible. But the widespread images gave an opening for skeptical claims that all talk of climate change was deceitful 'alarmism'. Most scientists felt they should say nothing that could be inter- preted as a doomsaying exaggeration, and worried about the minority who insisted on emphasizing hypothetical dangers. On the other hand, there did seem to be a chance (1 percent? 10 percent?) that feedbacks known and unknown could boost global tempera- ture six degrees if not more by the end of the twenty-first century. That would destroy agricultural systems and ecosystems vital to the maintenance of civilization (Lynas 2007). Yet policy makers, whether from choice or ignorance, preferred to concentrate on the median-probability scenario. That scenario did not seem to demand urgent attention.

To put the Kyoto Protocol into effect required ratification by nations with more than 55 percent of the world's CO_2 emissions. Absent the United States, the world's largest emitter, it was only in 2005 that the Kyoto Protocol went into effect with 141 signatory nations. Everyone understood that many would find it impossible to live up to their obligations, and even if they could, it would do little to forestall global warming. The treaty was meant as a first step to get people started on working out systems for monitoring and restricting emissions, and to stimulate the development of low-emission devices and practices. This experience would be needed for the next round of negotiations, aiming at a treaty that would take effect after the Kyoto Protocol expired in 2012.

In 2005 western Europeans adopted a 'cap-and-trade' scheme that required permits for carbon emissions and set up a market for trading the permits (see Jordan et al., this volume). The system was so badly designed that the price of the permits at first soared, then abruptly crashed to almost nil. In a perverse way, it was what the Kyoto negotiators had wanted: experiments to find how particular policies worked in practice. After some adjustments the mechanism began to function effectively. The Japanese, proud to be associated with the Kyoto Protocol, also made strong efforts to constrain emissions. Even the Chinese government, with no treaty obligations and with economic progress depending on the rapid deployment of coal-fired power plants, worked to limit automobile gas consumption and to promote wind, solar, and nuclear energy production.

In the United States, action proceeded more effectively at other levels of society (Rabe 2004). In 2007 several coalitions of states laid plans for mandatory regional systems to track and cut back their citizens' emissions. Practical steps were undertaken by hundreds of cities and thousands of smaller governmental and private entities as well as federal organs such as the armed forces. In fact, overall the United States was doing as much as most European nations. The net result of all these efforts, however, was at best to delay warming by a few years. While many experts had pointed to ways of reducing emissions that would be cheap, in some cases even a net benefit to society, powerful economic, political, and ideological interests continued to block strong action.

The fourth IPCC report, issued in 2007, warned that serious harm due to greenhouse warming had already become evident in some parts of the world. The panel's reports now included detailed social and economic analysis, which concluded that the cost of reducing emissions would be far less than the cost of the damage they would eventually cause. In a 'Synthesis' of the reports issued in November, the IPCC leaders spoke more plainly than ever. With CO_2 in the atmosphere rising a percent each year at an accelerating rate, they foresaw a variety of likely impacts, some 'abrupt or irreversible,' ranging from 'disruption of . . . societies' by storm floods to the extinction of up to a quarter of the world's species. If greenhouse gas levels kept rising unrestrained to well beyond twice the pre-industrial level, we were likely to see a radical impoverishment of many of the ecosystems that sustain modern society (IPCC 2007: 12–13).

A century of research had pinned down specific and disturbing predictions about what would happen if the mean global temperature climbed a few degrees. The experts admitted, however, that these predictions were only probable, not certain. They were also unable to say when the impacts would become truly severe, except that it was decades away. Most of the world's citizens were little inclined to confront the problem. After all, global warming was not clearly visible except to a few thousand scientists, and never seemed as urgent as the latest economic or political crisis. The realms of scientific research and public policy had become more entangled with one another than ever before, but they remained realms apart.

NOTES

1. Glacken (1967); Neumann (1985); Fleming (1990; 1998: chs. 2–4); Stehr et al. (1995). This essay is drawn from Weart (2009), especially the essays on impacts and international cooperation, where much fuller references may be found. A condensed version is Weart (2008).
2. Arrhenius (1896); this and other key scientific papers are available online at Fleming (2008).
3. On climate scientists and the US government see Hart (1992); Hart and Victor (1993); Fleagle (1992; 1994).
4. WMO (1989). On all these negotiations see Lanchbery and Victor (1995); O'Riordan and Jäger (1996); Franz (1997); Bolin (2007).
5. IPCC (1990). On the creation and processes of IPCC see Jäger (1992); Lanchbery and Victor (1995); Gelbspan (1997: ch. 5); Agrawala (1998a, 1998b, 1999); Leggett (1999: 9–28); Stevens (1999: ch. 13); Edwards and Schneider (2001); Bolin (2007), and for reports see <http://www.ipcc.ch/pub/reports.htm>.
6. Browne, speech at Stanford University, 19 May 1997. This and other recent matters are most easily documented by Web searches (note in particular the Google News Archive <http://news.google.com/archivesearch/>), and see references in the essays on US government, international cooperation, and public opinion in Weart (2009).
7. Schwartz and Randall (2003). See also CNA corporation (2007); Pumphrey (2008), and media reports.

REFERENCES

AGRAWALA, S. 1998a. Context and early origins of the intergovernmental panel for climate change. *Climatic Change* 39: 605–20.

AGRAWALA, S. 1998b. Structural and process history of the intergovernmental panel for climate change. *Climatic Change* 39: 621–42.

—— 1999. Early science-policy interactions in global climate change: Lessons from the Advisory Group on Greenhouse Gases. *Global Environmental Change* 9: 157–69.

ARRHENIUS, S. 1896. On the influence of carbonic acid in the air upon the temperature of the ground. *Philosophical Magazine* 41: 237–76.

—— 1908. *Worlds in the Making*. New York: Harper & Brothers.

BEUERMANN, C., and JÄGER, J. 1996. Climate change politics in Germany. Pp. 186–227 in T. O'Riordan and J. Jäger (eds.), *Politics of Climate Change: A European Perspective*. London: Routledge.

BODANSKY, D. 1997. *The History and Legal Structure of the Global Climate Change Regime*. Potsdam: PIK.

BOLIN, B. 2007. *A History of the Science and Politics of Climate Change. The Role of the Intergovernmental Panel on Climate Change*. Cambridge: Cambridge University Press.

—— DÖÖS, B. R., JÄGER, J., et al. (eds.) 1986. *The Greenhouse Effect, Climatic Change, and Ecosystems. SCOPE Report No. 29*. Chichester: John Wiley.

BOYKOFF, M., and BOYKOFF, J. 2004. Balance as bias: Global warming and the US prestige press. *Global Environmental Change* 14: 125–36.

BROECKER, W. S. 1966. Absolute dating and the astronomical theory of glaciation. *Science* 151: 299–304.

BRYSON, R. A. 1973. *Climatic Modification of Air Pollution, 2, The Sahelian Effect*. Madison Institute for Environmental Studies, Report 9. Madison, WI: University of Wisconsin—Madison Institute for Environmental Studies.

—— 1974. A perspective on climatic change. *Science* 184: 753–60.

CALLENDAR, G. S. 1938. The artificial production of carbon dioxide and its influence on climate. *Quarterly J. Royal Meteorological Society* 64: 223–40.

CNA Corporation, Military Advisory Board. 2007. *National Security and the Threat of Climate Change*. Alexandria, VA: CNA Corporation.

Conservation Foundation. 1963. *Implications of Rising Carbon Dioxide Content of the Atmosphere*. New York: The Conservation Foundation.

EDWARDS, P. N., and SCHNEIDER, S. H. 2001. Self-governance and peer review in science-for-policy: The case of the IPCC Second Assessment Report. Pp. 219–46 in C. A. Miller and P. N. Edwards (eds.), *Changing the Atmosphere. Expert Knowledge and Environmental Governance*. Cambridge, MA: MIT Press.

FLEAGLE, R. G. 1992. The U.S. government response to global change: Analysis and appraisal. *Climatic Change* 20: 57–81.

—— 1994. *Global Environmental Change: Interactions of Science, Policy, and Politics in the United States*. Westport, CT: Praeger.

FLEMING, J. R. 1990. *Meteorology in America, 1800–1870*. Baltimore: Johns Hopkins University Press.

—— 1998. *Historical Perspectives on Climate Change*. New York: Oxford University Press.

—— 2007. *The Callendar Effect. The Life and Work of Guy Stewart Callendar (1898–1964), the Scientist Who Established the Carbon Dioxide Theory of Climate Change*. Boston: American Meteorological Society.

—— 2008. *Climate Change and Anthropogenic Greenhouse Warming: A Selection of Key Articles, 1824–1995, with Interpretive Essays*. Online at <http://wiki.nsdl.org/index.php/PALE:ClassicArticles/GlobalWarming>.

FRANZ, W. E. 1997. *The Development of an International Agenda for Climate Change: Connecting Science to Policy* (IASA interim report IR-97-034) Austria: International Institute for Applied Systems Analysis. Online at <http://citeseer.ist.psu.edu/cache/papers/cs/6624/http:zSzzSzwww.iiasa.ac.atzSzPublicationszSzDocumentszSzIR-97-034.pdf/franz97development.pdf Laxenburg>.

GELBSPAN, R. 1997. *The Heat Is On: The High Stakes Battle over Earth's Threatened Climate.* Reading, MA: Addison-Wesley.

——— 2004. *Boiling Point. How Politicians, Big Oil and Coal, Journalists, and Activists Are Fueling the Climate Crisis—and What You Can Do to Avert Disaster.* New York: Basic Books.

GLACKEN, C. J. 1967. *Traces on the Rhodian Shore: Nature and Culture in Western Thought from Ancient Times to the End of the Eighteenth Century.* Berkeley: University of California Press.

GOODELL, J. 2006. *Big Coal: The Dirty Secret behind America's Energy Future.* Boston: Houghton Mifflin.

HANSEN, J. E., LACIS, A., RUEDY, R. A., et al. 1992. Potential climate impact of Mount Pinatubo eruption. *Geophysical Research Letters* 19: 142–58.

——— 2009. *Storms of my Grandchildren. The Truth about the Coming Climate Catastrophe and our Last Chance to Save Humanity.* New York: Bloomsbury USA.

HART, D. 1992. *Strategies of Research Policy Advocacy: Anthropogenic Climate Change Research, 1957–1974.* Cambridge, MA: Harvard University, John F. Kennedy School of Government.

——— and VICTOR, D. G. 1993. Scientific elites and the making of US policy for climate change research, 1957–74. *Social Studies of Science.* 23: 643–80.

HECHT, A.D., and TIRPAK, D. 1995. Framework agreement on climate change: A scientific and policy history. *Climatic Change* 29: 371–402.

HOFFERT, M. I., and COVEY, C. 1992. Deriving global climate sensitivity from palaeoclimate reconstructions. *Nature.* 360: 573–6.

HOFFMAN, J. S., KEYES, D., and TITUS, J. G. 1983. *Projecting Future Sea Level Rise. Report for the US Environmental Protection Agency.* Washington, DC: US Government Printing Office.

IPCC (Intergovernmental Panel on Climate Change). 1990. *Climate Change: The IPCC Scientific Assessment. Report Prepared for IPCC by Working Group I,* ed. J. T HOUGHTON, G.J. JENKINS, and J. J. EPHRAUM. Cambridge: Cambridge University Press.

——— 1996. *Climate Change 1995: The Science of Climate Change. Contribution of Working Group I to the Second Assessment Report of the Intergovernmental Panel on Climate Change.* Cambridge: Cambridge University Press. Online at <http://www.ipcc.ch/pub/reports.htm>.

——— 2001. *Climate Change 2001: The Scientific Basis. Contribution of Working Group I to the Third Assessment Report of the IPCC.* Cambridge: Cambridge University Press. Online at <http://www.ipcc.ch/pub/reports.htm>.

——— 2007. Summary for policymakers. Pp. 1–22 in *Climate Change 2007: Synthesis Report of the Intergovernmental Panel on Climate Change Fourth Assessment Report.* Cambridge and New York: Cambridge University Press. Online at <http://www.ipcc.ch/reports.htm>.

JÄGER, J. 1992. From conference to conference. *Climatic Change* 20: iii–vii.

KEELING, C. D. 1960. The concentration and isotopic abundances of carbon dioxide in the atmosphere. *Tellus* 12: 200–3.

——— 1970. Is carbon dioxide from fossil fuel changing man's environment? *Proceedings of the American Philosophical Society* 114: 10–17.

LANCHBERY, J., and VICTOR, D. 1995. The role of science in the global climate negotiations. Pp. 29–39 in H. BERGESEN and G. PARMANN (eds.), *Green Globe Yearbook of International Cooperation on Environment and Development 1995.* Oxford: Oxford University Press.

LEGGETT, J. 1999. *The Carbon War: Dispatches from the End of the Oil Century.* London: Allen Lane—Penguin.

LONG, M., and ILES, A. 1997. *Assessing Climate Change Impacts: Co-evolution of Knowledge, Communities and Methodologies. ENRP Discussion Paper E-97-09.* Cambridge, MA: Harvard University, John F. Kennedy School of Government. Online at <http://www.hks.harvard.edu/gea/pubs/e-97-09.pdf>.

LORIUS, C., JOUZEL, J., RAYNAUD, D., et al. 1990. The ice core record: Climate sensitivity and future greenhouse warming. *Nature* 347: 139–45.

LOVELOCK, J. 2009. *The Vanishing Face of Gaia. A Final Warming.* New York: Basic Books.

LYNAS, M. 2007. *Six Degrees: Our Future on a Hotter Planet.* London: Fourth Estate (HarperCollins UK).

MANABE, S., and WETHERALD, R. T. 1967. Thermal equilibrium of the atmosphere with a given distribution of relative humidity. *J. Atmospheric Sciences* 24: 241–59.

—— BRYAN, K., and SPELMAN, M. J. 1975. A global ocean-atmosphere climate model. Part I: the atmospheric circulation. *J. Physical Oceanography* 5: 3–29.

MINTZER, I., and LEONARD, J. A. (eds.) 1994. *Negotiating Climate Change: The Inside Story of the Rio Convention.* Cambridge: Cambridge University Press.

MOONEY, C. 2005. *The Republican War on Science.* New York: Basic Books.

National Academy of Sciences, Geophysics Research Board, Panel on Energy and Climate. 1977. *Energy and Climate: Studies in Geophysics.* Washington, DC: National Academy of Sciences.

—— Climate Research Board. 1979. *Carbon Dioxide and Climate: A Scientific Assessment (Jule Charney, Chair).* Washington, DC: National Academy of Sciences.

—— Carbon Dioxide Assessment Committee. 1983. *Changing Climate.* Washington, DC: National Academy of Sciences.

NEUMANN, J. 1985. Climatic change as a topic in the classical Greek and Roman literature. *Climatic Change* 7: 441–54.

NOLIN, J. 1999. Global policy and national research: the international shaping of climate research in four European Union countries. *Minerva* 37: 125–40.

OBERTHÜR, S., and OTT, H. E. 1999. *The Kyoto Protocol: International Climate Policy for the 21st Century.* Berlin: Springer.

O'RIORDAN, T., and JÄGER, J. 1996. The history of climate change science and politics. Pp. 1–31 in T. O'RIORDAN and J. JÄGER (eds.), *Politics of Climate Change: A European Perspective.* London: Routledge.

PETERSON, T. C., CONNOLLEY, W. M., and FLECK, J. 2008. The myth of the 1970s global cooling scientific consensus. *Bulletin of the American Meteorological Society* 89: 1325–37.

PLASS, G. N. 1956. The carbon dioxide theory of climatic change. *Tellus* 8: 140–54.

President's Science Advisory Committee. 1965. *Restoring the Quality of Our Environment. Report of the Environmental Pollution Panel.* Washington, DC: The White House.

PUMPHREY, C., ed. 2008. *Global Climate Change: National Security Implications.* Carlisle, PA: Strategic Studies Institute, U.S. Army War College. Online at <http://www.strategicstudiesinstitute.army.mil/pdffiles/PUB862.pdf>.

RABE, B. G. 2004. *Statehouse and Greenhouse: The Emerging Politics of American Climate Change Policy.* Washington, DC: Brookings Institution Press.

RAMANATHAN, V., CICERONE, R. J., SINGH, H. B., et al. 1985. Trace gas trends and their potential role in climate change. *J. Geophysical Research* 90: 5547–66.

REVELLE, R. 1957. Testimony to Committee on Appropriations, United States Congress (85:2), House of Representatives. *Report on the International Geophysical Year.* Washington, DC: Government Printing Office.

—— and SUESS, H. E. 1957. Carbon dioxide exchange between atmosphere and ocean and the question of an increase of atmospheric CO_2 during the past decades. *Tellus* 9: 18–27.

SANTER, B. D., TAYLOR, K. E., WIGLEY, T. M. L., et al. 1996. A search for human influences on the thermal structure of the atmosphere. *Nature* 382: 39–46.

SCEP (Study of Critical Environmental Problems). 1970. *Man's Impact on the Global Environment: Assessment and Recommendation for Action.* Cambridge, MA: MIT Press.

SCHNEIDER, S. H. 2009. *Science as a Contact Sport: Inside the Battle to Save the Earth's Climate.* Washington, DC: National Geographic.

SCHWARTZ, P., and RANDALL, D. 2003. *An Abrupt Climate Change Scenario and its Implications for United States National Security.* Emeryville, CA: Global Business Network. Online at <http://www.ems.org/climate/pentagon_climatechange.pdf>.

SEITZ, F., JASTROW, R., and NIERENBERG, W. A. (eds.) 1989. *Scientific Perspectives on the Greenhouse Problem.* Washington, DC: George C. Marshall Institute.

STEHR, N., STORCH, H. VON, and FLÜGEL, M. 1995. The 19th century discussion of climate variability and climate change: analogies for present debate? *World Resources Review* 7: 589–604.

STEVENS, W. K. 1999. *The Change in the Weather: People, Weather and the Science of Climate.* New York: Delacorte Press.

Time 1956. One big greenhouse. *Time* 67: 59. Online at <http://www.time.com/time/magazine/article/0,9171,937403,00.html>.

WATSON, R. T., ZINYOWERA, M. C., and MOSS, R. H. (eds.) 1997. *Summary for Policymakers. The Regional Impacts of Climate Change: An Assessment of Vulnerability. A Special Report of IPCC Working Group II.* Cambridge: Cambridge University Press. Online at <http://www.ipcc.ch/reports.htm>.

WEART, S. R. 2008. *The Discovery of Global Warming.* 2nd edn. Cambridge, MA: Harvard University Press.

—— 2009. *The Discovery of Global Warming.* rev. edn. Online at <http://www.aip.org/history/climate/>.

WILSON, C. L., and MATTHEWS, W. H. (eds.) 1971. *Inadvertent Climate Modification. Report of Conference, Study of Man's Impact on Climate (SMIC), Stockholm.* Cambridge, MA: MIT Press.

WMO (World Meteorological Organization). 1979. *Declaration of the World Climate Conference.* Geneva: World Meteorological Organization. Online at <http://www.dgvn.de/fileadmin/user_upload/DOKUMENTE/WCC-3/Declaration_WCC1.pdf>.

—— 1989. *The Changing Atmosphere: Implications for Global Security, Toronto, Canada, 27–30 June 1988: Conference Proceedings.* Geneva: Secretariat of the World Meteorological Organization.

CHAPTER 6

VOICES OF VULNERABILITY: THE RECONFIGURATION OF POLICY DISCOURSES

MAARTEN HAJER AND WYTSKE VERSTEEG

1 Introduction: Giving Voice to the Planet

'THE true problem for the climatologist to settle during the present century is not whether the climate has lately changed, but what our present climate is, what its well-defined features are, and how they can be most clearly expressed in numbers', thus Cleveland Abbe (1889). Abbe's statement testifies to an unlimited faith in science not only to settle the problem, but also to define the features of that problem and to express it in statistics. It suggests that the climate is out there, waiting to be described.

Although the planet is clearly expressive (think of storms, earthquakes, or droughts), it cannot speak for itself: it is the interpretation of these storms, earthquakes, and droughts that constitutes the societal phenomenon of climate change. Various interpreters tell us stories in different voices of how to understand what is going on. Some stories are directly political and speak about global injustices or, on the contrary, of preferred ways of addressing the problem (e.g. the promotion of market solutions to climate change policy); some are scientific and seek to order how to make sense of 'biospheric' relationships.

Given the complexities of the biosphere, experts hold a privileged position when it comes to giving voice to the planet. Drawing on their scientific expertise they end up working with representations of the planet, some of which are very powerful and are able to structure the global political debate over a decade or more. Here global climate modeling fulfills a crucial role. Much depends on the interpretation and underlying assumptions of the models in use. Speaking on behalf of the planet always comes with a particular emphasis, whether explicit or implicit, intended or unintended. Scientific knowledge and social order 'coproduce' each other in complicated ways (Jasanoff 2004). As scientists are but too aware: each and every model is value laden, which is where bias comes in. The question is how society deals with this value-ladenness: how do the features of a problem become well defined, which facts solidify in society and which do not, which facts develop into policy relevant facts, which do not?

Discourse analysis is a way to make sense of this process of solidifying, or of the biases in the way we address public problems. We aim to make sense of the reconfigurations of discourses by relating the way in which climate change is conceived to the institutional practices by means of which meaning is given, illuminating the way in which discursive dominance is produced. Crucial is how particular ways of talking about the climate reorder and re-relate political debates. In general we see how a 'new' debate is marked by traces of previous discussions. As climate change emerged more strongly as an issue in the late 1980s and early 1990s, the policy debate was marked by discussions on the diminishing ozone layer and the successful policy concept of 'sustainable development' that the Brundtland Commission had defined in 1986. In this article we trace the reconfiguration of climate change discourse paying particular attention to the influence of the institutional molds in which this discourse was produced. Analyzing thirty years of discourse we see how fragile discursive dominance can be.

We examined the period of intensive climate change discourse from the 1980s to 2010. We distinguish three phases and a hiccup: (1) the phase in which climate rose to the political agenda, (2) the classical-modernist response to 'first get the facts right' (the UN conference in Toronto 1988 to Kyoto), (3) the mature years of the IPCC and UNFCCC as fully-fledged global science-policy interface producing successive policy results, to the 'hiccup'; the moments of confusion of the UNFCCC COP-15 at Copenhagen (2009) and the 'errors' of the IPCC (2010).

2 CLIMATE CHANGE: WHY TALK ABOUT WORDS WHEN WE HAVE THINGS TO DO?

Discourse is here defined as an ensemble of ideas, notions, and categories through which meaning is given to social and physical phenomena, and which is produced and reproduced through an identifiable set of practices (Hajer 1995, 2009). Thus, the subject of analysis in this chapter is the way in which certain *natural* phenomena are conceptualized within *institutional* practices, such as the intermingling of politics and science in the IPCC reviewing process; a process that produces and reproduces definitions, notions, and categories that together give meaning to our current idea of what climate change 'actually' is.

The social construction of the meaning of climate change is not only a question for scientists of course. Policy makers and interest groups (such as lobby groups from industry, environmental NGOs) will try to influence the public image not only by using text, but also, and importantly, symbols and images. What is more, often scientists themselves can be found to be active in 'boundary work' in which they also employ symbols to communicate what they, as scientists, see happening, in an effective way. Discourse analysis thus helps us to understand how political actors actively position themselves to make 'nature' linguistically intelligible and influence what can and what cannot be thought, drawing on their particular conceptualization—indeed, it helps to see how 'nature' and 'environment' are, and have always been, contested notions (Hajer and Versteeg 2005).

A discourse analysis also opens for investigation how the definition of problems on the one hand and chosen policy solutions on the other are interrelated. Wildavsky (1979) aptly described policy analysis as the study of public problems that cannot be expressed until they are at least tentatively solved. Thus, the answer to the question whether or not we have 'things to do', let alone which things, is highly dependent upon the words and concepts we use to discuss the problem, the knowledge we choose to mobilize, the symbols that are found to get complex messages across. We can try to see how some arguments gain prominence and 'stick' while others don't, showing how some discourses become dominant in, first, structuring the way in which people conceptualize their world (*discourse structuration*) and, second, solidifying into institutions and organizational practices (*discourse institutionalization*). What is more, we can analyse to what extent these definitions are robust to attacks and counterclaims later on.

Which framings of the problem become dominant may influence the possibilities for action, or on the contrary, the likelihood of collective denial (see K. M. Norgaard in this volume). Moreover, a discursive analysis will help to see the bias in particular discourses—emphasizing only some aspects of the problem and thus the interests of some actor groups rather than others. Hence it is not about which discourse is 'true', it is more about how power is exercised through language and discourse.

3 'MORE EQUABLE AND BETTER CLIMATES': FROM THE EARLY SCIENTISTS TO THE IPCC PROCESS

The relationship between climate and human society has been a topic of debate for ages; theorists related climate to the character or culture of a nation (Aristoteles, Montesquieu), to health (Hippocrates), or, in a reversed causal relationship, to the cultivation of the land (Hume). Economic progress was a recurring theme in these reflections; for instance, climate pioneer Arrhenius confidently expressed his hope that the increasing percentage of carbonic acid would allow us 'to enjoy ages with more equable and better climates, especially as regards the colder regions of the earth, ages when the earth will bring forth much more abundant crops than at present, for the benefit of rapidly propagating mankind' (quoted in Fleming 2005: 74; cf. Weart, this volume).

However, the idea of 'climate' as a meaningful concept on its own was by no means self-evident. As Miller (2004: 51–3) notes, in the first half of the twentieth century, the distinction between 'weather' and 'climate' was still seen as an artificial one, with climate as an aggregation of local weather averages. This implied that there was no perceived need for international cooperation; whereas changes in weather and climate might be disastrous for local communities, there was no risk for global civilization. This would change radically between the 1960s and the 1980s. Pioneers such as Arrhenius might already have conceived of climate change as a global phenomenon, but in the second half of the twentieth century a dramatic change took place in the *representation* of climate: instead of signifying an aggregation of locally specific weather averages, the term was now conceived of as an ontological whole, an integrated system (Miller 2004: 53–4). Ironically, the global network-ing that was to become characteristic for the modern climate change regime would be greatly stimulated by new technical developments, such as telecommunication and cheap airplane travel, but most important: computers. Modern computer modeling played a crucial role in the 'experience' of the global environmental system, comparable to the role of satellite images for the idea of 'the blue planet' which came to be the icon of the seminal UN conference on the human environment held in Stockholm with the motto: 'Only One Earth' (cf. Hajer 1995: 8 ff.). The importance of computer modeling would make specialized scientists the authoritative interpreters of a non-tangible reality. As in other environmental domains, climate science would become increasingly entangled in the process of policy making as metaphors from systems ecology started to fulfill an increas-ingly important role in conceptualizing the world. In the end the scientist is the only arbiter to determine what levels of pollution nature can endure (Hajer 1995: 27).

In the second half of the twentieth century, scientific interest in climate change was organized and stimulated by a series of conferences, focusing not only on global warming, but also on global cooling (due to the effect of industrial aerosols) and climate variability (a theme that gained prominence due to the prolonged drought in the Sahel). The latter theme formed the impetus for the United Nations Conference on Desertification (Nairobi, 1977), one of the earliest North–South dialogues with, however, little follow-up as funding from the Northern countries was voluntary (Agrawala 1998: 607). Whereas climate change had first been discussed as one of the several 'Global Effects of Environmental Pollution' (Dallas, 1968) or 'Critical Environmental Problems' (MIT, 1970), it gradually acquired the status of a theme in its own right. In 1979, the World Meteorological Organization (WMO) organized the first World Climate Conference in Geneva, which subsequently led to the establishment of the World Climate Program (WCP).[1] The WCP was a cooperation between the International Council of Scientific Unions (ICSU) and the WMO, two organizations that had previously cooperated in the setting up of its predecessor, the Global Atmospheric Research Program (GARP, 1967). Together with the UNEP (United Nations Environment Program), the two bodies organized a series of workshops, culminating in the 1985 Villach conference. Experts reached the remarkably activist conclusion that *in the first half of the next century a rise of global mean temperature will occur which is greater than any in man's history*, calling on governments to incorporate the results of the Villach assess-ment in their social, economic, and environmental policies. Moreover, they argued that *the understanding of the greenhouse question is sufficiently developed that scientists and policy makers should begin an active collaboration to explore the effectiveness of alternative policies and adjustments.*[2]

In a phrase of Boehmer-Christiansen (1994), one could call this the creation of concern: a joint effort by scientists, NGOs, UN agencies, diplomats, and various other sponsors to place climate change on the political agenda and attract funds. This interweaving of the different social worlds of scientists and policy makers from various countries was a complicated endeavor; scientists were caught in a balancing act between their traditional role as distant knowledge providers and a felt concern about the threats to mankind only they could see.

The Villach conference was followed by the establishment of 'a small taskforce on greenhouse gases' in July 1986, the Advisory Group on Greenhouse Gases, a joint effort of the WMO, UNEP, and ICSU. In the same year, influential members of the US Congress began to pressure the White House to take action, persuaded that climate change required policy responses. In response to the Villach recommendations, UNEP Executive Director Tolba had also urged the US Secretary of State George Schultz to take appropriate policy action. However, whereas the US had the biggest cumulative expertise on climate change, it had also a huge stake in the problem and there was no consensus among US agencies, thus: '[a]t a time when it was difficult to get interagency agreement on any action, there was convergence (for different reasons) around the concept of an international scientific assessment' (Hecht and Tirpak 1995: 381). This convergence strongly influenced the WMO Executive Council resolution in which it was decided 'in coordination with the Executive Director of UNEP to establish an *intergovernmental* mechanism to carry out internationally coordinated scientific assessments of the magnitude, impact and potential timing of climate change' (Agrawala 1998). The famous intergovernmentality of the later IPCC—the most discerning characteristic of the organization—was initially the result of a compromise between those who hoped that the establishment of an intergovernmental assessment agency would further progress on the theme of climate change, and those who merely hoped for a delay.

While climate science was still a relatively small field of science, the number of scientific publications on climate change had started to grow exponentially (see Stanhill 2001). What is more, there was a slowly developing majority opinion that the risk of climate change was real enough to require policy action. George Bush Sr. inserted climate change into his campaign for the 1988 presidential election, assuring that 'those who think we're powerless to do anything about the greenhouse effect are forgetting about the White House effect. As president, I intend to do something about it.' For the lay public, the topic acquired a recognizable face by directly observable meteorological phenomena; in the 'endless summer' of 1988, the US experienced a record-breaking drought, in relation to which NASA scientist James Hansen testified before a Senate Committee, that: 'The greenhouse effect . . . has been detected and is changing our climate now.'[3] Hansen's testimony was particularly important as his explicit warning was in sharp contrast with the accepted neutral, cautious style of scientific rhetoric, thus attracting political and media attention. It also helped to establish a rhetorical figure that would remain central in media representations of the climate change debate; the lone voice of the early warner, a Cassandra who defies consensus to speak the truth but is not heard, with catastrophe as the result (Hamblyn 2009: 227). The following comment of Stephen Schneider (Schneider 1993: 57) is illustrative:

> In July [1988] alone, I had probably 100 phone calls from journalists, many of which asked, in effect, okay, you've been carrying a banner about global warming for the past 15 years. Are you finally ready to give us an 'I told you so'?

As he puts it (ibid.), in 1988 'The Environment' became a media event. The high temperatures also helped to attract attention for the Toronto World Conference on the Changing Atmosphere, a meeting nicknamed the 'Woodstock of CO_2,' that was organized in the same year. In Toronto, the invited scientific experts urged governments for the first time to set targets for greenhouse gas reductions—again following the example of the Montreal Protocol.

Whereas climate change had previously been just one of several environmental concerns, the discourse structuralized in the second half of the 1980s; climate change or 'the greenhouse effect' had become a central concept for the understanding of the world.

Apart from the felt concerns, climate change was also an *attractive* concept for several actors; for UNEP to prolong its ozone success, for scientists as a possible source of funding, for environmental NGOs as a theme that united and harmonized their concerns, for politicians to present themselves as 'green presidents', and for the media as a frame to 'explain' unexpected meteorological phenomena ('Heat waves and droughts are nothing new, of course. But on that stifling June day a top atmospheric scientist testifying on Capitol Hill had a disturbing message for his senatorial audience: Get used to it').[4]

But as we have seen in the previous paragraphs, the various actors that formed this discourse coalition did not necessarily agree with one another on what climate change meant, how serious it was or what kind of actions would be necessary. Central dilemmas in this regard were the appropriate politicization of the topic and whether climate change was to be seen as a local or a global problem. We will see in the following paragraphs how these questions recur when climate change discourse becomes institutionalized in the establishment of the IPCC, and treaties such as the UNFCCC and the Kyoto Protocol.

4 'FIRST GET THE FACTS RIGHT': THE IPCC AND THE DRAFTING OF A SCIENTO-POLITICAL CONSENSUS

The tasks for the newly established IPCC were to be divided between three working groups, which would prepare assessment reports on the available scientific information on climate change (WG1), environmental and socio-economic impacts (WG2), and the formulation of response strategies (WG3). The organizational arrangements reflected the conflicting desires of the actors involved; UNEP had wanted to replicate the ozone successes, the WMO had wanted to broaden the peer-review to prevent 'capture' by a relatively small group of scientists, whereas the US wanted to have a firm grip on the process to prevent the creation of a new political reality by scientific experts (Miller 2004: 56). In the end an *intergovernmental* structure was chosen: formal authority for the three working groups and their publications would be carried by a Plenary, consisting of formal government representatives of all member states. The Plenary established only a small bureau and relied on voluntary contributions from member countries to fund the actual assessment work.

Interestingly, the light 'network' organizational format was combined with the choice to do a *comprehensive* assessment. It was a major design decision—after all, the IPCC could also have confined itself to the less contested and more authoritative physical science, which would have been more in line with the reports of the 1970s and 1980s (Agrawala 1998: 615). Yet the IPCC had a double task: to get the facts right *and* to get member states to agree. Hence it had to also investigate the possible negative effects of climate change, in particular in developing countries.

The first IPCC assessment reports outlined the contours of the future climate change regime and would become the basis for the United Nations Framework Convention on Climate Change, discussed at the Earth Summit of Rio de Janeiro in 1992. Contrary to the UNEP's original request to convene under auspices of the Panel, an International

Negotiating Committee was established. The objective of the UNFCCC is defined in article 2, which states that:

> The ultimate objective of this Convention and any related legal instruments that the Conference of the Parties may adopt is to achieve, in accordance with the relevant provisions of the Convention, stabilization of greenhouse gas concentrations in the atmosphere at a level that would prevent dangerous anthropogenic interference with the climate system. Such a level should be achieved within a time frame sufficient to allow ecosystems to adapt naturally to climate change, to ensure that food production is not threatened and to enable economic development to proceed in a sustainable manner.

Guiding principles in attaining this should be equity, special needs, precaution, cost-effectiveness and comprehensiveness, sustainable development, and the (openness of the) international economic system. Whereas some of these principles are common in international law, we can also see the active alignment with the then prevailing neoliberal ideas on the global order.[5] Given the influence of this Framework Convention on the development of later treaties, such as the Kyoto Protocol, it is important to keep this connection of the UNFCCC to market principles in mind.

5 WHAT'S A SINK? THE ECONOMIZATION OF CLIMATE CHANGE DISCOURSE

The new climate change regime is a textbook example of ecological modernization: its framing of the environmental problem combines monetary units with discursive elements derived from natural science; portrays environmental protection as a positive sum-game (although the potentially enormous costs of climate change are considered, the UNFCCC draws on the discourse of sustainable development), and assumes that issues of environmental degradation are calculable (Hajer 1995: 26). The ecological metaphors—such as climate system, ecosystem, the idea of a dangerous level, and the possibility of natural adaptation—acquire a new function in this neoliberal discourse. As Shiva points out (in Sachs 1994), defining the environment in terms of the global instead of the local could be considered as a discursive strategy which advantages the more developed countries by facilitating their global access to environmental resources, while globally distributing the costs of environmental damage and regulation. Whereas climate politics tends to emphasize that 'we're all in the same boat', this metaphor should not be trusted given the markedly different consequences of climate change for different regions (Meyer-Abich ibid.).

The UNFCCC was relatively weak in terms of commitments and obligations. It formulated general bottom-up commitments for all countries, focused on national inventories; specific requirements for industrialized countries, the most important of which was to curb GHG emission to 1990 levels in 2000; and the need to transfer finances and technologies to developing countries. It was decided to periodically reassess the commitments (again a process copied from the ozone negotiations), thus creating possibilities for new science to feed the process and strengthening the role of the IPCC.

When the second IPCC assessment was published, the single most quoted sentence of the policy summary concluded that 'the balance of evidence suggests that there is a discernible human influence on global climate' (Houghton, Meira, et al. 1996: 39). In July 1996, US Under Secretary for Global Affairs Timothy Wirth said in his Remarks before the Second Conference of the Parties Framework Convention on Climate Change that 'The science is convincing; concern about global warming is real and we must continue to take steps to address this problem consistent with our long-term economic and environmental aspirations.'[6] His address was remarkable. The US had previously asked to delay policy action in order to do more research to diminish scientific uncertainty, a claim that had also been promoted by a vocal industrial lobbying group called the Global Climate Coalition. Now that the science was called convincing, it was no longer the climate scientist, but the industrial lobbyist or climate skeptic who was cast as the lonely voice.

6 FROM FACTS TO STRATEGIES

As the IPCC had achieved a basic consensus on the facts of climate change, the emphasis gradually shifted towards the options for action and the costs of action or inaction. In the mid-1990s the discourse became more economical (Hoffman 1998: 2–3). For instance, in February 1997, the environmental organization Redefining Progress initiated an *Economists' Statement on Climate Change*, signed by over 2,500 economists including nine Nobel laureates. The statement backs the conclusions of the IPCC, and states that, given the significant risks, preventive steps are justified. Casting both the problem of climate change and the solutions to it in neoliberal language, the economists proceed to argue that:

> II . . . For the United States in particular, sound economic analysis shows that there are policy options that would slow climate change without harming American living standards, and these measures may in fact improve U.S. productivity in the longer run.
>
> III. The most efficient approach to slowing climate change is through market-based policies . . . . The United States and other nations can most efficiently implement their climate policies through market mechanisms, such as carbon taxes or the auction of emissions permits. The revenues generated from such policies can effectively be used to reduce the deficit or to lower existing taxes.[7]

In the same year, the COP convened in Kyoto. The legally binding Kyoto Protocol commits thirty-seven industrialized, so-called 'Annex I countries' to cut their greenhouse gas emissions by an average of 5.2 percent against 1990 levels in the period 2008–12. The Protocol, which has not been ratified by the US, entered into force in February 2005. Kyoto also introduced a number of 'flexible mechanisms' to help countries to meet their commitments in an efficient way; emission trading, clean development mechanism (CDM), and joint implementation. These market driven instruments have been criticized as a new form of colonialism, discursively emptying the forests in developing countries from the indigenous, complex meanings of place, in order to commodify them as global carbon sinks or 'Kyoto lands' (Fogel 2004, 2005; Watson, Noble, et al. 2000; Boyd 2009). Moreover, the original UNFCC 1992 definition of a 'sink' as 'any process, activity or mechanism which removes a greenhouse gas, an aerosol or a precursor of a greenhouse gas from the atmosphere' had now been narrowed down to the removal of greenhouse gases by land use change and forestry activities (Lövbrand 2009: 404). This resulted in a discursive struggle to define the meaning of sinks—for instance: should the avoidance of deforestation be included?—and the IPCC was asked to prepare a special report on land use change and forestry. Political stakes were high; one negotiator described the report as 'probably the most immediately policy-relevant and therefore potentially the most sensitive' IPCC report ever (Fogel 2005: 193). Whereas the US insisted on a broad and 'comprehensive' accounting of sinks, the EU countries emphasized the responsibilities of industrialized countries, and developing countries were struggling to find their voice. All different sides attacked the IPCC in terms of its scientific integrity and credibility, which resulted in an intensive boundary struggle, among other things about the difference between 'policy relevant' and 'policy prescriptive'. Despite these pressures, the IPCC managed to produce a relatively transparent and accessible process; whereas the conditions of participation were by no means equal, the negotiated process of knowledge-generation forced actors to defend the validity of their definitions, claims, and strategy (Fogel 2005; Lövbrand 2009). The IPCC report was unable to prevent the failure of the COP-6 (The Hague, 2000), but conditions for the discussion changed when the US withdrew from the Kyoto Protocol in March 2001 and COP-7 institutionalized carbon sequestration as a GHG mitigation strategy.

7 BEATEN BY A HOCKEY STICK? PERCEIVING/ PERFORMING CLIMATE CHANGE

The casting of the political sink-discussion in terms of scientific integrity, shows the authoritative position of the IPCC as a scientific body able to demarcate its boundaries within a highly political process. This boundary, however, would remain contested in the years to come, when assessment report 3 (2001) and 4 (2007) published their increasingly confident conclusions about the anthropogenic warming of the climate system. After all, scientists are not only perceiving but also performing climate change; their visual and statistical models provide us with the contours of our idea of global climate change (cf. Demeritt 2001).

Boundary battles about scientific uncertainty do therefore not only attack the authority of the IPCC, but also the credibility of the issue as such. This makes it especially important for the organization to maintain both its critical peer-review process and an image of consensus for the outside world, for instance by the use of an anchoring concept such as 'climate sensitivity' (Sluijs, Eijndhoven, et al. 1998). Whereas the ambiguity of climate sensitivity allowed for different interpretations, thus promoting cohesion among different social worlds, visual representations and graphs 'spoke volumes' in their conciseness, entered the media, and sparked controversy. The hockey stick controversy is a prominent example of this. The 'hockey stick' refers to a graphic in the form of a hockey stick, representing Michael Mann's model of the temperature changes of the past 1,000 years, based on proxy data. Yet the assumptions of the model, and thus the steep form of the right end of the graph, have been criticized from various angles. This resulted in a widely publicized discussion.

MacKenzie (1990) has suggested that the understanding of uncertainty depends on the distance to knowledge. So, the best of scientists (closest to knowledge) are all too aware of the conditionality of their knowledge. Consequently, they would emphasize it. For instance, Swart et al. (Swart, Bernstein, et al. 2009: 25) concluded that even more attention should be given to the selection of graphs (or tables) and explaining why they have been selected, preferably always including different sources of information in the same graph (or table) (cf. Oude Lohuis 2009). Policy makers and politicians, standing further away, have difficulty handling uncertainty, which they see as standing in the way of action. The lay public, standing way apart from the practice of academic knowledge have yet another understanding: uncertainty delegitimizes the scientific basis for (costly) action.

The popular media—who have a literally mediating role in this regard—tend to draw a double portrait of science. While the authority of science as a whole is attacked in recurring affairs, gates, and controversies, scientists are also portrayed as committed, active witnesses, who:

> 'bear witness' of the crisis for the public, if not with their own eyes then through the screen or the reports on their desks. But secondly, they also act. They measure glaciers and peer into the mouths of polar bears. They study maps and aerial photographs, and they work in bustling offices and laboratories, where they are too busy engaging with the crisis to be interviewed anywhere other than at their desks. (Lester and Cottle 2009: 930)

The message is clear: scientists are the ones to overcome uncertainty and provide us with the truth. This, however, makes them all the more vulnerable to a scandal.

8 THE VULNERABILITY OF DISCOURSE: THE WINTER OF 2009–2010 AS A CRITICAL MOMENT

On 20 November 2009 a large batch of e-mails from the Climate Research Unit (CRU) of the University of East Anglia were made public on the internet. It led to a media hype focusing on the alleged inappropriate behavior of climate scientists. The suggested employment of a 'trick', the suggestion of an active attempt to keep critics out of representation in the IPCC assessment, and the general sense of a fight between different groups in science created a confusion in the public debate. Precisely at the time of the preparations of the COP-15 at Copenhagen, doubt was sewn on the climate science underlying the need for a global contract on climate abatement policies.

Within days the leaked e-mails were framed as 'Climategate', thus suggesting it revealed corruption and great social injustice (as was the case in the original 'Watergate' scandal). 'Climategate' led to a broader questioning of the practices of climate science, in particular the global institutional vehicle of the IPCC.

Then, in December 2009 the long-awaited 15th Conference of the Parties took place at Copenhagen, Denmark. Politicians, policy makers, NGOs, and media had framed it as the moment of the 'decisive act': the moment at which the world community, after years of preparation, would decide on joint action and a legally binding agreement on the reduction of greenhouse gas emissions in an inclusive follow-up to the Kyoto accord. When the conference closed no binding deal was on the table, the idea of united action of the COP was shattered, and new coalitions seemed to shape up. By the time people returned to work in January and February of 2010, errors had been found in the 3,000-page Fourth Assessment Report of the IPCC of 2007 and doubt proliferated. Suddenly the IPCC was questioned as the reliable foundation for climate policy. IPCC science, which had been the cornerstone for global politics, now was framed as a liability.

Time will tell what the meaning of the winter of 2009–10 will be for climate change policies. Yet right now it is safe to observe the fragility of (climate) discourses. Institutions may seem robust but in actual fact they need constant discursive reproduction. At the heart of it lies the question whether the particular institutional forms were flexible enough to cope with the changing political contexts. The UNFCCC officials worked relentlessly towards a global agreement, and science supported precisely this trajectory. The analysis of the dynamics at Copenhagen suggests, however, that the geopolitical situation changed. The joint treaty did not materialize. The eventual 'accord' was a 'deal' of the US with the new powerful nations such as China, India, Brazil, and South Africa. Arguably this represents the new reality.

The winter of 2009–10 calls for a rethinking of the role of the UNFCCC. The journalist Thomas Friedmann captured this when he suggested it was based on an 'earthday' format according to which everybody collaborates based on a joint moral agreement. As an alternative he suggested the frame of an 'earthrace': the idea that different countries, China and the US in particular, might be more motivated by a framing which constructs climate change and environmental degradation in terms of a 'race' for the invention of the new generation of green technologies with the prospect of competitive advantages, new jobs, and a coupling of economic growth to a serious greening of the economy.

The winter of 2009–10 also repositioned the IPCC. It had to manoeuvre between science and politics, gaining political legitimacy by scientific peer review but relying on consensus for its authority. Political struggles—such as the definition of sinks—were cast in terms of 'sound science', scientific integrity, and credibility, while assessment reports spoke with increasing certainty about human-induced climate change. Climate scientists thus acted as the perceivers, but were to a large extent also the performers of global climate change. The popular media reproduced this image of the scientist as an (en)acting witness, while at the same time vehemently attacking it.

The winter of 2009–10 was also the moment at which the power of the emerging 'blogosphere' was felt. Voices of skeptics found their way via websites into the official public sphere. Ironically, journalistic norms like balanced coverage then promoted a biased view on anthropogenic warming, overemphasizing uncertainties and providing the new 'lonely voices' of climate skeptics with a disproportionate amount of speaking time.

The winter of 2009–10 seems to illustrate not so much the power but the limits of the global climate change discourse. First, the COP-15 showed how the geopolitical order had changed. Not the general conference, but the negotiations behind closed doors of the newly developing countries China, India, Brazil with the United States were crucial. They did not aim at a binding treaty, but at a more open agreement. Secondly, it was not the physics of climate change that mattered so much as the question how to get to a financial deal, in particular, the ways in which the North would compensate the South for measures in the sphere of climate change adaptation. Thirdly, the public discussion ordered the politics back in. On the one hand through the accusation that the UNFCCC and IPCC processes were corrupted by politics, on the other hand by speaking out against particular ways of addressing climate change. Here new cleavages arise. By far the single biggest indicator of American's feelings about global warming is their party political affiliation (ACVS 2008), and reactions to communication about climate change differ markedly dependent upon previously existing attitudes. While IPCC might have brought us a firm understanding of the basics of climate science, it has not resolved the politics.

9 Conclusion

The science of climate change was performed and given shape by various actors. In the twentieth century, the climate came to be conceptualized as a global system instead of an aggregation of weather averages, with climate scientists as the only people able to 'perceive' and perform this global system. As the discourse structuralized, 'climate change' became an important frame to understand environmental and meteorological phenomena. What is more, during the 1990s it became the prime emblem of the ecological challenge facing the world.

New emblems carry over elements of the past. The newly developing climate discourse was influenced by the previous ozone debate and the discourse of sustainable development, and institutionalized in the IPCC, UNFCCC, and its various agreements, with CDM and carbon trading as its main, market-based instruments, and a general agreement or treaty as its vehicle.

Entering the second decade of the twenty-first century we can see how the particular configuration that dominated the debate on climate change was a product of its time. New geopolitical relationships erode the previously seemingly self-evident focus on the United Nations as forum for a global agreement. The emergence of the blogosphere eroded the stability that the IPCC had provided in providing the science base for policy making. The discursive construction of climate change as the emblematic environmental problem led to the latching on to the climate change negotiations of various other agendas. Climate change became a 'Christmas tree' in which a variety of other concerns found a place.

The discursive format, in the end, was one of an encompassing global accord. Precisely this striving to a unified globally agreed treaty based on unequivocal scientific knowledge contributed to the loss of momentum. It is too early to tell whether this 'UNification model' is merely going through a difficult phase or is out for good.

Notes

1. The WCP was divided into two research strands: the WCIRP (World Climate Impact Assessment and Response Strategies Programme, to be coordinated by the UNEP) and the much bigger WCRP (World Climate Research Programme (WCRP, joint responsibility of WMO and ICSU).

2. <http://www.icsu-scope.org/downloadpubs/scope29/statement.html>

3. <http://discovermagazine.com/1988/oct/23-special-report-endless-summer-living-with-the-greenhouse-effect/article_view?b_start:int=0&-C=>
4. <http://discovermagazine.com/1988/oct/23-special-report-endless-summer-living-with-the-greenhouse-effect/article_view?b_start:int=0&-C=>
5. For instance, article 3.4 on sustainable development argues that 'The Parties have a right to, and should, promote sustainable development. Policies and measures to protect the climate system against human-induced change should be appropriate for the specific conditions of each Party and should be integrated with national development programmes, taking into account that economic development is essential for adopting measures to address climate change,' and 3.5 adds to this that 'The Parties should cooperate to promote a supportive and open international economic system that would lead to sustainable economic growth and development in all Parties, particularly developing country Parties, thus enabling them better to address the problems of climate change. Measures taken to combat climate change, including unilateral ones, should not constitute a means of arbitrary or unjustifiable discrimination or a disguised restriction on international trade.'
6. Remarks before the Second Conference of the Parties to the Framework Convention on Climate Change, Geneva, Switzerland, 17 July 1996. Wirth also explicity addressed the allegations of deception, stating that: 'We are not swayed by and strongly object to the recent allegations about the integrity of the IPCC's conclusions. These allegations were raised not by the scientists involved in the IPCC, not by participating governments, but rather by nay-sayers and special interests bent on belittling, attacking, and obfuscating climate change science. We want to take this false issue off the table and reinforce our belief that the IPCC's findings meet the highest standards of scientific integrity.'
7. <http://www.rprogress.org/publications/1997/econstatement.htm>

REFERENCES

ABBE, C. 1889. Is our climate changing? *Forum* 6: 687–8.

ACVS 2008. *Research Summary October 2008: The American Climate Values Survey: Moving Towards a Tipping Point.*

AGRAWALA, S. 1998. Context and early origins of the intergovernmental panel on climate change. *Climatic Change* 39: 605–20.

BOEHMER-CHRISTIANSEN, S. 1994. Global climate protection policy: The limits of scientific advice. *Global Environmental Change* 4(2): 140–59.

BOYD, E. 2009. Governing the Clean Development Mechanism: global rhetoric versus local realities in carbon sequestration projects. *Environment and Planning A* 41: 2380–95.

DEMERITT, D. 2001. The construction of global warming and the politics of science. *Annals of the Association of American Geographers* 91(2): 307–37.

FLEMING, J. 2005. *Historical Perspectives on Climate Change.* Oxford: Oxford University Press.

FOGEL, C. 2004. The local, the global and the Kyoto Protocol. Pp. 103–26 in S. Jasanoff and M. Long Martello (eds.), *Earthly Politics: Local and Global in Environmental Governance.* Cambridge, MA: MIT Press.

—— 2005. Biotic carbon sequestration and the Kyoto Protocol: The construction of global knowledge by the Intergovernmental Panel on Climate Change. *International Environmental Agreements* 5: 191–10.

HAJER, M. 1995. *The Politics of Environmental Discourse: Ecological Modernization and the Policy Process*. Oxford: Oxford University Press.

—— 2009. *Authoritative Governance: Policy-Making in the Age of Mediatization*. Oxford: Oxford University Press.

—— and VERSTEEG, W. 2005. A decade of discourse analysis of environmental politics: Achievements, challenges, perspectives. *Journal of Environmental Policy and Planning* 7(3): 175–84.

HAMBLYN, R. 2009. The whistleblower and the canary: Rhetorical constructions of climate change. *Journal of Historical Geography* 35: 223–36.

HECHT, A., and TIRPAK, D. 1995. Framework agreement on climate change: A scientific and policy history. *Climatic Change* 29: 371–402.

HOFFMAN, A. 1998. *Global Climate Change: A Senior-Level Debate at the Intersection of Economics, Strategy, Technology, Science, Politics and International Negotiation*. San Francisco: New Lexington Press.

HOUGHTON, J., MEIRA, G., et al. (1996). *Climate Change 1995: The Science of Climate Change*. Cambridge: Cambridge University Press.

JASANOFF, S. 2004. *States of Knowledge*. London: Routledge.

LESTER, L., and COTTLE, S. 2009. Visualizing climate change: Television news and ecological citizenship. *International Journal of Communication* 3: 920–36.

LÖVBRAND, E. 2009. Revisiting the politics of expertise in light of the Kyoto negotiations on land use change and forestry. *Forest Policy and Economics* 11: 404–12.

MACKENZIE, D. 1990. *Inventing Accuracy: A Historical Sociology of Nuclear Missile Guidance*. Cambridge, MA: MIT.

MILLER, C. A. 2004. Climate science and the making of a global political order. Pp. 46–66 in S. Jasanoff (ed.), *States of Knowledge*. London: Routledge.

OUDE LOHUIS, J. 2009. De rol van wetenschappelijke gegevens in het klimaatdebat, het voorbeeld van de Hockeystick als historische reconstructie van het klimaat. Pp. 54–77 in A. De Kraker and J. van der Windt (eds.), *Jaarboek voor ecologische geschiedenis 2008. Klimaat en atmosfeer in beweging*. Gent: Academia Press.

SACHS, W. 1994. *Der Planet als Patient: ueber die Widersprueche globaler Umweltpolitik*. Basel: Birkhäuser Verlag.

SCHNEIDER, S. 1993. The greenhouse effect: Reality or media effect? Pp. 57–67 in *Preparing for Climate Change: Second North American Conference on Preparing for Climate Change*. Washington, DC: Climate Institute.

SLUIJS, J. V. D., EIJNDHOVEN, J. V., et al. 1998. Anchoring devices in science for policy: The case of consensus around climate sensitivity. *Social Studies of Science* 28(2): 291–323.

STANHILL, G. 2001. The growth of climate change science: A scientometric study. *Climatic Change* 48: 515–24.

SWART, R., BERNSTEIN, L., et al. 2009. Agreeing to disagree: Uncertainty management in assessing climate change, impacts and responses by the IPCC. *Climatic Change* 92(1–2).

WATSON, R., NOBLE, I., et al. 2000. *Special Report on Land Use, Land Use Change and Forestry*. The Hague: COP-6, IPCC.

WILDAVSKY, A. 1979. *Speaking Truth to Power: The Art and Craft of Policy Analysis*. Boston: Little Brown.

CHAPTER 7

....

ENVIRONMENTALITY

....

TIMOTHY W. LUKE

THIS chapter explores the utility of Foucault's analysis of governmentality, and subsequent elaborations of his project in work on 'environmentality,' for understanding global climate change and various social responses to mitigate and/or adapt to the disruptions associated with global warming. To accomplish this task, the chapter will lay out a brief theoretical account of governmentality as environmentality, and some different intellectual views of these eco-managerial practices. Next, it will stress how climate change has brought to light new issues to those viewing the world and the challenges of climate change tied to greenhouse gas (GHG) emissions through the theories of environmentality. In particular, it briefly will indicate how the theory of environmentality can shed new light on the practices of climate science, the responses of scientific experts and various agencies to global warming, and some attempts at international climate change governance. Finally, it concludes with a review of the dangers of environmentality in an era of climate change as well as the opportunities for resistance.

1 GOVERNMENTALITY AND ENVIRONMENTALITY
....

In Foucault's analysis of governance, 'the essential issue in the establishment of art of government' is quite clear, namely, the effective introduction of 'economy into political practice' (Foucault 1991: 92). As modernity unfolds, Foucault sees the influence of economy on political practices manifesting itself in the governmentalization of things, people, and their complex interrelations in territories as populations in motion. With governmentality, then, the finality of authority and expertise in government increasingly rests 'in the things it manages and in the pursuit of the perfection and intensification of processes which it directs; and the instruments of government, instead of being laws, now come to be a range of multiform tactics' (Foucault 1991: 95). The foci of such governance tactics, in turn, are populations and their welfare, since governmentality recognizes 'the population is the subject of needs, of aspirations, but it is also the object in the hands of government, aware, *vis-à-vis* the government, of what it wants, but ignorant of what is being done to it'

(Foucault 1991: 100) as the new techniques and tactics of politicized economy governmentalize the state, society, and self.

During the twentieth and twenty-first centuries, the rising pressure of human population growth on natural resource stocks, surviving wilderness areas, and terrestrial carrying capacities have created new anxieties as governmentality articulates 'new networks of continuous and multiple relations between population, territory and wealth' (Foucault 1991: 101). With the advent of political economy, a tacit by-product is the emergence of political ecology in its multiple forms, resulting, 'on the one hand, in the formation of a whole series of specific governmental apparatuses, and, on the other, in the development of a whole complex of *saviors*' (Foucault 1991: 103). At these sites and with these knowledges, the greening of governmentality in political ecology generates the practices of environmentality (Luke 1995). Environmentality marks efforts to bring governance of the state, society, and self into the ambit of 'ecoknowledges' and 'geopowers' as human and nonhuman populations are policed to provide and protect environmental biodiversity, resilience, and sustainability (Luke 1999a).

Agrawal (2005) and Braun (2003) more easily accede to a systematic reduction of the environment and biodiversity to natural capital assets to be artfully managed by well-disciplined green citizens, consumers, and collaborators in the economy of nature and society. Their relatively benign notions occlude how more expansive eco-managerialist engagements of natural resources from afar can have inequitable and/or exploitative effects by framing what resources will be managed where by whom, how, and for whose advantage (Luke 1999a, 1999b, 1997). Some regard environmentality as a more rational mode for the environmental management of forest resources (Li 2007), rural development (Agrawal 2005), resilient communities (Berkes et al. 2000), or green politics (Torgerson 1999). In these instances, environmentality operates as a more inclusive and exhaustive mode of government by experts. Hence, others regard it as another global ideology that merits popular resistance (Scott 2009; Escobar 2008; Luke 2005; Fischer 2000).

The project of environmentality, then, can be considered an evaluative index both of climate change, and of the social adaptations to it, inasmuch as this concept redirects Foucault's sense of governmentality into the continuous 24×7 management of Nature and Society through combating greenhouse gas emissions. Foucault's central insight in governmentality is how systems of knowledge and discourses of expertise have come to govern 'the conduct of conduct' (Foucault 1991: 95) in manifold relations for propounding 'the right disposition of things' (Foucault 1991: 94) by policing populations of humans and nonhumans alike up close and at a distance. Whatever complex systemic attributes that Nature is presumed to possess becomes predictable, and then protectable and exploitable as scientific experts systems (Jasanoff 1990) map and monitor its characteristics reduce its resources to manageable sites, stocks, and services (Luke 2009).

The professional papers and policy proposals of environmentalists and climatologists engaged in global warming research, therefore, are plainly significant sites for seeing environmentality at work (McNeill 2000; Berkes et al. 2000). Many evolving combinations of ecological research, technical governance, green entrepreneurialism, and futurological prediction now are a power/knowledge formation, which increasingly 'traverses and produces things. . . . It needs to be considered as a productive network which runs through the whole social body, much more than a negative instance whose function is repression' (Foucault 1980: 119).

As these discursive frames are placed around the planet's population, the practices of environmentality essentially interpose a fully ecomanagerial grid on the Earth in which there is a 'substitution of the "user" figure of everyday life, for the political figure of the "citizen"' (Lefebvre 1981: 71) to implement the resource managerialism of ecosystem management (Cortner and Moote 1999). The Earth is reduced again through these discursive frames into 'planetary infrastructures' in which humans are merely one more type of inhabitant, albeit the most central one, while each and every inhabitant also 'is reduced to a user, restricted to demanding the efficient operation of public services' (Lefebvre 1981: 79) that the networks of green experts are working to sustain, develop, and render more resilient (Berkes et al. 2000).

2 CLIMATE CHANGE AND ENVIRONMENTALITY

The conjuncture of climate change and society is a key link for the discourses and practices of environmentality to gain traction for its evolving multiform tactics of governance. Indeed, the challenges posed for society by rapid climate change, as these problems have been assessed by international governmental conferences and scientific panels since the Rio environmental summit in 1992, typifies the tendency in which 'received representations and commonly used words are insidious vehicles for a morality, an ethics and a aesthetics that are not declared to be such' (Lefebvre 1981: 71) by the policy makers and scientific experts that popularize them. Managing global climate change, on the one hand, is a central preoccupation for many environmentality networks, and the apparent urgency of acting decisively to reduce greenhouse gases, on the other hand, gives the scientific experts and green NGOs involved in this project a new collective morality, individual ethics, and common aesthetics in the production of new scientific studies without necessarily expressing that intent.

Some high-profile authority figures, like Al Gore, Jr., James Hansen, or Bill McKibben, do immediately propound a set of political imperatives from the conclusions they find in climate models, atmospheric science analyses, and global warming measurements, even though many scientists still keep their more guarded counsel in the domain of provisional hypotheses yet to be fully verified empirical analysis. Other scientific figures, however, like Rajendra Pachauri, chairman of the IPCC since 2002 (as well as the main proponent of keeping carbon dioxide concentrations less than 350 parts per million), frequently second Gore's, Hansen's, or McKibben's sense of political urgency by suggesting all precautionary interventions and anticipatory adaptations to climate change are worth accepting in coastal cities, island nations, and developing countries. Yet, until the current state-of-the-art scientific research is essentially confirmed or falsified by empirical analysis, in keeping with the dictates of experimental positivist science, one should not trust science more than provisionally.

Still, most approaches to global environmental governance largely appeal to 'technocratic expertise' (Fischer 1990) for answers that can anchor toward a more governmentalized terrestrial environment in more corporate globalist registers of power (Luke 2004). On the other hand, the greater contestedness of climate models and measurements, as both popular doubts and collective worries about rapid global warming have grown over the past two decades, highlights how the 'facts,' or 'findings,' or the 'formations' of global warming science are not firm, final, or fixed. They are instead the operational output of continuously evolving instrumental, professional, and technical practices in many fields of endeavor, ranging from atmospheric chemistry to paleoclimatology to tropical zoology. As the modes of modern climatological research, contemporary capitalist exchange, and global warming conference activity converge in green governmentality networks, one can see the formations of power/knowledge 'which shows how man in his being, can be concerned with the things he knows, and know the things that, in positivity, determine his mode of being' (Foucault 1994: 354).

Many climate change debates are evolving into the routinized operations of environmentality (Luke 2005, 1999b) as power/knowledge formations in which the few can exercise over the many in different approaches to 'green governance' (Torgerson 1999). Even for the professional-technical and intellectual classes (Hardt and Negri 2000), this globalist ideology-in-action maintains,

> The world market eliminates or supplants political action—that is, the ideology of rule by the world market, the ideology of neoliberalism. It proceeds monocausally and economistically, reducing the multidimensionality of globalization to a single, economic dimension that is itself conceived in a linear fashion. If it mentions at all the other dimensions of globalization—ecology, culture, politics, civil society—it does so only by placing them under the sway of the world-market system. (Beck 2000: 9)

In this vein, a green globalism motivates many to enact beliefs and practices that presume each state, society, and culture must be managed along the lines of a corporate capitalist enterprise. As Lefebvre observes, this turn to expert authority often leads to a negative outcome, or 'the debasement of civic life occurs in the everyday, facilitating the task of those who manage everyday life from above by means of institutions and services' (1981: 80).

Consequently, as one asks if political agendas are largely what global climate change science is all about, it is crucial to realize how 'the State and/or company must abandon the idealist and humanist narratives of legitimation in order to justify the new goal: in the discourse of today's financial backers of research, the only credible goal is power. Scientists, technicians, and instruments are purchased not to find truth, but to augment power' (Lyotard 1984: 46). Many dynamics of environmentality under today's neoliberal economic policies also have allowed 'sustainability' to be mobilized to give 'the mechanisms of competition' (Foucault 2008: 147) freer play as regulatory principles. While not discounting the centrality of commodity exchange and mass consumption, a neoliberal 'enterprise society' (Foucault 2008: 147) permits competition on energy efficiency, resource optimization, material reduction, and information intensification to serve as those mechanisms 'that should have the greatest possible surface and depth and should also occupy the greatest possible volume in society' to ensure green governmentality finds the most sustainable and developmental 'man of enterprise and production' (Foucault 2008: 147).

Global warming science in some sense becomes a field of studies about what Foucault suggests should be 'broadly called endemics, or in other words, the form, nature, extension, duration, and intensity of the illness of the population' (Foucault 2003: 243). As the biopower engagements of green governance evince themselves, they express the historic shift from power shifting from juridical logics tied to 'the right to *take* life or *let* live' to operational modalities working 'to *foster* life or *disallow* it to the point of death' (Foucault 1980: 136–8). What many regard as unsustainable fossil fuel use and excessive GHG emissions are now an embedded social endemic that characterizes the Earth's growing human populations, and environmentality seeks to reconfigure the endemic form, extent, and intensity of GHG to either halt or adapt to global warming and its many ill-effects for human populations.

The worldwide, but also a locally, nationally, and regionally articulated process of staging open contested debates about controlling climate change over the past two decades provide many splendid examples of environmentality at work. Despite hundreds of scientific meetings, thousands of new technical studies, and millions of careers blossoming up in many fields of scientific endeavor all devoted to halting the steady increase in average global temperatures since the 1992 Rio conference, the planet inexorably continues to warm. In becoming a constant concern of local authorities, average citizens, regional governments, big companies, nation-states, and international organizations, global warming successfully has changed the climate of debate about new social and environmental controls worldwide.

Not surprisingly, *Foreign Policy* magazine's December 2009 issue on '100 Top Global Thinkers of 2009' identified at least a dozen of its top hundred public intellectuals as strong voices raising questions—both positively and negatively—about how to police climate change. Global warming is perhaps the most significant engagement for what Foucault would regard as biopolitics in the twenty-first century. When viewing 'modernity at large' today, these climate change 'ecoscapes' are one of the main cultural fields of globalization (Appadurai 1996).

By leveraging greenhouse gas reduction as the cause of action, a tremendous range of new interventions into the conduct of conduct in everyday life have been made immediately more legitimate, more imperative, and more thorough to create a low-carbon economy. Voluntarily and involuntarily, the alignments of sovereignty, populations, and territory in the development of technology, use of energy, and consumption of resources are being governmentalized to advance green goals of managing climate change on the agendas of globalization (Beck 2000; French 2000; Tabb 2000; Briden and Downing 2002). From the discourse of planetary monitoring and management, with the discipline behind individual and collective carbon reduction, and in the biopolitics of optimizing ecological carrying capacity for human and nonhuman life's survival (Luke 2005), environmentality truly has come into its own by recasting the unfolding of 'Nature' and 'History' in these near totalizing terms.

The contemporary identification of 'the environment' as a focal engagement for science and government, which is worth spending the vast sums of money upon managing, transforms it into a legitimate pursuit for knowledge communities to study. As Lefebvre asserts,

> The concept of a scientific object, although convenient and easy, is deliberately simplistic and may conceal another intention: a strategy of fragmentation designed to promote a unitary and synthetic, and therefore authoritarian, model.... The sought-for-system constitutes its object by constituting itself. The constituted object then legitimates the system.... In other words, the 'real' sociological object is an image and an ideology. (2003: 56–7)

The state's and market's grudging endorsement of 'the environment' as the policy priority for green governance exemplifies these ideological tendencies toward operational analytical objectification. Yet, this object is unstable, elusive, and ideological. Many things coexist at the interface of Nature and Society, but global environmentality becomes real with certain preferred styles of examination, like the images of endangered environments and the ideologies of ecological emergency. By postulating the analytical constructs of global spatiality, the environment becomes another twist in scientists' and technologists' articulation of 'time-space,' or, if you prefer, the inscription of 'time in space,' becomes an object of knowledge (Lefebvre 2003: 73) in climatologies applied in climate policy making. Once again, however, the positivistic pretense and interventionist impulse behind scientific power and knowledge is quite problematic (Fischer 1990; Jasanoff 1990).

3 THE CHALLENGES OF ENVIRONMENTAL GOVERNANCE

With ecological footprint analysis and carbon-intensity calculators, monitoring the consumption patterns for citizens as 'users,' in turn, has become a job for many experts in climatological and ecological science (Darier 1999). Matching user demands with terrestrial services is a vitally significant assignment that an incipient environmental governance network has prepared itself to operate since Rio. In contemporary transnational capitalism, 'everyone knows how to live.' Yet, as other domains of disciplinary discourse illustrate, 'they know it thanks to a knowledge that does not originate with them, which they have assimilated, and which they apply to their own individual cases, managing their personal affairs—their everyday lives—in accordance with the models developed and diffused for them' (Lefebvre 1981: 81). Indeed, users are shaped to serve as critical receptacles of a particular culture, or 'a mixture of ideology, representations and positive knowledge. The enormous culture industry supplies specific products, commodities to which users have a "right," so that the output of this industrial sector no longer has the appearance of commodities but, rather, of objects valorized by them and destined exclusively for use' (Lefebvre 1981: 80). 'Use' here is clearly a mystification of many acts and artifacts in action, but so too is the role of the knowledge communities and sciences needed by the cultivation of environmental governance (Nowotny et al. 2001; Sklair 2001; Torgerson 1999) by transnational globalists (Ohmae 1990; Reich 1991; Fukuyama 1992; Mittelman 2000).

   The strategic spaces of the planetary environment are where conflicts over measuring and then mitigating global climate changes occur by socially creating Nature as the site of global warming (Evernden 1992). As Lefebvre asserts, any human space is social practice, but the political practices of highly visible and organized scientific networks and interest groups regard 'the Earth' now,

> as an ideology in action—an empty space, a space that is primordial, a container ready to receive fragmentary contents, a neutral medium into which disjointed things, people and habitats might be introduced. In other words: incoherence under the banner of coherence, a cohesion grounded in scission and disjointedness, fluctuation and the ephemeral masquerading as stability, conflictual relationships embedded within an appearance of logic and operating effectively in combination. (Lefebvre 1991: 308–9)

Regrettably, as the global climate change networks gel, their environmentality agendas find great merit in fragmenting the natural world into their respective domains of empty analytical space, as another site for their services as policy-making experts, to fill with new fragmentary and disjointed studies.

Environmentality arguably is another distinctive sign of 'the postmodernization of the world economy' through which 'the creation of wealth tends evermore toward what we call biopolitical production, the production of social life itself, in which the economic, the political, and the cultural increasingly overlap and invest one another' (Hardt and Negri 2000: xiii). How the economic, political, and cultural overlap and invest each other in the foci of green governmentality researchers approaching human and nonhuman, organic and inorganic, animate and machine populations 'as a datum, a field of intervention and as an objective of governmental techniques' (Foucault 1991: 102). Whether the environmentaliz-ing agency in question attends to natural recourses as a limited stock to be husbanded, an embedded system to be administered, or a continuous service to be optimized, one sees these ecological factors interacting with many populations 'in which are articulated the effects of a certain type of power and reference to a certain type of knowledge' enabling 'the machinery by which the power relations give rise to a possible corpus of knowledge, and knowledge extends and reinforces the effects of this power' (Foucault 1980: 69).

The 2009 Copenhagen summit provides another trace of the ecomanagerial project of green governmentality working through its precepts and practices on a global scale. At the root of climate change reduction, mitigation, or administration is a managerial impulse intent on working at the global scale. In accordance with Rio, Kyoto, or Bali summits, the process of planetarian resource managerialism is one of collaborative and collective governance. As the G-8 group of developed economies (a late Cold War label for a bloc of wealthy nations that *The Economist*, for example, now numbers with twenty nation-states) and G-77 group of developing economies (another tag from the 1980s for a group of poorer developing, underdeveloped, or even failed states that now number around 130) maneuver for marginal advantages in the struggle to reduce GHG emissions back to 1990 levels, hold global temperature increases to 1.5–2 °C, or cut all global GHG emissions 80 percent by 2050.

Without any legally binding accords, the more normalizing, disciplinary, and surveillant aspects of these international negotiations come to the fore as environmentality engage-ments for power to shape individual, group, national, and global behavior. The climate change conference cycle is resource managerialism brought to a more complete articulation by recognizing that the baneful by-products of fossil-fueled economies must be adminis-tered as closely and completely as their beneficial products. Moreover, much of collabora-tive governance activity now occurs in a register of accounts comparing national first-entrant, current high-emissions, large GHG legacy G-20 country profiles to late-comer, current lower/low-emissions, small GHG legacy G-77 country profiles in some effort to attain environmental justice for the latter group without abridging excessively the basic prosperity of the former group.

Environmentality relies upon these *dispositifs* (apparatuses) of resource managerialism as well as systemic surveillance to produce its governance effects. Sometimes described as 'machines for government,' *dispositifs* run best in atmospheres of crisis, emergency, or the urgent, such as those generated by Al Gore's climate change writings, to generate 'an effect that is more or less immediate' (Agamben 2009: 8). The *dispositif* fascinates Agamben, and he asserts all *dispositifs* must display the capacity 'to capture, orient, determine, intercept, model, control, or secure the gestures, behaviors, opinion, or discourses of living beings' (Agamben 2009: 14) to attain the new subjectifications needed for power to circulate effectively. Hence, as Foucault notes, 'individuals are vehicles of power, not its points of articulation' (1980: 98). The sustainability ethics of global climate change politics indicate how subjects/bodies/individuals act on themselves first and then pass such power effects on. The environmentality at the heart of greenhouse gas reduction is, once again, a green governmentality whose *dispositifs* leave everyone 'enjoined to take care and responsibility for our own lives, health, happiness, sexuality, and financial security' (Dean 1996: 211). As a result, the endemic of GHG is being addressed in a fashion that seeks to safeguard the vitality of the world's human, and nonhuman, population, but in different ways and varying rates defending on whether it is being done in China, the United States, Germany, or the Philippines.

Once in the grip of climate modelers, atmospheric science experts, and global complexity research foundations, an implied logic of granting experts limited provisional authority to command-and-control through analysis helps administer these ecosystems (Luke 2009). To track down greenhouse gas emissions, the modelers' maps soon become the terrain. In other words, with environmentality at work,

> models are presented as the product of objective analyses, described as 'systemic', which, on a supposedly empirical basis, identify systems of subsystems, partial 'logics', and so on. To name a few at random: the transportation system; the urban network; the tertiary sector; the school system; the work world with its attendant (labour market, organizations and institutions; and the money market with its banking-system. Thus, step by step, society in its entirety is reduced to an endless parade of systems and subsystems, and any social object whatsoever can pass for a coherent entity. (Lefebvre 1991: 311)

From such methodologies, the current global warming crisis and its ascription mostly to carbon dioxide emissions has evolved by fits and starts in the scientific communities at least since McKibben's popularization of the problem in *The End of Nature* (1989).

As 'the environmental' is reconfigured here as a coherent pacified entity ready to be managed beyond sovereign territoriality or accepted as a given facticity for 'governance' on a planetary scale, transnational 'moderating forces' do not wish to waffle over how often such climatology can be questioned inasmuch as 'all we have here is a tautology masquerading as science and an ideology masquerading as a specialized discipline. The success of all such "model-building," "simulation" and systemic analysis reposes upon an unstated postulate— that of a space underlying both the isolation of variables and the construction of systems. This space validates the models in question precisely because the models make the space functional' (Lefebvre 1991: 311–12). Nonetheless, 'daily life is where "we" must live; it is what has to be transformed' (Lefebvre 1981: 66). The prospects of attaining ecological sustainability, as well as the reality of it arriving as political domination through technical expertise, are becoming more social concerns for citizens as 'users' rather than autonomous agents (Fischer 2000). In fact, as users, environmentality practices assure that 'as they encounter more obstacles, barriers and blockages than ever in modernity, how is it that so many people have not realized that they were coming up against the boundaries of daily life, boundaries that are invisible, yet cannot be crossed because of the strength of daily life? They come up against these boundaries like insects against a window pane' (Lefebvre 1981: 66).

The United Nations Climate Change Conference, 7–18 December 2009, or COP-15, continued these trends in the many diverse and difficult practices of maintaining green governance patterns by permitting so many divergent clusters of conduct to converge with enough progress to permit new potential to be glimpsed as well as sufficient impasse to prevent any truly significant change in GHG emissions from being realized. Of course, the enormity of the threat was dramatically reaffirmed, the urgency for action was once again underscored, but the option for making a significant decision was pushed off until November 2010, the major choices to be made at Cancún in Mexico at the next UN climate change conference—COP-16. The utility of the Copenhagen proceedings, however, is undeniable, and it marks another of many more steps forward on subnational, local, and metropolitan levels of action (Luke 1994). Despite these efforts during 2009 and 2010, the general convocation of nation-states could not come to a mutually satisfactory resolution of how, and then how much, to regulate/reduce their GHG emissions.

Building on the Bali climate change accords, and in keeping with the Kyoto protocol, COP-15 again affirmed how the environmentalized 'government of men' must always 'think first of all and fundamentally of the nature of things' such that all ideas of a sustainable environmentalizing administration of things must 'think before all else of men's freedom, of what they want to do, of what they have a interest in doing, and of what they think about doing' (Foucault 2007: 49). Committing in principle to adapting to climate change as an anthropogenic reality, freedom remains central to mitigating average temperature rise around the world to 2 °C above preindustrial levels. Indeed, affirming the freedom of carbon markets, bunker fuel curtailments, GHG regulation, and national flexibility is foundational to the COP-15 accord from Copenhagen. Yet, as green governmentality expects, the Copenhagen Accord also commits all parties to more robust, rigorous, and reportable 'MRV' standards—or 'measurement, reporting, and verification' commitments to work with high professional objectivity and operation standards to oversee the integration of climate change mitigation into the free activities of all individuals and groups worldwide.

When environmentalized visions of space, which can be explicit or implicit, exalt 'conceptualized space,' or 'the space of scientists, planners, urbanists, technocratic subdividers and social engineers,' then it is not uncommon to find them eager to 'identify what is lived and what is perceived with what is conceived' (Lefebvre 1991: 38). Since they express the power/knowledge behind what is conceived, the fifth estate of science (Jasanoff 1990) often lives up to Lefebvre's fears about an eclipse of 'the lived' in their presumption of scientific authority, community knowledge, or environmental governance through their expert conceptual works-up of representational space. Indeed, Al Gore's fascination with NASA's earth imaging photography illustrates how ideology and knowledge can barely be divided as such representation 'supplants the concept of ideology and becomes a serviceable (operational) tool for the analysis of spaces' (Lefebvre 1991: 45) in environmentality.

4 ENVIRONMENTALITY AND OPTIONS FOR RESISTANCE

The fact that different versions of environmentalizing intervention all exist, compete, and angle for advantage goes without saying. At its heart, the politics of climate change mitigation too often is rooted in an adaptive environmentality intent upon only checking the rate of greater anthropogenic global warming for industrial society instead of returning to preindustrial temperature averages worldwide in order to safeguard what appears as more optimal atmospheric conditions for humans and nonhumans alike. A common style of resistance to this strategy is to contest the science of climate change per se. Partisans of most adaptive approaches to global warming, ranging from wait-and-see denialists to it-is-never-too-late geoengineers, stand ready to refashion Nature to suit their grand designs for its special transformation. These alternative environmentality networks all intend to guide individual and collective activities to meet their expert expectations for managing a continuous technologically remediated biosphere. As Foucault would observe, it is the case that the green governance regime of environmentality allows one to witness how 'there is only one true and fundamental social policy: economic growth' (Foucault 2008: 144). Hence, government rationalities soon overlap, and then begin to challenge each other, 'so that competitive mechanisms can play a regulative role at every moment and at every point in society' (Foucault 2008: 145) in the service of state sovereignty, collective safety, environmental sustainability, group security, or individual subjectivity. To interweave these sometimes complementary, but other times contradictory, goals together, 'government must not form a counterpoint or screen, as it were, between society and economic processes' (ibid.). Instead, it has to 'intervene on society as such, in its fabric and depth' (ibid.).

Climatology, then, is not a closed and certain system of discovery, and it is being challenged from many sides. Like any science, its theory and practice represent 'a will to knowledge that is anonymous, polymorphous, susceptible to regular transformations, and determined by the play of identifiable dependencies' (Foucault 1977: 200–1). By the same token, any expanse of expertly observed and administrated space, like our 'environment,' any 'locality,' or one's 'community,' always 'manifests itself as the realization of a general practical schema' rooted in orders of homogeneity, fragmentation, and hierarchy that give rise 'to multiple tactical operations directed towards an overall result' (Lefebvre 1981: 134). These problematic orderings result in many contradictory historical appearances, conceptual frameworks, or mental maps. As Lefebvre (1981: 135) would note, few thinkers, and especially those working on climatological projects of environmental governance, ever admit how fully, 'a representation of space—which is by no means innocent, since it involves and contains a strategy—is passed off as disinterested positive knowledge.'

Despite the well-meaning efforts tying together the deliberative projects of collaborative governance, collective self-management or communal administration as a disclosure of 'inconvenient truths,' many resistances soon become entangled in fairly stealthy globalist schematics of homogenized, fragmented, and hierarchical spatial practices as the endemic of global warming is addressed. At this juncture, then, it is imperative for any power/knowledge critique to investigate who sets the possibilities, what is the realm of the possible imagined to be, and how are they all to be realized?

To challenge environmentality, as it is being articulated through globalist climate change discourses, is to question today's infatuation with new systems of governance, while at the same time meeting Foucault's provisionally expressed 'first definition of critique, this general characterization: the art of not being governed too much' (Foucault 2007: 45). The efforts to formalize environmentality through local ordinances, national legislation, and international accords are expansive, growing, and thorough, as any survey of legislative, judicial, and executive outcomes in many nation-states since the 1970s affirms. Nevertheless, the green movements' relentless policing of everyday life—what light bulbs to buy, what foods to eat, which clothes to wear, how to commute, what sort of shelter to inhabit, and so forth—represents yet another proliferation of governance theories and practices beyond the political as such (Fischer 2000) in development itself.

Environmentality's embrace of 'sustainable development,' however, turns many peoples' energies towards a reinvention of deeply embedded machinic processes, complex systems, and social technics that make subtle new demands on individuals and groups to conform to flexible new sustainability criteria to sustain their social constructs of Nature (Eder 1996). As Foucault hints, this change keeps entangling them within the rationalization regime of that modernity created since the sixteenth century, but the addition of environmentality raises new worries about its green governance: 'what about this rationalization with its effects of constraint and maybe of obnubilation, of the never radically contested but still all massive and ever-growing establishment of a vast technical and scientific system' (Foucault 2007: 55). Rightly or wrongly, fresh conflicts will build here over: how green governance must work.

Projects of ecocritique, in part, draw from 'not accepting as true . . . what an authority tells you is true, or at least accepting it only if one considers valid the reasons for doing so' (Foucault 2007: 46). Environmentality appears as another 'movement through which individuals are subjugated in the reality of a social practice through mechanisms of power that adhere to a truth' (Foucault 2007: 47) as the sustainability revolution of climate change politics illustrates. While standing ready to accept these truths as valid, ecocritics assert 'the right to question truth on its effects of power and question power on its discourses of truth,' and thereby cultivate 'the art of voluntary insubordination, that of reflected intractability' (Foucault 2007: 47).

For example, some important avenues for resistance to large scale environmentality agendas are the development of 'transition towns' (Hodgson with Hopkins 2010), 'eco-villages' (Dawson 2006; Christian 2003), and 'permaculture' settlements (Holmgren 2002; Bell 1992). Anticipating the necessity of dealing with severe climatic disruptions caused by global warming and/or the chaotic difficulties of peak oil production in the near future, the proponents of these strategies are refusing to trust in green governmentality to create solutions to rapid climate change. Since environmentality tends to operate on a planetary scale of action, and depends upon international accords negotiated by large nation-states for most its policy responses, these more localist direct action strategies are taking the challenges of dealing with climate change into their hands within their own neighborhoods and towns linked together with other neighbors and towns around the world.

These efforts are more popular and collective, but locally focused campaigns. They work to create less energy-intensive, more labor-enriched, and less complex modes of economy under the control and ownership of those in the ecovillages, permacultures, and transition towns themselves. While these groups are not large, their numbers are considerable in many towns and villages around the world, and they are consciously working against the environmentality of global capitalist commerce, major nation-states, and big international organizations (Luke 1997, 1999b).

Ecocritiques can question today's historical confrontations of society with climate change by questioning the uses of still unstable climatological models as well as the purposes of rule by experts intent upon governing everyday life from afar. Voluntary simplicity, conscious frugality, and intentional community are new resistances that have been discussed during the same decades that governmentality has become green (Luke 1997; Torgerson 1999; Fischer 2000). Resistance to environmentality can easily find other sources of voluntary insubordination and reflected intractability to leverage immediately in concrete political actions.

Caution, then, is called for. Too often environmentality discourse scales up totalizing solutions off raw data, which read like the draft diktats of expert 'environmental governors.' By the same token, with a shared common faith in their own sense of good science, their expert community knowledge, and their warranted fears of impending crisis might erase discretionary space accorded to the lived, the local, and the living aspects of everyday life. Many global warming countermeasures advanced from Rio to Copenhagen are an increasingly institutionalized assembly of scientific experts, government bureau, and green NGOs. And, all together, they are intently coming together in a never-ending campaign to restructure everything artificial that extracts matter and energy from Nature by adding more information to more rightly dispose of things and conveniently arrange ends in pursuit of biopolitical efficiencies. Still, as the world's many ecovillages, permacultures, and transition towns all underscore, certain modes of resistance are not futile; and, indeed, these diverse experiments are become much more necessary as society at large confronts climate change.

References

AGAMBEN, G. 2009. *What is An Apparatus?* Stanford, CA: Stanford University Press.

AGRAWAL, A. 2005. *Environmentality: Technologies of Government and the Making of Subjects.* Durham: Duke University Press.

APPADURAI, A. 1996. *Modernity at Large: Cultural Dimensions of Globalization.* Minneapolis: University of Minnesota Press.

BECK, U. 2000. *What is Globalization?* Oxford: Blackwell.

BELL, G. 1992. *The Permaculture Way: Practical Steps to Create a Self-Sustaining World.* London: Thorsons.

BERKES, F., FOLKE C., and COLDING, J. (eds.) 2000. *Linking Social and Ecological Systems: Management Practices and Social Mechanisms for Building Resilience.* Cambridge: Cambridge University Press.

BRAUN, B. 2003. *The Intemperate Rainforest.* Minneapolis: University of Minnesota Press.

BRIDEN, J. C., and DOWNING, T. E. 2002. *Managing the Earth: The Lineacre Lectures 2001.* Oxford: Oxford University Press.

CHRISTIAN, D. 2003. *Creating a Life Together: Practical Tools to Grow Ecovillages and Intentional Communities.* Gabriola Island, BC: New Society.

CORTNER, H. J., and MOOTE, M. A. 1999. *The Politics of Ecosystem Management.* Washington, DC: Island Press.

DARIER, E. 1999. *Discourses of the Environment.* Oxford: Blackwell.

DEAN, M. 1996. Foucault, government, and the enfolding of authority. Pp. 209–29 in A. Barry, T. Osborne, and N. Rose (eds.), *Foucault and Political Reason: Liberalism, Neo-Liberalism, and the Rationalities of Government.* Chicago: University of Chicago Press.

DAWSON, J. 2006. *Ecovillages: New Frontiers in Sustainability.* Totnes: Green Books.

EDER, K. 1996. *The Social Construction of Nature: A Sociology of Ecological Enlightenment.* London: Sage.

ESCOBAR, A. 2008. *Territories of Difference: Place, Movements, Life*. Durham, NC: Duke University Press.

EVERNDEN, N. 1992. *The Social Creation of Nature*. Baltimore: Johns Hopkins University Press.

FISCHER, F. 1990. *Technocracy and the Politics of Expertise*. London: Sage.

—— 2000. *Citizens, Experts, and the Environment: The Politics of Local Knowledge*. Durham, NC: Duke University Press.

FOUCAULT, M. 1977. *Language, Counter-Memory, Practice*. Ithaca: Cornell University Press.

—— 1980. *The History of Sexuality, Vol. I*. New York: Vintage.

—— 1991. Governmentality. In G. BURCHELL, C. GORDON, and P. MILLER (eds.), *The Foucault Effect: Studies in Governmentality*. Cambridge: Cambridge University Press.

—— 1994. *The Order of Things: An Archaeology of the Human Sciences*. New York: Vintage.

—— 2003. Society must be defended. *Lectures at the Collège de France, 1977–1978*. New York: Palgrave.

—— 2007. *The Politics of Truth*, ed. S. LOTRINGER. New York: Semiotexte.

—— 2008. *The Birth of Biopolitics: Lectures at the Collège de France, 1977–1978*, ed. M. Senellart. New York: Palgrave Macmillan.

FRENCH, H. F. 2000. *Vanishing Borders: Protecting the Planet in the Age of Globalization*. New York: Norton.

FUKUYAMA, F. 1992. *The End of History and the Last Man*. New York: Free Press.

HARDT, M., and NEGRI, T. 2000. *Empire*. Cambridge, MA: Harvard University Press.

HODGSON, J., with Hopkins, R. 2010. *Transition in Action: Totnes and District 2030. An Energy Descent Action Plan*. Devon: Transition Town Totnes.

HOLMGREN, D. 2002. *Permaculture: Principles and Pathways beyond Sustainability*. Hepburn, VIC: Holmgren Design Services.

JASANOFF, S. 1990. *The Fifth Branch: Science Advisers as Policymakers*. Cambridge, MA: Harvard University Press.

LEFEBVRE, H. 1981. *The Critique of Everyday Life, Vol. 3: From Modernity Towards a Metaphilosophy of Daily Life*. London: Verso.

—— 1991. *The Production of Space*. Oxford: Blackwell.

—— 2003. *The Urban Revolution*. Minneapolis: University of Minnesota Press.

LI, T. M. 2007. *The Will to Improve: Governmentality, Development, and the Practice of Politics*. Durham, NC: Duke University Press.

LUKE, T. W. 1994. Placing powers, siting spaces: The politics of global and local in the new world order. *Environment and Planning A: Society and Space* 12: 613–28.

—— 1995. On environmentality: geo-power and eco-knowledge in the discourses of contemporary environmentalism. *Cultural Critique* 31: 57–81.

—— 1997. *Ecocritique: Contesting the Politics of Nature, Economy and Culture*. University of Minneapolis: Minnesota Press.

—— 1999a. Training eco-managerialists: Academic environmental studies as a power/knowledge formation. Pp. 103–20 in F. Fischer and M. Hajer (eds.), *Living with Nature: Environmental Discourse as Cultural Politics*. Oxford: Oxford University Press.

—— 1999b. Environmentality as green governmentality. Pp. 121–51 in ERIC DARIER (ed.), *Discourses of the Environment*. Oxford: Blackwell.

—— 2004. Ideology and globalization: From globalism and environmentalism to ecoglobalism. Pp. 67–77 in Manfred Steger (ed.), *Rethinking Globalism*. Lanham, MD: Rowman & Littlefield.

—— 2005. Environmentalism as globalization from above and below: Can world watchers truly represent the Earth? Pp. 154–71 in P. Hayden and C. El-Ojeili (eds.), *Confronting Globalization: Humanity, Justice and the Renewal of Politics*. New York: Palgrave Macmillan.

—— 2009. Developing planetarian accountancy: Fabricating nature as stock, service, and system for green governmentality. *Current Perspectives in Social Theory* 26: 129–59.

LYOTARD, J.-F. 1984. *The Postmodern Condition*. Minneapolis: University of Minnesota Press.

McKibben, B. 1989. *The End of Nature*. New York: Knopf.

McNeill, J. R. 2000. *Something New under the Sun: An Environmental History of the Twentieth-Century World*. New York: Norton.

Mittelman, J. 2000. *The Globalization Syndrome*. Princeton: Princeton University Press.

Nowotny, H., Scott, P., and Gibbons, M. 2001. *Re-thinking Science: Knowledge and the Public in an Age of Uncertainty*. Cambridge: Policy Press.

Ohmae, K. 1990. *The Borderless World: Power and Strategy in an Interlocked Economy*. New York: Harper & Row.

Reich, R. 1991. *The Work of Nations: Preparing Ourselves for 21st Century Capitalism*. New York: Knopf.

Scott, J. 2009. *The Art of Not Being Governed: An Anarchist History of Upland Southeast Asia*. New Haven: Yale University Press.

Sklair, L. 2001. *The Transnational Capitalist Class*. Oxford: Blackwell.

Tabb, W. 2000. *The Amoral Elephant: Globalization and the Struggle for Social Justice in the Twenty-First Century*. New York: Monthly Review Press.

Torgerson, D. 1999. *The Promise of Green Politics: Environmentalism and the Public Sphere*. Durham, NC: Duke University Press.

PART III

SCIENCE, SOCIETY, AND PUBLIC OPINION

....

CHAPTER 8

THE PHYSICAL SCIENCES AND CLIMATE POLITICS

....

HANS VON STORCH, ARMIN BUNDE,
AND NICO STEHR

1 ORIENTATION

....

IN the following sections, the physic-ness of climate science is discussed. One of the motivations for doing so is related to the observation that some scientists, trained as physicists, often play a very influential role as political actors, when interpreting and explaining the significance of scientific insights for policy implications using the 'linear' model, according to which knowledge leads directly to first political consensus and then decision making, while social and cultural processes related to preferences and values represent mostly invalid disturbances (e.g. Beck 2010; Curry and Webster 2010). Of course, the linear model means that those in control of the knowledge ought to be in control of the outcome of the political decision process.

We therefore thought it meaningful to examine to what extent this claim of political competence is warranted. We find that it is not. In order to become societally relevant, climate science has to become trans-disciplinary, by incorporating the social-cultural dimension.

We could also have done an analysis with fields related to ecology or economics. A similar phenomenon is also observed among some high-profile members of these groups; that scientists find it difficult to balance the authority of scientific competence, limitations, and integrity with the need to engage one's own values. We limit ourselves here to physics, first because this field is likely the most important.

Our chapter features three main sections.

In section 2, we discuss the historical development of the concept of climate leading us from an anthropocentric view to a strictly physical world-view, and one that is now moving once again towards a more anthropocentric view—this time concerning not only the impacts but also the drivers. This is not meant as a general review of the history of climate sciences, which is done competently by Weart in this volume. Instead we want to emphasize the circularity in the development, from an anthropocentric view, over an impassionate, distanced truly physical view, back to an anthropocentric view.

In section 3, a series of physical issues, from modeling, over parameterizations, the impossibility of experimentation, and data problems are discussed.

In section 4, the concept of 'post-normal' science is introduced, which is related to high uncertainties in the field of climate research, and the high stakes on the societal side. Here, at the boundary between science and policy, new dynamics emerge, which have little to do with physics; dynamics which depend on culture and history, on conflicting interests and world-views.

A brief concluding section argues for the need for a trans-disciplinary approach to climate in order to assist in developing policies consistent with physical insights and cultural and social constraints.

2 HISTORY OF CLIMATE SCIENCE

Historically, 'climate' was considered part of the human environment. Alexander von Humboldt ([1845] 1864: 323–4) in 1845 in his book *Cosmos: A Sketch of a Physical Description of the Universe* defined climate as the sum of physical influences, brought upon humans through the atmosphere:

> The term climate, taken in its most general sense, indicates all the changes in the atmosphere, which sensibly affect our organs, as temperature, humidity, variations in the barometrical pressure, the calm state of the air or the action of opposite winds, the amount of electric tension, the purity of the atmosphere or its admixture with more or less noxious gaseous exhalations, and, finally, the degree of ordinary transparency and clearness of the sky, which is not only important with respect to the increased radiation from the earth, the organic development of plants, and the ripening of fruits, but also with reference to its influence on the feelings and mental condition of men.

Thus, like astronomy, climate in much of nineteenth-century discourse was subject to an anthropocentric view. The global climate was little more than the sum of regional climates (cf. Hann 1903), and the challenge was to faithfully describe regional climates by measuring and mapping the statistics of their weather. Not surprisingly, a large body of information was generated, dealing with the impacts of climate on people and societies. It was the time of the prominence of the perspective of climatic determinism (Fleming 1998; Stehr and von Storch 1999, 2010). At the turn of the nineteenth and twentieth centuries, questions were formulated more in terms of climate as a physical system (e.g. Friedmann 1989; see also the systematic approach presented by Arrhenius 1908), and meteorology and oceanography became 'physics of the atmosphere' and 'physics of the ocean'. Climate was no longer primarily considered an issue of the field of geography, but of meteorology and oceanography, and climate science became 'physics of climate' (e.g. Peixoto and Oort 1992).

Since the 1970s the notion that unconstrained emissions of greenhouse gases into the atmosphere generated by human activities will lead to significant changes of climatic conditions—a theory first proposed by Svante Arrhenius (1896)—was supported by evidence of a broad warming and finally embraced by the majority of climate scientists. The series of Assessment Reports by the Intergovernmental Panel of Climate Change (IPCC) are central to and document this change. In the 1990s human-driven climate change became the absolutely dominant topic in climate sciences (Weart 1997, 2010). Climate research became to a large extent driven by concern with human-made climate and its impacts.

Unnoticed by most climate scientists, the developments in the last decades represent a return to the original but transformed *anthropocentric* view of the issue of climate (Stehr and von Storch 2010): In contrast to the perspective of 'climatic determinism', it was no longer the idea that climate *determines* the functioning and fate of societies, but that climate *conditions* human societies (Stehr and von Storch 1997).

3 METHODICAL CHALLENGES OF THE PHYSICS OF CLIMATE

In this section we outline, after a brief retrospect of the success of physics, several concepts in climate science, which are not normally met in conventional physics—and thus represent serious obstacles from a physics point of view. One of the obstacles is the absence of 'the equations' and the need for parameterizations; another is the difficulty to 'predict' and finally the issue of inhomogeneity of data.

The pillars of the success story of physics in the last two centuries are the unbiased observation and description of natural phenomena, the reproducibility of experimental data, and the mathematical description of the empirical results leading to a generalization of the experimental results and the elucidation of the underlying basic laws of nature. Perhaps the most prominent example is the Newtonian classical mechanics which Newton developed on the basis of Kepler's observations and Galileo's gravity experiments. In classical mechanics, the time evolution of a system (like the motion of the earth around the sun) follows Newton's equations. The important thing is that, when the state of the system is known for a certain time (for the earth sun system this is the position and velocity of the earth relative to the sun), then the time evolution of the system can be calculated rigorously and precise predictions can be made. By solving Newton's equations one can predict, for example, the trajectories of rockets, satellites, and space ships, which is the basis for modern space science.

Another example is electrodynamics which was established by Maxwell and based on the experimental and theoretical work by Coulomb, Volt, Ampere, Gauss, and others. Like Newton's equations, the celebrated Maxwell equations describe comprehensively all (classical) electrical and magnetic phenomena, and not only those that they aimed to describe initially. Among others, Maxwell's theory led to the recognition that light is an electrodynamical phenomenon.

Prerequisites of the success of physics were:

➤ the departure from the anthropocentric view of life that for the first time allowed an unbiased view onto the natural phenomena (like planetary motion);

➤ a new practice of publication: the protagonists did no longer (like the alchemists) hide their results but made them available to the public, allowing colleagues to reproduce (or falsify) them; and finally

> the norm of checking theoretical hypotheses experimentally. In the case of conflicting theories, an *experimentum crucis* is needed to decide which theory is correct. Perhaps the most important *experimentum crucis* is the Michelson experiment on the velocity of light, which forms the basis for Einstein's theory of relativity.

In climate science, at least two of these requisites do not exist. Climate science has become anthropocentric and *experimenta crucis* are not possible.

When approaching the subject of climate from a physics or mathematical point of view, the first question usually is—what is included, and how to describe it? What are its *equations*?

The climate system has different 'compartments', such as atmosphere, ocean, sea ice, land surface including river networks, glaciers, and ice sheets, but also vegetation and cycles of substances, in particular greenhouse gases (Figure 8.1). An important element of the dynamics is given by fluid dynamics of the atmosphere, ocean and ice, which are described by simplified Navier–Stokes equations. However, due to the unavoidable discrete description of the system, turbulence cannot be described in mathematical accuracy, and the equations need to be 'closed'—the effect of friction, in particular at the boundaries between land, atmosphere, and ocean, need to be '*parameterized*' (e.g. Washington and Parkinson 2005).

Additional equations describing the flow and transformations of energy are needed— part of this may be described by the first law of thermodynamics, the conservation of energy. In these equations we find source and sink terms, which are related to phase changes (condensation, for instance) and the interaction of cloud water and radiation. The sources and sinks often take place at the smallest scales and require additional state variables (such as the size spectrum of cloud droplets). Again, such processes cannot be taken care of explicitly—and need to be 'parameterized'.

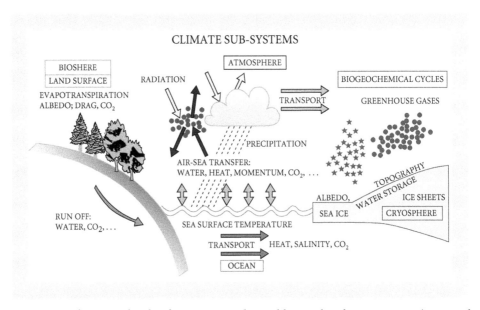

FIGURE **8.1** Schematic sketch of processes and variables in the climate system (reprinted with permission of Klaus Hasselmann)

This issue of parameterization is difficult to understand (Müller and von Storch 2004). The basic idea is that there is a set of 'state variables' $\{\Psi\}$ (among them the temperature field at a certain time t at certain *discrete* positions on the globe), which describe the system, and which dynamics is given by a differential equation $d\{\Psi\}/dt = F(\{\Psi\})$. The function F is nonlinear in Ψ and only approximately known. A rigorous analytical or numerical solution of the equations (as for Newton's equations) is impossible.

To simplify the equations, and make them tractable for a numerical treatment, one splits each of the state variables Ψ into a slowly and a rapidly varying component, $\Psi = \underline{\Psi} + \Psi'$.[1] $\underline{\Psi}$ represents that part of Ψ, which is well represented with the given spatial resolution (say 100 km), and Ψ' being the unresolved part of smaller spatial scale. The equations are then approximately written as $d\underline{\Psi}/dt = F(\underline{\Psi}) + G(\Psi')$. Here, $F(\underline{\Psi})$ describes the influence of the resolved part Ψ on the future development of $\underline{\Psi}$, whereas $G(\Psi')$ describes the influence of the non-resolved part, which is of course unknown. A conventional truncation of the equations would lead to $d\underline{\Psi}/dt = F(\underline{\Psi})$, and the non-resolved part would have no influence. This truncation is only valid in linear systems and thus unacceptable here, where the influence of the small-scale turbulence and the associated friction on the slowly varying state variable $\underline{\Psi}$ cannot be neglected. Another approximation is used, namely $G(\Psi') = H(\underline{\Psi})$. The latter is the 'parameterization'. The problem is to specify H.

The idea with the parameterization is that it would carry the information, which is to be expected from the small scales Ψ' when the resolved state is $\underline{\Psi}$. Or more precisely, $G(\Psi')$ is considered a random variable, which is conditioned by the resolved part $\underline{\Psi}$. Practically, the distribution of $G(\Psi')$ can be determined empirically—by observing the distribution of G (Ψ') when the large-scale state is $\underline{\Psi}$. The parameterization $H(\underline{\Psi})$ can then be a random realization of this distribution of $G(\Psi')$, or the $\underline{\Psi}$-conditional expectation of $G(\Psi')$.

The usage of parameterizations is normal practice in climate models, and they have turned out to make such models capable of describing the present climate, the ongoing change, and historical climates. It is plausible that the parameterizations are valid 'closures' also in a different climate (after all, in terms of physical (but not societal) magnitudes, any climate change would represent only a minor change), but the final evidence for this belief will be available only after the expected changes have taken place, have been observed and analyzed.

There are two important aspects of parameterizations.

One is a *linguistic* aspect, namely that in the language of climate modelers, parameterizations are named 'physics', a shorthand for 'unresolved physical processes'. For a person uncommon with the culture of climate sciences, this terminology may go with the false connotation that parameterizations would be derived from physical principles. While the functional form of the parameterization $H(\underline{\Psi})$ may be motivated by a physical plausibility argument, the specific parameters used are either guessed, fitted to campaign or laboratory data, or to make the model skilful in reproducing the large-scale climate $\underline{\Psi}$. Thus, the word 'physics' points to semi-empirical 'tricks'.

Another aspect of parameterizations is their *strong dependence on the spatial resolution*. When the model is changed to run on a higher resolution, the parameterizations need to be reformulated or respecified. There is no rule how to do that, when the spatial resolution is increased—which means that the difference equations do not converge towards a pre-specified set of differential equations, or, in other words: there is nothing like a set of differential equations describing the climate system per se, as is the case in most physical disciplines.

To summarize: In climate science, there are not 'the equations' but only useful approximations, which crucially depend on the spatial resolution of the system. This aspect causes many misunderstandings, in particular among mathematicians and physicists who often enough demand to see 'the equations'.

Unlike most physics disciplines climate cannot be associated with *spatial and temporal scales* in a certain limited range—instead climate varies on all spatial scales (on Earth) and extends across several magnitudes of timescales, from short-term events, measured in seconds, via timescales of decades and centuries, to geological timescales of millennia and more years. If we look at the relevant processes in the climate components atmosphere and ocean, we find a continuum of scales, as displayed in Figure 8.2. The implication is that there are hardly independent observations from different locations, and the temporal memory extends across many decades of years.

As a practical rule, the World Meteorological Organization (WMO) mandated a hundred or more years ago that thirty-year time intervals would represent 'normal climatic conditions'; every thirty years new normals are determined. If we accept this somewhat arbitrary number of thirty years, we have to wait about thirty years to get a new realization of the climate system, which is at least somewhat independent of previous states. Thus, tests of hypotheses, derived from historical data, using new data are hardly possible.

Real *experiments*, in the sense of paired configurations, which differ in a limited number of known details, are of course also not possible in the real world (as in any other

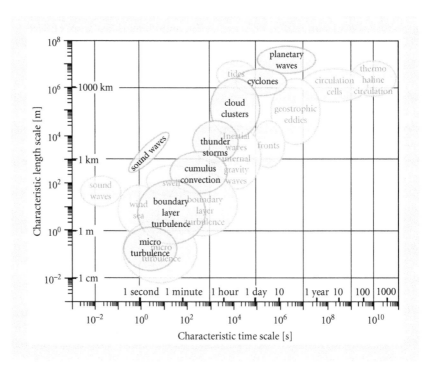

FIGURE **8.2** Spatial and temporal scales of processes in the climate components atmosphere and ocean.

geophysical set-up). However, with quasi-realistic models, which serve as a kind of virtual laboratory, it is possible to perform virtual experiments, for instance on the effect of different formations of clouds, different specifications of physiographic detail, but also on elevated greenhouse gas concentrations in the atmosphere. Independent realizations can be generated; extended long simulations are possible so that the weather noise may be reduced and the looked-after signals, caused by an imposed experimental change in the system, may be more easily isolated. The problem is, of course, that even if the models share indeed many properties with reality, it is unproven if the specific model response is realistic.

Real forecasts are also hardly possible: even if we are able to prepare a successful forecast for the coming ten or thirty years, we cannot claim the 'success' of our prediction scheme, because a single success may also have taken place by coincidence. Determining the skill of a forecast scheme needs many independent trials, as in case of weather forecasting. The long timescale in climate variability does not allow robust estimates of the skill of our methods to explore the future.

Indeed, in recent years sometimes real predictions are tried with dynamic climate models, with lead times of one or a few decades (e.g. Keenlyside 2011). The logic behind such forecasts is that the details of the emissions do not really matter for such a time horizon— as long as they exhibit some increase in the coming decades. A first attempt at forecasting the next ten years (in this case 2000–9) was in the year 2000 published by Allen et al. (2000)— now, ten years later, this prediction can be compared with the actual recent development. The scenario prepared by Hansen et al. in (1988) was retrospectively by Hargreaves (2010) considered a forecast, and compared with the recent development (Figure 8.3). In both cases (Figure 8.3), the future was well predicted. For the coming decades, a reduced warming has been predicted in an experimental forecast effort by Keenlyside et al. (2008).

Most outlooks of possible future climatic developments take the form of conditional predictions—assumed developments of greenhouse gas emissions/concentrations and other factors are used as external drivers in climate models (e.g. von Storch 2007). As such they are scenarios, namely possible future developments, and not predictions, namely most probable developments (cf. the discussion in Bray and von Storch 2009). Such scenarios are often falsely labeled as 'predictions' in the media, and even by some research institutions. They are prepared with *quasi-realistic climate models* (e.g. Müller and von Storch 2004), often abbreviated by GCM (which historically stands for General Circulation Models and not for Global Climate Models)

To summarize: most future outlooks available to the scientific community and presented to the general public are not descriptions of most probable futures (predictions) but plausibly consistent and possible futures (scenarios or projections). In a few cases, real predictions have been prepared for the nearer future, and they have turned out to point into the right direction.

While this is encouraging, such sporadic successes cannot be considered as significant evidence for the general validity of climate models. At the same time, evidence is not available that would positively disqualify such models for being valid tools to study man-made climate change.

The lack of the option to do experiments prevents many uncertainties from being resolved, as for instance the *climate sensitivity* (temperature increase after equilibration when CO_2 concentrations are doubled). Indirect evidence is used for improving the

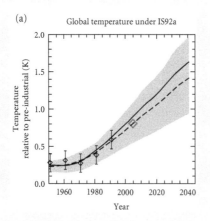

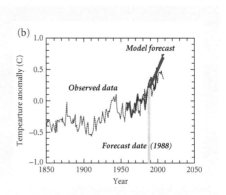

FIGURE **8.3** **Left:** Allen et al.'s (2000) forecast of global temperature made in 1999. Solid line shows original model projection. Dashed line shows prediction after reconciling climate model simulations with the HadCRUT temperature record, using data to August 1996. Grey band shows 5–95% uncertainty interval. Grey diamond shows observed decadal mean surface temperature for the period 1 January 2000 to 31 December 2009 referenced to the same baseline. **Right:** Hansen's scenario published in 1988 as a prediction up to 2010 (redrawn after Hargreaves 2010).

estimate of such uncertain quantities, but some uncertainty remains. This leaves a certain range for interpreting the policy implications differently.

Because of the long waiting time for getting a new realization of the climate system, climate science must rely on historical 'instrumental' data, data which have been measured for often quite different purposes, under different conditions, with different instruments and standards. Alternatively, proxy data may be used, for instance data on tree growth or ice accumulation, which may have 'recorded' aspects of the geophysical environment.

The 'instrumental' data usually suffer from 'inhomogeneities' (e.g. Jones 1995; Karl et al. 1993). An example is provided by Lindenberg et al. (2010), who examined statistics of wind speed recorded on islands along the German North Sea coast (Figure 8.4). The diagram shows periods, when the wind speeds co-varied to large extent, whereas at other times, marked by the dashed line, the statistics began to deviate strongly. These deviations are mostly related to the relocation of the instruments for a variety of reasons. Such effects are common in 'raw' data time series, and at least in modern data sets are well documented.

Indeed, it may be a good rule of thumb that almost all time series, extending across several decades of years, suffer from some inhomogeneities—the more easily detectable inhomogeneities are 'abrupt', such as those in Figure 8.4, but the more difficult to detect are continuous changes. An example is the effect of continuous urbanization, which can be separated from the natural variability only within large error bars (Lennartz and Bunde 2009).

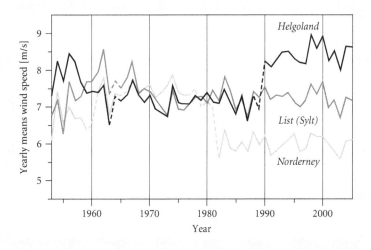

FIGURE 8.4 Yearly means of wind speed measurements from three synoptic island stations. Dashed lines label years with known station relocations (after Lindenberg et al. 2010).

Before using such data in climate analysis, the series have to be 'homogenized' (e.g. Peterson et al. 1998). For scientists and lay people, with insufficient insight into the contingencies of climate data, this significant hurdle is hardly recognized. Therefore, it happens every now and then that publications show surprising results, which in the end display changing data recording practices and not changes in the climate system. A nice example is the conjectured increase of the absolute number of deep cyclones in the last century, which is due to insufficient data knowledge (Schinke 1992). Also, contributors on weblogs often ask for 'raw data,' with the implicit suspicion that somebody may have tampered with the raw data in order to obtain preconceived results. This is in most cases not a wise approach—because of these invisible inhomogeneities (cf. Böhm 2010).

'Proxy'-data have other problems (Briffa 1995). The main problem is that the proxies, for instance growth rings in trees, or annual layers of sedimentary material, record not only some climate parameters, such as summer temperature, but also other influences. The fundamental problem is that only part of the variability in the proxies is related to climate variability, in particular temperature. The proportion of variability, which may be related to climate drivers, differs in time, and the empirically derived transfer functions may show different amplitudes for different timescales. The famous problem of the hockey stick-named temperature reconstruction, which was based mostly on tree rings, had much to do with the non-uniform representation of long-term and short-term variability. An interesting exchange about proxy-methodologies and robust claims making is provided by a series of papers, comment, and replies by Christiansen et al. (2009), Rutherford et al. (2010), and Christiansen et al. (2010).

To summarize: 'Data' entail complex issues in climate research. Historically collected 'instrumental' data often suffer from inhomogeneities, related to changing observational, archival, and analytical practices. Indirect proxy-data provide information about changing physical conditions, but compete with other unknown influences—so that stationarity and timescale dependence of the information content of such data are an issue. More expertise about the process of using instruments and of storing influences in indirect data is required.

Thus, there are a series of obstacles and uncertainties, which are uncommon in conventional physics, which represent special challenges when dealing with climate dynamics and impacts.

4 THE SOCIO-CULTURAL CONTEXT

There is another set of factors that makes the science of climate 'different' from other natural science fields—namely that climate research, the issues, results, and individuals—are firmly embedded into socio-historical, socio-cultural, and socio-economic contexts. This is already illustrated by virtue of the fact that most, perhaps almost all, of present climate research activities are related to the issue of anthropogenic climate change and its impact on the natural environment and society.

The main issue in the societal context concerns the statistics of weather (in atmosphere and ocean) and its changes, such as the frequency and intensity of extreme events such as storms, heatwaves, and flooding. Weather statistics are significant data for societies, its infrastructure, and inhabitants because they contain important information about possible impacts (and adaptive measures to deal with them) and options for keeping a check on the drivers (mitigation). Both strategies in response to a changing climate, that is, reducing emissions and reducing vulnerability, are subjects of a wide range of scientific fields including the engineering sciences, hydrology, law, geography, policy sciences, ecology, economy, and social sciences.

Thus, climate research has significant attributes beyond physics. We could now start to discuss the needed contributions from these other fields, but we do not. We instead concentrate more on the functioning of the science–society knowledge interaction.

This interaction of climate research with society in general and with policy making in particular is linked to the joint presence of two factors. One factor that we have just discussed above is the high uncertainty about the 'facts' of climate dynamics, ranging from the climate sensitivity, to regional specification, to the presence of other social drivers and to future options of dealing with emissions and impacts. The other factor concerns the societal response to climatic conditions, how we interpret and deal with climate-related processes. Our ordinary everyday understanding of climate is closely related to our way of life, mediated of course by the way in which the mass media shape climate issues according to media logic (Weingart et al. 2000; Boykoff and Timmons 2007; Carvalho 2010). References to climate and climate change in public communication may be employed for example as a tool to legitimate changes to our way of life, or, in the opposite sense, as a means to defend dominant world-views.

Under these circumstances, but not only because the nature of our understanding of climate is embedded in everyday life, climate science becomes along with other modern scientific fields 'postnormal' (Funtovicz and Ravetz 1985; Bray and von Storch 1999). A broad range of essentially contested terms and explanations enter the public arena and compete for attention, accounts that may also be brought in position to give credence to different world-views and legitimacy to political and economic interests.

There seem to be two major, contending classes of explanations of the climate and climate change (von Storch 2009). One, which we label as 'scientific construct' of human-made climate change, states that processes of human origin are influencing the climate—that human beings are changing the global climate. In almost all localities, at present and in the foreseeable future, the frequency distributions of the temperature continue to shift to higher values; sea level is rising; amounts of rainfall are changing. Some extremes such as heavy rainfall events will change. The driving force behind alterations beyond the range of natural variability is above all the emission of greenhouse gases, in particular carbon dioxide and methane, into the atmosphere, where they interfere with the radiative balance of the Earth system.

The scientific construct is widely supported within the relevant scientific communities, and has been comprehensively formulated particularly thanks to the collective and consensual efforts of the UNO climate council, the IPCC.[2] Of course, there is not a complete consensus on all aspects of the construct in the scientific community, so that speaking of 'the scientific construct' is somewhat of a simplification. What is consensual and enumerated in the previous paragraph is the core of the scientific construct.

A different conception of climate and climate change may be labeled the *social or cultural construct* (cf. Stehr and von Storch 2010). In the context of this concept, climate and weather patterns are also changing, the weather is less reliable than it was before, the seasons less regular, the storms more violent. Weather extremes are taking on catastrophic and previously unknown forms.

What causes these changes in weather patterns? A variety of economic reasons and psychological motives tend to be adduced, for example, sheer human greed and simple stupidity. The mechanism that is at work may be described as follows: Nature is retaliating and striking back. For large segments of the population, at least in central and northern Europe, this mechanism producing climate change is taken for granted. In older times, and even sometimes today, adverse weather patterns were the prompt response of the gods angered by human sins (e.g. Stehr and von Storch 2010).

The cultural construct of climate and changing weather patterns takes many different forms depending on the traditions in a society, its development and dominant aspirations—but what is described above as the everyday concept of climate and weather represents something like a standard core of such statements.

Obviously, the scientific construct is hardly consistent with such cultural constructs.

The position of so-called 'climate sceptics' is not discussed here because there is no consistent body of knowledge of 'sceptic' climate science but merely a collection of various, often highly contested issues that range from detailed matters to much more general assertions, e.g. that greenhouse gases would have no significant impact on climate. The absence of a consistent body of assertions does not imply in principle that the questions raised by 'sceptics' might not in one or the other instance be helpful to constructively move the science of climate forward.

In this postnormal situation where science cannot make concrete statements with high certainty, and in which the evidence of science is of considerable practical significance for formulating policies and decisions, this science is impelled less and less by the pure 'curiosity' that idealistic views glorify as the innermost driving force of science, and increasingly by the usefulness of the possible evidence for just such formulations of decisions and policy (Pielke 2007b). It is no longer being scientific that is of central importance, nor the methodical quality, nor Popper's dictum of falsification, nor Fleck's

(1980) idea of repairing outmoded systems of explanation; instead, it is social and political utility of knowledge claims that carry the day. Not correctness, nor objective falsifiability, occupies the foreground, but rather social acceptance and social utility.

In its postnormal phase, science thus lives on its claims, on its staging in the media, on its affinity and congruity with socio-cultural constructions. These knowledge claims are not only raised by established scientists, but also by other, self-appointed experts, who often are bound to special interests. Representatives of social interests seek out those knowledge claims that best support their own position. One need only recall the *Stern* report (see the critique by Pielke (2007a) or Yohe and Tol (2008)), or the press releases of US Senator Inhofe.

5 CONCLUSIONS

Two major conclusions about the science of climate, and the knowledge about climate may be drawn.

The scientific construct is mostly based on a physical analysis of climate and developed by natural scientists. It describes the left two blocks in Figure 8.5. In the 'linear model' (Beck 2010; Hasselmann 1990) the middle blocks, representing social and cultural dynamics, are not taken into account. Instead, once society has given a metric of determining 'good' and 'less good', it is simply a matter of understanding the 'physical' (including economic) system.

However, the climate scientists are also part of society and not immune to dominant societal conceptions of the nature and the impact of climate and climate change on human conduct, they tend to embed their analysis, especially in efforts to communicate their knowledge to policy makers and society at large in ways which are attentive to the socio-cultural construct of climate and climate change. It is not surprising that in this postnormal situation scientists concerned about the impact of the greenhouse gases, in their desire to

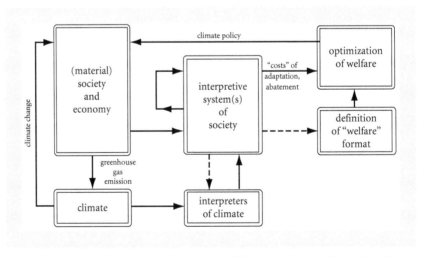

FIGURE **8.5** The perceived climate and society model (after Stehr and von Storch 2010).

save the world, may develop some bias towards an overdramatization. The discussion itself often resembles more a religious than a physics discussion where the non-believers (of the role of the greenhouse gases and their impact) are called 'deniers'.

One therefore is able to surmise that the transfer of the scientific construct into the societal realm goes along with a subtle transformation of the climate knowledge, by blending the scientific construct with the socio-cultural construct (the middle blocks in Figure 8.5). Obviously, in the model described by Figure 8.5, the basic assumption of physics, that there are given quantifiable laws (linear or nonlinear), is no longer valid. Understanding the interaction of climate and society is not only an issue of physical analysis (with laws) but of society/culture analysis (without laws) as well.

Obviously, the situation is not quite that straightforward, it is not easily deconstructed and the interrelations of scientific and everyday construct are difficult to dissemble. To comprehend and disentangle the multiple interactions of science and society in the case of our understanding of climate and climate change is nonetheless a real and worthy scientific and practical challenge. It needs a trans-disciplinary approach, bringing together scientists with a solid background in the physics, and scholars who understand societal and know-ledge dynamics (Pielke 2007b).

If this helps to implement a better climate policy, with an efficient constraining of climate change and socio-culturally acceptable measures of mitigation and adaptation, it needs to be developed. Summing up, climate science is and should be much more than just the physical analysis.

NOTES

1. In certain cases this can be done by expanding Ψ into a Fourier-series of trigonometric functions of spherical harmonics. Those with small wavenumbers then make up Ψ, and the rest Ψ'.
2. The support among climate scientists seems indeed very broad, when related to the key assertions just listed (cf. Bray and von Storch 2007). Whether the emergence of errors in Working Group (WG) 2, and possibly WG3, of 4th Assessment Report (AR4) of IPCC, which so far all point towards a dramatization, and after it became known that a key data set (CRU) could no longer be reproduced because of some original data having been 'lost', will have implications for this support within climate science and the general scientific community remains to be seen.

REFERENCES

ALLEN, M. R., STOTT, P., MITCHELL, J., SCHNUR, R., and DELWORTH, T. 2000. Quantifying the uncertainty in forecasts of anthropogenic climate change. *Nature* 407: 617–20, October.
ARRHENIUS, S. A. 1896. On the influence of carbonic acid in the air upon the temperature of the ground. *Philosophical Magazine and Journal of Science* 41: 237–76.

ARRHENIUS, S. A. 1908. *Das Werden der Welten.* Leipzig: Akademische Verlagsanstalt.

BECK, S. 2010. Moving beyond the linear model of expertise? IPCC and the test of adaptation. *Reg. Env. Change,* doi:10.1007/s10113–010–0136–2.

BÖHM, R. 2010. 'Faking versus adjusting'—why it is wise to sometimes hide 'original' data. <http://klimazwiebel. blogspot. com/2010/01/guest-contribution-from-reinhard-bohm. html> (as of 22 March 2010).

BOYKOFF, M. T., and TIMMONS, R. J. 2007. *Media Coverage of Climate Change: Current Trends, Strengths, Weaknesses.* Human Development Report 2007/2008.

BRAY, D., and VON STORCH, H. 1999. Climate science: An empirical example of postnormal science. *Bull. Amer. Met. Soc.* 80: 439–56.

—— —— 2007. *Climate Scientists' Perceptions of Climate Change Science.* GKSS-Report 11/2007.

—— —— 2009. 'Prediction' or 'projection'? The nomenclature of climate science. *Sci. Comm.* 30: 534–43, doi:10. 1177/1075547009333698.

BRIFFA, K. R. 1995. Interpreting high-resolution proxy climate data—the example of dendro-climatology. Pp. 77–84 in H. von Storch and A. Navarra (eds.), *Analysis of Climate Variability: Applications of Statistical Techniques.* Berlin: Springer Verlag.

CARVALHO, A. 2010. Media(ted) discourses and climate change: A focus on political subjectivity and (dis)engagement. *Climate Change* 1(2), doi:10.1002/wcc.13.

CHRISTIANSEN, B., SCHMITH, T., and THEJLL, P. 2009. A surrogate ensemble study of climate reconstruction methods: Stochasticity and robustness. *J. Climate* 22: 951–76.

—— —— —— 2010. Reply to comment on 'A surrogate mensemble study . . .' by Rutherford et al. *J. Climate* 23: 2839–44.

CURRY, J. A., and WEBSTER, P. J. 2010. Climate science and the uncertainty monster. *Bull Amer. Meteo. Soc.* forthcoming.

FLECK, L. [1935] 1980. Entstehung und Entwicklung einer wissenschaftlichen Tatsache: Einführung in die Lehre vom Denkstil und Denkkollektiv. Frankfurt am Main: Suhrkamp.

FLEMING, J. R. 1998. *Historical Perspectives on Climate Change.* Oxford: Oxford University Press.

FRIEDMAN, R. M. 1989. Appropriating the Weather: Vilhelm Bjerknes and the Construction of a Modern Meteorology. Ithaca, NY: Cornell University Press.

FUNTOWICZ, S. O., and RAVETZ, J. R. 1985. Three types of risk assessment: A methodological analysis. Pp. 217–31 in C. Whipple and V. T. Covello (eds.), *Risk Analysis in the Private Sector.* New York: Plenum.

HANN, J. 1903. *Handbook of Climatology.* Vol. i: *General Climatology.* New York: Macmillan.

HANSEN, J., FUNG, I., LACIS, A., RIND, D., LEBEDEFF, S., RUEDY, R., RUSSELL, G., and STONE, P. 1988. Global climate changes as forecast by Goddard Institute for Space Studies three-dimensional model. *J. Geophys. Res.—Atmospheres* 93(D8).

HARGREAVES, J. 2010. Skill and uncertainty in climate models. *Wileys Interdisplinary Reviews/ Climate Change,* in press.

HASSELMANN, K. 1990. How well can we predict the climate crisis? Pp. 165–83 in H. Siebert (ed.), *Environmental scarcity: The International Dimension.* Tübingen: JCB Mohr.

HUMBOLDT, ALEXANDER VON [1845] 1864. *Cosmos: Sketch of a Physical Description of the Universe.* Vol. i. London: Henry G. Bohn.

JONES, P. D. 1995. The instrumental data record: Its accuracy and use in attempts to identify the 'CO_2 Signal'. Pp. 53–76 in H. von Storch and A. Navarra (eds.), *Analysis of Climate Variability: Applications of Statistical Techniques.* Berlin: Springer Verlag.

KARL, T. R., QUAYLE, R. G., and GROISMAN, P. Y. 1993. Detecting climate variations and change: New challenges for observing and data management systems. *J. Climate* 6: 1481–94.

KEENLYSIDE, N. S. 2011. Prospects for decadal climate prediction. *Wiley Interdisciplinary Review/Climate Change*, in press.

—— LATIF, M., JUNGCLAUS, J., KORNBLUEH, L., and ROECKNER, E. 2008. Advancing decadal-scale climate prediction in the North Atlantic sector. *Nature* 453: 84–8.

LENNARTZ, S., and BUNDE, A. 2009. Trend evaluation in records with long-term memory: application to global warming. *Geophys. Res. Lett.* 36, L16706.

LINDENBERG, J., MENGELKAMP, H.-T., and ROSENHAGEN, G. 2010. Representativity of near surface wind measurements from coastal stations at the German Bight. Forthcoming.

MÜLLER, P., and VON STORCH, H. 2004. *Computer Modelling in Atmospheric and Oceanic Sciences—Building Knowledge*. Berlin: Springer Verlag.

PEIXOTO, J. P., and OORT, A. H. 1992. *Physics of Climate*. American Institute of Physics.

PETERSON, T. C., EASTERLING, D. R., KARL, T. R., GROISMAN, P., NICHOLLS, N., PLUMMER, N., TOROK, S., AUER, I., BOEHM, R., GULLETT, D., VINCENT, L., HEINO, R., TUOMENVIRTA, H., MESTRE, O., SZENTIMREY, T., SALINER, J., FØRLAND, E., HANSSEN-BAUER, I., ALEXANDERSSON, H., JONES, P., and PARKER, D. 1998. Homogeneity adjustments of in situ atmospheric climate data: A review. *Intern. J. Climatol.* 18: 1493–517.

PIELKE, R. A., JR. 2007a. Mistreatment of the economic impacts of extreme events in the Stern Review Report on the economics of climate change. *Global Environmental Change* 17: 302–10.

—— 2007b. *The Honest Broker*. Cambridge: Cambridge University Press.

RUTHERFORD, S. D., MANN, M. E., AMMANN, C. M., and WAHL, E. R. 2010. Comment on: 'A surrogate ensemble study of climate reconstruction methods: Stochasticity and robustness' by Christiansen, Schmith, and Thejll. *J. Climate* 23: 2832–8.

SCHINKE, H. 1992. Zum Auftreten von Zyklonen mit niedrigen Kerndrücken im atlantisch-europäischen Raum von 1930 bis 1991. *Wiss. Zeitschrift der Humboldt Universität zu Berlin, R. Mathematik/Naturwiss.* 41: 17–28.

SOLOMON, S., QIN, D., MANNING, M., MARQUIS, M., AVERYT, K. Tignor, M. M. B., Le Roy Miller, H. jr, and Chen, Z. (eds.) 2007. *Climate Change 2007: The Physical Basis*. Cambridge: Cambridge University Press.

STEHR, N., and VON STORCH, H. 1997. Rückkehr des Klimadeterminismus? *Merkur* 51: 560–2.

—— —— 1999. An anatomy of climate determinism. Pp. 137–85 in H. KAUPEN-HAAS (ed.), *Wissenschaftlicher Rassismus—Analysen einer Kontinuität in den Human- und Naturwissenschaften*. Frankfurt. a. M.: Campus-Verlag.

—— —— 2010. *Climate and Society: Climate as a Resource, Climate as a Risk*. Singapore: World Scientific.

VON STORCH, H. 2007. Climate change scenarios—purpose and construction. In H. von Storch, R. S. J. Tol, and G. Flöser (eds.), *Environmental Crises: Science and Policy*.

—— 2009. Climate research and policy advice: Scientific and cultural constructions of knowledge. *Env. Science Pol.* 12: 741–7.

—— and STEHR, N. 2000. Climate change in perspective: Our concerns about global warming have an age-old resonance. *Nature* 405: 615.

WASHINGTON, W. M., and PARKINSON, C. L. 2005. *An Introduction to Three-Dimensional Climate Modelling*. 2nd edn., Sausalito, CA: University Science Books.

WEART, S. R. 1997. The discovery of the risk of global warming. *Physics Today* January: 35–40.

—— 2010. The idea of anthropogenic global climate change in the 20th century. *Interdisciplinary Reviews/Climate Change*. Published online: 22 December 2009; doi:10.1002/wcc.6.

WEINGART, P., ENGELS, A., and PANSEGRAU, P. 2000. Risks of communication: discourses on climate change in science, politics and the mass media. *Public Understanding of Science* 9: 261–83.

YOHE, G. W., and TOL, R. S. J. 2008. The Stern review and the economies of climate change: An editorial essay. *Climatic Change* 89: 231–40.

CHAPTER 9

COSMOPOLITAN KNOWLEDGE: CLIMATE SCIENCE AND GLOBAL CIVIC EPISTEMOLOGY

SHEILA JASANOFF

1 INTRODUCTION

WORLD opinion divided on why the 15th Conference of the Parties (COP-15) under the United Nations Framework Convention on Climate Change (UNFCC), held in Copenhagen on 7–19 December 2010, failed. But one conclusion mustered almost universal agreement: COP-15 cut short any hope of vigorous forward movement to curb greenhouse gas emissions worldwide. Political reactions to COP-15 remained as fractured as the pre-COP-15 attempts to forge a climate policy consensus. Some seemed to give up on the very prospect of multilateralism, recommending instead a recourse to 'minilateralism' (Naím 2009), led by a small number of powerful, presumably like-minded, states that might build consensus more effectively than the unruly UN community. For analysts of climate politics, however, this turn to the pragmatics of negotiation seems premature. Much remains to be explored about the reasons for the lack of political action on what many view as *the* question for human survival in the twenty-first century. Analytically and normatively, a few steps backward may be necessary before plotting the next steps forward.

Among the unresolved questions exposed by the great Copenhagen fizzle of 2009 is a crucial one for science policy. Climate change should have presented, from one viewpoint, an easy case for global policy making. In the arena of environmental politics, it was long assumed that uncertainty about natural phenomena produces policy disarray; by the same token, scientific agreement was thought to facilitate action by positioning all actors on a baseline of shared knowledge. On this view, climate change should have moved by the turn of the century toward the low-conflict, strong-consensus end of the political action spectrum. Unlike many environmental problems that are hedged around with seemingly unresolvable unknowns, climate change appears to rest on a rock-solid consensus among those most qualified to judge (Anderegg et al. 2010; Oreskes 2007, 2004). There is remarkably little scientific dissent around the claim that the earth is warming dangerously as a consequence of human activity, specifically the release of gaseous carbon into the atmosphere. That consensus, moreover, is supported by one of the most inclusive and transparent exercises in international scientific consensus building the world has yet seen, through the successive assessment reports of the Intergovernmental Panel on Climate Change (IPCC). Most scientists who challenge the consensus position have well-documented ties to corporate interests that should have deprived them of all credibility (Oreskes and Conway 2010).

Why, then, has political consensus failed to follow where science leads? A popular line of argument blames poorly informed public opinion, fed by the petroleum industry's manufactured uncertainties and the news media's irresponsible exaggeration of dissent. Scientists' response to this analysis has been, on the one hand, to attack the integrity of the opposition, vehemently insisting on the strength of the consensus and the absence of any intellectually credible opposition, and, on the other, to bemoan the public's lack of scientific understanding, especially of probabilistic, temporally distant threats. In this self-defensive project, scientists enthusiastically embraced the work of historians of science such as Naomi Oreskes and Erik Conway, which not only provided quantitative evidence of agreement but joined the climate science elite in urging policy makers to take heed. As Oreskes (2004) put it: 'But there is a scientific consensus on the reality of anthropogenic climate change. Climate scientists have repeatedly tried to make this clear. It is time for the rest of us to listen.' Among influential politicians, former Vice President Al Gore notably did listen, repeatedly invoking Orseskes' work (though not citing her by name) to support his often-stated confidence in the strength of the science, as in his award-winning documentary *An Inconvenient Truth*.

The problem for social analysis, however, lies elsewhere. Decades of work in studies of scientific controversies have established that the credibility of science depends not only on the strength of internal consensus among scientists, as measured by objective and supposedly context-independent criteria, but as much or more on the persuasive power of the people and institutions who speak for science (Irwin and Wynne 1996). Indeed, this capacity to persuade appears to have been the foundation of scientific authority from the earliest times (Shapin 1994; Shapin and Schaffer 1985). If experts and institutions are trusted, then their claims may be accepted on faith even if contrary assertions circulate in public discourse; if such bodies lose public confidence, then the mere fact of strong internal agreement will not be enough to carry conviction for outsiders. Put differently, convergence in scientists' understanding of the facts is not the same thing as public assent to those understandings. Factors conditioning the reception and uptake of scientific claims by lay publics operate to some extent independently of the dynamics of knowledge production within scientific communities. Such predispositions, moreover, are especially likely to manifest themselves at moments of conflict or controversy.

These observations raise special difficulties for policy making on international and global problems such as climate change. Trust and credibility, as social achievements, are necessarily inflected by culture. Markers of trustworthiness depend on contexts of interpretation: a persuasive salesperson in one cultural setting may well come across as a hypocrite or charlatan in another. How to build scientific credibility across international public arenas emerges, then, as a conceptual and practical challenge in its own right. The mere fact that scientists are speaking as if with one voice on a particular issue may be highly relevant, but not dispositive, when it comes to persuading global publics of the need to act. Scientific consensus becomes one factor among many in the circulation of signs, symbols, and meanings that bring collectives together into anything resembling an 'epistemic community' (Haas 1990). As the noted sociologist Neil Smelser (1986: 26) observed in a National Academies report on the social and behavioral sciences:

> Social problems, then, can be defined by the presence of 'objective facts' only if there is consensus about the meaning and significance of those facts ... That kind of consensus rarely exists. We now know that social problems are not matters of objective fact but matters of an uncertain, disputed set of both facts and principles. Recognizing this, we can appreciate why such a large proportion of the debates about social problems are debates not about the existence of facts but about symbols, about the legitimacy of the competing sets of criteria by which a factual situation will or will not qualify as a genuine social problem.

In this chapter, I lay out an analytic framework that allows us to see why consensus building on responses to climate change cannot proceed through the institutions of science alone but requires a more differentiated and more culturally sensitive—indeed, more *cosmopolitan*—approach to confronting the climate phenomenon (Jasanoff 2010a).

Cosmopolitanism in its dictionary definition means the quality of being 'free from local, provincial, or national ideas, prejudices, or attachments,' and 'at home all over the world.'[1] This state of transcendence over local particularities has generated a large literature in philosophy and social theory, examining the pros and cons of seeking to eliminate or rise above difference, especially differences of nationality and culture (Beck 2006; Rorty 1991). Most writers recognize that the claim to cosmopolitanism may entail a false universalism, denying or denigrating differences that should be respected and that legitimately matter to others. Science, however, is taken by most of its practitioners to be unproblematically cosmopolitan, speaking the truths of nature in the same register, with equal force and conviction, to all people everywhere. Yet this assumption flies in the face of evidence from countless controversies surrounding policy-relevant knowledge claims. Science, I argue, risks becoming parochial, like a bumbling tourist in a strange land, when it is generated and put to use without attentiveness to contexts of interpretation. Global problems demand a more plastic approach, sensitized to the situated character of all knowledge claims and to the performative and persuasive demands of reasoning for culturally diverse audiences.

I begin by placing science itself in a changing historical context, in which the ideal of science as a detached, curiosity-driven inquiry, guided by truthfulness to nature, has gradually yielded to the social reality of sciences that are more problem driven and politically accountable. I then draw on comparative studies of three national science and decision-making cultures (US, UK, and Germany) to show how the credibility of public knowledge claims relates to long-established, culturally situated practices of interpretation and reasoning. I conclude with reflections on the institutional changes that will be needed to build robust cosmopolitan knowledge for collective action on climate change and other global problems.

2 PUBLIC SCIENCE: A CONCEPTUAL HISTORY

Theorizing the role of science in public life has long roots. It reaches back if one is philosophically inclined to Aristotle's reflections on oligarchy, or government by the few. Those few in ancient times comprised mainly rich and powerful families, entitled by birth to rule. In the modern world challenges to democracy come not only from the thin stratum of the super-rich but also from a much wider array of experts who command specialist knowledge in the technical fields that increasingly underwrite public policy (Price 1965). By what right do those experts shape policy, and how do we account for their legitimacy?

We need not return to ancient Greece in search of answers. A sea-change in the role of experts occurred in most technologically advanced nations following the Second World War, justifying a rather more historically compressed narrative. Since the mid-twentieth century, science has become an immensely more pervasive presence in political life, both as an eager recipient of public funds and as a crucial provider of technical knowledge to governments. Theoretical understandings of the science–society–politics relationship have also evolved, giving rise to three partially overlapping and partially distinct framings through which to examine the scientific debates around climate change.

Mid-century views about the scientific enterprise were influentially captured by the American sociologist Robert K. Merton (1942) in a well-known essay on the normative structure of science. Scientists, in Merton's view, are institutionally conditioned to abide by certain distinctive norms that become a part of the 'scientific conscience,' arousing collective 'moral indignation' when they are violated. The four elements of the scientific ethos that Merton identified—communalism, universalism, disinterestedness, and organized skepticism—ensure that scientists will share results, eschew parochialism, avoid contaminating political or financial ties, and hold each others' work to impartial critical scrutiny. Widely accepted by scientists, politicians, and the public as an accurate characterization of science, Merton's norms served as the basis for what we may call the 'separatist model' of science in society. On this view, science is a realm apart: it can be trusted to operate largely on its own, setting its own research priorities, and organizing or regulating its internal conflicts, held adequately in check by nature and peer criticism, without need of significant external supervision. Most important for our purposes, science in this model is always already cosmopolitan, because it purveys truths underwritten by a nature that is stable, unchanging, and identical for all. Science merely discloses that universal reality: when consensus is reached, in Oreskes' words, 'It is time for the rest of us to listen.'

Science's institutional relationships changed with the advent of major public funding programs, such as those of the National Science Foundation and the National Institutes of Health in the United States. Publics underwriting research with substantial tax revenues acquired a stake in what science produces, just as science acquired stakes in making its findings useful as a basis for continued public support. At the same time, the number and variety of scientific advisory bodies assisting governments multiplied.[2] With growing entanglement between science and politics came a further reevaluation of the science–policy relationship, leading in purest form to what has been called the 'linear model' (Pielke 2007). This view acknowledges that science has a useful role to play in society, by finding relevant facts and informing policy makers, but these are merely inputs, untouched by the later political tasks of balancing information with other relevant factors to arrive at public decisions. The governance implications of the linear model were articulated by the US National Research Council (NRC 1983) in a report on risk analysis that still exercises far-reaching influence on regulatory discourse and decisions. Briefly, the NRC recommended that the scientific aspects of assessing risk should be kept apart from the value-laden task of managing risk, an injunction that became almost a mantra for regulators seeking to shore up their legitimacy in areas of uncertain, contested knowledge. Indeed, one can discern the impact of the NRC's boundary drawing in the IPCC's repeated claim that its mission is to provide scientific, technical, and socio-economic information that is policy-relevant but policy-neutral, that is, not 'policy prescriptive.' On this view, science generated for policy can be cosmopolitan so long as parochial political biases are not allowed to enter into its production.

Both the separatist and the linear model of science in relation to policy and politics continue to enjoy wide circulation in public discourse. At the same time, a more complex and interactive account of that relationship has been recognized in the academic literature and partly reflected in policy practice. In one well-known articulation, contemporary science is seen as subject to an altogether different set of imperatives, producing a culture in which older Mertonian norms have at best attenuated significance. Labeled Mode 2 (Gibbons et al. 1994), this model characterizes large components of scientific production as more interdisciplinary, institutionally dispersed, context dependent, and problem oriented than in earlier times. Mode 2 science responds to social problems; its very reason for being originates outside the communities of curiosity-driven research. In these domains, claims to the virtues of communalism, universalism, and disinterestedness are offset by pressures to be practical, useful, goal directed, and accountable to publics, political sponsors and industrial patrons. Much policy-relevant science, including such second-order scientific activities as the IPCC's climate assessments,[3] can properly be characterized as Mode 2. Given the potentially high stakes associated with its production, the legitimacy of this kind of knowledge necessarily depends on engagements with audiences beyond tightly bounded, disciplinary scientific communities (Jasanoff 1990, 2010b).

In the twenty-first century, numerous states and supra-state entities, especially in the European region, recognized the need for wider public consultation, and undertook varied efforts to engage citizens in the steering of science. For the most part, such tactics were targeted to emerging sciences and technologies, in order to forestall the stubborn resistance that greeted crop biotechnology at the turn of the century and nuclear power a generation earlier. Democratization efforts, in other words, filtered upstream into the processes of scientific discovery and invention, chiefly for purposes of shoring up the legitimacy of public funding for science—by seeking citizen acquiescence well in advance of widespread commercial applications. Largely absent from these initiatives was an attention to the democratization of public reasoning based on scientific and technical expertise, let alone to the problems of expert credibility that have arisen in relation to transboundary problems such as AIDS, ozone depletion, and climate change. How can or should global entities such as the IPCC seek legitimation in the eyes of global publics? What are the obstacles to building cosmopolitan knowledge on problems of worldwide spread and significance? Some answers to these rarely articulated questions can be found in the comparative politics of health, safety, and environmental regulation.

3 Climate Science and National Civic Epistemologies

An important finding of the field of science and technology studies over the last several decades is that claims about nature, like claims about any aspect of human life, win social acceptance only as a result of intensive preparatory work. The credibility of scientific facts rests not only on their conformance with well-established research paradigms (Kuhn 1962), but also on steps taken to make those facts visible, tangible, and comprehensible to those not immediately engaged in a particular piece of research. Scientists use instruments, measurements, visualization techniques, specialized languages, and many other agreed-upon conventions to let nature speak, as it were, 'for itself' (Latour 1987). It may take years of conflict and negotiation to establish such conventions in emerging research areas, and specific assertions may generate persistent controversies or be overturned by later results. Yet, as long as scientific production remains a largely in-house affair, conducted by and for scientists themselves, such struggles remain mostly invisible except to insiders. The institutional authority of science as ultimate purveyor of truth is not affected by its internal conflicts.

The credibility landscape shifts considerably when science goes public, whether because the claims are extraordinary (as in the case of cold fusion in the 1980s), or because they challenge deeply acculturated views of reality and thereby arouse resistance (as in the well-known case of American creationists), or because the new reality calls on people to change prior expectations and behaviors, as with most modern findings of environmental risk. In all these situations, science faces a double bind—not only the hard problem of consensus building within (and, in Mode 2 science, also across) relevant disciplinary communities, but additionally the harder problem of gaining public acceptance for claims that have destabilized settled social relations and ways of living. For global problems such as climate change, the second hurdle looms especially large, because the processes of testing and accounting for policy-relevant scientific claims vary substantially across countries. In particular, modern polities have developed what I have elsewhere called 'civic epistemologies,' or publicly accepted and procedurally sanctioned ways of testing and absorbing the epistemic basis for decision making (Jasanoff 2005: 247–71). Even Western nation-states with shared or intertwined histories of cultural and political development diverge in their public knowledge ways. Thumbnail sketches of US, British, and German responses to climate science over the course of the IPCC assessments illustrate the point.

3.1 United States

Analyses grounded in international relations and negotiation theory have not been able to account for the prevalence or persistence of climate skepticism in the United States. The nearest approach to explanation, as already noted, is to blame powerful corporate interests, chiefly the oil industry, for sowing the seeds of doubt by funding scientists to question the growing scientific consensus on anthropogenic climate change (Oreskes and Conway 2010). In brief, so the story goes, this lobby manufactured uncertainty by buying, or bending, science to support its political interests. Many US scientists, historians, and activists have condemned this cynical politicization of science and the power of moneyed lobbies to buy political support with distorted facts (Schrag 2006; Schneider 2009). But granting the reality of lobbies and of science for hire, a question still hangs uneasily in the air. Why did the anti-climate consensus lobby gain such powerful adherents in the US Congress, the White House, and large segments of the American public? Why then such striking differences in the intensity and duration of public controversy? Multinational corporations, after all, spread their messages everywhere, and there are no glaring differences between the United States and other Western countries in either media coverage of scientific controversies or the public understanding of science.

A more complete answer requires us to look not only at the circumstances of knowledge production (e.g. partisan funding of science), but also at the institutions and practices that condition receptivity toward public knowledge claims of any kind—in short, the dominant forms of a nation's civic epistemology. Such public knowledge ways consist of established commitments to particular forms of evidence garnering, argument, and demonstrations of reasonableness. In America's decision-making culture, founded on the common law's adversary system, information is typically generated by interested parties and tested in public through overt confrontation between opposing, interest-laden points of view. Underlying this approach is a somewhat paradoxical epistemic theory that places science above values and preferences, in keeping with Mertonian norms, but presupposes that the best way to attain unvarnished or disinterested truths is to display, and then eliminate, bias through the adversarial give and take of competing expert assertions.

The theory 'works' in practice because the American system has also created numerous procedures and institutions designed specifically to mitigate or end lengthy epistemic conflicts.[4] These include expert advisory committees (Jasanoff 1990), regulatory agencies with power to evaluate competing arguments, and of course a multi-tiered judicial system that frequently acts as the arbiter of last resort. Controversies that scientists cannot solve by themselves may nevertheless end for all practical purposes through the combined operation of these legal or political closure mechanisms. A case in point is the US Supreme Court's important decision in *Massachusetts v EPA*, 549 US 497 (2007), holding that climate science was compelling enough to require the US Environmental Protection Agency to take note, and to explain why it had failed to regulate greenhouse gases as air pollutants under the Clean Air Act. The Court in effect partially overrode US climate skeptics' attempts to stall the regulation of CO_2, and it did so by declaring that going against the international consensus without explanation was irrational.

Returning, then, to the persistent public controversy over anthropogenic climate change, it becomes apparent that the IPCC's carefully constructed procedures for seeking peer review failed to answer adequately to US traditions of achieving knowledge closure—traditions constitutive of American civic epistemology. Two comparisons, with Britain and Germany, underscore this point and help to guide our thinking about the role of science in future climate negotiations.

3.2 Britain

In contrast to American public knowledge making, which privileges adversarial give and take, the British approach has historically been more consensual. Underlying Britain's construction of public reason is a long-standing commitment to empirical observation and commonsense proofs, attested to by persons with an authorized right to see on behalf of the people (Jasanoff 2005). Two rooted elements of British political culture thereby come into somewhat paradoxical alignment: on the one hand, a long history of government by elites, under the aegis of monarchical rule, and on the other validation of the facts of the matter through references to common, even populist, practices of seeing and witnessing (Shapin and Schaffer 1985). Ruling elites, in other words, are held accountable to popularly accepted criteria of epistemic validity when presenting public proofs. At the same time, the decision maker's capacity to speak on behalf of the people is never taken for granted, but must be demonstrated through accredited social markers, which importantly include a record of public service and problem solving. Britain's meritocratic culture of 'the great and the good' elevates just such well-tested people ('safe hands' as they are often called) to positions of high authority.

The defining features of British civic epistemology have manifested themselves repeatedly in developments surrounding climate science. To begin with, traditions of trusted epistemic consensus-building 'worked' in the sense that neither policy makers nor the public displayed the ugly rift between conviction and skepticism that marked the US political debate. On the rare occasions when questions of scientific reliability did spill into the open, they were resolved in a matter-of-fact fashion that upheld both the right to question and the basic sturdiness of the consensus. Thus, a British citizen's challenge to the UK government for distributing *An Inconvenient Truth* for showing in schools failed to convince a high court that the film was propaganda and should be suppressed.[5] Instead, in *Dimmock v Secretary of State for Education and Skills,* [2007] EWHC 2288 (Admin), Justice Michael Burton took evidence on several specific allegations of misrepresentation. His decision faulted the film for nine instances of factual inaccuracy (the 'nine errors') and ordered that these be flagged in guidance notes accompanying future educational showings of the film.

A bigger furor greeted the independent television station Channel 4's airing of the skeptical documentary, *The Great Global Warming Swindle,* in March 2007. Asked to investigate, Ofcom, the UK watchdog agency for telecommunications, upheld the principle of open debate on important public issues but also acknowledged the mainstream character of the anthropogenic climate change hypothesis.[6] Ofcom expressed overall confidence in the British public's ability to assess the statements made in the program in the light of already available information:

> The anthropogenic global warming theory is extremely well represented in the mainstream media. A large number of television programmes, news reports, press articles and, indeed, feature length films have adopted the premise that global warming is caused by man-made carbon dioxide. In light of this it is reasonable for the programme makers to assume that the likely audience would have a basic understanding of the mainstream man-made global warming theory, and would be able to assess the arguments presented in the programme in order to form their own opinion.[7]

Ofcom, in short, both responded to and reaffirmed a civic epistemology accustomed to weighing and balancing empirical facts in public circulation. In this respect, the Ofcom decision rejected the 'deficit model'—a model of public ignorance and unreason—historically embraced by segments of the UK government (Irwin and Wynne 1996).

Confidence in facts and their facticity, however, demands a corresponding confidence in the trustworthiness of fact makers, and this foundation stone of Britain's civic epistemology was severely tested in the episode dubbed 'climategate.' The trigger was the disclosure in November 2009, weeks before the Copenhagen COP-15 meeting, of some 1,000 e-mails from the computers of the respected Climatic Research Unit (CRU) at the University of East Anglia. The hacked messages showed scientists expressing disrespect for holders of dissenting opinions, refusing to share data with known opponents, and possibly even doctoring their data to overstate the case for global warming. Pressed by instant uproar on the internet and intense (though, in many scientists' view, unmerited) media scrutiny, the UK House of Commons undertook an inquiry into CRU's research practices to determine whether breaches of scientific integrity had taken place. The Select Committee on Science and Technology explicitly avoided assessing the science itself, focusing instead on the CRU researchers' motives and actions—and it acquitted the principals of wrong-doing: 'We believe that the focus on CRU and Professor Phil Jones, Director of CRU, in particular, has largely been misplaced' (UK House of Commons 2010: 3). That personal exoneration conformed well to a culture of public reasoning in which trust crucially depends on the capacity of the state and its advisors to see, and speak, honestly for the people (Jasanoff 2005). Though a gaunt and visibly strained Phil Jones was harshly grilled during the inquiry, the parliamentarians eventually pinned blame on sloppy institutional practices rather than on individual misbehavior.

Despite the absence of rampant, US-style public skepticism, an empiricist distrust of untested and overextended models remains alive in British climate discourse, together with an acute awareness of uncertainties—and the consequent interpenetration of scientific assessment and public values. This residual skepticism toward *all* model-based predictions may account for the acceptance of worst-case economic scenarios in the famous Stern Report (2005) prepared for the UK government by its chief economic advisor, Sir Nicholas Stern. Instructively, leading US economists criticized Stern's projections as speculative and bordering on philosophy rather than disciplined, scientific prediction (Nordhaus 2007). And contrasting with US climate scientists' insistence that the strength of the climate consensus is alone sufficient to drive urgent action, prominent UK experts advocated less science-centric approaches, looking instead toward alleviating human vulnerability as a problem on its own (Hulme 2009). This position crystallized in the Hartwell Paper (Prins et al. 2010). Coordinated through the London School of Economics and co-authored by well-known UK and US science policy analysts, this report recommended reframing climate policy as not being about carbon emissions at all, but as a question of human dignity, access to energy, and the Earth's survival.

3.3 Germany

The German case provides a counterpoint to the British and American ones, and accentuates from a non-Anglo-American cultural context the importance of civic epistemologies in shaping national responses to the science and policy of climate change. The climate issue had been brewing at various levels in Germany for decades, through concerns about above-ground nuclear tests, cloud seeding, and the oil shocks of the 1970s (Cavender and Jäger 1993). But it broke upon German consciousness in the mid-1980s as part of a wider wave of concern about technological risk and a world tilting toward unmanageable dangers. The 1986 Chernobyl nuclear accident provided a peg on which to hang these formless fears. A sign of the times was the runaway success of an academic sociological text that no one suspected would turn into a bestseller: Ulrich Beck's now famous *Risk Society* (1992 [1986]). Around the same time, on 11 August 1986, Germany's leading weekly news magazine, *Der Spiegel*, published a cover story that captured with an unforgettable image one dimension of the anxiety that Beck had also tapped into. It was a picture of the twin towers of Cologne's iconic cathedral rising above the waters of the North Sea that had already claimed the rest of the building. The story was starkly titled *Klimakatastrophe*, or 'climate catastrophe,' a term first coined by a subgroup of the German Physical Society (Beck 2004) that remains current in German but has not caught on in other languages or nations.

That postwar Germany remains economically, socially, and technologically risk averse needs little elaboration here. Important to note, though, is that the idea of catastrophic climate change took hold in the German context with virtually no public debate about the strength of the scientific evidence. It was as if the picture made famous by the *Spiegel* cover, of a building that stands for German strength in adversity, fed into a national consciousness already scarred and sensitized by similar images of disaster—which in the aftermath of the Second World War prominently featured broken and destroyed churches. Indeed, images of a devastated Cologne, with the silhouette of the bombed church rising above the city's ruins, were familiar to most Germans who came of age in that period. Sudden and complete wreckage of civilization as we know it was not something late twentieth-century Germans found hard to imagine. The bigger challenge was how to avoid falling into the vortex of destruction and overwhelming loss that the nation knew too well, and feared profoundly.

In Germany's national civic epistemology, building communally crafted expert rationales, capable of supporting a policy consensus, offers protection against a psychologically and politically debilitating risk consciousness. On climate change as on a host of regulatory issues, long-standing administrative practices came into play, along with newly elaborated traditions of constitutional government, to achieve just this result. A key mechanism for producing technically grounded policy consensus in Germany is the parliamentary inquiry commission (*Enquete Kommission*), a hybrid decision-making forum that combines scientific expertise and political representation within a single advisory body (see e.g. NRC 2007: 90–2). An inquiry commission on 'Preventive Measures to Protect the Earth's Atmosphere' (*Vorsorge zum Schutz der Erdatmosphäre*) was appointed in 1987 and renewed in the next parliamentary election period. One of its major political achievements was to keep anti-regulatory forces from upsetting the scientific consensus on climate. The commission's third report, issued in 1990, endorsed the theory of anthropogenic climate change, attributing already observed atmospheric warming and sea-level rises to that phenomenon. The report also recommended a 30 percent reduction in carbon dioxide emissions below 1987 levels by 2005, a goal the cabinet reduced to a more feasible 25 percent (Enquete Kommission 1991 [German version 1990]; Cavender and Jäger 1993: 13–14). Noteworthy, especially in the light of the US case, was the virtual absence of scientific or political conflict over the reality of human-induced global warming itself.

The 2009 brouhaha over the hacked University of East Anglia e-mails did not leave German climate science unscathed, but there was no public backing off by scientists or politicians from the consensus on the need for emission reductions. A few visible scientists, notably Hans von Storch of Hamburg University's Meteorological Institute, tried to frame the episode as a wake-up call to restore the IPCC's credibility through an independent review process that might even include skeptics (Evers et al. 2010). Von Storch's reaction is consistent with an epistemic and political culture that leans toward the inclusion of all relevant viewpoints, even if it means sacrificing strict scientific purity. Illustrating a sharp difference of cultures, a review of climate assessments by the US National Research Council faulted the German inquiry commission for precisely this tendency: selecting experts through parliament rather than through scientific bodies, the NRC noted, 'could have some significant ramifications in terms of the credibility and legitimacy of the process' (NRC 2007: 92). Yet the hybrid politico-scientific position produced in Germany proved in this case more resistant to challenge than the corresponding expert determinations in the United States or Britain.

Chancellor Angela Merkel's decision early in her second term to reverse Germany's planned nuclear phase-out, and the sharp nationwide protests triggered by that policy reversal, offer powerful evidence of the strength of the climate accord and of a felt rejection of important norms of German civic epistemology by the ruling party. Merkel's Christian Democrats believed that they could take advantage of the public's still solid belief in the climate crisis to (again) throw support behind the nuclear industry, which had been clamoring for a comeback throughout the Western world. But vociferous criticism and the sudden remarkable rise in popularity of the Green Party (which achieved parity in some 2010 polls with the Social Democrats) showed that Merkel may have miscalculated the strength of Germany's nuclear-free vision, as well as underestimated the need to include anti-nuclear advocates in any reforged consensus around nuclear energy. Her government was attacked for violating powerful norms of democratic accountability. At the dawn of the twenty-first century's second decade, then, not climate skeptics but nuclear skeptics sounded the loudest notes of dissonance in German environmental and energy policy. It was in energy policy, rather than climate per se, that German consensus-building efforts had failed to produce the customary socio-technical accord.

4 BUILDING COSMOPOLITAN KNOWLEDGE

When like cases pattern unlike, the divergence cries out for elucidation. The ups and downs of climate science in the public spheres of the three Western nations discussed above provide a telling example. The United States, Britain, and Germany seem at one level to share all of the epistemic, technological, and political characteristics that should have pushed them toward similar understandings of climate science and similar actions on climate policy. Each boasts a strong, homegrown research tradition on climate change, a highly educated citizenry, multiple pathways for civic engagement, vigorous news media with the capacity to report on science and technology, and—not least—ample representation of its scientists in the workings of the IPCC. Each national economy depends on high levels of energy use, producing greenhouse gas emissions, and each nation has invested with varying degrees of enthusiasm in renewable energy technologies. Nuclear power was embraced by policy makers and dogged by continual citizen protest in each. Nonetheless, debate on climate issues has diverged across even these 'like' countries, indicating that the passage from technical assessment to the public sphere to policy choice is anything but linear, predictable, or deterministic. Comparison of these three cases suggests instead that global environmental policy making—especially around the climate threat—needs to take serious account of deep-seated national ways of knowing and acting, in short, of multiple civic epistemologies (Jasanoff 2005).

Our examples show that public reasoning, a process that all democratic societies are committed to in avoidance of arbitrary power, depends on prior criteria of what counts as valid reason. In societies infused with science and technology, those criteria importantly relate to the reception and uptake of scientific evidence by citizens and policy makers, and this is where political culture steps in. Believing in something so complex and multidimensional as climate change, let alone believing strongly enough to undertake urgent action, requires considerable faith in what I have called the 'wisdom of strangers.' Others' judgment, not our own, lays the groundwork for many collective decisions in the modern world, and the conventions by which we evaluate those judgments are part and parcel of our political cultures. Schematically speaking, scientific and technical judgments are exercised at the level of the individual expert, the discipline or specialty, and the advisory bodies that help translate technical knowledge to policy domains (Jasanoff 2010b). The importance accorded to each layer of judgment, however, differs across countries, as the three cases discussed above illustrate in almost ideal-typical fashion. Disciplinary norms uncontaminated by overt interest count most in the United States; the probity of individual experts is the prime focus of concern in Britain; and the capacity to form inclusive consensus positions functions as a *sine qua non* of stability and closure in German policy making.

If three closely similar nation-states, sharing hundreds of years of Enlightenment history, can diverge so much in their modes of linking knowledge to action, with seemingly dismal consequences for policy coordination, what hope is there for collective, knowledge-based action at the far more heterogeneous, indeed cacophonous, global level? Is it not inevitable that the world will stall on endless 'debates not about the existence of facts but about symbols, about the legitimacy of the competing sets of criteria by which a factual situation [in this case the reality of climate change] will or will not qualify as a genuine social problem' (Smelser 1986: 26)? Isn't minilateralism, after all, the only realistic answer? These are weighty questions, and cautious optimism is the best that can be offered against a future of scientific and political uncertainty, but the bright arrows, those pointing toward possibilities for global cooperation and stewardship respectful of democratic values, are worth rehearsing in conclusion.

First, in the examples above, we note that each nation's political system has put methods in place for confronting and resolving technical disputes and for moving ahead with policy implementation even in situations of unresolved uncertainty. Second, though their accountability mechanisms vary cross-nationally, each nation accepts in principle that governmental decisions must be explained in ways deemed acceptable by the public. Third, no exception is made for technical decisions: whether through administrative rule making (US), consultation (UK), or parliamentary inquiries (Germany), each state has found ways to engage its citizens, directly or indirectly, in the substance of technically grounded decisions, such as policies for climate change. In these circumstances, perhaps the greatest threat to collective action is to strive for an artificial separation of scientific assessment, in keeping with the separatist and linear models, from considerations of credibility, accountability, and persuasiveness. Arguably, the InterAcademy Council

(2010) formed in the wake of 'climategate' represented such an inward-looking response to the crisis of faith on climate science. This was an effort to shore up science's internal accounting processes and to make them more transparent, which as we have seen is a necessary but not sufficient condition for the production of cosmopolitan policy-relevant knowledge. Minilateralism fares no better, judged by standards of enlightened democracy: indeed, one can regard it as the political counterpart to scientific parochialism.

The need disclosed by decades of debate on climate science is for stronger processes of mediation and translation woven into the processes of knowledge making itself. The history of the climate controversy illustrates the limitations of separatist and linear thinking about science in a time characterized by ever greater interpenetration of knowledge and values, in a world that deeply respects scientific knowledge but follows its implicit norma-tive dictates only when science is ratified by diverse civic epistemologies. The histories recounted above suggest that procedures akin to the German parliamentary inquiry model—though critiqued by adherents of the linear paradigm as too 'politicizing'—may hold more promise for a global world than approaches that stress the purified integrity of experts and disciplines to the detriment of plurality of opinion. But the German model did not spring from nowhere; it, too, was forged in a historical tradition including two traumatic world wars that sensitized a nation to the perils of excluding dissent. On climate change, one hopes that building cosmopolitan knowledge that recognizes and respects justifiable differences will come at lesser cost. It should not take a holocaust.

NOTES

1. See <http://dictionary.reference.com/browse/cosmopolitan>.
2. At the end of 1972, a possible high water mark, there were some 1,400 federal advisory committees in existence in the United States. *Federal Advisory Committees, First Annual Report of the President*, 29 March 1973, p. 1. By the end of 1998, following years of deregulation and belt tightening in the federal government, the number stood at a still significant 892. *Twenty-Seventh Annual Report of the President on Federal Advisory Committees*, 1 March 1999, p. 7.
3. IPCC assessments do not produce original research. Rather, they provide original and comprehensive syntheses of data already in the public scientific record. In this sense they can be referred to as 'second-order' science—consisting neither of basic research nor technological applications, but requiring sophisticated interdisciplinary scientific judgment for its production (Mitchell et al. 2006).
4. Without these closure mechanisms, public science tends to spiral into the infinite decon-struction of 'experimenters' regress', which involves opponents continually challenging the assumptions underlying each other's claims (Collins 1985). America's long-running science controversies, such as that over creationism, points to the absence of credible closure devices. Why some protracted controversies close (e.g. over the health effects of tobacco or asbestos) and others do not requires case-by-case analysis.
5. Although the lawsuit was nominally brought by Stewart Dimmock, a truck driver and member of a political party known for climate skepticism, his action was financially backed

by Christopher Walter, Third Viscount Monckton of Brenchley, a conservative hereditary peer and noted critic of the theory of anthropogenic climate change.
6. From the Ofcom decision: 'Ofcom considers it of paramount importance that broadcasters, such as Channel 4, continue to explore controversial subject matter. While such programmes can polarise opinion, they are essential to our understanding of the world around us and are amongst the most important content that broadcasters produce.' Broadcast Bulletin Issue number 114–21|07|08, <http://stakeholders.ofcom.org.uk/enforcement/broadcast-bulletins/obb114/>.
7. Broadcast Bulletin Issue number 114–21|07|08.

REFERENCES

ANDEREGG, W. R. L., PRALL, J. W., HAROLD, J. and SCHNEIDER, S. H. 2010. Expert credibility in climate change. *Proceedings of the National Academy of Sciences* 107(27): 12107–9.

BECK, S. 2004. Localizing global change in Germany. Pp. 173–94 in S. Jasanoff and M. Martello (eds.), *Earthly Politics: Local and Global in Environmental Governance*. Cambridge, MA: MIT Press.

BECK, U. 1992 [1986]. *Risk Society: Towards a New Modernity*. London: Sage Publications.

—— 2006. *The Cosmopolitan Vision*. Cambridge: Polity.

CAVENDER, J., and JÄGER, J. 1993. The history of Germany's response to climate change. *International Environmental Affairs* 5(1): 3–18.

COLLINS, H. M. 1985. *Changing Order: Replication and Induction in Scientific Practice*. London: Sage Publications.

Enquete Kommission. 1991. *Protecting the Earth—A Status Report with Recommendations for a New Energy Policy*. Third Report of the Enquete Commission of the German Bundestag Preventive Measures to Protect the Earth's Atmosphere. Bonn: Deutscher Bundestag.

EVERS, M., STAMPF, O., and TRAUFETTER, G. 2010. A superstorm for global warming research. *Spiegel Online International*, 27 September.

GIBBONS, M., LIMOGES, C., NOWOTNY, H., SCHWARTZMAN, S., SCOTT, P., and TROW, M. 1994. *The New Production of Knowledge*. London: Sage Publications.

HAAS, P. 1990. *Saving the Mediterranean: The Politics of International Environmental Cooperation*. New York: Columbia University Press.

HULME, M. 2009. *Why We Disagree about Climate Change*. Cambridge: Cambridge University Press.

InterAcademy Council. 2010. *Climate Change Assessments: Review of the Processes and Procedures of the IPCC*. <http://reviewipcc.interacademycouncil.net/report.html>.

IRWIN, A., and WYNNE, B. (eds.) 1996. *Misunderstanding Science? The Public Reconstruction of Science and Technology*. Cambridge: Cambridge University Press.

JASANOFF, S. 1990. *The Fifth Branch: Science Advisers as Policymakers*. Cambridge, MA: Harvard University Press.

—— 2005. *Designs on Nature: Science and Democracy in Europe and the United States*. Princeton: Princeton University Press.

—— 2010a. A new climate for society. *Theory, Culture & Society* 27(2–3): 233–53.

—— 2010b. Testing time for climate science. *Science* 328(5979): 695–6.

KUHN, T. S. 1962. *The Structure of Scientific Revolutions*. 1st edn., Chicago: University of Chicago Press.

LATOUR, B. 1987. *Science in Action: How to Follow Scientists and Engineers through Society*. Cambridge, MA: Harvard University Press.

MERTON, R. K. 1942. Science and technology in a democratic order. *Journal of Legal and Political Sociology* 1: 115–26.

MITCHELL, R. B., CLARK, W. C., CASH, D. W., and DICKSON, N. M. 2006. *Global Environmental Assessments: Information and Influence*. Cambridge, MA: MIT Press.

NAÍM, M. 2009. Minilateralism: The magic number needed to get real international action. *Foreign Policy*, 1 July.

National Research Council (NRC). 1983. *Risk Assessment in the Federal Government: Managing the Process*. Washington DC: National Academies Press.

—— 2007. *Analysis of Global Change Assessments: Lessons Learned*. Washington, DC: National Academies Press.

NORDHAUS, W. 2007. Critical assumptions in the stern review on climate change. *Science* 317 (5835): 201–2.

ORESKES, N. 2004. The scientific consensus on climate change. *Science* 306(5702): 1686.

—— 2007. The scientific consensus on climate change: How do we know we're not wrong? Pp. 65–99 in J. F. C. DiMENTO and P. DOUGHMAN (eds.), *Climate Change: What It Means for Us, Our Children, and Our Grandchildren*. Cambridge, MA: MIT Press.

—— and CONWAY, E. M. 2010. *Merchants of Doubt: How a Handful of Scientists Obscured the Truth on Issues from Tobacco Smoke to Global Warming*. New York: Bloomsbury Press.

PIELKE, R. A., JR. 2007. *The Honest Broker: Making Sense of Science in Policy and Politics*. Cambridge: Cambridge University Press.

PRICE, D. K. 1965. *The Scientific Estate*. Cambridge, MA: Harvard University Press.

PRINS, G., et al. 2010. *The Hartwell Paper: A New Direction for Climate Policy after the Crash of 2009*. London: London School of Economics.

RORTY, R. 1991. Cosmopolitanism without emancipation: A response to Jean-Francois Lyotard. Pp. 211–22 in R. Rorty, *Objectivism, Relativism and Truth*. New York: Cambridge University Press.

SCHNEIDER, S. H. 2009. *Science as a Contact Sport: Inside the Battle to Save Earth's Climate*. Washington, DC: National Geographic.

SCHRAG, D. P. 2006. On a swift boat to a warmer world. *Boston Globe*, 17 December.

SHAPIN, S. 1994. *A Social History of Truth*. Chicago: University of Chicago Press.

—— and SCHAFFER, S. 1985. *Leviathan and the Air-Pump: Hobbes, Boyle, and the Experimental Life*. Princeton: Princeton University Press.

SMELSER, N. J. 1986. The Ogburn vision fifty years later. Pp. 19–35 in National Academy of Sciences, Commission on Behavioral and Social Sciences and Education, *Behavioral and Social Science: 50 years of Discovery*. Washington, DC: National Academies Press.

STERN, N. 2005. *Stern Review on the Economics of Climate Change*. London. <http://www.hm-treasury.gov.uk/sternreview_index.htm>.

UK House of Commons Science and Technology Committee. 2010. *The Disclosure of Climate Data from the Climatic Research Unit at the University of East Anglia*. Select Committee Report, 31 March. London: The Stationery Office.

CHAPTER 10

ORGANIZED CLIMATE CHANGE DENIAL

RILEY E. DUNLAP AND AARON M. McCRIGHT

EVEN as the consensus over the reality and significance of anthropogenic climate change (ACC) becomes stronger within the scientific community, this global environmental problem is increasingly contested in the political arena and wider society. The spread of debate and contention over ACC from the scientific to socio-political realms has been detrimental to climate science, as reflected in significant declines in public belief in global warming in 2009 and 2010 (Leiserowitz et al. 2010). Contrarian scientists, fossil fuels corporations, conservative think tanks, and various front groups have assaulted mainstream climate science and scientists for over two decades. Their recently intensified denial campaign building on the manufactured 'Climategate' scandal (Fang 2009) and revelations of various relatively minor errors in the 2007 IPCC Fourth Assessment Report appears to have seriously damaged the credibility of climate science (Tollefson 2010). The blows have been struck by a well-funded, highly complex, and relatively coordinated 'denial machine' (Begley 2007).[1] It consists of the above actors as well as a bevy of amateur climate bloggers and self-designated experts, public relations firms, astroturf groups, conservative media and pundits, and conservative politicians.

The motivations of the various cogs of the denial machine vary considerably, from economic (obvious in the case of the fossil fuels industry) to personal (reflected in the celebrity status enjoyed by a few individuals), but the glue that holds most of them together is shared opposition to governmental regulatory efforts to ameliorate climate change, such as restrictions on carbon emissions. While the claims of these actors sometimes differ and evolve over time (there's no warming, it's not caused by humans, it won't be harmful, etc.), the theme of 'no need for regulations' remains constant (McCright and Dunlap 2000; Oreskes and Conway 2010). A staunch commitment to free markets and disdain of governmental regulations reflect the conservative political ideology that is almost universally shared by the climate change denial community.[2] This suggests how the diverse elements of the denial machine are able to work in a compatible and mutually reinforcing manner even when their efforts are not necessarily coordinated. By attacking climate science and individual scientists in various venues and fashions, the denial machine seeks to undermine the case for climate policy making by removing (in the eyes of the public and policy makers) the scientific basis for such policies—i.e. by challenging the reality and seriousness of climate change.

Viewed through a broader theoretical lens, climate change denial can be seen as part of a more sweeping effort to defend the modern Western social order (Jacques 2006), which has been built by an industrial capitalism powered by fossil fuels (Clark and York 2005). Since anthropogenic climate change is a major unintended consequence of fossil fuel use, simply acknowledging its reality poses a fundamental critique of the industrial capitalist economic system. European scholars such as Ulrich Beck and Anthony Giddens describe the current era as one of 'reflexive modernization,' in which advanced nations are undergoing critical self-confrontation with the unintended and unanticipated consequences of industrial capitalism—especially low-probability, high-consequence risks that are no longer circumscribed spatially or temporally such as genetic engineering, nuclear energy, and particularly climate change (Beck 1992; Beck et al. 1994; Giddens 1990). Reflexive modernization theorists like Beck and Giddens argue that a heightened level of reflexivity is a necessary precondition for dealing effectively with this new set of human-induced ecological and technological threats.

Crucial drivers of this reflexivity, or societal self-confrontation and examination, are citizen action/social movements (Beck's 'sub-politics') and science, most notably environmental activism and those scientific fields that examine ecological and human health impacts of technologies and economic activities. By directing societal attention to environmental disasters like massive oil spills and crescive problems like climate change that result from economic production, the forces of reflexivity draw the ire of defenders of the capitalist system who often mobilize against them (Beck 1997; Mol 2000). This has been particularly true in the United States, where a combination of corporate and conservative interests have long battled environmentalism (Helvarg 2004) and environmental science (Jacques et al. 2008). We have argued elsewhere that these interests are now mobilizing more broadly in opposition to reflexive modernization writ large and are becoming a source of 'anti-reflexivity' (McCright and Dunlap 2010). Nowhere is this anti-reflexive orientation—particularly the dismissal of scientific evidence and methodology—more apparent than in climate change denial.

This chapter provides an overview of organized climate change denial.[3] We begin by describing the growth of conservative-based opposition to environmentalism and environmental science in general, and then explain why climate change became the central focus of this opposition, which quickly evolved into a coordinated and well-funded machine or 'industry' (Monbiot 2007). We also examine denialists' rationale for attacking the scientific underpinnings of climate change policy and the crucial strategy of 'manufacturing uncertainty' they employ (Michaels 2008; Oreskes and Conway 2010). The remainder of the chapter describes the complex and evolving set of actors espousing climate change denial, touching on their tactics when appropriate and tracing their interconnections when possible. Describing the climate change denial machine is difficult, because it is both a complex and ever-evolving labyrinth *and* because many of its components intentionally mask their efforts and sources of support. We focus primarily on the US, where denial first took root and remains most active, but also include a brief look at its international diffusion. We conclude with observations about the dangers of growing anti-reflexivity in an era of profound ecological threats such as climate change.

1 History and Strategy of Climate Change Denial

Riding the wave of a conservative resurgence launched in reaction to the progressive gains of the 1960s and early 1970s (Lapham 2004), including an impressive set of environmental agencies and regulations, the Reagan Administration came into office promising to get government off the back of the private sector. However, the administration's efforts to curtail environmental protection created a backlash that forced it to moderate its anti-environmental rhetoric and actions, albeit not its objectives (Dunlap 1987). This experience taught conservatives (and industry) that it was more efficacious to question the *need* for environmental regulations by challenging evidence of environmental degradation, rather than the *goal* of environmental protection. Promoting 'environmental skepticism' which disputes the seriousness of environmental problems (Jacques 2006) has subsequently been heavily employed by conservative think tanks and their corporate allies, especially since the 1990s when the downfall of the Soviet Union and the rise of global environmentalism represented by the 1992 Rio 'Earth Summit' led conservatives to substitute a 'green threat' for the disappearing 'red threat' (Jacques et al. 2008). Perception of the Clinton-Gore Administration as receptive to environmental protection heightened conservatives' fears of increasing national *and* international environmental regulations.

These fears crystallized around climate change, as creation of the Intergovernmental Panel on Climate Change (IPCC) by the United Nations Environmental Program and the World Meteorological Organization represented an unprecedented international effort to develop a scientific basis for policy making.[4] This, combined with the encompassing nature and wide-ranging implications of climate change, turned ACC into a *cause célèbre* for conservatives. The mainstream conservative movement, embodied in leading foundations and think tanks, quickly joined forces with the fossil fuels industry (which recognized very early the threat posed by recognition of global warming and the role of carbon emissions) and wider sectors of corporate America to combat the threat posed by climate change—not as an ecological problem but as a problem for the pursuit of unbridled economic growth (Gelbspan 1997). In the process this coalition took the promotion of environmental skepticism to a new level, attacking the entire field of climate science as 'junk science' and launching attacks on such pillars of science as the importance of peer-reviewed publications (Jacques et al. 2008). The result has been an evolution of environmental skepticism into a full-blown anti-reflexivity in which the ability and utility of science for documenting the unintended consequences of economic growth are being undermined (McCright and Dunlap 2010).

The conservative movement/fossil fuels complex quickly adopted the strategy of 'manufacturing' uncertainty and doubt (perfected by the tobacco industry) as its preferred strategy for promoting skepticism regarding ACC (Union of Concerned Scientists 2007). Early on contrarian scientists—with considerable support from industry and conservative think tanks—stressed the 'uncertainty' concerning global warming and human contributions to it (Oreskes and Conway 2010). As the threat of international policy making increased, from the 1997 Kyoto Protocol to the 2009 COP in Copenhagen, the growing army of opponents to carbon emissions reduction policies has stepped up their attacks (Greenpeace 2010a; McCright and Dunlap 2003; Pooley 2010). They have also broadened their tactics well beyond manufacturing uncertainty, increasingly criticizing peer-review, refereed journals, governmental grant making, scientific institutions (American Association for the Advancement of Science, US National Academy of Sciences, etc.) and the expertise and ethics of scientists (Nature 2010a, 2010b; Sills 2010). Again, this assault on scientific practices, evidence, and institutions weakens a major mechanism of reflexive modernization. We now turn to an examination of the major actors in the denial machine, which are portrayed in Figure 10.1 to help readers readily identify them and visualize their interconnections.

Key Components of the Climate Change Denial Machine

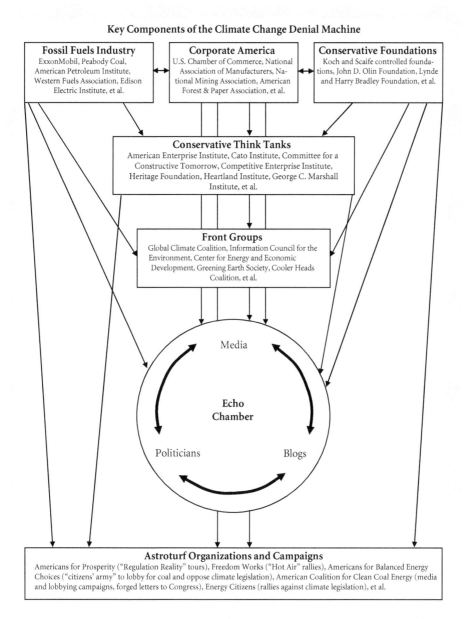

FIGURE 10.1 Key components of the climate change denial machine.

2 MAJOR ACTORS

2.1 Fossil Fuels Industry and Corporate America

Coal and oil corporations recognized the implications of global warming and efforts to combat it for their industries early on, as burning fossil fuels was quickly identified as a major source of greenhouse gas emissions. Not surprisingly, therefore, the fossil fuels industry pioneered the charge against climate science and policy making (Begley 2007; Gelbspan 1997; Goodell 2007). Both individual corporations such as ExxonMobil and Peabody Coal as well as industry associations such as the American Petroleum Institute, Western Fuels Association, and Edison Electric Institute provided funding for individual contrarian scientists, conservative think tanks active in climate change denial, and a host of front groups we discuss below. ExxonMobil[5] has long been the leading contributor to think tanks and front groups involved in climate change denial, although it cut back somewhat in recent years in response to negative publicity and severe criticism (Mooney 2005; Union of Concerned Scientists 2007).

The efforts of fossil fuels corporations and industry associations to combat climate science and policy making were quickly supplemented by those of numerous energy companies (e.g. Southern Company), other resource-based corporations in the steel, forestry, and mining industries as well as their associations (e.g. National Mining Association), numerous manufacturing companies such as automobile corporations (e.g. Chrysler, Ford, and General Motors), and large national associations such as the National Association of Manufacturers and the US Chamber of Commerce (Gelbspan 1997, 2004; Hoggan with Littlemore 2009; Layzer 2007). Thus, in the early 1990s it appeared that much of corporate America was lining up against climate science and policy making, with the IPCC being the crucial target.

The growing evidence of anthropogenic climate change reported in the IPCC's Second Assessment Report in 1995 and the adoption of the Kyoto Protocol at the 1997 Kyoto Conference led to some fracturing within the business community, and several corporations including BP announced that they no longer questioned the reality of ACC and were halting efforts to undermine climate science. Several oil companies and other major corporations joined with leading environmental organizations to form the US Climate Action Partnership, and it appeared that a major segment of corporate America was ready to accept the reality of climate change and the inevitability of carbon reduction policies (Kolk and Levy 2001; Layzer 2007). However, with the inauguration of the George W. Bush Administration, which institutionalized climate change denial in the federal government (McCright and Dunlap 2010), the fossil fuels industry in particular had little to fear.

The election of Barack Obama and a Democratic majority in both houses of Congress has made the reality of legislation to limit carbon emissions salient, and the result has been enormous corporate lobbying to oppose or weaken the various measures introduced in the House and Senate as well as international efforts such as COP-15 in Copenhagen (Goodell 2010; Pooley 2010). This lobbying has been accompanied by escalating attacks on climate science and scientists as well as the IPCC, with considerable support from corporations such as ExxonMobil and associations such as the US Chamber of Commerce (Greenpeace 2010a; Mashey 2010). Thus, while there are divisions within corporate America over policy proposals such as 'cap-and-trade,' it appears that significant portions of it remain active in climate change denial.

2.2 Conservative Philanthropists, Foundations, and Think Tanks

The earlier-mentioned conservative resurgence began when wealthy conservative philanthropists such as Joseph Coors began to fund, typically through their family foundations, the establishment of conservative think tanks (CTTs) such as the Heritage Foundation to wage a 'war of ideas' against the progressive gains of the 1960s (Himmelstein 1990; Lapham 2004). By the 1990s conservative foundations were funding a 'conservative labyrinth' designed to implant conservative values and goals in academic, media, governmental, legal, and religious institutions (Covington 1997: 3). Particularly important is the network of well-heeled and influential think tanks that churn out an endless flow of policy proposals credited with moving the US policy agenda significantly to the right (Krehely et al. 2004) and—since the 1990s—influencing climate policy (McCright and Dunlap 2003, 2010).

Major funders include foundations controlled by Richard Mellon Scaife and David and Charles Koch (both drawing upon family fortunes stemming in part from oil interests). Besides giving generously to a vast range of CTTs and conservative causes, they are responsible for establishing the Cato Institute (C. Koch), Citizens for a Sound Economy, now Americans for Prosperity (D. Koch) and Committee for a Constructive Tomorrow or CFACT (R. M. Scaife)—three particularly crucial elements of the denial machine. In fact, in recent years the Scaife and Koch families of funds may have exceeded ExxonMobil in terms of funding climate change denial actors and activities (Grandia 2009; Greenpeace 2010b; Mashey 2010).

CTTs represent 'social movement organizations' that typically serve as spokespersons and facilitators for conservative causes, and share a universal commitment to free enterprise, limited government, and the promotion of unfettered economic growth (Jacques et al. 2008; McCright and Dunlap 2000). While corporations like ExxonMobil have joined conservative foundations in providing generous funding for CTTs, many of the latter appear to oppose climate science and policy making for purely ideological reasons (McCright and Dunlap 2010: 109–11), and some of their leaders have criticized corporations for disengaging from climate change denial (Layzer 2007: 112). CTTs involved in climate change denial range from large, multi-issue ones (e.g. the Heritage Foundation and American Enterprise Institute), to medium ones with a strong interest in environmental/climate issues (e.g. George Marshall Institute and the Heartland Institute), to small shops currently dedicated to climate change denial (e.g. Fred Singer's Science and Environmental Policy Project and Republican operative Robert Ferguson's Science and Public Policy Institute) (Mashey 2010; McCright and Dunlap 2003).

CTTs are a fundamental and highly effective component of the denial machine, providing institutional bases for leading contrarians such as Patrick Michaels (a Cato Fellow), hosting anti-IPCC conferences (Heartland Institute), sponsoring 'educational events' for politicians (National Center for Policy Analysis), assisting the George W. Bush Administration's efforts to impede climate policy (Competitive Enterprise Institute), and producing and circulating a vast range of anti-climate change material via various forms of media (reports, press releases, press conferences, videos, radio and television interviews), among other activities (see e.g. Hoggan with Littlemore 2009; Lahsen 2008; McCright and Dunlap 2000, 2003; Oreskes and Conway 2010).

More generally, CTTs help shield the efforts of corporations and philanthropists to combat climate change policy, as for example ExxonMobil, the Koch brothers, and R. M. Scaife support contrarian scientists and denial campaigns effectively but 'discreetly' by funneling millions into think tanks that sponsor the contrarians and organize the campaigns (Greenpeace 2010b; Mashey 2010; Union of Concerned Scientists 2007).[6] Furthermore, CTTs have been successful in marketing themselves as objective sources of information, basically an alternate academia, and thus they have more credibility with much of the public, many media outlets, and some policy makers than do corporations (Jacques et al. 2008). They enhance their credibility by sponsoring contrarian scientists who are treated as 'experts' (regardless of the relevance or quality of their research records) by the media and public, and whose ideas are amplified considerably by CTTs' media access (Hoggan with Littlemore 2009; McCright and Dunlap 2000, 2003; Mooney 2005; Oreskes and Conway 2010). Finally, CTTs work carefully with corporate America to set up a maze of front groups and astroturf campaigns to combat climate science and policy making.

2.3 Front Groups

Most corporations prefer to shield their anti-environmental activities from public scrutiny, and creating front groups that act on their behalf is one way to do this. The Global Climate Coalition (GCC), formed in 1989 in reaction to establishment of the IPCC, was an early front group designed to combat evidence of climate change and climate policy making. Sponsored by oil companies (ExxonMobil, Texaco, and BP), automobile manufacturers (Chrysler, Ford, and GM) and industrial associations such as the American Petroleum Institute (API), US Chamber of Commerce, and the National Association of Manufacturers, it was originally led by William O'Keefe of API. The GCC was very active in opposing US ratification of the Kyoto Protocol, running television ads against it, and played a critical role in launching a vicious (and unfounded) attack on climate scientist Benjamin Santer for allegedly altering a chapter in the 1995 IPCC report in an effort to discredit the entire report and the IPCC (Gelbspan 2004: 78–80; Oreskes and Conway 2010: 207–13). The accumulating scientific evidence in support of climate change led BP, Shell, and other companies to leave the GCC in the late 1990s, presumably because they no longer wanted to be associated with its aggressive and highly visible opposition to climate science and policy. The GCC disbanded in 2002, confident that its goals were shared by the George W. Bush Administration (Gelbspan 2004; Greenpeace 2010a; Pooley 2010).

The Information Council on the Environment (ICE) was created in 1991 by coal and utility interests, including the National Coal Association, Western Fuels Association, and Edison Electric Institute, and launched a campaign to 'reposition global warming as a theory (not fact)' (Pooley 2010: 41). Assisted by contrarian scientists such as Patrick Michaels, Robert Balling, and Sherwood Idso, ICE ran a media campaign designed to denigrate the notion of global warming and campaigned against US agreement to mandatory greenhouse gas emissions at the 1992 Earth Summit in Rio. ICE folded up when its strategic plans were leaked to the press, but the Western Fuels Association subsequently established the Greening Earth Society in 1998 to promote the idea that CO_2 was good for the environment and thus global warming was to be welcomed. Besides an advertising campaign, it sponsored a quarterly *World Climate Review* edited by contrarian Patrick Michaels, which has been replaced by the *World Climate Report* blog also edited by Michaels but with unknown sources of support (Gelbspan 2004; Hoggan with Littlemore 2009; Pooley 2010).[7]

The Cooler Heads Coalition (CHC) is the final major US front group for climate change denial, and unlike its predecessors its membership consists primarily of CTTs including CFACT, the Marshall Institute, the Heartland Institute, and the Competitive Enterprise Institute (CEI)—who, of course, receive significant corporate and conservative foundation funding. It emerged in 1997 as a subgroup of the National Consumer Coalition, a project of Consumer Alert—an industry-funded entity founded in 1977 to oppose consumer protection regulations such as mandatory seatbelts. It is tied closely to CEI, which hosts its website <www.globalwarming.org>, where CHC is described as 'an informal and ad-hoc group focused on dispelling the myths of global warming.' CHC/CEI leaders Myron Ebell and Christopher Horner are central figures in the denial machine, and use both CHC and CEI to distribute a flood of denial material, host press conferences and Congressional briefings, and amplify the voices of contrarian scientists (Hoggan with Littlemore 2009; Mooney 2005; Pooley 2010). They have played a crucial role in promoting 'Climategate' and waging war on the IPCC, and often launch malicious attacks on individual climate scientists.

2.4 Contrarian Scientists

From the earliest stages of climate change denial the fossil fuels industry and conservative think tanks, and their fronts groups like GCC, recognized the importance of employing credentialed scientists to manufacture uncertainty concerning climate change (building on the tobacco industry's success with this strategy—Oreskes and Conway 2010), and they readily found scientists who were eager to assist (Gelbspan 1997; McCright 2007). Some had expertise relevant to climate science (e.g. Patrick Michaels and Fred Singer), but many did not. For instance, the George C. Marshall Institute was established by a trio of prominent physicists who, despite having no expertise in climate science per se, quickly made climate change denial a central mission of the Institute and created a magnet that eventually attracted several contrarians, such as Roy Spencer, who do have climate science expertise (Lahsen 2008; Oreskes and Conway 2010).

It is impossible to discern whether contrarian scientists sought affiliations with CTTs (and front groups) or were solicited by them, but at this point most of the highly visible contrarians have some form of affiliation with CTTs, such as: having formal appointments like Patrick Michaels at the Cato Institute; serving on boards, as scientific advisors, or as affiliated experts; giving talks at the CTTs and participating in CTT press conferences, political briefings, and public lectures; and especially publishing material for CTTs (McCright and Dunlap 2000, 2003).[8] Being affiliated with CTTs enables contrarians to avoid the 'stigma' of being directly linked to fossil fuels corporations (see e.g. Gelbspan 1997: 41), while still benefiting from the industry's largesse to many CTTs (Mashey 2010; Union of Concerned Scients 2007).

The strong bond between contrarian scientists and CTTs reflects the staunch conservative aversion to governmental regulations and commitment to free markets shared by nearly all leading contrarian scientists (Oreskes and Conway 2010). It may also reflect contrarians' realization that their marginal standing within mainstream climate science (Anderegg et al. 2010) can be offset by moving into the public and policy spheres where their messages are greatly amplified by their very influential CTT sponsors and often welcomed by journalists eager to provide 'balanced' reporting (Boykoff and Boykoff 2004; McCright 2007; McCright and Dunlap 2003).

As climate change denial has matured, the number of 'scientists' who promote it has grown both in size and diversity (as well as spread internationally). CTTs and fossil-fuels front groups, in particular, now sponsor a multitude of 'experts' who often have no discernible credibility as climate scientists. An increasing number of their spokespersons appear to lack any scientific training or expertise, such as the ubiquitous Christopher Walter Monckton (aka Lord Monckton) who is affiliated with the Science and Public Policy Institute in the US. However, manufacturing uncertainty is most successful when it is done by individuals that the media and public will accept as experts, and CTTs continue to find and support a number of credentialled scientists critical of climate science, giving them unprecedented visibility regardless of how poorly their typically non-peer-reviewed work fares among the scientific community (see e.g. Enting 2010 on one example). Indeed, Monbiot's (2007) characterization of the 'denial industry' reflects the fact that climate change denial now offers the possibility of a rewarding 'career' for contrarian scientists and others eager to work with CTTs, front groups, and conservative media.

2.5 Conservative Media

The influence of the conservative media or 'echo chamber' has been well documented and has been credited with helping move the US rightward in recent decades (e.g. Jamieson and Cappella 2008). For right-wing talk radio commentators, most notably Rush Limbaugh, attacks on 'environmental wackos' is standard fare, and climate change (and Al Gore) a favorite target (Nature 2010b; Wolcott 2007). Perhaps exceeding the impact of the right-wing dominance of talk radio is Rupert Murdoch's Fox News, as both its reporters and most popular commentators (Glenn Beck, Bill O'Reilly, and Sean Hannity) consistently denigrate climate change by, for example, highlighting 'Climategate' and critiques of the IPCC and providing frequent opportunities for contrarian scientists and CTT representatives to disparage climate change, the IPCC, and climate scientists.

The conservative media assault on climate science also occurs in print media, especially conservative newspapers such as the Murdoch-owned *Wall Street Journal* (whose editorial pages have become a regular forum for climate change denial, including columns by contrarian scientists) and the *New York Post* and the Reverend Moon's *Washington Times*. Climate change denial is also a regular feature in leading conservative magazines such as *The Weekly Standard*, *National Review*, and *The American Spectator* as well as online publications such as *The American Thinker*. Add in prominent conservative columnists like George Will and Charles Krauthammer (infamous for their erroneous statements about climate change—Dickinson 2010), who reach vast newspaper audiences via national syndication, and the result is a barrage of assaults on climate science (and, increasingly, climate scientists) that not only inundates committed conservative audiences but also reaches a large segment of the general public. Conservative media consistently present contrarian scientists and CTT representatives as 'objective' experts, in stark contrast to their portrayal of scientists working with the IPCC as self-interested and biased, further magnifying the influence of the former relative to the latter.

In recent years these conservative media outlets have been supplemented (and to some degree supplanted) by the conservative blogosphere, and numerous blogs now constitute a vital element of the denial machine. While a few are hosted by contrarian scientists (most notably Roy Spencer), the most popular North American blogs are run by a retired TV meteorologist (wattsupwiththat.com), a retired mining executive and dedicated critic of the 'hockey stick' model of historical climate trends (climateaudit.org), and a self-styled 'warrior' in the climate wars (climatedepot.com). The latter individual, Marc Morano, exemplifies the deep roots of climate change denial in conservative circles. Before setting up Climate Depot, which is modeled on the popular right-wing 'Drudge Report' and supported by R. M. Scaife's CFACT, Morano—who has a BA degree in political science—worked for Rush Limbaugh, right-wing Cybercast News Service (where he played a key role in the 'swift-boat' campaign against 2004 Democratic Presidential candidate John Kerry), and then for Republican Senator James Inhofe (Dickinson 2010; Harkinson 2009b).

Having this powerful, pervasive, and multifaceted media apparatus at its service provides the denial machine with a highly effective means of spreading its message, as reflected quite recently by its success in turning a tiny and highly unrepresentative sample of thirteen years worth of personal e-mails hacked from the Climate Research Unit at the University of East Anglia into a major scandal that has generated a decline in public belief in climate change and trust in climate scientists (see Leiserowitz et al. 2010 on public opinion and Greenpeace 2010b on the role of Koch-funded actors in publicizing Climategate)—despite the fact that several investigations have concluded that the e-mails neither demonstrate unethical behavior nor undermine climate science (Young 2010).

2.6 Conservative Politicians

Most conservative politicians have been highly skeptical of climate change from the outset, as accepting its reality challenges their faith in inevitable progress created by the free market and raises the specter of increased governmental regulations. Republicans in Congress have been eager hosts of contrarian scientists, CTT spokespersons, and a raft of other non-credentialed deniers from novelist Michael Crichton (whose *State of Fear* portrayed climate change as a contrived plot) to, most recently, Lord Monckton. They have also called hearings to rebut and in some instances harass mainstream climate scientists. The most notable include the 1994–5 House of Representatives hearings called by California Republican Dana Rohrbacher devised to portray evidence for dioxin, ozone depletion, and global warming as 'junk science' (McCright and Dunlap 2003: 361), and a 2005 House hearing held by Texas Republican Joe Barton designed to disprove the hockey stick model of Mann, Bradley, and Hughes and thereby discredit the IPCC (Mashey 2010).[9]

The single most prominent Republican when it comes to climate change denial is Oklahoma Senator James Inhofe, famous for claiming in a Senate speech that global warming is 'the greatest hoax ever perpetrated on the American people.' When Inhofe was Chair of the Committee on Environment and Public Works he turned it into a bastion of climate change denial via its website run by Marc Morano and his frequent invitations to contrarian scientists to testify at Committee hearings (McCright and Dunlap 2010). More recently he has called for a criminal investigation of leading climate scientists (Nature 2010a). The ease with which Inhofe and his Republican colleagues gain access to conservative media like Fox News provides yet another means for amplifying the messages of contrarian scientists in the conservative echo chamber.

The inauguration of George W. Bush institutionalized climate change denial throughout the most powerful branch of the US government, allowing representatives of the fossil fuels industry and CTTs to undermine climate science and policy from within the administration. For eight years the Bush administration used a variety of techniques, ranging from emphasizing the 'uncertainty' of climate science and calling for 'sound science' to suppressing the work of governmental scientists, to justify inaction on climate policy (McCright and Dunlap 2010). By the time it was replaced by the Obama Administration, most Republican politicians had followed its lead in questioning the seriousness of climate change. The predictable upsurge in denial activism and lobbying against climate policy that has occurred following the change in administrations, especially the embrace of denialism among the more extreme elements of the Right (e.g. Tea Party supporters), has turned climate change denial into a litmus test for Republicans (Johnson 2010). As a consequence, even one-time sponsors of bipartisan climate legislation like Republican Senators John McCain and Lindsey Graham have had to back-pedal to appease Republican interest groups and supporters.

2.7 Astroturf Groups and Campaigns

The defining feature of astroturf groups is that they are generated by an industry, think tank, or front group, but disguised to appear as a spontaneous, popular 'grassroots' effort. They are created to lobby or campaign on behalf of their sponsors, who hope to remain hidden from view (Beder 1998). Front groups and PR firms typically play key roles, and are often inseparable from the astroturf group/campaign itself, the key distinction being that the former tend to last longer while astroturf efforts come and go in response to specific events and policies. The use of astroturf groups has flourished in the Obama era, being used to oppose healthcare reform and other progressive goals of the President and Democratic Congress. Especially important are the roles played by the Koch-funded Americans for Prosperity and FreedomWorks front groups in generating a significant portion of the 'Tea Party' *and* encouraging it to focus on climate change (Dickinson 2010; Goodell 2010; Pooley 2010).

For example, Americans for Prosperity sponsored a multi-state 'Hot Air Tour' in 2008 with the slogan, 'Global Warming Alarmism: Lost Jobs, Higher Taxes, Less Freedom,' while FreedomWorks played a major role in promoting the 2009 rallies against climate legislation in about twenty states that were 'officially' sponsored by 'Energy Citizens'—an astroturf group created by the American Petroleum Institute (API). While its website proclaims that Energy Citizens 'is a movement made up of tens of thousands of Americans,' API President Jack Gerard's memo to API member corporations urged them to provide 'strong support for employee participation at the rallies' and asked that his (inevitably leaked) memo be treated as 'sensitive information' because 'we don't want critics to know our game plan' (Dickinson 2010; Goodell 2010).[10]

More generally, the success of Americans for Prosperity and Freedom Works along with CFACT and other conservative organizations and spokespersons (e.g. Glenn Beck) in melding climate change denial into the faux populist rage of the Tea Partiers has put climate science squarely in the sights of right-wing extremists, which has no doubt contributed to the escalating attacks against climate scientists (Hickman 2010). As Levy (2010: 4) states, 'Tea Party activism has elevated climate change to the status of a litmus test of cultural politics in the U.S., up there with abortion, guns, god, gays, immigration and taxes.' This raises the politicization of climate change/science (Dunlap and McCright 2008) to a new level and into a treacherous domain.[11] While the entire denial machine (but particularly the Kochs and their operatives) has contributed to this 'accomplishment,' it epitomizes successful astroturfing.

2.8 International Diffusion of Climate Change Denial

We have concentrated on the US because it is where climate change denial was born and continues to be most active, but denialism has spread to other nations—often with some degree of assistance from American actors. It tends to be strongest in nations that currently have or have recently had conservative governments and in which CTTs are firmly planted, notably the UK, Canada, and Australia, reinforcing our claim that free-market conservatism (with the strong support of the fossil fuels industry in the latter two countries) is the unifying force behind climate change denial. The UK's International Policy Network and its affiliate the Institute of Economic Affairs, Canada's Fraser Institute, and Australia's Institute of Public Affairs, for example, have provided early and continuing support for contrarian scientists and others active in climate change denial in their respective nations (Hamilton 2010; Hoggan with Littlemore 2009; Monbiot 2007).

In addition, one finds a similar emphasis on the creation of a web of front groups to act on behalf of industry and think tanks, perhaps best exemplified by Australia. There the Institute of Public Affairs (a free-market think tank) created in 2005 the Australian Environment Foundation (to mimic the pro-environmental Australian Conservation Foundation), which in turn set up the Australian Climate Science Coalition to promote climate change denial. These organizations are complemented by the Lavoisier Group, funded heavily by mining interests, which focuses specifically on climate change. Most leading Australian contrarian scientists such as Robert Carter, William Kininmonth, Garth Paltridge, and Ian Plimer are connected in some fashion to these organizations, which are also active in bringing American contrarians to Australia. In fact, the US denial machine was very active in helping establish its counterpart in Australia. Contrarians such as Fred Singer and Patrick Michaels visited there early on, and in the mid-1990s the Competitive Enterprise Institute recognized that Australia's Howard Government could become a valuable ally in opposing the Kyoto Protocol and began to coordinate efforts with the Institute for Public Affairs and mining interests (Hamilton 2007; Climate Action Network Australia 2010).

Climate change denial is now spreading far beyond the US, UK, Canada, and Australia, and once again this is directly due to the efforts of crucial CTTs to diffuse their goals and influence internationally. In particular, the Atlas Economic Research Foundation (established by Sir Anthony Fisher of the UK, but based in the US) serves as an 'incubator' for free-market think tanks around the world, and is credited with helping plant them in several dozen nations where they are frequently active in climate change denial. Canada's Fraser Institute (which receives funding from Koch and Scaife foundations) has a similar international reach with its Economic Freedom Network having affiliates in scores of nations, many helping spread climate change denial. And finally, in 2007 Fisher's UK-based International Policy Network created the 'Civil Society Coalition on Climate Change' which consists of 'independent civil society organizations' in forty nations committed to denying the reality of climate change (Harkinson 2009a).

As Harkinson (2009a: 1) puts it, 'With US-backed overseas think tanks parroting denier talking points in dozens of languages, the echo chamber is already up and running.' In sum, we are witnessing the globalization of organized climate change denial, and this does not bode well for the future of climate science and especially for effective international action and policy making to deal with the reality of climate change.

3 CONCLUSION

Many factors influence both national and international policy-making on environmental (and other) issues (Dryzek et al. 2002). We are definitely not suggesting that organized climate change denial has been the sole factor in undermining efforts to develop domestic climate policies in nations such as the US, Australia, and Canada where it has been especially prominent, nor at the international level where diverging national interests are obviously a major obstacle (Parks and Roberts 2010). Nonetheless, it is reasonable to conclude that climate change denial campaigns in the US have played a crucial role in blocking domestic legislation and contributing to the US becoming an impediment to international policy making (McCright and Dunlap 2003; Pooley 2010). The financial and organizational resources and political and public relations expertise available to and embodied in the major components of this machine, and the various actors' ability to coordinate efforts and reinforce one another's impacts, have certainly had a profound effect on the way in which climate change is perceived, discussed, and increasingly debated—particularly within the US.

We have argued that because of the perceived threat posed by climate change to their interests, actors in the denial machine have strived to undermine scientific evidence documenting its reality and seriousness. Over the past two decades they have engaged in an escalating assault on climate science and scientists, and in recent years on core scientific practices, institutions, and knowledge. Their success in these efforts not only threatens our capacity to understand and monitor human-induced ecological disruptions from the local to global levels (Hanson 2010), but it also weakens an essential component of societal reflexivity when the need for the latter is greater than ever.

NOTES

1. The actions of those who consistently seek to deny the seriousness of climate change make the terms 'denial' and 'denier' more accurate than 'skepticism' and 'skeptic' (Diethelm and McKee 2009), particularly since all scientists tend to be skeptics (Schneider 2010: 205). We will, however, refer to scientists involved in the denial machine as 'contrarians.' For an alternative but complementary use of 'denial' see Kari Norgaard's chapter in this volume.
2. This may be somewhat less true of contrarian scientists, but the few examples of self-professed liberals active in climate change denial such as Freeman Dyson are clearly exceptions to the rule (Larson and Keating 2010).

3. For overviews that provide clear time-lines for the historical evolution of climate change denial see Greenpeace (2010a) and Mashey (2010).

4. The explicit merger of science and policy making within the IPCC has contributed to climate science *and* climate scientists, along with the IPCC, becoming targets for those fearful that strong evidence of climate change will lead to national and international regulations on carbon emissions that create restrictions on corporate behavior, free markets, and economic growth (Corfee-Morlott et al. 2007).

5. Exxon merged with Mobil to become ExxonMobil in 1999, and other oil companies have merged and/or changed their names (e.g. British Petroleum became BP) in the past two decades. To avoid confusion, we will employ the current names even when describing activities undertaken by earlier versions of the contemporary corporations.

6. Only recently have the links between the Koch brothers and right-wing activities, including the Tea Party and climate change denial, been publicized (Greenpeace 2010b; Mayer 2010).

7. Another important coal-based front group, the Center for Energy and Economic Development, and its offshoots Americans for Balanced Energy Choices and American Coalition for Clean Coal Energy, have also supported climate change denial; however, their primary focus has been on lobbying against climate legislation by generating phony citizens' or astroturf (see below) campaigns (see Hoggan with Littlemore 2009 and especially Pooley 2010).

8. While the activities of a number of contrarian scientists are discussed in Begley (2007), Gelbspan (1997, 2004), Mooney (2005), and Oreskes and Conway (2010), individuals seeking detailed information on the CTT affiliations of leading contrarians should consult Greenpeace's website detailing connections between ExxonMobil and CTTs and contrarians (<http://www.exxonsecrets.org>), the data base created by James Hoggan and colleagues at their Desmogblog website (<http://www.desmogblog.com/global-warming-denier-database>), or John Mashey's highly detailed report (Mashey 2010).

9. Mashey (2010) provides evidence suggesting that the Competitive Enterprise Institute and the Marshall Institute played a role in stimulating Barton's hearing by promoting the efforts of Canadians Stephen McIntyre and Ross McKitrick to critique the work of Michael Mann and his colleagues.

10. The leaked memo is trivial compared to an earlier API embarrassment: a 1998 'Global Climate Science Communication Action Plan' developed at a meeting of leading figures in the denial machine hosted by API was made public by Greenpeace. The document laid out a detailed astroturfing strategy (involving contrarian scientists) and suggested that 'Victory will be achieved when average citizens understand (recognize) uncertainties in climate science . . . , media [does the same], and those promoting the Kyoto treaty on the basis of extant science appear to be out of touch with reality' (Greenpeace 2010a: 9; Hoggan with Littlemore 2009: 42–5).

11. Readers are encouraged to read the 'comments' on various denial websites particularly in response to posts about climate scientists to get a sense of the vitriol aimed at the latter.

REFERENCES

ANDEREGG, W. R. L., PRALL, J. W., HAROLD, J., and SCHNEIDER, S. H. 2010. Expert credibility in climate change. *PNAS (Proceedings of the National Academy of Sciences)* 107(27): 12107–9.

BECK, U. 1992. *Risk Society: Toward a New Modernity.* London: Sage.

—— 1997. *The Reinvention of Politics.* Cambridge: Polity.

—— GIDDENS, A., and LASH. S. (eds.) 1994. *Reflexive Modernization.* Stanford, CA: Stanford University Press.

BEDER, S. 1998. Public relations' role in manufacturing artificial grass roots coalitions. *Public Relations Quarterly* 43: 20–3.

BEGLEY, S. 2007. The truth about denial. *Newsweek* 150 (13 August): 20–9.

BOYKOFF, M. T., and BOYKOFF, J. M. 2004. Balance as bias. *Global Environmental Change* 14: 125–36.

CLARK, B., and YORK, R. 2005. Carbon metabolism: Global capitalism, climate change, and the biospheric rift. *Theory and Society* 34: 391–428.

Climate Action Network Australia. 2010. *Doubting Australia: The Roots of Australia's Climate Denial.* <http://www.cana.net.au/sites/default/files/DoubtingAustralia.pdf> (accessed 2 August 2010).

CORFEE-MORLOTT, J., MASLIN, M., and BURGESS, J. 2007. Global warming in the public sphere. *Philosophical Transactions of the Royal Society,* 365: 2741–76.

COVINGTON, S. 1997. *Moving a Public Policy Agenda: The Strategic Philanthropy of Conservative Foundations.* Washington, DC: National Committee for Responsive Philanthropy.

DICKINSON, T. 2010. The climate killers. *Rolling Stone* 1096 (21 January): 35–41.

DIELTHELM, P., and McKEE, M. 2009. Denialism: What is it and how should scientists respond? *European Journal of Public Health* 19(1): 2–4.

DRYZEK, J. S., HUNOLD, C., SCHLOSBERG, D., DOWNES, D., and HERNES, H. 2002. Environmental transformation of the state: The USA, Norway, Germany and the UK. *Political Studies* 50: 659–82.

DUNLAP, R. E. 1987. Polls, pollution, and politics revisited: Public opinion on the environment in the Reagan era. *Environment* 29 (July/August): 6–11, 32–7.

—— and McCRIGHT, A. M. 2008. A widening gap: Republican and Democratic views on climate change. *Environment* 50 (September/October): 26–35.

ENTING, I. G. 2010. *Ian Plimer's 'Heaven + Earth': Checking the Claims.* <http://www.complex.org.au/tiki-download_file.php?fileId=91> (accessed 17 August 2010).

FANG, L. 2009. *A Case of Classic Swiftboating: How the Right-Wing Noise Machine Manufactured 'Climategate.'* <http://wonkroom.thinkprogress.org/2009/12/09/climate-gate-time-line> (accessed 13 December 2009).

GELBSPAN, R. 1997. *The Heat Is On.* Reading, MA: Addison-Wesley Publishing.

—— 2004. *Boiling Point.* New York: Basic Books.

GIDDENS, A. 1990. *The Consequences of Modernity.* Oxford: Polity.

GOODELL, J. 2007. *Big Coal: The Dirty Secret behind America's Energy Future.* New York: Mariner.

—— 2010. As the world burns. *Rolling Stone* 1096: 30–3, 62.

GRANDIA, K. 2009. *Research on the 'Sponsors' behind the Heartland's New York Climate Change Conference.* <http://www.desmogblog.com/research-sponsors-behind-heartlands-new-york-climate-change-conference> (accessed 25 February 2009).

Greenpeace. 2010a. *Dealing in Doubt: The Climate Denial Industry and Climate Science.* <http://www.greenpeace.org/usa/Global/usa/report/2010/3/dealing-in-doubt.pdf> (accessed 5 April 2010).

—— 2010b. *Koch Industries Secretly Funding the Climate Denial Machine.* <http://www.greenpeace.org/usa/Global/usa/report/2010/3/executive-summary-koch-indus.pdf> (accessed 5 April 2010).

HAMILTON, C. 2007. *Scorcher: The Dirty Politics of Climate Change.* Melbourne: Black Inc. Agenda.

—— 2010. *Requiem for a Species.* London: Earthscan.

HANSON, B. 2010. Stepping back; moving forward. *Science* 328 (7 May): 667.

HARKINSON, J. 2009a. *Climate Change Deniers without Borders.* <http://motherjones.com/print/33941> (accessed 1 January 2010).

—— 2009b. *The Dirty Dozen of Climate Change Denial.* <http://motherjones.com/print/30676> (accessed 4 July 2010).

HELVARG, D. 2004. *The War against the Greens.* Rev. edn., Boulder, CO: Johnson Books.

HICKMAN, L. 2010. *US Climate Scientists Receive Hate Mail Barrage in Wake of UEA Scandal.* <http://www.guardian.co.uk/environment/2010/jul/05/hate-mail-climategate> (accessed 5 July 2010).

HIMMELSTEIN, J. L. 1990. *To The Right: The Transformation of American Conservatism.* Berkeley, CA: University of California Press.

HOGGAN, J., with Littlemore, R. 2009. *Climate Cover-Up: The Crusade to Deny Global Warming.* Vancouver: Greystone Books.

JACQUES, P. 2006. The rearguard of modernity: Environmental skepticism as a struggle of citizenship. *Global Environmental Politics* 6: 76–101.

—— DUNLAP, R. E., and FREEMAN, M. 2008. The organization of denial: Conservative think tanks and environmental scepticism. *Environmental Politics* 17: 349–85.

JAMIESON, K. H., and CAPPELLA, J. N. 2008. *Echo Chamber: Rush Limbaugh and the Conservative Media Establishment.* New York: Oxford.

JOHNSON, B. 2010. *Grand Old Deniers: Nearly All GOP Senate Candidates Deny Global Warming.* <http://thinkprogress.org/2010/09/13/warming-deniers-gop-caucus> (accessed 15 September 2010).

KOLK, A., and LEVY, D. 2001. Winds of change: Corporate strategy, climate change, and oil multinationals. *European Management Journal* 19: 501–9.

KREHELY, J., HOUSE, M., and KERNAN, E. 2004. *Axis of Ideology: Conservative Foundations and Public Policy.* Washington, DC: National Committee for Responsive Philanthropy.

LAHSEN, M. 2008. Experiences of modernity in the greenhouse. *Global Environmental Change* 18: 204–19.

LAPHAM, L. H. 2004. Tentacles of rage: The Republican propoganda mill, a brief history. *Harper's Magazine* 309 (September): 31–41.

LARSON, C., and KEATING, J. 2010. *The FP Guide to Climate Skeptics.* <http://www.foreignpolicy.com/articles/2010/02/25/the_fp_guide_to_climate_skeptics> (accessed 28 February 2010).

LAYZER, J. 2007. Deep freeze. Pp. 93–125 in M. E. Kraft and S. Kamieniecki (eds.), *Business and Environmental Policy.* Cambridge, MA: MIT Press.

LEISEROWITZ, A. A., MAIBACH, E. W., ROSER-RENOUF, C., SMITH, N., and DAWSON, E. 2010. *Climategate, Public Opinion, and the Loss of Trust.* <http://environment.yale.edu/climate/news/climategate-public-opinion-and-the-loss-of-trust> (accessed 4 July 2010).

LEVY, D. L. 2010. *It's the Real Thing: The Power of Koch.* <http://climateinc.org/2010/09/koch_climate> (accessed 9 September 2010).

McCRIGHT, A. M. 2007. Dealing with climate change contrarians. Pp. 200–12 in S. C. Moser and L. Dilling (eds.), *Creating a Climate for Change*. New York: Cambridge.

—— and DUNLAP, R. E. 2000. Challenging global warming as a social problem: An analysis of the conservative movement's counter-claims. *Social Problems* 47: 499–522.

—— —— 2003. Defeating Kyoto: The Conservative movement's impact on U.S. climate change policy. *Social Problems* 50: 348–73.

—— —— 2010. Anti-reflexivity: The American Conservative movement's success in undermining climate science and policy. *Theory, Culture and Society* 26: 100–33.

MASHEY, J. R. 2010. *Crescendo to Climategate Cacophony*. <http://www.desmogblog.com/sites/beta.desmogblog.com/files/crescendo%20climategate%20cacophony%20v1%200.pdf> (accessed 18 March 2010).

MAYER, J. 2010. *Covert Operations: The Billionarie Brothers Who are Waging a War against Obama*. <http://www.newyorker.com/reporting/2010/08/30/100830fa_fact_mayer> (accessed 24 August 2010).

MOL, A. P. J. 2000. The environmental movement in an era of ecological modernization. *Geoforum* 31: 45–56.

MONBIOT, G. 2007. *Heat*. Cambridge, MA: South End Press.

MOONEY, C. 2005. Some like it hot. *Mother Jones* 30 (May/June): 36–49.

Nature (editorial). 2010a. Climate of fear. *Nature* 464(7286): 141.

—— 2010b. Science scorned. *Nature* 467(312): 133.

ORESKES, N., and CONWAY, E. M. 2010. *Merchants of Doubt*. New York: Bloomsbury Press.

PARKS, B. C., and ROBERTS, J. T. 2010. Climate justice, social theory and justice. *Theory, Culture & Society* 27: 143–66.

POOLEY, E. 2010. *The Climate War*. New York: Hyperion.

SCHNEIDER, S. H. 2010. *Science as a Contact Sport: Inside the Battle to Save Earth's Climate*. Washington, DC: National Geographic Press.

SILLS, J. (ed.) 2010. Letters: Climate change and the integrity of science. *Science* 328 (7 May): 689–90.

TOLLEFSON, J. 2010. An erosion of trust? *Nature* 466 (30 June): 24–6.

Union of Concerned Scientists. 2007. *Smoke, Mirrors, and Hot Air*. Cambridge, MA: Union of Concerned Scientists.

WOLCOTT, J. 2007. Rush to judgment. *Vanity Fair* 561 (May): 100–6.

YOUNG, J. R. 2010. *British Panel Large Clears 'ClimateGate' Scientists of Misconduct Charges*. <http://chronicle.com/article/British-Panel-Largely-Clears/66163> (accessed 7 July 2010).

CHAPTER 11

COMMUNICATING CLIMATE CHANGE: CLOSING THE SCIENCE-ACTION GAP

SUSANNE C. MOSER AND LISA DILLING

1 INTRODUCTION

THE first decade of the twenty-first century was a big one for climate science and policy. Climate change steadily rose on the policy agenda of nations, regions, states, and cities. Ironically, it was also the decade in which public opinion vacillated on the reality of climate change and its human causation. Public opinion surveys dipped to new lows in the US, the UK, and Australia at the end of 2009 (Hanson 2009; Riddell and Webster 2009; Pew Research Center 2009b). Amidst what was called the greatest economic 'downturn' since the Great Depression, and a cooler, less extreme year in many regions across the globe, public concern about global warming dropped significantly even as scientific findings of accelerating impacts proliferated (e.g. Solomon et al. 2007). Commentators suggest the world is seeing a case of 'apocalypse fatigue' (Nordhaus and Shellenberger 2009).

This state of public opinion raises critical questions as to the effectiveness of twenty or more years of public education, outreach, and engagement approaches used to render a complex scientific issue meaningful and actionable for lay audiences. A growing body of literature on public attitudes, as well as on other aspects of the communication process, can help us understand not only the larger trends in public opinion, but also the challenges and opportunities for more effective approaches to climate change communication. While we are also concerned about the seeming lack of public engagement, we don't agree with the superficial diagnoses (and frequently implied disdain) that lay publics apparently don't care or don't 'get it.' Instead, we believe important insights can be gained from better understanding the way climate change has been communicated to date and how this communication has been received and interpreted. Based on these insights, we see something of a 'perfect storm' that has made communication of climate change so challenging, and thus limited its role in enabling public engagement and support for action: the characteristics of climate change itself (Jamieson, this volume; Steffen, this volume), the not-surprising politicization and its institutionalization (Dunlap and McCright, this volume; Hajer and Versteeg, this volume), the cognitive and psychological ways of processing information (K. M. Norgaard, this volume), and the structural challenges pertaining to the media used for communication. We focus on those aspects not discussed by others in this Handbook.

Before we detail some of the common approaches to climate change communication, we should make several assumptions and understandings explicit. First, what do we mean by 'engagement'? If communication succeeds in bridging the science-action gap, it does so by fostering public engagement with climate change. This involves a cognitive, an affective, and a behavioral dimension, i.e. people grapple mentally with and gain understanding of the issue; experience an emotional response, such as interest, concern, or worry; and actively respond by way of changes in climate-relevant behavior or political action (Lorenzoni et al. 2007; Moser 2009b; Moser and Dilling 2007; NRC 2002). This definition of engagement illustrates three normative assumptions on which our chapter rests. First, while science alone can never compel us to action (which instead rests on a value-driven interpretative judgment of the meaning of scientific findings), we believe science has sufficiently well established that climate change is underway, that the problem is mostly human-caused, and that there are significant threats to human and environmental systems (Solomon et al. 2007; Parry et al. 2007). This evidence warrants, in our view, immediate and significant actions to reduce climate change-related risks. Second, in a democracy policy action requires public input, support, or quiet consent, and implementation requires active engagement of the public. And third, communication is an essential means to link scientists, politicians, and the public, and thus can and should play an important and constructive role in enabling public engagement with climate change.

While we do not promote particular responses to climate change or advocate for particular policies here, our goal is to present insights from the multidisciplinary research literature on how communication can be shaped and carried out to assist in the task of engaging the public more effectively on climate change. While communication alone may not close the science-action gap, communication—cognizant of the pertinent literature that tells us what works and what doesn't—offers a better chance of reaching the goal of effective and meaningful public engagement. We claim that communication of climate change has been less effective than one might wish for four main reasons. Communicators have assumed that:

(1) a lack of information and understanding explains the lack of public engagement, and that therefore more information and explanation is needed to move people to action ('Inspiration with information');

(2) fear and visions of potential catastrophes as a result of inaction would motivate audiences to action ('Motivation by fear');

(3) the scientific framing of the issue would be most persuasive and relevant in moving lay audiences to action ('One size fits all'); and

(4) mass communication is the most effective way to reach audiences on this issue ('Mobilization through mass media').

Below we offer more detailed diagnoses of each and make suggestions for potential remedies based on the extant literature.

2 Opportunities and Challenges for Improving Communication

2.1 Inspiration with Information

Climate change is a challenging issue to convey. One reason lies in the fact that its principal culprits—carbon dioxide and other heat-trapping greenhouse gases—are colorless, odorless gases, and the long-term average changes set in motion by them have emerged only recently from the daily, seasonal, and interannual 'noise' of variability. Relative to human perceptual capacities, the problem is building far too slowly to be noticed by the lay eye, and the early environmental impacts of climate change have mostly occurred (and are clearly noticed) in regions far from where the majority of mid-latitudinal audiences live. The lack of direct experience makes climate change—at least for now—fundamentally a problem that requires signaling, illustrating, and explaining by those who have expert knowledge to those who don't. Communication is essential.

Unsurprisingly, communicators often assume that a lack of information and understanding explains the lack of public concern and engagement, and that therefore more information and explanation is needed to move people to action. This assumption has been studied widely and is known as the knowledge or information deficit model (Bak 2001; Sturgis and Allum 2004).

Evidence abounds that much communication is guided by this assumption: Former US Vice President Al Gore's movie *An Inconvenient Truth*, magnified by his books, inexhaustible lecture tour, and communication trainings (<www.theclimateproject.org>), are prominent examples, as are countless news and informational articles and presentations in which the science of climate change leads, features prominently, and is defended.

The information deficit model, however, is problematic in a number of ways, foremost because this deep-seated, if flawed belief assumes that information and understanding are *necessary and sufficient* conditions for behavioral or political engagement. Those guided by the deficit model believe that better problem understanding by way of more or 'better' information automatically eliminates any skepticism, raises concerns, and inevitably leads to the appropriate behavioral or policy response. At worst, it assumes people have to be 'little scientists' to make effective decisions.

Despite years of research showing the inaccuracy of this model, many still let it guide their communication efforts (Nisbet and Scheufele 2009). Moreover, this model can be detrimental in particular through the condescension that may emerge if the public is seen as or portrayed by communicators as irrational or ignorant (ibid.). Clearly, much can be said for broad public education in the principles and methods of science in general and in climate science specifically. A sturdier stand in science education may leave lay individuals less susceptible to misleading, factually untrue argumentation. But ignorance about the details of climate change is NOT what prevents greater concern and action. While knowledge about the causes of climate change is moderately correlated with appropriate behavioral responses (O'Connor et al. 2002), and deeper understanding of systems primes audiences to more readily understand the magnitude of the required response (Sterman 2008), there is also evidence that better knowledge about climate change does not necessarily raise concern, and even better understanding about possible solutions does not necessarily result in efficacious behavior (Gardner and Stern 2002).

Instead, deeply held pro-environmental values and beliefs, incentives, perceived benefits, skill and a sense of efficacy, social support, peer pressure, and practical assistance have been shown to foster behavior change (Downing and Ballantyne 2007; Gardner and Stern 2002; Semenza et al. 2008; Takahashi 2009). Even concerned individuals ready and willing to act on their conviction that climate change is a problem may encounter obstacles. For example, getting an energy audit for one's house but no help in prioritizing, how to select a contractor, finance the work and navigate other problems involved in retrofitting a home can thwart the intentions of even the most committed. Information has even been found to undermine concern and action: learning more about climate change may feel like 'having done something,' may feel overwhelming and disempowering, or even reinforce the desire to hold on to the status quo (Kahlor and Rosenthal 2009).

2.2 Mobilization by Fear

So maybe it is not that lay audiences need more information about climate change; maybe it's a matter of grabbing their attention? If people are not interested in the more complex or dry scientific questions about a global, difficult-to-visualize, seemingly distant issue, maybe a few pictures of fear-evoking weather-related catastrophes will do? Maybe what is needed are some hard-hitting fear appeals—images and projections of negative climate change impacts that evoke worry, feelings of loss, or dread—to bring the putatively distant risks of climate change closer to home and thus motivate people to take action.

In fact, there is plenty of evidence that communicators resort to these evocative means of communication: A *Time Magazine* cover in 2006 featured the now emblematic picture of the polar bear on melting ice and the headline warned in big bold letters, 'Be worried. Be *very* worried.' Countless news clips in papers, on television, or in ad campaigns use imagery of catastrophic flooding, drought, or people suffering in heat. Book titles speak of emergencies, catastrophes, and crisis (Kolbert 2006; Kunstler 2005). Words, images, tone of voice, and background music convey danger, darkness, pending doom. Increasingly, not just the messages of scared or passionate advocates but even serious scientific discussions of 'dangerous interference' in the climate system, tipping points, high-risk consequences, the 'runaway' greenhouse effect, or geoengineering as an 'emergency' response evoke apocalyptic fears in lay audiences.

Many climate change communicators believe that a sufficient dose of fear can serve as an effective motivator to increase the issue's urgency and get people to 'do something.' Certainly, fear appeals should help prioritize the issue as it keeps slipping behind more pressing matters such as jobs, the economy, terrorism, health, or any other issues dominating our daily lives (Nisbet 2009; Pew Research Center 2009a).

The principal problem with fear as the main message of climate change communication is that what grabs attention (dire predictions, extreme consequences) is often not what empowers action. Numerous studies have documented that audiences generally reject fear appeals (or their close cousin, guilt appeals) as manipulative (Moser 2007c; O'Neill and Nicholson-Cole 2009). Conservative audiences—at least on climate change—have been shown to be particularly resistant to them (Jost et al. 2007). Effective action motivators avoid being blatant and offer solutions that help audiences translate their concern into feasible and effective actions (Floyd et al. 2000). Fear appeals or images of overwhelmingly big problems without effective ways to counter them frequently result in denial, numbing, and apathy, i.e. reactions that control the unpleasant experience of fear rather than the actual threat (APA 2009; CRED 2009). This is particularly important in light of the fact that individuals have been shown to only have a 'finite pool of worry,' in which issues rise and fall (Weber 2006). An excessive focus on negative impacts (i.e. a severe 'diagnosis') without effective emphasis on solutions (a feasible 'treatment') typically results in turning audiences off rather than engaging them more actively.

Clearly these findings pose difficult dilemmas for communicators: Should we avoid telling what scientists have established as facts and reasonable outlooks about the serious-ness, pace, and long-term commitment of climate change? Should we instead only discuss energy- and money-saving actions and convey pictures of hope by focusing on the easy actions, the 'doability' of mitigation? Should we perpetuate the idea that there are fifty 'simple ways to save the planet,' just to spare lay publics rather appropriate anxiety? Existing research suggests otherwise. While neither alarmism nor Pollyannaism seem to yield desired results, wise integration of strategies may well result in greater engagement. First, communication that affirms rather than threatens the sense of self and basic world-views held by the audience has been shown to create a greater openness to risk information (Kahan and Braman 2008). Second, risk information and fear-evoking images should be limited and always be combined with messages and information that provide specific, pragmatic help in realizing doable solutions. These solutions must be reasonably effective in reducing the problem, especially together with other solutions being implemented. Importantly, communicators must establish a sense of *collective* response, especially by people in like social and cultural groups. Moreover, solutions should be broadly consistent with individuals' personal aspirations, desired social identity, and cultural biases (CRED 2009; Segnit and Ereaut 2007). Finally, given the ideological polarization around responses to climate change (discussed below), the legitimate experience of fear and being over-whelmed, and the deep and lasting societal changes required to address the problem, there is an important place for facilitated dialogue and structured deliberation of the issues as they emerge (Kahan and Braman 2006). Such deliberation has been shown to improve interpersonal knowledge and trust of people with very different values, provide critical social support and affirmation, increase openness to different opinions and risk informa-tion, and thus to enable decision making, rather than obstruct it (Nagda 2006).

In the end, communicators must temper their own temptation to persuade with fear by recognizing that issues have attention cycles (McKomas and Shanahan 1999). Climate change is not always going to be on the top of people's agendas. But communicators can make important gains by framing climate change and solutions in ways that link them to more salient (local) issues people consistently care about—the economy, their children, their health and safety.

2.3 One Size Fits All

The science and scientists have been central to climate change communication. As we discussed above, however, knowledge is not sufficient to motivate action; to the contrary, as the focus of much debate it has been used to delay action. Maybe even more problemat-ically, science has been assumed to be of central interest to most audiences. However, the science of climate change is not nearly as interesting to most audiences as some might wish. The big 'take-home' message why anyone should care is often lost in jargon and 'he said—she said' debates. At worst, technical information or scientific debates over minute details beyond the grasp of the non-inducted can be experienced as exclusionary or even condescending. Finally, when experts tell lay audiences about the technical aspects of climate change, typically in one-way communication, there is little room for dialogue, building a shared understanding of the problem and possible solutions. In short, science is not the most resonant grounding for a fruitful exchange on climate change. To rethink how communicators might reach audiences more effectively, the audience itself must become the first concern.

The importance of identifying the audience that has decision-making power or influence over an intended outreach goal cannot be overstated. Many communicators, however, are primarily concerned with the message or information they want to convey and rush over the critical question who they are trying to reach with it. Once the relevant audience is identified, communicators must try to understand what people care about, what they value and how they think about climate change and related matters (e.g. energy, environment). Audience segmentation studies can provide important insights and communication strat-egies, but may be too general to apply to a specific (e.g. local) audience (Downing and Ballantyne 2007; Leiserowitz et al. 2008).

Trying to better understand the audience will reveal what issues and language resonate with individuals and groups, which values are important, what aspirations they have (as parents, as professionals), as well as any preexisting knowledge of the climate change issue, common mental models, and possible misconceptions. Audiences also differ by the infor-mation channels they use, what messengers are credible to them, and what challenges they may face in implementing any desired action. Audience segmentation clearly indicates that a 'one size fits all' approach will almost certainly fail. Without solid audience knowledge, outreach campaigns may not generate more than fleeting attention, fail to meet the information needs people have, and generate values and world-view-based resistance to considering the information communicated (Dickinson 2009; Jost and Hunyady 2005).

Deeper understanding of the audience also helps identify the best ways to frame an issue. Framing—through words, images, tone of voice, messengers, and other signals—provides essential context for people to make sense of an issue; it triggers a cascade of responses and can prime an audience for action or not.

Extensive research in risk perception, cognitive psychology, and the influence of cultural world-views on both, suggests that individuals view incoming information through a 'cultural' lens, i.e. they understand and evaluate information through a filter that is colored by their general beliefs about society, the world, and right or wrong. According to Kahan and Braman (2006: 148), 'culture is prior to facts in the cognitive sense that what citizens believe about the empirical consequences of [certain actions or] policies *derives* from their cultural worldviews.' Incoming information—however framed—may be rejected upon very quick (intuitive) judgment if it evokes some kind of threat to the listener's sense of self, i.e. if it challenges his or her deeply held beliefs or those of the group he or she most identifies with (Kahan et al. 2007). By the same token, people tend to selectively hear and collect evidence that supports their beliefs and underlying values—a confirmation bias (CRED

2009; Kahan and Braman 2008). A number of deep-seated psychological needs explain these first and even more considered reactions to information about climate change risks and may be at the heart of audience segmentation (Jost and Hunyadi 2005). They may also explain why there is such enormous ideological polarization around the issue (Gastil et al. 2006; Jost et al. 2008), despite a very strong scientific consensus on the fundamentals of anthropogenic climate change (Doran and Zimmerman 2009; Oreskes 2004).

Messengers—those who convey a message—are part of the framing. It is thus important that the messenger is consistent with the way the message is framed or else the importance and credibility of the message is undermined (FrameWorks Institute 2002). Historically, climate change—predominantly framed as a scientific issue—has been communicated by experts, or else scientists were the primary sources and voices in media reports (Nisbet 2009). And while scientists working in relevant fields are generally credible and trusted as information sources (NSF 2009), they are not the most trusted or most appropriate source with every audience or with any message (Cvetkovich and Löfstedt 2000). If climate change is framed as a moral issue, religious leaders may have greater suasion (Wardekker et al. 2009). If taking action on climate change is seen as an economic issue, it may be most credibly conveyed by a business person who has done it (Arroyo and Preston 2007). If climate change is framed as a national security issue, spokesmen such as former CIA director James Woolsey can serve as a trusted messenger (Nisbet 2009). If the audience is a teenage crowd, 'hip' celebrities may be able to generate enthusiasm and engagement (Boykoff and Goodman 2009), but teenagers already active in the climate movement may do so even more effectively (Isham and Waage 2007). The clear message from these findings is that messengers must be trusted and trigger a frame that opens a door for the message.

Care must be taken, however, that trusted messengers are also knowledgeable messengers. For example, on the question how good, convenient, cheap, or easy to use an energy-efficient appliance is, one's friend or neighbor may be trusted the most, even if they do not possess the greatest expert knowledge on how effective the action is (Gilbert et al. 2009). Importantly, trust in individuals roots in positive (personal) experiences in specific contexts or in certain expectations based on 'proxies' (e.g. past performance, credentials, or expertise) and can translate to other areas, even if the expertise doesn't (Wolf et al. 2010).

Trust in the messenger is particularly important in the context of a problem like climate change that is invisible, uncertain, seemingly remote in time and space, scientifically and morally complex, and which may pose significant demands on citizens' scientific literacy and their behaviors (Marx et al. 2007). Individuals not inclined or able to systematically process a large amount of (sometimes conflicting) complex or difficult to understand information will use heuristics—mental shortcuts—to make up their minds about it (Kahneman et al. 1982). Thus, for example, trust can be based on a messenger belonging to one's own social or cultural group (Cialdini 1993; Agyeman et al. 2007). In other instances, trust in a message can be enhanced if it comes from an outsider—perceived to be more objective and credible in situations where conflict and distrust exist among 'insiders' (Fessenden-Raden et al. 1987). Whether or not a messenger is persuasive to a certain audience also depends on the type and intensity of interaction between them (Maibach et al. 2008). Finally, messengers must be chosen carefully as they can be polarizing in certain contexts (a good example is former Democratic US Vice President Al Gore) (Dunlap and McCright 2008; Kahan et al. 2007). In short, there is no simple guidance as to who will be trusted, who is credible, and who the 'best' messenger is—it all depends on the audience and context, and must fit the frame.

2.4 Mobilization through Mass Media

All too often, communication campaigns assume that mass media is the most effective way to reach the wide audiences needed for mobilization around climate change. The appeal of reaching large numbers through television, newspapers, and the internet glosses over the fact that information passed along through media-ted channels is often 'consumed' without great attention, quickly discarded or ignored. It also tempts us to ignore the structural changes occurring in the media landscape that affect the quality and diversity of news, and social divides (Moser 2009b, 2010). Because the assumed goal of most communication on climate change is not only to reach an audience, but to actively engage people, understanding the effectiveness of different channels is critically important.

Research in the health field, where public service announcements and behavior change campaigns promoting healthy behavior have been used for some time, is relevant here. While mass media channels can sometimes reach millions of people, persuasive media campaigns only evoke a change in health behavior in a small percentage of the audience (9 percent and slightly higher if the behavior is legally enforceable, Maibach et al. 2008). Health issues tend to be more personal than climate change, thus we might find climate change messages to be even less effective at engaging the audience through media channels. Not surprisingly, many advocates focus not on climate change at all, but on more tangible issues, such as household energy use, cost savings, or energy security, and employ social marketing campaigns to affect behavior change (Kollmuss and Agyeman 2002; McKenzie-Mohr 2000).

In general, while mass media are important for agenda-setting, face-to-face communication is more persuasive than mass media communication (Lee et al. 2002). Several aspects make face-to-face communication more salient and effective: first, it is more personal; second, non-verbal cues can allow the communicator to gauge how the information is being received in real time and respond accordingly; direct communication also allows for dialogue to emerge; and finally, the trust between individuals participating in a two-way exchange goes a long way toward engaging and convincing someone. Interactive communication, whether face to face or over the internet, improves health outcomes and behavior change (Abroms and Maibach 2008). Not surprisingly, some have suggested employing a two-step flow in communicating climate change—from a source to an influential to the ultimate audience—to bank on the benefits of face-to-face, more direct communication (Nisbet and Kotcher 2009).

One of the limitations of mass media channels is the inability to tailor messages to particular audiences, although certainly strategic choices can be made about the publication used, the timing and placement of ads, and so on. Several recent studies suggest that different audiences relate to distinct frames, goals, messages, and messengers and have preferences as to the communication channels they frequent (Agyeman et al. 2007; Featherstone et al. 2009; Leiserowitz et al. 2008). Thus, choosing the appropriate channel with a tailored message is more likely to reach and actively engage a specific audience (Moser 2007a, 2007b).

3 COMMUNICATION IN CONTEXT:
SOME CONCLUSIONS

In some sense, communication on climate change has been spectacularly successful. Across nations, nearly everyone in surveyed populations has at least heard of the issue and many can identify at least some important climate change impacts (Leiserowitz 2007). However, upon deeper exploration, we find that understanding is superficial, personal concern is relatively low and ever-susceptible to be overwhelmed by more immediate, salient threats and interests. And while a majority favors generic 'action,' support for those that affect people's pocketbook such as carbon taxes or increased gasoline prices declines sharply, especially during economically difficult times (Moser 2008).

Faced with these facts, communicators tend to resort to well-worn strategies to raise concern and elicit active engagement. First, they attempt to increase public understanding and provide more information on the assumption that knowledge is the major stumbling block to action. Second, they resort to fear tactics to motivate action, if often only to achieve the opposite effect. Third, in banking on the credibility and overall consensus of thousands of scientists, they insist on the scientific framing of climate change as the most compelling story, regardless of the differences among audiences. And finally, they try to reach the masses through traditional communication channels, while disregarding the power and advantages of different channels, especially interpersonal ones.

Clearly, communication on climate change is only part of the picture. Raising awareness and discussing an issue does not directly result in behavior change or policy action. Other factors, especially policy options, windows, and barriers, come into play. Thus, for communication to be effective in leading to active engagement, it must be supported by policy, economic, and infrastructure changes that allow concerns and good intentions to be realized (Moser and Dilling 2007; Ockwell et al. 2009). No matter how much communicators may exhort individuals to use less energy, for example, if people have no alternative to heat their homes or get to work in a timely manner, such efforts may fail. Educating about the benefits of energy-efficient appliances will not produce results if easy ways to implement these changes are not provided (Dilling and Farhar 2007). In short, communication for social change must consist of efforts to increase the motivation to make a change and help to lower the barriers to realizing it (Moser and Dilling 2007).

In this chapter, we articulate a role for communication that is broader than some communicators might assume: it is not just a way of conveying information to, or persuading, a passive receiver. Rather, we suggest that people in a democratic society are best served by actively engaging with an issue, making their voices and values heard, and contributing to the formulation of societal responses. Imposing a deluge of scientific facts and technocratic solutions on a populace without discussion and awareness of risks and choices is likely to lead to resistance and opposition (Moser 2009b). There is no easy answer to societal discussions of value-laden issues with big stakes for everyone involved (as heated debates over regulating carbon emissions and siting wind farms suggest). Thus, effective communication serves two-way engagement, which—ultimately—enables societal action.

Given limited resources and attention spans, climate change communicators should re-examine their strategies in light of the insights from communications research. We must challenge our assumptions about climate change communication and acknowledge its most useful modes and roles. Better understanding the audience will help identify the most appropriate framings, messengers, and messages that will most powerfully resonate with different people. Audience-specific use of communication channels for 'retail communication' may in the end be more cost-effective than mass communication that speaks to no one really. If creatively combined with engagement in forums for direct dialogue, communicators can take advantage of economies of scale, persuasive power, and social capital to achieve their goals.

References

Abroms, L. C., and Maibach, E. W. 2008. The effectiveness of mass communications to change public behavior. *Annual Review of Public Health* 29(1): 219–34.

Agyeman, J., Doppelt, B., Lynn, K., and Hatic, H. 2007. The climate-justice link: Communicating risk with low-income and minority audiences. Pp. 119–38 in S. C. Moser and L. Dilling (eds.), *Creating a Climate for Change: Communicating Climate Change and Facilitating Social Change.* Cambridge: Cambridge University Press.

American Psychological Association (APA). 2009. *Psychology and Global Climate Change: Addressing a Multi-faceted Phenomenon and Set of Challenges. A Report by the Task Force on the Interface between Psychology and Global Climate Change.* Washington, DC: APA.

Antilla, L. 2005. Climate of scepticism: US newspaper coverage of the science of climate change. *Global Environmental Change* 15(4): 338–52.

Arroyo, V., and Preston, B. L. 2007. Change in the marketplace: Business leadership and communication. Pp. 319–38 in S. C. Moser and L. Dilling (eds.), *Creating a Climate for Change: Communicating Climate Change and Facilitating Social Change.* Cambridge: Cambridge University Press.

Bak, H.-J. 2001. Education and public attitudes toward science: Implications for the 'deficit model' of education and support for science and technology. *Social Science Quarterly* 82(4): 779–95.

Boykoff, M. T. 2007a. Flogging a dead norm? Newspaper coverage of anthropogenic climate change in the United States and United Kingdom from 2003 to 2006. *Area* 39(4): 470–81.

—— 2007b. From convergence to contention: United States mass media representations of anthropogenic climate change science. *Transactions of the Institute of British Geographers* 32(4): 477–89.

—— and Boykoff, J. M. 2004. Balance as bias: Global warming and the US prestige press. *Global Environmental Change* 14(2): 125–36.

—— and Goodman, M. K. 2009. Conspicuous redemption? Reflections on the promises and perils of the 'celebritization' of climate change. *Geoforum* 40: 395–406.

Caruso, E. M., Gilbert, D. T., and Wilson, T. D. 2008. A wrinkle in time: Asymmetric valuation of past and future events. *Psychological Science* 19(8): 796–801.

Carvalho, A. 2007. Ideological cultures and media discourses on scientific knowledge: Re-reading news on climate change. *Public Understanding of Science* 16(2): 223–43.

—— and Burgess, J. 2005. Cultural circuits of climate change in U.K. broadsheet newspapers, 1985–2003. *Risk Analysis* 25(6): 1457–69.

Center for Research on Environmental Decisions (CRED). 2009. *The Psychology of Climate Change Communication: A Guide for Scientists, Journalists, Educators, Political Aides, and the Interested Public.* New York: Columbia University, CRED.

Cialdini, R. B. 1993. *Influence: The Psychology of Persuasion.* 2nd rev. edn., New York: Quill—William Morrow.

Cvetkovich G., and Löfstedt R. (eds.) 2000. *Social Trust and the Management of Risk: Advances in Social Science Theory and Research.* London: Earthscan.

Dickinson, J. L. 2009. The people paradox: Self-esteem striving, immortality ideologies, and human response to climate change. *Ecology and Society* 14(1): 34. Available at: <http://www.ecologyandsociety.org/vol14/iss31/art34/>.

Dilling, L., and Farhar, B. 2007. Making it easy: Establishing energy efficiency and renewable energy as routine best practice. Pp. 359–82 in S. C. Moser and L. Dilling (eds.),

Creating a Climate for Change: Communicating Climate Change and Facilitating Social Change. Cambridge: Cambridge University Press.

DORAN, P. T., and ZIMMERMAN, M. K. 2009. Examining the scientific consensus on climate change. *EOS, Transactions of the American Geophysical Union* 90(3): 22–3.

DOWNING, P., and BALLANTYNE, J. 2007. *Tipping Point or Turning Point? Social Marketing and Climate Change.* London: Ipsos MORI Social Research Institute.

DUNLAP, R. E., and McCRIGHT, A. M. 2008. A widening gap: Republican and Democratic views on climate change. *Environment* 50(5): 26–35.

FEATHERSTONE, H., WEITKAMP, E., LING, K., and BURNET, F. 2009. Defining issue-based publics for public engagement: Climate change as a case study. *Public Understanding of Science* 18(2): 214–28.

FESSENDEN-RADEN J., FITCHEN, J. M., and HEATH, J. S. 1987. Providing risk information in communities: Factors influencing what is heard and accepted. *Science, Technology, and Human Values* 12(3/4): 94–101.

FLOYD, D. L., PRENTICE-DUNN, S., and ROGERS, R. W. 2000. A meta-analysis of research on protection motivation theory. *Journal of Applied Social Psychology* 30(2): 407–29.

FrameWorks Institute. 2002. *Framing Public Issues.* Washington, DC: FrameWorks Institute.

GARDNER, G. T., and STERN, P. C. 2002. *Environmental Problems and Human Behavior.* 2nd edn., Boston: Pearson Custom Publishing.

GASTIL, J., KAHAN, D. M., and BRAMAN, D. 2006. Ending polarization: The good news about the culture wars. *Boston Review* 31(2), <http://bostonreview.net/BR31.32/gastilkahanbraman.php>.

GILBERT, D. T., KILLINGSWORTH, M. A., EYRE, R. N., and WILSON, T. D. 2009. The surprising power of neighborly advice. *Science* 323: 1617–19.

HANSON, F. 2009. Australia and the world: Public opinion and foreign policy. *The Lowy Institute Poll 2009.* Sydney: Lowy Institute for International Policy.

ISHAM, J., and WAAGE, S. (eds.) (2007). *Ignition: What You Can Do to Fight Global Warming and Spark a Movement.* Washington, DC: Island Press.

JOST, J. T., and HUNYADY, O. 2005. Antecedents and consequences of system-justifying ideologies. *Current Directions in Psychological Science* 14: 260–5.

—— NAPIER, J. L., THORISDOTTIR, H., GOSLING, S. D., PALFAI, T. P., and OSTAFIN, B. 2007. Are needs to manage uncertainty and threat associated with political conservatism or ideological extremity? *Personality and Social Psychology Bulletin* 33(7): 989–1007.

—— LEDGERWOOD, A., and HARDIN, C. D. 2008. Shared reality, system justification, and the relational basis of ideological beliefs. *Social and Personality Psychology Compass* 2: 171–86.

KAHAN, D. M., and BRAMAN, D. 2006. Cultural cognition and public policy. *Yale Law and Policy Review* 24: 147–70.

—— —— 2008. The self-defensive cognition of self-defense. *American Criminal Law Review* 45(1): 1–65.

—— BRAMAN, D., SLOVIC, P., GASTIL, J., and COHEN, G. L. 2007. The Second National Risk and Culture Study: Making Sense of—and Making Progress In—The American Culture War of Fact. *SSRN eLibrary*, <http://ssrn.com/paper=1017189>.

KAHLOR, L., and ROSENTHAL, S. 2009. If we seek, do we learn? Predicting knowledge of global warming. *Science Communication* 30(3): 380–414.

KAHNEMAN, D., SLOVIC, P., and TVERSKY, A. (eds.) (1982). *Judgment under Uncertainty: Heuristics and Biases.* New York: Cambridge University Press.

KOLBERT, E. 2006. *Field Notes from a Catastrophe: Man, Nature, and Climate Change.* London: Bloomsbury.

KOLLMUSS, A., and AGYEMAN, J. 2002. Mind the gap: Why do people act environmentally and what are the barriers to pro-environmental behavior? *Environmental Education Review* 8(3): 239–60.

KUNSTLER, J. H. 2005. *The Long Emergency: Surviving the End of Oil, Climate Change, and Other Converging Catastrophes of the Twenty-First Century.* New York: Grove Press.

LEE, E.-J., LEE, J., and SCHUMANN, D. W. 2002. The influence of communication source and mode on consumer adoption of technological innovations. *Journal of Consumer Affairs* 36(1): 1–27.

LEISEROWITZ, A. 2005. American risk perceptions: Is climate change dangerous? *Risk Analysis* 25(6): 1433–42.

—— 2006. Climate change risk perception and policy preferences: The role of affect, imagery, and values. *Climatic Change* 77(1): 45–72.

—— 2007. Communicating the risks of global warming: American risk perceptions, affective images, and interpretive communities. Pp. 44–63 in S. C. MOSER and L. DILLING (eds.), *Creating a Climate for Change: Communicating Climate Change and Facilitating Social Change.* Cambridge: Cambridge University Press.

—— MAIBACH, E., and ROSER-RENOUF, C. 2008. *Global Warming's 'Six Americas': An Audience Segmentation.* New Haven: Yale Project of Climate Change, School of Forestry and Environmental Studies, Yale University and Fairfax, VA: Center for Climate Change Communication, George Mason University.

LIMA, M. L., and CASTRO, P. 2005. Cultural theory meets the community: Worldviews and local issues. *Journal of Environmental Psychology* 25(1): 23–35.

LIU, X., VEDLITZ, A., and ALSTON, L. 2008. Regional news portrayals of global warming and climate change. *Environmental Science and Policy* 11(5): 379.

LORENZONI, I., PIDGEON, N. F., and O'Connor, R. E. 2005. Dangerous climate change: The role for risk research. *Risk Analysis* 25(6): 1387–98.

—— NICHOLSON-COLE, S., and WHITMARSH, L. 2007. Barriers perceived to engaging with climate change among the UK public and their policy implications. *Global Environmental Change* 17(3–4): 445--9.

LOWE, D. C. 2006. *Vicarious Experiences vs. Scientific Information in Climate Change Risk Perception and Behaviour: A Case Study of Undergraduate Students in Norwich, UK.* Technical Report 43. Norwich: Tyndall Centre for Climate Change Research.

McKENZIE-MOHR, D. 2000. New ways to promote proenvironmental behavior: Promoting sustainable behavior. An introduction to community-based social marketing. *Journal of Social Issues* 56(3): 543–54.

McCOMAS, K., and SHANAHAN, J. 1999. Telling stories about global climate change: Measuring the impact of narratives on issue cycles. *Communication Research* 26(1): 30–57.

MAIBACH, E., ROSER-RENOUF, C., WEBER, D., and TAYLOR, M. 2008. *What Are Americans Thinking and Doing about Global Warming? Results of a National Household Survey.* Washington, DC and Fairfax, VA: Porter Novelli and Center of Excellence in Climate Change Communication Research, George Mason University.

MARX, S. M., WEBER, E. U., ORLOVE, B. S., LEISEROWITZ, A., KRANTZ, D. H., RONCOLI, C., et al. 2007. Communication and mental processes: Experiential and analytic processing of uncertain climate information. *Global Environmental Change* 17(1): 47–58.

MOSER, S. C. 2007a. Communication strategies to mobilize the climate movement. Pp. 73–93 in J. Isham and S. Waage (eds.), *Ignition: What You Can Do to Fight Global Warming and Spark a Movement*. Washington, DC: Island Press.

—— 2007b. In the long shadows of inaction: The quiet building of a climate protection movement in the United States. *Global Environmental Politics* 7(2): 124–44.

—— 2007c. More bad news: The risk of neglecting emotional responses to climate change information. Pp. 64–80 in S. C. Moser and L. Dilling (eds.), *Creating a Climate for Change: Communicating Climate Change and Facilitating Social Change*. Cambridge: Cambridge University Press.

—— 2008. Toward a deeper engagement of the U.S. public on climate change: An open letter to the 44th president of the United States of America. *International Journal for Sustainability Communication* 3: 119–32.

—— 2009a. Communicating climate change and motivating civic action: Renewing, activating, and building democracies. Pp. 283–302 in H. Selin and S. D. VanDeveer (eds.), *Changing Climates in North American Politics: Institutions, Policymaking and Multilevel Governance*. Cambridge, MA: The MIT Press.

—— 2009b. Costly knowledge—unaffordable denial: The politics of public understanding and engagement on climate change. Pp. 161–87 in M. T. Boykoff (ed.), *The Politics of Climate Change*. Oxford: Routledge.

—— 2010. Communicating climate change: History, challenges, process and future directions. *Wiley Interdisciplinary Reviews: Climate Change* 1(1): 31–53.

—— and DILLING, L. (eds.) 2007. *Creating a Climate for Change: Communicating Climate Change and Facilitating Social Change*. Cambridge: Cambridge University Press.

NAGDA, B. R. A. 2006. Breaking barriers, crossing borders, building bridges: Communication processes in intergroup dialogues. *Journal of Social Issues* 62(3): 553–76.

National Research Council (NRC). 2002. *New Tools for Environmental Protection: Education, Information, and Voluntary Measures*. Washington, DC: National Academy Press.

National Science Foundation (NSF). 2009. *Science and Engineering Indicators 2008*. Arlington, VA: NSF.

NISBET, M. C. 2009. Communicating climate change: Why frames matter for public engagement. *Environment* 51(2): 12–23.

—— and KOTCHER, J. E. 2009. A two-step flow of influence? Opinion-leader campaigns on climate change. *Science Communication* 30(3): 328–54.

—— and SCHEUFELE, D. A. 2009. What's next for science communication? Promising directions and lingering distractions. *American Journal of Botany* 96(10): 1767–78.

NORDHAUS, T., and SHELLENBERGER, M. 2009. Apocalypse fatigue: Losing the public on climate change. *Yale Environment* 360 (16 November), <http://www.e360.yale.edu/content/feature.msp?id=2210>.

OCKWELL, D., WHITMARSH, L., and O'Neill, S. 2009. Reorienting climate change communication for effective mitigation: Forcing people to be green or fostering grass-roots engagement? *Science Communication* 30(3): 305–27.

O'CONNOR, R., BORD, R. J., YARNAL, B., and WIEFEK, N. 2002. Who wants to reduce greenhouse gas emissions? *Social Science Quarterly* 83(1) 1–17.

O'NEILL, S., and Nicholson-Cole, S. 2009. 'Fear won't do it': Promoting positive engagement with climate change through visual and iconic representations. *Science Communication* 30(3): 355–79.

ORESKES, N. 2004. The scientific consensus on climate change. *Science* 306 (3 December): 1686.

PARRY, M. L., CANZIANI, O. F., PALUTIKOF, J. P., LINDEN, P. J. V. D., and HANSON, C. E. (eds.). 2007. *Climate Change 2007: Impacts, Adaptation, and Vulnerability. Contribution of Working Group II to the Fourth Assessment Report of the Intergovernmental Panel on Climate Change.* Cambridge: Cambridge University Press.

Pew Research Center for the People and the Press. 2009a. *Economy, Jobs Trump All Other Policy Priorities in 2009: Environment, Immigration, Health Care Slip Down the List.* Washington, DC: The Pew Research Center for the People and the Press.

—— 2009b. *Fewer Americans See Solid Evidence of Global Warming: Modest Support for 'Cap and Trade' Policy.* Washington, DC: Pew Research Center for the People and the Press.

RIDDELL, P., and WEBSTER, B. 2009. Widespread scepticism on climate change undermines Copenhagen summit. *The Times Online*, 14 November, online at: <http://www.timesonline.co.uk/tol/news/environment/article6916510.ece>.

SEGNIT, N., and EREAUT, G. 2007. *Warm Words II: How the Climate Story is Evolving and the Lessons We Can Learn for Encouraging Public Action.* London: Institute for Public Policy Research.

SEMENZA, J. C., HALL, D. E., WILSOND, D. J., BONTEMPO, B. D., SAILOR, D. J., and GEORGE, L. A. 2008. Public perception of climate change: Voluntary mitigation and barriers to behavior change. *American Journal of Preventive Medicine* 35(5): 479–87.

SOLOMON, S., QIN, D., MANNING, M., CHEN, Z., MARQUIS, M., AVERYT, K. B., et al. (eds.) 2007. *Climate Change 2007: The Physical Science Basis. Contribution of Working Group I to the Intergovernmental Panel on Climate Change's Fourth Assessment Report* (Vol. i). Cambridge: Cambridge University Press.

STERMAN, J. D. 2008. Risk communication on climate: Mental models and mass balance. *Science* 322: 532–3.

STURGIS, P., and ALLUM, N. 2004. Science in society: Re-evaluating the deficit model of public attitudes. *Public Understanding of Science* 13(1): 55–74.

TAKAHASHI, B. 2009. Social marketing for the environment: An assessment of theory and practice. *Applied Environmental Education and Communication* 8(2): 135–45.

UZZELL, D. L. 2000. The psycho-spatial dimension of global environmental problems. *Journal of Environmental Psychology* 20(4): 307–18.

WARDEKKER, J. A., PETERSENA, A. C., and SLUIJS, J. P. V. D. 2009. Ethics and public perception of climate change: Exploring the Christian voices in the US public debate. *Global Environmental Change* 19(4): 512–21.

WEBER, E. U. 2006. Experience-based and description-based perceptions of long-term risk: Why global warming does not scare us (yet). *Climatic Change* 77(1–2): 103–20.

WOLF, J., W. N. ADGER, I. LORENZONI, V. ABRAHAMSON, and R. RAINE 2010. Social capital, individual responses to heat waves and climate change adaptation: An empirical study of two UK cities. *Global Environmental Change* 20(1): 44–52.

PART IV

··

SOCIAL IMPACTS

··

CHAPTER 12

ECONOMIC ESTIMATES OF THE DAMAGES CAUSED BY CLIMATE CHANGE

ROBERT MENDELSOHN

1 INTRODUCTION

ECONOMIC theory argues that the optimal policy to control greenhouse gases should minimize the present value of the sum of the damages from climate change plus the costs of mitigation (Nordhaus 1991). In order to convert this premise into policy, it is essential to measure both mitigation costs and climate damages. This chapter examines estimates of the damages of climate change in order to guide policies on greenhouse gases.

One reason why it is hard to compare mitigation costs to climate damages is that mitigation costs must begin immediately whereas many climate damages will occur in the distant future. Natural scientists tend to lean towards simply calculating a cumulative number regardless of when the cost occurs. However, time is money. That is, a dollar is worth more today than a year from now. A dollar today is in fact worth a dollar plus the interest rate (roughly 4 percent in the long run) in a year. This does not make a big difference if one is comparing today versus next year. But if one is comparing today versus a century from now, a dollar in a century is worth only two cents today. To put it another way, one can invest two cents today at the market rate of interest and it would be worth a dollar in a century. Including an appropriate value of time is one of the most important though often misunderstood (and therefore violated) concepts of rational public policy. Because of the long time horizon of greenhouse gases, it is one of the most important concepts to incorporate into climate change policy.

A second reason why climate damages are hard to measure is that we have little experience with climate change. Human beings did live through one ice age but recorded history has largely occurred in a period (the last 20,000 years) when climate has not changed. Although this warm interglacial period (our current climate) is not expected to last, our experience has been built on a period of constant climate. We therefore do not have a set of historical precedents that reveal how sensitive we might be to climate change. We consequently have to turn to controlled experiments in laboratories, to weather, or to comparisons across different climate zones (across space), to learn about how climate affects us.

A third reason that climate damages are hard to measure is that the bulk of damages may be to non-market sectors, especially the ecosystem. For example, ecosystems across the world are likely to be affected by climate change and will either migrate or change productivity. The social values to place on such widespread changes are not known.

The original United States Environmental Protection Administration (USEPA) studies of the damages from climate change were exclusively concerned with measuring effects in the United States (Smith and Tirpak 1989). The analyses examined the consequences of the equilibrium climate that would be caused by doubling carbon dioxide (CO_2) concentrations in the earth (550 ppm). At the time, global climate models predicted this level of greenhouse gases would increase temperatures by 2.5 °C. Analysts asked how would warming by 2.5 °C change the 1990 US economy? The analyses used comparative statistics, comparing one equilibrium (today) against another (2.5 °C warmer). The USEPA studies did not address the dynamics of impacts over time. For example, the coastal, forestry, and ecosystem studies involve sectors that take decades if not centuries to adjust. The studies did not capture how these costs evolved over time. The USEPA studies revealed that a limited number of economic sectors were vulnerable to climate change: agriculture, coastal, energy, forestry, infrastructure, and water. In addition, several non-market sectors were also vulnerable including recreation, ecosystems, endangered species, and health. Most of the studies in this effort measured physical changes such as inundated lands, lost lives, and additional energy in TWh. Only a few of the USEPA studies actually measured damages in dollars. Some of the studies concentrated only on the part of each sector that was damaged. For example, the energy study focused on the additional electricity needed for cooling in summer but not the reduction in heating fuels in winter. The recreation study focused on skiing in winter and not the bulk of outdoor recreation in the summer.

Subsequent economic studies attempted to value the US economic damages associated with these impacts in terms of dollars (Nordhaus 1991; Cline 1992; Titus 1992; Fankhauser 1995; Tol 1995). These economic results were summarized in the Second Assessment Report of the Intergovernmental Panel on Climate Change (Pearce et al. 1996). The aggregate damage estimates to the US for doubling greenhouse gases (550 ppm) range from 1.0 to 2.5 percent of GDP. This summary is the first and last time that the IPCC reported a comprehensive estimate of the economic damages of climate change.

The damage estimates varied widely across the different authors reviewed even though each author relied on the same original USEPA sectoral studies. The agricultural studies provided a range of answers depending on whether one took into account carbon fertilization or not. Cline disregarded carbon fertilization and therefore argued that agriculture damages would be very high whereas most other authors thought they would be modest. Fankhauser looked at studies predicting reductions in water supply and assumed that the impacts would be uniformly distributed across all users and therefore very costly. Most of the other authors assumed that water would be allocated to its best use so that only low valued uses would lose water leading to only modest impacts. Titus assumed that forest fires would consume most of the forests in the US leading to widespread losses. Most of the other authors assumed that ecosystem change would not necessarily be this harmful. Nordhaus saw the need to construct vast sea walls across the entire coastline of the US and predicted that sea-level rise was the most harmful effect (coastal impacts). Most other authors assumed that sea walls would only have to be built in critical locations and slowly over time. Tol predicted new diseases would lead to many lost lives in the US and therefore a very high impact. Most of the other authors realized that a modest public health program would keep disease under control. The fact that all five of the different authors reviewed in the IPCC felt that a different sector was going to be the most heavily damaged indicates how unstable the initial damage estimates really were.

Two studies went beyond the US and predicted impacts across the world (Fankhauser 1995; Tol 1995). Unfortunately, there was little evidence at the time to base this extrapolation upon other than population and income. They predicted global impacts from doubling CO_2 would range from 1.4 to 1.9 percent of Gross World Product (GWP). They predicted that the bulk of these damages would fall on the OECD (60 to 67 percent) because they assumed that damages were proportional to income. Only 20 to 37 percent of the damages were predicted to fall on low latitude countries, although this would amount to a higher fraction of their GDP (over 6 percent). Africa, southern Asia, and southeast Asia (not including China) were predicted to be the most sensitive to warming with losses over 8 percent of GDP (Tol 1995).

To place these damages in perspective, the doubling of CO_2 was expected to occur by mid-century, although the equilibrium temperature would take another thirty years to reach. The study predicted that greenhouse gases could easily increase beyond this point in the absence of mitigation. If temperatures could rise to 10 °C in future centuries, damages could rise to 6 percent of GWP (Cline 1992). However, this is based largely on just extrapolating the results of the doubling experiment rather than upon additional research concerning higher temperatures. The IPCC report also considered catastrophe. If temperatures were 6 °C warmer by 2090, 'experts' predicted an 18 percent chance that damages would be greater than 25 percent of GWP (Nordhaus 1994). In this case, experts included economists but also natural scientists unfamiliar with damage estimation. The three catastrophes identified in the IPCC report are a runaway greenhouse gas effect, disintegration of the West Antarctic ice sheet, and major changes in ocean currents (Pearce et al. 1996).

It should be understood that including these damage estimates in an economic model of climate change suggests that the optimal mitigation strategy is a moderate reduction of greenhouse gas emissions. The effects mentioned above suggest the current present value of a ton of carbon would lead to damages on the order of $5 to $12 per ton (Pearce et al. 1996). This is equivalent to $1.4 to $3.3 per ton of carbon dioxide. This social cost of carbon should rise over time at approximately a 2 percent rate to account for the rising marginal damages associated with accumulating greenhouse gases in the atmosphere. Such low prices will not stop greenhouse gases from accumulating over this century, they will simply slow them down (Nordhaus 1991). This is not a failure of the price mechanism from working effectively. It is an indication that the benefits of controlling greenhouse gases are quite low compared to the costs. Only a moderate mitigation program is justified according to the early damage estimates.

2 RECENT DAMAGE STUDIES

More recent analyses of the impacts of climate change have made four significant advances over the original studies. (1) Modern impact studies incorporate adaptation. Adaptation can dramatically reduce the 'potential' damages from climate change. (2) Modern studies have specifically conducted dynamic analyses of impacts in high capital sectors with high adjustment costs (forestry and coastal). (3) Modern studies measure the benefits of warming, not just the damages. (4) Modern studies examine the impact of climate change on the future economy, not the 1990 economy. Many of the climate-sensitive sectors of the economy are expected to grow more slowly than the rest of the economy, which increases absolute damages but shrinks damages as a fraction of GDP. (5) Impact studies have expanded from just analyzing the United States and have begun to measure impacts across the world.

2.1 Adaptation

'Private adaptation' refers to changes that a person will make for their own benefit (Mendelsohn 2000). Private adaptation is in the agents own self-interest to undertake. Following self-interest, an actor will not adopt options where the cost exceeds the benefit because such choices will make the person worse off. Similarly, the actor will tend to adopt all options that improve welfare because they have something to gain from the change. There is consequently reason to believe that private adaptation will be efficient.

There are now extensive examples of private adaptation that have been measured in agriculture as farmers have adapted to their current climate. They choose different crops depending on temperature and precipitation in Africa (Kurukulasuriya and Mendelsohn 2008a) and Latin America (Seo and Mendelsohn 2008a). They choose different animals depending on climate (Seo and Mendelsohn 2008b). They adjust their farm type (livestock, mixed, and crops) (Seo and Mendelsohn 2008c). These are all adjustments that farmers currently make to adapt to their local climate. Given that these adaptations are observed across the world, there is every reason to believe that farmers will adjust to future climates. Of course, that does not imply that farmers can adjust perfectly to weather in every year. Despite the findings of Deschenes and Greenstone (2007) that weather has no effect on farmers, it is well known that weather continues to plague farmers across the globe. Fluctuations in weather will continue to be a problem for farmers in the future. The adaptation research simply suggests that farmers should be able to keep up with decadal changes in observed climate (long-term weather) as it occurs.

There are also examples of private adaptation in other sectors. People increase insulation in their homes in either very hot or very cold environments. Firms and people increase cooling capacity in warmer climates (Morrison and Mendelsohn 1999). Firms and people shift fuels towards electricity in warmer climates (Mansur et al. 2008). Firms can harvest timber that is likely to be subject to dieback or forest fires caused by climate change (Sohngen and Mendelsohn 2002). Firms can plant short rotation species to counteract predicted near-term reductions in timber. Farmers rely on water-saving irrigation technologies as water becomes scarcer (Dinar et al. 1992; Mendelsohn and Dinar 2003). People can take more outdoor recreation trips as temperatures warm, lengthening the summer season.

'Public adaptations' involve many beneficiaries (Mendelsohn 2000). We label these as 'public' because they resemble public goods. With many beneficiaries, it is difficult to coordinate the provision of public adaptation. In general, markets are likely to underprovide public adaptations. Governments have an important role to play in encouraging public adaptation. The key issue for governments is to be efficient. That is, only engage in adaptations where the benefits outweigh the costs. Although there is every reason to believe that private adaptation will be done efficiently, the same cannot be said about public adaptation. In fact, it is distinctly possible that governments can make matters worse (maladaptation) by spending too much on adaptation or engaging in adaptations that increase damages. For example, they might spend billions of dollars building an expensive sea wall that protects property worth only a few hundred million dollars. Or they might encourage people to stay on vulnerable coastal islands by giving them subsidized insurance or disaster relief and thereby increasing coastal damages. Of course, the government may also prove to be inefficient because it fails to act and encourage a pubic adaptation that should occur.

Many public adaptations relate to the non-market impacts of climate change. If ecosystem shifts cause threats to endangered species, the protection of those species benefits many people. Measures that society can take to address this threat include reducing pressures on these species from other causes such as pollution, encroachment, and hunting. Society can also develop dynamic habitat programs that shift protected areas as climate shifts, enabling each species to have a minimum habitat at all times. Society can directly enhance the chance a species survives by helping breeding, increasing food supplies, or reducing natural predators. These measures are costly but they could neutralize the enhanced risk to species loss from climate change.

Another obvious public adaptation measure concerns public health measures taken to control disease. Although there have been extensive studies showing that certain diseases such as malaria and dengue fever could expand their territories with warming (IPCC 2007b), there have been no studies that address public health measures that can be taken to eliminate this potential threat. Many countries, including the US, have historically eliminated the threat of malaria with effective mosquito control, netting, and medical care. Public health expenditures are an effective measure against these diseases.

Some public adaptations are needed to protect market sectors. The allocation of water must be revisited if the supply of water drops or if the demand for water increases. Water must be moved from low-valued to high-valued users (Mendelsohn and Bennett 1997; Hurd et al. 1999). This will be far cheaper than building new dams or canals, although there may be circumstances when new hard structures are needed. The primary low-valued user of water is agriculture although there are some circumstances where mining may also be a low-valued user. Farmers may have to learn to use less water by adjusting irrigation to only high-valued crops and by using irrigation technologies that consume less water such as drip irrigation. A study of California revealed that farmers could reduce water use by 24 percent and lose only 6 percent of their net revenue by just eliminating low-valued crops (Howitt and Pienaar 2005).

Another very important public adaptation concerns the creation of private property rights in locations dominated by common property ownership (Mendelsohn 2006). Whether the resource is fisheries, grazing lands, farmland, or forests, there is now pervasive evidence that common property management underinvests in natural resource capital. Fisheries are overharvested, grazing lands have too many animals, forests are understocked, and farmlands are undercapitalized under common property ownership. As long as people share in the ownership of the resource, they do not sacrifice short-run gains for long-run profit. The conversion of economically active lands to private property would not only make them more efficient today, but this conversion would give incentives for people to engage in private adaptation in the future.

Sea-level rise potentially can inundate coastal lands. Sea walls are an efficient adaptation measure to prevent inundation of developed lands (Yohe et al. 1999; Neumann and Hudgens 2006). However, sea walls require coordination as they are ineffective if done one property at a time. An entire coastline has to be protected for a sea wall system to be effective. The timing of the construction of these walls must match with the advance of the seas. At the moment, there is considerable uncertainty about what sea-level rise will be in 2100 or later. However, if governments build walls to keep up with observed sea levels, the uncertainty will be resolved by the time that construction must begin. The only far advanced planning that must be considered is that planners must take into account the eventual height of walls. That is, they should plan to build walls that can be heightened as needed.

It may not be possible or desirable to build sea walls in all locations. Sea walls would not preserve beaches for example. To insure that beaches are preserved into the future, there are several options. For small changes in sea level, it may be sufficient to engage in beach nourishment to protect existing sand and to replace it (Neumann and Hudgens 2006). For larger changes in sea level, undersea containment walls can be built to raise the effective height of the beach (Ng and Mendelsohn 2006). However, care must be taken not to create dangerous currents and exposed structures in the process. In many locations, the ideal adaptation may be retreat. Marshes, mangroves, and beaches could all move inland in response to higher seas. Planning must be undertaken to allow these coastal ecosystems to have sufficient inland space to retreat.

Research and development is another widespread public adaptation. For example, new species could be developed that would be more suited for a warmer world. Both crops and animals could be developed that would flourish in a climate-changed world. Research and development could find new trees that would be better suited for future climates. Alternatively, new ways to cope with a warmer world could be developed. New shelter systems for crops, new housing for animals, and possibly new fertilizer and feed that would help future crops and animals in a warmer world. Research and development may also develop better methods of building sea walls and structures to withstand extreme events. Current analyses of climate change adaptation tend to rely only on existing technologies. However, there is every reason to believe that when adaptation is going to be widely needed, research and development will help lower the cost of adaptation by providing new opportunities to adjust.

2.2 Dynamic Modeling

The initial studies of climate damages were comparative static studies contrasting the current world with a world where greenhouse gases doubled. However, it is quite apparent that climate change is actually a dynamic process involving a path from the current world to this future world. Understanding what happens along this path is quite important, especially for sectors that might take many decades to adjust. Some global impact models have explicitly modeled this dynamic path (Tol 2002b; Mendelsohn and Williams 2004). However, what is especially important is that some studies have grappled with the difficulties of adjusting sectors with extensive capital (timber and coastal). Because the capital stocks in these sectors are slow and expensive to adjust, capturing the dynamics of climate change is necessary to measure actual impacts on these sectors.

The sea-level rise literature has revealed that sea walls are an effective adaptation for developed coastal lands (Yohe et al. 1999; Neumann and Hudgens 2006). However, it is critical that sea walls be built to match sea-level rise. If the walls are built too late, the coasts get inundated. If they are built too early, the costs escalate quickly. For example, a recent study of adaptation by the World Bank encouraged sea walls to be built fifty years in advance of rising seas. Such an aggressive policy would increase the present value cost of sea walls about sevenfold. This suggests a gradual wall-building program that raises walls every two to three decades to keep pace with sea-level rise. Note that one advantage of such a program is that uncertainty about sea-level rise is kept to a minimum since the walls merely must keep up with the increase expected over the next twenty to thirty years.

The timber literature has revealed that a forward-looking model of global timber can make a number of adjustments to the timber stock despite the long rotations of some trees (Sohngen and Mendelsohn 1998; Sohngen et al. 2002). For example, the industry can facilitate nature by helping to harvest trees undergoing transition from one ecosystem to another. Rather than allowing them to die gradually in place, they can be targeted for harvest. Similarly, the industry can facilitate the planting of new tree species in these transition locations to speed the rate that new ecosystems can expand. These dynamic adjustments reduce the damages from natural dieback and regeneration and increase the potential benefits of moving towards more productive forests.

2.3 Benefits of Warming

Although warming clearly does impose damages in some locations, there will also be substantial benefits. Using a multiplicity of methods, the agriculture literature confirms that crops have a hill-shaped relationship with temperature (Adams et al. 1990; Rosenzweig and Parry 1994; Mendelsohn et al. 1994). There is an ideal temperature range for each crop. Yields fall as temperatures fall below that ideal or rise above it. Warming consequently will not have the same effect on crops everywhere. Places that are already cool will benefit from warming as it improves local conditions. High latitude and high altitude farmland is expected to increase in productivity. Warming is also expected to have beneficial effects on forests, allowing them to spread (Haxeltine and Prentice 1996) and increase in productivity (Mellilo et al. 1993). This leads to benefits from an increase in forest supply and subsequent reductions in timber prices (Sohngen et al. 2002). A third important source of benefits is the reduction in heating costs associated with warming. Counting units of energy, the heating benefits offset the cooling damages (Rosenthal et al. 1995). However, cooling is more expensive than heating, so that there is a net damage in the energy sector but this damage is much smaller than simply the impacts on cooling (Morrison and Mendelsohn 1999; Mansur et al. 2008). A final important benefit concerns outdoor recreation. Although skiing will clearly be damaged by warming, most outdoor recreation is summer oriented. Warming will extend the season allowing for large annual outdoor recreation benefits (Mendelsohn and Markowski 1999; Loomis and Crespi 1999).

2.4 Predicting Future Impacts

As temperatures increase, they will have increasingly harmful effects. However, to measure these dynamic outcomes, it is important that analysts update the economy that will actually experience the climate changes. Several climate-sensitive sectors (agriculture, forestry, and water irrigation) are all expected to grow over time but at a slower rate than the rest of the economy. These three sectors are projected to have smaller shares of future GDP. This can be seen in long-range projections for the United States but especially in long-range projections for emerging and least developed nations. Damages associated with these sectors will grow in proportion to the sector, but fall in proportion to GDP. Early studies based on 1990 economies overestimated long-term damages as a fraction of GDP.

The one sector that is likely to grow as fast if not faster than the rest of the economy is the coastal sector. Coastal economies have tended to grow quite quickly because there are advantages to having access to ports and because of the amenities of the coastline. If these trends continue, damages from sea-level rise and coastal storms could increase substantially.

2.5 Global and International Studies

Although the extensive analyses of impacts in the United States provided useful method-ologies and insights, it is clear that the United States is not representative of the world. Impact studies would have to be done in other locations in order to support reliable global results. Even inside the United States, there was evidence that the impacts from warming were more harmful in the southern compared to the northern states (Mendelsohn 2001). The results hinted that high-latitude countries would benefit from warming whereas low-latitude countries would be distinctly damaged.

A series of studies have been conducted measuring damages around the world. For example, Ricardian studies have been done to select countries: Brazil and India (Sanghi and Mendelsohn 2008); Sri Lanka (Seo et al. 2005), Israel (Fleischer et al. 2008), Germany (Lippert et al. 2009) and China (Wang et al. 2009). The Ricardian studies examine farm performance (net revenue or land value) across climate zones. By comparing farms in one climate zone to another, one can see what effect climate has on farm outcomes. These studies confirmed that low-latitude agriculture was vulnerable to warming but high-latitude agriculture would benefit. Ricardian techniques have also been conducted at a continental scale in Africa and South America. Studying eleven countries in Africa revealed that climate change would have harmful effects on crops if future climates were hot and dry but not if they were mild and wet (Kurukulasuriya and Mendelsohn 2008b). Livestock at small farms in Africa would do well with climate change but livestock at large farms would be harshly hit (Seo and Mendelsohn 2008b). Large farms are more vulnerable than small farms because they specialize in high-productivity temperate animals. Studying seven countries in South America revealed that crops and livestock in South America are more vulnerable to warming than crops in Africa (Seo and Mendelsohn 2008d). The more modern farms of South America appear to have also specialized in crops and animals that are more suited for temperate climates.

The global studies also confirmed a hypothesis that rainfed farms may be more vulner-able to warming than irrigated farms (Schlenker et al. 2005). Irrigated farms in both Africa and China were found to be much less sensitive to warming than rainfed farms (Kuruku-lasuriya and Mendelsohn 2008b; Wang et al. 2008). In fact, irrigated farms in China were expected to benefit from more warming because it would extend their cropping to two seasons. Curiously, irrigated farms in Latin America were found to be more sensitive to warming than rainfed farms (Seo and Mendelsohn 2008d; Mendelsohn et al. 2010). This may have to do with the specific crops that are irrigated in Latin America.

Global studies of timber have also been done more recently (Perez-Garcia et al. 1997; Sohngen et al. 2002). Global studies reveal that forests in different regions of the world will respond differently to warming. Price effects predicted on the basis of regional studies not only could get the magnitude of changes wrong but even the sign.

Sea-level rise studies are beginning to be conducted around the world. Studies have been completed on both developed and undeveloped land in Singapore revealing that sea walls should be built to protect the developed parts of Singapore as well as the popular beaches but that other coastlines should be allowed to retreat (Ng and Mendelsohn 2005, 2006). Sea-level rise will likely flood extensive marshland acreage around the world (Nicholls 2004) and threaten coastal cities (IPCC 2007b).

3 CONCLUSION

The early studies of the global impacts of doubling carbon dioxide (CO_2) in the atmosphere predicted damages ranging from 1.4 to 1.9 percent of GDP (Pearce 1996). These studies predicted that the bulk of impacts would be felt by the OECD.

More recent studies have made a number of improvements on that literature. Adaptation is now integrated into impact analysis. Dynamic studies are done of capital-intensive sectors such as the forestry and coastal sectors (sea-level rise). Benefits are measured as well as damages. The economy is updated to when the climate changes actually occur. Finally, empirical studies have been conducted outside the United States in order to get a more representative sample. With all of these innovations in hand, estimates of economic damages have fallen. More recent estimates of the market impacts of climate change suggest much lower global damages of between 0.05 and 0.5 percent of GWP by 2100 (Mendelsohn et al. 2006; Tol 2002a). Further, it now is clear that the bulk of damages will be felt in low-latitude countries. Mid- to high-latitude countries will bear very little damages. Of course, the magnitude of damages depends upon the severity of the climate scenario. Milder scenarios on the low end of what is predicted (IPCC 2007a) will likely be beneficial. Mid-scenarios will be mildly harmful. The only scenarios that will generate large damages this century are the scenarios suggesting warming of 4–5 °C by 2100.

These results support the original results found by the IPCC in 1996 that only mild mitigation policies make any sense in the near future. The benefits of controlling greenhouse gases are currently less than $1–$4 per ton of carbon dioxide. Any mitigation effort that requires more than this amount should be postponed until the future. As greenhouse gases accumulate in the atmosphere, the damages will rise (at approximately 2 percent a year). Eventually, these higher marginal damages will justify more aggressive mitigation efforts.

A recent study promoted by the British government to support their aggressive near-term mitigation policies suggests a much higher near-term price for a ton of carbon dioxide (Stern 2006). However, to justify aggressive near-term policies, one must make a number of assumptions that have no empirical validity. First, one must assume that the real interest rate is near zero. There are no opportunities to invest in the future at a positive rate of return for either private gain or social progress. Second, that technical change will dramatically force down mitigation costs in the near term if a sufficiently high charge is levied today. Third, that the magnitude of damages is an order of magnitude higher than anything that has yet been measured. Fourth, that the probability of a large catastrophe is non-negligible this century. If all of these things are true, investing heavily in climate change mitigation would be attractive. However, since not one of these propositions has any empirical support, it is premature to commit the trillions of dollars an aggressive mitigation program would require. In fact, such a program could be one of the greatest follies mankind ever considered.

REFERENCES

ADAMS, R. M., ROSENZWEIG, C., PEART, R., RITCHIE, J., MCCARL, B., GLYER, J., CURRY, B., JONES, J., BOOTE, K., and ALLEN, L. 1990. Global climate change and US agriculture. *Nature* 345: 219–24.

CLINE, W. 1992. *The Economics of Global Warming.* Washington, DC: Institute of International Economics.

DASGUPTA, P. 2008. Discounting climate change. *Journal of Risk and Uncertainty* 37: 141–69.

DASGUPTA S., LAPLANTE, B., MEISNER, C., WHEELER, D. and YAN, J. 2009. The impact of sea-level rise on developing countries: A comparative analysis. *Climatic Change* 93: 379–88.

DESCHENES, O., and GREENSTONE, M. 2007. The economic impacts of climate change: Evidence from agricultural output and random fluctuations in weather. *American Economic Review* 97: 354–85.

DINAR, A., CAMPBELL, M. B., and ZILBERMAN, D. 1992. Adoption of improved irrigation and drainage reduction technologies under limiting environmental conditions. *Environmental & Resource Economics* 2: 373–98.

FANKHAUSER, S. 1995. *Valuing Climate Change: The Economics of the Greenhouse.* London: Earthscan.

FLEISCHER, A., LICHTMAN, I., and MENDELSOHN, R. 2008. Climate change, irrigation, and Israeli agriculture: Will warming be harmful? *Ecological Economics* 67: 109–16.

HAXELTINE, A., and PRENTICE, C. 1996. BIOME3: An equilibrium terrestrial biosphere model based on ecophysiological constraints, resource availability, and competition among plant functional types. *Global Biogeochemical Cycles* 10: 693–709.

HOWITT, R., and PIENAAR, E. 2005. Agricultural impacts. Pp. 188–207 in J. Smith and R. Mendelsohn (eds.), *The Impact of Climate Change on Regional Systems.* Cheltenham: Edward Elgar Publishing.

HURD, B., CALLAWAY, J., SMITH, J., and KIRSHEN, P. 1999. Economic effects of climate change on US water resources. Pp. 133–77 in R. Mendelsohn and J. Neumann (eds.), *The Impact of Climate Change on the United States Economy.* Cambridge: Cambridge University Press.

IPCC. 2007a. *Climate Change 2007: The Physical Science Basis.* Intergovernmental Panel on Climate Change, Cambridge: Cambridge University Press.

—— 2007b. *Climate Change 2007: Impacts, Adaptation, and Vulnerability.* Intergovernmental Panel on Climate Change, Cambridge: Cambridge University Press.

KURUKULASURIYA, P., and MENDELSOHN, R. 2008a. Crop switching as an adaptation strategy to climate change. *African Journal of Agriculture and Resource Economics* 2: 105–26.

—— —— 2008b. A Ricardian analysis of the impact of climate change on African cropland. *African Journal Agriculture and Resource Economics* 2:1–23.

LIPPERT, C., KRIMLY, T., and AURBACHER, J. 2009. A Ricardian analysis of the impact of climate change on agriculture in Germany. *Climatic Change* 97: 593–610.

LOOMIS, J., and CRESPI, J. 1999. Estimated effect of climate change on selected outdoor recreation activities in the United States. Pp. 55–74 in R. Mendelsohn and J. Neumann (eds.), *The Impact of Climate Change on the United States Economy*. Cambridge: Cambridge University Press.

LUND, J., ZHU, T., TUNAKA, S., and JENKINS, M. 2006. Water resource impacts. In J. Smith and R. Mendelsohn (eds.), *The Impact of Climate Change on Regional Systems: A Comprehensive Analysis of California*. Northampton, MA: Edward Elgar Publishing.

MADDISON, D. 2000. A hedonic analysis of agricultural land prices in England and Wales. *European Review of Agricultural Economics* 27: 519–32.

MANSUR, E., MENDELSOHN, R., and MORRISON W. 2008. A discrete continuous model of energy: Measuring climate change impacts on energy. *Journal of Environmental Economics and Management* 55: 175–93.

MELLILLO, J. MCGUIRE, A., KICKLIGHTER, D., MOORE, B., VOROSMARTY, J., and SCHLOSS, A. 1993. Global climate change and terrestrial net primary productivity. *Nature* 363: 234–40.

MENDELSOHN, R. 2000. Efficient adaptation to climate change. *Climatic Change* 45: 583–600.

—— (ed.) 2001. *Global Warming and the American Economy: A Regional Analysis*. Cheltenham: Edward Elgar Publishing.

—— 2006. The role of markets and governments in helping society adapt to a changing climate. *Climatic Change* Special Issue on *Climate, Economy, and Society: From Adaptation to Adaptive Management* 78: 203–15.

—— 2008. Is the Stern Review an economic analysis? *Review of Environmental Economics and Policy* 2: 45–60.

—— and BENNETT, L. 1997. Global warming and water management: Water allocation and project evaluation. *Climatic Change* 37: 271–90.

—— and DINAR, A. 2003. Climate, water, and agriculture. *Land Economics* 79: 328–41.

—— and MARKOWSKI, M. 1999. The impact of climate change on outdoor recreation. Pp. 55–74 in R. Mendelsohn and J. Neumann (eds.), *The Impact of Climate Change on the United States Economy*. Cambridge: Cambridge University Press.

—— and WILLIAMS, L. 2004. Comparing forecasts of the global impacts of climate change. *Mitigation and Adaptation Strategies for Global Change* 9: 315–33.

—— —— 2007. Dynamic forecasts of the sectoral impacts of climate change. Pp. 107–18 in M. Schlesinger et al. (eds.), *Human-Induced Climate Change: An Interdisciplinary Assessment*. Cambridge: Cambridge University Press.

—— NORDHAUS, W., and SHAW, D. 1994. Measuring the impact of global warming on agriculture. *American Economic Review* 84: 753–71.

—— DINAR, A., and WILLIAMS, L. 2006. The distributional impact of climate change on rich and poor countries. *Environment and Development Economics* 11: 1–20.

—— CHRISTENSEN, P., and ARELLANO-GONZALEZ, J. 2010. The impact of climate change on Mexican Agriculture: A Ricardian analysis. *Environment and Development Economics* 15: 153–71.

MORRISON, W., and MENDELSOHN, R. 1999. The impact of global warming on US energy expenditures. In R. Mendelsohn and J. Neumann (eds.), *The Impact of Climate Change on the United States Economy*. Cambridge: Cambridge University Press.

NEUMANN, J., and HUDGENS, D. 2006. Coastal impacts. In J. Smith and R. Mendelsohn (eds.), *The Impact of Climate Change on Regional Systems: A Comprehensive Analysis of California*. Northampton, MA: Edward Elgar Publishing.

NEUMANN, J., and LIVESAY, N. 2001. Coastal structures: Dynamic economic modeling. In R. Mendelsohn (ed.), *Global Warming and the American Economy: A Regional Analysis.* Cheltenham: Edward Elgar Publishing.

NG, W., and MENDELSOHN, R. 2005. The impact of sea-level rise on Singapore. *Environment and Development Economics* 10: 201–15.

———— 2006. The impact of sea-level rise on non-market lands in Singapore. *Ambio* 35: 289–96.

NICHOLLS, R. J. 2004. Coastal flooding and wetland loss in the 21st century: Changes under the SRES climate and socio-economic scenarios. *Global Environmental Change* 14: 69–86.

NORDHAUS, W. 1991. To slow or not to slow: The economics of the greenhouse effect. *Economic Journal* 101: 920–37.

—— 1992. An optimal transition path for controlling greenhouse gases. *Science* 258: 1315–19.

—— 1994. Expert opinion on climate change. *American Scientist* 82: 45–51.

—— 2007. Critical assumptions in the Stern Review on climate change. *Science* 317: 201–2.

—— 2008. *A Question of Balance: Economic Modeling of Global Warming.* New Haven: Yale Press.

—— 2010. The economics of hurricanes and implications of global warming. *Climate Change Economics* 1: 1–20.

PEARCE, D., CLINE, W., ACHANTA, A., FANKHAUSER, S., PACHAURI, R., TOL, R., and VELLINGA, P. 1996. The social cost of climate change: Greenhouse damage and the benefits of control. Pp. 179–224 in Intergovernmental Panel on Climate Change, *Climate Change 1995: Economic and Social Dimensions of Climate Change.* Cambridge: Cambridge University Press.

PEREZ-GARCIA, J., JOYCE, A., MCGUIRE, A., and BINCKLEY, C. 1997. The economic impact of climatic change on the forest sector. In R. Sedjo, R. Sampson, and J. Wisniewski (eds.), *Economics of Carbon Sequestration in Forestry.* Boca Raton, IL: Lewis Publishers.

REINSBOROUGH, M. J. 2003. A Ricardian model of climate change in Canada. *Canadian Journal of Economics/Revue canadienne d'économique* 36: 21–40.

ROSENTHAL, D., GRUENSPECHT, H., MORAN, E. 1995. Effects of global warming on energy use for space heating and cooling in the United States. *Energy Journal* 16: 77–96.

ROSENZWEIG, C., and PARRY, M. L. 1994. Potential impact of climate change on world food supply. *Nature* 367: 133–8.

SANGHI, A., and MENDELSOHN, R. 2008. The impacts of global warming on farmers in Brazil and India. *Global Environmental Change* 18: 655–65.

SCHLENKER, W., HANEMANN, W. M., and FISCHER, A. C. 2005. Will US agriculture really benefit from global warming? Accounting for irrigation in the hedonic approach. *American Economic Review* 95: 395–406.

—— —— —— 2006. The impact of global warming on US agriculture: An econometric analysis of optimal growing conditions. *Review of Economics and Statistics* 81: 113–25.

SEO, N., and MENDELSOHN, R. 2008a. An analysis of crop choice: Adapting to climate change in Latin American farms. *Ecological Economics* 67: 109–16.

—— —— 2008b. Measuring impacts and adaptation to climate change: A structural Ricardian model of African livestock management. *Agricultural Economics* 38: 150–65.

—— —— 2008c. Climate change impacts on Latin American farmland values: The role of farm type. *Revista de economia e agronegocio* 6: 159–76.

—— —— 2008d. A Ricardian analysis of the impact of climate change on South American farms. *Chilean Journal of Agricultural Research* 68: 69–79.

—— —— and MUNASINGHE, M. 2005. Climate change impacts on agriculture in Sri Lanka. *Environment and Development Economics* 10: 581–96.

SMITH, J., and TIRPAK, D. 1989. *Potential Effects of Global Climate Change on the United States*. Washington, DC: US Environmental Protection Agency.

SOHNGEN, B., and MENDELSOHN, R. 1998. Valuing the market impact of large-scale ecological change: The effect of climate change on US timber. *American Economic Review* 88: 686–710.

—— —— and SEDJO, R. 2002. A global model of climate change impacts on timber markets. *Journal of Agricultural and Resource Economics* 26: 326–43.

STERN, N. 2006. *Stern Review on the Economics of Climate Change*. London: Her Majesty's Treasury.

TITUS, J. G. 1992. The cost of climate change to the United States. In S. Majumdar, L. Kalkstein, B. Yarnal, E. Miller, and L. Rosenfeld (eds.), *Global Climate Change: Implications, Challenges, and Mitigation Measures*. Easton, PA: Pennsylvania Academy of Sciences.

TOL, R. 1995. The damage costs of climate change: Towards more comprehensive estimates. *Environmental and Resource Economics* 5: 353–74.

—— 2002a. New estimates of the damage costs of climate change, Part I: Benchmark estimates. *Environmental and Resource Economics* 21: 47–73.

—— 2002b. New estimates of the damage costs of climate change, Part II: Dynamic estimates. *Environmental and Resource Economics* 21: 135–60.

WANG, J., MENDELSOHN, R., DINAR, A., HUANG, J., ROZELLE, S., and ZHANG L. 2009. The impact of climate change on China's agriculture. *Agricultural Economics* 40: 323–37.

YOHE, G., NEUMANN, J., and MARSHALL, P. 1999. The economic damage induced by sea level rise in the United States. In R. Mendelsohn and J. Neumann (eds.), *The Impact of Climate Change on the United States Economy*. Cambridge: Cambridge University Press.

....

CHAPTER 13

WEIGHING CLIMATE FUTURES: A CRITICAL REVIEW OF THE APPLICATION OF ECONOMIC VALUATION

....

RICHARD B. NORGAARD

SHOULD we aggressively transition away from fossil hydrocarbon combustion to reduce the likely significant, but uncertain, damages from climate change or transition more slowly, let climate change unfold, and then adapt to and live with the problems climate change actually brings? Aggressively transitioning away from fossil fuels entails relatively known costs in the near term; transitioning more slowly entails less well-known costs in the more distant future. Economists undertake cost-benefit analyses using economic valuations of future consequences to address such basic questions. Until the Stern Review (Stern 2007), the dominant economic analyses indicated a gradual response to the threats of climate change made the most sense. To the consternation of the economists supporting a gradual response, the Stern Review encouraged a fairly aggressive response. But a few economists, and a larger number of natural scientists studying climate change, argue that even the Stern Review is too conservative. This chapter provides a historical and theoretical explication of how these differences arise.

Underneath the climate policy question looms a larger and unknowable issue. Is climate change simply a speed bump on the road of economic development and human progress or the most solid evidence yet that modern beliefs about economics and progress are fundamentally misleading humanity and need to be replaced? The idea of progress, once focused on moral progress, has essentially transformed into beliefs with respect to material progress and the role of science and technology in controlling nature to meet material desires (Nisbet 1980). The economic systems we have, economics as a formal discipline, the professional culture of economists, individual economists themselves, and publicly held and politically invoked beliefs about economics have all evolved within broadly held beliefs about material progress and the course of human history (Norgaard 1994). A new and more global scientific understanding developing within a portion of the scientific community is now challenging these beliefs. In this sense, the differences in how mainstream and alternative economists weigh climate futures reflect fundamental differences in understandings of the human condition.

The central argument of the chapter is that the historic analytical approaches and their rationales developed by economists to advise governments on choices between specific projects affecting a region are not adequate for a global policy problem. The question with respect to the rate of transition away from fossil fuel combustion is fraught with many general problems that cost-benefit analysis and the economic valuation methods that have evolved along with it are ill equipped to address. Numerous chapters in this handbook stress that 'we' consists of rich and poor people living in developed and developing nations (Kartha; Figueroa; Baer), present and future peoples (Gardiner; Howarth), and specific but as yet unknown people who will bear some unfortunate consequence of climate change by chance (Barnett; Doyle and Chaturvedi). The diversity of situations different peoples find themselves in makes agreement on a single analysis and valuation approach very difficult. Furthermore, climate futures, unlike past public decisions about specific projects around which economic practices evolved, are global. While different nations and people can choose different adaptation strategies, mitigation has global benefits for which rules for sharing the costs need to be devised. Uncertainties over climate futures are large and, with the problem being global, betting the planet and the future of humanity on how fast to transition to a greener economy cannot be hedged. It is especially difficult to address a problem analytically when that problem challenges dominant beliefs about the course of history, the role of nature, and the nature of science.

Section 1 describes the early history of cost-benefit analysis and the concerns of economists with distributional issues, who benefits and who loses from a project, in this development. Section 2 presents a brief history of climate economic analysis and some of the early controversies. A general equilibrium framework, described in section 3, helps us see the shortcomings of partial equilibrium analysis for a global problem such as climate change. Nevertheless, while critically important insights are highlighted through general equilibrium analysis, 'answers' to basic questions do not become clearer. The Stern Review and its impact on the economics of climate change are covered in section 4. The final section summarizes the key points while adding an institutional dimension to the arguments.

1 Historical Background to Current Methodological Problems

Economists were brought into government during the Great Depression to help design policies and select public works projects to increase employment and, after the Depression, to speed economic development through efficient public investments. Economists developed cost-benefit analysis in the process (Chakravarty 1987). In turn, cost-benefit analysis created further opportunities for economists to advise on public investment decisions nationally and internationally thereafter (Nelson 1987; Pechman 1989).

The criterion of cost-benefit analysis is to maximize net present value, NPV, defined as:

$$NPV = \sum_{t=0}^{n} \frac{B_t - C_t}{(1 + r)^t}$$

where B_t = benefits in year t, C_t, = costs in year t, r = the rate of interest. Future benefits and costs are divided by $(1+r)^t$, a phenomenon known as 'discounting' to the present. The rationale is that a dollar put in a savings account today equals $1.05 a year from now when the interest rate is 5 percent, and so the value of a future benefit or cost of $1.05 is equivalent to $1 today. Within this analytical framework, the controversies are over how benefits and costs are valued and what is the appropriate rate of interest, or rate of discount, for public investments.

Questions surrounding the investment in dams raised three broad issues that needed to be resolved early in the development of cost-benefit analysis and valuation. First, water projects entail major construction costs in the near term with benefits that stretch out over long time periods, raising critical questions about the appropriate interest, or discount, rate on public investments. Second, flood control, and later the environmental consequences of water development, required new, non-market valuation techniques. Third, the benefits and costs of water projects are not evenly distributed between people, and rationales had to be developed for addressing, or not as the case turned out, questions of equity.

Public water projects were typically too large for the private sector alone to undertake yet they were also small relative to any nation's economy as a whole. This meant that the 'general equilibrium' of a national economy as a whole was not being affected significantly, that the prices of goods and services would not change significantly at a national level, by a particular public investment decision. This meant that 'partial equilibrium' analysis could be used. The key assumption is that prices for goods other than those produced by the project would not be affected by the project, hence prices in other markets could be taken as 'given.' Similarly, the project would not affect financial markets and the rate of interest used in the project analysis should be based on the economy as it was, whether the project existed or not. Many of these assumptions are embedded in climate change valuations even though climate change is global in nature and, to the extent it is a big problem, will affect the whole economy.

While partial equilibrium analysis helps hide who benefits and who pays for a project, economists practicing cost-benefit analysis still had to address distributional questions. To illustrate why this is a problem, consider an economy consisting of people falling into two groups, X and Y, perhaps laborers and capitalists, but any two categories will do. By the 1880s, economists realized that if rights to capital, land, and education were redistributed between groups X and Y, the economy could still be perfectly efficient, but it would operate differently, markets would clear at different prices sending signals to producers to produce different quantities of goods, and people in group X or Y would be better or worse off (McKenzie 1987). This is the essence of general equilibrium theory, illustrated in Figure 13.1 by a 'possibility frontier' between the well-being of two groups. At every point along the frontier, the economy is efficient, for neither group can be made better off without the other being made worse off, and for this reason it is also referred to as an efficiency frontier. While prices are not visible in this diagram, the prices at which markets clear are different from one point on the frontier to the next.

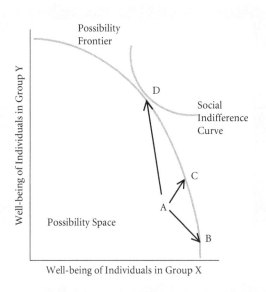

FIGURE **13.1** Who benefits and who loses from alternative decisions is clear in a general equilibrium model.

Consider the inefficient Point A, within the frontier and perhaps inefficient because private investors are unable to construct and manage a water project that the public could undertake. Now, imagine this economy can be made efficient by providing different water projects (B, C, and D) with different features affecting which group has the right to more water as well as financial characteristics affecting which group pays, moving the economy to Points, B, C, or D. All of these points are efficient, but the extent to which individuals in Groups X and Y benefit or not are different.

Economists have long recognized that a moral criterion, some sort of 'social welfare function' derived from reasoning beyond economics, is required to choose between the points on the possibility frontier. The social indifference curve of a hypothetical welfare function illustrated in Figure 13.1 indicates that Point D is the socially preferred alternative. Legislative bodies could be given the responsibility of choosing between the points on the efficiency frontier or could provide social welfare criteria to economists. Economists, however, were asked by legislative bodies, who knew nothing of general equilibrium theory, to indicate which projects were best from an economic perspective. Caught in this bind between economic theory and political expectations, economists reasoned that a project that made members of both groups better off must make society better off, a criterion labeled Pareto superior in honor of an early general equilibrium theorist. Point C fits this criterion. It is important to stress, however, that this criterion ignores historical phenomena, whether social injustices or pure luck, that may have made one group already better off than the other.

In spite of the theoretical shortcomings of the Pareto criterion, it too had to be compromised because projects could not be designed so that all people were winners. Economists then reasoned that winners should compensate losers. However, since economists could not enforce the payment of compensation, they next argued that, with many project decisions being made, winners in one project would likely be losers in another. Finally economists, at least for the purpose of the practice of cost-benefit analysis, simply accepted that worrying about the distributional effects was a matter they would simply leave to legislators and other policy makers. For the purposes of economic analysis, they reasoned that the existing distribution of income and wealth should be taken as a policy choice already made by the political process (Harberger 1971). Indeed, the original distributional concerns identified in welfare economics from a general equilibrium perspective have been overridden, at least in practice, by a new 'Harbergerian welfare economics' based in partial equilibrium economics (Jones 2005).[1] While these theoretical compromises may have made sense for the practice of advising on the selection of specific, regional projects, they make no sense for how climate change not only effectively affects everyone in the global economy but does so both differently and once and for all time.

In practice, ignoring distributional issues probably results in valuations and analyses that reproduce the status quo, or, more accurately, the existing dynamics. Had economic analysts looked into distributional issues, they would have also had to address the institutional changes required to redistribute rights to the poor. The historical evidence is that those who already were wealthy frequently got more. Those who already had rights to land, for example, received irrigation water in the summer and flood control in the winter. The underlying rationale was one of economic development, indeed human progress through the transformation of nature, wherein over time all would become better off. In the United States, cost-benefit analysis evolved in the context of spreading development from the industrialized east to poor regions such as the Tennessee Valley, delivering electricity to rural areas, and bringing water to the arid southwest. The association of cost-benefit analysis with development through environmental transformation was continued through its application to poor nations by international development agencies.

In summary, many of the problems associated with the application of cost-benefit analysis and valuation methods to climate change stem from their historical ties to a partial equilibrium framework. There are at least four important reasons for keeping a general equilibrium framework in mind, especially for climate valuation and analyses. First, the well-being of key actors rather than market quantities and prices are stressed, and the question of who benefits from a policy decision becomes more explicit. Second, the entire economy is considered; the *ceteris paribus*, or others things held equal, assumption is not necessary such that related changes are included in the analysis. Third, unlike partial equilibrium analysis or growth models, the general equilibrium model clearly illustrates limits in the form of a possibility curve and it is limits that force us to see trade-offs between parties. Fourth, it is clear that market and non-market prices change between policy choices. In section 3 we will see that the rate of interest also changes. While these advantages are critically important for climate analysis, general equilibrium analysis only illustrates these advantages in the abstract. It is more difficult to find solutions under the broad but weak light of general equilibrium analysis.

A complementary point is important. Economists were brought into policy analysis to offset pork-barrel politics and help assure that political decisions are economically efficient. This has led to a professional culture of virtuous skepticism among cost-benefit analysts wherein they think of themselves as guardians of the public good against wasteful, special-interest politics. As a consequence, they have tended to be as skeptical of climate scientists' warnings as they have been skeptical of dam builders' promises.

2 EARLY ECONOMIC ANALYSES AND VALUATIONS

Svante Arrhenius (1896) calculated the warming effect from a doubling of atmospheric carbon dioxide that has stood the test of time, but he mistakenly linearly extrapolated the then current rate of human emissions, estimating that it would take 600 or more years to double the stock in the atmosphere. For this reason, though not only this one, natural scientists brought economists into climate analysis during the 1980s to help provide reasonable economic scenarios for forecasting greenhouse gas emissions. But scientists also had good reason to presume economists would be solid allies in making the case for a strong policy response. When the ozone hole was discovered and its implications analyzed in the 1980s, economists showed that the damages to future generations from skin cancer were extremely high, bolstering the case for reducing the emission of ozone depleting substances as rapidly as possible (DeCanio 2003a). Climate change proved a much more complex issue for economists and many of their early valuations and policy analyses tended to support only a modest response.

Most of the economists who wrote on climate change in the early years were from the Northern developed countries who stressed that climate change was a complex problem with good and bad outcomes that demanded a careful accounting (Nordhaus 1991; Schelling 1992).[2] They saw benefits to cold regions and more rapid plant growth from higher concentrations of CO_2. Early valuations of climate change impacts looked at temperate agriculture where data on productivity, temperature, and rainfall were available, and in developed economies where the economic resources for adaptation were also available. Economists emphasized the possibilities for substituting crops as climate changed (Mendelsohn et al. 1994). Many early analyses portrayed the costs of climate change in terms of GDP as very small, lowering GDP by about 1 percent in any given year compared to a world without climate change (Repetto and Austin 1997). Given the importance of energy to economic growth, the early valuations determined that the starting value of reducing a ton of carbon at that time was in the neighborhood of US$10/ton (Nordhaus 1994) with later analyses increasing this to US$30/ton (Nordhaus 2008). Analyses then showed these values for reducing a ton of carbon rising steadily over time.

The early analyses supported a conservative response to climate change in part because many of the dynamics of climate change were poorly understood. The early analyses did not consider the possible acceleration of climate change due to ecological feedbacks, or the consequences of possible increased variability in year-to-year temperatures and precipitation as the climate system changed. Both the science and the empirical evidence now support more rapid and more variable change. Early economic studies made the typical *ceteris paribus* (other things being equal) assumption of partial equilibrium analyses, ignoring the complementary benefits of pollution reduction with reduced fossil fuel use as well as detrimental effects of other drivers of ecosystem change associated with the current economic course (Millennium Ecosystem Assessment 2005).

A few researchers from developed countries critiqued the analyses by Nordhaus and others supporting a 'go slow' response (Kaufmann 1997; Spash 2002; DeCanio 2003b). While the estimates of reducing carbon emissions in many climate policy models were typically costly, even at the beginning, there has long been good evidence that energy could be used more efficiently at a saving (Krause et al. 1992; Krause 1996).

Many researchers from developing countries argued that the early valuations from the developed countries ignored not only the costs that current greenhouse gas emissions from the developed countries imposed on the developing countries but past emissions as well. The impacts were not proportional to the emissions of developing countries (Srinivasan et al. 2008), and furthermore, the resources available for the developing countries to adapt to climate change were fewer. In addition, economists from the North were arguing, and continue to argue, that developed countries should sequester the carbon that developed countries emit in net forest growth in developing countries, the least expensive place because both land values and wages were lower. From the perspective of the South, this was simply another way for the rich North to continue emitting while exploiting the poor South. Clearly these critiques were rooted in distributional issues, both past injustices and development promises not yet fulfilled, as well as the fact that valuation needed to take a closer look at the impacts and possibilities for adaptation around the globe (Kartha, this volume).

Almost all of the critiques of the valuations and analyses pointed in the direction of a more rapid and larger response to climate change, especially by developed nations. Nevertheless, the arguments for a modest response made by key Northern economists continued to dominate within academic economics and the political-economic discourse of most developed nations until 2007.

3 CLIMATE CHANGE IN GENERAL EQUILIBRIUM AND THE NATURE OF THE VALUATION DEBATE

Many of the difficulties of climate change economics arose because the prevailing practices of valuation and analysis are embedded in partial equilibrium analysis, a framework suitable for project analysis but less suitable for questions of global change. The use of economic growth models to optimize between climate scenarios also compounded the problems. Neither approach highlights the distributional issues or the important issue of environmental constraints to economic development. Both of these problems can be highlighted in a general equilibrium framework.

Climate change is a special case of the generic problem of sustainable development, but the valuation methods and analytics of economics were never adjusted for sustainable development. Figure 13.2 is similar to the first figure except that now the groups on our axes are people of current and future generations. Sustainability is a matter of assuring that future generations are as well off as current generations (World Commission on Environment and Development 1987). The line at 45 degrees from the origin illustrates this criterion, and all those points above it are sustainable, whether they are efficient, i.e. on the frontier, or not. Rather than treat sustainability as a distributional problem in this fashion, to a large extent the economics profession has treated sustainability as an efficiency problem, as a matter of internalizing environmental externalities to make the economy more efficient (Norgaard 1992).

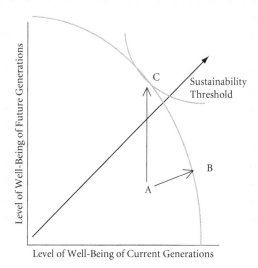

FIGURE 13.2 Sustainability as intergenerational equity.

If the economy is at an inefficient point such as A, existing economic valuation methods and analytical tools can help the economy move toward an efficient point such as Point B. Point B, however, is still below the sustainability criterion. To move to a sustainable and efficient point, such as Point C, a distributional decision has to be made. Rights, perhaps climate rights, need to be redistributed from current to future generations to assure the latter's well-being. To do this, current generations will have to relinquish rights and assume new responsibilities to assure those rights are available for future generations. But, this is the very type of decision that requires criteria from beyond economics. For the purposes of practically serving society's historical needs, economists have rationalized away the possibilities that such other points are within the purview of economics. Thus the methods economists have applied to improve sustainability over the past quarter-century have basically encouraged a movement toward Point B rather than Point C.

Recall that the prices that equilibrate markets efficiently, including the internalized prices of non-market environmental goods, are different along the possibility frontier, different, for example, at Points B and C. In a hypothetical example using an overlapping generations model, Howarth and Norgaard (1992) showed that when future generations are protected against climate change, the present value of the benefits of reducing greenhouse gas emissions go up because both the value in future generations is higher and the rate of discount goes down. The lower discount rate is illustrated in Figure 13.2 by the fact that well-being between current and future generations is being traded off at different rates as indicated by the different slopes at the points of tangency at Points B and C.

Economists tout how their methods of measuring non-market environmental values are empirical. What they ignore is that they are also rooted in the unsustainable economy that has existed, the economy at Point A, trying to move it to Point B, rather than the sustainable economy at Point C that most societies have long accepted to be their goal. If one thinks the existing economy merely needs a few minor adjustments to be sustainable, i.e. Point B and C are really quite close to the sustainability criterion, then this is not a significant problem. But if one thinks the economy is far from being sustainable, as pictured, then the concern with estimating values based on how people think and behave within the economy we have and treating sustainability as a matter of fine-tuning are significant. If there is a significant difference, then economists should model what an efficient, desirable sustainable economy would look like, ponder how people would live, value things, and behave in this significantly different economy, bring this projection into public debate and the democratic process, and then help societies set the course for a transition. This, however, is not what economists have been asked to do, nor is it how economists now see their role.[3]

While the concern with sustainability over the past quarter-century has brought little institutional change, or change in how economists carry out their role in the policy process, our scientific understanding of global ecological change and its implications for a sustainable economy has advanced considerably. Unfortunately, this advance is largely bad environmental news. While Arrhenius got the big picture pretty nearly correct with respect to moving from one equilibrium to another, the dynamics of how the climate system transitions are much more complicated. The more scientists have striven to fill out their understanding of the complexities of climate dynamics, the more uncertain the intermediate future has become. Reinforcing this new theoretical concern, the empirical evidence during the first decade of the twenty-first century indicates that climate change is occurring more rapidly than had been expected in only the decade prior. This documents that the reasons for concern were correct but that our understanding of climate dynamics—the details of when, where, and how the effects of climate change would appear—were not sophisticated enough to accurately predict the near future. Climate change, however, is not the only problem. The Millennium Ecosystem Assessment (2005) documented that ecosystems were already severely stressed due to other ongoing drivers in addition to climate change. Concerns over water scarcities have also increased, partly reinforced by what appears to be changes in climate that have increased the spatial and temporal uncertainties of water regimes (World Water Assessment Program 2009).

From an environmental science perspective, the future looks grimmer than it did a decade or two ago. The 'actual' possibility frontier in Figure 13.3 that is interior to the prior and now 'mistaken' possibility frontier illustrates this new perspective. Figure 13.3 is merely a way of illustrating the situation we appear to be in. As shown, Point C is not possible and, to achieve sustainability, society should now be moving toward Point D. This framing shows how the empirically based methods of deriving benefits and costs for weighing climate futures are not only based on an unsustainable economy but also based on information that was historically believed to be correct but is now thought not to be correct. If climate change were a small problem, it would make sense to consider people's stated preferences and observed behavior before the implications of climate change were understood and before institutions were established to guide individual behavior toward the public good. This 'small problem' assumption is standard for partial equilibrium analysis and its associated methods of looking at value and institutional design, but those who think climate change will have significant impacts are not at all comfortable with this assumption.

As illustrated in Figure 13.3, Point D is to the left of Point A, so moving to a sustainable economy requires a reduction in well-being by the current generation. This immediately raises the issue of whose well-being should be reduced. By most and certainly dominant

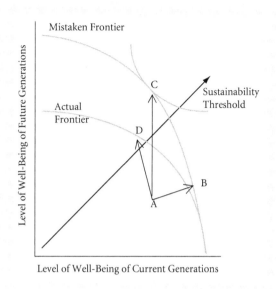

Level of Well-Being of Future Generations

Mistaken Frontier

Actual
Frontier

Sustainability
Threshold

C

D

B

A

Level of Well-Being of Current Generations

FIGURE 13.3 The implications of recent advances in global environmental science to economics.

moral criteria it should be the rich rather than the poor. Similarly, whose well-being is reduced probably should not be determined by where the next climate related environmental disaster happens to strike or who will have to move first due to sea-level rise. Thus Figure 13.3 provides a graphical portrayal rooted in general equilibrium economics of the equity and justice issues that drive global climate change policy debates as well.

4 THE STERN REVIEW OF CLIMATE ECONOMICS AND ITS CRITICS

The Stern Review of the economics of climate change (Stern 2007) was commissioned by the UK's Chancellor of the Exchequer in 2005 to provide an independent review of the full range of economic issues raised by climate science. The review was supervised by Sir Nicholas Stern, then UK's Chancellor of the Exchequer and formally Chief Economist and Senior Vice President at the World Bank, and prepared by a twenty-three-member team of economists and researchers with other expertise who were backed up by many more additional researchers assisting on specific parts. The Stern Review team ended up being much more concerned with the uncertainties of unmitigated climate change and the problems of equity than the analyses that had dominated before. Rather than ask what was the optimal response to a relatively well-known problem, the Stern Review asked what was the appropriate response, given the uncertainties in the science, to avoid, with a high probability of success, the imposition on future peoples of the more extreme dangers of climate change. Furthermore, the team worried about particular peoples including those living near sea level and those affected by extreme climate events in the near and medium term. In short, the team took the view that climate change was a big problem in a way that complements the issues that are highlighted by a general equilibrium perspective.

The Stern team directly addressed risks and ethics rather than hiding them within their assumptions. On account of this, the team openly pursued complex ethical issues eclectically, acknowledging the difficulties of taking a single economic perspective or sticking with the historic conventions of valuation and earlier rationales with respect to discount rates. In the end, they determined that the value of reducing the emission of carbon to the atmosphere today was US$390/ton, or 10 to 20 times greater than the value determined by most economists before them. Given the size of the difference between earlier policy conclusions and Stern's conclusions, there was a strong and highly critical backlash to the Stern report, especially by economists who had been covering the economics of climate change for a decade or more (see, for example, Mendelsohn 2008; Nordhaus 2007: Stern and Taylor 2007; Tol and Yohe 2006). The critics argue that their own work is empirical, consistent, and hence more scientific. Their bottom line is that the Stern Review pays too little attention to the development opportunities lost by investing in climate mitigation rather than investing in economic growth.

It is important to keep in mind, however, that the Stern team did not critique the use of economics as a policy instrument, though they did openly discuss whether past conventions in cost-benefit analysis were appropriate. The Stern Review accepted the importance of economic valuation; it did not argue for the intrinsic value of nature or that no trade-offs should be made between the well-being of different peoples. The team did not go so far as to argue, as presented in this chapter, that general equilibrium theory indicates that the appropriate discount rate depends on moral criteria beyond economics. To the consternation of climate scientists increasingly concerned about the ecological effects of even modest warming, the Stern Review argued, almost a priori, that, though 450 ppm CO_2 in the atmosphere was a reasonable long-term target, it was simply too difficult to stay below 450 ppm in the medium term. Thus while the Stern Review challenged the dominant economic policy recommendation, and in the process challenged the leading economists in the field, it is important to keep in mind that there are scientists and some economist who argue that a target of 350 ppm is morally and economically reasonable. For these scientists, the Stern Review is an improvement but still deficient (Hansen 2009; Ackerman et al. 2009).

In parallel with the Stern Review and the counter-response, Harvard professor of economics Martin Weitzman further expanded the valuation debate. He argued that the uncertainty of our knowledge of the dynamics of climate change is such that the critical information resides in the 'fat tails' of the distribution of the possible outcomes. The dominant economists had actually been looking at quite possible events and referring to them as extremes rather than taking the uncertainty seriously. According to Hanson, the uncertainties of climate dynamics vastly outweigh any of the differences in valuation methods or discounting being debated by economists (Weitzman 2009). Much in the same way that a general equilibrium view of the climate problem leads one to question the significance of valuation rooted in the current economy, Weitzman argues that benefits and costs would look a whole lot different if warming accelerated rapidly and a climate catastrophe unfolded. Weitzman complements other analyses that have looked at more disastrous climate outcomes, especially when combined with other resource scarcities, especially vulnerable populations, and global economic instabilities (Hall and Behl 2006; Hallegatte et al. 2007; Sterner and Persson 2008; Schlenker and Roberts 2009). In effect, the previously dominant analyses ignored outcomes that are as likely as the floods for which dams have long been built by the public sector.

5 CONCLUSIONS

Economics has multiple theories and related formal models that focus on different things yet overlap in their coverage. The multiple theories provide different, and often conflicting, insights. To meet the institutional needs of policy processes, i.e. to provide clear answers, economists agreed to conventions that rationalize away the conflicts, cover over complicating insights, and hide incongruent assumptions. The conventions, and in some cases simply selective collective denials, evolve to fit the niches of particular economic questions, institutions, and types of problems. The needs of different combinations evolve, noticeably so over periods of decades, the process of rationalization is always a discourse in progress. And economists from rich and poor nations have always leaned toward different rationalizations that favor the North or the South. Yet, throughout their gradual evolution, cost-benefit analysis and valuation techniques have morphed in the context of a larger story of how economic development contributes to human progress. Climate change challenges this larger story as well as the economic discourse that sustains historic conventions.

Ecology also has multiple theories and related formal models, but ecologists, or at least the top academic ecologists, are more dedicated than top economists to the science of ecology rather than to its practice in policy settings that require unique, simple answers. While ecologists typically contribute to one or two approaches in ecology, they are relatively more comfortable than economists with the contradictions that arise between patterns of thinking. The tensions between acknowledging scientific incongruities and unknowns and providing policy advice are rife in climate science and policy (von Storch in this volume; Schneider 2009).

This chapter documents that partial equilibrium analysis has been the dominant theory used in the practices of valuation and cost-benefit analysis. By looking at the practices from a general equilibrium perspective, the perspective that prompted the serous moral questions in the early years of public policy economics, we see why environmental valuation and cost-benefit analysis became largely a process of fine-tuning the existing economy. The practices have done little to reset the course of the economy to avoid further environmental problems and are thus also inappropriate for weighing alternative climate futures. At the same time, however, we also see how the practices, having evolved around the promotion of development, are propped up by the economic growth paradigm. Many ecological economists, conservation biologists, and natural scientists generally argue that material consumption cannot keep growing. Indeed, to the extent material consumption has been growing, it appears to have been doing so by depleting nature's stocks (Millennium Ecosystem Assessment 2005).

The challenges of climate change put the prospects for the poor becoming better off in a much deeper shadow. Economic practices evolved in a world where most people held to the faith that economic progress eventually would deliver net benefits to all. In the third millennium, partly because the benefits of economic development over the last half-century were so poorly distributed, partly because progressive economic philosophy has been replaced by neoliberal economic philosophy that rationalizes an unequal world, many people no longer hold expectations of progress for themselves, or for others, or they fear for their children and grandchildren. For many people on the globe, economics as practiced, whether progressive or neoliberal, did not deliver; for all people it has not delivered sustainable development. The Millennium Development Goals are faring no better. Thus there is little wonder that many participants in the global political discourse on climate question using the old conventions of expert economic practice to address the new problems of global environmental change. The old promises through the conventional practice of economics have not been met. Many of the new and stronger challenges, especially the distribution of climate impacts, are exactly where the old conventions are weakest.

Nor do new 'universal' conventions seem likely to become established. It seems unlikely that new rationales can develop that will allow economists to achieve a near consensus on how to value and analyze global environmental problems. Part of the problem is that there is only one globe on which to practice, and the experiments take a long time to play out. Part of the problem, however, is that the international institutional contexts in which global environmental economics is practiced are so amorphous.

The need for expert economic practice has changed from advising agencies and legislatures on whether one water project is better than another to how a very significant, but uncertain, global economic problem should be addressed through weak international institutions. While the UNFCC policy process was set up as if it would be driven by expert advice with equity issues resolved more or less democratically, it is becoming increasingly clear that this process is not working as hoped. The global institutional context within which climate will be managed, no doubt poorly from a progressive democratic perspective, will likely be eclectic, chaotic, and contradictory, probably more boisterous and roughshod like the maintenance of peace than like the formal capping of sulfur emissions and silent trading of permits in the United States. A heterogeneous economics of climate change will evolve within this amorphous context.

NOTES

1. The tendency to play down distributional issues, or believe they will be overcome through economic development, has been strongest in the United States. British economists, for example, have been more favorably inclined to give more weight to benefits and costs affecting the poor. Even in the United States, however, there was never an academic consensus. Rather, in the practice of cost-benefit analysis it has proven difficult to rationalize one set of weights over another and hence rarely used.

2. William Cline (1992) was a very important early exception. He argued for a low discount rate for a variety of reasons and, largely because of this, ended up arguing for a more aggressive response to climate change.

3. I have argued that the process of bringing scientists together from different disciplines and cultures from all around the world to work toward a consensus in understanding global environmental change is a model of discursive learning that suggests the possibility of working toward a discursive and relatively democratic political understanding across disparate interests (Norgaard 2008).

References

ACKERMAN, F. 2007. *Debating Climate Economics: The Stern Review vs. Its Critics* Report to Friends of the Earth-UK.

—— DeCANIO, S. J., HOWARTH, R. B., and SHEERAN, K. A. 2009. Limitations of integrated assessment models of climate change. *Climatic Change* 95: 297–315.

ARRHENIUS, S. 1896. On the influence of carbonic acid in the air upon the temperature of the ground. *Philosophical Magazine and Journal of Science*. Series 5, 41: 237–76.

CHAKRAVARTY, S. 1987. Cost-benefit analysis. Pp. 687–90 in *The New Palgrave: A Dictionary of Economics*. ed. J. Eatwell, M. Milgate, and P. Newman, vol. i. London: Macmillan Press.

CLINE, W. R. 1992. *The Economics of Global Warming*. Washington, DC: Institute for International Studies.

DeCANIO, S. J. 2003a. Economic analysis, environmental policy, and intergenerational justice in the Reagan administration: The case of the Montreal Protocol. *International Environmental Agreements: Politics, Law and Economics* 3: 299–321.

—— 2003b. *Economic Models of Climate Change: A Critique*. New York: Palgrave Macmillan.

HALL, D. C., and BEHL, R. J. 2006. Integrating economic analysis and the science of climate instability. *Ecological Economics* 57(3): 442–65.

HALLEGATTE, S., HOURCADE, J.-C., and DUMAS, P. 2007. Why economic dynamics matter in assessing climate change damages: Illustration on extreme events. *Ecological Economics* 62: 330–40.

HANSEN, J. 2009. *Storms of my Grandchildren: The Truth about the Coming Climate Catastrophe and our Last Chance to Save Humanity*. New York. Bloomsbury.

HARBERGER, A. C. 1971. Three postulates for applied welfare economics: an interpretive essay. *Journal of Economic Literature* 9(3): 785–97.

HOWARTH, R. B., and NORGAARD, R. B. 1992. Environmental valuation under sustainable development. *American Economic Review* 82(2): 473–7.

JONES, C. 2005. *Applied Welfare Economics*. Oxford: Oxford University Press.

KAUFMANN, R. K. 1997. Assessing the dice model: uncertainty associated with the emission and retention of greenhouse gases. *Climatic Change* 35: 435–48.

KRAUSE, F. 1996. The costs of mitigating carbon emissions: A review of methods and findings from European studies. *Energy Policy* 24(10/11): 899–915.

—— BACH, W., and KOOMEY, J. 1992. *Energy Policy in the Greenhouse*. New York: John Wiley, Sons.

McKENZIE, L. W. 1987. General equilibrium. Pp. 498–511 in *The New Palgrave: A Dictionary of Economics*. ed. J. Eatwell, M. Milgate, and P. Newman, vol. iv. London: Macmillan Press.

MENDELSOHN, R. 2008. Is the Stern Review an economic analysis? *Review of Environmental Economics and Policy* 2(1): 45–60.

—— NORDHAUS, W. D., and SHAW, D. 1994. The impact of global warming on agriculture: A Ricardian analysis. *American Economic Review* 84(4): 753–51.

Millennium Ecosystem Assessment. 2005. *Ecosystems and Human Well-Being*. Washington, DC: Island Press.

NELSON, R. H. 1987. The economics profession and public policy. *Journal of Economic Literature* 25(1): 49–91.

NISBET, R. 1980. *History of the Idea of Progress*. New York: Basic Books.

NORDHAUS, W. 1991. To slow or not to slow: The economics of the greenhouse effect. *Economic Journal* 101: 920–37.

NORDHAUS, W. 1994. *Managing the Global Commons: The Economics of Climate Change.* Cambridge, MA: MIT Press.

—— 2007. A review of the Stern Review of the economics of climate change. *Journal of Economic Literature* 45: 686–702.

—— 2008. *A Question of Balance: Economic Modeling of Global Warming.* New Haven: Yale University Press.

NORGAARD, R. B. 1992. Sustainability as intergenerational equity: Economic theory and environmental planning. *Environmental Impact Assessment Review* 12: 85–124.

—— 1994. *Development Betrayed: The End of Progress and a Coevolutionary Revisioning of the Future.* London: Routledge.

—— 2008. Finding hope in the millennium ecosystem assessment. *Conservation Biology* 22(4): 862–9

PECHMAN, J. A. (ed.) 1989. *The Role of the Economist in Government: An International Perspective.* New York: New York University Press.

REPETTO, R., and AUSTIN, D. 1997. *The Costs of Climate Protection: A Guide for the Perplexed.* Washington, DC: World Resources Institute.

SCHELLING, T. C. 1992. Some economics of global warming. *American Economic Review* 82(1): 1–14.

SCHLENKER, W., and ROBERTS, M. J. 2009. Nonlinear temperature effects indicate severe damage to U.S. crop yields under climate change. *Proceedings of the National Academy of Sciences* 106(37): 15594–8.

SCHNEIDER, S. H. 2009. *Science as a Contact Sport: Inside the Battle to Save Earth's Climate.* Washington, DC: National Geographic.

SPASH, C. L. 2002. *Greenhouse Economics: Value and Ethics.* London: Routledge.

SRINIVASAN, U. T. et al. 2008. The debt of nations and the distribution of ecological impacts from human activities. *Proceedings of the National Academy of Sciences.*

STERN, N. 2007. *The Economics of Climate Change: The Stern Review.* Cambridge: Cambridge University Press.

—— and TAYLOR, C. 2007. Climate change: Risk, ethics, and the Stern Review. *Science* 317 (5835): 203–4.

STERNER, T., and PERSSON, U. M. 2008. An even sterner review: introducing relative prices into the debate. *Review of Environmental Economics and Policy* 2(1): 61–76.

TOL, R. S. J., and YOHE, G. W. 2006. A review of the Stern Review. *World Economics* 7(4): 233–50.

WEITZMAN, M. L. 2009. On Modeling and Interpreting the Economics of Catastrophic Climate Change. *The Review of Economics and Statistics* 91(1): 1–19.

World Commission on Environment and Development. 1987. *Our Common Future.* Oxford: Oxford University Press.

World Water Assessment Programme. 2009. *Water in a Changing World.* London. United Nations Educational, Scientific, and Cultural Organization and Earthscan.

CHAPTER 14

GLOBAL CHANGE VULNERABILITY ASSESSMENTS: DEFINITIONS, CHALLENGES, AND OPPORTUNITIES

COLIN POLSKY AND HALLIE EAKIN

1 INTRODUCTION

CLIMATE-SOCIETY research has a long trajectory in both the physical and social sciences. By definition, this research area is inherently interdisciplinary. Yet despite what the name of the field might imply, to conduct a meaningful study of climate-society dynamics requires more than examining one feature of the climate system alongside one feature of the social system. For example, estimating the link between precipitation and crop yield is just the entry point, not the end point, for understanding the vulnerability of a country's agriculture sector to climate change. Crop yields are shaped by other biophysical and social factors. Indeed, each of the climate and society domains is multidimensional and the interactions across domains are complex, multi-scalar, and only beginning to be understood. An incomplete, yet still daunting, list of climate features to investigate includes the spatio-temporal patterns (e.g. mean, extent, frequency) of specific weather events (e.g. hurricanes, droughts, floods) or associated measures of energy and moisture fluxes (e.g. temperature, precipitation, albedo). Some social features to investigate include the perceptions, attitudes, impacts, and responses associated with climate variability/change—from numerous perspectives (e.g. health, income, political, prestige) at numerous levels of decision making (e.g. individual, household, community, state, institution). Even though this notion of interconnectedness in theory may appear self-evident, in practice a significant challenge of vulnerability assessments remains: knowing how many of the many other related factors to include in the study.

In recent years, a coherent, if still young, literature on global change vulnerability has emerged to structure this far-reaching line of inquiry. The vulnerability literature draws from ecological studies of resilience, social science studies of disasters, and from the 'Climate Impacts' tradition in geography and allied fields (Eakin and Luers 2006). This chapter begins by introducing Global Change Vulnerability Assessments (GCVAs), which achieved international prominence with the publication of the Third Assessment Report of the IPCC (McCarthy et al. 2001). Vulnerability in this context is defined as the product of processes along three dimensions: exposure to the climate and other relevant stresses; sensitivity to the exposure; and adaptive capacity to mitigate or prevent impacts generated by the sensitivities. The intellectual roots of, and the recent evolution of ideas behind, GCVAs have been widely reviewed (Dow 1992; Böhle et al. 1994; Ribot 1995; Kelly and Adger 2000; Golding 2001; Liverman 2001; Turner et al. 2003a; Schröter et al. 2005b; Adger 2006; Eakin and Luers 2006; Fussel and Klein 2006; O'Brien et al. 2007). We begin this chapter by summarizing the ground covered by these more extensive reviews, and outlining the primary criteria or principles that currently are guiding Global Change Vulnerability Assessments. We then look forward by outlining challenges and opportunities for future Vulnerability research.

2 Origins and Current Trajectories of GCVAs

2.1 Climate Impact Assessment

Early global climate change impact assessments focused principally on climate-society interactions drawing heavily from the tradition of natural hazards research (e.g. Burton et al. 1978). As outlined in a now-classic book edited by Kates et al. (1985), the key challenge of climate impact assessment is to isolate climate sensitivities in biophysical and social systems. The influence of any given climate stress is therefore traceable through direct influences on biophysical attributes, to sequential (and more indirect) impacts on social activities and attributes (Kates et al. 1985). This approach formed the basis of much of the research in the 1980s and 1990s by the climate change research community, and eventually evolved into a framework of 'vulnerability' assessment adopted by the IPCC (Carter et al. 1994).

Climate impact assessment was designed to be compatible with climate change modeling and simulation experiments, emphasizing the mechanisms and processes by which a specific climate signal translates into a measurable impact on the biophysical environment (e.g. crop yields, streamflow, forest growth) and, through the material and economic ties of society to a resource, into (second- or third-order) social and economic impacts (Carter et al. 1994). Integrated modeling of climate change impacts typically focused on specific geographic regions, scaled to a spatial resolution where climate parameters could be well understood and modeled, and often focused on a specific suite of biophysical or economic measures where the mechanism of climate influence could be directly argued (e.g. Rosenberg 1982; Rosenzweig 1985; Liverman et al. 1986). Considerable attention was given in these early assessment efforts to accurately characterizing the climatic parameters of interest and relevance to the region, sector, and decision makers who might make use of the assessment outcomes. Estimates of proxies for social processes such as demographic change and technological innovation were incorporated into these models to parameterize projections of greenhouse gas emissions and anticipate changes in the sensitivities of key system drivers. Serious critiques of this modeling approach emerged surrounding the realism of the underlying assumptions of human behavior (Dowlatabadi 1995; Berkhout and Hertin 2000; Kandlikar and Risbey 2000).

2.2 Social Vulnerability Assessment

Parallel to the development of climate impact modeling was the growth of another approach to vulnerability research, aimed less at process-based *assessment* for policy applications, but instead at the expansion of theory for understanding both the climate- and nonclimate-related causes of losses associated with climatic disturbance, and the differential consequences of such climate-society interactions. This line of research evolved from a response of research on 'natural' disasters, that, critics argued, paid inadequate attention to the social and political determinants of the damages from earthquakes, hurricanes, and the like (e.g. O'Keefe et al. 1976; Hewitt 1980). Sen's theory of entitlements as an explanation for famine in the developing world became particularly influential in the emergent critique (Sen 1981), highlighting distributional inequities in resource access associated with distinct social positions and the relation of resource access to capacities to cope with environmental disturbance. For example, scholars studying the causes and consequences of the Sahel famines in the 1970s and early 1980s argued that vulnerability was more the product of colonial legacies of social disruption than of climate variability (e.g. Watts 1983). The physical driver—in this case, drought—was interpreted as playing a secondary role in the famines of that period.

Attention to the details of social and political dynamics suggested that vulnerability assessments should be sensitive to the differential characteristics of households, to historical conditions and trajectories of resource distribution and management, and to global processes of institutional change and political interventions affecting the differential ways populations manage risks (see, for example, the 'Pressure and Release (PAR)' model of social vulnerability analysis in Wisner et al. (2004), originally published in 1994. When brought to bear on climate change vulnerability assessment, the lessons of this approach highlighted the importance of adding global and national socio-economic processes alongside climate as exogenous drivers of climate change vulnerability, and characterized vulnerability largely in terms of differential capacities for coping with shocks (rather than differential exposure to exogenous stress). Over the last decades the lessons from research in this domain has illustrated that differential social outcomes associated with climate stress may have as much (or more) to do with historical inequities and disparities in the social and institutional contexts of human activity than with differential exposure to climate shocks (Böhle et al. 1994; Ribot 1995). Attention to defining the complex nature of vulnerability at the local level also led to a proliferation of research aimed at defining appropriate socio-economic indicators as determinants of susceptibility to harm for different households and communities. These efforts in turn led to new conceptual developments in the area of social vulnerability analysis, such as the Social Vulnerability Index (SOVI; Cutter et al. 2003), and the Social Risk Management framework (Heltberg et al. 2009). The Third Assessment Report of the IPCC incorporated some of the contributions of social vulnerability research with a new emphasis on adaptive capacity as one of three dimensions of the concept of vulnerability, which were defined as exposure to climate (and other stresses), sensitivities to the exposures, and adaptive capacities to respond to the impacts from the exposures (McCarthy et al. 2001).

2.3 Assessment of Social-Ecological Systems

Whereas climate impact assessment tended to emphasize the biophysical drivers of vulnerability and meso-scale analyses (regions, sectors, countries), and social vulnerability oassessment tended to emphasize social and institutional drivers and focus on outcomes and processes occurring at the level of individual, households, and communities, over the last decade a demand for improved analysis of the *interaction* of biophysical and social processes in driving both vulnerability and structuring social responses has led to a concerted attempt at theoretical and methodological synthesis (Turner et al. 2003a, 2003b; Schröter et al. 2005b; Eakin and Luers 2006; Fussel 2007; Polsky et al. 2007). Vulnerability research is now being reframed in the context of understanding the sustainability of the coupled dynamics of human and environmental systems (Kates et al. 2001; Clark and Dickson 2003).

For vulnerability research, this new emphasis on system dynamics means that assessments initiate with the premise that social and biophysical processes are interdependent and co-evolving. As outlined by Turner et al. (2003a), approaching vulnerability through the lens of sustainability science requires understanding the mechanisms of perturbation, feedback, and response that produce change. Rather than focusing on outcomes (as in impacts analysis) or conditions (as in social vulnerability assessment), the emphasis in this perspective is on process, function, and interaction. Incorporating insights on what makes ecosystems resilient (or not) in face of disturbance (Holling 1973), vulnerability assessments of coupled systems leads to a focus on processes of learning, the exchange of knowledge, and the flexibility of institutions, communities, and ecosystems in response to change (Vogel et al. 2007). In this perspective, there is one primary unit of analysis—the coupled system—in which other subsystems are embedded (Gallopin 2006). Stressors should not be examined in isolation—e.g. modeling the effect of a doubling of CO_2 holding all other factors constant—but rather multiple, and stem from both biophysical and socio-economic processes. Assessments are focused in the temporal and spatial scales most pertinent to the decision-making process, while recognizing the temporal and spatial diversity of biophysical and social processes at work in shaping system disturbance and response (Turner et al. 2003a). Recognizing the important role of social institutions in regulating human-environment interactions in particular places, this approach highlights the ways in which interventions may exacerbate sensitivities, or create circumstances in which society and resource are buffered from disturbance in the short term only to face catastrophic losses in the longer term (Janssen et al. 2007).

3 GLOBAL CHANGE VULNERABILITY ASSESSMENTS: FIVE CRITERIA

The foregoing review suggests five specific criteria that assessments should satisfy to answer the various critiques of past scholarship (Schröter et al. 2005b). First, the knowledge base engaged for analysis should be varied and flexible. This criterion goes beyond the standard call for interdisciplinary scientific research. Vulnerability researchers should also collaborate with stakeholders to learn their perspectives and concerns. This imperative may require serious engagement with indigenous, or local, knowledge—despite difficulties in testing such information within a classical scientific framework. This approach furthermore argues that it may be valuable to engage with the stakeholders as co-equals when generating the research design (Cash et al. 2003). In this spirit we find, for example, Fox's (2002) inquiry into selected indigenous perspectives on recent environmental changes in two Inuit communities in Arctic Canada. Similarly, Patt and Gwata (2002) use participatory methods to learn from Zimbabwean subsistence farmers what factors matter most to farming success—rather than presupposing what factors should be most important based on externally defined agronomic principles. Finally, Brooks et al. (2005) validate their statistical correlations by engaging a panel of experts using Delphi elicitation techniques, on the assumption that there may be important variation in expert opinion about which factors matter most in the construction of vulnerability indices.

Second, the scale at which the coupled human-environment system is studied is generally 'place-based,' meaning that local-scale vulnerabilities are examined in the context of processes and outcomes at other scales. Most GCVAs examine the consequences of large-scale processes (e.g. climate change) at smaller scales, and are therefore multi-scale in scope, by definition. Yet few GCVAs have explicitly examined the roles of processes from multiple scales simultaneously, and instead have typically focused on processes and outcomes at the local scale. Exceptions to this trend include the work of Eakin (2006) on 1990s Mexican farming vulnerabilities associated with local-scale droughts in the context of rapid national-scale land reforms and international-scale economic liberalization programs. Similarly, Adger and Kelly (1999) and Kelly and Adger (2000) examine local-scale typhoon impacts on coastal Vietnamese aquaculture in the context of national-scale post-Communist political-economic restructuring (*Doi Moi*). O'Brien et al. (2004b) show that the picture of Norwegian vulnerability to climate change varies with the scale of analysis: at the national level, the country appears to exhibit low sensitivities and strong adaptive capacities, but at the regional and local levels, there emerges a more varied picture of undesirable exposures, sensitivities, and adaptive capacities.

Third, the potential drivers of vulnerability are understood to be possibly multiple and interacting. It would be a mistake to assume that outcomes associated with climate variability and change are necessarily the only issues of concern for a given population or human-environment system. In fact, stakeholders may rank daily matters associated with poverty, war, or health on a par or ahead of climate issues. O'Brien et al. (2004a) formalize this idea with the concept of 'Double Exposure' (see Leichenko and O'Brien (2008) for an expanded discussion), and apply it to the case of climate change and trade liberalization policies in Indian agriculture. The same point of departure is used in Klinenberg's (2002) examination of how elderly mortality during the 1995 Chicago heatwave was in part a product of the simultaneously unfolding municipal restructuring to reduce/outsource funding for social services for older populations.

Fourth, understanding adaptive capacity is a fundamental part of understanding vulnerability. Adaptive capacity is included as a vulnerability dimension to permit separation of human-environment systems that are exposed and sensitive to an external stress or perturbation but cannot successfully restructure or recover, from those systems that respond to the impacts in ways that give plausible hope for avoiding a future disaster repeat by limiting exposures, reducing sensitivities, and/or strengthening response options and adaptive capabilities. Yet vulnerability research has shown that adaptive capacity may vary within a human-environment system, depending on demographic, social, economic, or other factors. The dimension is framed as adaptive capacity rather than 'adaptations' because the adaptation options of some individuals or groups may be constrained by inadequate resources (including information), or political-institutional barriers. Moreover, in some cases it is difficult for researchers to know whether a given adaptation is good or bad; focusing on capacity, and capacity building, reframes the issue in terms of decision processes rather than outcomes. In this way a GCVA might identify that a given group has the knowledge and desire to reduce their vulnerabilities but is prevented from adapting by legal or cultural obstacles. Differential adaptation profiles can account for the possible combinations of adaptation constraints and opportunities for a given case, and how these factors may vary both between and within populations. In this spirit, Böhle et al. (1994) argue that, when studying agricultural vulnerability in Zimbabwe, even though female-headed households are uncommon, failing to account for the socially produced limited adaptive capacities of these households may produce a significantly misleading picture of vulnerabilities and the processes that produce the vulnerabilities. Alternatively, vulnerabilities can differ because feedback between ecological/climatic change to human activities is not occurring: the human system may in theory possess robust adaptation capacities, but not possess the leadership to trigger the responses.

Fifth, an understanding of historical vulnerabilities should be linked to scenarios of future social and environmental conditions. The primary objective of GCVAs is 'to inform the decision-making of specific stakeholders about options for adapting to the effects of global change' (Schröter et al. 2005b: 575). As such, some attempt at extrapolating past vulnerabilities, or projecting new vulnerabilities associated with new trends not observed in the historical record, is necessary. In this way we see Cutter and Finch's (2008) projection of the SOVI for 2010 on the basis of trends in the SOVI estimated for five points in time (1960, 1970, 1980, 1990, and 2000). Of course, a major challenge associated with projections is that no researchers have a 'crystal ball,' enabling a precise picture of the future, and stakeholders appear at least as aware of this fact as the researchers. Thus it is helpful to vet the projections with stakeholders during the production of the projections. Failing to engage stakeholders in this way risks having the stakeholders view the GCVA as neither credible, salient, nor legitimate, possibly resulting in the wholesale rejection of the GCVA, whatever the intellectual merit of the assessment (Cash et al. 2003). Alternatively, GCVAs should, in collaboration with stakeholders, attempt to produce and/or evaluate a range of plausible scenarios of future vulnerabilities, where each scenario is driven by projections of trends of important variables or combinations of variables, such as demographics, energy use, and decision-making. This approach is exemplified by Schröter et al. (2005a), who evaluate with stakeholders the outcomes of sixteen possible futures for Europe, driven by different combinations of four climate projections and four greenhouse gas emissions scenarios.

Admittedly, it is a tall order for a single GCVA to satisfy all five of the above criteria. Nonetheless, the five criteria provide a statement about the ideal case—what, ideally, a GCVA *should* look like, to capture all of the complexity and interactions identified beginning with the outline of the Climate Impacts approach by Kates et al. (1985). Conversely, as these five criteria collectively represent an ideal type, the criteria also provide a benchmark for where future research should be directed: if the literature on vulnerability for a given place or sector has consistently underemphasized one of these five criteria, then future research might benefit from emphasizing that criterion.

4 Two Major Outstanding Challenges

There are two integrating themes that connect the five criteria for GCVAs outlined above and that would benefit from immediate research attention. First, the shift in thinking described above—where vulnerability is seen as a process embedded in a complex coupled human-environment system—demands a more coherent methodological framework that can facilitate inter-case comparisons and cross-scale analysis. This need highlights the challenge of measuring, and combining, indicators of the three dimensions of vulnerability. In recent years there have been numerous attempts in the global change literature to produce vulnerability indicators (e.g. Clark et al. 1998; Hurd et al. 1999; Cutter et al. 2000; Downing et al. 2001; Moss et al. 2001; Wu et al. 2002; Cutter et al. 2003; Luers et al. 2003; O'Brien et al. 2004a; Brooks et al. 2005; Luers 2005; Rygel et al. 2006; Cutter and Finch 2008; Eakin and Bojorquez-Tapia 2008). Yet there remains a lack of consensus about how to combine the measures of the dimensions into a single index. Do all three of the commonly used dimensions of vulnerability deserve the same weights? If so, on what basis (subjective or objective) are the different weights assigned, and do those weights vary with the location, time, or unit of analysis? Are stakeholders engaged to help define the weights? Can the vulnerability index be readily decomposed by stakeholders into its constituent components to be useful for decision makers? Answers to these questions will provide several benefits: first, researchers will gain guidance on project-specific research design; second, the research community will gain a basis for comparing GCVA results across studies; and third, the likelihood will increase that vulnerability research will be more appropriate to user needs.

Furthering cross-study comparisons of GCVAs is particularly important. There is significant diversity of topics in the GCVA literature, which means that the measures relevant for one study may not be relevant for another study. Yet if the measures can be combined into a normalized index, the otherwise differing studies should be comparable, and thus contribute to both the development of more robust theory on the causes and consequences of vulnerability while also informing the development of policy to reduce vulnerabilities. There is a well-developed literature in social science (e.g. Wolf 1986; Ragin 1987) specifying how to pool results from independent studies into a larger dataset that permits more powerful inferences and generalizations—even when the variables measured, and the methods used to measure the variables, vary between studies. There are a few notable examples of meta-analyses in fields allied to vulnerability, including Geist and Lambin (2002), Misselhorn (2005), Keys and McConnell (2005), and Rudel (2005; 2008). Polsky et al. (2007) present an organizing tool designed to encourage between-study comparisons of GCVAs. Coupling insights from these recent developments in the vulnerability and allied literatures presents a ripe opportunity for advance in the coming decade.

The second integrating theme connecting the five criteria for GCVAs that would benefit from immediate research attention is, building on the first theme, improving our understanding of trade-offs. Recognizing the interconnectedness of climate and society means not only a need to examine in an integrated fashion many factors from many disciplines. Interconnectedness also means that decisions about reducing vulnerabilities— through some combination of reducing exposures and sensitivities, and increasing adaptive capacity—will have to be considered in the context of trade-offs, i.e. missed benefits from other uses for the resources required to reduce the vulnerabilities. For example, one means for reducing vulnerabilities associated with climate-related vulnerabilities linked to the urban heat island effect may be to increase use of water for evaporative cooling. Yet increasing water use, for whatever purpose, may be itself an action that elevates vulnerabilities in other parts of the coupled human-environment system, such as potentially limiting local agricultural productivity or increasing toxic emissions from energy production. Similarly, there may be social trade-offs entailed in the strategies adopted by different populations as they attempt to come to terms with changing risks. Decisions made to enhance resilience at one scale can have negative implications for vulnerability at other scales or for specific subsystems (Eakin et al. 2009). Effective management of the resilience of the entire Norfolk coast from sea-level rise, for example, may well entail significant losses of property and livelihood for many specific localities (Cole-Nicholson and O'Riordan 2009). Conversely, lack of attention to the social and ecological implications of the loss of smaller-scale mixed farms in regions of the Argentinean Pampas could enhance the risk the broader ecosystem faces to ecological disturbances (Eakin and Wehbe 2009). Even though the need to account for trade-offs is a necessary consequence of the coupled nature of the human-environment system being studied, the specific types of trade-offs in specific cases have been underemphasized to date.

5 CONCLUSION

Understanding vulnerabilities associated with climate-society relationships, including climate change, requires an interdisciplinary, multi-scaled approach. Recent scholarship has advanced the conceptual, and to a lesser degree, methodological frameworks for structuring vulnerability research. In this way, applied vulnerability research should account for not only relevant exposures of the human-environment system in question to climate and other stresses, and sensitivities to the exposures, but also adaptive capacities to reduce and/or prevent the impacts associated with the sensitivities. Even though this emerging conceptualization draws heavily from the well-known work on 'Climate Impacts,' vulnerability is understood to address three primary limitations of the Climate Impacts approach as commonly operationalized: an underemphasis of the social and ecological variables that often mediate climate-society relationships; the multi-scaled nature of the interactions among humans, ecosystems, and the atmosphere; and the responses of people, societies, ecosystems, and the atmosphere of longer duration than the initial climate event and manifestation of associated impacts. In this way the goal of global change vulnerability assessments—to inform the decision making of specific stakeholders about options for adapting to the effects of global change—is more readily achievable than in cases where only one or a small number of climate and social variables are linked. The vulnerability literature would benefit from advances in a number of areas, including the production of vulnerability indices (and weights for the constituent components), and the identification of the trade-offs involved in reducing vulnerability by enhancing adaptive capacity. These two topics present both theoretical and methodological challenges. Overcoming these, and other, challenges will represent an important advance in the larger agenda to improve the sustainability of contemporary societies and ecosystems.

REFERENCES

ADGER, N. 2006. Vulnerability. *Global Environmental Change* 16: 268–81.

ADGER, W. N., and KELLY, P. M. 1999. Social vulnerability to climate change and the architecture of entitlements. *Mitigation and Adaptation Strategies for Global Change* 4(3/4).

BERKHOUT, F., and HERTIN, J. 2000. Socio-economic scenarios for climate impact assessment. *Global Environmental Change* 10: 165–8.

BÖHLE, H. G., DOWNING, T. E., and WATTS, M. 1994. Climate change and social vulnerability: Toward a sociology and geography of food insecurity. *Global Environmental Change* 4(1): 37–48.

BROOKS, N., ADGER, W. N., and KELLY, P. M. 2005. The determinants of vulnerability and adaptive capacity at the national level and the implications for adaptation. *Global Environmental Change* 15(2): 151–63.

BURTON, I., KATES, R. W., and WHITE, G. F. (eds.) 1978. *The Environment as Hazard.* New York: Oxford University Press.

CARTER, T. R., PARRY, M. L., HARASAWA, H., and NISHIOKA, S. 1994. *IPCC Technical Guidelines for Assessing Climate Change Impacts and Adaptations,* 59. Department of Geography, University College London and Center for Global Environmental Research, National Institute for Environmental Studies, London, UK, and Tsukuba, Japan.

CASH, D. W., CLARK, W. C., ALCOCK, F., DICKSON, N. M., ECKLEY, N., GUSTON, D. H., JÄGER, J., and MITCHELL, R. B. 2003. Knowledge systems for sustainable development. *Proceedings, National Academy of Sciences* 100(14): 8086–91.

CLARK, G. E., MOSER, S., RATICK, S., DOW, K., MEYER, W. B., EMANI, S., JIN, W., KASPERSON, J. X., KASPERSON, R. E., and SCHWARZ, H. E. 1998. Assessing the vulnerability of coastal communities to extreme storm: The case of Revere, MA, USA. *Mitigation and Adaptation Strategies for Global Change* 3: 59–82.

CLARK, W. C., and DICKSON, N. M. 2003. Sustainability Science: The Emerging Research Program. *Proceedings, National Academy of Sciences,* 100(14): 8059–61.

CUTTER, S., and FINCH, C. 2008. Temporal and spatial changes in social vulnerability to natural hazards. *Proceedings, National Academy of Sciences* 105(7): 2301–6.

—— MITCHELL, J. T., and SCOTT, M. S. 2000. Revealing the vulnerability of people and places: A case study of Georgetown County, South Carolina. *Annals of the Association of American Geographers* 90(4): 713–37.

—— BORUFF, B. J., and SHIRLEY, W. L. 2003. Social vulnerability to environmental hazards. *Social Science Quarterly,* 84(2): 242–61.

DOW, K. 1992. Exploring differences in our common future(s): The meaning of vulnerability to global environmental change. *Geoforum* 23: 417–36.

DOWLATABADI, H. 1995. Integrated assessment models of climate change. *Energy Policy* 23 (4/5): 289–96.

DOWNING, T. E., BUTTERFIELD, R., COHEN, S., HUQ, S., MOSS, R., RAHMAN, A., SOKONA, Y., and STEPHEN, L. 2001. *Vulnerability Indices: Climate Change Impacts and Adaptation*. Policy Series 3. United Nations Environment Programme.

EAKIN, H. 2006. *Weathering Risk in Rural Mexico: Climatic, Institutional, and Economic Change*. Tucson, AZ: The University of Arizona Press.

—— and BOJORQUEZ-TAPIA, L. A. 2008. Insights into the composition of household vulnerability from multicriteria decision analysis. *Global Environmental Change* 18: 112–27.

—— and LUERS, A. 2006. Assessing the Vulnerability of Social-Environmental Systems. *Annual Review of Environment Resources* 31: 365–94.

—— and WEHBE, M. 2009. Linking local vulnerability to system sustainability in a resilience framework: Two cases from Latin America. *Climatic Change* 93: 355–77.

—— TOMPKINS, E., NELSON, D. R., and ANDERIES, J. M. 2009. Hidden costs and disparate uncertainties: Trade-offs involved in approaches to climate policy. Pp. 212–26 in W. N. Adger, I. Lorenzoni, and K. O'Brien (eds.), *Adapting to Climate Change: Thresholds, Values, Governance*. Cambridge: Cambridge University Press.

FOX, S. 2002. *When the Weather is Uggianaqtuq: Inuit Observations of Environmental Change*. Multi-media, Interactive CD-ROM, Cartography Lab, Department of Geography, University of Colorado at Boulder, Boulder, CO. Online.

FUSSEL, H.-M. 2007. Vulnerability: A generally applicable conceptual framework for climate change research. *Global Environmental Change* 17: 155–67.

—— and KLEIN, R. J. T. 2006. Climate change vulnerability assessments: An evolution of conceptual thinking. *Climatic Change* 75: 301–29.

GALLOPIN, G. 2006. Linkages between vulnerability, resilience, and adaptive capacity. *Global Environmental Change* 16: 293–303.

GEIST, H. J., and LAMBIN, E. F. 2002. Proximate causes and underlying driving forces of tropical deforestation. *Bioscience* 52(2): 143–50.

GOLDING, D. 2001. Vulnerability. In A. S. Goudie and D. J. Cuff (eds.), *Encyclopedia of Global Change: Environmental Change and Human Society*. Oxford: Oxford University Press.

HELTBERG, R., SIEGEL, P. B., and JORGENSEN, S. L. 2009. Addressing human vulnerability to climate change: towards a 'no-regrets' approach. *Global Environmental Change* 19: 89–99.

HEWITT, K. 1980. Book review: The Environment as Hazard, by I. Burton, R. Kates and G. White. *Annals of the Association of American Geographers* 70(2): 306–11.

HOLLING, C. S. 1973. Resilience and stability of ecological systems. *Annual Review of Ecological Systems* 4: 1–23.

HURD, B., LEARY, N. A., JONES, R., and SMITH, J. 1999. Relative regional vulnerability of water resources to climate change. *Journal of the American Water Resources Association* 35(6): 1399–409.

JANSSEN, M. A., ANDERIES, J. M., and OSTROM, E. 2007. Robustness of social-ecological systems to spatial and temporal variability. *Society & Natural Resources* 20(4): 307–22.

KANDLIKAR, M., and RISBEY, J. 2000. Agricultural Impacts of Climate Change: If Adaptation is the Answer, What is the Question? An Editorial Comment. *Climatic Change* 45: 529–39.

KATES, R. W., AUSUBEL, J. H., and BERBERIAN, M. (eds.) 1985. *Climate Impact Assessment: Studies of the Interaction of Climate and Society*. SCOPE 27. Chichester: Wiley.

—— CLARK, W. C., CORELL, R., HALL, J. M., JAEGER, C. C., LOWE, I., MCCARTHY, J. J., SCHELLNHUBER, H. J., BOLIN, B., DICKSON, N. M., FAUCHEUX, S., GALLOPIN, G. C., GRUEBLER, A., HUNTLEY, B., JÄGER, J., JODHA, N. S., KASPERSON, R. E., MABOGUNJE, A.,

MATSON, P., MOONEY, H., MOORE, B., III, O'RIORDAN, T., and SVEDIN, U., 2001. Sustainability science. *Science*, 292 (27 April): 641–2.

KELLY, P. M., and ADGER, W. N. 2000. Theory and practice in assessing vulnerability to climate change and facilitating adaptation. *Climatic Change* 47: 325–52.

KEYS, E., and MCCONNELL, W. 2005. Global change and the intensification of agriculture in the tropics. *Global Environmental Change* 15: 320–37.

KLINENBERG, E. 2002. *Heat Wave: A Social Autopsy of Disaster in Chicago*. Chicago: University Chicago Press.

LEICHENKO, R., and O'BRIEN, K. 2008. *Environmental Change and Globalization: Double Exposures*. Oxford: Oxford University Press.

LIVERMAN, D. 2001. Vulnerability to global environmental change. Pp. 201–16 in J. X. Kasperson and R. E. Kasperson (eds.), *Global Environmental Risk*. Tokyo: United Nations University Press.

—— TERJUNG, W. H., HAYES, J. T., and MEARNS, L. O. 1986. Climatic Change and Grain Corn Yields in the North American Great Plains. *Climatic Change* 9: 327–47.

LUERS, A. 2005. The surface of vulnerability: An analytical framework for examining environmental change. *Global Environmental Change* 15: 214–23.

—— LOBELL, D. B., SKLAR, L. S., ADDAMS, C. L., and MATSON, P. A. 2003. A method for quantifying vulnerability, applied to the agricultural system of the Yaqui Valley, Mexico. *Global Environmental Change* 13: 255–67.

MCCARTHY, J. J., CANZIANI, O. F., LEARY, N. A., DOKKEN, D. J., and WHITE, K. S. (eds.) 2001. *Climate Change 2001: Impacts, Adaptation, and Vulnerability*. Published for the Intergovernmental Panel on Climate Change. Cambridge: Cambridge University Press.

MISSELHORN, A. 2005. What drives food security in southern Africa? A meta-analysis of household economy studies. *Global Environmental Change* 15: 33–43.

MOSS, R. H., BRENKERT, A. L., and MALONE, E. L. 2001. *Vulnerability to Climate Change: A Quantitative Approach*. Pacific Northwest National Laboratory. PNNL-SA-33642. Online.

NICHOLSON-COLE, S., and O'RIORDAN, T. 2009. Adaptive governance for a changing coastline: Science, policy and publics in search of a sustainable future. Pp. 368–83 in W. N. Adger (ed.), *Adapting to Climate Change: Thresholds, Values, Governance*. Cambridge: Cambridge University Press.

O'BRIEN, K., LEICHENKO, R., KELKAR, U., VENEMA, H., AANDAHL, G., TOMPKINS, H., JAVED, A., BHADWAL, S., BARG, S., NYGAARD, L., and WEST, J. 2004a. Mapping vulnerability to multiple stressors: climate change and globalization in India. *Global Environmental Change* 14: 303–13.

—— SYGNA, L., and HAUGEN, J. E. 2004b. Vulnerable or resilient? A multi-scale assessment of climate impacts and vulnerability in Norway. *Climatic Change* 64: 193–225.

—— ERIKSEN, S., SCHJOLDEN, A., and NYGAARD, L. P. 2007. Why different interpretations of vulnerability matter in climate change discourses. *Climate Policy* 7: 73–88.

O'KEEFE, P., WESTGATE, K., and WISNER, B. 1976. Taking the naturalness out of natural disasters. *Nature* 260: 566–7.

PATT, A. G., and GWATA, C. 2002. Effective seasonal climate forecast applications: examining constraints for subsistence farmers in Zimbabwe. *Global Environmental Change* 12: 185–95.

POLSKY, C., NEFF, R., and YARNAL, B. 2007. Building comparable global change vulnerability assessments: The vulnerability scoping diagram. *Global Environmental Change* 17: 472–85.

RAGIN, C. C. 1987. *The Comparative Method. Moving beyond Qualitative and Quantitative Strategies*. Berkeley: Univ. of California Press.

RIBOT, J. C. 1995. The causal structure of vulnerability: Its application to climate impact analysis. *GeoJournal* 35(2): 119–22.

ROSENBERG, N. J. 1982. The increasing CO_2 concentration in the atmosphere and its implication on agricultural productivity, Part II: Effects through CO_2-induced climatic change. *Climatic Change* 4: 239–54.

ROSENZWEIG, C. 1985. Potential CO_2-induced climate effects on North American wheat-producing regions. *Climatic Change* 7: 367–89.

RUDEL, T. K. 2005. *Tropical Forests: Regional Paths of Destruction and Regeneration in the Late 20th Century*. New York: Columbia University Press.

——2008. Capturing regional effects through meta-analyses of case studies: An example from the global change literature. *Global Environmental Change*.

RYGEL, L., O'SULLIVAN, D., and YARNAL, B. 2006. A method for constructing a social vulnerability index: An application to hurricane storm surges in a developed country. *Mitigation and Adaptation Strategies for Global Change* 11: 741–64.

SCHRÖTER, D., CRAMER, W., LEEMANS, R., PRENTICE, I. C., ARAÚJO, M. B., ARNELL, N. W., BONDEAU, A., BUGMANN, H., CARTER, T. R., GRACIA, C. A., VEGA-LEINERT, A. C. D. L., ERHARD, M., EWERT, F., GLENDINING, M., HOUSE, J. I., KANKAANPÄÄ, S., KLEIN, R. J. T., LAVOREL, S., LINDNER, M., METZGER, M. J., MEYER, J., MITCHELL, T. D., REGINSTER, I., ROUNSEVELL, M., SABATÉ, S., SITCH, S., SMITH, B., SMITH, J., SMITH, P., SYKES, M. T., THONICKE, K., THUILLER, W., TUCK, G., ZAEHLE, S. and ZIERL, B. 2005a. Ecosystem service supply and vulnerability to global change in Europe. *Science* 310: 1333–7.

—— POLSKY, C. and PATT, A. 2005b. Assessing vulnerabilities to the effects of global change: An eight step approach. *Mitigation and Adaptation Strategies for Global Change* 10(4): 573–95.

SEN, A. 1981. *Poverty and Famines: An Essay on Entitlement and Deprivation*. New York: Oxford University Press.

TURNER, B. L., KASPERSON, R. E., MATSON, P., McCARTHY, J. J., CORELL, R. W., CHRISTEN-SEN, L., ECKLEY, N., KASPERSON, J. X., LUERS, A., MARTELLO, M. L., POLSKY, C., PULSI-PHER, A., and SCHILLER, A. 2003a. A framework for vulnerability analysis in sustainability science. *Proceedings, National Academy of Sciences* 100(14): 8074–9.

—— MATSON, P., McCARTHY, J. J., CORELL, R. W., CHRISTENSEN, L., ECKLEY, N., HOVELSRUD-BRODA, G., KASPERSON, J. X., KASPERSON, R. E., LUERS, A., MARTELLO, M. L., MATHIESEN, S., POLSKY, C., PULSIPHER, A., SCHILLER, A., and TYLER, N. 2003b. Illustrating the coupled human-environment system for vulnerability analysis: Three case studies. *Proceedings, National Academy of Sciences* 100(14): 8080–5.

VOGEL, C., MOSER, S., KASPERSON, R., and DABELKO, G. 2007. Linking vulnerability, adaptation, and resilience science to practice: Pathways, players, and partnerships. *Global Environmental Change* 17: 349–64.

WATTS, M. 1983. On the poverty of theory: Natural hazards research in context. In K. Hewitt (ed.), *Interpretations of Calamity from the Viewpoint of Human Ecology*. Winchester, MA: Allen & Unwin.

WISNER, B., BLAIKIE, P., CANNON, T., and DAVIS, I. 2004. *At Risk: Natural Hazards, People's Vulnerability, and Disasters*. London: Routledge.

WOLF, F. M. 1986. *Meta-analysis: Quantitative Methods for Research Synthesis. Quantitative Applications in the Social Sciences*. London: Sage.

WU, S.-Y., YARNAL, B., and FISHER, A. 2002. Vulnerability of coastal communities to sea-level rise: A case study of Cape May County, New Jersey. *Climate Research* 22: 255–70.

CHAPTER 15

....

HEALTH HAZARDS

....

ELIZABETH G. HANNA

1 INTRODUCTION
....

HUMANS are social creatures. Our choice to live in cooperative groups has proven to be an eminently successful survival tactic. It is also clearly a choice borne from preference, as we enjoy the enrichment provided by community living, and the interesting social complexities gained through membership of a social group. But perhaps even more fundamental than our social needs is the fact that we are also biophysical creatures. Dependence upon the healthy functioning of earth's biophysical systems is a matter of survival, not merely preference.

Our social systems and cultural practices evolved in response to our elected congregation into communities, whereas human physiology and physical needs evolved in response to the earth's environment. Accordingly, we are superbly adapted to this environment (Richardson, Steffen, et al. 2009). Our apparent preference, and indeed survival, in non-natural environments such as urban centers is a relatively recent phenomenon, made possible only by our exceptional adaptive capacity, application of technology, and social organization. However increasing evidence suggests these artificial environments carry health risks.

Widespread anthropogenic degradation of ecological systems and now interruptions to global climate are creating an environment that does not suit our physiology, and hence is challenging the survival of our species.

Projected average global warming by 4 to 6 °C by 2100 will transform the planet to a very different world to the one in which human physiology (and existing ecologies) evolved. Climate change is currently having, and perhaps more alarmingly, will increasingly have wide-ranging and mostly adverse impacts through multiple pathways. Warming poses direct health threats via exposure to extremes of heat beyond our physical tolerance. Ecological systems threatened by climate change concern food yields, infectious pathogens, and river health. These supply the food, air, and water we need to survive, so by interfering with these, we are disrupting the core building blocks for human health.

Developing countries, already experiencing initial impacts, are predicted to suffer the most dramatic impacts of continued warming. Unless we rapidly reverse the process, continued degradation and destruction of life support systems will create widespread deprivation, generate resource competition, and will ultimately disrupt the social structures responsible for peace, security, and stability. This chapter outlines how these changes threaten population health and the ultimate survival of our species.

2 CLIMATE CHANGES AND HUMAN HEALTH

A warming climate delivers significant changes to climate systems, which in turn, have profound impacts on ecosystems. The following sections describe the links between climate change interfering with ecosystems and social infrastructures and human health and well-being.

Figure 15.1 depicts the various pathways of impact. Climate change warms the planet and creates anomalies in precipitation patterns, increases in extreme weather events, heatwaves, and rising sea levels. Direct pathways include the acute and chronic stress of heatwaves, and the immediate trauma from increased bush fires, storms, flooding, and coastal inundation. Indirect pathways occur when heatwaves or rainfall shortages reduce crop yields, or when altered distributions of vectors and pathogens produce changes in the epidemiology of infectious diseases. Butler classifies these as primary and secondary impacts (Butler and Harley 2010). Butler's tertiary effects include famine, resource competition, social disruption, conflict, war, and significant population displacement, which carry serious widespread implications for governance, health, and health inequity.

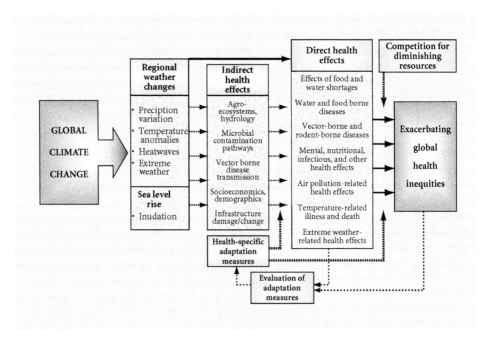

FIGURE 15.1 Pathways that climate change can affect human health and well-being. (Hanna et al. 2011; adapted from McMichael, Campbell-Lendrum, et al. 2003)

3 HEAT EXPOSURE

Temperatures are rising, and the first half of 2010 was the warmest on record, with heat extremes experienced on many continents (National Climatic Data Center 2010). There is an upper limit to human tolerance to heat exposure. Complex physiological and behavioral thermoregulatory systems maintain our core body temperature within the very narrow tolerance range of one degree around 37 °C. Eighty percent of the energy produced by exercising muscles is heat, which must be dissipated, and this heat transfer between the human body and the environment depends on climate and clothing. If air temperatures exceed 35 °C, the human body can only maintain normal core temperature by the mechanism of sweat evaporation. Sweating becomes ineffective in humid conditions and when wearing certain clothing, such as personal protective gear. The threat to human health from a warming climate is less from the rise in average global temperatures than increases in frequency and intensity of extreme heat events, beyond physiological comfort and capacity to work and function.

Under a medium-high emissions scenario, by 2020, 2050, and 2080, the number of hot days could increase by 2.1, 3.6, and 5.1 times relative to 1961–90 (Cueto, Martínez At, et al. 2010). In many urban populations, an average 2 °C rise in temperature would increase the annual death rate by an estimated doubling or more, due to hotter heatwaves. Heatwaves are already more frequent and more intense than in the past, and the health toll is apparent. The 2003 European heatwave was responsible for a total of over 60,000 premature deaths. Unprecedented heatwaves across the United States, Canada, Europe, Asia, and Russia in 2010 caused major disruptions to power, interrupting transport, air conditioning, and exacerbating heat deaths. A heatwave in India and Pakistan in June 2010 saw temperatures reaching 53.5 °C (128 °F) (Pakistan Meteorological Department 2010).

Communities unused to high temperatures are especially at risk, as they have not previously had need to develop behavioral or technical adaptive strategies via housing design and air conditioning, and they are not acclimatized. The European heatwave is a case in point. In all communities exposed to heat events, it is the elderly, the socially isolated or immobile, the very young, and people with chronic diseases, who are considered most vulnerable to extreme heat events. Cardiovascular disorders accentuate risk as extreme heat places additional load on the heart. Cognitive impairment, diabetes, cancer, and obesity also elevate susceptibility to heat stress. Prescribed medications such as anticholinergics, antiparkinsonian agents, and some antipsychotics and tranquilizers can interfere with the body's thermoregulatory system and fluid and electrolyte balance (Beggs 2000; Kwok and Chan 2005). Settings are also vitally important. For example, in urban environments, upper floor dwellings increase heat exposure risks where cooling and ventilation is limited, as they accumulate heat. High night-time minimum temperatures also exacerbate health risks, and the urban heat island effect can raise temperatures in highly urbanized areas by as much as 6 °C (Erell and Williamson 2007).

People working or needing to be outdoors, and those without access to cooling are another high risk group (Hanna, Kjellstrom, et al. 2010). Their risk is exceptional during extreme heat events, but they are also susceptible during hot days that are not extreme. Unless acclimatized and very fit, humans cannot withstand exercising in hot temperatures for any length of time, and the capacity to be productively active is halted. By the 2080s, population-based labor work capacity is predicted to fall by 11 percent to 27 percent due to rising heat (Kjellstrom, Kovats, et al. 2009). Additional warming to areas of the world that are already warm is deeply concerning. To achieve the same productive output, workers may need to work longer hours, or more workers must be employed. Employers will need to consider economic costs of lost production and/or implementation of occupational health interventions protective against heat exposures. People remunerated by output, and those not at liberty to self-pace may be pressured to continue working beyond their thermal tolerance limit. Chronic heat exposure takes a heavy toll where workers are chronically dehydrated, and the result is diminishing health, performance, and cognitive ability, which further impoverishes. Global warming therefore presents both economic and health threats.

Heat stress injuries range from mild through to organ failure and death. Heat is dangerous because the victims of heat stroke often do not notice the symptoms, and therefore do not take steps to alleviate the risk. This means that employers, colleagues, family, neighbors, and friends are essential for early recognition of the onset of the conditions. The affected individual's survival depends on others to identify symptoms, provide assistance, and to seek medical help. Early signs can include disorientation, confusion, euphoria, or an unaccountable irritability or complaints of a general malaise. The skin is moist, the pulse rate is typically rapid and weak, and breathing is rapid and shallow.

If left untreated, heat exhaustion can progress to heatstroke, a serious, life-threatening condition characterized by a high body temperature (>39.4 °C [>103 °F]). Signs and symptoms include skin that is red, hot, and dry (sweating has ceased); rapid, strong pulse; throbbing headache; dizziness; nausea; confusion; and eventually unconsciousness (Parsons 2003). Symptoms can progress to multiple organ system dysfunction and death. Prevention is the most effective treatment, by allowing the person to rest in a cool environment, with access to food and plenty of fluids (1–2 liters/hour).

In 2009, temperatures in the outer suburbs of Melbourne (Australia) reached 48 °C. Associated with winds over 100 km/hour, eucalyptus wildfires reached an intensity never witnessed before, burning over 2,000 homes, over 60 businesses, and causing the loss of 173 lives (Department of Human Services 2009). The pattern of intense fires following heat extremes and droughts is similarly affecting many other countries, and carries long-lasting effects on infrastructure and business, including tourism.

Severe mental health effects can follow the trauma of seeing charred landscapes that were once beautiful forests. The term 'solostalgia' has been coined to describe the distress caused by environmental change (Albrecht, Sartore, et al. 2007) which can arise following storms, floods, fires, or droughts. Loss of family, friends, and livelihoods, especially when there appears to be limited opportunities to recover, are predictors of mental health issues. Children witnessing these events and the impact on their supporting adult generation will inevitably be impacted as they ponder their uncertain future (Hanna, McCubbin, et al. 2010).

4 POLICY RESPONSES TO HEAT

With certain notable exceptions, such as the mining industry and the military, Occupational Health and Safety legislation and guidelines have given limited attention to protecting workers against excessive heat exposure. Consideration is particularly needed to protect the health of workers in the emergency and essential services during heatwaves (Hanna, Kjellstrom, et al. 2010). Public health responses depend on the ability of fire-fighting, police, ambulance, caring, and nursing sectors to function at maximum capacity and efficiency during the most adverse conditions for extended periods of time.

Many nations are rolling out heat warning systems (HWS) but they are not sufficient by themselves to prevent poor outcomes, especially among low-income, minority, and politically marginalized groups with fewer economic opportunities. For example, old and poor housing stock is less likely to have effective cooling. The poor do not have resources to modify their homes, nor incentives to adapt; they may also be less likely to work in cooled environments, and be less able to travel to cooled public buildings. Inability to escape the heat increases their exposure and exacerbates health inequities. HWS need to be associated with a platform of health protective actions across a range of sectors, which must be well considered, and locally feasible. Some actions involve national or state polices, such as national health promotion campaigns, and others can be enacted at the local scale, but national programs are often needed to spearhead their adoption.

Climate change will likely reinforce and amplify current as well as future socio-economic disparities leaving the disadvantaged with greater health burdens. The US CDC is focusing on increased heat events to reduce the impacts of heat events on vulnerable populations (CDC 2009). Technical responses such as evacuations to cooling centres are complex. Preconditions to development of proactive policies are, first, an acknowledgement that heat events are a current and increasingly future risk, that is, the will to act, and secondly, sufficient economic capacity to initiate strategies. Environmental justice movements can assist to motivate governments to invest in health protection, and public infrastructure and support services. Lessons can be learned from Australia when in the Victorian 2009 heatwave, train lines buckled in the heat and peak electricity demand led to power blackouts which interrupted air conditioning, caused traffic chaos with traffic lights affected, and compromised emergency services in their attempts to reach people in need. Hospitals and shopping centers were inundated (overwhelmed) by people seeking a cool environment. An outpouring of community reaction prompted government action.

5 EXTREME WEATHER EVENTS

Climate change is predicted to bring extreme rain events, and the trend is already obvious. Hot air absorbs more water and carries more kinetic energy, which increases the ferocity of storms. Insurance industries report that Category 5 storms are becoming more frequent globally. The greater destructive power of these stronger winds in cyclones, typhoons, and hurricanes generate significantly more injuries, widespread loss of homes, and greater damage to infrastructure and livelihoods. Economic costs of these events are high, and expenses such as repairing roads, bridges, and buildings can curtail government expenditure on proactive investments in capacity building and reducing inequities.

Heavy rainfall events create flooding and landslides. The hottest year to date, 2009–10, brought a series of catastrophes. Floods in Pakistan affected more than 21 million people, caused over 80,000 deaths, displaced 700,000 people homeless, destroyed crops, contaminated water supplies, and ruined food supplies. Russia's extreme heatwaves and peat fires near Moscow exceeded its response capacity, and generated dangerous air pollution. The destruction of over 20 percent of Russia's crops led to an export freeze which inflated global food prices, and diminish the amount of food that aid agencies can afford and distribute. Flooding and landslides in China in July 2010 affected over 29 million people and left one million people homeless. The reinsurance company, Munich Re reports that weather-related catastrophe losses have increased by 2 percent each year since the 1970s (Munich Re 2009). Typhoon Ketsana submerged 80 percent of the Philippine capital, affecting over 4.4 million people in Manila and neighboring provinces. In Vietnam, landslides and flash floods killed 162 people. In Cambodia and Laos, hundreds of families lost household items and food stocks to rapidly rising flood waters. Floods destroy crops and infrastructure, and by interrupting food sources and contaminating water supplies, also bring diseases such as cholera. Meanwhile women have special needs because they continue to give birth, and nurture children regardless of the dangers surrounding them.

6 FOOD AND WATER SECURITY

Increased food and water insecurity threatens millions with hunger and malnutrition. Long-term extreme drying trends are being observed in North and South America, Africa, the Middle East, China, and other parts of Asia. Sub-Saharan Africa has shown a marked decline in rainfall leading to an average decline in discharge of some watercourses in the range of 40–60 percent since the early 1970s (FAO 2010).

Water is also a health issue. Droughts diminish food yields, raise prices, and reduce water availability for drinking and sanitation. Climate change will exacerbate droughts, and yield more extreme precipitation events and more fierce storms, and will therefore interrupt food and water supplies. For millions of people these basic necessities for human life, food, water, and shelter are already being interrupted by increasing climate extremes.

Currently, almost 1 billion people live worldwide in arid lands; one-fifth of these are in Africa alone. Arid and dry semi-arid land are defined as less than 120 days length of growing period, where temperature and soil moisture conditions are favorable to crop cultivation. Arid and dry semi-arid areas in Africa are predicted to increase by about 11 percent, making it at risk of being severely constrained for crop agriculture.

More than 70 percent of current world population, 4.2 billion people live in the 80 poor food-insecure countries, and in 2010, over 1.2 billion people, more than one-sixth of the global population, are hungry and undernourished (FAO 2009, 2010). This hunger crisis is historically unprecedented, with several factors converging to make it particularly damaging to people at risk of food insecurity. These were partly climate/environment induced. Widespread droughts reduced global cereal yields and sparked the 2006–8 food crisis, which pushed the prices of basic staples beyond the reach of millions of poor people. Two years later, food commodity prices remain 17 percent higher. This protracted problem plunged an additional 172 million people into hunger, and forced many poor families to sell assets or sacrifice healthcare and education to purchase food (Myers 2010).

Among children and pregnant women, malnutrition has a multiplier effect, accounting for more than a third of the disease burden of children under age 5, and over 20 percent of maternal mortality. Diarrheal diseases arise from insufficient clean water or lack of sanitation, and diminish the body's capacity to absorb nutrients. People with insufficient access to nutritious food are more susceptible to diseases, and being deprived of sufficient clean water therefore doubly compromises their prospects for health.

The importance of the various dimensions and the overall impact of climate change on food security will differ across regions and over time. All 'normal' climate patterns are expected to change and alter local agricultural productivity, therefore every country will be exposed to climate change, and the net impact is expected to be negative. Perhaps most importantly, sensitivity to these impacts will depend on the overall socio-economic status of the country as the effects of climate change set in.

Some areas will have longer growing seasons, less snow, and more rainfall, however as farmers transition to novel production new pests and challenges will emerge. In some developed countries, notably parts of North America, the Former Soviet Union, and northern Europe (Fischer, Shah, et al. 2005) many reports suggest that some yields might increase. The United States produces 41 percent of the world's corn and 38 percent of the world's soybeans, two of the four largest sources of caloric energy produced and are thus critical for world food supply. In the US staples such as corn and soybeans can tolerate temperatures up to 30 °C, but temperatures above these thresholds are very harmful. Without shifting the current growing regions average yields are predicted to decrease by 30–46 percent under the slowest warming scenario (B1—International cooperation towards global sustainability), and decrease by 63–82 percent under the most rapid warming scenario (A1FI World markets and technology fail to deliver sustainable solutions, rapid economic growth fossil fuel intensive) before the end of the century (Schlenker and Roberts 2009). The immediate impact this will bring for wealthy countries is unknown. Inevitably, the poor will suffer most, especially farmers unable to transition to new crop varieties, or move to new growing areas. Communities depending on unsustainable agriculture will also be at risk.

An average global 2 °C rise in temperature could reduce cereal grain yields in South Asia and Sub-Saharan Africa by 5–20 percent. Southern Africa could lose more than 30 percent of its main crop, maize, in the next two decades, with possibly devastating implications for hunger in the region. Warming of more than 3 °C is expected to have negative effects on production in all regions (IPCC 2007a).

Elsewhere, different climate change impacts are also threatening food security. Some of the most productive agricultural areas, such as the Mekong Valley are delta regions which will become inundated by rising sea levels. Such regions covering 60–90 million hectares feed millions of people. Small Island States are also at extreme risk of losing their agricultural lands and water supplies, bringing into doubt their capacity to survive on their traditional homelands. The risk for them is livelihood, health, and culture.

The world's fish stocks are threatened by climate change. Ocean uptake of anthropogenic CO_2 alters ocean chemistry, leading to ocean acidification and reduction in calcium carbonate ($CaCO_3$) used to make the shells and skeletons of the marine food web (Doney 2010). Fishing is a major industry for coastal communities, and the absence of fish as a protein source risks the nutrition of millions of people worldwide. On a more positive note, fisheries do have a potential for autonomous adaptive capacity. Climate variability has occurred throughout history. Natural systems and fisheries have developed a capacity to adapt that may help them mitigate the impact of future changes. Marine life can move locations if conditions and food sources suit. Whether migration of sufficient elements of marine food webs can occur to sustain survival is yet to be seen.

Complex systems are not insular. Fish stocks may be challenged as climate change places extra pressure on marine food systems. Population growth and food shortages may drive more people to seek nourishment from the seas. Land-based food shortages may also drive more intensive farming patterns and chemical contamination of rivers, and interfere with water runoff. Multiple stresses can cause fish stocks to collapse.

Existing projections indicate that a future population of 9 billion people coupled with economic growth will require a doubling of current food production by 2050, including an increase from 2 billion to>4 billion tons of grains annually. This becomes problematic, as economic advancement has been accompanied by a shift in food consumption patterns from crop-based to livestock-based diets which are less effective in terms of land and water use per unit energy and protein produced. Against this rising demand, urban growth extends further into agricultural lands, and crop yields diminish through the effects of increased heat, extreme weather events, and water shortages, and world phosphate supplies are depleting. Irrigation, fertilizers, and pesticides, land-clearing, and new crop varieties addressed the world's last food crisis, but these strategies are now unadvisable, for they exacerbate environmental degradation. As the world continues to warm, the problem of feeding a growing population is expected to become a major humanitarian challenge.

7 INFECTIOUS DISEASES

The prospect that climate change could further escalate the global burden of disease due to infectious diseases is another source of growing concern.

Harm to health is less obvious via secondary effects from climate change, such as when diseases arise due to alterations in the ecology of vectors, parasites, and host animals (Butler and Harley 2010). Most studies show an increase in disease transmission results from alterations in environmental conditions. For example, vector biting rates are 300 times higher in deforested areas of Peruvian Amazon compared to intact forest. Building micro-dams and waterways are beneficial in securing water supplies, but also increase malaria vector habitat and infection rates by a factor of seven in nearby villages. Draining wetlands where mosquitoes breed can eliminate local sources of malaria but also reduces water quality and increases risk of diarrhea and water-borne diseases (Myers 2010). Increasing urbanization and the growth of urban slums that lack sanitation and clean water provide fertile ground for infections. Human forays into virgin areas of forests have brought us into close contact with animals and their pathogens. Pathogens can sometimes change hosts and infect humans who are not adapted to these zoonotic diseases, unlike the original animal host. Climate adaptive strategies therefore need careful planning.

Climate change is expected to impact vector-borne disease epidemiology. Temperature and rainfall patterns affect the distribution of disease vectors, and also reproductive cycles and biting frequencies. These determine vector capacity to transmit pathogens, such as those responsible for Malaria and dengue. Studies in the United States suggest that recent increasing temperatures contribute to the 35–83 percent higher incidence of reported West Nile Virus infection. With regard to future trends, the key factor for public health is the migration of vectors into areas previously free, where resident populations have yet to develop immunity, nor developed strategies to protect health and minimize harm.

However some conditions are reportedly less favorable to certain infectious disease vectors, so the area is attracting considerable debate in the literature with regard to net global impacts. For example there were concerns about a rise in malaria cases in the UK and the US, but there have been marked global declines in the disease and a substantial weakening of the global correlation between malaria endemicity and climate (Gething, Smith, et al. 2010). The Malaria Atlas Project funded by the Wellcome Trust study found preventative measures such as the widespread use of bed nets have outweighed the effects of climate warming on malaria. This example provides a positive demonstration of coordinated public health investments countering climate-induced health burdens.

In addition to affecting the health of individuals directly, infectious diseases impact whole societies, economies, and political systems. The loss of qualified personnel, most notably to human immunodeficiency virus (HIV)/acquired immune deficiency syndrome (AIDS), tuberculosis (TB), and malaria limits the capacity of crucial sectors involved in nation building and sustained development, such as health and education. These and other infectious agents not only take an enormous physical toll on humanity, but also cause significant economic losses both directly in the developing world and less directly in the developed world. AIDS, TB, and malaria are increasingly being acknowledged as important factors in the political and economic destabilization of the developing world.

8 COMBINED IMPACTS

Many of the impacts of climate change are interrelated. For example heat and droughts are often co-linked, along with fires and water shortages. Floods precipitate disease outbreaks such as cholera and other diarrheal disease, and also wreak damage to infrastructure and can disrupt food and water security. Regions will therefore be commonly exposed to multiple, consecutive, or sequential impacts. The interaction of these and associated effects may result in profound and unpredictable impacts. Reparation of buildings, bridges, agricultural lands, and breeding stock can take many years to be fully effective. In many instances, residual deficits to national infrastructure, livelihoods, social capital, and food supplies will linger. Responses to even quite gradual changes to climate may not be complete before another event unfolds. The risk, especially for communities with limited adaptive and recuperative capacity, is that climate change will exert constant downwards pressure on their ability to recover.

A consensus has emerged that developing countries are more vulnerable to climate change than developed countries, and despite their own negligible contribution to greenhouse gas emissions, their health outcomes are already decidedly worse (St Louis and Hess 2008). The most important risk factors for death and disability in poor countries differ strikingly from those in the developed world, because of the predominance of agriculture in their economies, the scarcity of capital for adaptation measures, their often warmer baseline climates, and their heightened exposure to extreme events. Poor communities have fewer resources to respond to climate change health threats such as increased natural disasters, food and water insecurity, and changing disease distribution. This existing disproportion in risk of being affected by weather-related natural disasters is almost 80 times higher in developing countries than in developed countries, and women are up to 14 times more likely than men to die from natural disasters (Neumayer and Pluemper 2007).

Developed countries cannot be complacent. They will not be spared, as indicated by extreme weather events such as the 2003 European heatwave, the drought, bushfires, and extreme heatwaves in southeast Australia, Hurricane Katrina, and many other events. Whereas it is impossible to say with certainty that any specific weather event is a result of climate change, the significant increase observed in both the frequency and intensity of extreme whether events—heatwaves, storms, and floods—is entirely consistent with climate change (IPCC 2007b).

9 CLIMATE CHANGE: DEEPENING GLOBAL INEQUITIES

The picture painted thus far focuses principally on the most vulnerable populations; many of these are already confronting the early effects of climate change. Unabated, global warming will significantly interrupt food supplies, water quality, and disease transmission to also significantly impact developed countries

Climate change is already bringing significant perturbations to human life support systems in many regions, notably among water and food security, in some cases with disastrous results. Post-disaster health effects persist long after the global attention moves on. Physical deterioration and chronic disease can launch a downward spiralling of health status, and family economic well-being, and produce intergenerational effects from childhood stunting and learning deficits. Mental health deterioration can also follow these events (Oxfam 2009). Combined, these can damage social functioning and group or family cohesion, which can then drive further deterioration of living conditions.

A healthy population is key to high agricultural productivity and hence food security. Reductions in food production and nutrition can lead to higher rates of malnutrition and susceptibility to other diseases such as HIV/AIDS, which in turn affects agriculture through loss of labor, knowledge, and assets (World Bank 2008). Hunger and malnourishment therefore creates a vicious circle of disadvantage. Their poverty trap acts as a *positive feedback loop*. This occurs when a system responds to a perturbation *in the same direction* as the perturbation, thereby exacerbating the original impact. A positive feedback loop can run out of control, and result in the collapse of the system (Hunter 2007). Climate change exacerbated hunger can tip populations into a state where they have not the human resources to grow the food they need to survive. As the majority of people in developing countries employ labor-intensive methods of food production, any changes in the health status of the community as a result of climate change is most likely to also affect future food security.

Climate change has therefore major repercussions for the social determinants of health. Inequalities in these determinants are the major cause of health inequities within countries and between countries. Socio-economic status determines where one lives, and therefore can influence the environmental determinants (Lin 2008). By acting on these social causes, climate change could further entrench a vicious circle of disadvantage by greatly exacerbating global health inequities among current generations and establishing a profound intergenerational inequity (Walpole, Rasanathan, et al. 2009).

Although vital for global health, strategies to reduce carbon emissions will not necessarily improve health equity. Ill-considered policies such as flat pricing mechanisms could easily be regressive in terms of income inequality by increasing the price of essential items. Low-income households, even those in developed countries, are struggling to afford basics. Policies that exclude consideration of impacts on the poor, by means such as subsidies, will result in them paying a proportionally greater financial burden, thereby increasing income inequality and worsening health inequities.

Climate change therefore has the potential to further widen the gap between rich and poor. Countries that are existing at the edge of survival have minimal resilience to withstand additional shocks. By the 2080s, the total population of *food-insecure* countries is projected to increase to 6.8 billion, or about 80 percent of the world population at that time (Fischer, Shah, et al. 2005). Such disparities have the capacity to destabilize social structures in rich and poor countries. Environmental stresses caused by climate change exacerbate competition for soil and water resources; they place high demands on disaster risk management and on planning and systematically prioritizing the allocation of limited public funds. Disrupted markets, shortages of goods, and strained social structures can exceed the capacities of struggling governments to maintain services and order within and beyond their localities. When a disaster occurs, there is the risk of collapse of public order, especially in megacities.

Underlying determinants of health inequity and environmental change are signs of an economic system predicated on asymmetric growth and competition, shaped by market forces that disregard health and environmental consequences or values of fairness and support (Friel, Marmot, et al. 2008). Climate-induced insecurity poses a population health threat. Adaptive health-maximizing strategies are needed that narrow the gap, and lessen the within country and between country disparities in access to resources, opportunities, and health.

10 ADAPTATION

Adaptation landscapes are fundamentally different in rich and poor countries, and this difference also has implications for equity and for global health policy (St Louis and Hess 2008).

The Least Developed Countries Fund (LDCF) was established under the United Nations Framework Convention on Climate Change to address the special needs of the 48 Least Developed Countries (LDCs), regarded as especially vulnerable to the adverse impacts of climate change. Support included preparing and implementing National Adaptation Programs of Action (NAPAs) to identify urgent and immediate needs of LDCs to adapt to climate change. While the NAPA process is aimed at achieving procedural justice, in practice often both procedural and distributive justice concerns are unresolved because of lack of trust, local or national political tensions, lack of capacity of locals, and corruption. The refusal of the US to contribute to the LDC Fund served to slow down the process of assisting vulnerable nations (Global Environment Facility 2008).

As many health determinants are social and environmental in origin, many health maintenance strategies lie outside the formal heath sector. For example the provision of safe food and safe water is a health necessity, and is therefore a key health issue, although not regarded as health sector activity. Climate change threatens human survival, initially primarily in the developing countries, and ultimately globally (Schneider 2009). The vulnerability of those suffering current or potential food, water, and shelter insecurity should therefore be encapsulated into all policies across government. For example, agricultural policies should be designed to develop capacity to ensure continuous affordable food supplies can be maintained during the immediate period post disaster events, as well as for the longer term. Strategies can include education and practical assistance, such as seeds, to produce food that can flourish in the 'new' local climate. This includes ensuring women have rights to land tenure, and to form cooperatives. In urban areas, establishment of community gardens can also assist food security, this also applies in developed countries (Dixon, Donati, et al. 2009).

Adaptation strategies within the health sector include augmenting public health infrastructure. The key elements of population health are promoting health and well-being; by providing the supports for people to self-manage their own health, and the health needs of their families. Water security must therefore take high priority in water scarce lands. Options to boost self-sufficiency and resilience therefore include installation of community water catchments, reuse and recycle infrastructure, community water storage capacity, and providing education on requirements to manage and maintain clean water systems. Renewable energy sources are necessary to support pro-health activities. Health services include bolstering disease surveillance and prevention and treatment of infectious diseases. Disease prevention includes strategies for families to avoid contact with contagion, building of sewerage systems, education to boil water during disease outbreaks, and provision of facilities to heat water.

Preparedness also includes capacity building to cater for surge in health care demands from heatwaves, fires, floods, and storms, and catering to the recovery phase, sheltering the homeless, and providing food, water, protection, and counselling services. The health sector was slow to embrace climate change, and recognize the health threats, so there is much yet to be done in terms of preparation, skills building, training the health professions, and educating the public. Vulnerability assessments and monitoring and surveillance systems will need to be modified to ensure programs remain effective under a changing climate (Ebi 2009).

Another category of positive adaptation is promoting strategies that bring *co-benefits*. Co-benefits occur when one action carries an additional secondary positive result. Shifting to bicycles as a primary urban transport mode reduces car mileage and reduces carbon footprint. Side benefits are numerous: a more amenable environment, cleaner air, and health benefits are gained from physical exercise, such as a reduction of cardiovascular disease, diabetes, and obesity.

The condom should be classified as a powerful climate change adaptive strategy, promoted as reproductive rights rather than as a form of population control. Family planning can be made available as a method of empowering families to control the number and timing of children and help reduce unwanted pregnancies. The health advances in childhood survival rates and maternal health gained from access to effective family planning are well established. Global availability of the condom could reduce population by 1–2 billion by 2100, thereby reducing demand for limited resources, and ultimately, greenhouse gases (AAAS 2009).

Failure to mitigate against and to adapt to climate change is certain to bring human misery on a grand scale. Oxfam and Global Humanitarian Fund reports testify to the millions of people alive today who are suffering through extreme shortage of the basic necessities for life (Global Humanitarian Forum (GHF) 2009; Oxfam 2009). This will generate millions of refugees as people move in search of opportunities to feed their families. The conflict in Darfur has been attributed in part to climate change and environmental degradation (see Gilman et al., this volume). Climate change could be the catalyst for widespread social unrest that ultimately brings this species to its knees. In conflict-prone areas, the most vulnerable group is the refugee community, the bulk of whom are women and children (FAO 2010).

11 CONCLUSION

The burden of ill-health attributable to climate change is likely to aggravate, and in some cases even provoke, further economic decay, social fragmentation, and political destabilization, especially in (but by no means restricted to) the developing world and countries with unstable governments.

In the face of little progress on mitigation strategies, it becomes increasingly imperative that focus is given to adaptation strategies on preparing for the health issues to come, and growing capacity to protect health, and provide safe environments, food, water, and shelter. Population health services focus on reducing inequities, and promoting and maintaining health. In essence, boosting climate change adaptive capacity therefore entails augmenting population health capacity; as these strategies build preparedness of communities to withstand health challenges such as climate stress. Such approaches also constitute 'no-regrets' polices, as they provide and promote basic human rights, and if strengthened, could deliver immediate benefits by minimizing today's health burden. Without such adaptation, community resilience will remain low, and human misery on a scale not yet seen will unfold.

REFERENCES

AAAS 2009. *Science and Technology Forum Panel: Addressing Climate Change Will Benefit Global Health*. American Association for the Advancement of Science.

ALBRECHT, G., SARTORE, G. M., et al. 2007. Solastalgia: The distress caused by environmental change. *Australasian Psychiatry: Publication of The Royal Australian and New Zealand College of Psychiatrists* 15(1 supp. 1): 95–8.

BEGGS, P. J. 2000. Impacts of climate and climate change on medications and human health. *Aust N Z J Public Health* 24: 630–2.

BUTLER, C. D., and HARLEY, D. 2010. Primary, secondary and tertiary effects of eco-climatic change: The medical response. *BMJ* 86: 230–4.

CDC 2009. Basic principles of healthy housing. Chapter 2 in *Healthy Housing Reference Manual*. Atlanta, GA: CDC.

CUETO, R. O. G., MARTÍNEZ AT, et al. 2010. Heat waves and heat days in an arid city in the northwest of México: current trends and in climate change scenarios. *Int J Biometeorol* 54(4): 335–45.

Department of Human Services. 2009. *January 2009 Heatwave in Victoria: An Assessment of Health Impacts*. Melbourne: Victorian Government.

DIXON, J. M., DONATI, K. J., et al. 2009. Functional foods and urban agriculture: Two responses to climate change-related food insecurity. *NSW Public Health Bulletin* 20(1–2): 14–18.

DONEY, S. C. 2010. The growing human footprint on coastal and open-ocean biogeochemistry. *Science* 328(5985): 1512–16.

EBI, K. L. 2009. Public health responses to the risks of climate variability and change in the United States. *J Occup Environ Med* 51(1): 4–12.

ERELL, E., and WILLIAMSON, T. 2007. The spatial variability of air temperature in the urban canopy layer. *2nd PALENC Conference and 28th AIVC Conference on Building Low Energy Cooling and Advanced Ventilation Technologies in the 21st Century*. Crete island, Greece.

FAO 2009. 1.02 billion people hungry. One sixth of humanity undernourished—more than ever before. *The State of Food Insecurity in the World, SOFI*. Rome: Food and Agriculture Organization of the United Nations.

——2010. Climate change implications for food security and natural resources management in Africa. Paper presented to Twenty-Sixth Regional Conference for Africa Luanda, Angola, 3–7 May 2010. Rome: Food and Agriculture Organization of the United Nations.

FISCHER, G., SHAH, M., et al. 2005. Socio-economic and climate change impacts on agriculture: An integrated assessment, 1990–2080. *Philosophical Transactions of the Royal Society B: Biological Sciences* 360(1463): 2067–83.

FRIEL, S., MARMOT, M., et al. 2008. Global health equity and climate stabilisation: A common agenda. *The Lancet* 372(9650): 1677–83.

GETHING, P. W., SMITH, D. L., et al. 2010. Climate change and the global malaria recession. *Nature* 465: 342–5.

Global Environment Facility. 2008. Least developed countries fund factsheet. Retrieved 12 May 2009 from <http://www.thegef.org/uploadedFiles/Publications/LDCF-factsheets.pdf>.

Global Humanitarian Forum (GHF). 2009. *Human Impact Report: Climate Change—The Anatomy of a Silent Crisis*. Geneva: Global Humanitarian Forum.

HANNA, E. G., KJELLSTROM, T., et al. 2010. Climate change and rising heat: population health implications for working people in Australia. *Asia Pacific Journal of Public Health*. In press.

——McCUBBIN, J., et al. 2010. Australia, lucky country or climate change canary: What future for her rural children? *International Journal of Public Health* 2(4): 501–12.

——McMICHAEL A. J., and BUTLER, C. D. 2011. Climate change and global public health: Impacts, research and actions. In R. Parker R and M. Sommer (eds.), *The Routledge International Handbook on Global Public Health*. Oxfordshire: Routledge.

HUNTER, B. 2007. Cumulative causation and the productivity commission's framework for overcoming indigenous disadvantage? Paper submitted for the Australian Social Policy Conference Social Policy through the Life Course: Building Community Capacity and Social Resilience.

IPCC. 2007a. Climate change 2007: Impacts, adaptation and vulnerability. Contribution of Working Group II to the Fourth Assessment Report of the Intergovernmental Panel on Climate Change. (Eds. M. L. Parry, O. F. Canziaani, J. P. Palutikof, P. J. van der Linden, and C. E. Hanson). Cambridge: IPCC.

——2007b. *Climate Change 2007: The Physical Science Basis. Contribution of Working Group I to the Fourth Assessment Report of the Intergovernmental Panel on Climate Change*. Cambridge and New York: IPCC.

KJELLSTROM, T., KOVATS, R. S., et al. 2009. The direct impact of climate change on regional labor productivity. *Archives of Environmental & Occupational Health* 64(4): 217–27.

KWOK, J. S., and CHAN, T. Y. 2005. Recurrent heat-related illnesses during antipsychotic treatment. *Ann Pharmacother* 39(11): 1940–2.

LIN, S. W. 2008. *Understanding Climate Change An Equitable Framework.* PolicyLink.

MCMICHAEL, A. J., Campbell-Lendrum, D. H., et al. (eds.) 2003. *Climate Change and Human Health: Risks and Responses.* Geneva: World Health Organization, World Meteorology Organization, UNEP.

Munich Re 2009. *Topics Geo. Natural catastrophes 2008: Analyses, Assessments, Positions.* Munich: Munich Re.

MYERS, S. S. 2010. *Global Environmental Change: The Threat to Human Health.* Worldwatch Report 181. Washington, DC: Worldwatch Institute.

National Climatic Data Center. 2010. *State of the Climate Global Hazards July 2010.* National Oceanic and Atmospheric Administration.

NEUMAYER, E., and PLUEMPER, T. 2007. The gendered nature of natural disasters: the impact of catastrophic events on the gender gap in life expectancy, 1981–2002. *Annals of the American Association of Geographers* 97(3): 551–66.

Oxfam. 2009. *Suffering the Science: Climate Change, People, and Poverty.* Copenhagen: Oxfam International.

Pakistan Meteorological Department. 2010. *Record breaking heat in Pakistan. Highest maximum temperature was recorded 53.5 °C in Mohenju Daro and 53 in Sibbi.* Islamabad: Government of Pakistan.

PARSONS, K. 2003. *Human Thermal Environment. The Effects of Hot, Moderate and Cold Temperatures on Human Health, Comfort and Performance.* New York: CRC Press.

RICHARDSON, K., STEFFEN, W., et al. 2009. *Climate Change: Global Risks, Challenges & Decisions.* Synthesis Report. Copenhagen, 10–12 March.

ST LOUIS, M. E., and HESS, J. 2008. Climate change: Impacts on and implications for global health. *American Journal of Preventive Medicine* 35(5): 527–38.

SCHLENKER, W., and ROBERTS, M. J. 2009. Nonlinear temperature effects indicate severe damages to U.S. crop yields under climate change. *Proceedings of the National Academy of Sciences* 106(37): 15594–8.

SCHNEIDER, S. 2009. The worst-case scenario. *Nature* 458(7242): 1104–5.

WALPOLE, S., RASANATHAN, K., et al. 2009. Natural and unnatural synergies: Climate change policy and health equity. *Bulletin of the World Health Organization* 87: 799–801.

World Bank. 2008. *World Development Report 2008: Agriculture for Development.* Washington, DC.

CHAPTER 16

INDIGENOUS PEOPLES AND CULTURAL LOSSES

ROBERT MELCHIOR FIGUEROA*

1 INTRODUCTION

In the international climate change discourse, distributive justice is often expressed through concern for intergenerational harms and measures needed to retrieve historical, present, and future compensation. Some common sentiment has identified First Nations, Indigenous Peoples, and a range of native populations[1] as the most vulnerable human communities to climate change. The Permanent Forum on Indigenous Issues and the Universal Declaration on Indigenous Rights echo what the United Nations Convention on Biological Diversity has made explicit: that 'Indigenous and local communities are among the first to face the direct adverse consequences of climate change, due to their dependence upon and close relationship with the environment and its resources (UNEP Convention on Biological Diversity 2007).' Of the distributive inequities in burdens and benefits, H. E. Miguel d'Escoto Brockmann, President of the United Nations General Assembly, admitted in his Statement to the 2009 Indigenous Peoples' Global Summit on Climate Change that 'climate change poses threats and dangers to the survival of Indigenous communities worldwide, even though they contribute the least to greenhouse emissions' (Galloway et al. 2009: 2). More extensively, the Preamble of the *Anchorage Declaration* drafted from the same indigenous summit elucidates, 'We are experiencing profound and disproportionate adverse impacts on our cultures, human and environmental health, human rights, well-being, traditional livelihoods, food systems and food sovereignty, local infrastructure, economic viability, and our very survival as Indigenous Peoples' (Galloway et al. 2009: 5).

Embedded in these climate justice accounts is the primary struggle of indigenous people to sustain their environmental identity and environmental heritage, in the face of threats to the physical resources that shape their living ecology and the threats to values, beliefs,

* I am deeply indebted to Kyle Powys Whyte for sharing his extensive knowledge and insights, as well as his invaluable suggestions offered throughout this chapter.

behaviors, histories, and languages. By *environmental identity*, I mean the amalgamation of cultural identities, ways of life, and self-perceptions that are connected to a given group's physical environment. And, my use of *environmental heritage* pertains to the meanings and symbols of the past that frame values, practices, and places peoples wish to preserve as members of a community. Environmental heritage is the expression of an environmental identity in relation to the community viewed over time (Figueroa 2006: 371–2). Both environmental identity and environmental heritage are concepts that can work in and across scales—local, national, global. These concepts are also unrestricted by ethnic identity, but they are most often interfused to cultural identities that track ethnic, historical, and other situation-specific identities. Indigenous environmental identity and heritage can be described and explored by virtue of the close, often inseparable, relationship that many indigenous communities have between environmental and cultural values and behaviors. Nonetheless, in the wake of climate change non-indigenous communities find commonalities with indigenous environmental identity and heritage.

In this chapter, a variety of experiences and philosophical reflections on cultural loss will be discussed under an environmental justice framework, wherein 'environmental justice' is broadly construed as the conceptual connections, causal relationships, and strong correlations that exist between environmental issues and social justice. Environmental justice frames social issues (including cultural contexts and political economies) as environmental issues. Social and environmental issues are inseparable, co-causally related, and always in a context that requires a political interpretation; in particular, such a consideration of justice accounts for power dynamics and socio-environmental practices that maintain historical relations, as well as the remedies for injustices. The primary argument of this chapter is that the environmental justice framework is a proper theoretical and practical approach to understanding the cultural loss among indigenous peoples caused by climate change. Specifically, I employ an interpretative lens through which several dimensions of justice—distributive, recognition, participatory, and restorative—can be better enjoined to handle the extent and complexities of justice as it pertains to the cultural losses of both indigenous and non-indigenous peoples across the globe. Under this framework, considerations of traditional environmental knowledge, international policy, transformations of knowledge in climate adaption and mitigation, relocation and loss of place, and general prescriptions for environmental justice are brought to bear on the trauma of cultural loss.

2 Cultural Loss and Environmental Justice

2.1 Indigenous Social-Ontology and Environmental Colonialism

The social-ontology of indigenous peoples is one historically linked to colonialism, and can be traced to the cultural losses of recent centuries driven by global exploration and knowledge transfer. As numerous indigenous scholars, such as Donald Fixico (1998), Donald A. Grinde and Bruce E. Johansen (1995), Winona LaDuke (1999), and Jace Weaver (1996), have described in explicit detail, the environmental resources, environmental knowledge, and subsequently the plight of environmental heritage for indigenous peoples was inspired by the designation of common heritage, which, ironically, invited the colonial exploitation and appropriation of natural resources found in indigenous lands. Hence, *environmental colonialism* best describes the environmental injustice common to most historical relations between indigenous and non-indigenous peoples since 1492 (Kleinman 2005).

In one sense, indigenous peoples have existed throughout human history, and in the abstract ontology of deep anthropological time, everyone links back to indigenous peoples. But does that make us all indigenous in the current social-ontology, the one in which climate change is an anthropogenic cause for cultural loss? No; indigeneity pertains to a dynamic set of non-essentialist, non-relativistic, historical, and self-identifying features that compose individual and community experiences. Cultural loss hangs in an odd balance for this social-ontology. On the one hand, cultural loss by assimilation and by agent-driven transformation is a fact of humanity's existence through deep anthropological time. Cultures survive by their abilities to change. Sometimes changing culture is an intended improvement and sometimes a more powerful society dominates another into changing its culture, perhaps intended by dominant actors as an improvement, albeit informed by prejudices against the subjected party. Alternatively cultural loss could be the aim for positive revolution. But, analogous to concepts of risk, when cultural loss is self-initiated it is quite different than other-initiated risk. Even apparently self-initiated loss could reveal a cascade of historical events in assimilation and cultural elimination that limits the survival of vital traditional lifeways. For instance, Australian aboriginal knowledge that has intimately connected environmental and genetic knowledge may be available only to elders who have gained such knowledge under strict rites of passage; in some cases the failure of non-elders to reach a passage causes the complete loss of this knowledge (Grasshoff et al. 1988). It is true that cultural loss is common in the history of humanity, and not always a bad thing, but there are important philosophical nuances that make the distinction between cultural injustice and cultural transformation. I am concentrating on the cultural loss that is compounded by overt and subversive injustices, as opposed to cultural transformation that encourages avenues for restoration of agency and environmental justice. More could be explored on the ontological conditions of indigenous peoples, in order to avoid any romanticizing or mythologizing the indigenous identity. My point is simply that we should avoid sliding the current social-ontology of indigenous peoples into abstractions of deep anthropological time. Instead, we should focus on the unwanted cultural losses faced by self-identified indigenous communities, especially given their extensive vulnerability to climate change.

2.2 An Environmental Justice Framework

Due to the dominance of the distributive justice paradigm in Western philosophical and legislative practice, especially regarding meanings of 'fairness' and legal interpretations of harms and punishment, it is nearly impossible to offer an appropriate definition of 'environmental justice' without some framing of commensuration between environmental burdens and benefits, or the distributive inequities of such burdens (United Church of Christ 1987, 2007; Bullard 2000; Figueroa 2003; Figueroa and Waitt 2008; Schlosberg 2007;

Shrader-Frechette 2005; Bryant 1995; Jamieson 1994). Climate justice has been wedded to environmental justice in accounts of intergenerational justice, distributive inequities, and active contribution to the causes of climate impacts. However, the embedded cultural dynamics of the above-mentioned UN conventions and global summits point to fundamental linkages between political participation and the cultural side of environmental justice. The references to 'traditional livelihoods' and 'dependence upon and close relationship with the environment' speak directly to issues concerning environmental identities and heritage; these, in turn, reflect upon the recognition justice paradigm. Thus, environmental justice is a 'bivalent' form of justice, requiring both distributive and recognition justice bridged by participatory forms of procedural justice. This is a widely accepted theoretical perspective from environmental justice scholars (Hunold and Young 1998; Peña 2005; Cole and Foster 2001; Figueroa 2003, 2006; Figueroa and Waitt 2008; Schlosberg 2007; Shrader-Frechette 2005).

In the myriad of environmental justice movements, this kind of bivalent framework has many representatives, from Lois Gibbs in Love Canal to Ken Saro-Wiwa in Nigeria, and from South Bronx to South Africa. Specific reasons for including the recognition paradigm vary according to specific theoretical architecture and historiographical conditions, but the explicit reasoning that is offered across the board is that conceptions of justice based solely in distribution and compensation are simply unable to provide a full description or full remedy for environmental injustices. This bivalence is particularly evident in indigenous peoples' claims requiring consideration of impacts upon the cultural and participatory features of justice that cannot be resolved by a strictly distributive framework. For instance, Native philosopher Dale Turner (2006) devotes his volume to documenting numerous indigenous writers who describe the critically fundamental aspects of the politics of recognition as part of the appropriate paradigm of justice. Identifying and repairing disparate distributive impacts according to a commonly shared human-environmental health metric is insufficient to *recognizing* both the community-specific losses that arise when traditional lifeways are dramatically disrupted and the need for solutions that flow from the community's agency decision making. Thus, the most promising environmental justice framework is one that simultaneously addresses distributive and recognition justice, including the subcategories of justice that fall under these paradigms.

2.3 Backgrounding Environmental Colonialism

A more useful discussion can be gleaned from Henry Shue's (1992) concept of *background injustices,* such as colonial practices of resource exploitation, relocation, land appropriation, and persistent economic exploitation, which compound an even greater moral crisis for indigenous peoples. For indigenous peoples, the legacy from environmental colonialism includes historical under-representation in environmental decision making and the gross historical distributive inequities in consumption and production. The argument continues that the terms distributive and procedural agreement are historically corrupted by the fact that nearly every sustaining indigenous population affected by climate change is affected by multiple magnitudes of background injustices. As a form of environmental colonialism, the background conditions capture the causal roots of precisely why indigenous groups are the most vulnerable and impacted by climate change. For instance, the background struggles over sovereignty and self-determination play into the environmental colonialism of climate change. Indigenous communities must be able to exercise sovereign jurisdiction over their territories in order to maintain their lifeways and to protect them from overt and inadvertent non-indigenous peoples' colonial presumptions. Environmental colonialism continues to compromise indigenous jurisdictional authority and exclude indigenous leaders from participating on policy making about the very environments upon which their communities depend (Robyn 2002).

In the United States, for example, tribes are subject to the plenary power of the US Congress to determine matters as intimate as their political status and membership (Cohen 1971). Tribes that require good water quality for religious, cultural, and subsistence uses previously had no control over the discharges that came from non-reservation point-sources, as that policy was determined in Washington, DC. In the 1980s, when amendments were made to the US Clean Water Act, the solution was to 'treat tribes as states,' which would allow them some control in setting standards that fit their cultural, religious, and subsistence uses. But, of course, whether a tribe qualifies as a state depends upon the US Environmental Protection Agency's contrived criteria for determining a tribe's fitness for such treatment. Subsequently, the state-status language is received suspiciously by Indian nations and tribes, since indigenous participation 'as states' has the circular obstacle of achieving self-determination: the direct and absolute political power and participation over tribal matters (Ranco and Fleder 2005; Tweedy 2005; Suagee 2005). Climate policy in the US reflects a similar problem. For instance, the 2009 Waxman-Markey Bill is designed to generate funds for improved energy efficiency. Section 202's program to retrofit existing buildings for energy efficiency mentions state governments but not tribal governments, which raises suspicions over whether the climate change policy (if passed) will arrive in Indian country and be administrable by tribes (Cordalis and Suagee 2008; Suagee 2009).

Thus, environmental colonialism, as it pertains to the legacy of environmental injustice, is fused to many climate change impacts, both in terms of negative impacts and any positive ones, since indigenous peoples are part of a global economy that will include carbon-emitting technology (cars, boats, snowmobiles, planes, etc.). Many indigenous communities have assimilated to or adopted the dominant colonial cultures and/or political economy of the modern global market. Indigenous economies range the entire spectrum, where casinos dominate the political economy in some and nomadic hunting-gathering economies drive indigenous practices in others. Nothing in this vast spectrum entails that indigenous peoples desire to completely abandon their environmental heritage or traditional lifeways.

2.4 Participatory Parity, Adaptation, and Mitigation

Direct and robust participation in the decisions that affect a people is a matter for participatory justice, or parity, which serves to bridge the distributive and recognition dimensions of bivalent environmental justice. Thus, we find in the *Anchorage Declaration* from the 2009 Indigenous Peoples' Global Summit on Climate Change stipulating obligations upon high carbon-emitting parties, responsible nations, corporate agents, and international agencies to proactively ensure recognition in participatory parity:

4. We call upon the UNFCCC's decision-making bodies to establish formal structures and mechanisms for and with the full and effective participation of Indigenous Peoples. Specifically we recommend that the UNFCCC:

> e. Take the necessary measures to ensure the full and effective participation of Indigenous and local communities in formulating, implementing, and monitoring activities, mitigation, and adaptation relating to impacts of climate change. (Galloway et al. 2009: 6)

And, to proactively engage the distributive dimensions of participatory parity:

> 7. We call for adequate and direct funding in developed and developing States and for a fund to be created to enable Indigenous Peoples' full and effective participation in all climate processes, including adaptation, mitigation, monitoring and transfer of appropriate technologies in order to foster our empowerment, capacity-building, and education. We strongly urge relevant United Nations bodies to facilitate and fund the participation, education, and capacity building of Indigenous youth and women to ensure engagement in all international and national processes related to climate change. (Ibid.)

Participatory parity is needed for fair agreements and appropriate epistemological representation between indigenous and non-indigenous, low carbon-emitting and high carbon-emitting, actors. Compensation, mitigation, and adaptive strategies may severely compromise cultural welfare and survival if parity in the decision-making process is ill-conceived. In climate science and policy, indigenous knowledge may have strong adaptive prospects and thus strong opportunities for indigenous participatory parity; but, mitigation still gets pared away as a technoscientific preference of dominant political economies. However, mitigation is more than justifying technological advances, since it simultaneously involves observation and prediction. Vital observations in weather changes, ice melt, sea level, and predictions for response are regularly made by local indigenous peoples (Leung 2005). Participatory parity in adaption and mitigation would resemble the Nunavik Research Centre, an organization that responds to climate change, whose strategies include involving indigenous residents, non-indigenous scientists, participatory research, a method for addressing issues raised by residents, and checks along the process that are reviewed by elders (Woodard 2005).

3   TRADITIONAL ENVIRONMENTAL KNOWLEDGE

3.1   Traditional Environmental Knowledge and Climate Change

Article 8(j) of the UN Convention on Biological Diversity focuses specifically on traditional knowledge and indigenous peoples:

> '[T]raditional knowledge' refers to the knowledge, innovations and practices of indigenous and local communities, developed and shared through experience gained over time and adapted to the local social structure, culture and environment. *Such knowledge tends to be collective in nature.* It is usually communicated through indigenous peoples' way of life, stories, songs, folklore, proverbs, cultural and religious values, beliefs, rituals, customary laws, practices and traditions, languages and other ways of transmission. This knowledge is normally of a practical nature, and covers areas such as traditional livelihoods, health, medicine, plants, animals, weather conditions, environment and climate conditions, and environmental management.

Moreover,

> Such knowledge is not merely a collection of facts and observations; it includes analysis and understanding of the subject matter from a practical perspective. Consequently, adverse external impacts on indigenous and local communities' way of life, social structures, culture and habitat will also affect their knowledge, innovations and practices. (UN Convention on Biological Diversity, Article 8(j), 2007)

The Convention's use of 'indigenous *and* local communities' is worthy of attention in discourse about cultural loss, yet many indigenous peoples would argue that these definitions of knowledge are not appropriate. Many Western understandings of traditional knowledge simply see it as a repository of information that can be accessed (Berkes 1999; Johnson 1992; Nakashima 1993; Callicott 1994).

Anishnabe scholar Deborah McGregor (2004, 2008) advances a more aboriginal understanding of traditional knowledge for non-indigenous environmentalists, which she refers to as *traditional ecological knowledge* (TEK):

> TEK is viewed as the process of *participating* (a *verb*) fully and responsibly in such relationships [between knowledge, people, all of Creation (the 'natural' world as well as the spiritual)], rather than specifically as the knowledge gained from such experiences. For aboriginal people, TEK is not just about understanding relationships, it *is* the relationship with Creation... Equally fundamental from an Aboriginal perspective is that TEK is *inseparable from the people who hold it... This* means that, at its most fundamental level, one cannot ever really 'acquire' or 'learn' TEK without having undergone the experiences originally involved in doing so. This being the case, the only way for TEK to be utilized in environmental management is to involve the people, the TEK holders... Once separated from its original holders, TEK loses much of its original value and meaning. (McGregor 2008: 145–6)

Here, TEK cannot be separated from environmental identity and environmental heritage and the performed lifeways (Hester et al. 2000; McGregor 2004; Berkes 1993). Even under the continuous threats of assimilation many indigenous peoples continue to sustain some significant ethos of environmental identity and heritage that remains embedded in the environmental imagination: the epistemological and phenomenological horizon of TEK. However, TEK is not to be equated with static knowledge. The lifetime of TEK depends upon factors of functionality, human relations, predictability, explanation, religious value, and a host of other applications. New challenges brought on by the remarkable speed of some climate impacts require that TEK must often undergo transformations or lose its most effective practicality for sustaining its peoples. The efforts to preserve TEK and address climate vary from the steadfast Traditional Seminoles of Florida (LaDuke 1999) to emerging programs, such as Oglala Lakota Community College's program for students to examine the effects (and future effects) of climate change on traditions of gathering medicinal herbs, vegetables, and berries; or the Confederated Tribes of the Umatilla Reservation's early efforts to anticipate climate change threats to traditional foods that have grown wild since time immemorial (LaDuke 1999; Melmer 2007; Associated Press 2008). There are at least six lessons that can be gleaned from this discussion: (1) Traditional indigenous knowledge is a lived experience rather than a storage of information; (2) Indigenous traditional knowledge is inherently wedded to the surrounding ecology and embedded in an ecological history; (3) Such knowledge is currently retrievable and capable of being sustained in present indigenous communities; (4) Climate change poses extensive damage to the ecological relationships that bind TEK, which entails that climate change threatens the ability to sustain the knowledge and culture; (5) Embedded within TEK are substantial insights for observation and prediction of climate change impacts—extreme weather events, protecting biodiversity, and maintaining cultural survival; and (6) TEK is therefore valuable to the cultural survival of non-indigenous peoples and should be used to forge a wider epistemological spectrum about climate change and political agency. Effective knowledge sharing will occur only if indigenous communities experience self-determination over what, how, where, and with whom this vital knowledge is shared.

To further explore effective observation and prediction, as well as the exceptional vulnerability of TEK, consider the devastation of the 2004 Tsunami in the South Pacific, which left over 100,000 dead, extensive casualties, and unimaginable property destruction. Yet, in the epicenter several groups of extensively long-existing indigenous communities were found to be with minor, if any, casualties. The only available explanation for the high survival rate in the tsunami's epicenter is the long-term accruing of TEK about the warning signs of extreme weather events (Leung 2005). Local news observations, confirmation by recovery efforts and census counts, and even network journalism, such as *60 Minutes*, reported that the Moken people, a nomadic indigenous people of Thailand, observed signs in wind and tides, as well as insects and other land and sea animals that offered reliable evidence for predicting the stories describing 'hungry seas.' The TEK response to move to forests and higher ground proved quite effective (Leung 2005). In the *60 Minutes* report,

> The Moken has a legend that is passed from generation to generation about the Laboon, the 'wave that eats people.' It is believed that the angry spirits of the ancestor brought the tsunami. The myth tells that, before the giant wave comes, the sea recedes. Then the waters flood the earth, destroy it, and make it clean again. On these islands the cicadas are usually loud, but suddenly went silent before the tsunami hit. Saleh Kalathalay (an interviewed Moken man) noticed the silence and warned everyone about the tsunami. The Moken started to flee toward higher ground long before the first wave struck and were saved. (Leung 2005)

Sustained knowledge of tsunami behavior in traditional explanation (that the sea is hungry) and the reliable reaction to TEK gives indigenous peoples an opportunity to reaffirm their heritage by adapting to climatic threats.

3.2 Sharing Knowledge

The *Anchorage Declaration* (2009) closes with a final clause,

> We offer to share with humanity our Traditional Knowledge, innovations, and practices relevant to climate change, provided our fundamental rights as intergenerational guardians of this knowledge are fully recognized and respected. We reiterate the urgent need for collective action. (Galloway et al. 2009: 6)

What are appropriate conditions of environmental justice for sharing knowledge? A fine line triangulates the dire need for shared knowledge, collective action, and vulnerability by assimilation. The reality of such vulnerability is exemplified by a comparison between the tsunami survivors and their tribal cousins on neighboring islands. The latter were much more assimilated to Western economies, religions, and tourism. Their populations were totally annihilated by the tsunami (Devraj 2005). Moreover, TEK, genetic and family group knowledge may not be easily pulled apart, since the peoples have been a vital element in the ecological web for a good long time (Robyn 2002). Meanwhile, the survivors living in secluded resistance are subject to the pursuit of those who desire to collect extensive TEK and experiential information—in any form (oral, written, experiential, empirical, historical, mythological) and about nearly any ecological connection (winds, tides, animals, foods, descendants, archival practices). At the same time, they are highly sought after by researchers who desire to study the genetic pool dating ancestry and knowledge at least 20,000 to 40,000 years back (Devraj 2005). If we are convinced that TEK is valuable to climate science and policy responses, we must remain cognizant of the grand cultural stakes that indigenous peoples face in this shared endeavor. Severe anthropogenic threats have confronted indigenous peoples over colonial histories and climate change is another anthropogenic threat caused largely by those former colonial powers. Self-determination and agency over the paths of traditional and ecological knowledge must be respected against the background of injustices wherein knowledge appropriation was a vital component of colonial oppression.

3.3 Endangered Languages

Language loss is another extraordinary threat to cultural sustainability and ensures a clear and direct move towards cultural assimilation. Comparatively many non-indigenous and dominant world languages already frame an environmental imagination in which relationships to plants and animals experienced by indigenous peoples are less significant, or non-existent (Kassam 2008). For the self-identifying indigenous peoples who desire to sustain environmental identity a language that no longer refers to an existing physical manifestation limits available cultural options: assimilate, lose a history, or sustain the cultural imaginary by readapting the place in cultural memory and environmental imagination.

Yet, even in the face of some language loss, other terms may reemerge in the environmental imaginary. Shari Gearhead (2005) discusses an oral history project led by Shari Fox and Ruby Irngaut in which, 'many Inuit have shared other interpretations and meanings of *Uggianaqtuq* with me such as a reference to people fighting, tension, extreme heat, something unseasonable or untimely, and the root of the word refers to a dog taking something in its mouth and shaking it' (Gearhead 2005: 1). Although not a very common expression, *Uggianaqtuq* is a term more widely found in the various media capturing the climatic impacts and cultural transformation. While indigenous survivors of the tsunami were able to draw appropriate responses from a fairly common environmental knowledge, these Inuit are drawing from a far less common term, though nonetheless appropriate for describing uncertain environmental future and impacts. In sum, TEK suffers vulnerability from several consequences of climate change. The background injustices and anthropogenic threats from colonial practices have had a history of its own to chisel away at sustaining traditional lifeways, vital environmental knowledge, cultural identity, and living languages. The acceleration of climate impacts subsequently exacerbates earlier injustices and threatens cultural loss on multiple fronts.

4 RESTORATIVE JUSTICE AND CLIMATE CHANGE

4.1 Climate Refugees

We are already witnessing one of the ultimate expressions of compounded climate impacts as cultural losses are seen in the total displacement of climate refugees. In the South Pacific, over 2,500 Carteret Islanders have been relocated to Papua New Guinea, and in the attempt to avoid complete loss of environmental identity residents of Tuvalu will be relocated to New Zealand. These relocation efforts dramatically compromise the cultural heritage of place and create remarkably different forms of socialization, environmental valuation, and environmental identity. Haulangi relates her current self-perception as a recently arrived climate refugee to New Zealand:

> What really concerns me—because, like, at the end of the day—is I may be a Kiwi now, call myself a Kiwi, 'cause I'm living in New Zealand. But hey, people will still look at my color and go 'hey, where are you from? Which island?' And I'll say, 'Oh, I'm from Tuvalu.' They'll say 'and where is that?' What shall I say, 'oh, it has disappeared or submerged under the sea because of global warming?' So, like that's our identity, our culture. Everything will disappear. We may get together here as a community and celebrate when it's Independence Day, our successes and things, but it's different. Definitely, it's going to be really hard for us to accept that we're no longer on the map. (Berzon 2006)

Where goes the environment, so goes the culture. Haulangi describes a complete loss of place, a loss of the referent for her culture, history, and environment. Ecopsychologists regard this complete shift of environmental identity by loss of place to be a form of post-traumatic stress disorder (Khanna 2010). The distress of place-based nostalgia or solastalgia (though, 'atoposalgia' may be more linguistically consistent), is an emotional and psychological trauma that can be applied to indigenous and non-indigenous people alike. For the Tuvalus, readapting place in cultural memory will occur against a background injustice of becoming the living example of a new, and soon to be undeniable, global environmental identity, the climate refugee.

Projecting into the next fifty years, or by 2060, there are estimates of up to 200 million environmental refugees, most of whom are expected to be relocated because of climate impacts (Myers 2002). At this scale, no easy formula of distributive justice will resolve the obligations of high carbon emitters to indigenous communities. Different background and current injustices mold nuances of justice on a case-to-case scalar basis. In another example, DeNeen Brown (2001) reports on the Inuvialuit people of Herschel Island in the Yukon, who witnessed the melting permafrost due to increasing temperatures over the past few decades. This has resulted in a spiritual, cultural, and environmental crisis. The background injustices include the denied recognition of the Inuvialuit peoples' heritage by changing the island's name, and the Western whalers who invaded the resources, generated cultural assaults through racism, violence, rape, and later introduced a devastating strain of influenza. Inuvialuit sacred tradition included raising the dead on platforms of honor, but since the epidemic, the deceased have been buried in the permafrost. These Inuvialuit believe that anyone who touches (and sometimes even is in the vicinity of) the possessions of the dead after they are buried will be cursed. This is a wise taboo, as some are concerned that frozen flu victims may have preserved the deadly strain (Brown 2001). The island is now a national heritage site, and its inhabitants relocated to the mainland of Canada.

These examples of indigenous climate refugees indicate that the available solutions to the ultimate loss of place may likely accelerate and intensify compromises to the whole web of TEK and cultural identity. The loss of whole identities for the gain of the new political identity of climate refugee and the loss of whole ways to communicate with and understand the surrounding ecology places further burden on sustaining both biological and cultural diversity. Moreover, indigenous climate refugees are much more likely to lose cultural sustainability than non-indigenous climate refugees who, though severely impacted in many similar ways, are often relocated within the dominant culture in which they originated. While climate refugees of Hurricane Katrina in the United States were severely impacted by great losses and inhumane conditions, many residents of the area were relocated to neighboring states in which the dominant culture remained intact. On the other hand, the local indigenous populations, in particular the Houma and Cajun communities, existed on the margins of dominant culture and either lose what they have of place or relocate to further marginalization from dominant culture.

4.2 Restorative Justice

The subcategory of recognition justice known as restorative justice may be the best procedural context for offsetting the cultural losses and the varieties of paternalism, such as debating the status of climate refugees, determining where to relocate refugees, and initiating services for refugees without their own active participation by those global citizens most responsible for climate change (Doyle and Chaturvedi in this volume). The primary virtues and function of restorative justice is that involved parties (victims, offenders, impacted relatives, and community members) voluntarily come 'face to face' in a participatory, mediation-based process of conflict resolution and healing, wherein the involved parties aim towards productive reintegration into the community. Rather than being punitive in the context of bureaucratic hierarchy of courtroom disputes, the exchanges and models of restorative justice bring much more self-realization and subjectivity exchange between parties (Johnstone and VanNess 2007). It pushes recognition justice from marginalized victim-offender relations, or adversarial self-other relations, to reconfigure the roles, histories, and subjective experiences (Figueroa 2006; Oliver 2001). In national contexts restorative justice is the context under which truth and reconciliation commissions exist. It has been used in a number of places to address procedures for conversation and full participation between the beneficiaries and recipients of historical harms, such as between Canada and First Nations Peoples, between the Australian Commonwealth and Aboriginal Peoples, and in the South African transition from apartheid (Chrunik 2009; Maepa 2005). Restorative justice has a variety of origins. It has been documented in religious traditions that favor forgiveness and community integration over brute retribution (Hadley 2001). It has other formal connections to juvenile criminal procedures, domestic abuse, and broader human rights claims (Johnstone and VanNess 2007). Truth and reconciliation proceedings are more than a blame-game for historical harms between people; the proceedings are representative of a form of justice historically utilized in a variety of indigenous traditions, international courts, and many communities (Mirsky 2004; Ulen 2010; Gibbs 2009).

However, background conditions must be heeded. Kelly Richards (2009) warns that the espoused virtues of restorative justice and the traditional indigenous connections that advocates point to should not underestimate the dominance of state criminal court systems. As

Richards argues, state-dominated courts are well entrenched and the integration of restorative procedures, or the presumption that contemporary indigenous peoples fundamentally endorse restorative justice according to some essentialist attribution, could easily underestimate the corrosive capacity state systems have upon attempts for restorative justice. Nonetheless, there are variations of indigenous restorative justice that can transform our understanding from strictly distributive responses that exclude indigenous peoples from full and effective participation in mitigation strategies to bivalent environmental justice in which TEK is legitimate for adaptation and mitigation strategies. Restorative justice, witnessing opportunities, inclusion and respect for TEK need to be brought in a transformative vision of justice in which we reassess the trajectory of the dominant environmental heritage.

In the context of cultural loss caused by climate change, restorative justice may be achieved in a variety of ways from mutual agreements for mitigation and adaptation to apologies, memorials, and bearing witness to a process of transformation between parties (Figueroa 2006; Figueroa and Waitt 2008; Oliver 2001). Moreover, restorative justice helps to better capture the ways in which all parties may mutually experience harms and mutually mend the individual and community wounds. For instance, restoring and reconciling the cultural wealth in TEK would aim at the recognition that epistemological diversity has benefits beyond the specific language community (Robyn 2002). As noted earlier, TEK can be seen as a climate adaptive strategy for whole regions, and reconciliation between the more-burdened populations and less-burdened populations can begin with the respect and acceptance of TEK in the epistemological framework.

5 CONCLUSION

We are witnessing a global phenomenon in which differently situated peoples are facing in the same direction of the traumatic worry for the future of the planet and struggle for human survival. The moral magnitude of cultural loss to be faced by the most vulnerable indigenous communities is akin to the loss of cultural opportunity, knowledge, and legacy available to humanity. An environmental justice framework that includes recognitional, participatory, and restorative justice for threatened indigenous cultures will demand that we change a causal trajectory in climate change by transforming not only the environmental colonialism long felt by such cultures, but also practices that contribute to the rise of climate change.

This chapter has discussed some of the vital dimensions in cultural loss for indigenous peoples, which is by no means comprehensive. Furthermore, the full breadth of connections to unique and specific biodiversity, cultural diversity, and language diversity that connects environmental knowledge important for adaptive and mitigation strategies could not be fully presented in any single chapter. Instead, this chapter takes the strategy of discussing cultural loss from climate change in an environmental justice framework. From this approach, the prescriptions I have suggested are both recognition and distributive, and I have emphasized restorative justice philosophies and procedures that can address the future consequences of cultural loss.

NOTES

1. Following the most common terminology in policies, declarations, documents, reports, literature, and self-referential citations and interviews, I will use the term 'indigenous people' in a general sense throughout this chapter. I will refer specifically to communities, civilizations, and societies by their name when case-specific points are made. I am aware that the general use of 'indigenous' may get over-extended or misrepresent peoples who do not self-identify as such. It is my intention to avoid any use or reference of 'indigenous' that is inappropriate, and for any failure on my part I humbly apologize in advance."

REFERENCES

Associated Press. 2008. Climate change and health: Oregon tribal health experts worry over native foods. *Indian Country Today*, 10 October <http://www.indiancountrytoday.com/living/health/30808764.html>.

BERKES, F. 1993. Traditional ecological knowledge in perspective. Pp. 1–9 in J. Inglis (ed.), *Traditional Ecological Knowledge: Concepts and Cases*. Ottawa: International Program on Traditional Ecological Knowledge and International Development Research.

——1999. *Sacred Ecology: Traditional Ecological Knowledge and Resource Management*. Philadelphia: Taylor & Francis.

BERZON, A. 2006. *Tuvalu is Drowning*. Living Earth Series, National Public Radio. Transcript. <http://www.salon.com/news/feature/2006/03/31/tuvalu/index.html>.

BROWN, D. 2001. Waking the dead, rousing taboo in Northwest Canada, thawing permafrost is unearthing ancestral graves. *Washington Post*. 17 October: A27.

BULLARD, R. D. 2000. *Dumping in Dixie: Race, Class, and Environmental Quality*. Boulder, CO: Westview Press.

BRYANT, B. (ed.) 1995. *Environmental Justice: Issues, Policies, Solutions*. Washington, DC: Island Press.

CALLICOTT, J. B. 1994. *Earth's Insights: A Survey of Ecological Ethics from the Mediterranean Basin to the Australian Outback*. Berkeley: University of California Press.

CHRUNIK, E. 2009. Stepping toward restorative justice. *The Sasquatch* 1: 4.

COHEN, F. S. 1971. *Handbook of Federal Indian Law*. Albuquerque: University of New Mexico Press.

COLE, L., and FOSTER, S. 2001. *From the Ground Up: The Rise of Environmental Racism in the United States*. New York: New York University Press.

CORDALIS, D., and SUAGEE, D. 2008. The Effects of Climate Change on American Indian and Alaska Native Tribes. *Natural Resources & Environment* 22: 45–9.

DEVRAJ, R. 2005. Tsunami Impact: Andaman Tribes Have Lessons to Teach Survivors. *Inter Press Service News*. 6 January 2005. <http://ipsnews.net/new_nota.asp?idnews=26926>.

FIGUEROA, R. M. 2003. Bivalent environmental justice and the culture of poverty. *Rutgers Journal of Law and Urban Policy* 1: 27–43.

——2006. Evaluating environmental justice claims. Pp. 360–76 in J. BAUER (ed.), *Forging Environmentalism: Justice, Livelihood, and Contested* Environments. Armonk, NY: E. M. Sharpe.

——and WAITT, G. 2008. Cracks in the mirror: (Un)covering the moral terrains of environmental justice at Uluru-Kata Tjuta National Park. *Ethics, Place, and Environment* 11: 327–48.

FIXICO, D. L. 1998. *The Invasion of Indian Country in the Twentieth Century: American Capitalism and Tribal Natural Resources.* Niwot, CO: University Press of Colorado.

GALLOWAY MCLEAN, K., RAMOS-CASTILLO, A., GROSS, T., JOHNSTON, S., VIERROS, M., and NOA, R. 2009. *Report of the Indigenous Peoples' Global Summit on Climate Change: 20–24 April 2009, Anchorage, Alaska.* Darwin: United Nations University—Traditional Knowledge Initiative.

GEARHEARD, S. 2005. Using interactive multimedia to document and communicate Inuit knowledge. *Études/Inuit/Studies* 29(1–2): 91–114.

GIBBS, M. 2009. Using restorative justice to resolve historical injustices of Indigenous peoples. *Contemporary Justice Review* 12: 45–57.

GRASSHOFF, A., BOYAJIAN, A., HAANSTRA, B., SELTZER, D., and KAUFMAN, J. 1988. *National Geographic's Australia's Aborigines.* VHS format. Washington, DC: National Geographic Video.

GRINDE, D. A., and JOHANSEN, B. E. 1995. *Ecocide of Native America: Environmental Destruction of Indian Lands and Peoples.* Sante Fe, NM: Clear Light.

HADLEY, M. L. (ed.) 2001. *The Spiritual Roots of Restorative Justice.* Albany, NY: State University of New York Press.

HESTER, L., MCPHERSON, D., BOOTH, A., and CHENEY, J. 2000. Indigenous worlds and Callicott's land ethic. *Environmental Ethics* 22: 273–90.

HUNOLD, C., and YOUNG, I. M. 1998. Justice, democracy, and hazardous siting. *Political Studies,* 46: 82–95.

JAMIESON, D. 1992. Ethics, public policy, and global warming. *Science, Technology, Human Values,* 17: 139–53.

——1994. Global environmental justice. Pp. 199–210 in R. ATFIELD and H. BELSEY (eds.), *Philosophy and the Natural Environment.* Cambridge: Cambridge University Press.

JOHNSON, M. (ed.) 1992. *Lore: Capturing Traditional Environmental Knowledge.* Ottawa: Dene Cultural Institute/IDRC.

JOHNSTONE, G., and VANNESS, D. W. (eds.) 2007. *Handbook of Restorative Justice.* Cullompton: Willan Publishing.

KASSAM, K.-A. 2008. Diversity as if nature and culture matter: Bio-cultural diversity and indigenous peoples. *International Journal of Diversity in Organizations, Communities and Nations* 8: 87–95.

KHANNA, S. 2010. What does climate change do to our heads? *Culture Change.* <http://www.culturechange.org/cms>. Generated: 27 November.

KLEINMAN, D. L. 2005. *Science and Technology in Society: From Biotechnology to the Internet.* Malden, MA: Blackwell Publishing.

LADUKE, W. 1999. *All Our Relations: Native Struggles for Land and Life.* Cambridge, MA: South End Press.

LEUNG, R. 2005. Sea gypsies saw signs in the waves: How Moken people in Asia saved themselves from deadly tsunami. *60 Minutes,* 20 March 2005, updated 10 June 2007. <http://www.cbsnews.com/stories/2005/03/18/60minutes>

McGREGOR, D. 2004. Traditional ecological knowledge and sustainable development: Towards coexistence. Pp. 72–91 in M. Blaser, H. A. Feit, and G. McRae (eds.), *In the Way of Development: Indigenous Peoples, Life Projects and Globalization*. Ottawa: Zed/IDRC.

——2008. Linking traditional ecological knowledge and western science: Aboriginal perspectives from the 2000 State of the Lakes Ecosystem Conference. *Canadian Journal of Native Studies* 28: 139–58.

MAEPA, T. (ed.) 2005. *Beyond Retribution: Prospects for Restorative Justice in South Africa*. Monograph no. 111. Pretoria, South Africa: Institute for Security Studies, with the Restorative Justice Centre.

MELMER, D. 2007. Project explores relationship between climate, cultural changes. *Indian Country Today*, 19 November <http://www.indiancountrytoday.com/archive/28142829.html>.

MIRSKY, L. 2004. Restorative justice practices of Native American, First Nation and other Indigenous People of North America: Part One. Restorative practices e-Forum. 27 April. International Institute of Restorative Practices. <http://fp.enter.net/restorativepractices/natjust1.pdf>.

MYERS, N. 2002. Environmental refugees: a growing phenomenon of the 21st century. *Philosophical Transactions of the Royal Society B: Biological Sciences* 357: 609–13.

NAKASHIMA, D. 1993. Astute observers on the sea ice edge: Inuit knowledge as a basis for Arctic co-management. Pp. 99–110 in J. Inglis (ed.), *Traditional Ecological Knoweldge: Concepts and Cases*, ed. J. Inglis. Ottawa: International Program on Traditional Ecological Knowledge and International Development Research Centre.

OLIVER, K. 2001. *Witnessing: Beyond Recognition*. Minneapolis: University of Minnesota Press.

PEÑA, D. 2005. *Mexican Americans and the Environment: Tierra y vida*. Tucson: University of Arizona Press.

RANCO, D., and FLEDER, A. 2005. Tribal Environmental Sovereignty: Cultural Appropriate Protection or Paternalism? *Journal of Natural Resources and Environmental Law* 19: 35–58.

RICHARDS, K. 2009. Returning to the practices of our ancestors? Reconsidering Indigenous justice and the emergence of restorative justice. Pp. 182–93 in M. Segrave (ed.), *Conference Proceedings Australian and New Zealand Criminology Conference*. School of Political and Social Inquiry, Monash University.

ROBYN, L. 2002. Indigenous knowledge and technology: Creating environmental justice in the twenty-first century. *American Indian Quarterly* 26: 198–220.

SCHLOSBERG, D. 2007. *Defining Environmental Justice: Theories, Movements, and Nature*. Oxford: Oxford University Press.

SHRADER-FRECHETTE, K. 2005. *Environmental Justice: Creating Equality, Reclaiming Democracy*. Oxford: Oxford University Press.

SHUE, H. 1992. The unavoidability of justice. Pp. 373–97 in A. Hurrell and B. Kingsbury (eds.), *The International Politics of the Environment: Actors, Interests, and Institutions*. New York: Oxford University Press.

SUAGEE, D. 2005. Indian tribes and the Clean Water Act. *A.B.A Trends*, 36 Jan.–Feb.

——2009. Tribal Sovereignty and the Green Energy Revolution. *Indian Country Today*, 19 September 2009. <http://www.indiancountrytoday.com/archive/44585422.html> accessed 7 June 2010.

TURNER, D. A. 2006. *This Is Not a Peace Pipe: Towards a Critical Indigenous Philosophy*. Toronto: University of Toronto Press.

TWEEDY, A. E. 2005. Using plenary power as a sword: Tribal Civil Regulatory Jurisdiction under the Clean Water Act after *United States v. Lara*. *Environmental Law* 35: 471–90.

ULEN E. N. 2010. No Word for 'Prison'. *The Defenders.* <http://www.thedefendersonline.com/2010/01/20/no-word-for-prison/> accessed 9 June 2010.

United Church of Christ. 1987. *Toxic Wastes and Race in the United States: A National Report on the Racial and Socio-economic Characteristics of Communities with Hazardous Waste Sites.* New York: Public Data Access.

United Church of Christ and Witness Ministries. 2007. *Toxic Waste and Race at Twenty: 1987–2007.* Cleveland: United Church of Christ.

United Nations Environmental Programme, Convention on Biological Diversity. 2007. *Report on Indigenous and Local Communities Highly Vulnerable to Climate Change Inter Alia of the Arctic, Small Island States and High Altitudes, with a Focus on Causes and Solutions,* Ad Hoc Open-Ended Inter-Sessional Working Group on Article 8(j) and Related Provisions of the Convention on Biological Diversity, Fifth Meeting. Montreal, Canada, 15–19 October.

WEAVER, J. 1996. *Defending Mother Earth: Native American Perspectives on Environmental Justice.* Maryknoll, NY: Orbis Books.

WOODARD, S. 2005. Blending science and tradition in the Arctic. *Indian Country Today.* 25 March. <http://www.indiancountry.com/content.cfm?id=1096410621>.

PART V

SECURITY

CHAPTER 17

....

CLIMATE CHANGE AND 'SECURITY'

....

NILS GILMAN, DOUG RANDALL,
AND PETER SCHWARTZ

CLIMATE change represents a unique and novel security threat: it has the capacity to devastate human civilization if not humanity as a biological species, yet it is not produced by enemies intending to do harm to a particular state; its impacts are direct and physical, yet it is not respectful of national boundaries; its consequences will be dire, yet it is a long-unfolding process with root causes and specific outcomes that are difficult to characterize objectively and scientifically. This complexity makes it challenging for policy makers to assess climate change threats and consequently to prioritize resources for countering them. And yet they must—because *the question is no longer whether to securitize the climate change debate, but how to do so properly*. The goal of this chapter is to provide a clear framework for analyzing the full range of security threats posed by climate change, with a view to determining appropriate governmental policy responses.

In proposing this assessment, we make two assumptions about climate change and national security. First, we believe that *over the long term, anthropogenic climate change is virtually certain to emerge as a security threat that is existential in scope*. If greenhouse gas (GHG) emissions continue unabated, then over the next several centuries (and possibly much sooner) the resulting climate change will kill or displace billions of people and destroy the majority of today's industrial infrastructure and seaside global cities, including Hong Kong, Shanghai, Tokyo, San Francisco, Washington, New York, Rio de Janeiro, London, Amsterdam, St Petersburg, Lagos, Karachi, Mumbai, and Sydney. Biologist James Lovelock has suggested that a continuation of our current GHG output will produce a shift as dramatic as the one that took place during the Eocene (55 million years ago, when palm trees shaded crocodile-like creatures on Alaska's North Slope), which could reduce the human carrying capacity of the planet by 90 percent or more (Lovelock 2006). Even organizations as sober as the Center for a New American Security and the Center for Strategic and International Studies suggest that, 'unchecked climate change equals the world depicted by Mad Max, only hotter, with no beaches, and perhaps with even more chaos' (Campbell et al. 2007). Short of global thermonuclear war, it is hard to imagine a more hair-raising threat, or one more worth prioritizing security resources against (Brand 2009).

Second, we believe that *there are policy options for addressing climate change-related security threats that go beyond traditional military action.* As we demonstrate below, determining the appropriate policy response to the threats posed by climate change requires properly identifying the nature of each threat, then mobilizing the right kind of security knowledge and technology to counter those threats. Critics of the 'securitization' of the climate change debate are not wrong that the debate itself, if misconceived, has the potential to misdirect resources away from the security arenas in most urgent need of shoring up in the face of climate change. The solution, however, is not to avoid discussing the security impact of climate change, but rather to ensure that the conceptualization(s) of security are appropriate to the diversity of climate change threats.

1 THREE MODES OF NATIONAL SECURITY THINKING

Throughout the last several hundred years, what states regard as a 'security' matter has continuously evolved and expanded. Security scholars Andrew Lakoff and Stephen Collier have documented this broadening, identifying three distinct ways of conceptualizing national security, each of which addresses a different type of threat and demands a different sort of preparation by governments (Collier and Lakoff 2008a, 2008b).

1.1 Sovereign State Security

In this mode of security thinking, the object of protection is the *territorial sovereignty* and *continuity* of the state itself. Threats center on *enemy attacks*, either foreign or internal, and the technology used to address these threats focuses on armed intervention. Sovereign state security is the classic conception of security as formalized in the work of Thomas Hobbes during the seventeenth century, and it is the one that remains paramount in the thinking of most military leaders today.

1.2 Population Security

In this mode, the object of protection is the *population* of the state. Threats are the 'normal "pathologies" of everyday life,' such as *disease*, *crime*, or *poverty*; the technology used to address these threats includes surveillance, statistics, and interventions such as public health measures or Keynesian macroeconomic management tools. Population security arose in the nineteenth century, when security thinkers came to believe that the capacity of a state to wage war, and thus defend its sovereignty, depended on the physical, economic, and social health and well-being of its population. Despite the dominance of liberal notions that government should be minimized, security concerns justified the expansion of government to take responsibility for public health and economic development. In practice, this meant much more detailed state monitoring of the population, as well as the construction of all sorts of infrastructure, from highways to hospitals, in the name of security.

1.3 Vital Systems Security

In this mode, the object of protection is the *critical infrastructure* of the state. Threats such as *nuclear war, terrorism,* and *natural disasters* are unpredictable in their specifics and potentially catastrophic; the technology used to address these threats focuses on 'preparedness' based on scenario and all-hazards planning. Vital systems security emerged in the 1940s, as strategists and planners began to consider the threats that total war (particularly nuclear war) posed to the very infrastructure that had been developed to promote population security. Government planners developed new kinds of conceptual and physical technology to address a host of unpredictable and undeterrable risks, ranging from terrorist attacks to pandemics to natural disasters like earthquakes and cyclones. Rather than try to predict and prevent specific events, planners bundled these potential catastrophes under a single rubric of 'hazards' that needed to be prepared for via scenario planning, simulations, wargaming, and other non-stochastic approaches to foresight (Tellmann 2009).

Each successive mode of national security thinking emerged as a way to address the limitations of the historically anterior mode. As such, all three are interconnected: threats to vital systems pose a second-order threat to the security of the populations they support, and threats to populations pose a second-order threat to the state itself. Appreciating the interconnectedness of these three different modes of security thinking is crucial to the analysis of any security threat—including those posed by climate change (Lakoff 2008b). But appreciating their distinctiveness is perhaps even more crucial, because different security threats demand divergent kinds of policy responses. Indeed, it is only by properly conceptualizing and categorizing this suite of threats that the most appropriate government responses to each can be realized.

1.4 Different Threats, Different Responses

Many debates among policy makers about how to respond to threats—from terrorism to financial disasters—in fact turn on how to classify those threats. During the 2004 US presidential race between George Bush and John Kerry, for example, a key point of debate was whether Al Qaeda constituted an existential threat to the US (sovereign state security) or mainly a threat to the lives and infrastructure of Americans (population and vital systems security). The former position, as promoted by Bush, implied the need for a forceful military-led response, the so-called global war on terror; the latter position, proposed by Kerry, suggested that policy should focus less on military action and more on better global policing and disaster preparedness. Not only was the distinction momentous from a policy perspective, it was also among the decisive political factors that determined the election's outcome.

Likewise, the right policy for regulating financial markets depends on the categorization of the threat that market malfunctions pose. Until recently, financial regulators had traditionally viewed economic security in population security terms. As Stephen Collier puts it, 'Risk managers looked at an archive of past events (for example, recent movements in bond prices or in defaults on mortgage-backed securities) and used this information to estimate the probable distribution of future fluctuations.' 2008's global financial meltdown, however, brought home the way in which the financial system as a whole is an essential piece of critical infrastructure, a vital system subject not just to stochastic risks but also to

Table 17.1 Three modes of security and the impact of climate change

	Sovereign State Security	Population Security	Vital Systems Security
Object of protection	Territorial sovereignty and regime continuity	The domestic human population	Vital systems and critical infrastructure
Nature of threat	Attacks by internal and external adversaries	Regularly occurring medical, social, economic pathologies	Unpredictable (and often undeterrable) catastrophes
Goals	Interdiction, deterrence	Prevention	Resilience, preparedness
Key security technologies and modes of intervention	Kinetic military platforms; armed force; military strategy	Surveillance and stochastic analysis of population, disease, sanitation and welfare; Keynesian economic interventions; policing	Scenario planning; imaginative (re) enactment; all-hazards planning; vulnerability reduction
Relevant climate change-related threats (examples)	Militarized resource competition; land losses due to sea-level rise, including lost military bases; breakdown and obsolescence of military kit	Shifting disease zones; lethal heatwaves; economic disruption; food price inflation; climate refugees; famine	Extreme weather events compromising transportation, water, and agricultural infrastructure
Appropriate governmental responses (examples)	Improved multilateral crisis response mechanisms; retooling of militaries to focus on relief operations	Achievement of Millennium Development Goals; geospatial mapping of flood/fire zones	Bioengineering of flood- and drought-resistant crops; investment in infrastructure redundancy and resiliency; long-term evacuation plans

Sources: Top portion of table adapted from Lakoff 2008a, Collier and Lakoff 2008a

unpredictable catastrophic events, what Nicholas Taleb refers to as 'black swans' (Taleb 2007). Viewing financial markets as a vital systems security rather than a population security matter carries huge policy implications, underscoring the need for a powerful new 'systemic risk regulator' with transnational authority, whose mission would be to mitigate the risk of endogenous financial catastrophe rather than maximize economic growth while limiting inflation.

As with terrorist and financial threats, so for climate security: it is crucial to correctly characterize the nature of the security threat in order to determine the proper policy responses. Insofar as climate change challenges sovereign state security, planning possible military responses is appropriate. At the same time, this framework underscores that the military is not the most effective vehicle for addressing climate threats to population security and vital systems security. Complicating this picture, however, is the realization that first-order (direct) threats to population security and vital systems security in fact pose second-order threats to sovereign state security. For example, the failure of one country to prepare for climate change's threats to their population can create—via a stream of refugees, for instance—a sovereign state security threat to a neighboring country. Rather than prepare for the cascading second-order effects to manifest themselves, however, it makes far more sense to prepare for the first-order effects of climate change on vital systems and population security with measures that are designed to address these matters *before* they cascade into military-grade crises. While mindful of these complexities, the discussion that follows is designed to allow policy makers to appreciate the diversity of security threats posed by climate change and the consequent breadth of policy responses that climate change demands.

2 CLIMATE CHANGE-RELATED THREATS TO VITAL SYSTEMS SECURITY

The science of climate change indicates a broad variety of geophysical impacts that may have implications for national security. On the one hand, some of climate change's impacts will be slow and mounting, producing deaths by a thousand cuts: increasing salinity in low-lying deltas that puts pressure on agriculture; shifting geographies of disease vectors; rising sea levels; and so on. On the other hand, climate change will also increase the frequency of extreme and unpredictable weather events, whose impacts will be rapid and immediate. To illustrate the various ways in which climate change will affect national security, we choose to focus on the latter category of climate change impacts, imagining a scenario from the year 2031.

> *Take 1: In early November 2031, the category-five Cyclone Shiva slammed into the Ganges-Brahmaputra delta. It was the second major cyclone of the season and the second category five in as many years to hit southern Bangladesh. The result was utter physical devastation. Half of what had been the country in 2010 was underwater. Most of the country's major roads were destroyed, making relief efforts far more difficult. Even worse than the disruption of food delivery was the sea surge's destruction of 10 percent of the country's arable land. The mobile telecommunications web was destroyed, and potable water was virtually unobtainable. Pundits from Dakka to D.C. declared that forty years of development had been destroyed overnight . . .*

2.1 Impacts of Climate Change on Vital Systems Security

The most direct and near-term effects of climate change will be its impact on the social and physical systems that societies rely on, including water supply infrastructure, agriculture, and public health systems. The primary vital systems that climate change will most directly affect include:

- *Energy production and delivery*. More frequent severe storms will compromise and destroy hydrocarbon wells, refinery capacity, and electric grids (Wong et al. 2009). Severe hurricanes in the Gulf of Mexico already regularly knock out offshore oil production and onshore refinery capacity for weeks or months at a time. In 2005, for example, Hurricane Katrina caused US oil output to shrink from 5.4 million barrels per day (MBD) to 4.4 MBD; one month later, Hurricane Rita further reduced output to 3.8 MBD. The combined hurricanes lowered refinery output by 28 percent, due to downed power lines, flooding, and so on (Kaufman 2005). Similar vulnerabilities exist for energy production and transmission networks in Europe and Asia, where heatwaves and storms generated by climate change will cause these systems to buckle.

- *Water supply.* Climate change increases 'the frequency of events that have always been at the heart of the concerns of water managers' (World Water Council 2006). There are two distinct impacts that climate change will have on water supply and delivery systems. First, water delivery infrastructure—pipes, canals, aqueducts, irrigation systems, etc.—will be vulnerable to damage from more frequent extreme weather events. Second, climate change will cause some water sources themselves to dry up— by receding glaciers, reducing snowpack, and altering rainfall patterns—thus rendering existing infrastructure obsolete or inadequate. For example, a warming climate means more water will fall as rain rather than snow; less snow accumulation in the winter and earlier snow-derived water runoff in the spring will challenge the capacities of existing water reservoirs in parts of the world reliant on snowmelt. The Western US and Central Asia are especially vulnerable to this effect. Facing even more serious consequences by 2050 will be regions dependent on glacial melting for water: once the glaciers have melted, there will be no replacement. Glaciers in Peru have experienced a 25 percent reduction in the past three decades, with further shrinking all but certain. The biggest impacts from glacial melting, however, will be felt in Asia, where hundreds of millions of people rely on waters from vanishing glaciers on the Himalayan plateau (Barnett et al. 2005; Pomeranz 2009; Gleick 2002).

- *Agriculture.* Food security has received considerable attention from climate change researchers, focusing in particular on food production and distribution systems, especially as mediated through world trade (Parry et al. 1999; Schmidhuber and Tubiello 2007). Climate change may have direct effects on crop production (e.g. changes in rainfall leading to drought or flooding, or warmer or cooler temperatures leading to changes in the length of growing season), but can also drive adverse changes in markets, food prices, and supply chain infrastructure (Gregory et al. 2005). Food-importing nations, such as Egypt and the Philippines, are most vulnerable, not least because in the event of a global crisis some governments will react by restricting food exports, as happened during the food price bubble of 2008. A widespread shift to biofuels would exacerbate the situation, crowding out at least some food production.

- *Transportation.* Transportation systems are vulnerable to climate change on numerous levels: not only will more severe and frequent storms damage roads and bridges, but rising sea levels will destroy many seaside roads (Mills and Andrey 2009). In addition, warming tundra will make arctic roadways impassable; one recent estimate for Alaska claims that melting permafrost will increase road maintenance costs by 10–20 percent (Larsen et al. 2008).

In addition to the risks specific to each of these vital systems, climate-related disruptions will also reverberate, causing cascading negative impacts from one system to the next. A climate-related crisis in one system, such as energy production, may domino into another system, such as transportation, which in turn may provoke a failure in a third system, such as food distribution. Moreover, climate change-vulnerable vital systems exist within a web of other systems—for example, political systems for generating collective decision making and legitimacy, and social systems for creating ethical norms, consensus, and integration—that collectively can be characterized as the 'macrosystem' (Gilman et al. 2007). The convergence of systemic stresses and climate change may produce simultaneous nonlinearities across multiple systems that risk overloading the macrosystem as a whole, creating the possibility of catastrophic failure.

Importantly, the globalization of financial, food, and energy markets means that states are vulnerable to the consequences of vital systems failures in distant geographies, even if they themselves remain insulated from the extreme weather and hydrological events associated with climate change. Analyzing these systemic interdependencies is crucial to appreciating the full scope of the security threat that climate change poses to vital systems. Finally, the threat to vital systems will become increasingly important as the planetary population continues its rapid urbanization. Cities depend on vast flows of goods to sustain their populations: food, water, fuel, and construction materials must be continuously imported; sewage and other waste must be exported; and fresh air must circulate. Although cities in developed countries appear to have solved the most pressing health and economic problems associated with hyperdense human settlement, rapid urban growth in the Global South has led to extreme shortages of basic social services, a glaring lack of infrastructure, and chronic air and water pollution. These problems will only worsen, as most of the oncoming growth will perforce take place on less desirable, more at-risk land: hillsides, gullies, swamps, and so on (Revkin 2009).

2.2 Addressing the Climate Change Threat to Vital Systems Security

In general, threats to vital systems security can be addressed by (1) improving the resiliency of a state's vital systems and (2) improving the ability of planners to foresee the low-probability, high-impact events that would disrupt these systems. In order to build resiliency to climate-related threats, countries and industries should increase the redundancy of critical infrastructure, improve the transparency of vital systems, and improve crisis response effectiveness. Improving hospitals and first-responder technology and personnel, building better communications and transportation infrastructure (preferably located further away from coastlines and flood zones), and developing reserve agricultural facilities will all improve resiliency. So will prophylactic removal of the communities and facilities at highest risk, such as those located in riparian and seaside flood zones, and rethinking agriculture that relies on regular rains. Finally, improving resiliency also means increasing local autonomy, both to decrease the risk that external authorities will impose measures inappropriate to local conditions and to avoid bottlenecks in simultaneous-crisis situations.

In order to improve the ability to anticipate in general the specific catastrophes that might impact vital systems, governments and other organizations can take several steps. The first is to engage in scenario planning exercises, both to develop detailed scenarios of specific catastrophes and to identify which systems in which countries are most vulnerable. Once the most vulnerable regions and systems have been identified, policy makers can 'wind-tunnel' the robustness of various contingency plans in response to specific crisis scenarios and run simulations to uncover stakeholder commitments and biases in high-pressure situations.

3 CLIMATE CHANGE-RELATED THREATS TO POPULATION SECURITY

Take 2: By late December 2031, the humanitarian situation in Bangladesh was beyond catastrophic. The death toll of Cyclone Shiva's initial impact had been estimated at a quarter million. But now far worse was unfolding: twin epidemics of typhus and cholera, combined with famine-like conditions, ripped across the country, devastating cities in particular. Reporting via satellite link on Al Jazeera on the misery in camps across the country, one NGO predicted that many of the displaced persons would never be able to return to their former homes. Exacerbating the situation was the international humanitarian relief community's failure to respond; all the groups that might normally have been expected to help had been deployed to the Philippines, to deal with the aftermath of an almost equally devastating cyclone that had flattened the island of Luzon . . .

3.2 Impacts of Climate Change on Population Security

Climate change threatens population security by impacting public health and welfare in a variety of consequential ways–but three central population security issues in particular must be considered:

- *Shifts in the geography and demography of death and disease.* A hotter, more volatile climate will exacerbate both morbidity and mortality almost everywhere (Parker and Shapiro 2008). More frequent and severe heatwaves will increase the risk of heat-related mortality, which will be particularly pronounced in cities. Because of their energy use, cities experience 'urban heat island' effects; temperatures are 1 to 5 °C warmer in a city than in surrounding countryside, with the gradient most pronounced at night. Overheated urban environments will experience increased energy demand for air conditioning and refrigeration, creating a positive feedback loop toward more air pollution. Meanwhile, vector-borne diseases, which are today largely confined to the tropics, will extend their range into higher latitudes and altitudes as the tropics expand, and make opportunistic advances in the wake of more regular heavy rainfall events (Githeko 2000). These effects will be most severe in the epicenter of global demographic growth, namely the cities of the Global South. While pandemics are likely to begin in the countryside, they will combust in urban slums.

- *Reduced economic productivity.* Climate change will have a deleterious impact on both individual economic well-being and the resources available to states and societies as a whole. Not only will storms hurt millions through direct hits on urban centers and disruption to the vital systems that support health and welfare—e.g. hospitals, food transport, water cleanliness—extreme weather will also disrupt livelihoods. Direct damage from storms alone is increasing exponentially: from US$1 bn. in the 1960s to US$24 bn. in the 1980s to US$273 bn. in just the first six years of the 2000s (Larsen 2006). These figures capture only the direct cost of physical storm damage, not the billions more that result from concomitant economic disruption. Business is lost as communities struggle to recover. Goods and services cannot be shipped to the necessary locations, and it could take years for cities to right themselves economically. As climate change produces bigger, more frequent storms across broader geographies, the damage will only increase. Exacerbating the situation is the fact that insurers are pulling out of high-risk regions, leaving governments as the insurers of last resort. Indeed, the retreat of insurers is already limiting development opportunities, creating a feedback loop of negative economic consequences (Mills 2005; Lenton et al. 2009). In addition to damage and disruption from storms, long-term droughts brought on by climate change will also have devastating economic consequences, rendering enormous swaths of farmland unusable in Africa, Australia, and Asia. The expansion of the subtropics is in some ways more fearsome than the widening of the tropical zone itself (Lu et al. 2009). The Mediterranean, Southern Australia, and the Western United States appear headed toward more or less permanent drought conditions, best characterized as a trend toward desertification, rendering agriculture in these regions untenable. This impact will grow far more severe as states complete their heedless drainage of major aquifers such as the Ogallala aquifer in North America, the Nubian and Saharan aquifers in Africa, the Great Artesian Basin in Australia, and the North China and Rum-saq aquifers in Asia. Beyond the economic damage caused by the loss of agricultural productivity, the dire implications for food security are obvious.

- *A profusion of climate refugees.* As a spokesman for the Office of the UN High Commissioner for Refugees recently put it, 'If we were a corporation, climate change is what you might call a "growth area" ' (Sanders 2009). Estimates vary enormously as to the scale of the forthcoming climate refugee crisis. At the high end, Christian Aid Week, the organ of a US-based evangelical organization, has forecast that by 2050 the world might have as many as 1 billion climate refugees (Christian Aid 2007). Even if the number is an order of magnitude less, it is still very large, and for virtually all these people it will be a personal disaster. Collectively, it will be a humanitarian disaster of unprecedented scale (Kakissis 2010). Most predictions forecast that the significant majority of climate refugees will not cross borders directly, instead becoming, in the jargon of the UN, internally displaced persons. UN General Secretary Ban Ki Moon courted controversy when he characterized refugees from the Darfur conflict as 'climate refugees.' But rather than exculpating the regime in Khartoum, he was simply underscoring that many climate refugees will also be refugees from political conflicts generated by climate change-related environmental stress, and that it is foolish to attempt to distinguish these two categories of refugees (Moon 2007; Faris 2007). (For more on climate refugees, see the chapter by Doyle and Chaturvedi in this volume.)

3.3 Addressing the Climate Change Threat to Population Security

In general, threats to population security are best addressed by improved government surveillance and measurement of the populations in question, establishing relevant baselines, and following with targeted improvements to public health, policing, and economic management. This presupposes that effective governance is available and that the historical data that the governments are building policy on is relevant to future conditions. The former assumption is dubious for most of the Global South (and for some of the developed world as well), while the latter condition is thrown into radical doubt by the uncertainties produced by climate change. Nonetheless, improved biosurveillance and public health measures, both national and international, are of paramount importance. Awareness of probable shifts in disease vectors and extreme weather severity and frequency will also help governments prepare for the health impacts of climate change on their populations.

While the economic threats posed by climate change are severe, the good news is that the needed responses are well established. In a word, what is required is development, and to a large extent the program for achieving global population security is already central to the agenda of the World Bank, most national governments in the Global South, and many NGOs, including the Bill and Melinda Gates Foundation. On the other hand, it should be underscored not only that climate change is making the achievement of developmental goals ever more difficult, but also that development itself, insofar as it increases carbon-intensivity, also tends to exacerbate the underlying driver of anthropogenic climate change.

Just as important as what governments *should* do is what they *should not* do. That states should avoid Sudan-style ethnic cleansing need not be emphasized, but other counterproductive and even pernicious policies are in fact widespread and should be avoided. For example, governments should not react to food crises by imposing export restrictions. Rather than subsidize water extraction, as they often do today, governments should penalize unsustainable water extraction. Likewise, governments should avoid passively becoming the insurer of last resort and instead accept that a withdrawal of private insurers means that the regions in question are no longer sustainable. Finally, governments should begin drawing up long-term plans for how to address, both nationally and internationally, the oncoming refugee crisis. Planned evacuations will be far more humane, and much less likely to produce interstate conflict.

4 CLIMATE CHANGE AS A THREAT TO SOVEREIGN STATE SECURITY

Take 3: The inability of Dakka or the international community to respond adequately to the humanitarian crisis sparked by Cyclone Shiva had two main effects. The first was to generate an upwelling of political rage, leading to the collapse of the Bangladeshi government and strengthening the hand of rebels in Chittagong, who were accused of attacking refugees from the delta. Then again, the refugee crisis in the Southeast paled in comparison with the ones to the North and West, where hundreds of thousands of people were attempting to cross the border into India. India's decision to respond to the flood of refugees by closing the border was hailed as necessary in some quarters of the West, though few saw fit to defend the slaughter of refugees at Bangaon. Across the Islamic world, India's action (and the international community's muted response) was denounced as yet another example of anti-Muslim discrimination, a charge rendered more plausible by the weakness of the international response in Bangladesh compared to the response in the Philippines. Pakistan in particular denounced the Indian actions, threatening a forceful response . . .

4.2 Impacts of Climate Change on Sovereign State Security

Threats to sovereign state security represent direct threats to the stability and continuity of governments and states, including their ability to defend themselves militarily. With respect to climate change, the question is: how could climate change compromise the operations of states and governments, possibly fatally?

- *Decreased military operational capacity.* Shifting disease boundaries will impact soldier health by exposing soldiers to a broader and more virulent array of pathogens. Drought conditions will require new logistical plans and equipment for moving water to US troops in war zones. Meanwhile, rising sea levels will threaten coastal bases at home and abroad. Military hardware will degrade more rapidly in harsher operational environments, and changing ocean salinity could require changes in sonar and submarine systems (CNA Corporation 2007).

- *Loss of territorial integrity and rising resource conflicts.* Climate change will change the territorial boundaries of some nations. Rising sea levels pose a direct threat to low-lying countries, particularly atoll states like the Maldives, Kiribati, and the Marshall Islands (Barnett and Adger 2003), as well as to countries with large populations located in flood-prone deltas, such as Bangladesh. Also, climate change will make some 'old' resources less available and create new ones to be competed for. An example of the former is the much-discussed (and perhaps exaggerated) possibility of 'water wars.' Potential adversaries include Israel and its enemies, China and India, India and Bangladesh, Angola and Namibia, and Egypt and Ethiopia (Reid 2006). An example of the latter would be the way that the warming of the Arctic and Antarctic is likely to instigate competition between states over resources and rights of passage in these regions (Weinberger 2007). Should climate change begin to take place abruptly (and some will certainly experience it as abrupt), it will reduce both global and local carrying capacities, raising the Malthusian specter of aggressive wars fought over food, water, and energy rather than ideology, religion, or national honor. Such shifting motivations for confrontation would alter which countries are most vulnerable and the existing warning signs for security threats.

- *Population security spillover.* Even absent armed violence, conflict between countries that share a river basin, for example, will hinder sustainable development, thus driving poverty, migration, and social instability and exacerbating non-water-related conflicts (OECD 2005). The extreme case is that governments unable to protect their populations from climate change may collapse into 'environmentally failed states.' Even short of that limit case, states will produce large numbers of environmental refugees that will at minimum cause localized violence. Some models suggest that by exacerbating resource stress, global warming will increase the risk of civil wars, particularly in already-weak states in Africa (Burke et al. 2009). Certainly if environmental refugees attempt to cross borders they are likely to be perceived by neighboring states as a threat to those states' own security (Homer-Dixon 1993).

- *Fundamental shifts to modern states' social contracts.* Because of the vast collective risks associated with climate change, a variety of new political forms, which do not fit neatly within contemporary political paradigms, may emerge in response to climate change. These new forms may threaten the internal stability of individual states or

increase tensions between states. Although speculative, the following possible political reactions to climate change are worth considering:

○ *Pro- and anti-risk sharing parties.* Divergent attitudes toward the social sharing of
risk form an important emerging fault-line in many modern polities (Beck 1992).
Climate change will accelerate this process, producing one characteristic faction that
demands a far more activist government to mitigate a variety of 'big' security risks
(military, terrorist, environmental), while another coalition forms around allowing
people to fend for themselves, with more private provision of security services.

○ *Dominion.* Religious leaders may come to be crucial voices in interpreting the
impacts of extreme weather events, which have traditionally been seen 'acts of
God(s).' Religiously oriented political parties may form coalitions with Green
parties or adopt environmentalism as part of their political agenda (Simon
2006). Or they could become a major force of climate change denial.

○ *Enviro-fascism.* Radical movements tend to flourish during times of economic and
social disturbance. Given that climate stress will form a major causal vector in
these disturbances, or may be perceived as such, the political valence of environmental politics could shift dramatically. Although environmentalism has generally
operated in the orbit of left-liberalism, there may emerge hard-right environmental parties that seek to restrict immigration and hoard resources for their own
ethnic group.

○ *World government.* The Anglo-American political right has long fulminated that
climate change is a hoax, the basis of a conspiracy designed to deliver world
government and possibly world socialism. While the conspiracy charge is silly, and
the claim that the climate data has been systematically falsified simply delusional
(Kleiner 2010), it is far from absurd to say that a serious appreciation of the climate
change threat might lead reasonable people to conclude that a single world
authority, capable of imposing the necessary and painful GHG emissions restrictions, may be the only hope for salvaging some version of modern civilization.

Any or all of these changes may arrive in different permutations in different locations, with
unpredictable geopolitical consequences. What is certain is that the transition to a new
global politics in a climate-compromised world will be bumpy.

4.3 Addressing the Climate Change Threat to Sovereign State Security

How can and should governments respond to these climate change threats to sovereign state
security? Most importantly, governments should do all they can to prevent the eventualities
laid out in this paper from transpiring in the first place. Instituting stringent GHG emissions
reduction programs should be a top priority for all states, particularly wealthy ones. Many of
the risks described in the preceding sections can only be headed off by removing the
underlying driver of change, namely GHG emissions. If these emissions continue unabated,
then the gravest forecasts laid out in this chapter will become all but inevitable.

Beyond this paramount prophylactic measure, states should redesign their militaries so
that they are well equipped to deal with environmentally failed states and provide security

in humanitarian situations. Relief operations and security details, more than direct combat engagements or counterinsurgency, will represent the key missions of the future. Governments should also build more formal partnership arrangements with both NGOs and corporations to provide relief in acute situations. Moreover, because the threat of climate change is supranational in scope, governments should work to build multilateral institutions, perhaps modeled after today's strategic alliances, and develop plans to deal with the security consequences of climate, so that premeditated responses, rather than panic or adhocracy, become the order of the day.

Just as important is what states should *not* do. Above all, states should not regard the climate change threat in adversarial, nation-centric terms—that is, as a source of potential competitive advantage against other states. Preemptive military action to address potential or actual refugee crises, or to force the hand of other governments to take action (including GHG emissions reduction programs), risks producing results as bad or worse than climate change itself (Glass 2009). In addition, the prospect of geoengineering, recently given prominent airing in the US (Victor et al. 2009), should be avoided. Not only will the delusion of an easy 'technical fix' distract governments from making painful but necessary GHG emissions reductions, but if enacted unilaterally it will likely be perceived as an act of war by some other states.

5 CONCLUSION

The foregoing analysis suggests that most climate change-related security threats pertain to vital systems and population security, and only secondarily (that is, more remotely in time) to sovereign state security. The most immediate security threats posed by climate change will involve acute insults to and chronic compromising of critical infrastructure, including energy production and delivery systems, transportation networks, agriculture, and water supplies. Nearly as proximate and even more severe are the threats to population security, above all to public health and economic well-being, some of which will happen because of the direct effects of a hotter, more volatile climate, and some of which will be a second-order result of the damaging effects of climate change on critical infrastructure.

Addressing these climate change-related threats to vital systems and population security should be the highest priorities for governments. This means redoubling investment in the processes and technologies designed to address these threats: improving preparedness, resiliency, and redundancy in the case of vital systems, and more effective development programs in the case of population security, including investment in biosurveillance, healthcare delivery programs, and programs to improve economic growth. In addition, a great deal of attention should be paid to ensuring that anticipatory adaptations, by both the private sector and governments, focus on delivering Pareto-efficient benefits rather than on merely redistributing the risks and threats associated with climate change.

In the longer run (toward the second half of this century) the threats to vital systems and population security may become so severe that they indeed begin to impact the sovereign state security of large, populous nations. Already we have a foretaste in the fate of small island nations: these pioneers of the brave new climate world show to the economically and technologically more advanced nations an image of their own future. Mass refugee crises and environmentally failed states, each of which for different reasons may seem to necessitate the intervention of armed forces, will become an increasingly pressing possibility as the century advances—and the environmental conditions under which these armed forces will be forced to operate will be increasingly harsh.

It is thus crucial that security analysts be able to correctly characterize the different threats posed by climate change, and above all not to assume that the military should be the primary vehicle for addressing these threats. Using soldiers to improve civilian preparedness and the resiliency of vital systems, or to address climate change's threats to population health and well-being, is not only inefficient but very likely to be ineffective (Lakoff 2006). Although armed forces are appropriate for dealing with climate change threats that have cascaded into sovereign state security threats, using the military to address vital systems and population security threats is as inefficient as using emergency rooms to provide primary care—and to invest in military systems as a way to deal with the climate change threat, at the expense of improving public health and the resiliency of critical infrastructure, is akin to neglecting to provide primary care, only to deal with the much more dire and expensive crises once they reach the emergency rooms. It is therefore crucial that security planners learn to correctly identify and differentiate the different sorts of threat posed by climate change, and design and fund threat-appropriate responses, rather than assume that a single class of preparation will be sufficient.

Above all, however, states should do everything they can to arrest the threat posed by climate change before it erupts into a threat to sovereign state security. In the final analysis this can only mean addressing the underlying driver of the threat, namely the massive ongoing emission of GHG. But doing this presents states and leaders with a dilemma: although increasing economic growth is the single best way to acquire the tools to defend oneself from the security impacts of climate change, the only way to stem the underlying drivers of climate change is to curtail GHG emissions, which will compromise the ability to sustain economic growth. Finding a geopolitical solution to this growth-climate conundrum is as urgent to the twenty-first century as finding a way to avoid global thermonuclear war was to the twentieth.

ACKNOWLEDGMENTS

Thanks to William Barnes, Jenny Johnston, Stephan Faris, Ceinwyn Karne, and Andrew Lakoff, who commented on various drafts of this chapter.

REFERENCES

BARNETT, J. 2000. Destabilizing the environment-conflict thesis. *Review of International Studies* 26.
—— and ADGER, N. W. 2003. Climate dangers and atoll countries. *Climatic Change* 61: 3.
BARNETT, T. P., et al. 2005. Potential impacts of a warming climate on water availability in snow-dominated regions. *Nature* 438.

BECK, U. 1992. *The Risk Society: Toward a New Modernity*. London: Sage Publications.

BRAND, S. 2009. *Whole Earth Discipline: An Ecopragmatist Manifesto*. New York: Viking Books.

BRAUCH, H. G. 2009. Securitizing global environmental change. In *Facing Global Environmental Change: Environmental, Human, Energy, Food, Health and Water Security Concepts*. Berlin: Springer.

BROCK, L. 1992. Security through defending the environment. In E. Boulding (ed.), *New Agendas for Peace Research: Conflict and Security Reexamined*. Boulder, CO: Lynne Reinner.

——2004. Vom erweiterten Sicherheitsbegriff zur globalen Konfliktintervention: Eine Zwischenbilanz der neuen Sicherheitsdiskurse. <http://web.uni-frankfurt.de/fb3/brock/mat/Brock_2004_erweiterter_Sicherheitsdiskurs.pdf>.

BRUGGEMEIER, F.-J. et al. (eds.) 2006. *How Green Were the Nazis?: Nature, Environment, and Nation in the Third Reich*. Athens: Ohio University Press.

BURKE, M. B., et al. 2009. Warming increases the risk of civil war in Africa. *PNAS* 106(49).

CAMPBELL, K. M. et al. 2007. *The Age of Consequences: The Foreign Policy and National Security Implications of Global Climate Change*. Center for Strategic and International Studies and the Center for a New American Security.

Christian Aid. 2007. *Human Tide: The Real Migration Crisis*. A Christian Aid Report.

CNA Corporation. 2007. *National Security and the Threat of Climate Change*.

COLLIER, S. J., and LAKOFF, A. 2008a. On vital systems security. Presentation at the University of Helsinki Collegium.

——2008b. The vulnerability of vital systems: How critical infrastructure became a security problem. In M. DUNN ET AL. (eds.), *Securing the Homeland: Critical Infrastructure, Risk, and (In)security*. New York: Routledge.

FARIS, S. 2007. The real roots of Darfur. *The Atlantic*.

GILMAN, N., et al. 2007. *Impacts of Climate Change: A Systems Vulnerability Approach*. Global Business Network white paper.

GITHEKO, A. K., et al. 2000. Climate change and vector-borne diseases: A regional analysis. *Bulletin of the World Health Organization* 78: 9.

GLASS, M. 2009. *Ultimatum*. New York: Atlantic Monthly Press.

GLEICK, P. H. 2002. *Dirty Water: Estimated Deaths from Water-Related Diseases 2000–2020*. Pacific Institute Report.

GREGORY, P. J., et al. 2005. Climate change and food security. *Philosophical Transactions of the Royal Society B* 360.

HOMER-DIXON, T. 1993. Environmental change and violent conflict. *Scientific American Magazine*.

JACKSON, W. D. 2003. *Homeland Security: Banking and Financial Infrastructure Continuity*. Congressional Research Service report RL31873.

KAKISSIS, J. 2010. Environmental refugees unable to return home. *New York Times*, 3 January.

KAUFMAN, R. K. 2005. The forecast for world oil market. *Project Link Oil Forecast*.

KLEINER, K. 2010. Climate science in 2009. *Nature Reports Climate Change*.

KNICKERBOCKER, B. 2004. A hostile takeover bid at the Sierra Club. *The Christian Science Monitor*, 20 February.

LAKOFF, A. 2006. From disaster to catastrophe: the limits of preparedness. Social Science Research Council Forum on Understanding Katrina: Perspectives from the Social Sciences.

——2008a. From population to vital system: National security and the changing object of public health. In A. LAKOFF and S. J. COLLIER (eds.), *Biosecurity Interventions: Global Health and Security in Question*. New York: Columbia University Press.

LAKOFF, A. 2008b. The generic biothreat, or how we became unprepared. *Cultural Anthropology* 23: 3.

LARSEN, J. 2006. Hurricane Damages Soar to New Levels: Insurance Companies Abandoning Homeowners in High Risk Coastal Areas. <http://www.earth-policy.org/index.php?/plan_b_updates/2006/update58>.

LARSEN, P. H., et al. 2008. Estimating future costs for Alaska public infrastructure at risk from climate change. *Global Environmental Change* 18: 3.

LENTON, T., et al. 2009. *Major Tipping Points in the Earths Climate System and Consequences for the Insurance Sector*. A report commissioned by Allianz and the World Wildlife Fund.

LOVELOCK, J. 2006. *The Revenge of Gaia: Earths Climate Crisis and the Fate of Humanity*. New York: Basic Books.

LU, J., et al. 2009. Cause of the widening of the tropical belt since 1958. *Geophysical Research Letters* 36.

MILLS, B. and ANDREY, J. 2009. *Climate Change and Transportation: Potential Interactions and Impacts*. US Department of Transportation.

MILLS, E. 2005. Insurance in a climate of change. *Science* 309.

MOON, B. K. 2007. A climate culprit in Darfur. *Washington Post,* 16 June.

OECD. 2005. *Water and Violent Conflict*.

PARKER, C., and SHAPIRO, S. M. 2008. *Climate Chaos: What You Can Do to Protect Yourself and Your Family*. New York: Praeger.

PARRY, M., et al. 1999. Climate change and world food security: A new assessment. *Global Environmental Change* 9: 1.

POMERANZ, K. 2009. The Great Himalayan watershed: Agrarian crisis, mega-dams, and the environment. *New Left Review* 58.

REID, J. 2006. Water wars: Climate change may spark conflict. *The Independent,* 28 February.

REVKIN, A. C. 2009. Disaster hot spots on a growing planet. *New York Times,* 17 May.

SANDERS, E. 2009. Fleeing drought in the Horn of Africa. *Los Angeles Times,* 25 October.

SCHMIDHUBER, J., and TUBIELLO, F. N. 2007. Global food security under climate change. *Proceedings of the National Academy of Sciences* 104: 50.

SIMON, S. 2006. Evangelicals ally with Democrats on environment. *Los Angeles Times,* 19 October.

TALEB, N. N. 2007. *The Black Swan: The Impact of the Highly Improbable*. New York: Random House, 2007.

TELLMANN, U. 2009. Imagining catastrophe: Scenario planning and the striving for epistemic security. *economic sociology: the european electronic newsletter* 10: 2.

VICTOR, D. G., et al. 2009. The geoengineering option: A last resort against global warming? *Foreign Affairs*.

WEINBERGER, S. 2007. Arctic warfare heats up: Canadians step in. *Wired.com,* 10 August.

WONG, P. C. et al. 2009. Predicting the impact of climate change on the u.s. power grids and its wider implications on national security. Paper delivered at Spring symposium for Association for the Advancement of Artificial Intelligence.

World Water Council. 2006. *Synthesis of the 4th World Water Forum*.

WÆVER, OLE. 2009. All dressed up and nowhere to go? Securitization of climate change. Paper presented at the annual meeting of the ISAs 50th Annual Convention Exploring the Past, Anticipating the Future.

CHAPTER 18

....

HUMAN SECURITY

....

JON BARNETT

1 INTRODUCTION

....

CLIMATE change poses risks to the basic needs, human rights, and core values of individuals and communities. These risks are increasingly being described as risks to human security, which contrast with the more abstract notion that climate change poses risks to national security.

That climate change is increasingly being understood as a security problem is not surprising given recent evidence of trends in greenhouse emissions, which suggest that there will very likely be mean global warming of more than 4 °C above pre-industrial levels by the end of this century (Anderson and Bows 2008). At 2 °C of warming above pre-industrial levels millions of people will be exposed to increased water stress, the supply of some ecosystem goods and services will decline, the productivity of crops in low latitudes will fall, coastal areas will increasingly flood, and there will be significant increases in morbidity and mortality from diseases and extreme events (IPCC 2007). Thus, even the optimistic climate change scenarios suggest that social systems will have to deal with significant changes in environmental conditions.

Concern about the impacts of these changes on social systems is justified given that at the global scale the degree and rate of change is arguably unprecedented in the history of human civilization. Thus, climate change is increasingly being identified as a risk to security (however defined). Intergovernmental and international organizations such as the European Union (EU 2008), and the World Bank (2010) have identified climate change as a security issue, with a report by the UN Secretary General arguing that climate change 'can impact on security, namely by: increasing human vulnerability; retarding economic and social development; triggering responses that may increase risks of conflict, such as migration and resource competition; causing statelessness; and straining mechanisms of international cooperation' (UNGA 2009: 8).

A number of governments are also concerned about the security implications of climate change, including those in Europe (Beckett 2007; Carius et al. 2008) and the United States (Blair 2009). Leaders of Pacific Island Countries have long identified climate change as a security issue, stating most recently 'for Pacific Island states, climate change is the great challenge of our time. It threatens not only our livelihoods and living standards, but the very viability of some of our communities' (PIF 2009: 12). Feeding off these governmental concerns, think tanks and consulting firms such as the Centre for Strategic and International Studies (2007), and the CNA Corporation (2007) have produced reports on the strategic implications of climate change. Non-governmental organizations such as Christian Aid (2006) and International Alert (Smith and Vivekananda 2007) have also picked up on the topic, as have numerous scholars (for example: Busby 2008; Dupont and Pearman 2006; Gleditsch et al. 2007; Nordås and Gleditsch 2007; Scheffran 2008).

To some extent this heightened concern about climate change and security is not new (Dabelko 2009). Research on environmental security has always noted the potential for climate change to be a major cause of social problems, and in the mid-1990s this research began to produce some important if tentative findings about the connections between environmental change and security. The lessons from this inform contemporary research on the security implications of climate change—including its implications for human security—which began in earnest after the 2001 Third Assessment Report of the Intergovernmental Panel on Climate Change (IPCC).

There are nevertheless considerable differences in interpretation of the connections between climate change and security. These are largely the function of different interpretations of security which, as explained below, can refer to different kinds of risks as well as different referent objects (that is, things to be secured). Thus, some accounts of climate change and security focus more on issues of conflict between states (Dupont 2008), some are mostly concerned with the stability of national political and economic systems (Busby 2007), others focus on the risks climate change poses to the well-being of people (Smith and Vivekananda 2007). Almost all reports refer to some degree to the same three key drivers of climate insecurity, namely the effects of climate change on human security, on migration, and on conflict. These three issues very clearly define the core content of research and policy on climate change and security. There are causal relationships between them, meaning that if climate change stimulates negative change in one or more of these issues then a process of mutually reinforcing increases in human insecurity, migration, and conflict may ensue (Barnett and Adger 2007).

This chapter focuses on the first of these commonly agreed climate change and security issues—that is the linkages between climate change and human security. The chapter has two major parts. First, it defines some of the key concepts of vulnerability, adaptation, and human security, which all come together in research and policy on climate change and human security. The relationship between these concepts is explained, as is the distinction between security as is it has traditionally been understood within the discipline of Political Science, and human security, which is a more interdisciplinary and less orthodox field of research and policy. Following this, the chapter provides an overview of the critical and applied uses of the concept of human security as it relates to climate change.

2 HUMAN SECURITY, VULNERABILITY, AND ADAPTATION

As explained in the Gilman et al. chapter in this volume, climate change poses risks to national security, and as alluded to above, this has become a new facet of both the climate change regime and the institutions of the United Nations system that concern themselves with security. Historically, national and human security have been treated as somewhat distinct categories of research and policy. For the most part national security has focused on a particular level of political action—that is what transpires between nation-states in the international system—to the exclusion of lower levels of political action. This leads to what Agnew (1994) calls the 'territorial trap' of imagining the world as a series of relatively simple disembodied rational political actors, which ignores the messy reality of global politics, and the complex processes that lead states to make seemingly irrational decisions. National security has also for the most part focused on one major risk (to the state, if not always the population it purports to govern), which is that of armed conflict.

Yet security need not mean 'national' security. In a broader sense security means safety and certainty, and it can refer to different kinds of risks such as hunger (food security), employment (job security), or income (economic security), as well as to different kinds of objects to be secured, such as people, communities, economies, nation-states, regions, and species. Therefore national security is a particular (if the most discursively and institutionally powerful) account of security. So too is human security, which is defined here as the state under which 'people and communities have the capacity to manage stresses to their needs, rights, and values' (Barnett et al. 2009: 18).

Indeed, human security has been used as a vehicle to critique the prevailing idea that security is something that states do to protect themselves from violence (Booth 1991; Shaw 1993). Such critique is both of an analytical and normative kind. By highlighting that security applies to people as well as states, and that people's insecurity arises from different processes, including from the security practices of states, human security has played a key role in the broadening and deepening of security theory and analysis. In this respect human security also has a strong normative element—it strongly implies that the security of people is more important than the security of states.

Human security has more practical application in the field of international development, where it is used to synthesize concerns about basic needs, violent conflict, and human rights (Gasper 2005). The 1994 *Human Development Report* launched the inclusion of human security in the field of development. It defined human security as a 'concern with human life and dignity' (UNDP 1994: 22), and it identified economic, food, health, environmental, personal, community, and political components to human security. A later definition of human security defined it as a process, 'to protect the vital core of all human lives in ways that enhance human freedoms and human fulfilment' (Commission on Human Security 2003: 4). This definition builds on Amartya Sen's (1999) work on the importance of freedoms and capabilities. The idea of a 'vital core' recognizes that there are some non-instrumental values (such as love, a sense of community, and identity) that are critical to a meaningful life, and which are neither basic needs (in the development sense of this term) nor necessarily human rights (in the legal sense of the term).

Like its use in international relations, human security has a critical function in development discourses, shifting the focus of development thinking away from large-scale issues such as macro-economic growth and infrastructure planning, towards the potential benefits and risks of these activities on people's needs, rights, and values. Yet human security also has more practical applications in that it identifies for the purposes of research and policy a scale of focus—people and communities—and clear metrics for measuring outcomes of processes (their effects on basic needs, human rights, and important social values).

The aforementioned definition of human security as the state under which people and communities have the capacity to manage stresses to their needs, rights, and values implies that there may be 'strong' and 'weak' criteria for deciding when climate change is a human security problem (as there is for the discussion of sustainability). In the strong case, climate change could not be said to be a human security problem unless it impacts on more than one of a person or group's needs, rights, and values. In most cases where climate change impacts on a person or group this is likely to be the case: in many developing countries it is hard to think of an impact of environmental change of any significance that does not impinge on at least people's basic needs and core values. A strong interpretation of human security denies that impacts on things that are valued is a sufficient criteria for calling something a human security problem. The pursuit of some values, such as driving big cars and buying lots of goods, is a cause of greenhouse gas emissions. Such activities are likely to be curtailed by any significant effort to reduce emissions of greenhouse gases. It is very likely that there will be arguments that restrictions on superfluous consumption would be an impact on human security, although there are compelling arguments from ethicists about the injustices (e.g. Shue 1999) of such consumption, from economists about the erosion of happiness that can arise from such consumption (e.g. Common 1995), and from sociologists about the psycho-social benefits of moving beyond unconscious consumption (e.g. Giddens 1991) (see Szasz in this volume).

Put most simply, vulnerability means the potential for loss (Cutter 1996). Although the concept is by no means unique to climate change research, it is its application in this context that is most relevant for this discussion. The IPCC defines vulnerability as 'the degree to which a system is susceptible to, and unable to cope with, adverse effects of climate change, including climate variability and extremes' (Parry et al. 2007: 883). Vulnerability is therefore a latent state in that it concerns potential outcomes, although too rarely are the potential outcomes specified in much detail. It is generally understood that vulnerability to climate change is the product of three factors: the degree to which natural and social entities are exposed to changes in climate, the degree to which they are sensitive to changes in climate, and their capacity to act to avoid or minimize the negative consequences of changes in climate.

Adaptation, too, has its applications outside of climate change research, but in climate change terms means actions that are taken to reduce vulnerability. The IPCC define it as 'adjustment in natural or human systems in response to actual or expected climatic stimuli or their effects, which moderates harm or exploits beneficial opportunities' (Parry et al. 2007: 869). The ability to act is called adaptive capacity, and it is seen to be a function of social opportunities and resources such as money, social networks, access to information, and influence in decision-making processes. It is not the case, however, that knowing what to do and having the ability to do it means that adaptation will actually transpire (Repetto 2009). The precise goals of adaptation are rarely articulated, which is to say that it is rare to see an academic or policy discussion about adaptation which identifies the specific risks that adaptation seeks to avoid, and to whom those risks most apply (such as, for example, the risk of increased malnutrition among women in the highlands of Papua New Guinea). For this reason, among others, adaptation will not necessarily be effective, efficient, or equitable (Adger and Barnett 2009).

Thus human security identifies a condition to be sustained, and broad sets of policy goals (that is, increases in the satisfaction of basic needs, human rights, and core values). It therefore offers something that is otherwise lacking in existing research and policy on climate change: it specifies clearly that climate change poses risks to core needs, rights, and values of people and local communities, and in so doing establishes some clearer goals for and metrics to measure the success of decisions about mitigation and adaptation. Thus it is helpful to talk about human security being vulnerable to climate change, and adaptation to sustain human security.

3 THE CRITICAL AND APPLIED DIMENSIONS OF CLIMATE CHANGE AND HUMAN SECURITY

Given that the concept of human security has critical, and applied functions, it is not surprising to find that these aspects of the larger literature are reflected in approaches to climate change and human security. Critical in this sense means approaches that seek to question established orders of knowledge, and applied means approaches that offer frameworks that can guide research and inform policy. These dimensions of climate change and human security are now discussed in turn.

3.1 Critical Dimensions

There is an underpinning normative argument to much of the literature that links climate change and human security, which is that climate change is a manifestation of inequities in political and economic processes, and will in turn magnify these inequities further. This position is most clearly advanced by O'Brien (2006), who uses the concept of human security to highlight the hitherto largely understated risks that climate change poses to people in both developed and developing countries. This has been a powerful argument, substantiated through various research projects (e.g. Leichenko and O'Brien 2008) and the work of the Global Environmental Change and Human Security Project (see Matthew et al. 2009). It has been well received by researchers and governments alike, almost all of whom have come to accept that climate change is as much a social problem as it is an environmental one. Thus in October 2009 the 31st session of the IPCC plenary and Working Group sessions approved for the first time the inclusion of a chapter on human security in the fifth assessment report.

There is embedded in these arguments of O'Brien and others a critical analysis of climate change research, which they imply is excessively driven by models of environmental processes, and measures of risk that exclude people and their needs, rights, and values (O'Brien et al. 2010). Yet the critical function of research on climate change and human security goes beyond this to offer a more profound critique of climate change policy. In particular, it highlights the limits to decision making about the mitigation of greenhouse gases, which remains concerned with the distribution of benefits and costs of decision making to economies, rather than people's needs and rights and values, and is constrained by decision-making institutions that do not give sufficient voice to local people (Adger et al. 2010).

A focus on human security also gives rise to a critique of the emerging logics and processes for deciding on and implementing adaptation, which are focused on engineering and planning responses that seek to reduce the exposure of things to risk (such as moving people, and building things like sea-walls) that are themselves likely to impact on human security (Adger and Barnett 2009). When the focus is on people's needs, rights, and values, alternative no-regrets strategies that seek to reduce sensitivity (such as changing design codes or better soil management) or increase adaptive capacity (such as improving access to healthcare, or the provision of microfinance) become preferred adaptation options.

Central to these human-security-based critiques of climate change research and policy is not just the idea that people matter, but also the idea that not all people are equally at risk from climate change. It is abundantly clear that because people have different levels of exposure, sensitivity, and adaptive capacity, some are far more at risk from climate change than others. There are therefore numerous arguments that the measure of dangerous climate change that climate mitigation should seek to avoid should be impacts on the most vulnerable people (Adger et al. 2006). This inequality in the distribution of risk becomes a significant problem of injustice when it is recognized that people are not equally responsible for climate change, and that the wealthy people who seem to be least vulnerable to climate change are those who are the most responsible for the problem (see the chapters on Justice later in this book). As Baer (2006) shows, the wealthiest people in developed countries produce 155 times more greenhouse gases than the poorest 10 percent of people in developing countries.

3.2 Applied Dimensions

Human security also offers a guide to research on the human dimensions of climate change. Such research has its antecedents in studies of human security and environmental change more broadly. Key lessons have been learned from studies of famine and food security (Sen 1981; Bohle et al. 1994), social vulnerability to disasters and slow-onset environmental changes such as desertification (Adger 1999; Blaikie and Brookfield 1987; Mortimore 1989), and the study of sustainable livelihoods (Bebbington 1999; Scoones 1998). Recent studies on environmental change and human security have narrowed their focus to concentrate on particular groups, such as women (Goldsworthy 2009) and the urban poor (Cocklin and Keen 2000), and particular outcomes such as morbidity (McDonald 2009), and violent conflict (Messer et al. 2001). Studies such as these show that environmental change interacts with other social and environmental processes to create impacts upon people's basic needs (such as food and shelter), human rights (such as the right to health), and social and cultural values (such as justice, identity, and belonging).

These and other studies have shown that people and groups that are already insecure are the most at risk of harm from environmental changes. The poor tend to be highly exposed to environmental risks, and have few assets with which to manage them (Adger 1999). The more people are dependent on natural resources and the less they rely on economic or social forms of capital, the more they are at risk from environmental change (Peluso et al. 1994). By virtue of their lower incomes and the gendered division of labor, women tend to suffer more and for longer during times of environmental change (Enarson and Morrow 1998). Elderly people with mobility constraints and preexisting health conditions tend to be more susceptible to harm (Hewitt 1997). Marginalized groups tend to have low incomes, and minimal access to the social services that help reduce sensitivity and increase the capacity to manage environmental changes (Narayan et al. 2000). People with insecure access to land are also at risk of loss (Liverman 1990).

The emerging wave of research on climate change and human security has learned from these earlier studies. In terms of basic needs, studies show that poverty and hunger are likely in places where multiple stressors—such as trade liberalization—interact with climate change (Eakin 2005; Leary et al. 2006; Leichenko and O'Brien 2008; Ziervogel et al. 2006). Others show that people's ability to avoid poverty and hunger is reduced where there is or has been violent conflict (Lind and Eriksen 2006). Denton (2000) argues that socio-economic factors disproportionately limit women's capacity to adapt to avoid poverty. Others show that human rights, such as the right to a means of subsistence, are at risk (Barnett 2009; Sachs 2006; Slade 2007). Studies from the South Pacific and the Arctic show that traditional practices that are important to cultures and identities are very sensitive to climate change (Adger et al. 2010; Ford et al. 2006; Tyler et al. 2007). There is a growing realization that human security can also be undermined by efforts to reduce emissions and promote adaptation (Adger and Barnett 2009; Bumpus and Liverman 2008). Yet despite the findings of these and other studies, climate change policy rarely recognizes the risks posed to people's needs, rights, and values (O'Brien 2006).

There are also linkages between human insecurity caused in part by climate change and the two other climate security problems that most concern the international community (large-scale migration, and violent conflict). Increasing migration does seem to be a likely outcome of climate change as people choose to move to minimize the impacts of climate change on their needs, rights, and values. Whether or not these moves are an impact of environmental change (in the sense that it is a response that migrants might rather have avoided) or an adaptation (a not necessarily unwelcome response to avoid or adjust to an even more undesirable outcome), and a problem or a benefit to the places the migrants move to, depends very much on the degree to which adaptation policies accept and plan for migration as an adaptation strategy (Barnett and Webber 2009). The links between climate change, human insecurity, and violent conflict are more tenuous, except to note that in times of livelihood shocks such as seem increasingly likely due to climate change, recruitment into armed groups tends to increase, and this, combined with weak states (which may also be an outcome of climate change) tends to increase the risk of violent conflict (Esty et al. 1999; Goodhand 2003).

Finally, a focus on climate change as a human security issue gives rise to new metrics for screening and monitoring policies. Policies that increase human insecurity by undermining people's access to enjoyment of basic needs, human rights, and core values should be considered to maladaptive, and all potential policies for both climate change mitigation and adaptation should be screened for these effects. Adaptation policies that seek to sustain and enhance access to and enjoyment of needs, rights, and core values in the face of climate risks are likely to have numerous co-benefits and synergize well with policy approaches that promote basic needs (such as those concerned with water and sanitation, for example) and human rights (such as the International Covenants on Economic, Social and Cultural Rights, and on Civil and Political rights).

4 CONCLUSION

As the social sciences become progressively more engaged with the problem of climate change the concept of security is increasingly being used. Human security has a particular place within the ambit of studies of security and development. It emphasizes the outcomes of power on the lives of people, it critiques research and decision-making processes that ignore local conditions, and it offers a framework for analysis and policy development. It has similar functions when applied to the problem of climate change: it highlights the social consequences of climate change, critiques the instrumental reasoning behind and constrained institutions for decision making about climate change, and it offers a structured basis for research on vulnerability and adaptation and synergies between policies and processes oriented towards basic needs and human rights. Human security, then, is a useful vehicle for understanding and acting on climate change.

REFERENCES

ADGER, W. N. 1999. Social vulnerability to climate change and extremes in coastal Vietnam. *World Development* 27(2): 249–69.

—— and BARNETT, J. 2009. Four reasons for concern about adaptation to climate change. *Environment and Planning A* 41(12): 2800–5.

—— PAAVOLA, J., MACE, M., and HUQ, S. (eds.) 2006. *Fairness in Adaptation to Climate Change*. Cambridge, MA: MIT Press.

—— BARNETT, J., and ELLEMOR, H. 2010. Unique and valued places at risk. Pp. 131–8 in S. Schneider, A. Rosencranz, and M. Mastrandrea (eds.), *Climate Change Science and Policy*. Washington, DC: Island Press.

AGNEW, J. 1994. The territorial trap: the geographical assumptions of international relations theory. *Review of International Political Economy* 1(1): 53–80.

ANDERSON, K., and BOWS, A. 2008. Reframing the climate change challenge in light of post-2000 emission trends. *Philosophical Transactions of the Royal Society A: Mathematical, Physical and Engineering Sciences* 366: 3863–82.

BAER, P. 2006. Adaptation: who pays whom? Pp. 131–54 in N. Adger, J. Paavola, M. Mace, and S. Huq (eds.), *Fairness in Adaptation to Climate Change*. Cambridge, MA: MIT Press.

BARNETT, J. 2009. Human rights and vulnerability to climate change. Pp. 257–71 in S. Humphreys (ed.), *Human Rights and Climate Change*. Cambridge: Cambridge University Press.

—— and ADGER, N. 2007. Climate change, human security and violent conflict. *Political Geography* 26(6): 639–55.

—— and WEBBER, M. 2009. Accommodating migration to promote adaptation to climate change. A policy brief prepared for the Secretariat of the Swedish Commission on Climate Change and Development and the World Bank World Development Report 2010 team. World Bank, Washington, and SCCCD, Stockholm.

—— MATTHEW, R., and O'BRIEN, K. 2009. Global environmental change and human security: an introduction. Pp. 3–32 in R. Matthew, J. Barnett, B. McDonald, and K. O'Brien (eds.), *Global Environmental Change and Human Security*. Cambridge, MA: MIT Press.

BEBBINGTON, A. 1999. Capitals and capabilities: A framework for analyzing peasant viability, rural livelihoods and poverty. *World Development* 27(12): 2021–44.

BECKETT, M. 2007. *The Case for Climate Security*. Lecture by the Foreign Secretary, the Rt. Hon. Margaret Beckett MP, to the Royal United Services Institute, 10 May 2007, available at <http://www.rusi.org/events/past/>.

BLAIKIE, P., and BROOKFIELD, H. 1987. *Land Degradation and Society.* New York: Methuen.

BLAIR, D. C. 2009. *Annual Threat Assessment of the Intelligence Community for the Senate Select Committee on Intelligence.* Available at <http://intelligence.senate.gov/090212/blair.pdf last accessed Nov 25 2009>.

BOHLE, H., DOWNING, T., and WATTS, M. 1994. Climate change and social vulnerability: Toward a sociology and geography of food insecurity. *Global Environmental Change*, 4(1): 37–48.

BOOTH, K. 1991. Security and emancipation. *Review of International Studies* 17(4): 313–26.

BUMPUS, A., and LIVERMAN, D. 2008. Accumulation by decarbonisation and the governance of carbon offsets. *Economic Geography* 84(2): 127–55.

BUSBY, J. 2007. *Climate Change and National Security: An Agenda for Action.* New York: Council on Foreign Relations.

—— 2008. Who cares about the weather? Climate change and U.S. national security. *Security Studies* 17(3): 468–504.

CARIUS, A., TANZLER, D., and MAAS, A. 2008. *Climate Change and Security: Challenges for German Development Cooperation.* Eschborn: GTZ.

Christian Aid. 2006. *The Climate of Poverty: Facts, Fears and Hopes.* London: Christian Aid.

CNA. 2007. *National Security and the Threat of Climate Change.* Alexandria, VA: CNA Corporation.

COCKLIN, C., and KEEN, M. 2000. Urbanization in the pacific: Environmental change, vulnerability and human security. *Environmental Conservation*, 27: 392–403.

Commission on Human Security. 2003. *Human Security Now.* New York.

COMMON, M. 1995. *Sustainability and Policy: Limits to Economics.* Cambridge: Cambridge University Press.

CSIS (Centre for Strategic and International Studies). 2007. *The Age of Consequences: The Foreign Policy and National Security Implications of Global Climate Change.* Washington, DC: CSIS.

CUTTER, S. 1996. Vulnerability to environmental hazards. *Progress in Human Geography* 20: 529–39.

DABELKO, G. 2009. Planning for climate change: The security community's precautionary principle. *Climatic Change* 96(1–2): 13–21.

DENTON, F. 2000. Gendered impacts of climate change: a human security dimension. *Energia News* 3(3): 13–14.

DUPONT, A. 2008. The strategic implications of climate change. *Survival* 50(3): 29–54.

——and PEARMAN, G. 2006. *Heating Up the Planet: Climate Change and Security.* Lowy Institute Paper 12. Double Bay: The Lowy Institute.

EAKIN, H. 2005. Institutional change, climate risk and rural vulnerability: Cases from central Mexico. *World Development* 33(11): 1923–38.

ENARSON, E., and MORROW, B. 1998. *The Gendered Terrain of Disaster.* New York: Praeger.

ESTY, D., GOLDSTONE, J., GURR, T., HARFF, B., LEVY, M., DABELKO, G., SURKO P., and UNGER, A. 1999. State failure task force report: phase II findings. *Environmental Change and Security Project Report* 5: 49–72.

European Union. 2008. *Climate Change and International Security.* Paper from the High Representative and the European Commission to the European Council, 14 March 2008, S113/08, available at: <http://www.consilium.europa.eu/uedocs/cms_data/docs/pressdata/en/reports/99387.pdf>.

FORD, J., SMIT, B., and WANDEL, J. 2006. Vulnerability to climate change in the arctic: A case study from arctic bay, Canada. *Global Environmental Change* 16(2): 145–60.

GASPER, D. 2005. Securing humanity: Situating 'human security' as concept and discourse. *Journal of Human Development* 6(2): 221–45.

GIDDENS, A. 1991. *Modernity and Self Identity: Self and Society in the Late Modern Age.* Cambridge: Polity Press.

GLEDITSCH, N., NORDAS, R., and SALEHYAN, I. 2007. *Climate Change and Conflict: The Migration Link.* International Peace Academy Coping with Crisis Working Paper, New York, International Peace Academy.

GOLDSWORTHY, H. 2009. Woman, global environmental change, and human security. Pp. 215–36 in R. Matthew, J. Barnett, B. McDonald, and K. O'Brien (eds.), *Global Environmental Change and Human Security.* Cambridge, MA: MIT Press.

GOODHAND, J. 2003. Enduring disorder and persistent poverty: A review of linkages between war and chronic poverty. *World Development* 31(3): 629–46.

HEWITT, K. 1997. *Regions of Risk: A Geographical Introduction to Disasters.* Essex: Longman.

IPCC. 2007. *Climate Change 2007.* Synthesis Report, Contribution of Working Groups I, II and III to the Fourth Assessment Report of the Intergovernmental Panel on Climate Change, IPCC, Geneva.

LEARY, N., ADEJUWON, J., BAILEY, W., BARROS, V., CAFFERA, M., CHINVANNO, S., CONDE, C., DE COMARMOND, A., DE SHERBININ, A., DOWNING, T., EAKIN, H., NYONG, A., OPONDO, M., OSMAN, B., PAYET, R., PULHIN, F., PULHIN, J., RATNASIRI, J., SANJAK, E., VON MALTITZ, G., WEHBE, M., YIN, Y., and ZIERVOGEL, G. 2006. *For Whom The Bell Tolls: Vulnerability in a Changing Climate.* A Synthesis from the AIACC project, AIACC Working Paper No. 21, International START Secretariat, Florida.

LEICHENKO, R., and O'BRIEN, K. 2008. *Environmental Change and Globalisation: Double Exposures.* Oxford: Oxford University Press.

LIND, J., and ERIKSEN, S. 2006. The impacts of conflict on household coping strategies: Evidence from Turkana and Kitui districts in Kenya. *Die Erde* 137(3): 249–70.

LIVERMAN, D. 1990. Drought impacts in Mexico: Climate, agriculture, technology, and land tenure in Sonora and Puebla. *Association of American Geographers* 80(1): 49–72.

McDONALD, B. 2009. Global health and human security: addressing impacts from globalisation and environmental change. Pp. 35–76 in R. Matthew, J. Barnett, B. McDonald and K. O'Brien (eds.), *Global Environmental Change and Human Security.* Cambridge, MA: MIT Press.

MATTHEW, R., McDONALD, B., BARNETT, J., and O'BRIEN, K. (eds.) 2009. *Global Environmental Change and Human Security.* Cambridge, MA: MIT Press.

MESSER, E., COHEN, M., and MARCHIONE, T. 2001. Conflict: A cause and effect of hunger. *Environmental Change and Security Project Report* 7: 1–16.

MORTIMORE, M. 1989. *Adapting to Drought: Farmers, Famines, and Desertification in West Africa.* Cambridge: Cambridge University Press.

NARAYAN, D., PATEL, R., SCHAFFT, K., RADEMACHER, A., and KOCH-SCHULTE, S. 2000. *Voices of the Poor: Can Anyone Hear Us?* Oxford: Oxford University Press.

NORDÅS, R., and GLEDITSCH, N. 2007. Climate conflict: Common sense or nonsense? *Political Geography* 26(6): 627–38.

O'BRIEN, K. 2006. Are we missing the point? Global environmental change as an issue of human security. *Global Environmental Change* 16(1): 1–3.

——ST CLAIR, A., and KRISTOFFERSEN B. (eds.) 2010. *Climate Change, Ethics, and Human Security.* Cambridge: Cambridge University Press.

PARRY, M. L., CANZIANI, O. F., PALUTIKOF, J. P., VAN DER LINDEN P. J., and HANSON, C. E. (eds.) 2007. *Climate Change 2007:* Impacts, Adaptation and Vulnerability. Contribution of Working Group II to the Fourth Assessment Report of the Intergovernmental Panel on Climate Change. Cambridge: Cambridge University Press.

PELUSO, N., HUMPHREY, C., and FORTMANN, L. 1994. The rock, the beach and the tidal pool: People and poverty in natural resource dependent areas. *Society and Natural Resources* 7 (1): 23–38.

PIF (Pacific Islands Forum). 2009. *Forum Communique 2009.* Fourtieth Pacific Islands Forum, Cairns, Australia, 5–6 August. Available at <http://www.pif2009.org.au/> last accessed 25 November 2009.

REPETTO, R. 2009. *The Climate Crisis and the Adaptation Myth.* WP 13, School of Forestry and Environmental Studies, Yale University, New Haven.

SACHS, W. 2006. Climate change and human rights. *The Pontifical Academy of Sciences* 106: 349–57.

SCHEFFRAN, J. 2008. Climate change and security. *Bulletin of the Atomic Scientists* 64(2): 19–25, 59–60.

SCOONES, I. 1998. *Sustainable Rural Livelihoods: A Framework for Analysis.* Working Paper 72, Institute for Development Studies, Brighton.

SEN, A. 1981. *Poverty and Famines: An Essay on Entitlement and Deprivation.* Oxford: Clarendon Press.

——1999. *Development as Freedom.* New York: Anchor Books.

SHAW, M. 1993. There is no such thing as society: Beyond individualism and statism in international security studies. *Review of International Studies* 19(2): 159–75.

SHUE, H. 1999. Bequeathing hazards: Security rights and property rights of future humans. Pp. 38–53 in M. Dore and T. Mount (eds.), *Global Environmental Economics: Equity and the Limits to Markets.* Oxford: Blackwell.

SLADE, N. 2007. Climate change: The human rights implications for small island developing states. *Environmental Law and Policy* 37(2–3): 215–19.

SMITH, D., and VIVEKANANDA, J. 2007. *A Climate of Conflict: The Links between Climate Change, Peace and War.* London: International Alert.

TYLER, N., TURI, J., SUNDSET, M., STRØMBULLD, K., SARA., M, REINERT, E., OSKAL, N., NELLEMANN, C., MCCARTHY, J., MATHIESEN, S., MARTELLO, M., MAGGA, O., HOVELSRUD, G., HANSSEN-BAUER, I., EIRA, N., EIRA, I., and CORELL, R. 2007. Saami reindeer pastoralism under climate change: Applying a generalized framework for vulnerability studies to a sub-arctic social–ecological system. *Global Environmental Change* 17(2): 191–206.

UNDP (United Nations Development Program). 1994. *Human Development Report 1994.* New York: Oxford University Press.

UNGA (United Nations General Assembly). 2009. *Climate Change and its Possible Security Implications.* Report of the Secretary-General, 11 September 2009, A/64/350, available at <http://www.un.org>.

World Bank. 2010. *World Development Report 2010: Development and Climate Change.* Washington, DC: World Bank.

ZIERVOGEL, G., BHARWANI, S., and DOWNING, T. 2006. Adapting to climate variability: Pumpkins, people and policy. *Natural Resources Forum* 30(4): 294–305.

CHAPTER 19

....

CLIMATE REFUGEES AND SECURITY: CONCEPTUALIZATIONS, CATEGORIES, AND CONTESTATIONS

....

TIMOTHY DOYLE AND SANJAY CHATURVEDI

1 INTRODUCTION

....

ADDING yet another complex layer to the ongoing, earth system science driven, climate change debate, various narratives of 'security' within the discipline of international relations are also in competition with one another for greater geopolitical salience and moral authority. This discursive battlefield is visibly marked by fast multiplying, at times highly imaginative, 'mega-solutions' to the seemingly 'mega-problem' of anthropogenic global warming. This chapter concerns itself with one subcategory of the broader climate debate: the *climate refugee*. First, we use the IPCC's three climate change-related categories of human migration or population displacement—drought incidence, increased cyclone (hurricanes and typhoons) intensity and sea-level rise—to initially describe how this phenomenon is usually presented and categorized. However, climate refugees—and climate change— not only exist (in terms of physicality) but are also socially constructed (and understood) using different world-views. As succinctly pointed out by Mike Hulme, 'we won't understand climate change by focusing only on its physicality. We need to understand the ways in which we talk about climate change, the variety of myths we construct about climate change and through which we reveal to ourselves what climate change means to us' (Hulme 2009: 355). Furthermore, 'the *idea* of climate change is now to be found active across the full parade of human endeavors, institutions, practices and stories. The idea that humans are altering the physical climate of the planet through their collective actions, an idea captured in the simple linguistic compound "climate change", is an idea as ubiquitous and as powerful in today's social discourse as are the ideas of democracy, terrorism or nationalism' (ibid. 322).

Understanding the substance of these different world-views and frames and, in turn, viewing the phenomenon through the lenses of these disparate political prisms, will alter our understandings of what a climate refugee actually *is*. In this light then, next, we introduce three theoretical frames—derived from the language of security which, in turn, is imbedded within the academic discourses of the discipline of international relations: realist, liberal, and critical frameworks. The very concept of *climate refugee* is one which is firmly entrenched in the literature of international relations; centered at the heart of discussions pertaining to nation-states, and those citizens who are forced—for whatever reasons—to migrate from them.

Our third task is to use these three theoretical frames as a means of coming to terms with diverse *prescriptive* positions, advocating very different approaches to both mitigation and adaption strategies to alleviate climate displacement and migration.

Finally, we offer a somewhat critical response ourselves. We question the validity of the climate refugee category, arguing that far from providing succor and solace to the most vulnerable communities within the global South, the climate refugee is a *subject of securitization*. The most dominant perspective remains a realist (state-centric), militarist narrative, and the climate refugee is constructed, at best, as a victim of a global polity with no human agency—a political entity *outside sovereignty*—or, even worse, as an environmental criminal or terrorist.

2 Describing the Problem: Mapping Climate Migrations and Displacements

Using the aforementioned IPCC categories of climate refugees, let us begin by describing several examples of the growing reporting and incidences of climate refugees across the globe.

2.1 Drought Incidence

The incidences of climate change-induced drought will increase in many regions across the planet: lands which were once deemed marginal, in agricultural terms, will increasingly become incapable of supporting both human and nonhuman species and ecosystems. Already vulnerable communities in semi-arid regions will be worse hit, most usually from the continents of Africa, Asia, and Australia. Looking at the case of Ethiopia is instructive here. Morrissey argues that drought should be considered as the primary environmental stressor in the highlands of Ethiopia. This is because one-off major droughts encourage 'both temporary-distress migration and permanent migration,' and that permanent migration is used as a means of allowing 'people to escape livelihoods which depend on the availability of water but is also a strategy for managing drought' (Morrissey 2008: 28). Also, in Africa, the case of Kenya is particularly salient. There have been four major droughts in Kenya in the past ten years, which have had a devastating impact on people's lives and livelihoods, in particular on the three million pastoralists of northern Kenya. Adow claims that 'with the increasing frequency and severity of the droughts, pastoralists' land can no longer sustain them and people have been forced to migrate' and that 'the way of life that has supported them for thousands of years is falling prey to the impact of climate change' (Adow 2008: 34).

And climate-induced changes to landscape, producing long-term droughts or other such degrading processes which cripple agriculture, are not the exclusive problematic domain of people with fixed addresses. In fact, impacts upon mobile peoples, such as the Bedouins, who are often forced to live within parcels of land which are most marginal, experience these tensions at an even greater level. Due to the increasing impacts of climate change, these 'nomadic (mobile) peoples are already deriving their livelihoods from marginal and extreme landscapes; changes in physical and biological resources—and the impacts of increasingly severe weather and climate change—are therefore of particular concern to them' (Sternberg and Chatty 2008: 25). Namely, the mobile indigenous peoples in question are from places such as Gabon, Kenya, Tanzania, Senegal, Iran, India, Australia, Jordan, and Mongolia. Extreme weather events due to climate change are threatening the viability of their livelihoods and limiting the effectiveness of their traditional adaptive strategies.

2.2 Increased Cyclone (Hurricanes and Typhoons) Intensity

The IPCC argues that climate change will lead to an increasing number of extreme weather events and, in turn, these events will increase in their intensity. The most vulnerable communities, in this case, will be those inhabiting the regions closest to the oceans and/ or those living in low-lying areas. Obviously, many of the planet's people dwell in these areas but, again, there are certain cases which are most often utilized as exemplars of this type of situation. For example, the case of Bangladesh is an oft-quoted prime example of this type of climate-induced displacement. As a consequence of the increasing sea-level rise in Bangladesh, cyclones may become more frequent as well as more powerful. Pender maintains that 'a higher sea level means that storm surges that accompany cyclones will drive sea water even further inland.' Additionally, the cyclone which occurred in 1991 killed 138,000 and affected over 13 million, with a surge 7.2 meters high. Pender also adds that it is highly likely that surges in future may surpass 10 meters in height, 'penetrating far inland in this country of which two-thirds is lower than five meters above sea level' (Pender 2008: 54).

Of course, climate-induced displacement and migration can also be experienced in the countries of the affluent world but, again, is usually depicted as manifesting itself in the least affluent communities. For example, on 29 August 2005, Hurricane Katrina, with winds of up to 200 kilometers per hour, lashed out at communities living on the Gulf Coast of the United States. Overall, due to the severity of the storm, coupled with poorly designed and maintained levees, there were 1,300 deaths and Hurricane Katrina's impact on human displacement was equally devastating. As Smith (2007: 617–18) writes:

> Stimulated by various offers of assistance from federal, state and private organizations, tens of thousands of Katrina 'migrants' resettled to over 1,042 shelters in 26 states and the District of Columbia. Overall, between 100,000 to 300,000 Louisiana residents alone may have been displaced permanently as a result of Hurricane Katrina and, to a lesser degree, Hurricane Rita, which struck about a month later.

According to Smith (2007: 617), 'The displacement of thousands of U.S. Gulf Coast residents in the aftermath of Hurricane Katrina is emblematic of a human migration challenge that will likely become more severe in the years and decades ahead.'

2.3 Sea-Level Rise

It is this third IPCC category of climate change impacts upon mass human movements which is most ubiquitous. In both academic papers, as well as reports in mass media, images of people being forced to evacuate their homeland are most prevalent. In this media bite, the image depicts people up to their knees in water, as if their habitat was a sinking boat unable to resist the rising waters any longer. Many of these media images have emerged from small island states and other islands such as Tuvalu and the Catarets in the Pacific Ocean, and the Maldives and the Sunderbans in the Indian Ocean.

For example, due to the rapidly increasing rise in sea level, the people from the group of islands which are known as the Catarets are often mentioned as the world's first refugees from the effects of global warming: 'Such is the effect of rising seas that tidal surges have cut one of the small islands in the Carteret group—1,000 miles north east of Australia—in half, the salt water slicing through the low centre' (Shears 2007: 1). In the Indian Ocean region, these rises in sea level are also most prevalent. For example, the Sunderbans are also affected by rising sea level and severe land erosion attributed to global warming. This has caused the concept of climate refugees to be a significant concern for these small islands located on the delta region of the Bay of Bengal. Ghoramara, in the Sunderbans, is predicted to disappear like its neighboring island, Lohachura. 'Over the past 25 years, Ghoramara's land mass is being eaten away by the advancing sea. It has forced more than 7000 people to relocate' (Roy 2009: 6).

In conclusion, of course, the three IPCC categories of climate change that are likely to result in climate refugees are intermeshed: the distinct categories really being for the purposes of structuring description rather than for describing separate phenomena. Others will be impacted by various other manifestations of climate change to the point of making their environments unsustainable: increased incidences of drought, flooding, sea-level rise, extreme weather events, wild-fire outbreaks, etc., are inextricably interwoven. As we began this descriptive section with a note on the impacts of climate change on migration in Africa, let us finish with another African example which illustrates succinctly the complexity of climate-induced population movements. As a result of climate change, Nigeria is experiencing 'increasing incidence of disease, declining agricultural productivity, increasing number of heat waves, unreliable or erratic weather patterns, flooding, declining rainfall in already desert prone areas in the north causing increasing desertification, decreasing food production in central regions, and destruction of livelihoods by rising waters in coastal areas where people depend on fishing and farming' (Chin 2008: 18–36). With such complex ecological processes leading to these migrations and displacements, it must now be obvious that the very concept of a climate refugee is equally complex, with no agreed and precise definition available.

3 Securitizing Climate Refugees

We utilize the term 'securitized' to explain how the dominant forms of debate pertaining to climate refugees have been 'pegged out' as a security issue, when it could just as easily have been framed, for example, as a 'human rights' or as a 'development' issue. As touched upon, this security frame emerges from mainstream discourses of international relations. We will now briefly visit three major schools of thought which divide the discipline, each providing a different interpretative lens. Obviously, climate-induced *migration* is not just a problem for international relations as there are many internal environmentally induced population movements *within* the boundaries of nation-states. These migrants, however, are not *refugees*, which always denote movements across national borders and boundaries. In this manner, it is further evident that any discussion of the *climate refugee* lies firmly under the rubric of international relations.

We then examine the manner in which advocates of these distinct approaches have interpreted environmental and climate change problems (as these, in turn, include the subset of problems oriented around climate refugees) and the ways in which they are imagined to be addressed. These theoretical categories are far from perfect, and often both descriptions and prescriptive policies to avert climate change (and by doing so, climate refugees) blur across these conceptual boundaries.

3.1 Realist Descriptions and Prescriptions

Paul J. Smith (2007) has pointed out that a growing number of countries in various parts of the world have begun to consider international migration as a threat to territorial integrity and national security. This securitization and often militarization, of state responses to migration partially reflects a paradigm shift in how international migration is being perceived. What was once (and ought to be still) a social or labor issue related to human security is now often transformed into a threat to national security. The print as well as visual media have also been responsible in giving far more (mostly sensational) attention to sudden 'disruptive' migration—such as the previously mentioned cases of rapid sea-level rises and subsequent flooding in the Cataracts and the Sundabans—rather than routine migration caused by various push and pull factors, including socio-economic dislocations and discriminations. In short, once immigration is discursively transformed from a law enforcement 'low politics' issue into a 'high politics' security matter, several geopolitical implications follow (Smith 2007). In the context of climate change, in our view, it is useful to be reminded that humans have long chosen to move (migrate) in order to adapt to natural-social calamities and environmental change and the process leading to such a decision is rather complex.

First and foremost, the 'traditionalists' tend to come from the 'realist' school. Not only do they see climate issues through the prism of 'national interests' within the context of an anarchical world system; they see climate as just another issue pertaining to the struggle for power amongst nation-states. In addition, climate, as a form of environmental security, is usually seen as a 'threat-multiplier'; rather than a base or fundamental threat. In this vein, climate can exacerbate tensions but, as an 'alternative' form of security (and, therefore, not as a 'fundamental' one of race, religion, ethnicity, finances, etc.), it acts as an 'accelerator' or 'catalyst' for existing tensions between nation-states (for examples of this dominant realist—and usually militarist—approach, see Myers 1993; Edwards 1999; Salehyan 2005; Reuveny 2007; and Chin 2008). Interestingly, not only does it imagine an anarchical world system; it views natural processes themselves as anarchical. In this world-view, humanity has not only declared a war against itself; but is also locked into mortal combat with the earth itself—Nature as enemy.

In this view of climate security, nation-states are seen as having to protect their borders from climate refugees driven from the global (and particularly global South periphery); protecting their 'natural comparative advantage' (in Ricardo's terms) of coal, uranium, and other markets. Ever since the concept of 'global climate change' rose to prominence in the 1980s, a series of metaphors have been deployed at the service of imaginative geographies of chaotic and catastrophic consequences of climate change, including 'mass devastation', 'violent weather', 'ruined' national economies, 'terror', 'danger', 'extinction', and 'collapse'. A number of 'security' experts and analysts are convinced that the United States will be the 'first responder' to numerous 'national security' threats generated by climate change (see Podesta and Ogden 2007–8). In April 2007, the CNA Corporation (2007), a think tank funded by the US Navy, released a report on climate change and national security by a panel of retired US generals and admirals that concluded: 'Climate Change can act as a threat multiplier for instability in some of the most volatile regions of the world, and it presents significant national security challenges for the United States.'

One of the specific and key pressure points in the realist and largely militarist reconstructions of environmental issues (using climate as a political metaphor) relates to *the blood-dimmed tides*, the washing up climate refugees—and even climate terrorists—upon the shore of the affluent world, due, in part, to rising sea levels. If anything, this imagined ecological Armageddon sees the global North re-engaging with the global South, not through choice, but through necessity (Doyle and Chaturvedi 2010: 532–3).

But realist responses to climate refugees are not always militarist, nor necessarily viewed from the perspective of the affluent, more *climate safe* countries. Indeed, some countries at most immediate risk from climate change sometimes equally endorse concepts of national sovereignty. This time, however, it is not to keep the flood *out,* but rather to affirm the existence and survival of national societies, rather than being seen as stateless victims. Indeed, as in the aforementioned cases of the peoples of the west Pacific Islands or in the Indian Ocean, the international label of climate refugee is often seen as quite offensive by these citizens, as it depicts their societies on the edge of oblivion, as post-event victims, while many in these communities prefer the language of resilience, unwilling to write off their sovereignty, their sense of place and history, their identity, and direct their energies to mitigation strategies, both locally and in international forums (McNamara and Gibson 2009: 481). In this more resilient and *victimless* tone, Thakur writes of certain islands in the Indian Ocean region. Before Ghoramara Island in the Sunderbans submerges and its inhabitants become environmental refugees, Thakur argues, 'they would have shown the world how an entire village can run on solar power at a time when cities are failing to grapple with the threat of global warming and climate change' (Thakur 2009: 4).

3.2 Liberal Interpretations

More liberal notions of international relations re-emerged after the so-called 'victory of capitalism' and the breakup of the communist-inspired USSR in the late 1980s, and world orders which had existed since the Second World War were called into question. During this time of uncertainty, there emerged a global policy-shaping concept embracing a shared plurality of interests which crossed nation-state borders, commonly referred to as *multilateralism*. The multilateralist decade of the 1990s, which ended as the current phase of US unilateralism emerged forcefully in the new millennium, was an era when new boundaries and borders were drawn in the sand, as alternative, more liberal concepts of identity and collectivity were imagined. One such idea which evolved at this time was that of *environmental security* (Doyle and Risely 2008).

A more liberal concept of environmental and climate security which is more inclusive of the interests of the majority of people in the global South is one that moves away 'from viewing environmental stress as an additional threat within the (traditional) conflictual, statist framework, to placing environmental change at the centre of cooperative models of global security' (Dabelko and Dabelko 1995: 4). In these terms, the Kyoto protocol is just one powerful example of how the affluent North seeks to re-engage with the less affluent world, through more cooperative, bi- and multilateral negotiations. Until very recently, of course, countries such as the United States and Australia positioned themselves well outside these cooperative ventures, maintaining a nation-state centric regional and global stance.

A good example of a liberal and more cooperative version of the securitization of climate lies in the political realm of non-state actors or organizations. John Ashton (CEO of E3G, a new organization that aims to convert environmental goals into accessible choices) and Tom Burke would argue that:

> Climate change is not just another environmental issue to be dealt with when time and resources permit. A stable climate, like, national security, is a public good without which economic prosperity and personal fulfillment are impossible. It is a prime duty of a government to secure such goods for their citizens. The current level of investment of political will and financial resources addresses climate change as an environmental rather than as a national security issue. Without a fundamental change in this mind-set governments will remain unable to discharge their duty to their citizens. (Burke and Ashton 2004: 6)

Two major liberal discourses have emerged particularly in relation to climate refugees. The first is legalistic and largely adaptive. There is an extensive legal debate pertaining to the definition of environmental refugee. Much of this literature aims to challenge traditional definitions of refugee under international law (e.g. 1951 Convention), arguing to include international climate migration within this traditional rubric, and then extending it. Other debates within this category seek legal recognition through others means, more directly connected to separate climate protocols.

This strand of liberal discourse relates to humanitarian responses dealing with the provision of sanctuary for refugees and internally displaced persons after climate catastrophe has occurred, and as such is largely adaptive.

The final strand focuses on mitigation, which includes programs and policies that concentrate on reducing ecological footprints (today). Through promoting and working within cooperative international conventions and protocols, like Kyoto, there are obvious efforts to mitigate against climate change: to avoid catastrophe (e.g. Cartaret islands and Indian Ocean Islands setting up solar and wind turbines etc.).

3.3 Examples of Liberal/Legal Arguments Regarding Environmental/Climate Refugees

Who is a refugee? Where and why does (or should) the category of 'economic refugee' end and the category of 'climate refugee' begin? What are the ethical and geopolitical concerns and considerations that define the category termed 'climate refugee' and differentiate it from 'climate migrant'?

There is much debate amongst liberal thinkers pertaining to climate change and refugees. Much of this debate focuses around the very concept of environmental and climate refugee in the strict terms of international and domestic laws. The question is asked: Should the concept of refugee, traditionally used to provide salve for those seeking migration on the basis of political or religious persecution, be extended to cover *climate* refugees? Some argue that these rights, endorsed by the UN, would provide additional protection for those seeking to migrate across national borders, and being granted asylum accordingly by nation-states operating in liberal cooperative fashion. Of course, the other consequence of such actions would be to provide further validity to climate securitization discourses. This is probably the major kind of *adaptation* discourse within the liberal frame, after-the-fact—as it were—of climate-induced migration.

In this vein, Bell, for example, refers to the standard definition of 'environmental refugees,' which comes from El-Hinnawi's United Nations Environment Program report as follows:

> Environmental refugees are defined as those people who have been forced to leave their traditional habitat, temporarily or permanently, because of a marked environmental disruption (natural and/or triggered by people) that jeopardized their existence and/or seriously affected the quality of their life. By 'environmental disruption' in this definition is meant any physical, chemical and/or biological changes in the ecosystem (or the resource base) that render it, temporarily or permanently, unsuitable to support human life. (Bell 2004: 137)

The legal status, then of environmental or climate migrants is very unclear. As Gemenne writes, 'not being prosecuted for their belonging to a particular group and not always crossing an international border', they do not qualify for the status of 'refugee' as spelled out under the Geneva Convention (Gemenne 2006: 13).

Bell refers to the common criticisms of this definition and, in many ways, his article encapsulates the different positions used to demarcate this category of debate. First, he spells out the idea of environmental refugees that has been criticized as, 'unhelpful and unsound intellectually, and unnecessary in practical terms'. The critics' main complaints are that: (1) the label 'environmental' oversimplifies the causes of forced migration; (2) there is no hard, legal evidence of very large numbers of people being displaced by environmental disruptions (particularly desertification and rising sea levels); and, (3) it is a strategic mistake to use the label 'environmental refugees' because it may 'encourage receiving states to treat [refugees] in the same way as "economic migrants" to reduce their responsibility to protect and assist' (Bell 2004: 1–4).

Bierman and Boas (2008) agree that some of the possible reform options—'extending the definition of refugees under the 1951 Geneva Convention Relating to the Status of Refugees or giving responsibilities to the UN Security Council—are less promising and might even be counterproductive.' They contend that a more plausible option lies with a new legal instrument constructed specifically for the needs of climate refugees—'a Protocol on the Recognition, Protection, and Resettlement of Climate Refugees to the United Nations Framework Convention on Climate Change,' supported by a separate funding mechanism, the 'Climate Refugee Protection and Resettlement Fund.'

What is different and interesting about Bierman and Boas' (2008) approach is that whereas much of the legal wrangling over the legitimacy of the climate refugee within this liberal discourse is largely about the responsibility of nation-states in relation to adaptation—after the fact of climate change—they also utilize the category of climate refugee in terms of mitigation. They write:

> It is crucial, then, that this protocol not be framed in terms of emergency response and disaster relief but in planned and organized voluntary resettlement programs. There is no need to wait for extreme weather events to strike and islands and coastal regions to be flooded. (Bierman and Boas 2008: 10)

Another use of this liberal approach is to allow costs of mitigation and adaptation incurred by migration to be met by nation-states or international regimes, and combinations thereof. As Gemenne (2006: 13) suggests, regardless of whether or not climate-displaced persons ever gain recognition as fully-fledged refugees, 'the issue of financial compensation will remain'. He argues: 'government that needs to expropriate residents for the completion of an infrastructure project will offer them a financial compensation. Shouldn't the same mechanism be applied to climate change forced displacements?' (Ibid. 15.) Therefore, Gemenne (2006: 15) suggests a twofold system of burden sharing: 'People burden-sharing on a regional basis, if possible within the framework of an international agreement on the status of these migrants;' and, 'Costs of burden-sharing on the international level, through an adaptation fund or a similar scheme.'

Of course, these liberal voices do not question those political acts *before* climate change became an item on the global environmental agenda. They merely deal with the questions: How do we mitigate the problem in order to stop it? And, how do we adapt to these changes? More critical responses, instead, amongst other approaches, use the category of climate refugee to address the climate *sins of the past*.

3.4 Critical Perspectives of the Climate Refugee

Finally, there are subservient, but more critical traditions within the rhetoric of environmental security and climate refugees. We have time to investigate only two categories of argumentation here. Advocates of the first type of critical position do not concentrate, as their liberal counterparts do, on ecological footprints, but rather upon ecological debts incurred over centuries of exploitation of the North over the South. This position is still crafted within discourses of security, focusing on the *causes* of environmental *in*securities (Barnett 2003; see also Barnett in this volume). For example, in the Asian region afflicted by the Tsunami in 2004, both realist and liberal responses were used as justifications to deploy military personnel to the worst affected areas; or to construct cooperative 'early warning systems' across the oceanic region. On the other hand, a critical response reviews and responds to the causative factors leading to environmental instabilities and insecurities, seeking, amongst other things, reparations for climate-vulnerable communities.

The approach of transnational green NGO Friends of the Earth International (FoEI) is of interest here. In a report made by the organization, Davissen and Long make FoEI's critical position clear:

> The global North, as the major greenhouse polluters, bears a significant responsibility for this disruption. Accordingly, we believe that the North must make reparations. In practical terms, this will mean we must make room for environmental refugees, as well as changing policies that contribute to the creation of more refugees. (Davissen and Long 2003: 8)

Climate debt is the special case of environmental justice—where industrialized countries have over-exploited their 'environmental space' in the past, having to borrow and steal from developing countries in order to accumulate wealth, and accruing ecological debts as a result of this historic over-consumption (ibid.). Friends of the Earth maintain that climate refugees add to the 'climate debt,' owed by the global North to the global South due to the 'unsustainable extraction and consumption of fossil fuels.' Furthermore, because refugees are among the world's most vulnerable people, the protection of their rights must be the principal concern in responses to climate change. FoEI argues that the responsibility for climate change and accommodating past earthly indiscretions rests firmly on the more affluent, minority world. Using this same type of critical approach, the Bangladeshi Environment Minister, Mrs Sajeeda Choudhury, commented that 'it is up to the developed countries of the world to rethink their immigration policies' (Kirby 2000: 1–2).

Whilst some commentators and organizations like FoEI seek to introduce critical arguments *within* the dominant climate change and refugee frame, another critical approach is to question the validity of the environmental security framework itself (Dalby 2009). As mentioned, this latter critical approach largely informs our own stance on the matter (Chaturvedi and Doyle 2012). From this position, due to the fact that climate change and refugee discourses have already been populated—and ultimately dominated—by more realist (and even liberal) discourses, climate-displaced persons do not gain through being included in this narrative. Despite FoEI's best emancipatory intentions, their organization will, at best, achieve little within this frame and, at worst, legitimate a climate discourse which further securitizes the peoples of the global South within a social and environmental agenda initiated, shaped, and controlled by the global North.

Within this latter type of critical approach, writers such as Simon Dalby also argue that the very act of securitization, with its implicit agenda set by military-industrial complexes, ultimately disenfranchises the majority, stripping environmental 'speech' from its more emancipatory projects (Dalby 2002). Whether it be climate or environmental refugees, or migrations in general, Buonfino, also, endorses this approach, when he contends:

> Nowadays, securitizing migration is a frequent response of European governments to people flows. However, securitization of migration creates more instability than it does security. It enhances fears of the Other and exacerbates difference, thus endangering peaceful coexistence. It is more than ever important for the intellectual and academic community to attempt to give accounts of immigration which are fair and balanced, or at least whose arguments are based on reliable material . . . Above all, it means understanding that human rights are global and not confined to small territories. Seeing immigration as an expression of globalization

which we can live with and which we can help improve, is not only a political challenge. It is a question affecting our quality of life. (Buonfino 2004: 4–5)

But, it is all very well in getting into *theoretical hair-splitting* of categories, premises, and contestations. The fact remains that the concept of the climate refugee seems to be universally embraced, although for very different reasons, within all three major schools of thought in international relations. Indeed, this is the strength and power of the climate change/climate refugee phenomenon. Not only does it cross the boundaries of nation-states, it also has the ability to waltz across theoretical academic boundaries which are usually guarded by scholars inhabiting these territories with a separatist ferocity. Perhaps climate change is simply a subset of environmental security, and in turn, environmental security is just a subset of non-traditional security. Indeed, these new categories of security and securization fit in nicely with Ulrich Beck's globalizing and homogenizing concept of 'risk society.' In this seminal work, Beck was referring to an 'era of modern society that no longer merely casts off traditional ways of life but rather wrestles with the side effects of successful modernization—with precarious biographies and inscrutable threats that affect everybody and against which nobody can adequately insure' (Beck 2009: 8). Beck has further argued in his more recent work that

> in the realist view [of world risk-society] the consequences of and dangers of developed industrial production 'are' now global. This 'are' is supported by natural scientific data ... this dynamic is reflected at the beginning of the twenty-first century in global warming, which is an ideal-typical illustration of the fact that environmental destruction 'knows no boundaries'. (Beck 2009: 85)

4 CONCLUSIONS: RELOCATING 'REFUGEES' IN HUMANITARIAN SPACES? IMAGINED GEOGRAPHIES OF EXCLUSION, INVISIBILITY, AND VIOLENCE

The different ways of understanding climate refugees can be most usefully seen as a battle (and sometimes a peace) between different lenses, frames, and discourses of international relations. Climate refugees, by their very definition, are seen as an international 'security problem,' as they cross the boundaries of nation-states. In this vein, the category of 'climate refugees' stands out in terms of a calculated ambiguity deployed by certain actors and agencies, especially in the global North, to mark the boundaries between 'our space' and 'their space' in pursuit of diverse agendas of control and domination (Gregory 2009: 369–70). In our own view, what lies at the core of dominant representations of the 'Climate Refugee' (be they the so-called 'realist' or 'liberal' or 'critical) is the fear and cartographic anxieties of the affluent and the privileged that their 'orderly' spaces are going to be invaded *en masse* by the unruly lives of refugees from global South (Chaturvedi and Doyle 2010a).

The questions then become: Where does it takes us? What does this provide for climate-displaced persons as they wrestle with the everyday realities of their predicaments? Jon

Barnett attempts to provide some way out of what can become a critical 'dead-end.' Like other critical geopolitical thinkers, he does acknowledge that, 'The crux of the problem is that national security discourse and practice tends to appropriate all alternative security discourses no matter how antithetical' (Barnett 2003: 14). But he also attempts to provide real policy choices: he proposes that the IPCC scientists should downplay such climate change militarist discourses—being 'cautious on the issue of violent conflict and refugees'—and, instead, focus on climate justice issues. This approach, he argues 'might helpfully integrate science and policy and usefully elucidate the nature of the "danger" that the UNFCCC ultimately seeks to avoid' (Barnett 2003: 14). Also, in desecuritizing the climate discourse, reparations for displacements may be pursued.

In addition, it is important to ask the people of the global South themselves, as to what *they* think of the climate refugee frame. The fact remains that many in the global South resent their depiction as 'victims' within climate refugee discourses. For example, McNamara and Gibson, as mentioned previously, state that the very definition of 'climate refugee' has encountered considerable resistance in the Pacific for various reasons. They argue that in part, this resistance is due to not wanting to be portrayed as 'weak and passive victims of climate change' (McNamara and Gibson 2009: 481). They argue that 'victimhood and vulnerability' have regularly been 'subtexts to the depiction of "climate refugees" by NGOs and the news media.' In addition, in extensive interviews they conducted between ambassadors of Small Island States, it is clear that any mention of being classified as 'climate refugees' or 'helpless victims' has been strongly opposed and is often considered offensive to these diplomats.

More importantly, a critical social science perspective on the 'climate refugee' would suggest that this particular category of climate change is likely to be defined not so much by the 'well-founded fear' of the 'victim' of climate change as it is going to be dictated and driven by largely 'ill-founded fear' of those who suspect that their 'borders' are going to be violated by thousands and millions of helpless climate migrants seeking protection and care (Chaturvedi and Doyle 2010b: 208). This geopolitics of fear is in some ways integral to what Ulrich Beck described as 'risk society' (Beck 1992).

Finally, the UNHCR has certain reservations about establishing a new legal category of 'climate refugees'. In our view, this is quite understandable as well as appropriate. Any attempt to broaden and deepen the 'humanitarian space' through the introduction of the category of 'climate refugees'—despite best possible intentions on the part of various actors, agencies, or social forces—is likely to be regressive rather than progressive, oppressive rather than emancipatory, and exclusive rather than inclusive. In the absence of a serious, sincere, and critical engagement with the ethics and the politics of representation and human rights of refugees in general (Limbu 2009), the securitization of climate-displaced persons remains problematic.

References

Adow, M. 2008. Pastoralists in Kenya. *Forced Migration Review* 31: 34.
Barnett, J. 2003. Security and climate change. *Global Environmental Change* 13(1): 7–17.

BECK, U. 1992. *Risk Society: Towards a New Modernity*. London: Sage.

—— 2009. *World at Risk*, trans. C. CRONIN. Cambridge: Polity Press.

BELL, D. 2004. Environmental refugees: What rights? Which duties? *Res Publica* 10(2): 135.

BRONEN, R. 2008. Forced migration of Alaskan indigenous communities to climate change: Creating a human rights response. Available online at <http://www.iom.int/jahia/webdav/site/myjahiasite/shared/shared/mainsite/events/docs/abstract.pdf>. Accessed 10 November 2009.

BIERMANN, F., and BOAS, I. 2008. Protecting climate refugees: A case for a global protocol. *Environment*, 50(6): 8–17.

BUONFINO, A. 2004. Securitizing migration. Available online at <http://www.opendemocracy.net/ . . . migrationeurope/article_1734.jsp>. Accessed on 15 October 2009.

BURKE, T., and ASHTON, J. 2004. The geopolitics of climate change. Unpublished paper given as a presentation by the authors at an SWP roundtable on climate change and foreign policy on 4 February 2004. Available online at <http://www.envirosecurity.org/conference/background/ClimateChangeGeopolitics.pdf>. Accessed on 29 October 2009.

CHATURVEDI, S., and DOYLE, T. 2010. Geopolitics of climate change and Australia's 'reengagement' with Asia: Discourses of fear and cartographic anxieties. *Australian Journal of Political Science*, special issue ed. Carol Johnson and Juanita Elias (forthcoming 2010).

—— 2010b. Geopolitics of fear and the emergence of 'climate refugees': imaginative geographies of climate change and displacements in Bangladesh, *Journal of the Indian Ocean Region*, vol 6, no. 2: 206–22.

—— Forthcoming 2011. *Climate Terror: A Critical Geopolitics of Climate Change*. Basingstoke and New York: Palgrave Macmillan.

CHIN, J. 2008. Coping with chaos: The national and international security aspects of global climate change. *The Journal of International and Policy Solutions* 9.

CNA Corporation. 2007. National security and the threat of climate change. Available online at <http://securityandclimate.cna.org/report/National%20Security%20and%20the%20Threat%20of%20Climate%20Change.pdf>. Accessed on 5 October 2009.

COULDREY, M., and HERSON, M. 2008. Climate change and displacement. *Forced Migration Review* 31.

DABELKO, G. D., and DABELKO, D. D. 1995. Environmental security: Issues of conflict and redefinition. *Woodrow Wilson Environmental Change and Security Project Report* 1: 3–12.

DALBY, S. 2002. *Environmental Security*. Minneapolis: University of Minneapolis Press.

—— 2009. *Security and Environmental Change*. Cambridge: Polity Press.

DAVISSEN, J., and LONG, S. 2003. The impact of climate change on small island states. *Friends of the Earth Australia*. Available online at <http://www.foei.org/search?SearchableText=climate+change+refugees> June 2003. Accessed on 7 October 2009.

DEMERITT, D. 2001. The construction of global warming and the politics of science. *Annals of the Association of American Geographers* 91(2): 307–37.

DOYLE, T., and CHATURVEDI, S. 2010. Climate territories: A global soul for the global South? *Geopolitics* special edition 15(3): 516–35.

—— and RISELY. M. 2008. *Crucible for Survival: Environmental Security and Justice in the Indian Ocean Region*. New Jersey: Rutgers University Press.

EDWARDS, M. J. 1999. Security implications of a worst-case scenario of climate change in the South-west Pacific. *Australian Geographer* 30(3): 311–30.

GEMENNE, F. 2006. Climate change and forced displacements: Towards a global environmental responsibility? The case of the Small Island Developing States (SIDS) in the South

Pacific Ocean. *Paper presented at the 47th Annual Convention of the International Studies Association (ISA)*, San Diego, 22–5 March 2006.

GREGORY, D. 2009. Imaginative geographies. Pp. 369–371 in D. Gregory, R. Johnston, G. Pratt, M. J. Watts, and S. Whatomore (eds.), *The Dictionary of Human Geography.* 5th edn., Chichester: Wiley-Blackwell.

HULME, M. 2009. *Why We Disagree about Climate Change: Understanding Controversy, Inaction and Opportunity.* Cambridge: Cambridge University Press.

KIRBY, A. 2000. West warned on climate refugees. *BBC News,* 24 January.

KIBREAB, G. 1997. Environmental causes and impact of refugee movements: A critique of the current debate. *Disasters* 21(1): 20–38.

LIMBU, B. 2009. Illegible humanity: The refugee, human rights, and the question of representation. *Journal of Refugee Studies* 22(3): 257–82.

McNAMARA, E., and GIBSON, C. 2009. We do not want to leave our land: Pacific ambassadors at the United Nations resist the category of 'climate refugees'. *Geoforum* 40(3): 475.

MORRISSEY, J. 2008. Rural-urban migration in Ethiopia. *Forced Migration Review* 31: 28.

MYERS, N. 1993. Environmental refugees in a globally warmed world. *Bioscience* 43(11): 752–61.

PENDER, J. 2008. Community-led adaptation in Bangladesh. *Forced Migration Review* 31: 54.

PODESTA. J., and OGDEN, P. 2007–8. The security implications of climate change. *The Washington Quarterly* 31(1): 115–38.

REUVENY, R. 2007. Climate change: induced migration and violent conflict. *Political Geography* 26: 656–73.

ROY. S. 2009. Film captures Sunderbans plight. *Statesman,* Kolkata, 26 February.

SALEHYAN, I. 2005. Refugees, climate change and instability. Presented at Human Security and Climate Change: International Workshop, Oslo, 21–23 June (conference).

SHEARS, R. 2007. The world's first climate change refugees to leave island due to rising sea levels. *Daily Mail Online.* Available online at <http://www.dailymail.co.uk/news/article-503228/The-worlds-climate-change-refugees-leave-island-rising-sea-levels.html>. Accessed on 7 October 2009.

SMITH, P. J. 2007. Climate change, mass migration and the military response. *Orbis: A Journal of World Affairs* 51(4): 617.

STERNBERG, T., and CHATTY, D. 2008. Mobile indigenous peoples. *Forced Migration Review* 31: 25.

TERI. 2009. The Energy and Resources Institute, New Delhi, press releases <http://www.teriin.org/index.php?option=com_pressrelease&task=details&sid=156>

THAKUR, J. 2009. Guv buoy for sinking island. *Hindustan Times,* Kolkata, 25 February.

PART VI

JUSTICE

....

CHAPTER 20

FROM EFFICIENCY TO JUSTICE: UTILITY AS THE INFORMATIONAL BASIS OF CLIMATE STRATEGIES, AND SOME ALTERNATIVES

....

SIMON DIETZ*

1 INTRODUCTION

....

THE aim of this chapter is to consider, from an ethical point of view, the role that economics should play in evaluating climate change strategies. Economics has been a prominent player in the intellectual and political debate about how to respond to climate change, and frequently a controversial one. Much of the controversy apparently surrounds how the consumption of individuals living in different places, at different times, and in different states of nature is weighted. On the temporal dimension, this issue is represented by the infamous discount rate. However, I argue that an equally, if not more, important ethical judgment made by economics comes earlier, when the informational basis of evaluation is accepted to be nothing more and nothing less than the 'utility' of individuals.

This is by no means an original insight, but on its basis I try to make constructive suggestions as to how evaluation of climate change can move forward, drawing on the strengths of economics, but compensating for its weaknesses. I draw on the work of John Broome (1999) and Amartya Sen (1987), among others, to argue that the strength of the economic approach lies in its emphasis on interdependence and comparability of changes

* I am grateful to Luc Bovens, Cameron Hepburn, Alec Morton, and one anonymous referee for comments, as well as to the editors. My research has been supported by the Grantham Foundation for the Protection of the Environment, as well as the Centre for Climate Change Economics and Policy, which is funded by the UK's Economic and Social Research Council (ESRC) and by Munich Re. The usual disclaimers apply.

to human well-being. That is, any climate strategy, be it adaptation, mitigation, or 'business as usual', consists in changes to human well-being that are linked across time, space, and states of nature. Any convincing evaluation of such strategies must be equipped to think about the comparisons entailed.

The weakness, however, lies in seeing human well-being solely through a 'utility' lens. Thus what would constitute real progress would be a systematic evaluation of the positive and negative changes in human well-being as a result of climate change strategies, on multiple dimensions of that well-being. It might be argued that existing assessments, such as those of the Intergovernmental Panel on Climate Change (IPCC), fit the bill, but I argue that, because they are informal, we miss an opportunity to bring to bear formal methods of comparison that might prove decisive in the climate debate.

2 EFFICIENT CLIMATE POLICY

To begin with, it is worth briefly summarizing how economics evaluates climate change strategies, the commonalities, and major debates within the tradition. Economists have been investigating the properties of efficient responses to climate change for over thirty years,[1] and have been a prominent voice in the climate debate since the early 1990s (Nordhaus 1991; Cline 1992). While differences in approach have from time to time become all too evident—William Nordhaus and William Cline in particular clashed over the opposing conclusions of their aforementioned studies, and Nordhaus and others have recently been involved in a very similar-looking disagreement with the findings of Nicholas Stern's (2007) study—there is a common ethical core to the great majority of analyses, which is derived from a particular mathematical model of social welfare.

In this model, the sole objective is to maximize the weighted sum of the utilities of individuals over time, space, and in different states of nature (with associated probabilities). The weights are given by the particular social welfare function chosen, which is almost always of the 'classical utilitarian' form. In a classical utilitarian social welfare function, individual utilities are initially unweighted, although a positive utility discount rate is usually applied, which means that the utility of future individuals is ultimately given less weight than the utility of present individuals. It is commonly assumed that each individual's utility can be estimated on the basis of their aggregate consumption of goods and services, using a utility function that is common to all of them. The impact of climate change, and of responses to it (i.e. adaptation and mitigation), is measured as a change in this consumption.

If the foregoing description appears a little dry, that is because care needs to be taken in ascribing ethical meaning to this model—building a 'straw man' would serve little purpose in any meaningful discussion of the pros and cons of the economic approach. 'Utility' in contemporary economics is perhaps most accurately understood as 'that which represents a person's preferences' (Broome 1991: 3) (see also Harsanyi 1977). This is 'axiomatic utility theory', in which the concept of utility means nothing more and nothing less than the value taken by a function describing a person's preferences over a set of alternative goods and services. These preferences must conform to a set of axioms; hence axiomatic utility theory. In this way, the particular meaning of utility has subtly changed since the time it was coined by Jeremy Bentham to denote the tendency to produce happiness: in contemporary economics utility simply means 'that which represents a person's preferences,' whether or not this has anything in particular to do with an individual's happiness or good, or society's happiness or good (and happiness and good do not, of course, necessarily come to the same thing).

However, there is a paradox. Axiomatic utility theory was constructed as a response to the charge, led by Lionel Robbins (1935), that interpersonal comparisons of utility were impossible. Yet, readers will already have noted, the welfare-economic model described above is based on a social welfare function, which enables the utilities of different individuals to be aggregated. Furthermore, for the sake of tractability, each individual is assumed to have the same utility function. The most obvious interpretation of this model is that it does indeed make interpersonal comparisons of (cardinal) utility, where utility is the satisfaction of preferences. Thus it is no longer clear precisely how to interpret the meaning of utility in the economic model of climate change. It can still be interpreted as the satisfaction of preferences, where each individual has identical preferences over a composite good, but there is a renewed sense in which by doing so we are taking a view on individuals' underlying well-being.

In order to apply the economic approach to climate change empirically, it is necessary to construct a coupled model of the economy and climate system, now widely known as an Integrated Assessment Model (hereafter IAM: see Hope 2005 for a review).[2] The sorts of IAM used for economic evaluation must be 'full-scale' (Weyant, Davidson, et al. 1996), in the sense that they must have some representation of every link in the chain from anthropogenic emissions of greenhouse gases to economic impacts of climate change, which, to fully optimize social welfare, must eventually work their way through economic activity back to emissions again.

What are the typical results of running models like this? Insofar as one can draw a caricature, economic evaluations of climate change have tended to conclude, on the one hand, that both adaptation and mitigation are efficient strategies (thus in respect of mitigation they contradict the climate 'deniers' or 'skeptics'), but, on the other hand, that the optimal rate of emissions reductions is rather modest (e.g. Nordhaus 1991; Maddison 1995; Manne and Richels 1995; Tol 1997; Nordhaus and Boyer 2000; Nordhaus 2008). Perhaps this should not surprise us, given President Truman's famous rhetorical call for 'one-handed' economists.[3] By modest is meant both gradual in the first instance and lower in the long run than many environmentalists and climate scientists would recommend. Thus, taking a representative sweep across the economic literature, one finds little support for a strategy that limits global warming to below 2 °C relative to pre-industrial levels, despite the recent political and scientific focus on just such a long-run target.

However, readers will no doubt already be complaining that such a caricature masks some well-known exceptions. These include the studies of Cline (1992) and Stern (2007), both of whom recommended more aggressive action to mitigate climate change in the short and long run (although even Stern's study does not firmly advocate a commitment to avoiding 2 °C warming). The reasons for disagreement from *within* the economic tradition are essentially twofold. First, studies have varied in the predictions they have made about the (undiscounted) costs and benefits of adaptation and mitigation. Some studies are essentially optimistic that the economic impacts of climate change will be relatively low, due to a moderate climatic response to emissions, and easy adaptation to environmental changes. Others are more pessimistic. Similarly, some studies are pessimistic about the economic cost of mitigation. Others are optimistic. Different researchers thus make different predictions in the face of uncertainty, illustrating in one way why uncertainty is relevant to ethical discussions of climate change (Gardiner 2004).[4]

Second, studies have varied in the way in which the standard welfare-economic model, described above, has been parameterized. In particular, studies have used different utility discount rates, which, to recall, changes the weight placed on the utility of future generations in the social welfare function. They have also made different assumptions about the rate at which the marginal utility of consumption (i.e. the additional utility obtained from one more unit of consumption) diminishes as one becomes richer (I will return to this specific issue in section 4). This rate is an argument in the social discount rate, and also describes the decision maker's attitude to risk and to inequality across individuals living at a particular point in time.

Navigating debates about the discount rate, risk, and inequality aversion would require a paper in itself, and they are indeed picked up elsewhere in this book. Therefore I will not focus on them here, turning instead in section 4 to the ethical implications of utility as the informational basis for a climate strategy. First, however, I want to examine what is compelling about the economic approach.

3 Interdependence as the Strength of the Economic Approach

In thinking about what there might be to commend the economic approach from an ethical point of view, particularly in the context of climate change, it is worth reflecting on some of the more important general contributions to the subject. Amartya Sen's *On Ethics and Economics* (Sen 1987) and John Broome's *Ethics out of Economics* (1999) draw similar conclusions. For them, the strength of the economic approach is its emphasis, via an elaborate set of analytical tools, on interdependence.

Sen uses as his main example the study of the causes of famine:

> The fact that famines can be caused even in situations of high and increasing availability of food can be better understood by bringing in patterns of interdependence which general equilibrium theory has emphasized and focused on. (1987: 9)

This argument stresses the contribution of so-called 'positive' economics. Indeed, Sen describes his example as one where the contribution comes from what he calls the 'engineering approach' to economics. In this vein, economics has enhanced our under-standing of effective climate strategies in many ways. Prominent examples of how the study of interdependent supply and demand relations has generated insights for climate strategies include the way in which decentralized incentives such as tradable carbon permits might bring about cost-effective emissions reductions across space and time, the so-called 're-bound effect', whereby increases in the efficiency with which we use energy lead to increases in energy use itself as prices fall (i.e. a cautionary tale), and the potential for changing relative prices to bring about adaptation to climate change.

But Sen's argument does not have anything to do with ethics per se. Rather, it highlights the contribution of positive economics to the design and implementation of effective climate strategies, which is of indirect ethical significance only (effective responses to climate change are presumably ethically desirable in the vast majority of cases). Broome (1999) takes on the contribution of economics to ethics more directly.

In the previous section I noted that economists' recommendations on optimal greenhouse gas emissions have tended to be of the 'on the one hand —— on the other hand ——' variety that (in a quite different context) so frustrated Harry Truman. This is even true of those economic studies that most look like a call to arms, such as the Stern Review (Stern 2007), for even Stern was careful to present his reasoning on optimal emissions as a broad comparison of benefits and costs (see in particular his chapter 13). One of the main reasons why economists tend to draw these conclusions is that responses to climate change are generally thought to involve weighing the interests of individuals in different places and times (and, less intuitively, in different states of nature). It is this aspect of interdependence that interested Broome:

> we have to balance the interests of future people against the interests of presently living people, fun in retirement against fun in youth, the wellbeing of the deprived against the wellbeing of the successful or lucky . . . These are places where the scarcity of resources forces a society to weigh up alternative possible uses for these resources, and economics claims to be the science of scarcity. (Broome 1999: 1–2)

How can economics help us to think about these problems? For Broome, the answer generally lies in the way in which economics forces us to think comparatively (where he thinks philosophers face some room for improvement), and specifically in the axioms of utility theory, which provide useful theorems that can, by analogy, be used to analyze the structure of good. However, we must remember not to equate utility with good—Broome was careful to argue that economics has useful things to say about the structure of good only, not the substance.

Again there are numerous examples of the power of economics in this regard. Indeed, it can be argued that the discounting debate has done as much as any other intellectual exchange to highlight the intergenerational balancing act we are required to resolve in the face of anthropogenic climate change, even though some economists involved would not like to admit the ethical issues involved (Dietz, Hepburn, et al. 2008). Similarly, while controversial in many quarters, Bjorn Lomborg has undoubtedly pursued a powerful line of questioning in the so-called Copenhagen Consensus (see Lomborg 2004), namely that scarce resources should be targeted at whichever public policy problem has the highest social rate of return. In this way he pits investment in climate change mitigation against, for instance, direct spending on public health in low-income countries, and he tends to find that climate change mitigation fares rather poorly. Many have objected to the concept of the Copenhagen Consensus (e.g. Sachs 2004)—and it is undoubtedly flawed even from an economic perspective[5]—but the general premise of comparing the return to investment in climate change with returns to other public investments such as health and education has rather a lot of appeal.

4 UTILITY AS AN INFORMATIONAL BASIS FOR CLIMATE CHANGE STRATEGIES

Climate change has the potential to bring about wide-ranging effects on human well-being, positive and negative. For example, in its Fourth Assessment Report, the IPCC provides a summary classification of five types of effect: on water supply, natural ecosystems, food supply, land in coastal zones, and health (2007b). In turn, this broad range of changes has the potential to affect human lives in a variety of ways, including directly, and indirectly by changing people's economic, social, and political circumstances. Mitigation of climate change will also affect human well-being. In this case, many of the relevant changes are likely to work their way through economic circumstances, as the cost of purchasing goods and services with embodied carbon rises. However, broader changes will also occur, such as the social impacts of unemployment.

Any such diversity of impact presents challenges for evaluation. The economic approach, however, responds in a singular way, attempting to measure all relevant changes as changes in utility. From an ethical point of view, this is likely to be the single most important precommitment, at least as important as particular choices about how utilities are weighted, which is essentially what much of the discounting controversy is about. As Sen put it:

> Each evaluative approach can, to a great extent, be characterised by its informational basis: the information that is needed for making judgments using that approach and—no less important—the information that is 'excluded' from a direct evaluative role in that approach. (1999: 56)

It is again worth taking some care to explain precisely how this transformation into the metric of utility is achieved in practice. As discussed, the basis of utility is an aggregate measure of the consumption of goods and services by individuals. Aggregate consumption per capita in empirical studies is simply derived from a future prediction of economic output (i.e. gross domestic product or GDP) per capita, by netting out investment. The impacts of climate change, and responses to it, are then estimated as equivalent changes in consumption, and added to this baseline flow of consumption per capita. Thus every effect of climate change, of adaptation, and of mitigation must be priced.

Consumption is transformed into utility by means of a utility function. This function is always assumed to be curved, specifically to be concave (Figure 20.1 helps visualize this property). What this means is that the marginal utility of consumption diminishes as one becomes richer. The effect of this is to place less (more) weight on the impacts of climate change and response strategies on rich (poor) individuals. Since consumption happens to be distributed unequally across time, space, and states of nature, this is how the curvature of the utility function affects the social discount rate, risk and inequality aversion.

There are undoubtedly things to commend about such an approach. Certainly any approach that takes seriously (i) the consequences of policy choices for (ii) human well-being can be insightful (Sen 1999). However, the narrowness of the approach also gives rise to some serious concerns, which have been expressed more generally about welfare economics in numerous other settings.

Perhaps the most obvious one is that the approach apparently ignores several factors that contribute to human well-being. In particular, what role do changes in environmental, political and social circumstances play? The answer is that if they can be estimated as equivalent changes in consumption—monetized—then they can be included in the estimation of utility. However, there are two problems. First, the baseline for utility is, as mentioned, taken from aggregate consumption per capita, or in other words essentially individual income. Thus the baseline is certainly a narrow measure of human well-being, and does ignore other non-monetary constituents. So we do not have a broad view of where people are starting from in terms of well-being. Second, it is in practice very difficult to place money values on many of the effects of climate change, and it is well known that the

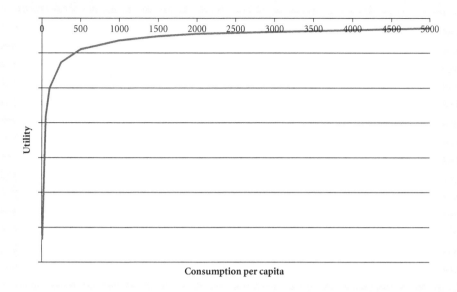

FIGURE **20.1** Utility as a function of consumption.

IAMs used to conduct economic evaluation omit some potentially important changes in environmental, political and social conditions (Watkiss and Downing 2008).

Third, the approach does not pay nearly enough attention to the distinction that is suggested to exist between the things that human beings vitally need, and the things that they merely desire (e.g. O'Neill, Holland, et al. 2008).[6] David Wiggins (2005) makes this point in a quite different, but relevant, context, reflecting on plans to demolish significant tracts of urban London in the 1960s and 1970s to make way for a system of motorways (a plan that, incidentally, failed to materialize). One argument for doing so was the result of a cost-benefit analysis (CBA), which compared the cost of land, resettlement, and construction with the benefit of travel-time savings. According to the CBA, the net present value of the project was positive, but for Wiggins this ran contrary to an intuitive sense in which the capacity of local communities to meet their vital needs was being diminished:

> These were places in which ordinary human lives of passable urban contentment were already being lived, and in which it was possible to satisfy after some fashion a huge variety of human needs [at least some of which were 'vital']. The disvalue of the destruction was swamped, however, by the simple numerosity of a vast sum of time savings. (Wiggins 2005: 27)

It does seem on the basis of Wiggins's account that the CBA of the London urban motorway system was very narrowly construed, such that the only costs of the project that were actually measured were of acquiring land, resettling residents, and construction. The broader social costs of dislocation were thus ignored, not to mention the environmental costs. But the theory and practice of CBA has moved on a long way since then, and in particular due to its increasing application to environmental problems, a great deal of work has been done on monetizing impacts that are not valued in markets (Pearce, Atkinson, et al. 2006). As a result, and indeed as a consequence of changing other assumptions, it is quite conceivable that a broader CBA would have suggested the project costs exceeded the benefits. It is also important to point out that the assumption of diminishing marginal utility of consumption does capture the fact that those on low incomes obtain (lose) far more utility from an increase (decrease) in that income than those on high incomes, which might be interpreted as capturing some notion of need.

However, these qualifications are rather beside the point, which is that there is no *guarantee* that the failure of some individuals to meet their basic needs is not overridden by an increase to other individuals of things that they merely desire (or in other words that they merely instrumentally need, as opposed to vitally needing).[7] Similar points have been raised in connection with the economic evaluation of climate change, where it has for instance been pointed out that it is possible for recreational benefits of warming in high-latitude, high-income regions to outweigh loss of life in low-income countries (Spash 1998).

In any case, the focus on diminishing marginal utility is also beside the point, since an approach founded on vital needs tends to point towards a variety of such needs, none of which are substitutable, either by instrumental needs, or by other vital needs (O'Neill, Holland, et al. 2008). The 'monistic' method of collapsing all relevant changes in human well-being to changes in utility does not reflect that. I shall return to notions of multiple criteria and pluralism below.

Closely related to the emphasis on vital needs are notions of liberty and in particular of rights. A vast amount has been written about liberties and rights, ranging from extensive systems such as Robert Nozick's (1974) to less extensive ones such as John Rawls's (1971), and I cannot hope to cover them here. What generally unites such approaches, however, is the idea that there is some set of rights (i.e. to some things) that must not be violated under any circumstances, or at least only under the most extreme of circumstances.[8] They are, in effect, side constraints, and like vital needs they would constitute the sort of absolute 'bottom line' for evaluation that utility analysis cannot represent.

Several have suggested that future generations have certain rights that might be compromised by climate change (Spash 1998; Shue 1999; Adger 2004; Caney 2008). Let us look at Simon Caney's formulation. According to it, climate change threatens several fundamental interests of future people, including their food supply, their access to land, and their health. These interests are weighty enough to impose obligations on others, which by definition allows corresponding rights to be established. These rights must not be violated, which leads to a presumption in favor of mitigation of climate change, until we can be sufficiently confident that they will be protected.

The difficulty with this argument, which is non-consequentialist, is the following. If it turned out that mitigation of climate change would also violate human rights of one form or another, it would no longer be clear what course of action to take. One would in effect have conflicting side constraints. In fact, Caney (2008) does address this matter, but his assumption is that the cost of mitigation (in a general sense) is sufficiently low to negate the potential conflict:

> Were the costs excessive [which they are judged not to be] then one might conclude that the interests in avoiding dangerous climate change are not fundamental enough to impose obligations on others. (Caney 2008: 539)

However this is not entirely reassuring, since Caney did not have at his disposal empirical evidence of the sort required to carefully evaluate the claim. What Caney was able to rely on were well-known studies of the overall economic cost of reducing emissions to a low level, expressed as a change in global GDP (e.g. IPCC 2007a). While informative, these studies, or at least their summary results, do not give us the evidence we need to carefully evaluate whether some human rights, somewhere, are violated due to mitigation.

This reveals a general problem with a rights-based approach, where rights might be in conflict. In particular, it strongly recalls the need for a framework for evaluation that places emphasis on comparison and interdependence. I praised economics for this in section 3, but opened the discussion to criticism of the utility metric in this section. Broome (1999), then, appears to be right—what we should take into the evaluation of climate change from economics is structure, but not substance, or at least not unreservedly so. This does beg the question, however, of what to do instead, and it is to this that I turn in the next section.

5 JUST KEEP MAXIMIZING, BUT...?

It seems possible to identify two ways forward. The first, which has been suggested by Alan Randall (2007), is to forge ahead with CBA, but to add side constraints to the maximization of utility. In Randall's formulation, which is expressly intended for use in environmental management, the side constraint takes the form of a Safe Minimum Standard of conservation (Ciriacy-Wantrup 1968), which is invoked when the moral consequences of human modification of the environment are deemed unacceptable. Unacceptability, according to Randall's pluralist ethical approach, can be defined with respect to various ethical theories, such as those that identify non-consequentialist duties and obligations as above. The key point here is that the Safe Minimum Standard is a commonsense decision heuristic, which represents, and can put into operation, these concerns. One might understand a limit of 2 °C warming above pre-industrial levels as such a Safe Minimum Standard for climate change. There is certainly no overwhelming evidence that unacceptable moral consequences begin at 2 °C above pre-industrial levels, as opposed to say 2.2 °C, so it seems to serve a heuristic purpose. CBA still has a role, Randall argues, because 'a society of thoughtful moral agents would agree to take seriously an account of benefits and costs, within some more complete set of principles' (Randall 2007: 111).

This certainly seems like an improvement on unalloyed utility maximization, and indeed it can be argued to be consistent with the disclaimer, which most economists would add to any of their evaluations, that CBA is but one input into the decision-making process. In fact, Randall goes further than the standard disclaimer, which, as he correctly points out, leaves entirely unanswered the questions (i) how much attention should be paid to economic evaluation and under what circumstances, and (ii) if CBA is systematically deficient, then why not try to fix it? Responding to (i), Randall seeks to establish the principle that CBA should be carried out 'where no overriding moral concerns are threatened' (2007: 112), and where those moral concerns are given by basic rights or similar—i.e. 'don't do anything disgusting' (ibid.).

The problem, however, is that maximization with side constraints does not really address the concern, expressed at the end of previous section, that the side constraints themselves could be in conflict. Indeed, this was recognized early on to be a problem with the notion of a Safe Minimum Standard of conservation, so that the ad hoc concept of 'intolerable cost' was later added (Bishop 1978). In effect, the method only works if there is an envelope of possible strategies that do not violate constraints either on the side of environmental change, or on the side of mitigating actions. Another problem, in practice, is that this approach seemingly does little to bring the wider set of non-utilitarian constituents of human well-being directly into the evaluation of climate change. Rather, they are cast outside the framework, giving rise to side constraints that may or may not be carefully justified with evidence. For an example of how careful evaluation may not underpin side constraints, see Tol's (2007) coruscating review of the adoption by the European Union of the 2 °C target. In any case, we still have not done anything to 'fix' economic evaluation, even though it does indeed seem to be systematically deficient from a moral point of view (point (ii) above).

What is being argued for, then, is an evaluation framework that makes best use of the structure of CBA, with all of its emphasis on interdependence, but with the 'benefits' and 'costs' under examination consisting in a broader set of determinants of human well-being. These determinants would need to be carefully specified, but there is much supporting work in ethics and in policy to inform them. Examples of the former include Rawls's (1971) notion of primary goods, and Sen's (1999) capabilities framework. Examples of the latter include the United Nations Development Program's *Human Development Reports* (see Barnett, in this volume, who also connects such an approach with climate change), the *World Development Indicators* of the World Bank, and the Millennium Development Goals. What all share is a conception of human well-being, whereby it consists in several factors, such as education, environmental quality, health, political and social opportunities, and of course income.

It might be said that such a pluralistic evaluation of climate change already exists. Take the IPCC's *Assessment Reports* for example. As indicated above, it is an established practice in the IPCC's Reports to evaluate the impacts of climate change and of adaptation to climate change on multiple dimensions (IPCC 2007b). Elsewhere in the Reports, the impacts of mitigation are considered (IPCC 2007b). However, the evaluation lacks formality and the sort of unifying framework that might shed more light on the choices we face. Impacts/adaptation are presented in a separate report to mitigation, and the treatment of the two issues, while similar, lacks some consistency.

Doing so, however, raises very challenging questions of aggregation. These are apparently much less at issue in approaches such as economic evaluation, which are interested in maximizing (or at least increasing) the value of a single, homogeneous quantity such as utility. Following such approaches, it is more straightforward to construct a complete ordering over social states (provided one rides over concerns about interpersonal comparisons). Working with multiple dimensions, complete orderings are less easily obtained.

Yet the process of doing so is likely to be very insightful. For example, could it be that a particular climate strategy dominates all of the alternatives, because it is better on one dimension, and at least as good on all others? If the dimensions of well-being consist mainly in meeting vital needs, it might be the case that a strategy to aggressively reduce carbon emissions, with the vast majority of that burden borne by the developed world, would dominate any other strategy. This might be true if the impacts of climate change threaten the vital needs of future generations of people in either the developed or currently developing world, *and* that mitigation in the developed world does not threaten to diminish the capacity of people to meet their vital needs. This is often argued, but to my knowledge the proposition has rarely if ever been put to careful empirical test. Even if it is not possible to identify a dominant strategy, it might be possible to identify a range of weights on the dimensions of human well-being, over which a particular strategy is best (Sen 1970). That it may not be necessary to come to agreement on unique weights for each dimension is significant.

This pluralistic, but formal, evaluation can be considered a *direct* response to the weaknesses identified with the economic approach. CBA with side constraints can be considered a *supplementary* response (Sen 1999). Both are likely to be informative, but the former may be more so in terms of the choices we face to respond to climate change.

6 CONCLUSION

This chapter has set out a view of the strengths and weaknesses of economic evaluation of climate change strategies. The main strength of the economic approach is argued to be the formal framework through which it is able to compare human well-being across time, space, and states of nature, under alternative courses of action. But its main weakness is the substance of that comparison—utility, as the satisfaction of preferences for the aggregate consumption of goods and services. It follows that the most celebrated debates within the economic profession about how to weight individual utility (e.g. about the discount rate) are somewhat (though certainly not entirely) beside the point.

In order to advance our understanding of the implications of different climate change strategies for human well-being, I then suggest that the formal aggregative framework be retained, but that the monistic measure of utility be replaced by a pluralistic 'vector' of factors affecting human well-being, which might well include income, but also education, health, environmental quality, and so on. Such a vector is capable of being sensitive to moral considerations such as basic needs and rights. This move introduces new difficulties to the task of aggregation, but much could be learned in the process of tackling them.

NOTES

1. In the early 1970s, the United States Department of Transportation carried out a study into, *inter alia*, the economic effects of climatic changes resulting from stratospheric flight by supersonic aircraft (d'Arge 1975).
2. Note, however, that while all economic evaluations of climate change require an IAM to be employed, not all IAMs are economic—some restrict their analysis to linking a model of the climate system with environmental changes in their natural units.
3. Truman complained that his economic advisers always said 'on the one hand —— and on the other ——'.
4. More generally, the reason for this is that, given the large uncertainties affecting responses to climate change, any viable ethical approach will need to yield sensible answers in a range of future states of nature.

5. In terms of welfare-economic theory, the problem with the question posed by the Consensus is that it posits an arbitrary set of policy problems and an arbitrary budget constraint for spending on them. From an economic point of view, climate change is a market failure, such that it should be adapted to and mitigated up to the point where the costs and benefits of doing so are all equal at the margin. This should be the case for any and all public expenditures, but the Consensus constrains the problem such that social welfare cannot be maximized in this way.

6. See also the Royal Institute of Philosophy Supplement 57 on *The Philosophy of Need*, edited by Soran Reader (volume 80, 2005). The concept of vital need—those things essential for a flourishing life (or, as per Aristotle, without which it is impossible to live)—must be distinguished from merely instrumental needs, since in everyday language the concept of need is used very broadly to cover both contexts.

7. From a technical point of view, the utility function is smooth and continuous, so there is no sharp distinction between a certain income that might be vitally needed (to live some minimally acceptable life) and income in addition to that. If the elasticity of the marginal utility of consumption were high, there would be a range of consumption over which the function would begin to *resemble* that sharp distinction (see how the utility function in Figure 20.1 becomes increasingly concave, and imagine increasing the elasticity still further; soon enough the function will resemble a right angle). However, in practice empirical studies have tended to use lower values in the region of 1.

8. See Nozick's (1974) 'catastrophic moral horrors'.

REFERENCES

ADGER, W. N. 2004. The right to keep cold. *Environment and Planning A* 36(10): 1711–15.

BISHOP, R. 1978. Endangered species and uncertainty: The economics of a safe minimum standard. *American Journal of Agricultural Economics* 60(1): 10–18.

BROOME, J. 1991. Utility. *Economics and Philosophy* 7: 1–12.

—— 1999. *Ethics out of Economics*. Cambridge: Cambridge University Press.

CANEY, S. 2008. Human rights, climate change, and discounting. *Environmental Politics* 17(4): 536–55.

CIRIACY-WANTRUP, S. 1968. *Resource Conservation: Economics and Policies*. Berkeley: University of California, Division of Agricultural Science.

CLINE, W. R. 1992. *The Economics of Global Warming*. Washington, DC: Institute for International Economics.

D'ARGE, R. C. (ed.) 1975. *Economic and Social Measures of Biologic and Climatic Change*. Climate Impact Assessment Program Volume 6. US Department of Transportation.

DIETZ, S., HEPBURN, C., et al. 2008. Economics, ethics and climate change. Pp. 365–86 in K. Basu and R. Kanbur (eds.), *Arguments for a Better World: Essays in Honour of Amartya Sen, ii: Society, Institutions and Development*. Oxford, Oxford University Press.

GARDINER, S. 2004. Ethics and global climate change. *Ethics* 114: 555–600.

HARSANYI, J. C. 1977. *Rational Behavior and Bargaining Equilibrium in Games and Social Situations*. Cambridge: Cambridge University Press.

HOPE, C. 2005. Integrated assessment models. Pp. 77–98 in D. Helm (ed.), *Climate-Change Policy*. Oxford: Oxford University Press.

IPCC. 2007a. Summary for policymakers. In B. Metz, O. R. Davidson, P. R. Bosch, R. Dave, and L. A. Meyer, *Climate Change 2007: Mitigation. Contribution of Working Group III to the Fourth Assessment Report of the Intergovernmental Panel on Climate Change.* Cambridge: Cambridge University Press.

IPCC. 2007b. Summary for policymakers. Pp. 7–22 in M. L. Parry, O. Canziani, J. P. Palutikof, P. J. van der Linden, and C. E. Hanson, *Climate Change 2007: Impacts, Adaptation and Vulnerability. Contribution of Working Group II to the Fourth Assessment Report of the Intergovernmental Panel on Climate Change.* Cambridge: Cambridge University Press.

LOMBORG, B. 2004. *Global Crises, Global Solutions.* Cambridge: Cambridge University Press.

MADDISON, D. J. 1995. A cost-benefit analysis of slowing climate change. *Energy Policy* 23(4/5): 337–46.

MANNE, A., and RICHELS, R. 1995. The greenhouse debate: Economic efficiency, burden sharing and hedging strategies. *Energy Journal* 16(4): 1–37.

NORDHAUS, W. D. 1991. To slow or not to slow: The economics of the greenhouse effect. *Economic Journal* 101: 920–37.

——2008. *A Question of Balance: Weighing the Options on Global Warming Policies.* New Haven: Yale University Press.

——and BOYER, J. 2000. *Warming the World: Economic Models of Global Warming.* Cambridge, Mass.: MIT Press.

NOZICK, R. 1974. *Anarchy, State, and Utopia.* Oxford: Blackwell.

O'NEILL, J., HOLLAND, A., et al. 2008. *Environmental Values.* London: Routledge.

PEARCE, D., ATKINSON, G., et al. 2006. *Cost-Benefit Analysis and the Environment.* Paris: OECD.

RAMSEY, F. P. 1928. A mathematical theory of saving. *Economic Journal* 38: 543–59.

RANDALL, A. 2007. Benefit-cost analysis and a safe minimum standard of conservation. In G. Atkinson, S. Dietz, and E. Neumayer (eds.), *Handbook of Sustainable Development.* Cheltenham: Edward Elgar.

RAWLS, J. 1971. *A Theory of Justice.* Oxford: Oxford University Press.

ROBBINS, L. 1935. *An Essay on the Nature and Significance of Economic Science.* London: MacMillan.

SACHS, J. D. 2004. Seeking a global solution. *Nature* 430: 725–6.

SEN, A. K. 1970. *Collective Choice and Social Welfare.* San Francisco: Holden-Day.

——1977. Social choice theory: a re-examination. *Econometrica* 45: 53–89.

——1987. *On Ethics and Economics.* Oxford: Basil Blackwell.

——1999. *Development as Freedom.* New York: Knopf.

SHUE, H. 1999. Bequeathing hazards: Security rights and property rights of future humans. Pp. 38–53 in M. Dore and T. Mount (eds.), *Global Environmental Economics: Equity and the Limits to Markets.* Oxford: Blackwell.

SPASH, C. L. 1998. *Greenhouse Economics: Values and Ethics.* London: Routledge.

STERN, N. 2007. *The Economics of Climate Change: The Stern Review.* Cambridge: Cambridge University Press.

TOL, R. S. J. 1997. On the optimal control of carbon dioxide emissions: An application of FUND. *Environmental Modelling and Assessment* 2: 151–63.

——2007. Europe's long term climate target: a critical evaluation. *Energy Policy* 35(1): 424–34.

WATKISS, P., and DOWNING, T. E. 2008. The social cost of carbon: Valuation estimates and their use in UK policy. *Integrated Assessment* 8(1): 85–105.

WEYANT, J. P., DAVIDSON, O. R., et al. 1996. Integrated assessment of climate change: An overview and comparison of approaches and results. *Climate Change 1995—Economic and Social Dimensions of Climate Change: Contribution of Working Group III to the Second Assessment Report of the Intergovernmental Panel on Climate Change*. IPCC. Cambridge: Cambridge University Press.

WIGGINS, D. 2005. An idea we cannot do without: What difference will it make (e.g. to moral, political and environmental philosophy) to recognize and put to use a substantial conception of need? *Royal Institute of Philosophy Supplement* 80(Supplement 57): 25–50.

CHAPTER 21

....

CLIMATE JUSTICE

....

STEPHEN M. GARDINER

THIS chapter presents a brief introduction to the emerging area of climate justice for an interdisciplinary audience. While other articles in this collection go into more detail about specific concerns, my aim is to provide a preliminary sense of the broader landscape. Section 1 considers the initial relevance of justice to climate affairs. Section 2 discusses why climate policy poses a strong challenge for ethical action. Section 3 considers how justice might guide us in making progress from within the constraints of current institutions and their immediately likely successors. Section 4 briefly considers the implications of relaxing such constraints.

1 THE RELEVANCE OF JUSTICE
....

Climate change is an ethical issue (Gardiner et al. 2010). Although justice is only one part of ethics, it is often understood to occupy an exalted position as 'the first virtue of social institutions' (Rawls 1999), where this implies that unjust institutions ought not to be tolerated except to avoid greater injustice. Although some have doubted that justice deserves quite such preeminence, almost all agree that it is a central concern.

1.1 The Marginalization of Justice

From this point of view, much conventional analysis of climate change is likely to seem puzzling, since it is dominated by prudential and economic terms. Typically, policies are said to be good or bad 'for us,' where the 'us' is either humanity as such, or particular countries or industries, and goodness and badness are measured by projections of economic costs and benefits in the present and future. Explicit discussion of justice is often marginalized or simply absent.

Some defend this approach by saying, 'When the problem is so urgent and the threat so large we may have to accept unjust solutions, since these are better than no solution at all.' Justice, in other words, runs a distant second to other values. But this response seems too quick. First, there is a major difference between being a distant second and not being in the race at all, and some of the contemporary discourse veers closer to this extreme. Second, even traditions that have other values at their heart typically show justice more consideration. For example, utilitarianism takes happiness as its primary value. Yet most contemporary utilitarians bend over backwards to accommodate justice, and for good reason. Claims of justice and rights have long been recognized to be strong promoters of utility, in ways that justify their having an exalted status in policy. On such views, the relegation of justice in climate policy will be highly worrying. Third, the claim that other values have priority is likely to mislead. In my view, if we do not highlight the harms and inequities inflicted upon the poor, current people more generally, future generations especially, and the rest of nature as well, then we are in danger of missing the central reasons for action, and so obscuring what is at stake.[1]

1.2 Varieties of Justice

Justice comes in a number of varieties, many of which seem immediately applicable to climate change. Most obviously, worries about procedural justice pervade the current approach of trying to address climate change primarily through international negotiations between nation-states. For example, it is widely thought that existing international structures have emerged under and reflect background circumstances of serious injustice (e.g. colonialism, and current trade rules (Pogge 2002)). Given this, many question the ability of such institutions to deliver climate justice. Indeed, some claim that they are likely to deliver 'compound' injustices: new wrongs that feed off the old to make the overall situation even worse (Shue 1990).

Almost as obviously, questions of distributive justice abound. Currently, the world is trying to move from a situation where emissions of the main greenhouse gas—carbon dioxide—are almost completely unconstrained to one that imposes strict limits. This shift raises questions about who should now be allowed to emit, to what extent, and with what justification. Since many actors will be required to change their behavior, and potentially take on significant burdens as a result, there is strong pressure for robust justifications of the distributive scheme.

Less obviously, climate change raises issues of corrective justice. Not only is there the matter of how to treat past emissions, but some actors may owe restitution because of their roles in past failures. For example, some nations have made public commitments to goals, or even specific targets, that they have then failed to meet, and some actors stand accused of actively facilitating procrastination and delay (e.g. through misinformation). Arguably, such agents should be held accountable.

1.3 Domains of Justice

Concerns of justice manifest themselves in many domains of human life—including the international, domestic, environmental, and ecological—to which climate change is clearly important. At the international level writers on justice often point out the sharp differences in national emissions levels. For example, the USA and China each have total carbon emissions that are roughly four times those of India, and more than eighteen times those of Bangladesh. Similarly, the average American's emissions are roughly equal to those of nearly five Chinese, fifteen Indians, and sixty-six Bangladeshis. Moreover, since, at the present time, such differences appear to be strongly correlated with economic prosperity, much is at stake in deciding how to distribute future emissions, at least in the near term.

Significant issues of justice also arise at the domestic level. One sign of this is that there are sharp differences in individual emissions within countries, and in ways that coincide with wider issues of economic inequality, class, gender, and race. For example, it has recently been noted that in India the top 10 percent of the urban class has collective emissions that are more than 24 times those of the bottom 10 percent of the rural class (Parikh et al. 2009), that poor women in developing countries are likely to bear a disproportionate burden of the impacts of climate change (UNPF 2009), and that although average Chinese per capita emissions are low by international standards, millions of Chinese emit at higher levels that are much more comparable to those in developed countries, so that there is there is 'a Germany in China' (Harris 2010).

Climate change also raises issues of intergenerational justice. Most prominently, climate impacts are subject to substantial time lags. Emissions of carbon dioxide, for example, typically remain in the atmosphere contributing to climate change for anywhere up to 300 years, and around 10 percent remains for thousands of years (IPCC 2007; Archer 2009). In addition, the processes triggered by the presence of elevated concentrations of greenhouse gases—such as sea-level rise—play out over centuries and millennia. Thus, the full consequences of any given generation's emissions will not be realized until long after they are dead. If these emissions impose significant risks of large costs or serious harms on future generations, then, other things being equal, this seems to constitute a serious injustice.

Finally, climate change raises issues of justice in relation to nature. On the one hand, 'environmental injustice' involves the propagation of harms or unfair relationships between humans as carried out through the medium of the natural environment (Shrader-Frechette 2002; Schlosberg 2007). Such injustice is already visible in existing domestic and international practices involving pollution, as the poor and ethnic minorities are disproportionately located in areas with much higher than average environmental risks. But climate change can be expected to increase such issues. Internationally, the disadvantaged are much more vulnerable to the negative impacts of climate change, both because these are expected to occur more often and more severely in areas of the world where they live (at least in the short term), and because they have poorer support mechanisms with which to cope (IPCC 2007). Domestically, there are similar issues. For example, in the United States poor and minority populations were affected more seriously by Hurricane Katrina than their rich and white neighbors.

On the other hand, there is the possibility of 'ecological injustice', or injustice between species. Climate change is almost certain to mean increasing negative impacts on many ecosystems and habitats, to the detriment (and perhaps extinction) of a wide range of animals and plants. This is part of a wider pattern of human exploitation of nature that seems to be accompanying our recent expansion in numbers and technological capacity. Though some would contest the use of the term 'injustice' to characterize such concerns, this is clearly an important area for discussion whatever the heading. Moreover, one need not be a radical environmentalist to take it seriously. Pretheoretically, most people seem to believe that if we were to protect humanity but nevertheless render the surface of the planet uninhabitable for most other forms of life, this would be a serious ethical problem. Few would be comfortable with the prospect of humanity living in vast domes atop the surface of an otherwise barren Earth.

The fourth and last point about the relevance of justice concerns the intersection of climate justice with more general issues of justice and ethics. For one thing, it seems clear that ultimately climate must be seen within the context of wider issues of justice, such as historical injustice (e.g. the legacies of slavery and unjust wars), international justice (e.g. trade and immigration), global justice (e.g. poverty, human rights), and the looming ecological crisis (e.g. species extinction, ocean acidification). For another, climate issues intersect with other ethical matters not most naturally seen as ones of justice. For example, perhaps earlier generations should be negatively evaluated for their approach to creating future people and the world they will inhabit. It may be that much of this evaluation could be done in terms of justice. For instance, perhaps a generation might be criticized for being unjust in their appropriation of resources, or their excessive projection of their own values into the future, or in their failure to respect some of the values and intergenerational projects of the past. Still, some issues might be better addressed in other ways, by appealing to values such as virtue or community. Perhaps we should also be asking what kind of people, communities, nations, or even species we aspire to be. For example, perhaps we should worry not just about being unjust, but also about being reckless, cruel, callous, or even morally weak or pathetic. After all, injustice is not the only way in which we may go wrong.

2 A CLIMATE OF INJUSTICE

So far, the world's attempts to address climate change seem to have fallen far short of taking justice seriously. Despite the fact that the foundational legal document (United Nations Framework Convention on Climate Change 1992) is full of the language of justice (Gardiner 2010a), most actual discussion of climate policy has proceeded on the basis of political horse trading. Nation-states encourage one another to make offers of particular cuts (or, in some cases, increases) in future carbon emissions, and then try to negotiate an agreement based on claims of reciprocity and (often nonclimate) side deals.

If political horse trading involves concern for justice at all, it suggests a focus on pure procedural justice, where the implicit idea is that informed bargaining offers a reasonable mechanism for producing just outcomes. But this idea faces significant challenges. For one thing, as mentioned above, it is not at all clear how one would set about negotiating a fair international agreement in a situation where there are sharp inequalities of power and resources, and where these reflect serious historical injustices. For another, any bargaining approach faces a very basic challenge in working out how to represent the interests of those thought to be most vulnerable to climate impacts, the poor, future generations, and nonhuman life. On the face of it, these groups have little or no voice in negotiations, and would in any case lack much to bargain with. After all, there is little that they could offer that the other bargainers could not simply take from them in any case (Gardiner 2009).

In light of such concerns, it is not surprising that substantive progress since the Framework Convention has been elusive. Globally, emissions have risen dramatically (by over 30 percent at this point), and (prior to the recent financial crisis) were increasing at a rate beyond the highest projections of twenty years ago (at 2–3 percent per year). Domestically, growth has been occurring at substantial rates in most of the large polluters.

For example, China has recently surpassed the United States in absolute annual emissions, and is up by more than 150 percent since 1990. Moreover, the US itself is up by almost 20 percent over the same period, despite starting from almost double China's 1990 total (Boden et al. 2009).

Elsewhere I have called this failure, 'the problem of political inertia', and argued that one of its central causes is that global climate change constitutes 'a perfect moral storm': the convergence of three nasty challenges (or 'storms') that threaten our ability to behave ethically (Gardiner 2006, 2011a). These challenges are global, intergenerational, and theoretical.

The global challenge is familiar. Both the sources and the effects of anthropogenic emissions are spread throughout the world, across local, national, and regional boundaries. According to many writers, this creates a tragedy of the commons situation, because the global system is not currently set up to govern this kind of commons. In addition, there are skewed vulnerabilities: those who are most vulnerable and least responsible will probably bear the brunt, at least in the short to medium term. Whereas the developed nations are, by and large, responsible for the bulk of emissions to this point, they appear much less vulnerable to the more immediate impacts than the less developed countries, where most of the world's poor reside. This mismatch of vulnerability and responsibility is exacerbated by the fact that the developed countries are more powerful politically, and so more capable of bringing about a solution, but the less developed are poorly placed to call them to account.

The intergenerational challenge is less familiar. As we have seen, the impacts of climate change are subject to major time lags, implying that a large part of the problem is passed on to the future. This suggests that each generation faces the temptation of intergenerational buck passing: it can benefit from passing on the costs and/or harms of its behavior to future people, even when this is morally unjustified. Moreover, if its behavior is primarily driven by concerns about what happens during its own lifetime, overconsumption is likely.

The third challenge is theoretical. We currently lack theoretically deep and compelling approaches to many of the ethical issues at stake in global warming policy, such as scientific uncertainty, international justice, intergenerational justice, and the appropriate form of human relationships to animals and the rest of nature. This causes special difficulties given the presence of the other storms. Most notably, given the intergenerational storm and the problem of skewed vulnerabilities, each generation of the affluent is susceptible to arguments for inaction (or inappropriate action) that shroud themselves in moral language but are actually weak, and self-deceptive. Such corruption is easily facilitated by the theoretical storm, and obscured by other features of the global storm.

The perfect moral storm not only provides a central explanation of past inertia, but also casts light on the threat of future injustice. Specifically, the climate tragedy is an evolving one. While procrastination and delay are plausible manifestations of the storm in its early stages, future manifestations need not take this form. For example, as the climate situation worsens, we might expect a generation focused on its own concerns to favor limited (and selective) action on mitigation and adaptation, and also investment in short-term 'solutions' to defer catastrophe (such as certain kinds of geoengineering). Such a generation may well act, and in ways that initially represent real progress over complete inertia. However, this kind of action might constitute only a 'shadow solution'—one that reflects only the limited set of concerns of those with the power to act—rather than a genuine and full response to the ethical challenge (Gardiner 2011a).

3 ETHICS FOR THE TRANSITION

Although buck passing is a major temptation, I see no reason to concede that it is irresistible. Still, we must acknowledge the large gap between recognizing the possibility of success and actually achieving it. One major issue is whether people and institutions will be able to muster the necessary motivation to act ethically. Another is that of working out what overcoming the storm would look like in practice. It is to this problem that I now turn.

Given the theoretical storm and the problem of moral corruption, the challenge of determining how to act is considerable. Still, I shall now argue that the basic direction of a just policy is clear enough, and we have some initially appropriate framings of the issues. Nevertheless, I shall also concede that much more difficult territory lies ahead.

In the rest of this section, I will be operating under several highly restrictive assumptions. First, my focus will primarily be on concerns of distributive justice; and I will make no attempt to cover all of the relevant domains. Second, even within this narrow area, my aim is simply to offer a preliminary sketch of one line of reasoning rather than to give a full, or even adequate defense. Third, I will be assuming for the purposes of discussion that, at least for the immediately foreseeable future, policy must be made largely within the framework of existing national and international institutions. Fourth, I shall assume that we must (at least for now, and to some extent) treat the issue of climate justice in isolation from wider concerns of international justice and ethics. These last two assumptions mark out what I have to say here as part of what I call 'the ethics of the transition'. In Section 4, I shall make a few remarks about the connections between this domain and that of ideal justice.

Three basic questions of climate policy are: (1) where to set a global ceiling on emissions, understood as a long-term trajectory consisting of a set of constraints on total net emissions at particular times; (2) how to distribute the emissions allowed under that ceiling at a particular time; and (3) what to do about unavoided impacts, especially negative impacts. Let us call these 'the trajectory question', 'the allocation question', and the 'impacts question', respectively.

3.1 Allocation

I begin with the allocation question, since this has received the most theoretical attention so far from writers concerned with justice. Many different proposals have been made. But there seems to be a broad ethical consensus that developed countries should shoulder most of the burden of action (at least initially), and so should have fewer emissions. Moreover, there is the sense that the agreements between rival views here are more politically important than their disagreements, because (it is assumed) almost any ethically guided policy will take us in the same general direction. This is so even though the physical constraints on action have increased dramatically over the last decade, so that it is now clear that the developing nations will have to constrain their own emissions quickly and significantly, even if the developed nations do take the lead (Baer et al. 2007). From the point of view of justice, the consensus appears to be that this only increases the burdens on developing countries to assist in other ways.

The basic grounds for the consensus are readily apparent. Following Peter Singer, we might say that, at least at first glance, the mainstream views of fairness all appear to support the idea that developing countries should take the lead (Singer 2002). First, historical theories do so because the developed countries are responsible for the majority of cumulative emissions. For example, the USA is responsible for 29 percent of global emissions since the onset of the industrial revolution (from 1850 to 2003), and the nations of the EU 26 percent; by contrast, China and India are responsible for 8 percent and 2 percent respectively. Second, theories based on moral equality support the consensus because the developed countries consume many more emissions per person than developing countries. For example, in 2005 average global emissions per capita were 1.23 metric tons of carbon. But the US average stood at 5.32 tons, the UK was at 2.47, China at 1.16, India at 0.35, and Bangladesh at 0.08 (Boden et al. 2009). Third, theories that prioritize the interests of the least well-off endorse the consensus because the developing countries are much poorer than the developed countries. Internationally poverty and inequality remain profound. In 2007, average per capita income in 2007 in the United States and United Kingdom was above $45,000 per year; in China it was $2,604, in India $976, and in Bangladesh $428 (United Nations 2009). Moreover, these averages conceal some of the worst problems. In 2005, more than 10 percent of the world's population lived in absolute poverty, on less than $1 per day, unable to meet their basic needs. Finally, utilitarian theories appear to support the consensus because taking the previous considerations seriously seems likely to promote happiness. For example, Singer argues that 'polluter pays' principles help to internalize incentives, principles of equality help to reduce conflict, and resources are very much more productive of well-being in the hands of those with little (Singer 2002).

This consensus has, of course, been challenged. Most prominently, some have rejected historical theories on the grounds that many past emitters were ignorant that their activities were dangerous, and are now dead (Caney 2005; Posner and Sunstein 2008). Because of this it is claimed that their descendants should not be held liable for past emissions. This argument might be extended to current infrastructure. Perhaps it can be argued that it is not the fault of the current populations in developed countries that they are stuck with carbon-intensive economies; instead, they should be pitied by those who can take a more direct path towards 'green' energy, and either assisted through the transition, or at least not punished.

This argument can seem impressive at first glance, but becomes less so under scrutiny (Gardiner 2010a). For one thing, humanity has been aware of the threat for some time. Not only was the first major IPCC report published in 1990, but discussion has been ongoing since at least the Johnson Administration (President's Science Advisory Committee 1965). For another, ignorance and death are not obviously complete excuses. Consider two quick analogies. First, suppose that I see a pile of bricks in your front yard with a 'free' sign in front of them. I take the bricks, but subsequently discover that the sign was blown there by the wind. While it is true that I shouldn't be blamed for 'stealing' your bricks in the normal sense, nevertheless, other things being equal, I should return them. If I cannot return them—for example, because I have already used them to build my own house—then I should compensate you in some other way. Ignorance does not neutralize the obligation. Second, suppose we amend the example so that it was not me but my (now dead) father who took your bricks. Assume also that he used them to build a hotel that has now made me very rich. Again, it is not obvious that I have no special responsibilities to assist you. (This is especially so if, as a result of losing the bricks, you have been left with only straw to make your house, and I am reckless with fireworks. (See also Gardiner 2011a, 2011b.))

A more pressing worry for the ethical consensus concerns its political stability moving forward. Presumably, as we do so, the differences between (and within) the distinct approaches to justice mentioned above will start to have significant practical implications for specific actors, and so complicate efforts to make progress. Arguably, then, as well as an ethical consensus on the general tendency of allocation policy, we will ultimately need a further convergence on what policies would count as 'fair enough' to particular stake-holders to provide a stable basis for ongoing agreement.

3.2 Unavoided Impacts

Much of the ethical consensus on allocation seems to carry over to unavoided impacts, since many of the same facts (e.g. concerning historical responsibility and current emissions levels) seem relevant. Nevertheless, there are complications, especially about how to understand the scope of the problem. Consider just two examples.

First, in climate policy, unavoided impacts are usually discussed in terms of assistance for 'adaptation', understood as 'adjustment in natural or human systems . . . which moderates harm or exploits beneficial opportunities' (IPCC 2001: 365). But this focus is liable to mislead, since adaptation efforts 'will not prevent all damages' (IPCC 2001: 226).

On the one hand, some unavoided impacts will simply have to be endured. This raises distinct issues of justice. Most obviously, there is a case for compensation, and perhaps in forms such as financial resources and immigration rights, rather than technical assistance, the usual focus of 'adaptation' measures. Less obviously, since some losses cannot be compensated, and since compensation is not the whole of justice in any case, matters of restitution, such as recognition and reconciliation, may also become prominent over time. On reflection, this should not be surprising. For example, the loss of indigenous homelands facing small island states appears to have similarities with the historical grievances of indigenous people elsewhere, where matters of recognition and reconciliation loom large.

On the other hand, there is the possibility of catastrophic changes than can neither be adapted to, nor endured. For example, if the earth really experiences a warming comparable in magnitude to an ice age shift (i.e. 5 °C), but over the course of only a century or so, or if climate change triggers dramatic threshold events, then the impacts on humanity might transcend historical experience. In such scenarios, the whole idea that we should address unavoided impacts through 'adaptation' may end up seeming 'quaint at best' (Jamieson 2008).

Second, much depends on what we are willing to call a climate impact. Not only will no one's death certificate ever read 'climate change', but many actual deaths will result from the interplay of climate with institutional failures caused by other moral and political problems (Jamieson 2005). As an illustration of this general problem, we might note that while it is often said that we can avoid 'dangerous climate change' if the global temperature rise can be limited to 2 °C, some claim that climate change is already responsible for around 300,000 deaths per year (Global Humanitarian Forum 2009).

3.3 Trajectory

Let us turn now to the trajectory question. Conventional climate policy implicitly involves envisioning a long-term aim, and then deciding how quickly to achieve that aim. On the first issue, a number of very different long-term targets for climate have been proposed. Some claim that we should prevent a temperature rise of above 2 degrees, some say that we should aim at a specific atmospheric concentration of carbon dioxide (or the equivalent) such as 350, 450, or 550 ppm, and others advocate that we should not exceed a given total of human emissions, such as one trillion tons of carbon (Allen at al. 2009).

The differences between these targets are not much discussed. One reason for this is presumably (again) that since all the targets actually offered are far from 'business as usual' projections, advocates assume that a move towards any would be substantially in the right direction, and so are disinclined to highlight disagreements on the specifics. A second reason is that there appears to be substantial agreement on the speed at which we should try to reach these long-term goals. Currently, many scientists and activists have converged on the claim that global reductions of 20–40 percent by 2020, and 50–80 percent by 2050, are roughly appropriate.

This political consensus is encouraging. Nevertheless, we should be careful. Such quantitative pronouncements tend to obscure the underlying ethical issues. Most prominently, the question of how quickly to reduce global emissions implicitly requires making a decision on how to balance the interests of the present and the future. Roughly speaking, on the face of it, it is probably better for the further future if we reduce emissions very rapidly, and so minimize the risks of climate catastrophe; but it is probably better for the present and closer future to proceed more slowly, and so minimize economic and social disruption. Given this, the pursuit of any given trajectory implicitly involves balancing competing concerns, and ultimately requires a moral judgment. More specifically, though much talk of specific percentage reductions is carried out in the language of 'feasibility', and so seems technical, this is a mistake. Presumably, it would be perfectly *technically* feasible for us all to reduce our emissions by 50–80 percent tomorrow, or even to eliminate them completely. We could, after all, just turn off our electricity for a large portion of the day, refuse to drive, and so on. The problem here is not that this cannot be done; it is rather that, given our current infrastructure, we assume that a very rapid reduction would cause social and economic chaos, and a humanitarian disaster for the current generation. If this assumption is correct, we are justified in not considering such drastic measures. But the justification is moral: a policy that demanded them of us would be profoundly unjust.

This move away from the 'feasibility' rationale makes a difference. Even if emissions cuts are disruptive at most levels, presumably at some point the risks imposed on future generations are severe enough to outweigh them. Perhaps the current proposals—such as 20 percent by 2020—capture the appropriate trade-off point here. Nevertheless, it would be nice to see some argument for this claim, especially since an issue of intergenerational justice is at stake, and since we are likely—given the perfect storm—to be biased in our own favor.

In addition, the essential rationale for the current generation's continuing with relatively high levels of emissions in the near term seems to be one of self-defense (Traxler 2002), and we should be aware that this has further implications. Rights of self-defense usually come with sharp limits, especially when directed towards the morally innocent. For example, one is normally required to use other (nonharmful) means of escaping the threat if possible. Moreover, if it is not possible, one is permitted only to use the minimum force necessary, and may be required to provide compensation if the victim is innocent.[2] Interestingly, such stringent restrictions seem to play very little role in current discussions of the trajectory question. Instead, the focus is on how the current generation may preserve its own expectations into the future by implementing a policy that allows as much as possible to go on exactly as before. It is not clear that this is a morally defensible policy; yet the perfect storm easily explains it.

There is, of course, an alternative to the line of argument that I have just offered. Some claim that the climate issue boils down to the question of whether to help the current or the future poor, and that the correct answer is to aid the poor now, because they are likely to be poorer than the future poor, and because they are more efficiently helped. I address this argument at greater length elsewhere (Gardiner 2010a, 2011a). But it may be worth just mentioning three initial points.

First, the attempt to reduce the climate issue to a trade-off between the current and future poor seems not to take mainstream scientific projections seriously. The IPCC envisions a temperature rise of 1.1–6.4 °C by 2100, with a best estimate of 1.8–4.0 °C (IPCC 2007). There is wide agreement that anything beyond 2 °C is very dangerous, and that 4–6 °C may be catastrophic (e.g. because a 5 °C shift in the opposite direction would be sufficient to bring on another ice age, but in a century rather than over millenia). If there is really a serious prospect of these changes, then the poor will not be the only ones affected.

Second, in the face of such threats the suggestion that a few billion dollars of aid to the current poor is our best strategy seems dubious. I do not doubt that we should provide such aid, and have compelling (other) reasons to do so. But is it really credible to claim that digging wells now in Africa is a better way to deal with the threat of climate catastrophe than reducing emissions? If profound droughts hit Africa, and the world's food supply, then wells may not be much help, either to Africans or anyone else.

Third, proponents of the above argument do not really mean to address climate change through direct aid, but rather claim that sustained economic growth will make the future much richer than the present, and so much more able to cope. But this move is also too quick. To begin with, the argument relies on an optimistic claim about the ability of global and local economies to flourish amid deteriorating climate conditions, especially in poor countries. This seems a little complacent in light of the science. In addition, the assumption that a few decades of growth would be enough to *offset* the impacts of temperature shift comparable to an ice age should be questioned, as should the idea that we are entitled to assume it in assessing our own obligations. Moreover, even if we could be very confident that the future will turn out to be richer, we should not concede too readily that it should therefore be forced to pay for climate damages. If we have a real obligation to pass on a sustainable planet to our successors, it is not clear that we can discharge it merely by assuming that they will have sufficient resources to clean up our mess.

In conclusion, though the ethical consensus surrounding strong action led by the developing nations looks robust, this conceals a range of deeper issues moving forward. For example, on the allocation question, differences in the rationale are likely to have significant implications for specific allocations, which may make a large difference to particular actors; on the impacts question, the issue of what to count as an unavoided climate impact will have profound distributive implications; and on the trajectory question, the current consensus on medium-term objectives masks important ethical assumptions about what is owed to the future. Such issues put pressure on existing theoretical approaches, especially as mediated through the ethics of the transition. So, let us conclude with some brief remarks about more ideal forms of theorizing.

4 IDEAL THEORY

The ethics of the transition aims to influence policy through existing institutional constraints and gradual attempts to modify those constraints. But it is reasonable to ask whether this is a feasible project. Early signs suggest that it may not be. After all, recent history implies that existing institutions have both allowed the threat to arise, but are (at best) reluctant to address it. Hence, the ethics of the transition is haunted by two more radical thoughts. First, current institutions might be seriously, and perhaps fatally, flawed so that they should be rejected (Dryzek 1987; Gardiner 2010b). Second, perhaps 'you can't get there from here.' If existing institutions and theories must be radically reconceptualized to reflect new global and ecological realities, perhaps these moves overwhelm the logic of a climate-focused account.[3]

In the face of such worries, some concessions seem inevitable. For example, most political philosophers working today believe that the current world order is seriously unjust, and it seems wise to acknowledge that climate change involves issues which current political institutions and theories do not seem designed for, nor obviously well equipped to handle. Hence, whatever one thinks about the ethics of the transition, it seems clear that ideal theory matters. Most prominently, climate change is one of a number of contemporary global problems that casts doubt on the traditional philosophical strategy of constructing basic justice on the model of a single self-sufficient nation-state. If we have truly entered a new epoch on the earth, a geological era dominated by humanity—the 'anthropocene' (Crutzen and Stoermer 2000)—then such a model seems at least seriously incomplete, and perhaps hopelessly outdated. Theorists should ask whether this requires revising their grand visions of ethics and justice.

Given these things, the project of ideal theory seems pressing.[4] Nevertheless, we should not be too quick to dismiss the ethics of the transition. Even if existing institutions and theories are hopelessly inadequate, we can hardly expect a transformation to better overnight; so, there remains a place for intermediate theorizing. In addition, we should recognize that such theorizing might play a number of different roles. At the extremes, some will conceive of climate ethics as operating completely in isolation of other, non-climate concerns (the isolation model); while others will see climate change as opening the door to a dramatically new world order (the vanguard model). But there are more moderate conceptions. For example, perhaps transitional climate policy should merely aim for modest improvement in other areas, insofar as it intersects with them (the mild rectification model); or perhaps it should remain content with not making wider injustice worse (the neutrality model). Importantly, discussions of the merits of these rival models seem part of the ethics of the transition rather than an obstacle to it.

More generally, it is important to note that what is at stake here is likely to depend as much on background beliefs about political reality as anything else, and so raise serious questions about the boundaries of the ideal. Practical 'political reality' is, of course, a treacherous notion, as geopolitical events of the last fifty years (e.g. the fall of the Berlin Wall, the end of apartheid in South Africa) have shown. But such worries infect ideal theory as well. Rawls, for example, claims to found his own political philosophy on a notion of 'realistic utopia' that aims to reconcile the real constraints of human nature and the world with the (equally treacherous) concept of 'utopia'. But how are we to decide what the 'real constraints' on ideal theory are? Given this problem, perhaps the differences between ideal and nonideal cases are more a matter of degree than of kind.

In closing, I want to make one final point about how to think about the roles of both ideal theory and the ethics of the transition. Even if we were pessimistic about the ability of current institutions and their likely successors to deal with climate in anything like a just way, there might still be a point to work on climate justice. While it is true that a central purpose of ethics is to guide change, it can also have other roles. In my view, prominent among these is the task of *bearing witness* to serious wrongs even when there is little hope of change. Ideal theory is central to this task. However, the ethics of transition can also play a part. Though we may not yet know either what a just climate would look like, or how to get there in the long term, visions of what might count as justice in the near term are still of some value in holding us accountable. This is so even if all they do is remind us that what we do now falls far short of any morally defensible goal.

NOTES

1. For example, suppose the rival value were the survival of the species. As it turns out, few scientists actually believe that climate change will lead to human extinction. The more extreme views are rather that it may cause a radical collapse in human societies, leaving only a few hundred thousand humans struggling to survive at the poles (Lovelock 2007). Thus, if we focus on survival alone, even severe climate change may seem tolerable. Justice, however, tells a different story.
2. According to some sources, the right does not apply if one is deemed the aggressor. This creates interesting complications in the present case.
3. Perhaps a conventionally unfair climate deal even leads to less injustice overall.
4. Of course, some of these concerns manifest themselves in the ethics of the transition. In particular, the basic moral logic of the situation may also drive us away from the status quo. For example, considering the allocation problem, no one cares much about carbon emissions for their own sake, but only about the role they play in human lives. Hence, some have advocated moving away from the focus on national emissions targets towards metrics such as development rights (Baer et al. 2007), human rights against environmental harm (Caney 2005; Vanderheiden 2008), or basic capabilities (Holland 2008; Schlosberg 2009). Such a shift may be morally justified. But it does suggest a substantial departure from current political norms and institutional structures. Consider, for example, that if there is a Germany in China, there must be something like a Pakistan or Bangladesh too (in order to generate China's low average per capita emissions). But China could address this by pursuing greater internal equality if it wished. If we insist that international policy must be adjusted instead—to ensure that different classes of Chinese emitters are treated differently—we seem to be saying that the international community should exert some authority over China's internal affairs.

REFERENCES

ALLEN, M. R., FRAME, D. J., HUNTINGFORD, C. et al. 2009. Warming caused by cumulative carbon emissions towards the trillionth tonne. *Nature* 458 (30 April): 1163–6.

ARCHER, D. 2009. *The Long Thaw*. Princeton: Princeton University Press.

BAER, P., ATHANASIOU, T., and SIVAN, K. 2007. *The Right to Development in a Climate Constrained World: The Greenhouse Development Rights Framework*. London: Christian Aid.

BODEN, T., MARLAND, G., and ANDREAS, R. J. 2009. *Global CO₂ Emissions from Fossil-Fuel Burning, Cement Manufacture, and Gas Flaring: 1751–2006*. Carbon Dioxide Information Center, Oak Ridge Laboratory. Available at <http://cdiac.ornl.gov/trends/emis/overview_2007.html>.

CANEY, S. 2005. Cosmopolitan justice, responsibility and global climate change. *Leiden Journal of International Law* 747–75.

CRUTZEN, P., and STOERMER, E. F. 2000. The Anthropocene. *Global Change Newsletter* 41: 17–18.

DRYZEK, J. 1987. *Rational Ecology*. Oxford: Blackwell.

GARDINER, S. M. 2006. A perfect moral storm: Climate change, intergenerational ethics and the problem of moral corruption. *Environmental Values* 15: 397–413.

—— 2009. A Contract on future generations? Pp. 77–119 in A. GOSSERIES and L. MEYER (eds.), *Intergenerational Justice*. Oxford: Oxford University Press.

—— 2010a. Ethics and climate change: An introduction. *Wiley Interdisciplinary Reviews: Climate Change*. 1: 54–66

—— 2010b. Climate change as a global test for contemporary political institutions and theories. Pp. 131–53 in K. O'BRIEN, A. L. ST CLAIR, and B. KRISTOFFERSEN (eds.), *Climate Change, Ethics and Human Security*. Cambridge: Cambridge University Press.

—— 2011a. *A Perfect Moral Storm*. Oxford: Oxford University Press.

—— 2011b. Is No One Responsible for Global Environmental Tragedy? Climate Change as a Challenge to Our Ethical Concepts. In D. ARNOLD (ed.) *The Ethics of Climate Change*. Cambridge: Cambridge University Press. (In press).

—— CANEY, S., JAMIESON, D., and SHUE, H. (eds.) 2010. *Climate Ethics: Essential Readings*. Oxford: Oxford University Press.

Global Humanitarian Forum. 2009. *Human Impacts Report*. Geneva.

HARRIS, P. 2010. *World Ethics and Climate Change*. Edinburgh: Edinburgh University Press.

HOLLAND, B. 2008. Justice and the environment in Nussbaum's 'capabilities approach': Why sustainable ecological capacity is a meta-capability. *Political Research Quarterly*.

Intergovernmental Panel on Climate Change (IPCC). 2001. *Climate Change 2001: The Synthesis Report*. Cambridge: Cambridge University Press.

—— 2007. *Climate Change 2007: The Physical Science Basis*. Cambridge: Cambridge University Press.

JAMIESON, D. 2005. Adaptation, mitigation, and justice. Pp. 221–53 in W. Sinnott-Armstrong and R. Howarth (eds.), *Perspectives on Climate Change*. Amsterdam: Elsevier.

—— 2008. The moral and political challenges of climate change. Pp. 475–82 in S. MOSER and L. DILLING (eds.), *Creating a Climate for Change: Communicating Climate Change and Facilitating Social Change*. New York: Cambridge University Press.

LOVELOCK, J. 2007. *The Revenge of Gaia: Why the Earth is Fighting Back and How We Can Still Save Humanity*. London: Penguin.

PARIKH, J., et al. 2009. CO₂ emissions structure of the indian economy. *Energy*. doi:10.1016/j.energy.2009.02.014

POGGE, T. 2002. *World Hunger and Human Rights*. Bristol: Policy.

POSNER, E., and SUNSTEIN, C. 2008. Climate change justice. *Georgetown Law Journal* 96: 1565–612.

President's Science Advisory Committee. 1965. *Restoring the Quality of our Environment*. Washington, DC: Government Printing Office.

RAWLS, J. 1999. *A Theory of Justice*. Rev. edn., Cambridge, MA: Harvard University Press.

SCHLOSBERG, D. 2007. *Defining Environmental Justice*. Oxford: Oxford University Press.

—— 2009. Rethinking climate justice: Capabilities and the flourishing of human and non-human communities. Paper presented at APSA.

SHRADER-FRECHETTE, K. 2002. *Environmental Justice: Creating Equality, Reclaiming Democracy*. Oxford: Oxford University Press.

SHUE, H. 1990. The Unavoidability of Justice. Pp. 373–97 in A. HURRELL and B. KINGSBURY (eds.) *The International Politics of the Environment*. Oxford: Oxford University Press.

SINGER, P. 2002. *One World: The Ethics of Globalization*. New Haven: Yale University Press.

TRAXLER, M. 2002. Fair chore division for climate change. *Social Theory and Practice* 28: 101–34.

United Nations. 2009. Social Indicators. Online at: <http://unstats.un.org/unsd/demographic/products/socind/inc-eco.htm>.

United Nations Framework Convention on Climate Change. 1992. *Framework Convention on Climate Change*. Online at <http://unfccc.int/essential_background/convention/background/items/1349.php>.

United Nations Population Fund (UNPF). 2009. State of the World Population. Available at <http://www.unfpa.org/swp/2009/en/pdf/EN_SOWP09.pdf>.

VANDERHEIDEN, S. 2008. *Atmospheric Justice: A Political Theory of Climate Change*. Oxford: Oxford University Press.

CHAPTER 22

....

INTERNATIONAL JUSTICE

....

PAUL BAER

1 INTRODUCTION

....

THE problem of anthropogenic climate change has given rise to numerous discussions of *justice*.[1] The focus of this chapter is *international* justice, which encompasses questions about duties and obligations between nations and between people of different nations. My approach will be organized around two themes; first, the broad distinction between cosmopolitan theories of justice and the alternatives (often classified as 'communitarian'); and secondly, the application of theories of justice (cosmopolitan theories in particular) to the primary ethical questions raised by the threat of anthropogenic climate change.

There are three ethical questions that lie at the heart of the climate problem, all of which have an international dimension. First, the question of *the target*: what limit should be set on greenhouse gas emissions, in order to limit the harm caused by climate change? Second, the question of *allocation*: how should the costs of meeting the target be distributed? Third, the question of *liability*: what is owed, and by whom, to those who will be put at risk or harmed by climate change—either in advance (adaptation) or after the fact (compensation)? The allocation question has been the focus of the majority of the debates about justice (or 'equity') and climate policy, but all three questions depend on reasoning about justice for their answers.

Clearly these questions raise issues of *international* justice. Emissions in any country contribute to the aggregate of harms that will be experienced in all countries, and thus protection from climate harm in one country requires emissions reductions in many, if not most or all, countries. Therefore if someone in one country has a right to protection from climate harm, the corresponding duties (or some part of them) must fall on people in other countries. Similarly, if there are costs to meeting a global reduction target, or a global bill for adaptation, those costs must be somehow be distributed across all countries.

What brings to the foreground the questions of international justice is the large asymmetry between those who benefit the most from GHG-producing activities and those who will be most vulnerable to climate change (Agarwal and Narain 1991; Sagar and Baer 2009; Srinivasan 2010). This asymmetry largely follows the structural patterns of global inequality, with roughly half of current and much more than half of historical GHG emissions coming from the roughly one-sixth of the world's population in the historically 'developed' countries. The vulnerability of the poor in developing countries due to their high dependence on agriculture, lack of basic health infrastructure, and social marginality is also well documented.[2] These facts give rise to many of the prominent features of the debates about climate justice, notably its polarization into developed vs. developing countries—'North vs. South'.

In practice there are at least four distinct discourses on ethics and justice in climate change. First, there is a philosophical discourse, to which the main contributors are academic philosophers and quasi-philosophers.[3] Second, there is a policy discourse on 'equity,' focused primarily on the question of fair allocation between countries of the costs of climate mitigation or the right to emit GHG pollution. Third, there is a public discourse on 'climate justice,' exemplified by the work of the 'Tck Tck Tck, Time for Climate Justice' campaign,[4] which focuses on the poor as victims of climate change, and appeals to a clearly privileged 'we' to take action to protect 'them' from the harms that will occur. Finally, there is a more radical discourse put forward by civil society coalitions including Climate Justice Now and Climate Justice Action. The language of these coalitions identifies directly with the global South, and criticizes northern governments and corporations as de facto perpetrators of injustice.[5]

In what follows I focus first on the philosophical discourse, which addresses the *reasons* for supporting or opposing various policy choices in the three domains I named (targets, allocation, and liability), and secondarily on the policy discourses; the civil society discourses are covered elsewhere in this volume.

2 BACKGROUND

The climate crisis has revealed the existence of both a kind of global interdependence and a level of environmental unsustainability to which the dominant norms and institutions of sovereign states are ill-suited to respond. The industrialized nations have long been able to operate as if the developing nations could do them no harm. The threat of climate change has made clear, however, that we share a global *life support commons* in which no country acting alone can preserve its own climate stability. Even the wealthiest countries are at substantial risk from climate impacts if emissions rise along 'business as usual' pathways (Stern 2006; Parry et al. 2007).

The risks from even the continuation of current global emissions levels, no less their increase, implies that those levels are unsustainable, and that emissions need to be reduced quickly (Hare and Meinshausen 2006; Hansen et al. 2008). But the established global model of industrial (and, worrisomely, agricultural) development depends on *growth* in energy use. The majority of the world's population live in countries that are still 'poor' by the global standard, and that have expected to be able to follow the same development path (and energy path) as the currently industrialized countries. Unsurprisingly, while fossil fuels remain by far the cheapest and easiest source of energy for economic growth, poor countries see any restrictions on growth in emissions as unfair constraints on their development.

Yet the climate problem is also strongly shaped by the dramatic inequalities *within* countries, both North and South. Particularly since the bursting of the housing bubble led to the deepest global recession in decades, most Northern consumers feel anything but wealthy, and as workers, have found themselves competing (often unsuccessfully) with lower-wage labor in developing countries. Thus opponents of action to reduce emissions in industrialized countries have been able to mobilize support by emphasizing the increase in energy costs and potential loss of jobs that are widely predicted to follow from steep emissions cuts.

Equally importantly, it is now widely recognized that the developing countries—particularly China and India, which accurately or not have become proxies for the entire developing world—have substantial populations of middle-class and wealthy consumers, many as wealthy or wealthier than middle-class citizens in the developed countries (Harris 2010). So the extension of the exemption from binding emissions reductions targets given to all developing countries in the Kyoto Protocol is easy to portray as fundamentally unfair, even if the average income and emissions of a Chinese or Indian citizen remain much lower than those of an average American or European. We will return to this issue below.

3 PHILOSOPHY AND CLIMATE JUSTICE

The problem of climate change has emerged in the wake of several decades of growing debate among philosophers on the topic of *cosmopolitanism*—that is, whether the rights and duties that obtain between people within the same nation or other community also obtain between people in different countries and communities. These debates have focused primarily on questions of distributive justice, and in particular on whether rich countries and their citizens have obligations of economic justice towards poor countries and their citizens. Supporters of cosmopolitanism claim that principles of justice are universal and that nationality, like race or gender, should not be a relevant factor in determining one's rights or one's obligations to others. This latter position has been used to justify claims for individual obligations (e.g. Singer 1972), for minimum standards of 'basic rights' (e.g. Shue 1996) and for more generalized egalitarian global distribution of resources (e.g. Beitz 1979; Pogge 2002).

Importantly, it is no longer credible to deny a certain kind of universal equality; one cannot in polite company argue that non-citizens are inferior and can be exploited, enslaved, or exterminated with no moral compunction. This principle of equal moral worth is foundational to existing international norms both of human rights and national sovereignty. Similarly, it is now standard to express concern with the well-being of persons wherever they live; however, whether those concerns reflect considerations of justice or of charity is fundamentally disputed. The cosmopolitan approach argues that obligations of justice apply across borders as well as within, deriving these obligations in some cases from actual causal interrelationships (e.g. Pogge 2002), sometimes from universal extension of the same egalitarian premises that ground nationally centered theories of justice (Singer 1972; Caney 2005b), or both (Beitz 1979; Moellendorf 2002). Rejection of such obligation typically relies on the argument that either a shared culture (e.g. Rawls 1999), shared nationality (Miller 1995) or a shared sovereignty (Nagel 2005) is required for the existence of obligations of justice, though proponents of such views—'communitarians' taken loosely—do usually advocate beneficence towards persons outside the community.

While not all the philosophers who have written on climate change are otherwise engaged in debates about cosmopolitanism, they are nearly uniformly united around some form of liberal egalitarianism, with a commitment to equality of dignity and respect. Starting from this commitment, most conclude that claims of justice do apply to the international distribution of the costs and benefits of climate policy. However, many authors have still found it necessary to engage John Rawls's famous dismissal of cosmopolitan justice claims (Rawls 1999); most conclude that the inescapable causal relationships of climate change render Rawls's position no longer supportable, if it ever was (see e.g. Vanderheiden 2008).

Rawls argued that the social ties required to make possible a contractarian theory of justice, like that of 'original position' bargaining, do not exist across national borders, and that only 'well-ordered societies'—in practice, nations—could plausibly be bound by such a standard of justice. The theories of Nagel and Miller make similar claims on slightly different grounds. However, such an argument fails when the status quo involves the imposition of significant cross-border harms, as from greenhouse pollution; any rejection of international justice claims therefore becomes a de facto endorsement of the right to do harm to non-citizens, which Rawls did not in fact endorse. And of course the fact that, considered as a resource, the atmosphere is not territorially bound (that is, it is a globally 'open access' resource) implies a need for global cooperation, providing further justification for cosmopolitan obligations (Vanderheiden 2008; Moellendorf 2009).

Even before the climate problem rose to the top of the international agenda, the decades-old establishment of universal human rights as both an aspirational vision and an international legal framework had weakened any arguments against cosmopolitan or 'universalist' ethics (Vanderheiden 2008). Indeed, if one accepts that climate change will cause harms of a type and scale that would clearly count as human rights violations if imposed by direct physical violence (e.g. Caney 2005a, 2009), it is difficult to reject the ethical obligation not to cause such harms. The harms cross borders, therefore the obligations cause borders. While one might argue that harms caused indirectly, as a byproduct of other activities, are not as morally repugnant as physical violence, the burden of proof would nevertheless appear to be shifted to those who would defend the lesser moral worth of citizens of other countries. And the general acceptance of a standard of universal human rights, however partially implemented, makes such a defense difficult.

Precisely because GHG emissions cause harm that crosses borders, and because of the asymmetry between the highest emitters and the most vulnerable, there is a near consensus among the philosophers who have written on the topic that considerations of justice do in fact justify the obligation of rich and high-emitting countries to reduce their emissions, pay for emissions reductions in poor countries, and aid poor countries in adapting to climate change. Both Henry Shue (1993, 1995) and Peter Singer (2002) exemplify this approach, arguing that on all plausible moral accounts, one reaches this general interpretation of the obligations of the wealthy and the rights of the poor. The few scholarly efforts to rebut these arguments—not from philosophers—rely on a variety of counter-strategies, arguing for example that if the rich have any obligations to the poor, preventing climate change is a very inefficient way to fulfill them (e.g. Beckerman and Pasek 2005; Lomborg 2006), or that nation-to-nation obligations unjustly permit the poor in the North to have obligations to the non-poor in the South (Posner and Sunstein 2008).

This latter argument in particular—highlighting the problems with nation-based 'collective responsibility'—is beginning to be addressed seriously by philosophers and others who explicitly apply cosmopolitan theories of justice to the climate problem. Some (Shue 1993; Neumayer 2000) have defended broad 'historical accountability', by which nations as a whole have obligations proportional to their historical emissions of greenhouse gases. Others (Caney 2009; Baer et al. 2010; Harris 2010) have argued that such collective, historical accounts are problematic (especially for emissions prior to the recognition of the risks of global warming[6]) and that obligations should also or instead be based on ability to pay. These 'ability to pay' arguments also focus on individuals rather than countries, which is consistent with the fundamental principles of a cosmopolitan approach.

Next I will consider the application of these principles to the three policy domains I outlined above: targets, allocation, and liability.

4 TARGETS

What do considerations of justice tell us about the appropriate goal for emissions reductions—the 'target' question? Philosophers have discussed the principles involved at some length, though drawing conclusions in terms of temperature, concentration, or emissions targets is made much harder due to the chain of scientific uncertainties involved.

Some including myself have argued that climate targets should be governed by a straightforward principle that it's wrong to cause avoidable harm to others.[7] Indeed, international 'soft law', including the Stockholm Declaration and the Preamble to the United Nations Framework Convention on Climate Change (UNFCCC), notes that nations have 'responsibility to ensure that activities within their jurisdiction or control do not cause damage to the environment of other States.' Since the likely harm from climate change includes substantial loss of life, generally considered the most grievous harm possible, while the cost of mitigation is primarily reduced consumption (and could be limited to reductions in luxury consumption), extremely rapid mitigation is warranted (Baer and Sagar 2009). Steven Gardiner (2006) frames the same argument in terms of the 'core precautionary principle,' arguing that the interests at risk from potentially catastrophic climate change vastly outweigh considerations of reduced economic growth. Others, notably Simon Caney (2005a, 2009), have argued that the right to a stable climate should be considered a fundamental human right, because the basic interests of life, health, subsistence, and security of place, all of which are endangered by climate change, are the foundations of both moral and legal human rights. Neither Gardiner nor Caney endorse particular targets, but their arguments would seem to support the most stringent targets currently entertained in the policy debates (e.g. reduction of CO_2 concentrations to 350 ppm, well below today's levels).

One counterargument is that loss of life is commonplace and should simply be treated as one more economic cost; otherwise resources will be wasted on climate mitigation that could save more lives through other means, such as the reduction of malaria (Schelling 1997; Lomborg 2006). Indeed it would seem odd to spend a billion dollars to prevent one more death from climate change if a thousand could be prevented with the same money spent differently. Yet it also seems wrong to say that we'll let millions of people die from pollution because we can spend part of the savings preventing harms to others more cheaply. There is, it seems, a fundamental tension between the utilitarian intuition that the sum of all suffering is what matters and the intuition about rights that what matters is precisely who is exposed to harm or risk and why (Baer and Sagar 2009).

Importantly, the links between the moral arguments and the policy conclusions depend on projections of climate impacts at various levels of emissions, concentrations and temperature increase, as well as on the costs of reducing emissions. A consensus has gradually emerged[8] that we should aim to keep temperature increase below 2 °C above pre-industrial; yet many of the least developed countries and small island states now argue that the objective should be 1.5 °C. However, the emissions reduction pledges made by various nations through June of 2010 seem to fall far short of meeting even a 2 °C objective, suggesting that, whatever the rhetoric, national economic interests still take precedence over global justice concerns.

5 ALLOCATION

The problem of equitably allocating mitigation responsibilities or emissions rights has been a primary focus of the climate policy literature since the late eighties (e.g. Krause et al. 1989; Grubb 1989), starting well before any philosophers got involved. The structure of the problem—dramatic differences in the emissions of rich and poor countries—made it clear that traditional negotiating solutions to burden-sharing among equals, such as equal percentage reductions from a baseline, would be patently implausible. Yet it was also already clear that, in the absence of emissions limits on developing countries, the problem could not be solved in the long run. Fortunately at least two solutions to these problems had already been developed; first, tradable emissions permits (pioneered in the US in the 1980s) and second, a multilateral fund with contributors from rich countries paying for reductions in poor countries, as implemented in the ozone regime.[9] Crucially, both solutions separate the question of where reductions are made from the question of who pays for them.

A thorough review of the literature on 'burden sharing' would be a chapter in itself, if not an entire book (see for example Gardiner 2004; Vanderheiden 2008; Klinsky and Dowlatabadi 2009). Several points stand out, however. First, it has been recognized since the beginning that principles of distributive justice were involved, and a number of non-philosophers made early contributions explicitly analyzing traditional theories of justice such as utilitarianism, Rawlsianism, and Kantianism and their applicability to mitigation policy (Solomon and Ahuja 1991; Ghosh 1993; Paterson 1996). Second, the idea of equal per capita emissions rights has consistently been a focal point for advocates of just climate policy, who have often argued that it is a necessary response to engender cooperation from developing countries (Meyer 2000; Athanasiou and Baer 2002) as well as a fair solution in its own right. Third, there have consistently been debates about how to incorporate historical emissions, which raises issues of measurement and modeling (Smith 1991; Müller et al. 2009) as well as questions of justice and responsibility (Neumayer 2000; Caney 2005a).

Although the term 'burden sharing' often subsumes all approaches to assigning responsibility for paying for mitigation or adaptation, a contrast is often made between 'burden-sharing' and 'resource-sharing' approaches to mitigation. In this sense, 'burden sharing' (like Kyoto-style percentage reduction targets) focuses on dividing the total costs or total amount of emissions reductions, whereas 'resource sharing' (like equal per capita allocations) focuses on the right to make use of global carbon sinks as an economic resource, and how to share those rights. Although one type of calculation can usually be converted to the other, the per capita approach was initially seen as an *alternative* to 'burden-sharing' approaches. In particular, an equal per capita approach could give many low-emitting countries a surplus of emissions rights that could be sold for cash like other natural resources. However, resource-sharing formulae are not necessarily more favorable to poor countries than burden-sharing formulae; for example, a resource-sharing formula which transitions from grandfathering to equal per capita allocations over time[10] can be significantly less generous to many developing countries than (for example) a burden-sharing formula like 'Greenhouse Development Rights' (Baer et al. 2008, 2010) which in fact assigns *negative* emissions allocations to most industrialized countries by shortly after 2020.

Crucially, resource-sharing formulae are unable to easily deal with two fundamental problems. First, they offer no good solution to the problem of funding adaptation or liability for climate damages. Under equal per capita allocations, revenue from surplus permit sales would (for at least some countries) provide a source of funds for adaptation activities, but there is no reason to think that this would be sufficient in total or appropriately distributed. Furthermore, using these funds for adaptation would reduce the ability to use them to supply low-carbon energy sources needed in the future when permit allocations become scarce. Thus even if a resource-sharing formula can solve the allocation question, additional principles are still required to address adaptation.

A second key problem, though, is that resource-sharing approaches do not usually address the wide variation in income levels across parties with similar levels of emissions. Since some high emitters are poor and some low emitters are rich, equal per capita allocations can be criticized as 'treating the unequal equally', and thus as unfair. This criticism also follows from the fact that differing 'national circumstances' such as climatic conditions and resource endowments mean that different levels of emissions are necessary to reach the same level of welfare.

In response to this, there is a significant convergence among philosophers and policy analysts towards 'hybrid' accounts which combine principles of responsibility, capacity, and need.[11] These types of proposals have appeared periodically (e.g. Smith et al. 1993; Sagar 2000; Ott et al. 2004; Baer et al. 2008, 2010; Caney 2009), and include a wide range of numerical formulae for allocating obligations or emissions rights. One key difficulty with all such proposals is that there is no obviously 'correct' way to combine multiple justifications like wealth and historical emissions; and since any specific formula creates predictable winners and losers, one can expect debates over such a formula to be contentious and difficult to resolve. And there are not in this case any obvious precedents for such a formula.

Another relatively recent development is the emergence of allocation principles which take the cosmopolitan approach even further by focusing on individuals rather than countries. Certainly the 'equal per capita' approach has always had this element, even though permits were usually assumed to be allocated to countries; now there are even proposals that would allocate tradable permits directly to individuals globally.[12] Two recent proposals which look at inequality within countries (and thus implicitly at individuals) are the Greenhouse Development Rights framework (Baer et al. 2008, 2010) and the proposal of Chakravarty et al. (2009). The former allocates obligations on the basis of a combined indicator of responsibility and capacity, exempting both income and emissions below a threshold; the latter focuses only on emissions and would require high-emitting individuals to reduce while low emitters are allowed to increase emissions. Both are thus compatible with the long-standing recognition of the distinction between 'luxury' and 'subsistence' emissions (Agarwal and Narain 1991; Shue 1993), but both still conclude that obligations, even if derived from looking at individuals, would still be incumbent on countries. In a recent book, however, Paul Harris (2010) develops the cosmopolitan focus on individuals in much greater detail, focusing at length on the wealth and emissions of the middle and wealthy classes in developing countries, and arguing for a more direct assessment of obligations on individuals. This approach is implicitly backed up by Simon Caney (2009), who argues against collective notions of responsibility but includes capacity and recent (e.g. since 1990) historical emissions in his account of individually based obligations.

It's painfully clear that a critical dimension of the failure of global climate policy to date has been the evident gap between what is fair and what is economically acceptable to key countries, notably the United States. Without here giving the detailed justifications for equal per capita allocations, suffice it to say that most analysts who have studied the problem from an ethical perspective have concluded that equal per capita is a *minimum* standard of distributive justice, inasmuch as it ignores disproportionate historical emissions, even if one limits the period of 'responsibility' to (say) 1990, by which time knowledge of the risks of GHG emissions were widespread (Jamieson 2001; Singer 2002). Yet it is also commonplace to simply state that 'the financial transfers that would be associated with equal per capita allocations would be unacceptable to the wealthy countries' (e.g. Ashton and Wang 2003; Posner and Sunstein 2008), which is not an argument about justice, but rather about power.

It is worth examining the public arguments given for rejecting the Kyoto Protocol by the United States, for it remains the case that international negotiating positions are typically framed as assertions of fairness rather than simple-self interest—a point which attests to the normative force of justice in international negotiations, however qualified it may be. Put in its simplest terms, the argument was that if the US has caps on its emissions and other countries with which it competes economically do not, its industries and thus the country as a whole will be unfairly harmed. This argument is prima facie reasonable, and thus very politically powerful in the US, as demonstrated by the unanimous vote for the Byrd-Hagel Resolution in 1997 in the US Senate[13] and the continuing concern over industrial competition expressed in current US political debates and proposed legislation. The counter-argument depends on a different assessment of the background conditions. In particular, the underlying rationale for the entire UNFCCC framework of 'common but differentiated responsibilities,' reflected in the Kyoto Protocol and other aspects of the convention, is that the extent of poverty in developing countries justifies exempting them from emissions reduction obligations that might reduce their rate of economic growth. Moreover, under any circumstances it is hard to ethically justify the complete (prior to 2009 in any case) refusal of the US to reduce its emissions (Brown 2002; Shue 1994).

Certainly a good case can be made that there should be no blanket exemption from emissions reductions for the non-Annex I countries like Singapore, South Korea, and many of the OPEC countries that are as rich as many Annex I countries. However, the real argument about developing country commitments has always focused in particular on China and India, which are substantial economic competitors with developed-country industries. Yet per capita emissions in China even today are still roughly a quarter of US levels and Indian per capita emissions are barely a twentieth, and per capita income levels are at least as unequal. A country-based assessment can hardly lead to a conclusion other than that the rich countries still need to 'go first,' as they pledged in the UNFCCC (Brown et al. 2006). In spite of this, the climate negotiations in Copenhagen in December of 2009 were still dominated by efforts of the developed countries to increase the emissions reductions commitments of the developing countries, and the resistance of the developing countries to those efforts. While a shift to an individually based approach to obligations might seem to offer some hope to transcend this continuing conflict, such a shift does not seem imminent.

6 ADAPTATION

While the allocation of mitigation obligations has certainly received the lion's share of attention in the debates about justice and climate change, there is a growing body of literature addressing the problem of adaptation in particular and liability for harm more generally (Sagar and Banuri 1999; Müller 2002; Tol and Verheyen 2004; Adger et al. 2006). Obviously the widely supported 'polluter pays' principle addresses damages that have been caused or will be caused by climate change (Baer 2006; Caney 2009), but there has remained substantial resistance to giving more than rhetorical attention to the problem in the climate negotiations themselves.

The logic of adaptation as a matter of justice is, at one level, quite straightforward: if a person or community has been put at risk from anthropogenic climate change, and that risk can be reduced or eliminated by some kind of proactive investment supported by those who have caused the risk, there is a moral obligation to make that investment. Arguably the obligations implied by 'adaptation' could be justified by an 'ability to help' principle, even in the absence of causal responsibility for the imposition of risk (Jamieson 1998; Caney 2009). If one accepts the causal chain from GHG emissions to risk of climate damages, the case is even stronger—support for adaptation becomes a kind of *corrective justice*, righting a wrong, rather than merely an example of *distributive justice*.

However, because the populations most at risk from anthropogenic climate change are by and large the same populations most at risk from 'normal' climatic variability and extreme weather events, it is conceptually very difficult if not impossible to separate investments that address only the 'additional' risk from anthropogenic change. Further-more, where it is possible to separate those who are most vulnerable to anthropogenic climate change, offering them protection through pro-active adaptation implies directing assistance not to those who are most vulnerable to climate-mediated harm, but to those whose vulnerability is increased by anthropogenic warming.

It is widely recognized that 'adaptation' looks very similar to 'development'. Indeed, it is one of the ironies that the logic of effective adaptation implies integrating adaptation funding and activities with existing development planning, policies, and projects (so-called 'mainstreaming—e.g. Huq et al. 2004; Yamin 2005), while the desire of developing countries to increase the level of transfers and their control over it has led to demands for adaptation assistance to be separated from and demonstrably 'additional' to existing development aid, and under the control of new, UNFCCC-directed institutions.

The relatively few writers who have engaged the questions of justice and adaptation have typically highlighted three obvious points: first, the need for aid to be directed to the most vulnerable persons, communities, and nations; second, the importance of fair participation by those who will be most affected; and third, the responsibility of those who have caused the risk and are most able to pay for adaptation to provide the necessary funds.[14] The question of how funding for adaptation to anthropogenic climate change could be made demonstrably separate from other development assistance is still awaiting a good answer; it seems safe to say, however, that if and when there is any significant funding available for implementing adaptation activities, there will be a dramatic increase in the attention paid to precisely this question.

7 CONCLUSION

As I suggested at the beginning, there is in fact a broad (though not universal) consensus about what should constitute a just (or fair or equitable) response to the threat of anthropogenic climate change. As noted in the UNFCCC itself, by virtue of their greater emissions and wealth, the industrialized countries bear primary responsibility to pay for emissions reductions and adaptation, and the requirement to reduce emissions should not be allowed to substantially impede the efforts of developing countries to raise their populations' standard of living.

Within the realm of philosophy, a similar point holds: although there is in fact a very wide disagreement even among broadly egalitarian scholars about the nature of cross-border obligations, even the philosophers who oppose cosmopolitan approaches in other domains have not tried to justify national self-interest as the basis of climate policy. Even those who deny a general obligation for rich countries to transfer wealth to poor countries find it hard not to assign emissions reductions obligations on egalitarian grounds, with capacity, responsibility, and need as crucial determinants of 'who pays.'

The existence of this consensus, however, should not lead us to ignore the very substantial differences on views of international justice that still remain, and their implications for policy choices. Crucially, a fully cosmopolitan view, which takes seriously the commitment to the primacy of individuals rather than countries as the basis of moral obligation even in international affairs, suggests a substantially different set of priorities. Such an approach would necessarily require that the necessary international transfers be based on a highly progressive assessment, and that developing-world elites accept climate obligations and some form of domestic redistribution. Arguably this type of fully cosmopolitan focus, in which the bulk of transfers went from the truly rich to the truly poor, is essential if we are to develop and sustain a regime with such redistributive implications (Baer et al. 2010; Harris 2010).

At the moment, however, perceptions of fairness seem primarily to be acting as a brake on ambitious emissions reductions—no country wants to do more than their 'fair share,' and there is very limited willingness to pay for reductions. It is widely agreed, including in the UNFCCC, that the disproportional historical and current emissions of the developed countries require that they accept the largest share of mitigation and adaptation effort. But inasmuch as this implies either explicit monetary transfers or a shifting of competitive advantage, considerations of justice are still outweighed by short-term economic concerns. One might imagine that as climate impacts become more and more visible, the search for a fair sharing of the costs of a more precautionary target will become more important (Victor 1999). Whether the current debates about international justice will make such negotiations easier in the future remains an open question.

Notes

1. In debates about climate change, as elsewhere, the terms 'justice,' 'fairness', and 'equity' are used somewhat interchangeably. Indeed, it is hard to imagine that an institution or an outcome could be considered fair but not just, or just but not equitable.
2. The reports of Working Group Two of the Intergovernmental Panel on Climate Change (e.g. Parry et al. 2007) provide a comprehensive review of this literature; see also Roberts and Parks (2007).
3. 'Quasi-philosophers' include a wide range of political scientists, economists, lawyers, and others who address the ethical questions raised by climate change.
4. See <http://www.tcktcktck.org>.
5. See <http://www.climate-justice-now.org> and <http://www.climate-justice-action.org>.
6. Typically placed around the first assessment report of the IPCC in 1990, although earlier dates are defensible.
7. See also Traxler (2002), Vanderheiden (2008), and Shue (1997). There are of course many justifications, such as self-defense or preventing a greater harm, which allow harming others in some circumstances.
8. As reflected for example in the Copenhagen Accord, <http://unfccc.int/resource/docs/2009/cop15/eng/l07.pdf>.
9. For the SO_2 trading experience in the US, see for example Solomon 1995; for the Montreal Protocol's Multilateral Fund, see <http://www.multilateralfund.org/>.
10. The Global Commons Institute has popularized this kind of transition to equal per capita rights as 'Contraction and Convergence.' See <http://www.gci.org.uk>.
11. It is not uncommon to also include elements of 'mitigation opportunity' (e.g. Claussen and McNeilly 1998).
12. Notably the 'Cap and Share' proposal—<http://www.capandshare.org>—though like other proposals which explicitly focus on allocating permits or refunds to individuals, it is primarily a nationally focused architecture with a sort of 'what if' extension to individuals globally).
13. The Byrd-Hagel Resolution, passed shortly before the Kyoto Protocol was negotiated, was a non-binding resolution stating that the US should not agree to any treaty in which developing countries did not have similar commitments to limit or reduce their emissions. S. 98, 105th Congress, 1997, Available at <http://thomas.loc.gov/cgi-bin/bdquery/z?d105:SE00098:>.
14. The various papers collected in Adger et al. 2006 make these points clearly.

REFERENCES

ADGER, W. N., PAAVOLA, J., HUQ, S., and MACE, M. J. (eds.) 2006. *Fairness in Adaptation to Climate Change*. Cambridge, MA: MIT Press.

AGARWAL, A., and S. NARAIN, S. 1991. *Global Warming in an Unequal World: A Case of Environmental Colonialism*. New Delhi: Centre for Science and Environment.

ASHTON, J., and WANG, X. 2003. Equity and climate: In principle and practice. Pp. 61–84 in J. E. Aldy, J. Ashton, R. Baron, et al. (eds.), *Beyond Kyoto: Advancing the International Effort Against Climate Change*. Washington, DC: Pew Center on Global Climate Change.

ATHANASIOU, T., and BAER, P. 2002. *Dead Heat: Global Justice and Global Warming*. New York: Seven Stories Press.

BAER, P. 2006. Adaptation to climate change: Who pays whom? Pp. 131–53 in W. N. Adger, J. Paavola, S. Huq, and M. J. Mace (eds.), *Fairness in Adaptation to Climate Change*. Cambridge, MA: MIT Press.

—— and SAGAR, A. 2009. Ethics, rights and responsibilities. Pp. 262–9 in S. H. Schneider, A. Rosencranz, and M. D. Mastrandrea (eds.), *Climate Change Science and Policy*. Washington, DC: Island Press.

—— ATHANASIOU, T., KARTHA, S., and KEMP-BENEDICT, E. 2008. The Greenhouse Development Rights Framework. 2nd edn., Heinrich Böll Stiftung, EcoEquity, Stockholm Environment Institute and Christian Aid. Available at <http://gdrights.org/wp-content/uploads/2009/01/thegdrsframework.pdf>.

—— —— —— —— 2010. Greenhouse development rights: A framework for climate protection that is 'more fair' than equal per capita emissions rights. Pp. 215–30 in S. M. GARDINER, S. CANEY, D. JAMIESON, and H. SHUE (eds.), *Climate Ethics: Essential Readings*. Oxford: Oxford University Press.

BECKERMAN, W., and PASEK. J. 1995. The equitable international allocation of tradable carbon permits. *Global Environmental Change* 5: 405–13.

BEITZ, C. R. 1979. *Political Theory and International Relations*. Princeton: Princeton University Press.

BROWN, D., TUANA, N., AVERILL, M., BAER, P., BORN, R., LESSA BRANDAO, C. E., FRODEMAN, R., HOGENHUIS, C., HEYD, T., LEMONS, J., MCKINSTRY, R., LUTES, M., MÜLLER, B., GONZALEZ MIGUEZ, J. D., MUNASINGHE, M., MUYLAERT DE ARAUJO, M. S., NOBRE, C., OTT, K., PAAVOLA, J., DE CAMPOS, C. P., PINGUELLI ROSA, L., ROSALES, J., ROSE, A., WELLS, E., and WESTRA, L. 2006. *White Paper on the Ethical Dimensions of Climate Change*. Available at <http://www.psu.edu/dept/rockethics/climate/whitepaper/edcc-whitepaper.pdf>

BROWN, D. A. 2002. *American Heat: Ethical Problems with the United States' Response to Global Warming*. Lanham, MD: Rowman & Littlefield Publishers.

CANEY, S. 2005a. Cosmopolitan justice, responsibility and climate change. *Leiden Journal of International Law* 18: 747–75.

—— 2005b. *Justice beyond Borders: A Global Political Theory*. Oxford: Oxford University Press.

—— 2009. Human rights, responsibilities and climate change. In C. R. BEITZ and R. E. GOODIN (eds.), *Global Basic Rights*. Oxford: Oxford University Press.

CHAKRAVARTY, S., CHIKKATUR, A., DE CONINCK, H., PACALA, S., SOCOLOW, R., and TAVONI, M. 2009. Sharing global CO_2 emission reductions among one billion high emitters. *Proceedings of the National Academy of Sciences of the United States of America* 106:11885–8.

CLAUSSEN, E., and McNEILLY, L. 1998. *Equity and Global Climate Change: The Complex Elements of Global Fairness.* Pew Center on Global Climate Change. Available at <http://www.pewclimate.org/publications/report/equity-global-climate-change-complex-elements-global-fairness>.

GARDINER, S. M. 2004. Ethics and global climate change. *Ethics* 114: 555–600.

—— 2006. A core precautionary principle. *Journal of Political Philosophy* 14: 33–60.

GHOSH, P. 1993. Structuring the equity issue in climate change. Pp. 267–74 in A. N. Achanta (ed.), *The Climate Change Agenda: An Indian Perspective.* Tata Energy Research Institute, New Delhi.

GRUBB, M. 1989. *The Greenhouse Effect: Negotiating Targets.* London: Royal Institute of International Affairs.

HANSEN, J., SATO, M., KHARECHA, P., BEERLING, D., BERNER, R., MASSON-DELMOTTE, V., PAGANI, M., RAYMO, M., ROYER, D. L., and ZACHOS J. C. 2008. Target atmospheric CO_2: Where should humanity aim? *Open Atmospheric Science Journal* 2: 217–31.

HARE, B., and MEINSHAUSEN, M. 2006. How much warming are we committed to and how much can be avoided? *Climatic Change* 75:111–49.

HARRIS, P. G. 2010. *World Ethics and Climate Change: From International to Global Justice.* Edinburgh: Edinburgh University Press.

HUQ, S., REID, H., KONATE, M., RAHMAN, A., SOKONA, Y., and CRICK, F. 2004. Mainstreaming adaptation to climate change in Least Developed Countries (LDCs). *Climate Policy* 4(1): 25–43.

JAMIESON, D. 1992. Ethics, Public Policy and Global Warming. *Science, Technology and Human Values* 17: 139–53.

—— 1998. Global responsibilities: Ethics, public health and global environmental change. *Indiana Journal of Global Legal Studies* 5: 99–120.

—— 2001. Climate change and global environmental justice. Pp. 287–308 in C. A. MILLER and P. N. EDWARDS (eds.), *Changing the Atmosphere: Expert Knowledge and Environmental Governance.* Cambridge, MA: MIT Press.

KLINSKY, S., and DOWLATABADI, H. 2009. Conceptualisations of justice in climate policy. *Climate Policy* 9: 88–108.

KRAUSE, F., BACH, W., and KOOMEY, J. 1989. *Energy Policy in the Greenhouse*, vol. i. El Cerrito, CA: International Project for Sustainable Energy Paths.

LOMBORG, B. (ed.) 2006. *How to Spend $50 Billion to Make the World a Better Place.* Cambridge: Cambridge University Press.

MEYER, A. 2000. *Contraction and Convergence: The Global Solution to Climate Change.* Totnes: Green Books.

MILLER, D. 1995. *On Nationality.* Oxford: Oxford University Press.

MOELLENDORF, D. 2002. *Cosmopolitan Justice.* Boulder, CO: Westview Press.

—— 2009. *Global Inequality Matters.* New York: Palgrave MacMillan.

MÜLLER, B. 2002. Equity in climate change: The great divide. September 2002. Available at <http://www.oxfordenergy.org/pdfs/EV31.pdf>.

—— HOHNE, N., and ELLERMANN, C. 2009. Differentiating (historic) responsibilities for climate change. *Climate Policy* 9: 593–611.

NAGEL, T. 1991. *Equality and Partiality.* Oxford: Oxford University Press.

—— 2005. The problem of global justice. *Philosophy & Public Affairs* 33: 113–47.

NEUMAYER, E. 2000. In defence of historical accountability for greenhouse gas emissions. *Ecological Economics* 33: 185–92.

OTT, H. E., WINKLER, H., BROUNS, B., KARTHA, S., MACE, M. J., HUQ, S., KAMEYAMA, Y., SARI, A. P., PAN, J., SOKONA, Y., BHANDARI, P. M., KASSENBERG, A., LA ROVERE, E. L., and RAHMAN, A. A. 2004. *South-North Dialogue on Equity in the Greenhouse.* Available at <http://www.gtz.de/de/dokumente/en-climate-south-north-dialogue-equity-greenhouse.pdf>.

PARRY, M. L., CANZIANI, O. F., PALUTIKOF, J. P., VAN DER LINDEN, P. J., and HANSON, C. E. (eds.) 2007. *Contribution of Working Group II to the Fourth Assessment Report of the Intergovernmental Panel on Climate Change, 2007.* Cambridge: Cambridge University Press.

PATERSON, M. 1996. International justice and global warming. Pp. 181–201 in B. Holden (ed.), *The Ethical Dimensions of Global Change.* New York: St Martin's Press.

POGGE, T. 2002. *World Poverty and Human Rights.* Cambridge: Polity Press.

POSNER, E. A., and SUNSTEIN, C. R. 2008. Climate change justice. *Georgetown Law Journal* 96: 1565–612.

—— —— 2009. Should greenhouse gas permits be allocated on a per capita basis? *California Law Review* 97: 51–93.

RAWLS, J. 1999. *The Law of Peoples.* Cambridge, MA: Harvard University Press.

ROBERTS, J. T. 2001. Global inequality and climate change. *Society & Natural Resources* 14: 501–9.

—— and PARKS, B. C. 2006. *A Climate of Injustice.* Cambridge, MA: MIT Press.

—— —— 2007. *A Climate of Injustice.* Cambridge, MA: MIT Press.

SAGAR, A. 2000. Wealth, responsibility and equity: Exploring an allocation framework for global GHG emissions. *Climatic Change* 45: 511–27.

—— and BAER, P. 2009. Inequities and imbalances. Pp. 251–61 in S. H. Schneider, A. Rosencranz, and M. D. Mastrandrea (eds.), *Climate Change Science and Policy.* Washington, DC: Island Press.

—— and T. BANURI. 1999. In fairness to current generations: Lost voices in the climate debate. *Energy Policy* 27: 509–14.

SCHELLING, T. C. 1997. The cost of combating global warming: Facing the tradeoffs. *Foreign Affairs* 76: 8–14.

SHUE, H. 1993. Subsistence emissions and luxury emissions. *Law and Policy* 15: 39–59.

—— 1994. After you: May action by the rich be contingent on action by the poor? *Indiana Journal of Global Legal Studies* 1: 343–66.

—— 1995. Ethics, the environment and the changing international order. *International Affairs* 71: 453–61.

—— 1996. *Basic Rights: Subsistence, Affluence and U.S. Foreign Policy.* 2nd edn., Princeton: Princeton University Press.

—— 1997. Eroding sovereignty: The advance of principle. Pp. 340–59 in R. Mckim and J. McMahon (eds.), *The Morality of Nationalism.* New York: Oxford University Press.

SINGER, P. 1972. Famine, affluence, and morality. *Philosophy & Public Affairs* 1: 229–43.

—— 2002. *One World: The Ethics of Globalization.* New Haven: Yale University Press.

SMITH, K. 1991. Allocating responsibility for global warming: The Natural Debt Index. *Ambio* 20: 95–6.

—— SWISHER, J., and AHUJA, D. R. 1993. Who pays (to solve the problem and how much)? Pp. 70–98 in P. Hayes and K. Smith (eds.), *The Global Greenhouse Regime: Who Pays?* London: Earthscan, for the United Nations University Press.

SOLOMON, B. 1995. Global CO_2 emissions trading: Early lessons from the U.S. Acid Rain Program. *Climatic Change* 30: 75–96.

—— and AHUJA, D. R. 1991. International reductions in greenhouse-gas emissions: An equitable and efficient approach. *Global Environmental Change* 1: 343–50.

SRINIVASAN, U. T. 2010. Economics of climate change: Risk and responsibility by world region. *Climate Policy* 10(3).

STERN, N. 2006. *The Stern Review on the Economics of Climate Change.* Available at <http://www.sternreview.org.uk>.

TOL, R. S. J., and VERHEYEN, R. 2004. State responsibility and compensation for climate change damages: A legal and economic assessment. *Energy Policy* 32: 1109–30.

TRAXLER, M. 2002. Fair chore division for climate change. *Social Theory and Practice* 28: 101–34.

VANDERHEIDEN, S. 2008. *Atmospheric Justice: A Political Theory of Climate Change.* Oxford: Oxford University Press.

VICTOR, D. 1999. The regulation of greenhouse gases: Does fairness matter? In F. Tóth (ed.), *Fair Weather? Equity Concerns in Climate Change.* London: Earthscan Publications Ltd.

YAMIN, F. 2005. The European Union and future climate policy: Is mainstreaming adaptation a distraction or part of the solution? *Climate Policy* 5: 349–61.

CHAPTER 23

....

INTERGENERATIONAL JUSTICE

....

RICHARD B. HOWARTH

1 INTRODUCTION

....

ANTHROPOGENIC climate change presents core issues of intergenerational justice. This chapter will delve into these issues with an emphasis on three distinct lines of moral reasoning: *presentism*, *utilitarianism*, and *rights-based ethics*. In brief, presentism is the view that the preferences of the present generation should play a dominant role in the formulation and evaluation of public policies. In this framework, the interests of future generations are pertinent only to the extent that the present generation holds an altruistic concern for its children, grandchildren, and subsequent descendants. Utilitarianism and right-based ethics, in contrast, assert that equal weight should be attached to the *welfare* or *rights* of both present and future human beings. Unsurprisingly, these theories have sharply different implications for climate stabilization policy.

In framing the distinction between these views, it is useful to begin by describing some factual premises concerning climate change and its moral implications. On the one hand, climate change is driven by activities that provide comfort, mobility, and a high material standard of living to members of the present generation. These benefits accrue in the short run and may be readily understood using the language and methods of economic analysis. A landmark study by Stern (2007), for example, found that implementing aggressive policies to stabilize climate would impose costs equivalent to a 1 percent reduction in economic output as the world made the transition towards low-emission technologies.

On the other hand, greenhouse gas emissions pose a threat to the long-term sustainability of ecological systems and the services they provide to human societies. Over the next century, standard estimates suggest that mean global temperature may increase by up to 6.4 °C in the absence of climate stabilization measures (IPCC 2007). Rising temperatures would give rise to pervasive impacts including sea-level rise, the spread of tropical diseases such as malaria, the intensification of tropical storms, exacerbated floods and droughts, the disruption of water supplies and agricultural production, and biodiversity loss. Hansen et al. (2006) note that a 2–3 °C increase relative to current temperatures would return the earth's climate to conditions last experienced some 3 million years ago, a time when sea level was many meters higher than it is today. In parallel, Thomas et al. (2004) conclude that a temperature increase of this magnitude would lead to the extinction of up to 37 percent of all terrestrial species. The upshot is that seemingly small changes in climate are likely to cause the wide-scale restructuring of environmental systems.

Focusing on impacts such as crop yields, sea-level rise, storm damages, and human health, economists have sought to assign a monetary value to the future damages imposed by climate change. Stern (2007), for example, gauged that a business-as-usual scenario in which greenhouse gas emissions remained unregulated would impose costs equivalent to a permanent 5–20 percent reduction in the level of economic activity. These costs would be concentrated in the twenty-second century and beyond, illustrating the need to consider intergenerational time horizons in understanding the impacts and implications of climate change.

More controversially, Weitzman (2009; see also Woodward and Bishop 1997; Gerst et al. 2010) argues that unregulated emissions might (with a low but positive probability) lead to a catastrophic collapse of the future economy, harkening back to early predictions associated with the Limits-to-Growth model (Meadows et al. 1972). To be clear, Weitzman's point is *not* that a climate catastrophe is in any sense likely to occur based on the hard logic of statistical analysis. But as Page (1978) noted some three decades ago, rational decision makers have good reason to mitigate environmental impacts that involve a low (near-zero) probability of imposing catastrophic (nearly infinite) costs. Unfortunately, the facts and uncertainties pertaining to climate science suggest that climate change constitutes what Page terms a 'zero-infinity dilemma.'

The question, then, is whether society should bear a significant and well-characterized short-term cost to avert long-term climate damages that—although uncertain—are potentially much greater in magnitude. As we shall see, presentists, utilitarians, and rights-based theorists approach this question from different perspectives that provide different insights and suggested solutions. Although this is an area where reasonable people can and do disagree, careful analysis is nonetheless useful in clarifying the plurality of values that can be applied in judging the justice and efficacy of policy alternatives.

2 PRESENTISM

Presentism is a moral framework that is implicitly adopted by climate economists such as Manne (1995), Nordhaus (1992, 2008), and Anthoff et al. (2009b). In this perspective, policy decisions should be based strictly on the *preferences* of the current generation with no explicit moral standing afforded to members of future generations. Taken to its logical extreme, presentism might imply an implausibly strong version of egoism in which present decision makers would never take actions that imposed short-run costs to provide benefits that accrued over intergenerational timescales. In fact, however, both introspection and a substantial body of evidence support a broader and more sensitive interpretation of this approach. Because people hold *altruistic preferences* regarding the welfare of their children and grandchildren, it is appropriate for policy makers to balance the interests of present and future generations (Passmore 1974). The rub is that presentism implies that the weight attached to the welfare of future generations should be based strictly on the degree of altruism that people exhibit through their private decisions (Arrow et al. 1996).

Advocates of presentism attach special importance to the market rate of return on capital investment, which they argue reveals people's willingness to give up present economic benefits for the sake of their children and grandchildren (Goulder and Stavins 2002). Suppose, for example, that households demanded a 6 percent annual return on investment, a figure that is consistent with long-run data from global stock markets. Then given rational decision making, a typical household would be willing to bear a short-run cost of one dollar if and only if it provided benefits of at least 1.06^t dollars to family members living t years from the present. This represents the compound return that could be achieved by investing on the market. Interestingly, this framework implies that the weight attached to the interests of future generations falls geometrically over time. Given a time horizon of 100 years, for example, a future benefit of at least 339 dollars is needed to justify bearing one dollar of costs today. This figure rises to 115,000 and 39 million dollars as the time frame shifts from 200 to 300 years.

The implications of this approach for the analysis of climate stabilization are well established in the literature. Nordhaus (1992, 2008), for example, has long advocated a presentist approach in which major reductions in greenhouse gas emissions should be deferred into the long-run future. In Nordhaus's analysis, the future benefits provided by climate stabilization are too small to justify imposing significant short-run costs given the degree of intergenerational altruism people reveal through their private decisions. In this perspective, people are simply too impatient to care especially about benefits and costs that accrue to their distant descendants.

The presentist approach to environmental policy analysis has been criticized on a variety of grounds. One line of critique argues that the market return on capital investment reveals the preferences that people hold regarding their own present and future well-being, not the conceptually distinct values they hold regarding the appropriate resolution of intergenerational conflicts (Burton 1993). In the economic models employed by presentists, these two behavioral motives are typically reduced to a single parameter for the sake of tractability and simplicity. Authors such as Howarth and Norgaard (1992), however, argue that this modeling approach is theoretically unsound and that fresh insights arise through the use of models that distinguish between personal time preference and intergenerational ethics. Gerlagh and van der Zwaan (2000) show how such models can be effectively applied in the economics of climate change.

As we shall see, critics also charge that presentism involves the unjust treatment of posterity because it denies the principle that all human beings—including members of future generations—should have full and equal moral standing (Broome 2008). Along these lines, Singer (2002: 26) argues that the moral salience of impacts such as 'suffering and death, or the extinction of species' does not diminish with the passage of time. In a similar vein, Ramsey (1928) argues that favoring the interests of present over future generations is 'a practice which is ethically indefensible and arises merely from the weakness of the imagination.' In this perspective, a lack of empathy cannot be used to justify actions that would inflict harms on future generations.

Advocates of presentism, however, counter that the strength of intergenerational altruism has been sufficient to ensure that the quality of life has steadily improved in the centuries following the industrial revolution. If one assumes that economic growth will continue for some time into the future, it follows that our descendants in future generations are likely to be substantially more wealthy than we are today. According to Schelling (2000), the relatively poor people who are alive today are under no compulsion to sacrifice their own interests to provide incremental benefits to the presumably richer people who will populate future society. One response to this is that the presentist stance abstracts away from the catastrophic risks that climate change poses to future generations—climatic impacts may be severe enough to threaten the sustainability and productivity of economic activity (Hoel and Sterner 2007). This point of view is supported by the findings of Woodward and Bishop (1997), Weitzman (2009), and Gerst et al. (2010).

More radically, authors such as Parfit (1983a) question the notion that present decision makers have any obligations to future generations aside from ensuring that future persons have lives that are minimally worth living. The reason is that present decisions will determine not only the welfare but also the identities of future human beings. To understand this point, suppose that wholly different sets of potential persons would live in: (a) a low-income future characterized by a degraded natural environment; and (b) a high-income future characterized by a flourishing environment. Parfit's argument is that the individuals living in the degraded state would be thankful for the fact that present decisions fostered the conditions necessary for *them* to come into being. Steps to stabilize climate would in no way benefit them and would lead to a different world in which they would never be born.

Parfit's argument has attracted considerable attention in the philosophical literature, casting an interesting light on the analysis of intergenerational social choice. De-Shalit (1995: 14–15), for example, argues that communitarianism provides an approach to questions of intergenerational justice that escapes the web of Parfit's reasoning:

> Our obligations to future generations derive from a sense of a community that stretches and extends over generations and into the future . . . If one accepts the idea of a community in one generation, including the principle that this entails certain obligations to other members, then one should accept the idea of a transgenerational community extending into the future, hence recognizing obligations to future generations.

De-Shalit's approach is related to the recent work of Norton (2005: 335), who emphasizes the importance of deliberation and social learning in the articulation of environmental values. For Norton, our duties towards posterity must be constructed based on 'processes by which communities can democratically, through the voices of their members, explore their common values and their differences and choose which stuff should be saved'—i.e. what set of social, economic, technological, and environmental assets should be conserved over time.

Alternatively, Gosseries (2008) notes that Parfit's argument abstracts away from a key fact of human demographics: At each point in time, the current generation of adults overlaps with its children and grandchildren whose existence and identities are fully determined. If one accepts the plausible premise that each generation of adults holds binding duties to its flesh-and-blood progeny, a 'chain of obligation' is then established between present decision makers and the unborn members of more distant generations (Howarth 1992). In particular, the principle of equal opportunity between contemporaries imposes a duty to ensure that human life opportunities are maintained from each generation to the next (Page 1983; see also Vanderheiden 2006). This moral claim is seemingly denied by presentism, which reduces the analysis of intergenerational trade-offs to the presence or absence of intergenerational altruism.

The implications of this 'opportunities' approach to intergenerational fairness are in line with the prescriptions of rights-based ethics as described later in this chapter. For the moment, however, it is pertinent to note that maintaining the structure, functioning, and integrity of natural systems is a secure means of leaving options open to future generations. Authors such as Norton (2005) and Sneddon et al. (2006) therefore stress the importance of environmental conservation based on perceived duties to posterity. Sneddon et al. relate this approach to Sen's (1999) writings on functionings and capabilities, which strongly emphasizes the importance of enhancing people's opportunities and effective freedoms in a broadly Aristotelian conception of justice.

3 UTILITARIANISM

Conferring moral status on members of future generations undercuts the foundations of presentism. While present decision makers might *prefer* to attach more weight to their own interests than to the welfare of future generations, this seems to conflict with a defining characteristic of moral reasoning—that moral subjects must be treated equally and impartially in the definition and pursuit of justice. One prominent alternative to presentism is *classical utilitarianism*, an ethical framework that dates to the seminal work of Jeremy Bentham (1823). According to utilitarians, social institutions and public policies should be designed to maximize total utility or well-being in society with equal weight attached to the welfare of each and every person. Mill (1863) termed this criterion the 'greatest happiness principle.'

Singer (2002: 42) discusses the implications of utilitarianism for climate stabilization policy. On the one hand, utilitarians favor an approach that balances the costs and benefits of greenhouse gas emissions. On the other hand, they also attach special importance to the interests of people suffering material deprivation:

> [W]hen you already have a lot, giving you more does not increase your utility as much as when you have only a little. One of the 1.2 billion people in the world living on $1 per day will get much more utility out of an additional $100 than will someone living on $60,000 per year. Similarly, if we have to take $100 from someone, we will cause much less suffering if we take it from the person earning $60,000 per year than if we take it from the person earning $365 a year. This is known as 'diminishing marginal utility.'

According to Singer, this reasoning implies that the costs of climate change mitigation should be borne disproportionately by the wealthiest members of the international community since a dollar of net benefits provides less utility to a rich person than to a poor person. For this same reason, utilitarians are especially concerned about the potential threat that climate change poses to incomes and livelihoods in low-income, developing countries that are resource dependent and therefore especially vulnerable to changes in environmental conditions (Anthoff et al. 2009a).

As a moral philosophy, utilitarianism provides no basis for attaching different weights to the welfare of present and future generations. On the contrary, utility is viewed as equally valuable regardless of who experiences it in either space or time (Broome 2008).[1] This point of view has a long and rich history in the development of economic thought. Ramsey (1928), for example, considered a theoretical model in which maximizing total utility over time required short-run sacrifices so that the economy would converge to a long-run state of 'bliss,' characterized as the highest degree of happiness that is psychologically achievable. In empirical applications, this approach implies higher rates of saving and economic growth than is typically observed in real-world economies.

Authors such as Cline (1992) and Stern (2007) have explored the consequences of classical utilitarianism in fully specified mathematical models of climate-economy inter-actions (see also Howarth 1998). These authors begin by gauging the monetary costs and benefits of greenhouse gas mitigation measures. Monetary costs and benefits are then converted to units of utility or well-being based on empirically plausible assumptions concerning the relationship between income and human flourishing. In contrast with Singer (2002), these authors limit their analyses to a concern for human welfare but attach equal weight to both present and future well-being. This analytical approach has strong policy implications: It implies that greenhouse gas emissions should be substantially if prudently reduced since the future welfare costs of unmitigated climate change would far exceed the short-run costs of making the transition towards more sustainable agricultural and energy technologies.

While utilitarianism supports aggressive steps to stabilize the earth's climate, this approach to policy analysis is controversial for both theoretical and practical reasons. The theoretical objections to utilitarianism are nicely summarized in Sen and Williams's edited volume *Utilitarianism and Beyond* (1982). In short, much of moral philosophy is concerned with understanding and managing the conflicts that exist between the pursuit of self-interest and the performance of one's moral duties. Utilitarianism approaches this problem by asserting that people's decisions should aim to maximize total utility in society without attaching special weight to personal needs and concerns.

Critics charge that this criterion seems psychologically implausible and inconsistent with our moral intuitions. It seems to suggest that it is morally wrong to spend resources to promote one's own happiness in a world of inequality in which transferring one's wealth to the poor would provide greater social utility. One response is that utilitarianism might be viewed as a criterion for *collective* decisions that is consistent with a framework in which individuals legitimately pursue their private preferences. Harsanyi (1955), for example, explores the circumstances under which utilitarian social choice rules can be derived from a situation in which free and equal persons negotiate constitutional arrangements behind a veil of ignorance.

A related concern arises especially in the economics of climate change, where as we have seen authors such as Schelling (2000) argue that phasing out greenhouse gas emissions would redistribute wealth from the poor of today to comparatively rich members of future generations. This runs afoul of what Parfit (1983b) terms the 'argument from excessive sacrifice.' The problem is that incurring one dollar of costs in the present would provide many more dollars of benefits accruing to future generations. Given realistic empirical assumptions, undertaking deep cuts in greenhouse gas emissions is therefore necessary to maximize total utility. But is it reasonable to demand self-sacrifice by the poor to provide increased benefits to people who are or will be considerably more affluent?

This issue signals an important and quite general objection to utilitarian ethics. In certain circumstances, utilitarianism appears to justify inflicting major hardships on the few as long as doing so would provide minor yet widely shared benefits for a sufficiently large number of people. This conflicts with the notion that individuals are entitled to protection against serious, uncompensated harms resulting from actions that provide benefits to third parties. As Shue (1999: 39) frames this point:

> One can try to imagine, say, a 'state of nature' in which assault, beating, rape, torture, and mayhem violate no rights and break no rules, because there are no such rights or underlying rules.

That said, a reasonable person:

> has no doubt whatsoever that it is unacceptable for a person's body to be damaged. It is simply not possible for a sane person to act in practice as if he or she believes that his or her body is not entitled to the kind of special protection against the depredations of others that a right constitutes.

In this sense, utilitarianism is in tension with the moral principles that support liberal-democratic political, economic, and legal institutions, which attach paramount importance to the extension and preservation of individual rights and freedoms.

4 RIGHTS-BASED ETHICS

Shue's analysis invites the question of what rights and entitlements should be afforded to members of future generations. Shue's own answer is that, at a minimum, future persons are entitled to protection against bodily harm. More broadly, Shue (2005: 276) imagines a future dystopia in which the present generation's unwillingness to reduce greenhouse gas emissions has inflicted suffering and immiseration on posterity. He reasons that 'If I were a desperate member of that later generation, I think I would be furious at our generation' for failing to take action. 'This is not how I was hoping to be remembered: as a good-for-nothing great-great-grandfather who wallowed in comfort and convenience.'

In developing this theme, Vanderheiden (2006: 343) notes that 'neither spatial nor temporal distance between agents and their victims can excuse acts of intentional or predictable harm.' Since compelling evidence suggests that climate change would inflict harms on future generations, this line of reasoning implies a duty to stabilize climate by cutting greenhouse gas emissions (see Vanderheiden 2008).

A related interpretation is provided by Caney (2008: 538), who argues that climate stabilization is necessary to secure and defend at least three kinds of fundamental human rights. In particular, Caney argues that climate change:

1. Violates people's right to *subsistence* by imposing risks of 'widespread malnutrition' that are well documented by the scientific literature.

2. Threatens people's capacity to 'attain a *decent standard of living*' (emphasis added), a point that resonates with the economic arguments advanced by Weitzman (2009).

3. Poses unacceptable risks to *human health* due to a range of mechanisms that include heat stress and the increased incidence of tropical diseases.

Further insights arise by locating Shue's (1999: 43) 'no-harm principle' in the history of Anglo-American political thought. In a letter to James Madison written in 1789, the American statesmen Thomas Jefferson reasoned that 'the earth belongs in usufruct to the living' (see Ball 2000). Jefferson's letter focused on the argument that the United States Constitution should include a provision preventing the federal government from accumulating unpaid financial debts that would be passed on from one generation to the next. His argument, however, built on the premise that the *earth*—i.e. the land and, by extension, the full suite of environmental resources—is the shared patrimony of present and future society. In this perspective, the present generation holds usufruct rights—i.e. an entitlement to reap the sustained flow of benefits provided by biophysical systems.[2] These rights, however, come with a correlative duty to conserve and protect resources for the benefit of future generations. To deplete natural resources would inflict uncompensated harm on posterity, thereby invading the rights and entitlements of future persons.

Jefferson's approach to questions of intergenerational justice builds on the rights-based ethical theories advanced by authors such as Locke (1690) and Kant (1963). Locke, for example, wrote that 'the earth, and all that is therein, is given to men for the support and comfort of their being' and that 'all the fruits it naturally produces . . . belong to mankind in common.' Although Locke famously argued that people could legitimately establish private property rights by mixing their labor with the land, he also held that the enjoyment of private property was thus limited by the proviso that there be 'enough and as good left in common for others' (see Singer 2002: 27–8). In this perspective, the legitimacy of private property seems to depend on the existence of institutions that ensure that different members of society have equal access to livelihoods and opportunities. Over intergenerational timescales, this may require policies that specifically protect environmental resources for the benefit of future generations.

Jefferson's concept of usufruct rights strongly anticipates more recent developments in conservationist thought and its applications to environmental governance. The founding Chief of the US Forest Service, for example, advanced the following principle that became institutionalized in the structures of US forest, land, and fisheries management (Pinchot 1910: 80)—publicly owned resources should be managed based on an approach that:

> recognizes fully the right of the present generation to use what it needs and all it needs of the natural resources now available, but [also] recognizes equally our obligation so to use what we need that our descendants shall not be deprived of what they need.

In close parallel, the Brundtland Commission's definition of *sustainable development* emphasizes the importance of '[meeting] present needs without compromising the ability of future generations to meet their own needs' (WCED 1987: 43). Quite explicitly, the Commission argued that this language entailed a responsibility to conserve natural resources and environmental quality.

The application of this approach to climate stabilization policy is described in detail by Brown's (1998, 2007) writings on 'stewardship.' If future generations hold a moral right to enjoy the benefits of a stable and non-degraded natural environment, then the present generation holds a corresponding trusteeship duty to limit greenhouse gas emissions to sustainable levels. This normative standard provides an important basis for the United Nations Framework Convention on Climate Change, which calls for:

[the] stabilization of greenhouse gas concentrations in the atmosphere at a level that would prevent dangerous anthropogenic interference with the climate system. Such a level should be achieved within a time frame sufficient to allow ecosystems to adapt naturally to climate change, to ensure that food production is not threatened and to enable economic development to proceed in a sustainable manner.

The language of this text echoes the 'safe minimum standards' criterion proposed by Ciriacy-Wantrup (1952), under which ecological resources should be managed in a manner that minimizes potential adverse impacts under conditions of uncertainty. Bishop (1993) interprets this approach as implying that the present generation should refrain from actions that threaten the stability and functioning of natural systems unless the costs are 'intolerable.'

More recently, authors such as Turner (1993) and Dobson (1998) have called for the conservation of 'critical natural capital'—those features of biophysical systems that provide potentially indispensable and irreplaceable ecosystem services—as a way of securing the interests of future generations under conditions of scientific uncertainty. Both safe minimum standards and the critical natural capital approach can be seen as ways of interpreting and applying the 'precautionary principle,' which endorses 'the commitment of resources now to safeguard against the potentially adverse future outcomes of some decision' (Perrings 1991; see also O'Riordan and Cameron 1994; Howarth 2001).

Paavola (2008: 657) presents an interesting discussion that links this precautionary, rights-based reasoning with a pluralistic, participatory approach to global environmental governance. According to Paavola:

[F]rom a social justice viewpoint, it is necessary to adopt atmospheric targets for GHGs, to adopt instruments such as a carbon tax to raise funds for assisting adaptation, and to establish procedural solutions that address inequities in participation in planning and decisionmaking on adaptation to climate change.

Paavola's point is that a concern for intergenerational justice mandates reducing greenhouse gas emissions to limit the future harms caused by climate change. In parallel, however, the design of just policies requires institutions that insure that burdens and benefits are fairly shared between members of society. For Paavola, the question of equitable burden sharing is best addressed through the mechanisms of deliberative democracy (see Dryzek 2000).

Critics of rights-based approaches to framing intertemporal decisions emphasize several lines of argument that are worthy of careful consideration. First is Parfit's (1983a) claim that future generations lack moral standing. As we have discussed, this view abstracts away from the well-recognized duties that each generation holds towards its children and grandchildren (Page 1983; Howarth 1992; Gosseries 2008).

Second is the concern that taking aggressive steps to stabilize climate would slow the rate of economic growth in ways that would reduce the welfare of both present and future generations. As we noted in the introduction, a substantial body of empirical research casts doubt on this argument (Stern 2007). On the one hand, the short-run costs of climate stabilization are thought to be relatively small—perhaps 1 percent of economic output, or less than one year's worth of economic growth over the course of several decades. On the other hand, the projected costs of unmitigated climate change are believed to be much larger in magnitude—a full 5–20 percent of future economic output. In short, climatic stability is a valuable resource that would support and sustain future prosperity and human flourishing. Stabilizing climate is therefore instrumental in securing the life opportunities of future generations.

Critics also advance the argument from excessive sacrifice (see above) as a reason to defer or delay reductions in greenhouse gas emissions. Here too, however, the evidence is equivocal. Given well-designed policies, cutting greenhouse gas emissions to stabilize climate would impose costs that would be unnoticeable by most members of society. Climate stabilization, however, would reduce the risk of imposing uncompensated and potentially catastrophic harms on members of future generations. With a low but nonetheless positive probability, unmitigated climate change might lead to a long-run collapse of the ecosystem services needed to support human welfare and economic activity (Weitzman 2009). In statistical terms, this risk can be reduced to effectively zero by stabilizing temperatures at a level no more than 2 °C above the pre-industrial norm (Gerst et al. 2010).

As Bromley (1989) argues, rights-based approaches to environmental management may be especially appropriate when: (a) the costs of environmental protection are comparatively low; and (b) the projected impacts of environmental degradation are uncertain, irreversible, and potentially catastrophic. This line of reasoning builds on Page's (1978) discussion of 'zero-infinity dilemmas' as outlined in the introduction of this chapter. Problems of this nature involve key asymmetries that provide a potential rationale to refrain from imposing prospective harms.

5 CONCLUSIONS

This chapter has reviewed three distinct approaches to addressing the important issues of intergenerational justice that arise in the evaluation of climate stabilization policies. The discussion supports the following findings and conclusions.

First, *presentism* is an ethical framework that emphasizes the interests of present generations while denying that future generations have full moral standing. Presentists note that people hold altruistic preferences concerning the welfare of their children and grandchildren and that those preferences provide the most appropriate basis for balancing short-run costs and long-run benefits in environmental policy analysis. On empirical grounds, authors such as Nordhaus (1992, 2008) argue that the degree of intergenerational altruism is too weak to justify aggressive steps to reduce greenhouse gas emissions. In this sense, presentism seems to suggest that it is better to endure the future costs of climate change than the short-run costs of climate stabilization.

Proponents argue that presentism is supported by the futurity or contingent status of future generations (Parfit 1983a) and by the 'argument from excessive sacrifice,' which reasons that present decision makers have no obligation to make sacrifices for the benefit of future generations that (in a world of economic growth) are likely to enjoy far higher levels of prosperity and well-being (Parfit 1983b; Schelling 2000). In the main body of this chapter we made the case that these arguments are morally and empirically unsound. The concern is that presentism unreasonably abstracts away from both: (a) the duties that the current generation holds towards its children and grandchildren; and (b) from the serious risks that climate change poses to future welfare.

Second, advocates of *classical utilitarianism* argue that equal weight should be attached to the welfare of each and every member of society including members of future generations (Broome 2008). Authors such as Cline (1992) and Stern (2007) show that utilitarianism supports aggressive policies to stabilize climate under plausible empirical assumptions (see also Howarth 1998). The key point is that stabilizing climate would reduce short-run economic output by roughly 1 percent while conferring gains of 5–20 percent on members of future generations (Stern 2007). Translated into units of experienced well-being, these figures imply that the benefits of climate stabilization considerably exceed the costs given the equal weighing criterion adopted by utilitarians.

As noted above, utilitarianism is vulnerable to the argument from excessive sacrifice. This critique, however, is less compelling than a superficial analysis might suggest. A consistent utilitarian, for example, would argue that the burden of climate stabilization should be borne disproportionately by the wealthiest members of the present generation. Indeed, climate change impacts are likely to fall hardest on poor communities that lack the resources and capabilities needed to adapt to changing environmental conditions. Preventing such impacts would arguably yield particularly large welfare gains.

A more serious objection is that utilitarian ethics leaves little room for the premise that people have fundamental rights and that the protection of rights—not simply the pursuit of aggregate social welfare—should play a key role in the design of public policies. In the context of climate change policy, advocates of *rights-based ethics* argue that (a) future generations are entitled to protection from harm (Shue 1999) or that (b) the natural environment is the shared property of both present and future generations (see Ball 2000). Rights-based theories accept the premise that the present generation has a right to derive benefit from the sustainable use of environmental resources. But they also imply a correlative duty to conserve the environment based on the rights and interests of future persons (Brown 1998). Caney (2008), for example, reasons that members of future generations have a right to be protected from the uncompensated and potentially catastrophic harms that climate change might impose on their livelihoods and life opportunities.

Advocates of the rights-based view would argue that it is only by protecting the rights and entitlements of future generations that we can be confident that future generations will enjoy a quality of life that is undiminished relative to the present (Norton 2005). This is one way of responding to the argument from excessive sacrifice. Moreover, they would argue that the present generation holds a duty to ensure that life opportunities are maintained from each generation to the next (Page 1983; Howarth 1992; Gosseries 2008). This argument extends the notion of equality of opportunity between each generation of adults and its children and grandchildren.

As noted in the introduction, this field of research and praxis is an area where reasonable people can and do hold strongly contrasting points of view. The author of this chapter, for example, is a critic of presentism, has written sympathetically though skeptically concerning the application of utilitarianism to climate change policy (Howarth 1998), and has endorsed the argument that rights-based ethics provides the most convincing approach to issues of sustainability and intergenerational justice (see Howarth 2001). Other contributors reach their own conclusions based on different value judgments and factual assumptions.

That said, both classical utilitarianism and rights-based ethics support the stated objective of the Framework Convention on Climate Change, which calls for the prevention of 'dangerous anthropogenic interference' with the earth's climate to protect the interests of future generations. We can say, then, that the simple move of conferring full moral standing on future generations seems to favor aggressive climate change policies in a way that brings together concepts of intra- and intergenerational justice. One could imagine a plausible if pragmatic policy approach that emphasized both the protection of basic rights and the pursuit of higher social welfare. In this sense, resolving the principled disagreement between utilitarians and right-based theorists may be unnecessary to reach agreement on just and effective climate change policies.

NOTES

1. In fact, Bentham (1823) suggested that the flourishing of all sentient beings might be pertinent to applications of the moral calculus. Developing this point, Singer's (1975) theory of animal liberation confers moral standing on nonhuman animals.
2. The Oxford English Dictionary (2nd edition, 1989) defines usufruct as 'the right of temporary possession, use, or enjoyment of the advantages of property belonging to another, so far as may be had without causing damage or prejudice to this.'

REFERENCES

ANTHOFF, D., HEPBURN, C., and TOL, R. S. J. 2009a. Equity weighting and the marginal damage costs of climate change. *Ecological Economics* 68: 836–49.
—— TOL, R. S. J. and YOHE, G. W. 2009b. Risk aversion, time preference, and the social cost of carbon. *Environmental Research Letters* 4: 1–7.
ARROW, K. J., CLINE, W. R., MÄLER, K. G., MUNASINGHE, R., SQUITIERI, R., and STIGLITZ, J. E. 1996. Intertemporal equity, discounting, and economic efficiency. In J. P. Bruce, H. Lee, and E. F. Haites (eds.), *Climate Change 1995: Economic and Social Dimensions of Climate Change*. Cambridge: Cambridge University Press.
BALL, T. 2000. The earth belongs to the living: Thomas Jefferson and the problem of intergenerational relations. *Environmental Politics* 9: 61–77.
BENTHAM, J. 1823. *An Introduction to the Principles of Morals and Legislation*. London: W. Pickering.
BISHOP, R. C. 1993. Economic efficiency, sustainability, and biodiversity. *Ambio* 22: 69–73.
BROMLEY, D. W. 1989. Entitlements, missing markets, and environmental uncertainty. *Journal of Environmental Economics and Management* 17: 181–94.
BROOME, J. 2008. The ethics of climate change. *Scientific American* 298: 97–102.
BROWN, P. G. 1998. Toward an economics of stewardship: The case of climate. *Ecological Economics* 26: 11–21.
—— 2007. *The Commonwealth of Life: Economics for a Flourishing Earth*. 2nd edn., Montreal: Black Rose Books.
BURTON, P. S. 1993. Intertemporal preferences and intergenerational equity considerations in optimal resource harvesting. *Journal of Environmental Economics and Management* 24: 119–32.

CANEY, S. 2008. Human rights, climate change, and discounting. *Environmental Politics* 17: 536–55.

CIRIACY-WANTRUP, S. V. 1952. *Resource Conservation: Economics and Policies*. Berkeley: University of California Press.

CLINE, W. R. 1992. *The Economics of Global Warming*. Washington, DC: Institute for International Economics.

DE-SHALIT, A. 1995. *Why Posterity Matters: Environmental Policies and Future Generations*. London: Routledge.

DOBSON, A. 1998. *Justice and the Environment: Conceptions of Environmental Sustainability and Dimensions of Social Justice*. Oxford: Oxford University Press.

DRYZEK, J. S. 2000. *Deliberative Democracy and Beyond: Liberals, Critics, Contestations*. Oxford: Oxford University Press.

GERLAGH R., and VAN DER ZWAAN, B. C. C. 2000. Overlapping generations versus infinitely-lived agent: the case of global warming. In R. B. Howarth and D. Hall (eds.), *The Long-Term Economics of Climate Change*. Stamford, CT: JAI Press.

GERST, M., HOWARTH, R. B., and BORSUK, M. E. 2010. Accounting for the risk of extreme outcomes in an integrated assessment of climate change. *Energy Policy* 38: 4540–8.

GOSSERIES, A. 2008. On future generations' rights. *Journal of Political Philosophy* 16: 446–74.

GOULDER, L. H., and STAVINS, R. N. 2002. An eye on the future. *Nature* 419: 673–4.

HANSEN, J., SATO, M., RUEDY, R., LO, K., LEA, D. W., and MEDINA-ELIZADE, M. 2006. Global temperature change. *Proceedings of the National Academy of Sciences* 103: 14288–93.

HARSANYI, J. C. 1955. Cardinal welfare, individualistic ethics, and interpersonal comparisons of utility. *Journal of Political Economy* 63: 309–21.

HOEL, M., and STERNER, T. 2007. Discounting and relative prices. *Climatic Change* 84: 265–80.

HOWARTH, R. B. 1992. Intergenerational justice and the chain of obligation. *Environmental Values* 1: 133–40.

—— 1998. An overlapping generations model of climate-economy interactions. *Scandinavian Journal of Economics* 100: 575–91.

—— 2001. Intertemporal social choice and climate stabilization. *International Journal of Environment and Pollution* 15: 386–405.

—— and NORGAARD, R. B. 1992. Environmental valuation under sustainable development. *American Economic Review* 80: 473–7.

Intergovernmental Panel on Climate Change (IPCC). 2007. *Climate Change 2007: Synthesis Report*. New York: Cambridge University Press.

KANT, I. 1963 edn. *Lectures on Ethics*, trans. L. Infield. Indianapolis: Hackett.

LOCKE, J. 1690. *Two Treatises of Government*. London: Awnsham Churchill.

MANNE, A. S. 1995. The rate of time preference: Implications for the greenhouse debate. *Energy Policy* 23: 391–4.

MEADOWS, D. H., MEADOWS, D. L., RANDERS, J., and BEHRENS, W. W. 1972. *The Limits to Growth*. New York: Universe Books.

MILL, J. S. 1863. *Utilitarianism*. London: Parker, Son & Bourn.

NORDHAUS, W. D. 1992. An optimal transition path for controlling greenhouse gases. *Science* 258: 1315–19.

—— 2008. *A Question of Balance: Weighting the Options on Global Warming Policies*. New Haven: Yale University Press.

NORTON, B. G. 2005. *Sustainability: A Philosophy of Adaptive Ecosystem Management*. Chicago: University of Chicago Press.

O'RIORDAN, T., and CAMERON, J. 1994. *Interpreting the Precautionary Principle*. London: Earthscan.

PAAVOLA, J. 2008. Science and social justice in the governance of adaptation to climate change. *Environmental Politics* 17: 644–59.

PAGE, T. 1978. A generic view of toxic chemicals and similar risks. *Ecology Law Quarterly* 7: 207–44.

—— 1983. Intergenerational justice as opportunity. In D. MacLean and P. G. Brown (eds.), *Energy and the Future*. Totowa, NJ: Rowman & Littlefield.

PARFIT, D. 1983a. Energy policy and the further future: The identity problem. In D. MacLean and P. G. Brown (eds.), *Energy and the Future*. Totowa, NJ: Rowman & Littlefield. Pp. 166–179.

—— 1983b. Energy policy and the further future: The social discount rate. In D. MacLean and P. G. Brown (eds.), *Energy and the Future*. Totowa, NJ: Rowman & Littlefield.

PASSMORE, J. 1974. *Man's Responsibility for Nature*. New York: Scribner Press.

PERRINGS, C. 1991. Reserved rationality and the precautionary principle: Technological change, time and uncertainty in environmental decision making. In R. C. Costanza (ed.), *Ecological Economics: The Science and Management of Sustainability*. New York: Columbia University Press.

PINCHOT, G. 1910. *The Fight for Conservation*. New York: Doubleday, Page & Company.

RAMSEY, F. 1928. A mathematical theory of saving. *Economic Journal* 38: 543–59.

SCHELLING, T. C. 2000. Intergenerational and international discounting. *Risk Analysis* 20: 833–7.

SEN, A. 1999. *Development as Freedom*. New York: Anchor Books.

—— and WILLIAMS, B. A. O. 1982. *Utilitarianism and Beyond*. New York: Cambridge University Press.

SHUE, H. 1999. Bequeathing hazards. In M. H. I. Dore and T. D. Mount (eds.), *Global Environmental Economics: Equity and the Limits to Markets*. Oxford: Blackwell.

—— 2005. Responsibility to future generations and the technological transition. In W. SINNOTT-ARMSTRONG and R. B. HOWARTH (eds.), *Perspectives on Climate Change: Science, Economics, Politics, Ethics*. Amsterdam: Elsevier.

SINGER, P. 1975. *Animal Liberation*. New York: Random House.

—— 2002. *One World: The Ethics of Globalization*. New Haven: Yale University Press.

SNEDDON, C., HOWARTH, R. B., and NORGAARD, R. B. 2006. Sustainable development in a post-Brundtland world. *Ecological Economics* 57: 253–68.

STERN, N. 2007. *The Economics of Climate Change: The Stern Review*. Cambridge: Cambridge University Press.

THOMAS C. D., CAMERON, A., GREEN, R. E., BAKKENES, M., BEAUMONT, L. J., COLLINGHAM, Y. C., ERASMUS, B. F., DE SIQUEIRA, M. F., GRAINGER, A., HANNAH, L., HUGHES, L., HUNTLEY, B., VAN JAARSVELD, A. S., MIDGLEY, G. F., MILES, L., ORTEGA-HUERTA, M. A., PETERSON, A. T., PHILLIPS, O. L., and WILLIAMS, S. E. 2004. Extinction risk from climate change. *Nature* 427: 145–8.

TURNER, R. K. 1993. Sustainability: principles and practice. In R. K. Turner (ed.), *Sustainable Environmental Economics and Management*. London: Belhaven Press.

VANDERHEIDEN, S. 2006. Conservation, foresight, and the future generations problem. *Inquiry* 49: 337–52.

—— 2008. *Atmospheric Justice: A Political Theory of Climate Change*. Oxford: Oxford University Press.

WEITZMAN, M. L. 2009. On modeling and interpreting the economics of catastrophic climate change. *Review of Economics and Statistics* 91: 1–19.

WOODWARD, R. T., and BISHOP, R. C. 1997. How to decide when experts disagree: Uncertainty-based choice rules in environmental policy. *Land Economics* 73: 492–507.

World Commission on Environment and Development (WCED). 1987. *Our Common Future*. New York: Oxford University Press.

PART VII

....

PUBLICS AND MOVEMENTS

....

CHAPTER 24

....

PUBLIC OPINION AND PARTICIPATION

....

MATTHEW C. NISBET

1 PUBLIC OPINION AND POLITICAL BEHAVIOR
....

IN political discourse and news coverage of climate change, nationally representative opinion surveys have come to dominate how we talk about the relationship between climate change and the public. The unfortunate tendency, however, is for survey research to be interpreted somewhat simplistically, with scant consideration for a respondent's social context or background and without regard to important communication behaviors and areas of knowledge. Instead, surveys are frequently referenced as if the public were comprised of relatively anonymous, geographically dispersed individuals who have very little or no shared interaction, common interests, or identity.

Across countries, this imagined public relative to climate change remains a source of ever growing anxiety among scientists and advocates for climate action. The focus typically is on how much the imagined public does not understand or know about climate change and the perceived 'gap' or 'divide' between aggregated survey results and expert views. To close this gap, communication is similarly imagined as a process of technical translation and popularization from experts to the mass public, with facts assumed to speak for themselves and to be interpreted by all individuals in similar ways. The difference between expert opinion and mass opinion is blamed on biases in news coverage, 'irrational' beliefs, the work of climate skeptics, or a combination of these three factors (Nisbet and Scheufele 2009).

Yet, instead of reducing public opinion formation to the aggregation of individual responses in nationally representative surveys, public opinion needs to be studied, understood, and discussed as a process that emerges from social context, interaction, and communication. It is this complex process that accounts for the difference between expert views and the subjective perceptions of a diversity of publics.

Examining the case of the United States, this chapter opens by describing the tail ends of public perspectives on climate change, examining the nature of an 'issue public' working to mobilize concern and a climate denial movement organized against policy action. These tail-end segments dominate popular discussion about public opinion, yet between these proportionally small segments, research shows a socially diverse and mostly ambivalent public. Constituting unique 'interpretative communities,' these middle-range segments vary in their size and demographic attributes; their levels of news consumption, attention, and forms of knowledge; the mental frameworks, values, and influences that guide their judgments and behaviors; and the strength and direction of their preferences, opinions, and participation. Importantly, research is being used to identify and develop specifically tailored communication initiatives that empower and enable these publics to reach decisions and to participate in societal debates over climate change.

2 THE MEDIA, THE 'ISSUE PUBLIC,' AND WIDER MOBILIZATION

In an era of digital and online media, the communication playing field has been leveled between expert institutions, traditional journalists as gatekeepers of information, and users of information. The balance of control has shifted in the direction of the people formerly known as the audience, with an engaged segment of media users participating as active contributors, collaborators, creators, disseminators, recommenders, and at times, critics in the climate change debate. These participatory individuals—empowered over the past decade by the many changes in the media system—are what communication researchers have traditionally defined and tracked as the 'issue' public (Krosnick et al. 2000; Kim 2009). Research on the connection between policy making and public opinion concludes that on most policy issues, decisions reflect the preferences of the small issue public surrounding a debate, since this segment is the most participatory and the voice that is heard loudest and most frequently among elected officials (Krosnick 1990; Manza and Cook 2002).

Studies find that the size of the issue public on climate change has increased over the past decade and is likely to continue to shift in marginal ways in reaction to focusing events, levels of news attention, and the efforts of advocates to intensify public concern and broaden involvement. Consider, for example, that in 1997 during the build-up to the Kyoto climate treaty meetings, the issue public on climate change grew from 9 to 11 percent over just a few months, an increase that translated into 5 million more Americans engaged and potentially involved on the issue (Krosnick et al. 2000).

Today, the issue public on climate change is estimated to be approximately 15 percent of Americans, a segment equal to the active public on issues such as abortion, gun control, and foreign policy. This proportion translates into approximately 35 million individuals—with more than 80 percent accepting the human-causes of climate change and supportive of policy action to reduce the threat (Krosnick 2010b). (As will be discussed later, other analyses depict the current proportion of Americans 'alarmed' and involved on the issue also at 15 percent (Leiserowitz et al. 2010). This figure also compares to the 15–20 percent of Americans who self-identify as 'active' environmentalists (Dunlap 2010)).

By working with others, members of the issue public have made climate change a major part of the agenda and criteria by which many organizations, companies, cities, and states reach decisions and interact across the government, business, and civic sectors. Through digital and face-to-face interactions, key members of the issue public are also serving as informal opinion leaders. More than just attentive and individually active on climate change, these opinion leaders also serve as influential go-betweens, receiving and passing on to their peers information, news, resources, and requests to get involved. In this 'two-step flow of information,' opinion leaders do not necessarily hold formal positions of power or prestige, but rather serve as the connective communication tissue that alerts their peers to what matters among political events, social issues, and consumer choices (Leiserowitz et al. 2010; Nisbet and Kotcher 2009).

Yet despite local impacts and interpersonal influence, members of the issue public in the US have yet to be able to create the public opinion conditions necessary to pass national climate change legislation. Climate change is one of a handful of enduring social problems such as immigration, social security, or healthcare that require non-incremental policy formulation and adoption. Previous studies of factors that have led to non-incremental, systemic policy change in Congress, such as 1990s welfare reform, find that pressure from an issue public is not enough. Instead, these studies find that widespread and intense public concern is a key factor in the success or failure of legislation. Consider that when welfare reform was passed in 1996, 27 percent of Americans considered the issue to be the most important issue facing the country and more than 80 percent supported President Clinton signing the bill into law (Nisbet 2009; Soss and Schram 2007).

In the US these public opinion conditions have yet to be met on climate change. In polls, typically few, if any Americans name climate change as the country's most important problem and in a ranking of 21 national issues, climate change ranks among the lowest in perceived priority (Pew 2010). Symptomatic of the still missing opinion intensity, polling suggests that majorities of Americans accept the science of climate change and support curbing greenhouse gas emissions (Nisbet and Meyers 2007; Krosnick 2010b), but when policy proposals are presented in the context of costs, support diminishes (Nisbet and Meyers 2007). In short, while Americans are concerned by climate change, only a small proportion possess the type of opinion intensity that motivates direct participation and contacts to elected officials (Leiserowitz et al. 2010).

Absent an increase in opinion intensity and wider public mobilization, no matter the policy proposal, national elected officials will have little incentive to take on the political risks needed to pass major legislation. As Bill McKibben expressed in 2009 following the failure of environmental advocates to gain US Senate support for Cap and Trade: 'We weren't able to credibly promise political reward or punishment. The fact is, scientists have been saying for the past few years the world might come to an end. But clearly that's insufficient motivation. Clearly, we must communicate that their careers might come to an end. That's going to take a few years' (Samuelson 2010).

Though digital media serve as a major resource for the issue public on climate change, the same dimensions of the contemporary media system also present barriers to building the wider public will necessary to exert pressure on national elected officials. This reflects in part the problem of limited attention in an age of digital media: Via the Web, individuals have more quality sources of information and opportunities to participate on climate change than at any time in history, but the availability of information does not mean that the wider public will use it. In a media world of many choices, if an individual lacks a preference or need for climate change-related information, they can avoid such content almost altogether (Prior 2005). This tendency is magnified by the multi-tasking nature of contemporary media use. While opinion leaders on climate change can take advantage of hand-held devices for news and social media influence (Nisbet and Kotcher 2009), as an average tendency, studies find that the multi-tasking facilitated by hand-held devices is negatively related to learning and recall, thereby amplifying the problem of choice in gaining the attention of the wider public (Ophir et al. 2009).

Yet when motivated—such as at times of a major relevant focusing event—otherwise inattentive or distracted individuals will turn to the news media and in particular Web sources for information (Pew 2006). A leading example is the Gulf oil spill. Through the spring and summer of 2010, the unfolding disaster had emerged as one of the top five issues covered across the news media with half of Americans saying that they were following news of the disaster 'very closely' (Pew 2010). Within this coverage, audiences have the potential to be exposed to discussion and news of the relevance of the oil spill to the climate change and energy debate. At other times, in the absence of a focusing event or direct personal need, wider audiences may simply 'bump' into climate change-related information while consuming entertainment or political media (Feldman et al. 2010). As will be discussed later, whether direct connections between a focusing event such as the oil spill and the relevance of climate change are effectively conveyed to the wider public, can be understood via past research on framing.

3 THE CLIMATE DENIAL MOVEMENT

Ambivalence on the part of the wider public—and intense opposition among a small segment of Americans—is also attributable to the organized activities of industry members, conservative think tanks, commentators, and elected officials. Applying a strategy first used to dispute the linkages between smoking and cancer; this 'climate denial' movement disputes the reality of man-made climate change and exaggerates the economic costs of action (Oreskes and Conway 2010). Studies have tracked the disproportionate number of appearances of a handful of contrarian scientists in Congressional hearings, in news reports, and as book authors, documenting the linkages with conservative think tanks and industry funders (Jacques et al. 2008; McCright and Dunlap 2003, 2010). The arguments of contrarians are echoed and magnified at conservative talk radio, cable news, and by conservative commentators, some who like syndicated columnist George Will contribute to traditional news outlets (Nisbet 2009). Other research has shown historically the tendency for even mainstream news reporters to falsely balance—i.e. portray as equivalent—the evidence for and against man-made climate change (Boykoff and Boykoff 2004).

There is little doubt that the climate denial movement has had an impact on policy debate, and these studies offer valuable details on the origins, strategies, and arguments of the movement. Yet in order to clearly understand the influence of the movement, the activities of climate deniers need to be placed within the context of the broader communication ecosystem surrounding the issue of climate change. In particular, few systematic studies and comparisons have turned the focus in the opposite direction, evaluating the communication resources, initiatives, strategies, successes, and failures of environmental groups, their funders, and political allies. Nor have the efforts of the denial movement been compared against the communication resources and activities of government agencies, universities, museums, popular science media, and scientific societies. To date, there exists not a single comprehensive evaluation of the communication activities of the US environ-mental movement or scientific community (Akerlof and Maibach 2008).

Moreover, while conservative media continue to dispute the reality of man-made climate change, research shows that since 2005, mainstream reporting reflects the strength of scientific agreement on this question (Boykoff 2007). This mainstream coverage reached record levels of attention in 2007 with a heavy emphasis at the time on the views of Al Gore and the dire nature of environmental impacts (Boykoff and Mansfield 2008; Nisbet 2009). As will be discussed later, as past research would have predicted, even the most high-profile arguments of the denial movement—such as those surrounding the 2009 'Climategate' event—were attended to and accepted by the small proportion of the public already deeply dismissive of climate change (Krosnick 2010b; Leiserowitz et al. 2010). In sum, the climate denier movement is only one—perhaps even a lesser—factor among several that make up the puzzle of lingering wider public ambivalence about climate change in the United States.

Separate from scholarly research, the focus in popular discussion on the climate denier movement also sometimes confuses the difference between political actors who reject the reality of the problem and others such as Bjorn Lomborg (2009) who accept the findings of climate science but who argue for different policy priorities or approaches. There is also an important difference between industry and think-tank coordinated efforts and the emerging online activities of a small segment of the issue public who are deeply dismissive of climate change and/or environmental problems generally.

At blogs and elsewhere online, this segment of the issue public are asking for greater transparency in climate science data and findings along with new participatory mechan-isms of scientific review. To date, studies have yet to examine this specific group of online activists, but based on her personal involvement engaging the users of these blogs, Georgia Institute of Technology scientist Judith Curry (2010) makes the following obser-vation:

> So who are the climate auditors? They are technically educated people, mostly outside of academia. Several individuals have developed substantial expertise in aspects of climate science, although they mainly audit rather than produce original scientific research. They tend to be watchdogs rather than deniers; many of them classify themselves as 'lukewarmers.' They are independent of oil industry influence. They have found a collective voice in the blogosphere and their posts are often picked up by the mainstream media. They are demanding greater accountability and transparency of climate research and assessment reports.

4 FORMING JUDGMENTS AND MAKING DECISIONS ABOUT CLIMATE CHANGE

Whether a member of the issue public or the inattentive public, an opinion leader, an elected official, a journalist, or even a scientist, it is impossible for any individual to be fully informed about climate change and it is rare that when faced with complexity, uncertainty, and limited time and attention, an individual will engage in active deliberation, weighing and assessing many sides and sources of information. Instead, as an average tendency, individuals are 'cognitive misers,' relying on personal experience, values, social influences such as friends or colleagues, personal identity, and the most readily available information about climate change in the media to make sense of an issue and to form judgments (Downs 1957; Popkin 1991).

In this section, I describe several major areas of research findings relative to how individuals reach judgments and form opinions about climate change examining specifically the influence of schema, values, knowledge, and framing. I then discuss how these factors relate to the strong proportion of the US public who fall between the poles of the 'issue public' working to mobilize concern and a denial movement opposed to policy action. Understanding these basic mental and social processes should inform strategies for effectively engaging various publics and for shifting individuals out of a default 'miserly' mode into a more active processing and participatory mode on climate change.

4.1 Schema and mental models

Studies in social psychology and communication point in particular to the role of cognitive and affective 'schema' as mental organizers that shape public judgments. A schema is the metaphorical term for an inferred system of related ideas about a concept or issue. Once activated, schema provide short cuts for reaching an opinion about a complex topic, serve as a basis for inference, and operate as a mechanism for storing and retrieving information from memory (Price 1992). People have multiple schema relevant to climate change which can be triggered by conversations, personal observation, and direct experience, or by way of news or entertainment (Maibach et al. 2008; Moser 2009; Weber 2010).

Examples of relevant climate change schema identified in past research include perceptions of the weather; lay models of how the climate works (Leiserowitz 2004); perceived overlapping issues such as the ozone hole; direct experience with the impacts of climate change such as flooding or hurricanes (Whitmarsh 2008); and vivid, affective imagery often cultivated or reinforced through media presentations such as depictions of melting ice, floods, climate 'alarmists' or 'naysayers' (Leiserowitz 2006). Research across national contexts suggests that tailoring climate change communication to these mental models can improve the ability of individuals and groups to reach decisions and to take actions, especially when statistical information is paired with affective, personally relevant images such as disease-related scenarios and discussed among like-minded peers (Marx et al. 2007; Weber 2010).

4.2 Values

Similar in function to schema, values serve as standards for evaluating personal behavior, societal actions or governance, and proposed policies (Price 1992). These socialized predispositions provide guidance on making sense of a desired end state for a problem such as climate change and the proposed actions for dealing with the issue.

Commonly referenced value predispositions, especially in assessments of US public opinion, are partisanship and political ideology. Survey analyses find that climate change has joined gun control, taxes, and abortion as a form of social identity marker (Hart and Nisbet 2010), one of a few issues that have come to define what it means to be a partisan in the United States (Nisbet 2009). Over the past decade the difference between self-identifying Democrats and Republicans' views on the reality of climate change has widened to a 30 to 50 percent gap depending on question wording (Dunlap and McCright 2008).

As discussed later in this section, these partisan differences can be explained in part by the framing strategies of political leaders, but partisanship and ideology also map onto deeper, more latent value predispositions that span national settings and cultures. In this research, individuals scoring high in terms of hierarchical and individualist values tend to reject the risks of climate change and proposed actions. Hierarchists view proposed climate policy solutions as threats to those they respect in power, to established order in society, and to status quo practices in the economy or their personal lives. Individualists, alternatively, view climate policy actions as unwise restrictions on markets, enterprise, and personal freedom. In contrast, for individuals scoring high in terms of egalitarian and communitarian values, arguments for action on climate change align easily with more generalized views about the need to manage markets and industry in favor of the collective good and to protect the most vulnerable (Leiserowitz 2006; Kahan et al. 2010). Of note, following from this research, a suggested communication strategy to engage individualists and hierarchists is to propose climate solutions that are market based and to promote those solutions using business leaders and national security experts as spokespeople (National Public Radio 2010).

4.3 Framing and news media portrayals

Framing—as a concept and an area of research—spans several social science disciplines. 'Frames' are the conceptual term for interpretative storylines that selectively emphasize specific dimensions of a complex issue over others, setting a train of thought in motion for audiences about who or what might be the cause of a problem, the relevance or importance of the issue, and what should done in terms of policy or personal actions (Gamson and Modigliani 1989). Framing research as applied to the news media offers a rich explanation for how various actors, including experts, define issues in strategic ways, how journalists from various beats selectively cover these issues, and how diverse publics differentially perceive, understand, and participate on climate change (Scheufele 1999).

To make sense of climate change, individuals integrate frames provided by media presentations with their preexisting schema and values. As a consequence, a specific media frame is only influential if it is relevant—or applicable—to the audience's preexisting interpretations and schema (Scheufele and Tewksbury 2007). For example, in the US, climate change has historically been either narrowly defined in news coverage as a looming and impending environmental problem with disastrous consequences and/or as a matter of holding industry accountable. These interpretative packages likely resonate with egalitarians and communitarians (values held more strongly among Democrats and liberals), but are likely ignored by individualists and hierarchists (values held more strongly by Republicans and conservatives). Selective acceptance of these frames of reference is reinforced by the climate denial movement who have emphasized in the news media and in direct messaging opposing frames of scientific uncertainty and negative economic consequences from any greenhouse gas controls (McCright and Dunlap 2003; Nisbet 2009).

Framing research is currently being applied to inform effective communication initiatives about climate change. For example, to date the public health risks of climate change have received limited attention in the US news media, mentioned in fewer than 5 percent of climate change-related stories (Nisbet et al. under review). Yet framing climate change in terms of public health not only reflects scientifically well-understood risks but also holds the potential to make climate change more personally relevant by drawing connections to already familiar problems such as asthma, allergies, and infectious disease. The emphasis also shifts the visualization of the issue away from remote arctic regions, peoples, and animals to more socially proximate neighbors and places such as suburbs and cities. In addition, the public health focus is also inclusive of the need for not just mitigation but also adaptation actions, while also bringing additional trusted communication partners into the fold on climate change, notably public health officials and leaders from minority and low-income communities who are the most at risk and the most vulnerable (Nisbet 2009).

Research involving in-depth interviews with representative segments of Americans finds that when climate change is introduced as a health problem with information then provided about specific mitigation-related policy actions that benefit health and well-being, this reframing of the issue is compelling and positively responded to by a broad cross-section of respondents even by segments otherwise skeptical of climate science (Maibach et al. 2010). Other frames of reference, such as an emphasis on national security or religious and moral teachings, may have similarly engaging influences across a diversity of publics (Nisbet 2009).

4.4 Knowledge

Given the central role of schema, values, and frames in guiding opinion formation, few studies have explored the relationship between knowledge and perceptions. Despite the popular assumption discussed at the opening of this chapter that the two are strongly linked, i.e. if the imagined mass public only understood the science better, they would see the urgency of climate change as most experts do, past studies find only a weak correlation between technical knowledge and perceptions (Achterberg et al. 2010; Allum et al. 2008).

Instead, opinion researchers view 'procedural' knowledge—understanding how to take actions or to get involved on an issue—as generally more important to decision making and behavior than 'declarative' knowledge, defined as a familiarity with the scientific and technical causes of a problem such as climate change (Kaiser and Fuhrer 2003; Roser-Renouf and Nisbet 2008). This finding parallels similar research on civic participation generally, with 'mobilizing information' on who are the key decision makers, where to vote, and how to get involved combining with perceived importance of the issue to be among the strongest predictors of political participation and activism (Eveland and Scheufele 2000; Goidel and Nisbet 2006).

Most survey research on climate change continues to assess general perceptions of expert agreement or awareness of the causes of climate change, yet survey measures should also explore respondent knowledge of the behavioral and policy changes needed to mitigate and adapt to climate change; the skills and resources needed to pursue these changes; the institutions, political actors, organizations, and decision makers involved in the debate; the skills to effectively engage with these decision makers and stakeholders; and how each of these dimensions of knowledge specifically apply to their local community (Maibach et al. 2008; Roser-Renouf and Nisbet 2008).

4.5 Interpretative Communities

As the discussion so far highlights, a complexity of factors shapes opinion formation and personal decisions relative to climate change. Recent analyses in the US have started to map how these factors and processes vary over time across distinct 'interpretative communities' of individuals, improving our understanding of why different segments of the public accept or reject certain arguments, risks, and dimensions of the climate debate (Leiserowitz 2007).

An interpretative community is a group of individuals who share common risk perceptions about climate change, reflect shared schema, mental models, values, and hold a common sociodemographic background. Not only do these interpretative communities share a common identity and world-view, but the fragmented nature of the media system helps reinforce, define, and shape a common shared outlook relative to climate change. Different interpretative communities tend to prefer their own ideologically like-minded news and opinion media; or alternatively, members of some communities tend to avoid most news coverage and instead pay attention mostly to entertain and popular culture (Mutz 2006).

Analyzing nationally representative US survey data, this research has identified six distinct interpretative communities on climate change, profiling their demographic characteristics, risk perceptions, affective reactions, levels of trust, forms of knowledge, political and personal behaviors, and media use patterns (Leiserowitz et al. 2010; Leiserowitz et al. 2009). These six interpretative communities include the Alarmed (approx. 18 percent of the adult population), the Concerned (33 percent), the Cautious (19 percent), the Disengaged (12 percent), the Doubtful (11 percent), and the Dismissive (7 percent).

The audience segments range along a continuum of knowledge, attitudes, and behavior from the Alarmed who accept climate change as a problem, are concerned, and who are looking for opportunities to take personal and political action to the Dismissive who reject the reality of climate change and strongly oppose action. Individuals in the four middle interpretative communities are less certain in their views on climate change, more ambivalent about the risks and relative importance of the issue, and disengaged personally and politically. In terms of public engagement and communication, for individuals between the two poles of perspectives on climate change, the challenge is to identify which frames of reference best enable and help them accurately understand and perceive the relevance of climate change, the personal choices and policy options available, and the common interests they share with others (see Maibach et al. 2008).

5 Structuring Opinion Formation via Organized Deliberation

The tendency for many individuals to be either highly selective—or alternatively inattentive—to news and information about climate change, and to reach decisions quickly relying on preexisting schema and values, leads to an important question: If individuals from different interpretative communities came together to learn about, discuss, and deliberate climate change, what judgments, preferences, and conclusions would they collectively voice? How would participation in such an event shape their subsequent attitudes and behaviors?

Over the past decade, on science issues generally, this question has inspired a number of consensus conferences, deliberative forums, and town meetings, initiatives designed to motivate and enable individual members of the public to voice collective opinions. In these initiatives, recruited lay participants receive background materials in advance, provide input on the types of questions they would like addressed at the meeting, and then provide direct input or recommendations about what should be done in terms of policy. Each initiative, however, varies by how participants are asked for feedback, and how much their feedback matters (Einsiedel 2008; Nisbet and Scheufele 2009).

Evaluation of these initiatives finds that participants not only learn directly about the technical aspects of the science involved, but perhaps more importantly, they also learn about the social, ethical, and economic implications of the scientific topic. Participants also feel more confident and efficacious about their ability to participate in science decisions, perceive relevant institutions as more responsive to their concerns, and say that they are motivated to become active on the issue if provided a future opportunity to do so (Besley et al. 2008; Powell and Kleinmann 2008). Research also finds that if carefully organized, these types of initiatives can shape perceptions of sponsoring institutions such as universities or government agencies as open to feedback and respectful of public concerns, perceptions that predict eventual acceptance and satisfaction with a policy outcome, even if the decision is contrary to an individual's original preference (Besley and McComas 2005; Borchelt and Hudson 2008).

On climate change, these forms of public engagement initiatives have been identified as promising tools for risk communication. As a National Academies (2010: 116) report concludes:

> What most risk researchers consider the ideal approach for communicating uncertainty and risk focuses on establishing an iterative dialogue between stakeholders and experts, where the experts can explain uncertainty and the ways it is likely to be misinterpreted; the stakeholders in turn can explain their decision-making criteria as well as their own local knowledge in the area of concern; and the various parties can work together to design a risk management strategy, answering each others' questions and concerns in an iterative fashion.

In 2009, leading up to the Copenhagen meetings, researchers and sponsoring organizations in more than thirty countries applied these principles to the design of deliberative forums on climate change. At each site, the initiative recruited 100 nationally or regionally representative citizens to spend a weekend discussing, deliberating, and voting on key policy issues related to climate change. The results of the meetings were aggregated by country and released via the project's website and at the Copenhagen meetings. Participants were provided informational materials and videos before the meetings, and had reference materials at their discussion tables.

Of note, the meetings did not feature climate change experts. Instead, careful planning was done in using a meeting facilitator and then trained discussion moderators at each table. The content of the meeting was the social interaction and discussion rather than an expert presentation or lecture. Across countries and meetings, following a weekend of discussion and reflection, when asked to vote on agreed concerns and preferences, widespread consensus was expressed, with strong majorities (80 to 90 percent) perceiving climate change as urgent and similar majorities favoring strong policy actions (World Wide Views on Global Warming 2010).

6 Conclusion

The studies reviewed in this chapter along with others from the growing literature in the area reveal a diversity of factors that shape individual perceptions and behavior relative to climate change. Major influences include media use, interpersonal discussion, schema, and values. Continued research in this area not only offers valuable insight into the dynamics that drive the trajectory of the climate debate in society but also can be applied to the design and implementation of public communication and engagement initiatives.

In particular, two key questions should be addressed in future research. First, more attention needs to be paid to putting into context the influence of the climate denial movement, comparing the movement to analyses of the resources and impacts of environmental organizations and their allies among think tanks, government agencies, scientific societies, science media organizations, and museums. Are advocates and institutions seeking to increase public engagement with climate science and policy solutions out-resourced and out-communicated by the climate denial movement? Conventional wisdom aside, what is the true relative impact of the climate denial movement on news coverage, public opinion, and societal decisions? Among the efforts of environmental community and their allies, what assumptions, practices, and strategies appear to be effective and which appear to be dead ends?

Second, to date, the diverse middle segments of the continuum of public opinion on climate change have been largely overlooked in political debate and in communication efforts. For these unique interpretative communities, who remain relatively ambivalent about the reality and urgency of the problem but are open to learning more, how can an understanding of the schema, values, and trusted information sources among members of these interpretative communities inform initiatives that empower these publics to reach personal decisions and participate in societal debate?

References

Achterberg, P., Houtman, D., Bohemen, S. Van, and Manevska, K. 2010. Unknowing but Supportive? Predispositions, knowledge, and support for hydrogen technology in the Netherlands. *International Journal of Hydrogen Energy* 25(12): 6075–83.

Akerlof, K., and Maibach, E. W. 2008. 'Sermons' as a climate change policy tool: Do they work? Evidence from the international community. *Global Studies Review* 4(3): 4–6.

Allum, N., Sturgis, P., Tabourazi, D., and Brunton-Smith, I. 2008. Science knowledge and attitudes across cultures: A meta-analysis. *Public Understanding of Science* 17(1): 35–54.

Besley, J. C., and McComas, K. A. 2005. Framing justice: Using the concept of procedural justice to advance political communication research. *Communication Theory* 4: 414–36.

—— Kramer, V. L., Yao, Q., and Toumey, C. P. 2008. Interpersonal discussion following citizen engagement on emerging technology. *Science Communication* 30(4): 209–35.

Borchelt, R., and Hudson, K. 2008. *Sci. Prog.* Spring/Summer: 78–81.

Boykoff, M. T. 2007. Flogging a dead norm? Newspaper coverage of anthropogenic climate change in the United States and United Kingdom from 2003 to 2006. *Area* 39(4): 470–81.

BOYKOFF, M. T. and BOYKOFF, J. 2004. Bias as balance: Global warming and the U.S. prestige press. *Global Environmental Change* 14(2): 125–36.

—— and MANSFIELD, M. 2008. 'Ye Olde Hot Aire': Reporting on human contributions to climate change in the UK tabloid press. *Environmental Research Letters* 3(2).

CURRY, J. 2010,. Opinion: Can scientists rebuild the public trust in climate science? *Physics Today*. (published online 24 Feb 2011). Available at <http://blogs.physicstoday.org/politics/2010/02/opinion-can-scientists-rebuild.html> accessed 24 May 2011.

DOWNS, A. 1957. *An Economic Theory of Democracy*. New York: Harper.

DUNLAP, R. E. 2010. At 40, environmental movement endures, with less consensus. Available at (published online 22 Apr 2010) <http://www.gallup.com/poll/127487/Environmental-Movement-Endures-Less-Consensus.apx> accessed 24 May 2011.

—— and MCCRIGHT, A. M. 2008. A widening gap: Republican and Democratic views on climate change. *Environment* 50(5): 26–35.

EINSIEDEL, E. 2008. Public engagement and dialogue: A research review. Pp. 173–84 in M. Bucchi and B. Smart (eds.), *Handbook of Public Communication on Science and Technology*. London: Routledge.

EVELAND, W. P., JR., and SCHEUFELE, D. A. 2000. Connecting news media use with gaps in knowledge and participation. *Political Communication* 17: 215–37.

FELDMAN, L., LEISEROWITZ, A., and MAIBACH, E. Forthcoming. The science of satire: The Daily Show and The Colbert Report as sources of public attention to science and the environment. In A. Amarasingam (ed.), *Perspectives on Fake News: The Social Significance of Jon Stewart and Stephen Colbert*. Jefferson, NC: McFarland & Company.

FISKE, S. T., and TAYLOR, S. E. 1991. *Social Cognition*. 2nd edn., New York: McGraw-Hill.

GAMSON, W. A., and MODIGLIANI, A. 1989. Media discourse and public opinion on nuclear power: A constructionist approach. *American Journal of Sociology* 95: 1–37.

GOIDEL, K., and NISBET, M. C. 2006. Exploring the roots of public participation in the controversy over stem cell research and cloning. *Political Behavior* 28(2): 175–92.

HART, S., and NISBET, E. C. 2010. *Boomerang Effects in Science Communication*. Working Paper. School of Communication, American University. Washington, DC.

JACQUES, P., DUNLAP, R. E., and FREEMAN, M. 2008. The organization of denial: Conservative think tanks and environmental scepticism. *Environmental Politics* 17: 349–85.

KAHAN, D. M., JENKINS-SMITH, H., and BRAMAN, D. 2010. *Cultural Cognition of Scientific Consensus*. Cultural Cognition Project Working Paper No. 77. Yale University School of Law. (published online 7 Feb 2010) Available at <http://papers.ssrn.com/sol3/papers.cfm?abstract_id=1549444> accessed 25 May 2011.

KAISER, F. G., and FUHRER, U. 2003. Ecological behavior's dependency on different forms of knowledge. *Applied Psychology: An International Review* 52(4): 598–613.

KIM, Y. M. 2009. Issue publics in the new information environment: Selectivity, domain-specificity, and extremity. *Communication Research* 36: 254–84.

KROSNICK, J. A. 1990. Government policy and citizen passion: A study of issue publics in contemporary America. *Political Behavior* 12: 59–92.

—— 2010a. The climate majority. *New York Times* (published online 8 June 2010). Available at <http://www.nytimes.com/2010/06/09/opinion/09krosnick.html> accessed 25 May 2011.

—— 2010b. Large majority of Americans support government solutions to address global warming. Woods Institute for the Environment. Stanford University. (published online 9 June 2010) Available at <http://woods.stanford.edu/research/americans-support-govt-solutions-global-warming.html> accessed 25 May 2011.

—— HOLBROOK, A. L., and VISSER, P. S. 2000. The impact of the fall 1997 debate about global warming on American public opinion. *Public Understanding of Science* 9: 239–60.

LAZARSFELD, P. F., BERELSON, B. R., and GAUDET, H. 1948. *The People's Choice: How the Voter Makes up his Mind in a Presidential Campaign.* New York: Duell, Sloan & Pierce.

LEISEROWITZ A. 2004. Surveying the impact of The Day After Tomorrow. *Environment* 46: 23–44.

—— 2006. Climate change risk perception and policy preferences: The role of affect, imagery, and values. *Climatic Change* 77: 45–77.

—— 2007. Communicating the risks of global warming: American risk perceptions, affective images and interpretive communities. In S. MOSER and L. DILLING (eds.), *Communication and Social Change: Strategies for Dealing with the Climate Crisis.* Cambridge: Cambridge University Press.

—— MAIBACH, E., and ROSER-RENOUF, C. 2009. *Global Warming's Six Americas: An Audience Segmentation.* New Haven, CT: Yale Project on Climate Change. Available at <http://environment.yale.edu/uploads/SixAmericas.pdf>.

—— —— —— 2010. *Global Warming's Six Americas: An Audience Segmentation.* New Haven, CT: Yale Project on Climate Change (published online June 2010). Available at <http://environment.yale.edu/climate/files/SixAmericasJune2010.pdf> accessed 25 May 2011.

LOMBORG, B. 2008. *Cool It: The Skeptical Environmentalist's Guide to Global Warming.* New York: Viking.

McCRIGHT, A. M., and DUNLAP, R. E. 2003. Defeating Kyoto: The Conservative Movement's Impact on U.S. Climate Change Policy. *Social Problems* 50(3): 348–73.

—— —— 2010. Anti-reflexivity: The American conservative movement's success in undermining climate science and policy. *Theory, Culture, and Society* 27(2–3): 100–33.

MAIBACH, E., ROSER-RENOUF, C., and LEISEROWITZ, A. 2008. Communication and marketing as climate change intervention assets: A public health perspective. *American Journal of Preventive Medicine* 35(5): 488–500.

—— NISBET, M. C., BALDWIN, P., AKERLOF, K., and GIAO, G. 2010. Reframing climate change as a public health issue: An exploratory study of public reactions. *BMC Public Health* 10: 299.

MANZA, J., and COOK, F. L. 2002. The impact of public opinion on policy: The state of the debate. In J. Manza, F. L. Cook, and B. J. Page (eds.), *Navigating Public Opinion: Polls, Policy, and the Future of American Democracy.* New York: Oxford University Press.

MARX, S. M, WEBER, E. U., ORLOVE, B. S., LEISEROWITZ, A., and KRANTZ, D. H. 2007. Communication and mental processes: Experiential and analytic processing of uncertain climate information. *Global Environmental Change,* 17: 47–58.

MOSER, S. 2009. Communicating climate change: History, challenges, process and future directions. *Wiley Interdisciplinary Reviews: Climate Change* 1(1): 31–53.

MUTZ, D. 2006. How the mass media divide us. Pp. 223–63 in P. Nivola and D. W. Brady (eds.), *Red and Blue Nation?,* vol. i. Washington, DC: The Brookings Institution.

National Academies. 2010. *Adapting to the Impacts of Climate Change.* Washington, DC: National Academies Press. Available at <http://www.nap.edu/catalog.php?record_id=12783> accessed 25 May 2011.

National Public Radio. 2010. Belief in climate change hinges on world view. 23 Feb. Available at <http://www.npr.org/templates/story/story.php?storyId=124008307>.

NISBET, M. C. 2009. Communicating climate change: Why frames matter to public engagement. *Environment* 51(2): 12–23.

NISBET, M. C., and KOTCHER, J. 2009. A two step flow of influence? Opinion-leader campaigns on climate change. *Science Communication* 30: 328–58.

NISBET, M. C. and SCHEUFELE, D. A. 2009. What's next for science communication? Promising directions and lingering distractions. *American Journal of Botany* 96(10): 1767–78.

—— et al. Under review. Communicating the public health relevance of climate change: A news agenda-building analysis. *Science Communication.*

OPHIR, E., NASS, C., and WAGNER, A. 2009. Cognitive control in media multitaskers. *Proceedings of the National Academy of Sciences.*

ORESKES, N., and CONWAY, E. M. 2010. *Merchants of Doubt.* New York: Bloomsbury Press.

PATTERSON, T. E. 2005. Of polls, mountains: US journalists and their use of election surveys. *Public Opinion Quarterly* 69(5): 716–24.

Pew Internet and American Life Project. 2006. The internet as a resource for news and information about science. Pew Internet and American Life Project and the Exploratorium science center (published online 20 Nov 2006). Available at <http://www.pewtrusts.org/uploadedFiles/wwwpewtrustsorg/News/Press_Releases/Society_and_the_Internet/PIP_Exploratorium_Science1106.pdf> accessed 15 May 2009.

Pew Project for Excellence in Journalism. 2010. The Gulf disaster becomes a beltway story. 14 June. Available at <http://www.journalism.org/index_report/pej_news_coverage_index_june_1420_2010>.

PRICE, V. 1992. *Public Opinion.* Newbury Park, CA: Sage.

POPKIN, S. L. 1991. *The Reasoning Voter.* Chicago: Univ. of Chicago Press.

POWELL, M., and KLEINMAN, D. 2008. Building citizen capacities for participation in nanotechnology decision-making. *Public Understanding of Science* 17(3): 329–48.

PRIOR, M. 2005. News v. entertainment: How increasing media choice widens gaps in political knowledge and turnout. *American Journal of Political Science* 49: 577.

ROSENSTIEL, T. 2005. Political polling and the new media culture: A case of more being less. *Public Opinion Quarterly* 69: 698–715.

Roser-RENOUF, C., and NISBET, M. C. 2008. The measure of key behavioral science constructs in climate change research. *International Journal of Sustainability Communication* 3: 37–95.

SAMUELSOHN, D. 2010. Greens defend climate decisions. *Politico* (published online 5 August 2010). Available at <http://www.politico.com/news/stories/0810/40680.html> accessed 25 May 2011.

SCHEUFELE, D. A. 1999. Framing as a theory of media effects. *Journal of Communication* 49(1): 103–22.

—— and TEWKSBURY, D. 2007. Framing, agenda setting, and priming: The evolution of three media effects models. *Journal of Communication* 57(1): 9–20.

SOSS, J., and SCHRAM, S. F. 2007. A public transformed? Welfare reform as policy feedback. *American Political Science Review* 101(1): 111–27.

WEBER, E. U. 2010. What shapes perceptions of climate change? *Wiley Interdisciplinary Reviews: Climate Change* 1(3): 332–42.

WHITMARSH, L. 2008. Are flood victims more concerned about climate change than other people? The role of direct experience in risk perception and behavioural response. *Journal of Risk Research* 11: 351–74.

World Wide Views of Global Warming. 2010. Project website and Report. Available at <http://www.wwviews.org/> accessed 25 May 2011.

....

CHAPTER 25

SOCIAL MOVEMENTS AND GLOBAL CIVIL SOCIETY

....

RONNIE D. LIPSCHUTZ AND CORINA MCKENDRY

EVEN as international climate change negotiations became bogged down over the decade leading up to the Conference of the Parties in Copenhagen in 2009, the global public has shown a growing interest in addressing the issue, with activism arising in many places and among many thousands of non-governmental (NGOs) and social movement organizations (SMOs). What cannot be denied is that civil society action and activism addressing climate change and its ostensible sources run the gamut from very localized home- and community-based efforts all the way to participation in a seemingly endless string of international meetings. These actors and their activities have become the focus of a considerable body of scholarly research (see Lipschutz 2006), much of which details what NGOs and SMOs *do*, and *how* they do it while paying rather less attention being paid to the *why*—that is, how to account for the efflorescence of such non-state activism?[1] Is this simply a consequence of individual and group concern or are there more fundamental driving processes at work? And, if the latter is the case, what are those drivers and how might we account for them? While much of the relevant scholarship focuses on norms, science, and self-interest as motivators of social activism, we propose that, at a structural level, the efflorescence of NGOs and SMOs has much to do with the emergence of global *governmentality*, a form of global governance that depends, in part, on the regulatory role filled by civil society, complementing the more conventional government of political economy by the state. In other words, social activism is one element in a newly emerging regime of global regulation, not quite politics, not quite economics, not quite opposition and resistance (Lipschutz with Rowe 2005).

In this chapter, we examine both what such actors *do* and offer some speculative ideas about *why* they are doing it, beyond the usual appeals to self-interest and norms. We begin with a general discussion of non-state environmental activism focused on climate change, as it has grown in scope and quantity since the 1970s. We then turn to theoretical explanations of this growth, briefly addressing both instrumental and structural accounts. In the third section of the chapter, we describe several examples of NGO activism in the realm of climate change, focusing on the Climate Action Network (CAN), which is an organization in its own right as well as a transnational/global networked alliance of environmental groups. As we shall see, CAN's activism or, more specifically, the activism of the organizations that constitute the network, runs the gamut from the very local to the very global. We conclude the chapter with a brief assessment of the effect of these activities on both climate change policies and outcomes.

1 THE ORIGINS OF SOCIAL ACTIVISM ADDRESSING CLIMATE CHANGE

The history of the environmental movement is, by now, fairly well documented, and we do not repeat any of it here (see McCormick 1991; Shabecoff 2003; Gottlieb 2005). What is somewhat more difficult is tracing the early involvement of environmental organizations in activism addressing climate change. While a few NGOs included climate change in their early social agendas, it was not until the 1980s that the issue became a major object of social activism; before that, most NGOs and SMOs were more concerned with local and national environmental problems, such as air and water pollution, endangered species, and exposure to toxics (see e.g. Hays 1959, 1987; Shabecoff 2003). Although the publics of industrialized states were vaguely aware of global environmental issues such as climate change, there was only limited hard evidence and almost no personal experience to confirm that it was an imminent problem. Not until the latter half of the 1980s, as nuclear tensions and protests diminished, along with the 'nuclear winter' debate (Williams and Parker 2008), did environmental organizations begin to turn their attention to other matters.

As scientific data accumulated indicating significant ozone depletion over the polar regions, a growing number of major environmental NGOs began to focus on degradation of the Earth's atmosphere, for several reasons. First, this was an issue with clear import not only for the global environment but also, in the longer run, public health and species survival. Second, it offered an opportunity to generate organizational credibility through provision of expert knowledge to the public, media, and policy makers (including those from science-poor countries) in public ways that atmospheric scientists tended to eschew.[2] Third, it offered a new issue for restoring depleted membership levels and fundraising, as well as developing links with corporate interests and funding sources. Fourth, climate change appeared to provide a threat that, following the end of the 'resource wars' of the 1970s, might facilitate North–South collaboration and redistribution.[3] Finally, as the pace of international meetings and conferences picked up during the late 1980s and early 1990s, NGOs found it possible to send staff to them and, sometimes, participate directly in the negotiation process, thereby influencing the shape of the resulting conventions and protocols (Betsill and Corell 2008). Note that in these early years, NGOs and SMOs could sometimes be found protesting outside of corporate offices and conference halls even as they were inside, too; at that point, they were engaged in public consciousness raising about but not yet in localized project building to address climate change.

The influence of NGOs and SMOs inside and outside of international conference halls was demonstrated during the 1992 Earth Summit in Rio de Janeiro, at which the UN Framework Convention on Climate Change was presented, debated, and approved. In the run-up to that meeting, representatives of major environmental NGOs from North

America and Europe attended the dozen-odd preparatory conferences, lobbying national representatives in coffee shops and hallways, organizing informational sessions for attendees, and offering expert advice to delegates from those countries lacking scientists and atmospheric research facilities.[4] A major source of information about the goings-on at these meetings, both public and otherwise, was *ECO*, a daily newsletter published by the Climate Action Network (see below).[5] Eventually, NGO representatives were even allowed to give short presentations in climate conference plenaries, an opportunity that was often used to scold delegates for their inability or unwillingness to agree on any substantive procedures or policies.

Notwithstanding the conventional wisdom that global climate change could only be dealt with through interstate agreements, ten years of largely fruitless meetings began to take their toll on efforts to influence the shape of conventions and protocols. NGOs continued to participate in international meetings, but there was a growing sense that action and activism might have to be refocused. In this regard in particular, and notwithstanding the supposed environmental credentials of Vice President Al Gore, the Clinton Administration was a major disappointment. Blocked by a Republican-controlled Congress, it showed itself to be largely unwilling to present for ratification any international convention that seemed to disadvantage the United States. Furthermore, widespread skepticism about the data and models that purported to predict or even suggest the onset of global climate change required NGO holding actions rather than gains in US federal policy (DePledge 2005). The European Union was much more open to international agreements and, indeed, came to be regarded as a leader on the issue of climate change, but, in any case, was hardly called on to do very much so long as the United States refused to act. It is pure speculation to think that, had the US Supreme Court elected Al Gore, the vast expansion of local activism during the first decade of the 21st century might have well not materialized. The refusal of the Bush Administration to do much of anything where climate change was concerned not only generated intense interest (and debate) among the public and the media, it also motivated NGOs and SMOs to seek other means of and locales for policy making and activism. Indeed, it could be argued that it was the Bush Administration's very intransigence, in concert with American military and economic policies, that motivated both widespread concern about climate change and the search for ways to address it (Bäckstrand and Lovbrand 2007; Betsill and Corell 2008). Barack Obama's promises to lead the search for a new climate agreement have yet to bear fruit and, notwithstanding a considerable presence at the December 2009 conference of the UN Framework Convention on Climate Change (COP-15) in Copenhagen, the environmental movement has not had much success in moving countries toward a new agreement to reduce greenhouse gas emissions.

2 WHO ACTS AND WHY?

International relations theory focuses primarily on interactions among nation-states (Tétreault and Lipschutz 2009). Depending on the epistemological approach one adopts—realism, liberalism, or critical social theory—the specific concerns of states are said to vary. For example, realists tend to put interstate conflict and the possibility of war at the center of their analysis; liberals, interstate bargaining and markets (Waltz 1979;

Keohane 1984). Critical social theories are more ecumenical about how state interests and interactions are constructed (Wendt 1999; but see also Jabri 1996). Our approach is more sociological and critical, insofar as we treat the 'global' as a realm of intense social activity, as much a construction as a material reality, in which the state system is only one of a number of what Pierre Bordieu called 'fields' (1984) and 'habitus' (2002), and Michel Foucault, *dispositifs* (1980: 194). A *dispositif* is

> firstly, a thoroughly heterogeneous ensemble consisting of discourses, institutions, architectural forms, regulatory decisions, laws, administrative measures, scientific statements, philosophical, moral and philanthropic propositions–in short, the said as much as the unsaid... The apparatus itself is the system of relations that can be established between these elements. (Foucault 1980: 194)

There is no comparable term or concept in the British or American literatures, but the term has been translated as 'assemblage.' The automotive 'system' is such an assemblage (Urry 2006; Paterson 2007). It includes not just the car but also the material infrastructure (highways, gas stations, parking lots, streets, pipelines), the production system (mining and manufacture of raw materials, shipping, parts production, assembly plants, tire plants, gasoline refining), auto-related labor, tourism, advertising, the arrangement of cities and suburbs, patterns of mass transit within and without major urban areas, and individuals' and people's subjectivities and mentalities regarding both car and system. In the United States, mobility of those lacking cars is highly constrained and getting around can be expensive and time consuming. Life is much easier with an automobile and possession of one constructs and reinforces 'normality' and *habitus*, as do the other elements of the automobility assemblage. The result is that practices associated with automobility are, for the most part, assumed, unquestioned, and regarded positively. When the assemblage imposes externalities on society, these are either treated as a problem of individual agency (e.g. fastening seat belts) or to be addressed instrumentally by technological and economic fixes (e.g. air bags). The assemblage, as a whole, is not subject to deliberate and directed transformation or conversion. Indeed, notwithstanding a host of such externalities, the consumer of automobility is pressured to sustain and support the assemblage through a variety of instruments focused on status, freedom, and the mobile imaginary.

Nevertheless, much analysis of the politics and economics of global climate change has remained bogged down in the state-centric paradigm, and a combination of realist-liberal theorizing has dominated scholarship. Practice largely tracks this understanding: states (like any buyer in a market) are rational actors, have interests and preferences, and calculate the costs and benefits of a menu of actions (or non-actions) from which they might select. Given the market-like structure of this environment, they bargain with each other over desires and goals, make trade-offs and exchanges that appear affordable and reject those that are too costly, and seek to contribute as little as possible to the public goods that collective action is intended to provide (Olson 1965). We will not belabor the complexity of ongoing negotiations over global climate change or the difficulties in arriving at strategies that can be ratified by the world's 200-odd states and, more importantly, the major greenhouse gas emitters. Suffice it to say that, over the twenty years of international activity, there has been little actual reduction in greenhouse gas emissions as a result of international agreements (IPCC 2007).

As suggested above, the logjam in dealing with global climate change and growing indications and evidence that climate change is taking place, and more rapidly than anticipated, has gradually raised anxieties among those members of the public who hold to environmentalist norms and beliefs. Whether such concern is a result of experiencing climate change is unclear; most of the world's population has yet to experience specific impacts that can be linked unequivocally to global warming, the melting of the Arctic ice cap, catastrophic storms, and droughts notwithstanding.[6] We might better explain the growth of activism as a combination of 'moral panic' (Ungar 2003), on the one hand, and some element of enlightened self-interest and concern for future generations, on the other. In the former instance, actions that contribute to environmental damage are increasingly regarded as both violating ethical codes and reflecting a lack of 'civic virtue': people (consumers) must make proper ethical choices or humanity is doomed. In the latter instance, people imagine how they and their children might be adversely affected by the physical and economic impacts of climate change. On the whole, to the extent that climate change activism reflects the emergence of a social movement—about which we will say more, below—it is difficult to specify any single cause or driver (unlike, for example, the cases of water and air pollution affecting health and aesthetics or threats to 'charismatic megafauna').

Much of the relevant literature on NGOs and SMOs regards them as discrete, autonomous agents, rationally seeking to inform and influence those actors—states, international agencies, corporations—who are believed to exercise forms of power that can be applied to change practices and trajectories of climate change (Wapner 1996; Keck and Sikkink 1997; Fogel 2007). Although civil society groups are assumed to be normatively motivated—the 'collective action problem' seems to suggest that self-interest alone would be an insufficient motivation for such groups (Olson 1965)—they are nonetheless embedded in a global capitalist economy and have quite specific material requirements that must be fulfilled in order to operate successfully. At the end of the day, the usual panoply of social movement concepts, such as 'opportunity structures,' 'framing,' and 'resource mobilization' all deal with organizational production for institutional continuity as much as problem solving (Tarrow 2007; Meyer 2004; Cox 1987). To be successful, an organization must survive and, in a market-based environment, this means finding ways to generate the funds necessary to sustain operations. We will not detail these behavioral models here, except to note that they largely follow the liberal convention of rational choice and organizational self-interest.[7]

We find Michel Foucault's (1991) concept of *governmentality* to be more useful here. On the one hand, it does not obviate the more instrumental motivations discussed above; on the other, it permits us to recognize both the ways in which everyone participates in global regulation and government under conditions of advanced liberalism and it elides the conventional distinction among states, societies, and markets. Moreover, as new matters and concerns rise into broad consciousness and fail to be addressed through institutionalized political processes, there is a growing tendency for the emergence of movements and actors seeking to governmentalize them and turn them into phenomena to be regulated and managed. This is not a new phenomenon by any means, but there is almost certainly a qualitative change under way in how the world is 'governed.' Inasmuch as climate change is a quintessential global 'problem,' it is hardly surprising that we should see high levels of social activity around it (see Luke, this volume).

But what is governmentality? It is about management, about ensuring and maintaining the 'right disposition of things' of that which is being governed or ruled, and bringing those being managed into the process of governing themselves. As Foucault put it, governmentality is 'the ensemble formed by institutions, procedures, analyses and reflections, the calculations and tactics that allow the exercise of this very specific albeit complex form of power, which has as its target populations, as its principal form of knowledge, political economy, and as its essential technical means apparatuses of security' (Foucault 1991: 102). This 'right disposition' has as its purpose not the action of government itself, but the 'welfare of the population, the improvement of its condition, the increase of its wealth, longevity, health, etc.,' which we might assume also contributes to the maintenance of administrative apparatuses as well as the well-being and productivity of the population (Foucault 1991: 100; see also Dean 1999: chapter 1).

There is more to governmentality than governing, however; it also rests on 'responsibilization' (Hier 2008; Hester 2009), which amounts to a form of self-government through the shaping of individual behaviors and comportment according to expert standards, societal values and peer pressure, and norms and regulations, as reflected in and propagated through a variety of social institutions. For example, the 'responsible' environmental activist not only engages in visible actions and demonstrations of her virtuous beliefs and concerns but also hews to a set of practices that have been vetted, validated, and approved by appropriate experts and disseminated via education, law, media, and publicity. Thus, not only should one recycle cans, bottles, boxes, and containers after use, one should also choose what to consume with pollution and waste reduction, as well as health and well-being, in mind. Driving the 'right' car, wearing the 'correct' clothing, and using 'appropriate' renewable energy sources, and standing as an example to others who might not be as advanced are all elements of governing the self while setting standards for the government of society. Thus, whereas the conventional view of environmental activism is that it seeks to influence various parties to devise new policies or curtail offending practices, governmentality regards that activism as part of the governing process itself.

Why use this concept rather than some other, more neutral-sounding one such as 'global management' or 'governance?' Foucault coined the term for reasons both normative and analytical. First, he sought to reveal the extent to which power was ubiquitous in human political and social relations, and governmentality reflected this presence, rather than regarding it as something oppressive and requiring elimination. Second, he recognized that complex social arrangements rely on much more than simple coercion or fear to function, and that those who are being 'managed' must participate, or be complicit in, such management. This is not the stuff of conspiracy or central planning, however; it is a simple necessity.[8] Hence, the question to ask is not 'governmentality or not?' Rather, it is 'what form of governmentality and how can we participate in its shaping?'

To return to an earlier point, climate can be regarded as a complex assemblage or field characterized by material phenomena, technological processes, social practices, deeply held values regarding processes and practices that, directly or indirectly, interact with the Earth's atmosphere and climate. Inasmuch as virtually everything that human beings do is linked, somehow, to the 'burning' of carbon—there is no possibility of doing away with that set of normalized practices, institutions, and infrastructures that are most deeply implicated in the production and reproduction of industrialized, capitalist society and its effects on the climate assemblage. Nonetheless, efforts to address and redress these effects will take place through new and different forms of governmentality, which will require more than simply raising the effective price of carbon or deploying new, greener technologies and processes. It is in working to transform individual and collective consciousness, beliefs, norms, practices, and habits, and associated forms of governmentality that civil society can be most effective. Some of these efforts can be seen in the projects of the Climate Action Network, discussed below.

3 THE CLIMATE ACTION NETWORK: TRANSFORMING THE CLIMATE ASSEMBLAGE

We now turn to an examination of a specific manifestation of environmental governmentality, the Climate Action Network (CAN), a transnational/global organization and network of organizations operating at the local, national, and international levels (Lipschutz with Mayer 1996). At a local level, individual member groups and organizations engage in projects and educational efforts intended to engage various publics adopting behaviors and practices that will, ultimately, reduce greenhouse gas emissions. Some of these 'local' projects are elements of larger transnational ones, and some of the local groups provide funds and assistance to communities in other parts of the world. At the national level, CAN member organizations provide information and education, conduct research and analyses, and try to influence governments and their agencies to adopt 'climate-friendly' policies. Some engage, too, with corporations and corporate associations in the effort to alter business practices and attitudes. Finally, there are the NGOs, and CAN itself, which operate in multiple countries and send representatives to international conferences in an effort to shape international law relating to climate change.

The overarching goal of the Climate Action Network is 'to keep global warming as far below 2 °C as possible.'[9] It was founded in 1989 to coordinate the lobbying efforts of environmental organizations during the Second World Climate Conference in 1990 (Duwe 2001). Throughout two decades of international climate negotiations since that meeting, CAN has worked to bridge tensions between European and American environmental organizations and between groups from the North and the South in order to put forward a strong and consistent message from the environmental community and to create a 'single, dominant environmental voice in the climate debates' (Pulver 2005: 28). Though this ability to bridge differences between the quite different environmental organizations present at the international negotiations has arguably been CAN's most important achievement, other pillars of its centrality in the climate negotiations have been its presence at every UN climate-related negotiation since 1990 and its role as the liaison between environmental NGOs and the UN Climate Change Secretariat, which has given it significant access to delegates (ibid.).

Though it maintains a small staff, CAN International is primarily a global network of over 450 environmental organizations from 85 countries, all of which have climate change mitigation as an organizational priority and work 'to promote government, private sector and individual action to limit human-induced climate change to ecologically sustainable levels.'[10] The international network is further composed of regional networks including

CAN-Europe, CAN-Latin America, CAN-United States, and CAN-South Asia, among others. These regional networks consist, in turn, of member organizations, ranging from large environmental organizations that focus on a variety of issues, such as the Sierra Club, the Union of Concerned Scientists, and the World Wildlife Fund in the United States, to a wide variety of smaller chapters and organizations, including those specifically focused on climate change as well as other groups interested in issues of environmental justice, renewable energy, and related concerns.

Beyond this, CAN serves another crucial role. Through online resources, including a blog and an electronic version of the newsletter *ECO*, CAN provides the concerned public with consistent, timely updates on the progress of climate negotiations. Although it is difficult to know precisely the impact of these resources in terms of increased public pressure on policy makers, it is widely acknowledged that CAN's efforts have helped to make the negotiating process more transparent and to force a certain amount of accountability on states and their representatives through 'shaming' (Keck and Sikkink 1997). Furthermore, in order to stay abreast of environmentalist perceptions of the negotiations and, indeed, to garner news and information not otherwise available or easily accessed, many delegates make it a point to read the information and analysis provided in *ECO* on a daily basis (Duwe 2001). One of *ECO*'s more colorful lobbying tools is the 'Fossil of the Day Award,' first presented at the climate talks in Bonn, Germany, in 1999. This prize goes to the country that has, in the judgment of CAN members, 'performed "best" at blocking progress in the negotiations' each day.[11] According to one notable analysis, the daily announcement of the award 'receives remarkable on-site attention' (Duwe 2001: 180). As a result, CAN International has been the key organization for coordinating ENGO climate-related activism at international negotiations. With the launch of post-Kyoto international climate negotiations in Copenhagen, it is clear that CAN will continue in its efforts to influence rules, procedures, agreements, policies, and outcomes through lobbying, protest, public shaming of recalcitrant nations, and a certain amount of 'speaking truth to power.' Taken together, these activities constitute a good deal more than activism as it is generally understood.

As argued above, however, years of such negotiations have had little real impact on greenhouse gas emissions and it remains to be seen whether future rounds are any different. As a result, environmental organizations have begun to shift their efforts increasingly toward local projects and actions designed to influence, if not transform, those norms, beliefs, behaviors, and practices that sustain the assemblage underlying the current climate crisis. These efforts, as suggested earlier, aim to change what, how, and why people engage with the atmosphere and to focus on 'what can be done' through reciprocal changes in *habitus* and the structures of governmentality than enable and support such practices, rather than (or at least in addition to) passage of legislation and regulations.

We now examine a few of CAN's member groups. The network's breadth makes it difficult to address the many different types of members, a task well beyond the scope of this chapter. A brief accounting of a few CAN-associated groups can, however, begin to illuminate the broader and more localized efforts through which they address climate change. We look at three different kinds of environmental organizations: a local project of the Birmingham (UK) chapter of Friends of the Earth, two corporate campaigns by the Rainforest Action Network, and a local climate action group in Parramatta, Australia. While these choices show a selection bias towards Northern NGOs, they are nonetheless typical of CAN members and their programs and projects.

Friends of the Earth (FOE) is a member of CAN as well as a global environmental network in its own right, with national and local chapters around the world, each with significant autonomy (see Doherty 2006). The network members campaign broadly on both environmental and social issues and 'promote solutions that will help to create environmentally sustainable and socially just societies.'[12] The Birmingham chapter of FOE is one of the largest and best-established in the UK, working on a variety of local and national issues. Recently, climate change has become one of Birmingham FOE's major priorities, and its Faith and Climate Change Project (FCCP) is especially noteworthy. This initiative focuses on members of the diverse faith communities in the city, supporting and encouraging activities that address climate change and 'bring faiths together to work towards a shared vision of a greener more sustainable future.'[13] Through the coordinating efforts of one full-time staff person and a number of volunteers, the FCCP offers training for people who want to initiate sustainability projects in their communities and places of worship, organizes regional events to bring together faith leaders who are interested in climate change and sustainability, and provides a forum for the sharing of climate mitigation resources and best practices between faith groups that might otherwise have limited contact with each other.[14] Activities include training people to provide energy audits to community members, and helping places of worship begin and expand recycling programs, install solar hot water heaters, improve their energy efficiency, and find other ways to reduce their carbon footprint.[15]

While FCCP's goals include more traditional climate change priorities such as strong, binding national and international greenhouse gas reduction policies, it emphasizes relationships among individuals, communities, and activities that impact the atmosphere. It also attempts to reject the fear and hopelessness of much climate discussion and strives to encourage 'people to act because they love the planet and they want their communities to thrive.'[16] The hope is that a positive vision that includes questions of equity and climate justice, combined with the provision of resources and skills training for people in the community, will result in sustained climate action that, as an example, will be disseminated to and reproduced in other locations.

The Rainforest Action Network (RAN) is another CAN member working on climate change outside of international negotiations. Though more centralized than FOE, RAN also addresses a wide variety of environmental issues, with particular emphasis on protecting the world's forests. Putting pressure on corporations to change their environmentally destructive activities has long been an important part of RAN's efforts (see Asmus 1998; Cashore 2002). Recently the organization has turned these efforts towards climate change which is, of course, linked to forest destruction. Indeed, an increasingly important tactic pursued by RAN and other environmental organizations involves winning changes in corporate behaviors related to climate change, even when the mandates of local, national, and international law are vague or elusive (on Corporate Social Responsibility, see Lipschutz with Rowe 2005).

Two examples of such activism illustrate RAN's approach, a campaign to change bank investment patterns and another focused on the fashion industry. For the past decade, RAN has pressured large banking institutions to avoid investing in companies that engage in environmentally harmful practices. Combining protests outside of banks with discussions with officials on the inside, RAN has successfully convinced a number of the world's largest investors to adopt environmental policies to guide their lending practices. These efforts have recently been focused on getting banks to divest from industries that contribute most heavily to climate change. In 2006, for example, RAN played an important part in a campaign to convince more than twenty global banks not to finance construction of new coal-fired power stations in Texas, leading to a considerable scaling back in the proposed plants.[17] The second example strikes even more closely at the *dispositif* of climate change, particularly to the extent that modern identities are defined by consumption, a relationship that growing numbers of environmental groups seek to alter as part of the effort to address climate change. Highlighting the climate impact of deforestation, RAN has convinced major players in the fashion industry, including Gucci and Tiffany & Co., to stop using paper made from trees cut down in Indonesian rainforests. These policies cover everything from office copy paper to the companies' branded shopping bags (The Independent 2009). During New York's annual Fashion Week in September 2009, RAN sponsored a 'Green-Shows EcoFashion Week,' which highlighted ecological designs by numerous well-known fashion designers as well as steps designers could take to reduce deforestation and associated climate change (Pinson 2009). That a rather radical environmental organization hosts a highly visible fashion event in which well-known designers promote their eco-credentials begins to hint at the effectiveness of efforts to transform people's thinking about climate change and what they might do to change their practices.

Our final example is the Parramatta Climate Action Network (ParraCAN), a member of the Climate Action Network of Australia (CANA), which is itself a member of CAN International (Bond 2009). Like the United States, the Australian government has been very resistant to adopting binding policies aimed at reducing the country's greenhouse gas emissions. Consequently, and like their American counterparts, local organizations have emerged in Australia to protest the government's inaction as well as to encourage local, business, and individual efforts to address the issue. Founded in 2007, ParraCAN is a local climate organization based in Parramatta, part of metropolitan Sydney, whose mission is 'to gather together community effort on climate action..., to support our political and business leadership in taking corrective action,'[18] and 'to raise awareness of how climate change will affect western Sydney and what Parramatta residents can do to help stop [it]' (ParraCAN 2008). By contrast with RAN and FOE, ParraCAN is a small, local group formed specifically to address climate change. It is involved in a number of activities, some of which are a part of lobbying the national government and others that emphasize actions individuals can take to mitigate climate change.

For example, within a month of ParraCAN's founding, the organization hosted a public forum in which candidates running for federal office discussed what their party would do to reduce greenhouse gas emissions if elected (ibid.). The group has also organized lobbying of the local MP around the issue of climate change, both in face-to-face meetings and in nationally organized protests outside Parliament. With the goal of engaging in immediate, direct actions to impact climate change, the business section of ParraCAN's 2009 annual meeting was followed by the planting of sedge along the river bank at the park where the meeting was held. While the explicit goals of this activity were to create more plant cover to absorb greenhouse gases and to help prevent soil erosion,[19] it also allowed participants to feel that they were quite literally 'getting their hands dirty' in the effort to stop climate change. ParraCAN has also played an important role in increasing the breadth and visibility of local climate events. On the International Day of Climate Action in October 2009, for example, ParraCAN was key to the day's local events. Activities for the day included letter writing to the Australian government, making tiles for a community sculpture about climate change, and hosting an educational forum about the issue.[20]

For ParraCAN, as with other environmental organizations in Australia and elsewhere, 'general awareness-raising [is] undertaken to increase understanding and profile of the issue of climate change in the community, and the NGOs then encourage community members to transfer this heightened awareness into politically-visible actions' (Hall and Taplin 2007: 326; see also Bond 2009). Though relatively small in scope, local climate initiatives such as ParraCAN's 'can potentially provide positive impacts on local sustainability through the transformation of consumption and production practices' (Bond 2009: 3) in addition to contributing to the pressure for legislative change on the national and international levels through their membership in CANA and CAN International.

4 Impacts and Effectiveness

What can be said about the impacts and effectiveness of such climate-oriented activism? If we take as our benchmark progress, or the lack thereof, in international negotiations, we might conclude that the influence of the multitude of NGOs and SMOs which have taken up the issue in recent years have been quite limited. Were we to try to quantify effectiveness in terms of tons of avoided carbon emissions, not only would we be hard put to do the calculation, we would also find few, if any, reductions in greenhouse gas emissions. As we have proposed in this chapter, the metric for assessing impacts and effectiveness is not so much numbers or laws or technologies; rather, it is transformation in the larger field that we have called the climate change *dispositif* through the participation of civil society in global governmentality. Although we cannot wholly attribute the significant rise in public awareness of global warming or the burgeoning environmental consciousness, manifest most clearly in the 'greening of markets,' to NGOs and SMOs, it is fairly clear that various forms of environmental action and activism have played a significant role in setting the stage for greater sustainability and, perhaps, some kind of climate mitigation. The key here is, as suggested earlier, changes in normalized practices and their significance for everyday life.

The object of environmental NGO and SMO activism is, consequently, not only persuasion, education, and lobbying, it is also *transformation* of those normalized behaviors that sustain a particular form of the *dispositif*. As carbon-based beings, we cannot disengage from the atmosphere; we can, however, engage with it in ways that change what we do, how we do it, and why. Without the kind of social activism characteristic of civil society, this change will not happen. With it, there is at least some chance that it will.

Notes

1. Because we have been tasked with discussing only movements and organizations, we do not include business and corporations. To the extent that they engage in lobbying and action, they are also elements of civil society (Lipschutz with Rowe 2005).

2. Two exceptions were Stephen Schneider and James Hansen, both of whom were criticized for 'politicizing' science. On the role of so-called epistemic communities of scientists, see Haas (1992). On science and knowledge, see Kütting and Lipschutz (2009), as well as the chapters addressing science in this Handbook.

3. 'Resource wars' refers to the general notion that countries go to war over scarce minerals and energy supplies; see Lipschutz (1989) and Klare (2002).

4. The senior author on this article attended several days of a prep com in Geneva, as an accredited delegate from the low-lying island country of Nauru. The senior delegate at that session was Jackson Davis, at the time a Professor of Biology at the University of California, Santa Cruz.

5. Issues of *ECO* are archived at <http://www.climatenetwork.org/eco> and provide an indispensable source of information about a broad range of environmental conferences.

6. We stand here, perhaps, on perilously thin ice, although it will likely take a crisis, such as significant and sudden sea-level rise, to fully convince the public that climate change is really taking place.

7. To be sure, many SMOs lack both strong organization and an evidently self-interested need to produce and reproduce. It is difficult to sustain such groups on any scale larger than a very local one, with the result that many organizations which began life as social movements find it necessary to bureaucratize (Moser 2007; Tarrow 2007).

8. This might remind some readers of Antonio Gramsci's concept of 'hegemony,' and it does bear some resemblance to that notion, insofar as the forms and practices of governmentality do reflect and shape domination and subordination within societies.

9. Available at <http://www.climatenetwork.org/about-can/three-track-approach>, accessed October 2009.

10. Ibid.

11. Available at <http://www.climatenetwork.org/fossil/barcelona-climate-talks-2009/BarcelonaFossilAwards3Nov09.pdf>, accessed November 2009.

12. Available at <http://www.foei.org/en/what-we-do>, accessed October 2009.

13. Available at http://faithandclimatechange.wordpress.com/about/, accessed November 2009.

14. Available at http://www.biggreenchallenge.org.uk/home/finalists/faith-and-climate-change-in-birmingham/, accessed November 2009.

15. Note, however, that such activities amount to a form of 'green consumerism' if they are not linked to deeper structural changes in the political economy; see Szasz, this volume.

16. Available at http://faithandclimatechange.wordpress.com/2009/09/28/interesting-article-in-the-guardian/, accessed November 2009.

17. Available at http://ran.org/campaigns/global_finance/about_the_campaign/, accessed October 2009.

18. Available at http://www.parracan.org/, accessed November 2009.

19. Available at http://www.parracan.org/page9.htm, accessed November 2009.

20. Available at http://www.parracan.org/Sept.pdf, accessed October 2009.

REFERENCES

Asmus, P. 1998. A template for transition: how Mitsubishi and the Rainforest Action Network found the Natural Step. *Corporate Environmental Strategy* 5: 50–9.

BÄCKSTRAND, K., and LOVBRAND, E. 2007. Climate governance beyond 2012: Competing discourses of green governmentality, ecological modernization and civic environmentalism. Pp. 126–7 in M. E. Pettenger (ed.), *The Social Construction of Climate Change*. Burlington, VT: Ashgate Publishing Company.

BETSILL, M. M., and CORELL, E. 2008. *NGO Diplomacy: The Influence of Nongovernmental Organizations in International Environmental Negotiations*. Cambridge, MA: MIT Press.

BOND, M. 2009. Localizing climate change: Stepping UP local climate action. Paper presented at the Proceedings of the Environmental Research Event, Noosa, QLD.

BORDIEU, P. 1984. *The Field of Cultural Production: Essays on Art and Literature*. New York: Columbia University Press.

—— 2002. *Habitus: A Sense of Place*. Aldershot: Ashgate Publishing Company.

CASHORE, B. 2002. Legitimacy and privatization of environmental governance: How non-state market-driven (NSMD) governance systems gain rule-making authority. *Governance: An International Journal of Policy, Administration, and Institutions* (15)4: 503–29.

COX, R. W. 1987. *Production, Power, and World Order: Social Forces in the Making of History*. New York: Columbia University Press.

DEAN, M. 1999. *Governmentality: Power and Rule in Modern Society*. London: Sage Publications.

DEPLEDGE, J. 2005. Against the grain: the United States and the global climate change regime. *Global Change, Peace & Security* (17)1: 11–27.

DOHERTY, B. 2006. Friends of the Earth: Negotiating a transnational identity. *Environmental Politics* 15(5): 697–712.

DUWE, M. 2001. The Climate Action Network: A glance behind the curtains of a transnational NGO network. *Reciel* 10(2): 177–89.

FOGEL, C. 2007. Constructing progressive climate change norms: The US in the early 2000s. Pp. 99–122 in M. E. Pettenger (ed.), *The Social Construction of Climate Change*. Burlington, VT: Ashgate Publishing Company.

FOUCAULT, M. 1980. *Power/Knowledge: Selected Interviews and Other Writings 1972–1977*, trans. C. Gordon. New York: Pantheon.

—— 1991. Governmentality. Pp. 87–104 in *The Foucault Effect: Studies in Governmentality*, ed. G. BURCHELL, C. GORDON, and P. MILLER. Chicago: University of Chicago Press.

GOTTLEIB, R. 2005. *Forcing the Spring: The Transformation of the American Environmental Movement*. Washington, DC: Island Press.

HAAS, P. M. 1992. Introduction: Epistemic communities and international policy coordination. *International Organization* 46(1): 1–35.

HALL, N. L., and TAPLIN, R. 2007. Solar festivals and climate bills: comparing NGO climate change campaigns in the UK and Australia. *Voluntas* 18: 317–38.

HAYS, S. P. 1959. *Conservation and the Gospel of Efficiency: The Progressive Conservation Movement, 1890–1920*. Cambridge, Mass.: Harvard University Press.

—— 1987. *Beauty, Health, and Permanence: Environmental Politics in the United States, 1955–1985*. Cambridge: Cambridge University Press.

HESTER, R. 2009. Embodied politics: Health promotion in indigenous Mexican migrant communities in California. Ph.D. Dissertation, Department of Politics, University of California, Santa Cruz.

HIER, S. P. 2008. Thinking beyond moral panic: Risk, responsibility, and the politics of moralization. *Theoretical Criminology* 12(2): 173–90.

The Independent (UK). 2009. Gucci joins other fashion players in committing to protect rainforests. 5 November.

Intergovernmental Panel on Climate Change (IPCC). 2007. *Climate Change 2007: Synthesis Report.* IPCC Secretariat, World Meteorological Organization, Geneva. At <http://www.ipcc.ch/pdf/assessment-report/ar4/syr/ar4_syr.pdf>, accessed June 2009.

JABRI, V. 1996. *Discourses of Violence.* Manchester: Manchester University Press.

KECK, M. E., and SIKKINK, K. 1997. *Activists beyond Borders: Advocacy Networks in International Politics.* Ithaca, NY: Cornell University Press.

KEOHANE, R. 1984. *After Hegemony: Cooperation and Discord in the World Political Economy.* Princeton, NJ: Princeton University Press.

KLARE. M. 2002. Resource Wars: The New Landscape of Global Conflict. New York: Holt.

KÜTTING, G., and LIPSCHUTZ. R. D. (eds.) 2009. *Global Environmental Governance—Power and Knowledge in a Local-Global World.* London: Routledge.

LIPSCHUTZ, R. D. 1989. *When Nations Clash: Raw Materials, Ideology and Foreign Policy.* New York: Ballinger/Harper & Row.

—— (ed.). 2006. *Civil Societies and Social Movements: Domestic, Transnational, Global.* Aldershot: Ashgate Publishing Company.

—— with MAYER, J. 1996. *Global Civil Society and Global Environmental Governance: The Politics of Nature from Place to Planet.* Albany, NY: State University of New York Press.

—— with ROWE, J. K. 2005. *Globalization, Governmentality and Global Politics: Regulation for the Rest of Us?* New York: Routledge.

McCORMICK, J. 1991. *Reclaiming Paradise: The Global Environmental Movement.* Bloomington, IN: Indiana University Press.

MEYER, D. S. 2004. Protest and political opportunities. *Annual Review of Sociology* 30: 125–45.

MOSER, S. C. 2007. In the long shadows of inaction: The quiet building of a climate protection movement in the United States. *Global Environmental Politics* 7(2): 124–44.

OLSON, M. 1965. *The Logic of Collective Action.* Cambridge, MA: Harvard University Press.

ParraCAN. 2008. Press Release: ParraCAN Receives Prestigious Award. Available at <http://www.parracan.org/userimages/Pressreleaseawardfinal.doc>, accessed October 2009.

PATERSON, M. 2007. *Automobile Politics: Ecology and Cultural Political Economy.* Cambridge: Cambridge University Press.

PINSON, L. 2009. The GreenShows brings eco to Fashion Week. *NBCNewYork.com* 31 August.

PULVER, S. 2005. A public sphere in international environmental politics: The case of the Kyoto Protocol negotiations. Paper presented at the 2005 Berlin Conference on the Human Dimensions of Global Environmental Change: International Organizations and Global Environmental Governance.

SHABECOFF, P. 2003. *A Fierce Green Fire: The American Environmental Movement.* Washington, DC: Island Press.

SKLAIR, L. 1991. *Sociology of the Global System: Social Change in Global Perspective.* Hemel Hempsted: Harvester Wheatsheaf.

TARROW, S. 2007. *The New Transnational Activism.* Cambridge: Cambridge University Press.

TÉTREAULT, M. A., and LIPSCHUTZ, R. D. 2009. *Global Politics as if People Mattered.* 2nd edn., Lanham, MD: Rowman & Littlefield.

UNGAR, S. 2003. Moral panic versus the risk society: The implications of the changing sites of social anxiety. *British Journal of Sociology* 54(1): 3–20.

URRY, J. 2006. Inhabiting the car. *Sociological Review* (54)1: 17–31.

WALTZ, K. 1979. *Theory of International Politics.* Reading, MA: Addison-Wesley.

WAPNER, P. 1996. *Environmental Activism and World Civic Politics.* Albany, NY: State University of New York.

WENDT, A. 1999. *Social Theory of International Politics.* Cambridge: Cambridge University Press.

WILLIAMS, A., and PARKER, M. 2008. In 1988, nuclear war was 'undoubtedly the gravest' threat facing the environment. *Environment,* 50(3): 34–45.

CHAPTER 26

....

TRANSLOCAL CLIMATE
JUSTICE SOLIDARITIES

....

PAUL ROUTLEDGE*

1 COPENHAGEN CONVERGENCE

....

While the UN Conference of Parties (COP-15) meeting in Copenhagen in December 2009 saw no meaningful decision on carbon emission reductions reached by the world's governments, the city did witness a series of mass mobilizations by a range of civil society organizations, and an alternative conference of activists and environmental practioners called the *Klimaforum* which articulated a radical climate justice agenda. Such a convergence of different civil society actors—e.g. social movements, non-government organizations (NGOs), autonomist groups—represented the culmination of many months of collaborative action focusing upon the collective demand of climate justice. For example, 24 October 2009, a global day of action organized by the 350 campaign, saw 5,200 actions in 181 countries unite in a call for an equitable and meaningful solution to the climate crisis (White 2009). Such mobilizations led some to argue in Copenhagen that a 'climate justice movement of movements' is emerging[1] (see also Agyeman et al. 2007).

This chapter is concerned with how solidarities might be forged between different communities and civil society actors (such as social movements) within the context of how injustices associated with climate change are experienced differently in different places. In particular, this chapter will argue that a spatialized understanding of both particular placed-based struggles as well as how such struggles attempt to forge solidarities beyond the local are crucial in order to construct meaningful translocal alliances.[2] After a brief discussion of 'climate justice' as interpreted by civil society actors, I will ground my discussion within the context of recent fieldwork on Bangladesh with a landless peasants' movement concerned with issues of food sovereignty and climate change. I will then consider the potentials for constructing what I term 'translocal climate justice solidarities'.

* This research was supported by a grant from the Carnegie Trust for the Universities of Scotland.

2 CLIMATE JUSTICE AND TRANSLOCAL SOLIDARITY

2.1 Climate Justice Action

A substantial academic literature exists concerning issues of justice (e.g. Young 1990; Fraser 1997; Burczak 2006; Heynen 2006; Pickerill and Chatterton 2006); the scope and content of justice in the context of anthropogenic climate change (e.g. Dobson 1998; Beckerman and Pasek 2001; Page 2006; Roberts and Parks 2006; Beckman and Page 2008); and environmental justice (e.g. Schlosberg 2004, 2007; Agyeman et al. 2007; Wolch 2007; Walker 2009). See also Simon Dietz, Stephen Gardiner, Paul Baer, and Richard Howarth in this volume.

The concept of 'climate justice' emerged from the Global South referring to attempts to conceptualize the interrelationships between, and address the roots causes of, the social injustice, ecological destruction, and economic domination perpetrated by the underlying logics of capitalism that has seen industrialized countries reap the benefits of fossil fuel-intensive development (e.g. see the Bali Principles of Climate Justice 2002). Such conceptualizations acknowledge that capitalism, as a social and ecological relation, is implicated in anthropogenic-induced climate change. This is entangled with the increasing pressure on, and conflict over, resources induced by what David Harvey (2003) terms 'accumulation by dispossession' i.e. capital accumulation by (trans)national corporations through the appropriation (dispossession) of key resources such as land, water, and public utilities from others (e.g. peasants, workers, and indigenous peoples). Such processes are especially pertinent for the poor of the Global South who face multiple injustices. They are the victims of resource conflicts generated by capitalism's search for profits; they are located at the frontline of the effects of climate change; they possess small carbon footprints; and they do not have the resources to mitigate against its effects.

Therefore, climate justice implies both the direct participation of those most affected by economic and climate injustices, and collective struggle. Climate justice activism is currently focused around a coalition of grassroots social movements, including the US-based Indigenous Environment Network, the international small and landless farmers' network *La Via Campesina*, and autonomous activist groups such as the UK's Camp for Climate Action, and the Climate Justice Action and Rising Tide networks (Wolf and Mueller 2009).

Rather than being indicative of a coherent climate justice 'movement' emerging I would argue that we are witnessing a range of overlapping, interacting, competing, and differentially placed and resourced networks concerned with issues of climate change and justice (e.g. see Juris 2004; Routledge and Cumbers 2009). Such 'climate justice networks', building upon those developed during the alter-globalization mobilizations, involve the prosecution of political action both within and beyond the scale of the state. This enjoins us to be attentive to a geographical, or spatial, imaginary.

2.2 The Spatiality of Struggle

The production of space manifests various forms of injustice but also produces and reproduces them (Dikeç 2001). For example, responsibility for the production of outcomes, consumption, and impacts of environmental injustice (e.g. carbon emitters; state policies; etc.) has distinct spatialities (e.g. scales of transmission and effect) (Walker 2009). The politicized inequality of flows (e.g. of carbon emissions, toxic waste), movements between places, and responsibilities intersect with inequalities of population proximity to environmental events (e.g. floods; cyclones). The distribution of vulnerabilities among bodies, households, neighborhoods, etc. are unequally experienced by men and women; rich and poor etc. (Walker 2009).

Clearly, those vested with the power to produce the physical spaces we inhabit through capitalist development, investment, and planning—as well as through grassroots-embodied activisms are likewise vested with the power to perpetuate injustices and/or create just spaces. Justice is thus a shared responsibility of engaged actors in the socio-spatial systems they inhabit and (re)produce (Bromberg et al. 2007). Moreover, 'social struggle remains crucial to the actual structuring and shaping of social justice' (Mitchell 2003: 235).

Spatially, Dikeç (2001) argues that struggles for justice claims require that people act spatially in four interrelated ways to secure such rights and claims. First, people must act from space, politically mobilizing from the material conditions of their (local) spaces and seek alternative spatializations. Second people must also act on space—to appropriate or dominate it with a group identity. Third, people must act in space, such as taking to the streets for protests, or occupying land. Fourth, people must make space: creating conditions to expand public political involvement, for example through the creation of solidaristic alliances. In the pursuit of climate justice these will require action locally as well as translocally.

2.3 Translocal Solidarities

Solidarities are forged out of the collective articulations of different place-based struggles, and constituted as the varied interconnections, relations, and practices between participants (Featherstone 2005). They are part of the ongoing connections, social and material relations, articulations and negotiations between places and place-based struggles. In particular what requires negotiation is the politics of extension and translation of place-based interests and experiences (Katz 2001), in order for common ground (e.g. shared antagonisms) and productive connections to be generated between different places and organizations (e.g. see Featherstone 2003, 2008).

Certainly, shared notions of climate (in)justice can inform the practice of solidarity, potentially creating a common ground that enables different themes to be interconnected, and different political actors from different struggles and cultural contexts to join together in common struggle (della Porta et al. 2006). However, this must negotiate the problem of how militant particularisms (Harvey 1996) concerning climate change transcend local concerns to form common ground upon which solidarities can be constructed. An initial requirement for the construction of such solidarities has been the construction of 'convergence spaces' (Routledge 2003; Cumbers et al. 2008) where groups can meet one another, exchange experiences, and plan collective strategies such as the alternative conferences and global days of action that have attended international summits such as those that occurred at Copenhagen (e.g. see Routledge and Cumbers 2009).

More sustained (and sustainable) 'climate justice solidarities,' will need to negotiate how shared translocal antagonisms, vulnerabilities,[3] insecurities, and aspirations of geographically, culturally, economically, and politically different and distant peoples are constructed that enable connections to be drawn that extend beyond the local and particular (Olesen 2005). Solidarity here is less altruistic (i.e. based upon the worthiness of, and sympathy towards, the suffering of distant others) than reciprocal (i.e. when activists in different groups draw connections between the suffering of others and their own plights or claims) and based upon shared threats or harm suffered as a consequence of common identities between activists (where identity is dynamic, contingent, contested, and socially constructed) (Reitan 2007: 20–1). Solidarity also includes feelings and emotions (e.g. anger; passion; empathy; fear) that are amplified into senses of collective solidarity (see Juris 2008), particularly when faced with the threats and uncertainties of climate change.

The creation of such climate justice solidarities must face at least two important and related issues: (i) the different strategies deployed, and concerns and constraints faced, by place-based social movements as activists make, and act from, on, and in space; and (ii) the networking problems that movements face when engaged with other place-based struggles. I will ground such a spatialized understanding of climate justice solidarity with reference to land struggles in Bangladesh, a country that is at the frontline of climate change impacts, focusing upon my critical engagement with the Bangladesh Krishok (peasant) Federation (as a co-facilitator of People's Global Action (Asia), an international network that they participated in, between 2002 and 2009).

3 THE BANGLADESH KRISHOK FEDERATION

Bangladesh is considered to be one of the most vulnerable countries in the world to climate change and sea-level rise (IPCC 2008). Despite the establishment by the national government of a Climate Change Strategy and Action Plan in 2008, little has as yet been initiated in terms of policy (Ayers and Huq 2009). The coping capacity in Bangladesh remains limited due to the relatively poor physical infrastructure and the political structure (Chowdhury 2009). Land degradation and scarcity have been growing in Bangladesh since the 1950s, and the majority of the population are poor and dependent on agriculture, and are thus more vulnerable to the tropical cyclones, storm surges, floods, and droughts to which the country has been historically prone (Reuveny 2007; see also Karim and Mimura 2008). Moreover, food security for peasants has been undermined by unequal land distribution (especially landlessness) and lack of credit (interviews, Bangladesh, 2009). In such a context, climate change is emerging as an important challenge for Bangladeshi social movements that are already involved in conflicts over access to key resources. One such movement is the Bangladesh Krishok Federation.

The Bangladesh Krishok Federation (BKF), is the largest rural-based peasant movement in the country, and was established in 1976. Currently, the BKF has 700,000 members and it belongs to the Aaht Sangathan (the Eight Organizations),[4] whose total membership is now close to two million members. Since its inception the BKF has been actively involved in land occupation struggles. From 1977 until 1991 the BKF conducted various types of nonviolent struggles to compel local government officials, at different times, to make commitments about the distribution of land amongst landless men and women. In 1987 the national government introduced a land law which enabled landless people to occupy and farm fallow (*khas*) land (interviews, Dhaka, 2004, 2009).

Since 1987 the BKF has demanded the distribution of *khas* land among landless men and women as stated in the land law. Because of ongoing government refusal to take any initiative on behalf of the landless, since 1992 the BKF has organized landless people to occupy *khas* land. For example, during the early 1990s the movement organized the occupation of over 22,000 acres of land on 4 *chars* (small islands) in the southern coastal belt of Bangladesh. During the occupation movement, the BKF has encountered many impediments from the local big landowners and their *goondas* (armed thugs) who have made several attacks on the landless people's settlements on the *chars* (interviews, Dhaka, 2004, 2009).

In remaining in their settlements, the people have built their homes, cultivated their land, and grown different indigenous crops (e.g. rice, vegetables, and fruits). Since 1992, the land occupation movement has continued, and so far, under the leadership of the BKF, the landless people have been able to occupy approximately 80,000 acres of *khas* land, across Bangladesh. Most of the occupations are concentrated in the south of the country (i.e. that part of Bangladesh most vulnerable to climatic events) and land has been distributed to more than 107,000 of the poorest men and women living in the countryside. Currently twenty-seven *chars* are occupied throughout Bangladesh (interviews, Dhaka, 2004, 2009).

In addition to struggles over land rights within Bangladesh, the BKF has been active in a range of international networks, concerned with collective action around a range of justice issues associated with neoliberal globalization, land rights, and, more recently, climate change. It is to the forging of solidarity around issues of climate justice that I now turn, and to the place-based, as well as networking, problems and potentials faced by the BKF.

4 FORGING CLIMATE JUSTICE SOLIDARITIES
FROM BANGLADESH

4.1 Occupation

In this section, following Dikeç (2001), I consider the particular conditions under which the BKF operates, particularly with regard to land occupations and the emerging threat of climate change. I will subsequently consider the particular networking problems faced by the BKF when working within translocal alliances. Concerning the emerging threat of climate change, a prominent BKF activist remarked:

> Climate change is a global concern but it is true that Bangladesh is going to be the most affected because it is low lying. Many people have migrated to the cities and towns because of [Cyclones] Sidr and Aila, so we already have climate refugees. Climate change is now one of the focal issues of BKF and the *Aaht Sangathan* since 2005 because it is related to life and livelihood, particularly for peasants. (Interview, Dhaka, Bangladesh, 2009)

For the BKF, place-specific conditions of local organizing—including a paucity of resources; the time taken up with the struggles to occupy land; the necessities of maintaining the occupation; and the role of (local and national) government actors and policy—act to constrain their capacity to organize their struggles and participate within solidarity networks.

First, the BKF acts from space politically mobilizing landless peasants from the material conditions of their (local) villages and towns and seeks to reconfigure both the political economy, and the geography, of peasant livelihoods in order to construct alternative spatializations, i.e. land controlled and used by peasants, not least for the growing of food. Activist organizers must negotiate both a relatively poor transport network that includes their travelling by public buses, motor and bicycle rickshaws, as well as by foot under often difficult weather conditions. More important is the lack of funds available to the movement to prosecute land occupations, as another BKF activist commented:

> The movement has not got enough funds to really expand, organize demonstrations, or train organizers at the thana[5] or district level, and to fight the legal cases brought by landlords in the local courts in an attempt to stop the occupation. We are only keeping the movement alive at present. We are not really developing. (Interview, Dhaka, Bangladesh, 2009)

In addition, the BKF faces ongoing problems from local and national government officials:

> When we attempt to establish land for the landless, the district authority gives a response to the thana authority. They make a committee of social workers, development workers, local government members, and local political party members who decide if the claim can proceed. But the committee has no representatives from the movement. It is corrupt. They choose their own people to receive land. (Interview, Kurigram District, Bangladesh, 2009)

Second, the movement, through its occupation of land, acts on space, appropriating it with movement identity. The BKF seeks to remake particular rural spaces in more humane, egalitarian ways—not least through appropriating *khas* land intended for the landless, as an activist commented:

> Then we occupy the land. We build makeshift shelters for the occupying families, and provide food relief until the peasants can sow *padi* (rice). The peasants must drink river water, and many get sick, until we have dug tube wells. (Interview, Kurigram District, Bangaldesh, 2009)

Peasant communities continually confront physical hardships associated with the climate. For example, in Satkhira district in southern Bangladesh (where 12,000 families have occupied land since 1998) peasants have faced Cyclone Sidr, in 2007, which caused 3,500 deaths (Karim and Mimura 2008), and Cyclone Aila, in 2009, which damaged homes and inundated the land of Satkhira with salt water. As a result women have to travel for four hours a day to collect fresh water (interviews, Kaliganj, Bangladesh, 2009).

A further problem encountered is landlord and *goonda* violence and harassment against peasant communities. False criminal charges against peasants keep them in courts and unable to farm their land:

> Peasant activists are attacked, beaten, burned, jailed, and their homes are burned. That is the reality that we face. (Interview, BKF activist, Kurigram District, Bangladesh, 2009)

Third, the movement acts in space, through its occupation of land, conducting demonstrations, and organizing local and national meetings with landless villagers. A key organizer in the BKF explained the material conditions of organizing land occupations thus:

> Depending on the local context the BKF will use strikes, and sit ins, *gheraos* (surrounding the district government offices and local politicians), hunger strikes and mass actions prior to the invasion of the land. (Interview, BKF activist, Kurigram District, Bangladesh, 2009)

Finally, the movement makes space, creating the conditions to expand the landless peasants' political involvement, and through the creation of translocal solidarities (discussed in the next section). The BKF painstakingly organizes communities of the landless over a period of six to twelve months to occupy land:

> For a successful land occupation, the movement needs a strong occupation committee, whose leaders can withstand attacks by the landlords' *goondas*; a strong mass mobilization; a medical team who can provide medical treatment to those who suffer physical attacks; and a legal team to fight the legal cases brought by landlords in the local courts in an attempt to stop the occupation. This takes time and money. (Interview, BKF activist, Kurigram District, Bangladesh, 2009)

However, the movement is increasingly encountering new problems associated with a changing climate. Agricultural practices are being disrupted because of the changing frequency and character of the Monsoon, which in turn impacts peasants' ability to fully participate in the movement, as a local peasant organizer told me:

> The frequency of the Monsoon has also changed. Before, it used to start in June or July. Now it is beginning in August. I have noticed a change in the Monsoon since 2002. It is affecting planting practices and we are seeing an increase in pests in the dry weather. Also the character of the monsoon has changed. It is increasingly unpredictable. Our planting of *padi* is being disturbed by the Monsoon changes. (Interview, Barguna District, Bangladesh, 2009)

What is apparent from the preceding discussion is that any attempt by the BKF to forge translocal solidarities with movements beyond Bangladesh must constantly negotiate the geographical dilemmas inherent in their embeddedness in particular spatial (local, regional, national) realities (Chesters 2003). The formation of translocal solidarities continues to be compromised by a range of place-specific issues concerning the realities of everyday struggle and resource availability. Certainly translocal alliances are more likely to be facilitated and sustained when movements possess significant mobilization capacities already; when they have the capacity for regular communication with other movements; and when each organization's members take some responsibility for brokering bonds of solidarity (Bandy and Smith 2005). In addition, the ability of movements to participate in solidaristic alliances is also shaped by the actions, policies, limitations, and challenges posed by the governments of the states in which they are located and their relationship to the state (Burawoy et al. 2000; Glassman 2001; Rai 2003). In these ways, networks are both influenced by and replicate the existing 'power geometries' (Massey 1999) that distinguish connections between places under economic globalization. Moreover, the desire for translocal solidarities notwithstanding, for the BKF the national scale remains important for the mobilization of resources (interviews, Bangladesh, 2009). However, where networks are fashioned and sustained, a range of other issues come to the fore.

4.2 Networking

The BKF is enmeshed in a variety of translocal networks, including La Via Campesina (the peasant way, LVC), the international peasant network established in 1993 (Desmarais 2007); People's Global Action Asia (PGA Asia), an international network of peasant movements (Routledge 2003); the Asia Peasants' Coalition, the South Asia Peasants' Coalition, and the People's Coalition on Food Sovereignty (interviews, Dhaka, 2009); and has also been engaged in the World Social Forum process. Such networks have been concerned with issues of food sovereignty; resistance to neoliberal capitalism and more recently to issues of climate change. As a BKF activist commented:

> We prioritize direct action in trying to stop carbon emissions. Our strategy is to continue to struggle here against the government but also internationally to push our demands. We are fighting for alternatives, for land, but we need to build networks. (Interview, Dhaka, Bangladesh, 2009)

In my work as a facilitator of the PGA Asia network, in which I worked closely with the BKF, certain problems with the networking process became apparent. Much networking devolved to a small number of key actors who sustained networks through their own interaction and personal communication—termed imagineers (Routledge et al. 2007; Cumbers et al. 2008; Routledge and Cumbers 2009; see also Juris 2004). They conducted much of the routine (international) organizational work of networks helping to organize conferences, mobilize resources, and facilitate communication and information flows between movements and between movement offices and grassroots communities. They also attempted to 'ground' the concept or imaginary of the network (what it is, how it works, what it is attempting to achieve) within grassroots communities who comprised the membership of the participant movements. Because of their structural positions, communication skills, and experience in activism and meeting facilitation these imagineers tended to wield disproportionate power and influence within networks such as PGA Asia. There was an over-reliance upon the imagineers to instigate events, raise funds, etc. (Routledge and Cumbers 2009).

While solidarities are forged to challenge unequal power relations (manifested locally, nationally, globally), in practice the operational processes of networks such as PGA Asia frequently involved significant power differences due to differences in resource access between activists (Rai 2003; Eschle and Maiguashca 2007), exemplified by the role of the imagineers. This vitiated against the full involvement of grassroots communities within the networks. Hence, for a Thai PGA Asia activist, the operational logic of the network was underpinned by 'literate' and conceptual communicational forms (e.g. the writing of e-mails and documents, the analysis of how networks function), whereas the operational logic of most grassroots movements continues to be based upon oral communication:

> There is a real limitation to the capacity of grassroots movements to take ownership of the process. Movements do not know each other very well and . . . do not really know the PGA process at all. Thus participation is limited and language affects this too. Most movements are based on oral communication, whereas the PGA process is more literate and concept-based, thus it is difficult for grassroots movements to understand. (Interview, Bangkok, Thailand, 2004)

The importance of getting grassroots communities more involved in networks such as PGA Asia was affirmed by an Indian PGA Asia activist:

> Movements in South Asia have a limited resource capacity to fully engage in global solidarity, things like time, money, language skills and computer skills. Hence most Indian movements are not really ready to fully participate in a global movement, to commit to it full time, or to fully involve and engage the grassroots in it. Most movements in India are leader based and many of these leaders have neither computer skills nor English language skills and thus they profess to be uninterested in global organizing since they do not possess the necessary skills

for it. Most folk who do global organizing primarily like to travel and enjoy the benefits of conference hotels—they aren't serious about global solidarity. The language of many movement leaders is influenced by NGO discourse and not by the language of the grassroots. We need to return to the grassroots since most global work is too much in the air. (Interview, Kathmandu, Nepal, 2006)

Therefore, a key issue concerning the forging of meaningful solidarities is how the network's 'imaginary' is visualized and developed at the grassroots: how to construct senses of shared (or 'tolerant') identities (della Porta 2005) concerning climate justice amongst very different place-based communities. This will require the co-recognition and internalization of others' struggles in a 'global' community. In part this must be based on shared values and principles (common ground) concerning economic and political justice and ecological sustainability. For climate justice solidarity, this will require the grounding of network imaginaries in grassroots communities, a process that will need to be attentive to the place-specificity of each movement (and its struggle) that comprise a particular translocal network. Moreover, this must continually negotiate the spatially specific problems, potentials, and constraints faced by grassroots movements both in their everyday organizing practices as well as their attempts to engage with solidaristic alliances with others.

One area of common ground for peasant movements in the Global South that has emerged during the past decade is that of food sovereignty. Food sovereignty has been recognized by peasant movements as one of the most important practices that enable peasant communities to both mitigate, and adapt to, the effects of climate change. Food sovereignty implies control over territory and biodiversity (commons); self-governance; ecological sustainability; the articulation of cultural difference, etc. and has acted as a point of encounter, common interest, and solidarity. Moreover, it is a matter of survival, as Junya Yimprasert (2009: 18) argues:

> The cities of the South are now jam-packed with people that a few years ago were self-sufficient. There is no way that the authorities in our mega-cities can ensure the physical and mental well-being of these migrants. The exodus from the land must be halted . . . Extinguishing and substituting small-scale farming with mono-culture agri-industry commanded by multi-national corporations has no logical connection to the eradication of poverty. On the contrary, all evidence points in the opposite direction—to increasing poverty, slum conditions, risk of epidemics and a whole bunch of negative impacts reflected in negative climate change.

LVC introduced the concept of food sovereignty in 1996 and now promote the issue internationally. Many of the struggles (in the Global South as well as Global North) concerning privatization of the commons, displacement, the construction of mega-dams, genetically modified crops, pesticide use, etc. are connected to agriculture and those who produce food (Desmarais 2007). As Amory Starr (2000: 224) notes: 'Centering food in economic and community analysis is an important way to get people to deal with environmental and economic issues.'

As a participant in LVC, the BKF has also taken up the demand for food sovereignty. Moreover, the connections between climate change and the movement's goal of food sovereignty were recognized by the activists who informed me:

> Our own vision is based on food sovereignty. We are trying to link up with other organizations with similar climate change concerns. We are trying to build a network regionally, both

in South Asia and Asia. We need links with movements in Asia where climate change effects will be greatest. We are also working to pressure the Ministry of Agriculture and the Ministry of the Environment to take a clear position on climate change and incorporate food sovereignty into the National agricultural policy. We need popular pressure from the grassroots. The government is concerned with food security, but this is different from food sovereignty since with food security you do not have to grow your own food; you can import food. We need culturally appropriate foods. (Interview, BKF activist, Dhaka, Bangladesh, 2009)

As part of their commitment to this vision, the BKF has embarked on popular education on food sovereignty to peasant communities, participated in an Asian Food Sovereignty Caravan (in 2004), and participated in conferences on food sovereignty and peasant rights in Nepal and Bangladesh in 2007; and an LVC-organized conference in Dhaka, 2008, on climate change and food sovereignty (interviews, Dhaka, Bangladesh, 2009). Such events point to an emerging common ground between different movements that potentially provides the basis for shared principles concerning a politics of climate change and justice—what Laclau and Mouffe (1985) termed a chain of equivalence.

Importantly, at the climate justice mobilizations in Copenhagen (in which the BKF President participated), the *Klimaforum* issued a declaration that, as of January 2010, had been signed by 488 civil society organizations. This declaration included a series of principles around which different movements concerned with issues of climate justice, and located in different local and national realities, could forge common ground as the basis for translocal solidarity and cooperation between them.[6] Indeed, such principles articulated a common agenda concerning climate justice that required systemic structural economic changes. As such, this formed one of two strategic goals for emerging climate justice activism, the other being the forging of translocal solidarities that can sustain climate justice networks.

Such principles and solidarities based upon common identities (as farmers/peasants), as well as shared hopes, fears, and threats (concerning climate change and justice) provide potential currents of equivalence that do not require groups or movements to sacrifice diversity and autonomy for the unity of the whole, and also provide the ground for reciprocity between interconnected struggles (Reitan 2007).

However there are considerable differences of experience between activists and movements from the Global South and Global North that have the potential for undermining the possibilities of such translocal solidarities. For example, different class and class-fractional positions exist within the various constituencies of Global Northern and Global Southern movements, such as the differential powers of certain Global Northern trade unions in key industries as compared to Global Southern peasant farmers' organizations. Moreover, there are profound situational differences of different militant groups' relationships to the regional and national movements in which they are embedded (Glassman 2001).

In addition, tensions continue to arise between the more horizontalist politics practiced by many climate justice groups in the Global North, and the more verticalist politics practiced by many Global Southern peasant movements, that may be shot through with class, caste, and gender inequalities (interviews, Copenhagen, 2009; see also Routledge and Cumbers 2009). Indeed, for Yimprasert greater communication, knowledge, and understanding between activists from such different geographical locations (and political, cultural, and economic realities) is required:

> Campaigning for global justice in the South remains North-led . . . [t]o be able to work in a true spirit of solidarity with the struggle of workers in the South, activists need at least some first-hand experience of living under oppression in the South. (2009: 6)

In networks that the BKF has participated in, such as PGA Asia, one of the ways that this problem has begun to be addressed is through regional conferences (such as one hosted by the BKF in Dhaka in 2004[7]). These have provided important spaces within which representatives of different movements can meet face to face, communicate with one another, and exchange experience, strategies, and ideas, and generate collective energy and sense of identity, and develop deeper interpersonal ties between activists from different struggles in order to nurture solidarity (interviews, Bangladesh, Nepal, 2004).

5 TRANSLOCAL CLIMATE JUSTICE SOLIDARITIES

Solidarities are co-produced with the constitution of networks, forged out of the collective articulations of different place-based struggles, and constituted as (often messy, problematic, always negotiated) interconnections that mediate between the participants (Featherstone 2005, 2008). They are part of the ongoing connections, social and material relations, and articulations between places.

Hence, the formation of a sustainable politics of climate justice solidarity involves understanding the way that the local is enmeshed in wider spatial relations, as well as how particular struggles act from, on, and in, as well as make, space in their everyday realities. Clearly, many movements in the Global South, such as the BKF, see defense of their local spaces and opposition to national governments as their most appropriate scales of political action (Mertes 2002; Tarrow 2005). Even where international campaigns are organized, local and national scales of action have continued to be important (Herod 2001). Indeed, though global events (such as the Copenhagen mobilizations) and networks are important, arguably more time and resources should be spent on networking locally and nationally e.g. constructing more effective *grounded* resistance to injustice and responses to climate change.

This is because climate justice networks and solidarities can only flourish as a consequence of capacity built at the local level. Translocal communication and coordination is only possible to the extent that there are active place-based struggles from which resistance tactics and strategies are developed. When translocal solidarities are manifested in networks and convergences, these can reinforce place-based movements and campaigns (Nunes 2009).

Hence, any discussion of possibilities of translocal climate justice solidarity must confront, and then negotiate, a range of messy place-specific 'ground realities'. In a recent article Harvey (2009) has argued that the terrain of (anti-capitalist) political struggle and of political possibilities includes NGOs (state and private funded); anarchist and autonomist grassroots organizations (e.g. Camp for Climate Action); traditional labor unions and left political parties; social movements resisting displacement and dispossession (e.g. LVC); and emancipatory movements around questions of identity (e.g. women, children, gays, racial, ethnic, and religious minorities). Clearly such constituencies are neither mutually exclusive nor exhaust possibilities for political action. However, certain constituencies such as trade unions and emancipatory movements around questions of identity were largely (if not completely) absent from Copenhagen's climate justice mobilizations. The articulation of common ground across such differences will require an engagement with the entangled character of capitalism and climate change and the crises inherent within each. Solidarities across diversity will require 'imaginative geographies of connection, composed of sympathies and affinities' (Featherstone et al. 2007: 388). For example, the DIY Education Collective (<http://diyeducation.wordpress.com>) in the UK is attempting to bring together climate activists with anti-poverty activists (in Glasgow) and anti-racist activists in London, through a series of collaborations that explore differences as well as shared concerns. Another example is that of the campaign since July 2009 to protect jobs at the Vestas wind-turbine plant in Newport, Isle of Wight, which has seen collaboration between trade unionists and green activists.

Solidarities must also negotiate the dangers of any emerging parochialisms (and an associated fetishism of the local) within climate justice activism. In attempting to build upon those connections and networks established during the earlier alter-globalization mobilizations, it is important to rigorously analyze how and why solidarities between different groups compose, recompose, and decompose. For climate justice solidarity to inspire local struggles in local communities requires not so much a reactive, defensive, or parochial politics but rather a politics of hope (see Solnit 2004) and shared notions of justice embodied in proactive, positive interventions and initiatives that produce what Matt Sparke (2007) terms 'geographies of repossession' of those resources and spaces colonized by capitalism. It remains a 'work in progress;' something that remains to be fought for, constructed, defended, sustained, and expanded.

NOTES

1. Such as Naomi Klein at a meeting organized by the Reclaim Power network in Cristiania during the week of protests.
2. I use the term translocal solidarity to refer to the connections, relations, and campaigns between different placed-based (but not place-restricted) social movements and other grassroots actors. This is in preference to the term 'transnational solidarity' which elides the specificity of the particular places in, and from which, collective action emerges and operates.
3. Vulnerability includes (i) socio-economic factors (e.g. economic resources; distribution of power, etc.); and (ii) biophysical factors (e.g. environmental conditions) acting across a range of scales. These are related to coping and adapative capacities of communities (Fussel 2007). Because socio-economic factors are intimately linked to issues of poverty; state practice; access to economic opportunities, health, and education, etc., they are directly related to issues of justice (Barnet and Adger 2007).
4. The *Aaht Sangathan* also includes the Bangladesh Kisani Sabha (BKS, peasant women's organization, 800,000 members); Floating Labour Union (100,000 members; Floating Women's Labour Union (150,000 members); Bangladesh Adivasi Samiti (indigenous committee, 50,000 members); Rural Intellectual Front (5,000 members); Ganasaya Cultural Centre (200 members); and the Revolutionary Youth Association (5,000 members).

5. Sub-district.
6. These common principles included: leaving fossil fuels in the ground; reasserting peoples' and community control over production; re-localizing food production; massively reducing overconsumption, particularly in the Global North; respecting indigenous and forest people's rights; and recognizing the ecological and climate debt owed to the peoples in the Global South by the societies of the Global North necessitating the making of reparations. (See <http://www.klimaforum09.org/Declaration>. Accessed 18.01.10.)
7. The six-day conference saw the convergence of 150 delegates from Bangladesh, India, Nepal, Thailand, Malaysia, Philippines, and Vietnam representing 46 grassroots peasant movements (of farmers, fisherfolk, indigenous people, women, and labourers). Forty percent of the delegates were women (see Routledge and Cumbers 2009).

References

AGYEMAN, J., and EVANS, B. 2004. 'Just sustainability': The emerging discourse of environmental justice in Britain? *Geographical Journal* 170(2): 155–64.
—— BULKELEY, H., and NOCHUR, A. 2007. Climate justice. Pp. 135–44 in J. Isham and S. Waage (eds.), *Ignition: What You Can Do to Fight Global Warming and Spark a Movement*. Washington, DC: Island Press.
AYERS, J., and HUQ, S. 2009. Leading the way. *Himal Southasian* October–November <http://www.himalmag.com/Leading-the-way_nw3581.html> accessed on 6/10/09.
Bali Principles of Climate Justice. 2002. See <http://www.ejnet.org/ej/bali.pdf>. Accessed 18.01.10.
BANDY, J., and SMITH, J. (eds.) 2005. *Coalitions across Borders*. Oxford: Rowman & Littlefield.
BARNET, J., and ADGER, W. N. 2007. Climate change, human security and violent conflict. *Political Geography* 26: 639–55.
BECKERMAN, W., and PASEK, J. 2001. *Justice, Posterity and the Environment*. Oxford: Oxford University Press.
BECKMAN, L., and PAGE, E. A. 2008. Perspectives on justice, democracy and global climate change. *Environmental Politics* 17(4): 527–35.
BROMBERG, A., MORROW, G. D., and PFEIFFER, D. 2007. Editorial note: Why spatial justice? *Critical Planning* 14 (summer).
BURAWOY, M., BLUM, J. A., GEORGE, S., GILLE, Z., GOWAN, T., HANEY, L., KLAWITER, M., LOPEZ, S. H., ÓRIAIN, S., and THAYER, M. (eds.) 2000. *Global Ethnography: Forces, Connections and Imaginations in a Postmodern World*. London: University of California Press.
BURCZAK, T. A. 2006. *Socialism after Hayek*. Ann Arbor: University of Michigan Press.
CHESTERS, G. 2003. Shape shifting: Civil Society, complexity and social movements. *Anarchist Studies* 11(1): 42–65.
CHOWDHURY, A. 2009. The coming crisis. *Himal Southasian* October–November <http://www.himalmag.com/The-coming-crisis_nw3575.html> accessed on 6/10/09.
CUMBERS, A., ROUTLEDGE, P., and NATIVEL, C. 2008. The entangled geographies of global justice networks. *Progress in Human Geography* 32(2): 183–201.
DELLA PORTA, D. 2005. Multiple belongings, tolerant identities, and the construction of 'another politics': Between the European social forum and the local social fora'. Pp. 175–

202 in D. Della Porta and S. Tarrow (eds.), *Transnational Protest and Global Activism.* Lanham, MD: Rowman & Littlefield Publishers.

—— ANDRETTA, M., MOSCA, L., and REITER, H. 2006. *Globalization from Below.* London: University of Minnesota Press.

DESMARAIS, A. 2007. *La Via Campesina: Globalization and the Power of Peasants.* London: Pluto Press.

DIKEÇ, M. 2001. Justice and the spatial imagination. *Environment and Planning A* 33: 1785–805.

DOBSON, A. 1998. *Justice and the Environment.* Oxford: Oxford University Press.

ESCHLE, C., and MAIGUASHCA, B. 2007. Rethinking globalised resistance: Feminist activism and critical theorising in international relations. *BJPIR* 9: 284–301.

FEATHERSTONE, D. 2003. Spatialities of transnational resistance to globalization: The maps of grievance of the Inter-Continental Caravan. *Transaction of the Institute of British Geographers* 28(4): 404–21.

—— 2005. Towards the relational construction of militant particularisms: Or why the geographies of past struggles matter for resistance to neoliberal globalisation. *Antipode* 37(2): 250–71.

—— 2008. *Resistance, Space and Political Identities: The Making of Counter-global Networks.* Oxford: Wiley-Blackwell.

—— PHILLIPS, R., and WATERS, J. 2007. Introduction: Spatialities of transnational networks. *Global Networks* 7(4): 383–91.

FRASER, N. 1997. *Justice Interruptus: Critical Reflections on the 'Postsocialist' Condition.* New York: Routledge.

FUSSEL, H.-M. 2007. Vulnerability: A generally applicable conceptual framework for climate change research. *Global Environmental Change* 17: 155–67.

GLASSMAN, J. 2001. 'From Seattle (and Ubon) to Bangkok: The scales of resistance to corporate globalization. *Environment and Planning D: Society and Space*, 20(5): 513–33.

HARVEY, D. 1996. *Justice, Nature and the Geography of Difference.* Oxford: Blackwell.

—— 2003. *The New Imperialism.* Oxford: Oxford University Press.

—— 2009. *Organizing for the Anti-Capitalist Transition.* <http://davidharvey.org/2009/12/organizing-for-the-anti-capitalist-transition/>. Accessed 12.01.10.

HEROD, A. 2001. *Labor Geographies.* New York: Guildford Press.

HEYNEN, N. 2006. 'But it's alright, ma, it's life, and life only': Radicalism as survival. *Antipode* 38(5): 916–29.

IPCC. 2008. *Climate Change 2007: Synthesis Report.* Sweden: Teri Press.

JURIS, J. 2004. Digital age activism: Anti-corporate globalization and the cultural politics of transnational networking. Unpublished Ph.D. dissertation, Department of Anthropology, University of California, Berkeley.

—— 2008. Performing politics: image, embodiment, and affective solidarity during anti-corporate globalization protests. *Ethnography* 9(1): 61–97.

KARIM, M. F., and MIMURA, N. 2008. Impacts of climate change and sea level rise on cyclonic storm surge floods in Bangladesh. *Global Environmental Change* 18: 490–500.

KATZ, C. 2001. On the grounds of globalization: A topography for feminist political engagement. *Signs* 26(4): 1213–29.

LACLAU, E., and MOUFFE, C. 1985. *Hegemony and Socialist Strategy: Towards a Radical Democratic Politics.* London: Verso.

MASSEY, D. 1999. Imagining globalization: Power-geometries of time-space. Pp. 27–44 in A. Brah, M. J. Hickman, and M. Mac an Ghaill (eds.), *Global Futures: Migration, Environment and Globalization.* Basingstoke: Macmillan.

MERTES, T. 2002. Grass-roots globalism. *New Left Review* 17: 101–10.

MITCHELL, D. 2003. *The Right to the City: Social Justice and the Fight for Public Space*. New York: Guilford Press.

MORALES, A. L. 1998. *Medicine Stories*. Boston: South End Press.

NUNES, R. 2009. What were you wrong about ten years ago? *Turbulence* 5: 38–9.

OLESEN, T. 2005. *International Zapatismo*. London: Zed Books.

PAGE, E. A. 2006. *Climate Change, Justice, and Future Generations*. Cheltenham: Edward Elgar.

PICKERILL, J., and CHATTERTON, P. 2006. Notes towards autonomous geographies: Creation, resistance and self-management as survival tactics. *Progress in Human Geography* 30: 730–46.

RAI, S. M. 2003. Networking across borders: The South Asian Research Network on Gender, Law and Governance. *Global Networks* 3(1): 59–74.

REITAN, R. 2007. *Global Activism*. London: Routledge.

REUVENY, R. 2007. Climate change-induced migration and violent conduct. *Political Geography* 26: 656–73.

ROBERTS, R. T., and PARKS, B. C. 2006. *A climate of injustice: global inequality, north-south politics and climate Policy*. Boston: MIT Press.

ROUTLEDGE, P. 2003. Convergence space: Process geographies of grassroots globalization networks. *Transactions of the Institute of British Geographers* 28: 333–49.

—— and CUMBERS, A. 2009. *Global Justice Networks: Geographies of Transnational Solidarity*. Manchester: Manchester University Press.

—— —— and NATIVEL, C. 2007. Grassrooting network imaginaries: Relationality, power, and mutual solidarity in global justice networks. *Environment and Planning A* 39: 2575–92.

SCHLOSBERG, D. 2004. Reconceiving environmental justice: Global movements and political theories. *Environmental Politics* 13(3): 517–40.

—— 2007. *Defining Environmental Justice: Theories, Movements and Nature*. Oxford: Oxford University Press.

SOLNIT, R. 2004. *Hope in the Dark: Untold Histories, Wild Possibilities*. New York: Nation Books.

SPARKE, M. 2007. Geopolitical fears, geoeconomic hopes, and the responsibilities of geography. *Annals of the Association of American Geographers* 97(2): 338–49.

STARR, A. 2000. *Naming the Enemy: Anti-corporate Movements against Globalization*. London: Zed Books.

STERN, N. 2006. *Stern Review on the Economics of Climate Change*. London: New Economic Foundation.

TARROW, S. 2005. *The New Transnational Activism*. Cambridge: Cambridge University Press.

WALKER, G. 2009. Beyond distribution and proximity: Exploring the multiple spatialties of environmental justice. *Antipode* 41(4): 614–36.

WHITE, A. 2009. *The Movement of Movements: From Resistance to Climate Justice*. <http://www.stwr.org/the-un-people-politics/copenhagen-the-global-justice-movement-comes-of-age.html>. Accessed 4/12/09.

WOLCH, J. 2007. Green urban worlds. *Annals of the Association of American Geographers* 97 (2): 373–84.

WOLF, F. O., and MUELLER, T. 2009. Green new deal: Dead end or pathway beyond capitalism? *Turbulence* 5: 12–16.

YIMPRASERT, J. 2009. *A Critical Report on Solidarity Building and Networking with Europe*. Bangkok: Thai Labour Campaign.

YOUNG, I. M. 1990. *Justice and the Politics of Difference*. Princeton: Princeton University Press.

CHAPTER 27

....

CLIMATE DENIAL: EMOTION, PSYCHOLOGY, CULTURE, AND POLITICAL ECONOMY

....

KARI MARIE NORGAARD

GLOBAL climate change is not only the single most significant environmental issue of our time, widespread and potentially catastrophic social impacts are predicted from sea-level rise and changing patterns of precipitation and disease. Climate change will likely jeopardize state economic resources, exacerbate social inequality, alter community structures, and generate new patterns of economic and social conflict. Yet we see remarkably little public reaction for a phenomenon of this magnitude. By 'reaction' we can think of the widest possible range of responses from planning by federal and state officials, to social movement activity from concerned citizens, individual behavioral changes, even the extent to which individuals ac-knowledge climate change by talking about it with others.[1] Instead climate change is like a proverbial 'elephant in the room.' Climate scientists may have identified global warming as the most important issue of our time, but it has taken over twenty years for the problem to penetrate the public discourse in even the most superficial manner. Although public concern is beginning to arise, climate change remains low on the public list of priorities worldwide (Brechin 2008; Pew Research Center for the People and the Press 2009; Poortinga and Pidgeon 2003).

While 'apathy' in the United States is particularly notable, this gap between the severity of the problem and its lack of public salience is visible in most Western nations (Lorenzoni et al. 2007; Poortinga and Pigeon 2003). Especially for urban dwellers in the rich and powerful Northern countries climate change is seen as 'no more than background noise' (Brechin 2008; Lorenzoni and Pidgeon 2006). Indeed, no nation has a base of public citizens that are sufficiently socially and politically engaged to effect the level of change that predictions of climate science would seem to warrant. Instead we are confronted with a series of paradoxes: as scientific evidence for climate change pours in, public urgency and even interest in the issue fails to correspond. In a number of cases, public interest has actually declined at the same time as scientific consensus on the problem has increased. What can explain the misfit between scientific information and public concern? Are people just uninformed of the facts? Are they inherently greedy and self-interested?

Given the seriousness of what is at stake, these gaps between information, concern, and social response have been the subject of much scientific study (see e.g. Lorenzoni et al. 2007; Moser and Dilling in this volume; Moser and Dilling 2007). Survey researchers repeatedly demonstrate the minimal public interest in climate change. Psychologists conduct experiments outlining conceptually flawed mental models and apply theories of cognitive dissonance and emergency helping behavior to climate change. Sociologists note relationships between oil company executives and federal governments, analyze climate skeptic campaigns, and describe how media framing skews public understanding. Each of these efforts points to important answers. However few of their findings support either the theory that people fail to respond because they are uninformed (the 'information deficit model'), or the notion that people have stopped caring about the environment, future generations, or people living in poor nations. Yet if our collective passivity comes from neither ignorance nor greed, it would seem even more irrational.

This chapter outlines the phenomenon of climate denial, that is, the active resistance to information on a collective level. I begin with a review of existing explanations for the public failure to respond to climate change from psychology and sociology. I then use ethnographic data to introduce the framework of socially organized denial. This view from the ground up builds upon many of the above explanations, and highlights the intersecting role of emotions, culture, social structure, and inequality in people's lived experience. This research concerns not the outright rejection of climate science by so-called climate skeptics, but the more pervasive and everyday problem of how and why people who purport to be concerned about climate change, manage to ignore it. The term 'denial' is sometimes used to describe the phenomenon of outright rejection of information as true, in this case, the reaction of climate skeptics mentioned above. But this is a very different, more literal use of the term 'denial' than I will describe. Instead, people actually work to *avoid acknowledging disturbing information* in order to avoid emotions of fear, guilt, and helplessness, follow cultural norms, and maintain positive conceptions of individual and national identity. As a result of this kind of denial, people describe a sense of 'knowing and not knowing' about climate change, of having information but not thinking about it in their everyday lives. Information from climate science is known in the abstract, but disconnected from, and invisible within political, social, or private life. As troubling as the success of climate skeptic campaigns may be to our sense of how a rational democracy works, the paradox of apathy in the face of knowledge and concern about climate change poses an even larger barrier to our collective response.

1 Current Psychological and Sociological Explanations

For nearly twenty years the majority of research on climate change from both disciplines presumed information was the limiting factor in public non-response. The thinking was that, 'if people only knew the facts,' they would act differently. These studies emphasized either the complexity of climate science or political economic corruption as reasons people do not adequately understand what is at stake. Given the complexity of climate change, it is not surprising that these researchers found evidence of conceptual misunderstanding.

Systematic reviews of surveys and polling data by Nisbet and Myers (2007) and Brewer (2005) describe widespread misunderstanding regarding climate science extending back into the 1980s. Researchers have lamented the confusion between global warming and the ozone hole (e.g. Bell 1994; Bostrom et al. 1994; Read et al. 1994), investigated the role of media framing (Bell 1994; Ungar 1992; Brossard et al. 2004; Dispensa and Brulle 2003; Weiskel 2005; Carvalho 2007), and described how understanding global warming requires a complex grasp of scientific knowledge in many fields (Moser and Dilling 2007). Recent work by Sterman and Sweeney (2007) examines public misperceptions of climate models as a cause for inaction. Similarly, working from the assumption that information limits present engagement, psychologists Grame Halford and Peter Sheehan write, 'With better mental models and more appropriate analogies for global change issues, it is likely that more people, including more opinion leaders, will make the decision to implement some positive coping action of a precautionary nature' (1991: 606).

Yet as Read (et al. 1994) pointed out more than a decade ago, only two simple facts are essential to understanding climate change: global warming is the result of an increase in the concentration of carbon dioxide in the earth's atmosphere, and the single most important source of carbon dioxide is the combustion of fossil fuels, most notably coal and oil. So how can it be that people around the world fail to understand these basic facts? And while such 'information deficit' explanations are indispensable, they do not account for the behavior of the significant number of people who know about global warming and express concern, yet still fail to take any action.

A second body of scholarship points to relationships between political economy and public perception. Here scholars have identified the fossil fuel industry influence on government policy (the US holds prominent examples), the tactics of climate skeptic campaigns (Jacques 2009; Dunlap and McCright in this volume; Jacques et al. 2008; McCright and Dunlap 2000, 2003), how corporate control of media limits and molds available information about global warming (Dispensa and Brulle 2003), and even the 'normal' distortion of climate science through the 'balance as bias phenomenon' in journalism (Boykoff 2008). Presumably such political economic barriers have far-reaching and interactive effects with the other factors discussed above. Yet note that explanations for public non-response that highlight corporate media and climate skeptic campaigns, also implicitly direct our attention to a lack of information as the biggest barrier to engagement, though for different reasons. Certainly there are cases when the public may either lack information or be outright misinformed, but are these issues the limiting factor behind greater public interest, concern, or political participation? Clearly knowledge is necessary to generate public response (e.g. O'Connor et al. 2002), but is knowledge sufficient (Bord et al. 2000)?

A third body of scholarship applies psychological theories on cognitive dissonance, efficacy, and helping behavior to climate change (see e.g. Stoll-Kleeman et al. 2001; Lorenzoni et al. 2007; Kollmuss and Agyeman 2002). Festinger's (1957) concept of cognitive dissonance describes 'dissonance' as a condition which emerges when an actor has two thoughts (cognitions) that are inconsistent. This dissonance is an unpleasant condition which people seek to resolve, often through changing one of their cognitions. Studies drawing upon these frameworks point to multiple factors that would seem to 'complicate' how people process information on climate change. For example, Paul Kellstedt and colleagues (2008) have found that increased levels of information about global warming have a negative effect on concern and sense of personal responsibility. In particular, respondents who are better informed about climate change feel less rather than more responsible for it. Furthermore, they find that 'in sharp contrast with the knowledge-deficit hypothesis, respondents with higher levels of information about global warming show less concern' (120). Note that these findings are in accordance with cognitive dissonance because people with low self-efficacy will be likely to deny responsibility and concern since unless they feel able to do something about the problem, an awareness of concern or responsibility would be conflicting cognitions. Similarly, Krosnic et al. (2006) observe that people stopped paying attention to global climate change when they realized that there is no easy solution for it. Instead they note that many people judge as serious only those problems for which they think action can be taken. In a third highly relevant application, Cynthia Frantz and Stephan Mayer (2009) apply a classic model of helping behavior to the public response to climate change. Based on the criteria of this model, the authors note that climate change is difficult to notice, is marked by a diffusion of responsibility, and there are psychological costs of acting, each of which inhibit the likelihood of individual response.

While emphasizing many important factors, the above exclude either the emotional and psychological complexity of our response to climate change, or the significance of political economy in shaping that response. Yet interesting results emerge when these two are integrated. Norwegian sociologist Hanno Sandvik (2008) reports a negative association between concern for climate change and national wealth, and a 'marginally significant' tendency that nations' per capita carbon dioxide emissions are negatively correlated to public concern. Sandvik writes, 'these findings suggest that the willingness of a nation to contribute to reductions in greenhouse gas emissions decreases with its share of these emissions' (333). Although Sandvik is the first to explicitly test a relationship between wealth and concern, his findings are in accordance with earlier work. For example, Zahran et al. (2006) found that citizens residing in US states with higher emissions of climate gases are somewhat less likely to support climate change policies. O'Connor et al. (2002) found that higher income negatively affected participants' willingness to take actions such as driving less. Similarly, an inverse relationship between wealth and concern is also reported in Dunlap's 1998 cross-national research, but with a smaller sample of nations. Further-more, there are no examples of the reverse relationship, in which higher income is positively correlated with concern or support for climate protection policy. Note that these studies contradict Inglehart's (1990) theory of post-materialism in which moderniza-tion and wealth promote greater environmental concern amongst citizens.

2 INTRODUCING THE CLIMATE ELEPHANT

Surveys are excellent for documenting large-scale patterns in human response, but what is going on behind the numbers? Qualitative and ethnographic research allow us to look into the details of people's lived experience. We move now to examine ethnographic data on how people make sense of climate change in a nation with some of the highest levels of education, political activity, and environmental concern in the world (Dryzek et al. 2003). What follows is a view into how people experience the reality of global climate change in their everyday lives. It is not an exhaustive attempt. I use the voices of people in one community in Norway during a recent very dry and warm winter in order to make visible the narratives and cultural constructions that can inform a larger story behind worldwide public paralysis in the face of predictions from climate scientists.

As it happened, in this rural community there was unusually warm weather during my stay. November brought severe flooding across the entire region. The first snowfall did not come until late January—some two months later than usual. As of January 2001, the winter of 2000 for Norway was recorded as the second warmest in the past 130 years. This fact was highly publicized. Regional and national newspapers carried headlines like, 'Warmer, Wetter and Wilder,' 'Green Winters—Here to Stay?' and, 'Year 2000 Is One of the Warmest in History.' As a result of these conditions, the local ski area only opened in late December with the aid of 100 percent artificial snow—a completely unprecedented event with dramatic recreational effects and measurable economic impacts on the community. The local lake failed to freeze sufficiently to allow for ice-fishing. Casual comments about the weather, a long-accepted form of small talk, commonly included references to unusual weather, shaking of heads, and the phrase 'climate change.'

It was not just the weather that was unusual that winter. As a sociologist, I was perplexed by the behavior of the people as well. Despite clear social and economic impacts on the community, no social action was taking place. People could have reacted differently to that strange winter. The shortened ski season affected everyone in the community. In the words of one taxi driver: 'It makes a difference if we move from five months of winter tourism to only three. It affects all of us, you know, not just those up on the mountain. It affects the hotels, the shops in town, us taxi drivers, we notice it too.' Why didn't this awareness translate into social action? Community members could have written letters to the local paper, brought the issue up in one of the many public forums that took place that winter, made attempts to plan for the local effects of climate change, put pressure on local and national leaders to develop long-term climate plans or short-term economic relief, de-creased their automobile use, or at the least, engaged their neighbors, children, and political leaders in discussions about what climate change might mean for their community in the next ten and twenty years. The residents of this town could have rallied around the problem of the lack of snow and its economic and cultural impacts. But they did not. Whether or not the warm weather and lack of snow in town were actually a result of global warming cannot be determined for certain. But, among competing explanations for it, the unusual weather *was* widely linked to global warming in both the media and in the minds of citizens. What perplexed me was that despite the fact that people were clearly aware of global warming as a phenomenon, everyday life went on as though it did not exist. Mothers listened to news of unusual flooding as they drove their children to school. Families watched evening news coverage of the failing climate talks in The Hague, followed by American sit-coms. Few people even seemed to spend much time thinking about global warming.

2.1 'We Don't Really Want To Know'

That winter and spring I spent a lot of time attending public meetings, reading the newspapers, talking with people on the street, and generally watching and listening to what was going on. I conducted forty-six interviews with a range of community members. Global warming was frequently mentioned and people in the community seemed to be both informed and concerned about it. Yet at the same time I noticed that it was an uncomfortable issue. People were aware that climate change could radically alter life within the next decades, yet they did not go about their days wondering what life would be like for their children, whether farming practices would change or whether their grandchildren would be able to ski on real snow. They spent their days thinking about more local, manageable topics. Vigdis, a college age student told me that she was afraid of global warming, but that it didn't enter her everyday life:

> I often get afraid, like—it goes very much up and down, then, with how much I think about it. But if I sit myself down and think about it, it could actually happen, I thought about how if this here continues we could come to have no difference between winter and spring and summer, like—and lots of stuff about the ice that is melting and that there will be flooding, like, and that is depressing, the way I see it.

In the words of one person who held his hands in front of his eyes as he spoke, 'people want to protect themselves a bit.' Other community members in Norway described this sense of knowing and not knowing, of having information but not thinking about it in their everyday lives. As one young woman told me, 'In the every day I don't think so much about it, but I know that environmental protection is very important.' As a topic that was troubling, it was an issue that many people preferred to avoid. Thus community members describe climate change as an issue that they have to 'sit themselves down and think about,' 'don't think about in the everyday,' 'but which in between is discouraging and an emotional weight.' Since members of the community did know about global warming but did not integrate this knowledge into everyday life, they experienced what Robert Lifton (1982) calls the *absurdity of the double life*, a phrase I adapt in coining the term *double reality*. In one reality was the collectively constructed sense of normal everyday life. In the other reality existed the troubling knowledge of increasing automobile use, polar ice caps melting, and the predictions for future weather scenarios. In the words of Kjersti, a teacher at the local agricultural school in her early 30s: 'We live in one way and we think in another. We learn to think in parallel. It's a skill, an art of living.'

What was happening in that community, and indeed what we can all observe in the public silence on climate change in the United States and elsewhere, was not a rejection of information per se, but the failure to integrate this knowledge into everyday life or transform it into social action. British sociologist Stanley Cohen calls this implicatory denial: 'the facts of children starving to death in Somalia, mass rape of women in Bosnia, a massacre in East Timor, homeless people in our streets are recognized, but are not seen as psychologically disturbing or as carrying a moral imperative to act . . . Unlike literal or interpretive denial, knowledge itself is not at issue, but doing the "right" thing with the knowledge' (Cohen 2001: 9).

3 THE SOCIAL ORGANIZATION OF DENIAL: WEAVING EMOTION, CULTURE, AND POLITICAL ECONOMY

The denial metaphor of the elephant in the room is useful because it reminds us that ignoring a serious problem is not easy to do. Ignoring the obvious can be a lot of work. In her work on apathy in the United States, sociologist Nina Eliasoph observes, 'We often assume that political activism requires an explanation, while inactivity is the normal state of affairs. But it can be as difficult to ignore a problem as to try to solve it, to curtail feelings of empathy as to extend them . . . If there is no exit from the political world then political silence must be as active and colorful as a bright summer shadow' (Eliasoph 1998: 6). How did people manage to outwardly ignore what was happening in the community? Did they manage to ignore it inwardly as well?

How we respond to disturbing information is a complex process. Individuals may block out certain information in order to maintain coherent meaning systems (e.g. cognitive dissonance see e.g. Festinger 1957; Gecas and Burke 1995), desirable emotional states (Rosenberg 1991), a sense of self-efficacy (Gecas and Burke 1995), and in order to follow norms of attention, emotion (Hochschild 1983), and conversation (Eliasoph 1998). Society organizes patterns of perception, memory, and organizational aspects of thinking (Zerubavel 1997). These cultural norms are in turn attuned to specific political economic relations. Thus, alongside the serious threat to democracy posed by capital's control of the production and dissemination of knowledge—e.g. the fact that increased corporate control of media limits and molds available information about global warming (Carvahlo 2007; Dispensa and Brulle 2003) and corporate funded research centers generate conflicting knowledge (McCright and Dunlap 2000, 2003; Jacques et al. 2008; Jacques 2009) is another phenomenon that reinforces public non-response: how people cope with information which *does* become available. Overt and more readily identifiable processes such as manipulation and control of information set the stage for the less visible (and to date less studied) process of socially organized denial which I describe here.

The concept of denial is generally considered the domain of psychology. But the information individuals find disturbing, and the mechanisms they employ to protect themselves from such information, may also be analyzed within the context of both social interaction and the broader political economy. Social context itself can be a significant part of what makes it difficult to respond to climate change. Sociologists remind us that notions of what is normal to think and talk about are not given, but are socially structured. It is by paying simultaneous attention to individual responses and social context that we can begin to analyze people's reactions to global warming in reference to the larger political economy. Drawing next from my ethnographic data from Norway, I will describe how people use a variety of methods for normalizing or minimizing disturbing information, what can be called 'strategies of denial.' I placed the strategies I observed into two broad categories: *interpretative* and *cultural*.

4 INTERPRETATIVE DENIAL: COMBATING GLOBAL WARMING BY INCREASING CARBON DIOXIDE

On the one hand, residents structured their relationship to information on global warming through narrative interpretation. Community members used a variety of social narratives, some produced by the national government, to deflect responsibility for and legitimate Norwegian climate and petroleum policy. I observed three types of narratives: selective interpretation, perspectival selectivity, and claims to virtue.

In the case of selective interpretation, to the extent that they are able, 'people tend to assign those meanings to events that will produce the desired emotions' (Rosenberg 1991: 135). In this case, community members had a set of 'stock stories' about who they were. By portraying Norwegians as close to nature, egalitarian, simple and humble, these narratives of national identity served to counter the criticism and doubt Norwegians face with regards to climate and petroleum policies. Notions of Mythic Norway were portrayed in official government images, and drawn upon by advertisers and everyday people in the town.

People also normalized information about global warming using what Morris Rosenberg calls 'perspectival selectivity' (Rosenberg 1991). Perspectival selectivity 'refers to the angle of vision that one brings to bear on certain events' (ibid. 134). For example, people may manage unpleasant emotions by searching for and repeatedly telling stories of others who are worse off than they are. Three narratives in this category—'Amerika as a Tension Point,' 'We Have Suffered,' and 'Norway is a Little Land'—served to minimize Norwegian responsibility for the problem of global warming by pointing to the larger impact of the United States on carbon dioxide emissions, stressing that Norway has been a relatively poor nation until quite recently, and emphasizing the nation's small population size. For example, multiple newspaper articles in the national papers in the winter and spring of 2001 listed the figure that the United States emits 25 percent of total greenhouse gas emissions, while accounting for only 4 percent of the global population, visibly in their articles. While obviously the US must be held accountable for our emissions, framing the figure in terms of total emissions and population makes the difference between the US and 'little Norway' appear greatest. When looking at per capita emissions in each country the contrasts are not so large. Perspectival selectivity was used to create what social psychologists Susan Opotow and Leah Weiss call 'denial of self-involvement' (2000). Examples of these narratives are discussed in more detail elsewhere (Norgaard 2006a, 2006b, forthcoming 2011).

A third interpretative strategy is in the vein of what historical psychologist Robert J. Lifton calls 'claim to virtue.' He coined the phrase to describe how the Nazi doctors in concentration camps who gave Jews lethal injections interpreted their genocidal actions in terms of compassion. From the doctor's perspective, their acts were compassionate because, by killing people who were ill (or who might become ill) they were able to prevent the spread of disease in the camps. Through the claim that unjust acts are actually working towards the opposite end as they appear (in the case of the doctors, saving the Jews rather than killing them), these actions are made acceptable. Two such claims to virtue were in use that winter with respect to climate change. Although the Norwegian government speaks urgently of the need to reduce emissions of climate gases, they were at the time involved in two projects that do exactly the opposite: the building of two new natural gas facilities and expansion of the petroleum sector by increasing oil development. Both actions have been justified by switching the focus from national targets and measures (as specified under the Kyoto Protocol), to emphasizing climate change as an *international* problem and attempting to meet Norwegian climate commitments through the *trading* of climate gas emissions rather than reduction of actual output.

4.1 'Gas Plants Are Better Than Coal'

Beginning in the early 1990s, the Norwegian government in combination with oil and gas companies began presenting a series of justifications for the development of new natural gas facilities: as natural gas produced less carbon dioxide than coal, Norway could sell this excess energy to other nations and actually be helping overall global emissions. Thus, although the government acknowledges that Norway's emissions of climate gases must decrease, it has used a claim to virtue to argue that the building two new natural gas plants—thereby *increasing* Norway's contribution to climate gases—was actually helping to solve the problem of global warming. However, as Norwegian researches Hovden and Lindseth (2002: 158) point out: 'While it is claimed that these would be off-set by reductions elsewhere, this does not change the fact that emissions from Norwegian gas-based power would increase the CO_2 emission reductions that Norway would have to complete in order to fulfil its international obligations'.

4.2 'Increasing Production of Norwegian Oil Will Help the Climate'

A second example, the justification for increasing national oil production, follows a similar pattern. Norway had increased production of oil and gas threefold in the preceding ten years, dropped its plan of a national carbon dioxide emissions stabilization target, and shifted from a focus on national strategies (mandated under the Kyoto Protocol) to a focus on international efforts. Within the new international perspective, the government has argued that 'since Norwegian petroleum products are not the dirtiest in the international market, Norwegian oil and gas production is good climate policy internationally' (Hovden and Lindseth 2002: 153). Hovden and Lindseth describe how

> Miljkosok, an environmental cooperative forum consisting of the petroleum industry, the government and various interest groups and organizations produced a report in 1996 that in effect, concluded that Norwegian oil production was environmentally benign. The arguments were a) that a cut in Norwegian production would increase the price of oil on the world market, which would make coal more competitive, and, most importantly, b) that as Norwegian petroleum production has fewer emissions per unit oil produced, it was environmentally preferable to the oil produced by other countries. The unavoidable conclusion was that Norway should increase its Continental Shelf activity, as this would, in sum, be beneficial with respect to the global emissions of CO_2 and No_x. (Ibid. 152)

Thus, by shifting attention from the national level (on which Norway is retreating from the Kyoto Protocol and other earlier reduction goals) to the international (in which Norway produces 'cleaner' oil than other nations), the Norwegian government claims that increasing oil production is the best thing it can do for the global climate, even though these activities increase carbon dioxide emissions and are in direct opposition to their agreement under the Kyoto Protocol!

The interpretative strategies of selective interpretation, perspectival selectivity, and claims to virtue worked together to reinforce one another. For example, selective interpretation and perspectival selectivity gave a background picture of Norwegian environmentalism and innocence, whereas claims to virtue were linked to particular, contested climate and petroleum activities such as the expansion of oil and gas production or plans of carbon trading.

5 CULTURAL DENIAL

In addition to the more identifiable strategy of interpretation, people collectively held information about global warming at arm's length by following established cultural norms about what to pay attention to, feel, talk, and think about in different contexts. I categorize these as 'cultural denial.' From the perspective of sociology of cognition, people learn to think through socialization into different 'thought communities' (Zerubavel 1997). At the same time as they feel 'just like everyday life,' these culturally prescribed norms of attention reflect a particularly insidious form of social control akin to Steven Lukes's third dimension of power. While outright coercion is a serious matter, it is also more easily recognized, identified and, in (so-called) democratic societies, condemned. As Cohen notes 'Without being told what to think about (or what not to think about), and without being punished for "knowing" the wrong things, societies arrive at unwritten agreements about what can be publically remembered and acknowledged' (Cohen 2001: 10–11).

Thus information about climate change disappeared into daily life for reasons that were more culturally diffuse. For example, simply upholding norms of attention with respect to space made the lack of snow and warm temperatures seem less significant (depoliticized in part because connections to unusual weather events elsewhere were not made), while following norms of attention with respect to time encouraged community members to not think too far ahead into the future, hence minimizing the extent to which the implications of immediate events are forecast. Cultural norms of emotion limited the extent to which community members could bring strong feelings they privately held regarding climate change into the public political process, which in turn served to reinforce the sense that everything was fine. Mechanisms of cultural denial are however more complex. Elsewhere I describe other cultural aspects of denial such as how community members used an available repertoire of conversational tactics, emotion management strategies, and techniques of shifting attention in order to follow local norms (Norgaard 2006a, 2006b, 2009, forthcoming 2011).

6 CONCLUSION

Were Norwegians in this one community the only people to normalize their behavior through selective interpretation and claims to virtue? The rather puzzling behavior of people that winter in this community is related to larger questions about social and environmental action in Norway, the United States, and around the world: How are the citizens of wealthy industrialized nations responding to global warming? Why are so few people taking any sort of action? Why do some social and environmental problems result in people rising up when others do not? And, given that many do know the grim facts, how do people manage to produce an everyday reality in which this urgent social and ecological problem is invisible? These are critical questions facing citizens of all the wealthier nations of the world today. Climate change is not unique to Norway, nor are its present and future impacts. Nor, unfortunately, is the failure of response unique to this small community.

Despite the extreme seriousness of this global environmental problem, the pattern of meager public response—in terms of social movement activity, behavioral changes, or public pressure on governments—exists worldwide.

While I know of no other ethnographic or sociological studies of climate denial, a handful of research speaks to the salience of these circumstances for people around the world. For example, Stoll-Kleemann and co-authors found that participants in Swiss focus groups experienced fears about future climate scenarios, dissonance with respect to their contribution to the problem, and developed justifications for their own inaction: 'To overcome the dissonance created in their minds' they 'created a number of socio-psychological denial mechanisms. Such mechanisms heightened the costs of shifting away from comfortable lifestyles, set blame on the inaction of others, including governments, and emphasized doubts regarding the immediacy of personal action when the effects of climate change seemed uncertain and far away' (2001: 107). Both unique and universal mechanisms of denial from interpretative narratives to cultural practices surely exist, especially in wealthy nations around the world. Indeed the phenomenon of denial, our collective resistance to disturbing information, poses a new challenge for our modern society that is increasingly relevant in our globalized information age, even beyond the issue of climate change.

Until recently denial has been studied almost exclusively as a psychological phenomenon. Yet even the briefest examination of Norwegian political economy illustrates the relevance of linking psychological material on interactions and culture with macro-level political economy in order to make sense of why people don't want to know about global warming. The notion that well-educated, wealthy people in the Northern hemisphere do not respond to climate change because they are poorly informed fails to capture how, in the present global context, 'knowing' or 'not knowing' is itself a political act. All nations emit carbon dioxide and other climate gases into the common atmosphere. While in the perhaps distant future climate change may have drastic consequences for Norwegians, in the immediate sense Norwegian wealth comes directly from the production of oil and their economy flourishes with their current level of carbon dioxide emissions. Given that Norwegian economic prosperity and way of life are intimately tied to the production of oil, denial of the issue of climate change serves to maintain Norwegian global economic interests and perpetuate global environmental injustice. It is easy to see power operating when key political and economic decision makers negotiate contracts with Shell, British Petroleum, and Exxon, or representatives of nation-states negotiate emissions-trading strategies. Yet everyday people play a critical role in legitimizing the status quo by not talking about global warming even in the face of late winter snow and a lake that never froze. The absence of these conversations worked to hold 'normal' reality in place.

Citizens of wealthy nations who fail to respond to the issue of climate change benefit from their denial in economic terms. They also benefit by avoiding the emotional and psychological entanglement and identity conflicts that may arise from knowing that one is doing 'the wrong thing.' Socially organized denial is thus connected to studies of privilege and has important implications for environmental justice. Most environmental justice research has focused on the experience of less powerful groups who have disproportionate exposure to environmental problems. While important, this approach passes over the role of citizens in wealthy nations, who as we turn a blind eye to the impacts of our high carbon lifestyles and lead a comfortable life, perpetuate environmental problems such as global warming (see also Baer in this volume; Gardiner this volume).

The conditions for denial are supported by the dynamics of global capitalism. Ongoing changes in social organization, especially the twin forces of globalization and increasing inequality create a situation in which, for privileged people, environmental and social justice problems are increasingly distant in time or space or both. Social inequality helps to perpetuate environmental degradation making it easier to displace visible outcomes and costs across borders of time and space, out of the way of those citizens with the potential time, energy, cultural capital, and political clout to generate moral outrage and take action in a variety of ways. The issue of climate change will deeply affect nations with less infrastructure long before it will significantly touch the lives of Norwegians or other wealthy people in the Global North. As a result, ecological collapse seems a fanciful issue to people living in 'safe' and 'stable' societies. And with the dynamics of global capitalism in which gaps between rich and poor increase, the problem of denial will become increasingly salient for those who have the economic resources and incentive to build physical, mental, and cultural walls around our daily lives.

To be 'in denial' has a negative connotation—associated with stupidity or ineptitude. Yet a key point in labeling this phenomenon *denial* is to highlight the fact that our non-response is not a reflection of our greed or inhumanity. Indeed, if information on climate change is repelled because it is too disturbing this is the very opposite of an inhumane interpretation. Instead it is my hope that the perspective of denial will draw attention to a new psychological predicament for privileged people. At the same time as it poses individual challenges, our capacity for denial in the face of problems that feel too large to tackle threatens to dangerously erode the critical democratic role of the public sphere at a time when we would seem to need it more than ever.

NOTE

1. This failure of public response is the subject of a large body of literature some of which is addressed in this chapter. In their work on the topic, Lorenzoni et al. 2007 define 'engagement' as consisting of three parts: cognitive (knowledge), affective (concern and motivation), and behavioral (taking an action). See also Moser and Dilling this volume. However, I am interested in response more broadly, including for example the presence or absence of conversations that could generate future democratic outcomes.

REFERENCES

BELL, A. 1994. Climate of opinion: Public and media discourse on the global environment. *Discourse and Society* 5(1): 33–64.

BORD, R., O'CONNOR, R. E., and FISCHER, A. 2000. In what sense does the public need to understand global climate change? *Public Understanding of Science* 9(3): 205–18.

BOSTROM, A., GRANGER MORGAN, M., FISCHOFF, B., and READ, D. 1994. What do people know about global climate change? I Mental models. *Risk Analysis* 14(6): 959–70.

BOYKOFF, M. 2008. Lost in translation? The United States television news coverage of anthropogenic climate change 1995–2004. *Climate Change* 86: 1–11.

BRECHIN, S. R. 2008. Ostriches and change: A response to 'Global Warming and Sociology.' *Current Sociology* 56: 467–74.

BREWER, T. 2005. U.S. public opinion on climate change issues: Implications for consensus building and policymaking. *Climate Policy* 4: 359–76.

BROSSARD, D., SHANAHAN, J., and McCOMAS, K. 2004. Are issue-cycles culturally constructed? A comparison of French and American coverage of global climate change. *Mass Communication and Society* 7(3): 359–77.

CARVALHO, A. 2007. Communicating global responsibility? Discourses on climate change and citizenship. *International Journal of Media and Cultural Politics* 3(2):180–3.

COHEN, S. 2001. *States of Denial: Knowing about Atrocities and Suffering.* New York: Polity Press.

DISPENSA, J. M., and BRULLE, R. J. 2003. Media's social construction of environmental issues: Focus on global warming—a comparative study. *International Journal of Sociology and Social Policy* 23(10): 74–105.

DRYZEK, J., HERNES, H. K., HUNOLD, C., and SCHLOSBERG, D. 2003. *Green States and Social Movements: Environmentalism in the United States, United Kingdom and Norway.* Oxford: Oxford University Press.

DUNLAP, R. E. 1998. Lay perceptions of global risk: Public views of global warming in cross-national context. *International Sociology* 13: 473–98.

ELIASOPH, N. 1998. *Avoiding Politics: How Americans Produce Apathy in Everyday Life.* Cambridge: Cambridge University Press.

FESTINGER, L. 1957. *A Theory of Cognitive Dissonance.* Stanford, CA: Stanford University Press.

FRANTZ, C., and MAYER, S. 2009. The emergency of climate change: Why are we failing to take action? *Analyses of Social Issues and Public Policy* 9(1): 205–22.

GECAS, V., and BURKE, P. 1995. Self and identity. Pp. 41–67 in K. Cook, G. A. Fine, and J. House (eds.), *Sociological Perspective on Social Psychology.* Boston: Allyn and Bacon.

HALFORD, G., and SHEEHAN, P. 1991. Human responses to environmental changes. *International Journal of Psychology* 269(5): 599–611.

HOCHSCHILD, A. 1983. *The Managed Heart: Commercialization of Human Feeling.* Berkeley: UC Press.

HOVDEN, E., and LINDSETH, G. 2002. Norwegian climate policy 1989–2002. Pp. 143–68 in W. Lafferty, M. Nordskog, and H. A. Aakre (eds.), *Realizing Rio in Norway: Evaluative Studies of Sustainable Development.* Oslo: Program for Research and Documentation for a Sustainable Society (Prosus), University of Oslo.

INGLEHART, R. 1990. *Culture Shift in Advanced Industrial Society.* Princeton: Princeton University Press.

JACQUES, P. 2009. *Environmental Skepticism: Ecology, Power and Public Life.* Burlington VT: Ashgate.

—— DUNLAP, R., and FREEMAN, M. 2008. The organisation of denial: Conservative think tanks and environmental skepticism. *Environmental Politics* 17(3): 349–85.

KELLSTEDT, P., ZAHRAN, S., and VEDLITZ, A. 2008. Personal efficacy, the information environment, and attitudes toward global warming and climate change in the United States. *Risk Analysis* 28(1): 113–26.

KOLLMUS, A., and AGYEMAN, J. 2002. Mind the gap: Why do people act environmentally and what are the barriers to pro-environmental behavior? *Environmental Education Research* 8 (3): 239–60.

KROSNIC, J., HOLBROOK, A., LOWE, L., and VISSER, P. 2006. The origins and consequences of democratic citizen's policy agendas: A study of popular concern about global warming. *Climate Change* 77: 7–43.

LIFTON, R. 1982. *Indefensible Weapons*. New York: Basic Books.

LORENZONI, I., and PIDGEON, N. F. 2006. Public views on climate change: European and USA perspectives. *Climatic Change* 77: 73–95.

—— NICHOLSON-COLE, S., and WHITMARSH, L. 2007. Barriers perceived to engaging with climate change among the UK public and their policy implications. *Global Environmental Change* 17(3–4): 445–59.

McCRIGHT, A. M., and DUNLAP, R. E. 2000. Challenging global warming as a social problem: An Analysis of the Conservative movement's counter-claims. *Social Problems* 47: 499–522.

—— —— 2003. Defeating Kyoto: The Conservative movement's impact on U.S. climate change policy. *Social Problems* 50: 348–73.

MOSER, S. C., and DILLING, L. 2007. *Creating a Climate for Change: Communicating Climate Change and Facilitating Social Change*. New York: Cambridge University Press.

NISBET, M., and MYERS, T. 2007. The polls—trends: Twenty years of public opinion about global warming. *Public Opinion Quarterly* 71(3): 444–70.

NORGAARD, K. M. 2006a. 'People want to protect themselves a little bit': Emotions, denial, and social movement nonparticipation. *Sociological Inquiry* 76: 372–96.

—— 2006b. 'We don't really want to know'. The social experience of global warming: Dimensions of denial and environmental justice. *Organization and Environment* 19(3): 347–470.

—— 2009. *Cognitive and Behavioral Challenges in Responding to Climate Change*. World Bank, World Bank Policy Research Working Paper no. WPS 4940, May.

—— Forthcoming 2011. *Living in Denial: Climate Change, Emotions and Everyday Life*. Cambridge, MA: MIT Press.

O'CONNOR, R., BORD, R. J., YARNAL, B., and WIEFEK, N. 2002. Who wants to reduce greenhouse gas emissions? *Social Science Quarterly* 83(1): 1–17.

OPOTOW, S., and WEISS, L. 2000. New ways of thinking about environmentalism: Denial and the process of moral exclusion in environmental conflict. *Journal of Social Issues* 56(3): 475–90.

Pew Research Center for the People and the Press. 2009. *Economy, Jobs Trump All Other Policy Priorities in 2009: Environment, Immigration, Health Care Slip Down the List*. Washington, DC: Pew Research Center for the People and the Press.

POORTINGA, W., and PIDGEON, N. 2003. *Public Perceptions of Risk, Science and Governance: Main Findings of a British Survey of Five Risk Cases*. University of East Anglia and MORI, Norwich.

READ, D., BOSTROM, A., GRANGER MORGAN, M., FISCHOFF, B., and SMUTS, T. 1994. What do people know about global climate change? II. Survey studies of educated lay people. *Risk Analysis* 14 (6): 971–82.

ROBERTS, J. T., and PARKS, B. C. 2007. *A Climate of Injustice: Global Inequality, North-South Politics, and Climate Policy*. Cambridge, MA: MIT Press.

ROSENBERG, M. 1991. Self-processes and emotional experiences. Pp. 123–42 in J. Howard and P. Callero (eds.), *The Self-Society Dynamic: Cognition, Emotion and Action*. Cambridge: Cambridge University Press.

SANDVIK, H. 2008. Public concern over global warming correlates negatively with national wealth. *Climatic Change* 90(3): 333–41.

STERMAN, J., and SWEENEY, L. 2007. Understanding public complacency about climate change: Adults mental models of climate change violate conservation of matter. *Climate Change* 80: 213–38.

STERN, P., DIETZ, T., and GUAGNANO, G. 1995. The new ecological paradigm in social-psychological context. *Environment and Behavior* 27(6): 723–43.

STOLL-KLEEMANN, S., O'RIORDAN, T., and JAEGER, C. 2001. The psychology of denial concerning climate mitigation measures: Evidence from Swiss focus groups. *Global Environmental Change* 11: 107–17.

UNGAR, S.. 1992. The rise and (relative) decline of global warming as a social problem. *The Sociological Quarterly* 33(4): 483–501.

WEISKEL, T. 2005. From sidekick to sideshow: Celebrity, entertainment, and the politics of distraction. Why Americans are 'sleepwalking toward the end of the Earth'. *American Behavioral Scientist* 49(3): 393–409.

ZAHRAN, S., BRODY, S., GROVER, H., and VEDLITZ, A. 2006. Climate change vulnerability and policy support. *Society & Natural Resources* 19: 1–19.

ZERUBAVEL, E. 1997. *Social Mindscapes: An Invitation to Cognitive Sociology*. Cambridge, MA: Harvard University Press.

—— 2006. *The Elephant in the Room: Silence and Denial in Everyday Life*. New York: Oxford University Press.

CHAPTER 28

....

THE ROLE OF RELIGIONS
IN ACTIVISM

....

LAUREL KEARNS

> Nobody was ever moved to change the way they live by a pie chart, but they are
> moved by a story...Nobody actually changes what they do unless they are
> inspired, touched, given hope.

Martin Palmer (Secretary General, Alliance of Religions and Conservation)

Such are the responses of Palmer in a November 2009 interview to the frequently heard questions: What has religion to do with the environment? Why would religious groups go to the 2009 Copenhagen international climate talks? What difference does it make if religious groups are involved? Such questions might be asked even by many in this volume, as very few mention religion. Palmer went on to comment that scientists present facts, and in relation to climate change, often depressing, alarming ones, but it is religious traditions and leaders that can motivate people to change and can offer hope in the face of such facts. This cooperation, or tension, between science and religion is a major aspect of all religious environmentalism, but particularly so with action regarding climate change, the subject of this chapter. Religious leaders and organizations, as illustrated below, can make a difference by mobilizing followers and framing the issue in moral terms; however, they can also stymie action on climate change through their vigorous rejection of the science or opposition to the implied economics. In the case of the United States, which will be the primary focus of this chapter, one cannot examine political action/inaction on climate change without understanding the role of religious groups. Thus my first assertion is that to understand the nature of the problem, as Dale Jamieson suggests in this volume, one must also understand the various roles of religion.

Palmer, and many scholars, asserts that stories motivate people toward social action (see Hajer and Versteeg in this volume).[1] *The Psychology of Climate Change Communication*, a forty-three-page booklet released by the Center for Research on Environmental Decisions (2009), also suggests that stories, which can help make something as enormous and complicated as climate change real, are part of a successful strategy. Social scientists increasingly recognize the importance of narratives in social movement success (Davis

2002). Religious groups can take the ineffable, the hard to grasp, which Jamieson explains is part of the problem of responding to climate change, and relate it to concrete moral action. Religious groups can embed prescribed action in a story of living one's faith, following one's tradition, and being a moral exemplar. Once religious institutions, local congregations, mosques, and temples try something new such as installing solar power, or CFL bulbs, etc., and incorporate the action into the story of their mission and lived practice, it becomes easier for members to follow suit. The group has taken the risk of new behavior, and its success becomes the story to be told to others. But it also matters how the 'story' is framed, as Matthew Nisbet, both in his chapter in this volume and in his widely read 2009 article on 'Communicating climate change' illustrates—'There is no such thing as unframed information.' Certainly, just following the news coverage leading up to the Copenhagen COP-15 talks makes the point evident. Nisbet (2009) further explains that 'frames are interpretive storylines that set a specific train of thought in motion, communicating why an issue might be a problem, who or what might be responsible for it, and what should be done about it.' Thus frames provide a shorthand for journalists, policy makers, religious believers, and others, that count on the reader/listener making associations that can be left unsaid (Snow and Benford 2000). Nisbet concludes what religious groups on the issue have long known: that it is important to frame environmental issues in moral and ethical terms. This is a central contribution of religious groups in responses to climate change, but what those moral and ethical issues are varies greatly.

This chapter will explore the myriad ways that religious groups are taking the story told by the mounting scientific consensus on the reality of climate change (see Moser and Dilling in this volume for more on this), and grounding it in the moral and ethical imperative of their faith traditions. Religious traditions can ask one to think about the effect of one's actions and about what is the right action, and then to imagine something greater and more important than what is immediately apparent. As more and more people of different faiths have responded to climate change as a religious/ethical/justice imperative, climate skeptics and deniers (see Dunlap and McCright in this volume for information on secular groups) are striving to tell a different story about the data, or to point out perceived inconsistencies (and leaked e-mails), or to come up with competing data, so that the story of climate change has a different plot line and conclusion, a different moral to the story, so to speak.

It is not just the ethical framing of environmental issues as part of the narrative of what it means to be religious that is crucial; it is also the sheer number of listeners and actors. As Palmer (2009) points out, 'Faiths reach out to 85 percent of the world's population, own seven to eight percent of the habitable land on the planet and five percent of commercial forests. They also have huge financial investments[2] and they are involved in half of all schools worldwide. So what they do with their assets and their influence matters a lot.' Palmer's opening remarks were about the significance of the actions pledged by 200 representatives of nine faith traditions from all over the globe—Bahai, Buddhism, Christianity, Daoism, Hinduism, Islam, Judaism, Shintoism, and Sikhism—at the Alliance of Religions and Conservation (ARC) and the United National Development Program (UNDP) award ceremony and forum in November 2009, as a lead-up to the Copenhagen climate talks. Leaders of these faiths, as well as nineteen secular environmental organizations in partnership with them, came ready to announce major initiatives to mitigate climate change. The Grand Mufti of Egypt announced a seven-year plan by Muslims to green the Hajj and to print the Koran (over fifteen million copies/year) on recycled paper. Nigel Savage, of the Jewish environmental organization Hazon, announced a seven-year plan that included reducing meat eating by 50 percent and linking synagogues with local farms. Many faith groups announced education plans to foster environmental practices such as recycling, reducing energy usage, using recycled paper, etc. Other announced projects included solar power for 26,000 Daoist temples in China, or the reforestation of sacred lands, such as Buddhist efforts toward reforestation in Mongolia and China, or the planting of 8.5 million trees in Tanzania. Leaders of the Shinto religion in Japan, in partnership with the Church of Sweden, presented the development of a religious forestry standard for such replanting efforts and the management of the millions of acres of forested land owned by various faiths (ARC 2007). The ARC website lists many other projects similar to those proposed above (ARC N.D.b.). Obviously, like Palmer, the participants think that religions can make a big difference.

This insight has been heard for a while from both sides of the aisle—religious and scientific leaders. In 1990, Pope John Paul II stated that the environmental crisis is a moral issue; his remarks were followed by many religious leaders around the globe. The 1991 Joint Appeal in Religion and Science was a response to an 'Open Letter' by thirty-four internationally known scientists calling for scientists and political leaders to work with religious leaders to address the looming environmental crisis. The Orthodox Ecumenical Patriarch Bartholomew I, since his election in 1991, has been an extraordinary leader, repeatedly calling attention to environmental degradations as sins. Despite these clear calls from prominent figures (many more could be cited, such as former US Vice President Al Gore), it is not as simple as just adding religion. When examining the subsequent religious responses, it is how the scientific and economic issues are framed, even more so than the ethical issues or theological/religious differences between groups, that is key to how the message is perceived, and central in recognizing important differences between religious groups, even within the same theological/faith tradition, such as Christian evangelicalism. Thus it is important, when looking at the role of religious environmental groups in stimulating and prohibiting action on climate change, to recognize all the frames being employed. Before analyzing some illustrative discourse on climate change by religious groups, it is helpful to provide a short history, which by definition will leave out many important aspects, figures, and groups of religious environmentalism of the past fifty-plus years (see Kearns 2004 for more detail) but will demonstrate the need to consider the role of religion in the overall discussion of responding to climate change.

1 History of Religious Environmentalism

Perhaps the fifty-year time frame took you a bit by surprise? Religious environmentalism, while making news headlines more frequently, is not as new as it may seem, and scholars in many traditions have been hard at work to uncover ecological 'roots' and 'saints' in their traditions or to add an ecological turn, as Lynn White (1967) suggested in his infamous essay on the 'Historical Roots of our Ecological Crisis.' White is widely credited with blaming Christianity, charging it carried 'a huge burden of guilt' for the ecologic crisis, but his argument is more complex and subtle than it is often portrayed. But long before

White, Walter Lowdermilk, while in Jerusalem in 1939, penned an 11th Commandment concerning stewardship and conservation for future generations (Nash 1989: 97–8) and, Lutheran eco-theologian Joseph Sittler first began writing in the 1950s on topics related to his 'theology of the Earth.' Other key eco-theologians, such as Jurgen Moltmann, John Cobb, and Rosemary Radford Ruether, also began writing on environmental concern in the 1970s, and in the US mainline Protestant denominations issued statements in the 1970s, and in the 1980s, established offices with staff working on education and lobbying on environmental issues. Internationally, the World Council of Churches has played a key role since the 1970s. In 1986, the World Wildlife Fund International called together leaders of world religions in Assisi, Italy who issued statements, leading to the formation of the Alliance of Religion and Conservation, which now consists of eleven faiths (ARC N.D.a). In the 1990s, the focus of these groups and a wide array of others, such as the National Religion Partnership for the Environment, increasingly turned toward climate change and the related issues of justice (1988 for the WCC, see WCC N.D. for full history). Scholars have also been very active— perhaps the most monumental undertaking was the Forum on Religion and Ecology's series of conferences regarding a range of religions held from 1996 to 1998. More than 800 scholars and environmental activists participated and ten edited volumes on ecology and the different religious traditions were published (Forum 2004). The theological, ethical, and how-to literature for religious traditions has dramatically increased since then.

Religious groups have also been very active at all the United Nations (UN) conferences on the environment or climate change and participated in all of the sessions of the UN Commission on Sustainable Development and the UN Framework Convention on Climate Change (UNFCCC). The UN does its own religious outreach, issuing a 'Spiritual Declaration on Climate Change' at the 2005 United Nations Climate Change Conference (referred to as UNFCCC-COP-11 meeting) in Montreal (Hallman 2005). The Secretary of the UN, Ban Ki-moon, presented the awards at ARC's Windsor Castle ceremony preceding the Copenhagen talks.

As a way to make sense of all the activity briefly described above, in earlier work, I identified three emerging 'ideal types' in the 1990s that captured the spectrum of Christian environmentalism in the US—Christian stewardship, eco-justice, and creation spirituality (Kearns 1996)—as well as the ideological perspectives that they embodied regarding scripture, science, and economy. Although these types still are useful to identify dominant themes, ranging from the more bible-oriented stance of stewardship to the more liberal or post-Christian creation spirituality, there is now far less clear separation as the spectrum has broadened, and coalitions, individuals, and groups that embody or reflect aspects of all three are easily found. For example, with increased awareness of climate change, concerns for justice are now a central aspect of any of these religious positions, and thus the designation of only one strand as eco-justice can be misleading. And divisions have emerged within those ideal types—the evangelical world that I previously labeled simply Christian Stewardship is now more complex, and in the following section I will analyze how understanding that complexity demonstrates the tensions and hurdles that US faith groups who are active regarding climate change must face. Further, these positions could be used to describe the spectrum in other faiths, such as Judaism and Islam, and the interfaith efforts in which they participate (religious environmentalism on the whole is one of the most interfaith movements, as the concern for a common earth often transcends religious differences).

2 US Religious Activism on Climate Change

As this brief history shows, religious groups in the US and abroad have been working on climate change almost since it became a broad scientific concern. My research has focused primarily on Christianity in the US because it is a major actor both in its refusal to sign the Kyoto Treaty and all subsequent negotiations and in its imbalanced contribution of roughly 25 percent of global warming emissions, and because Christian religious groups have an enormous influence on public policy issues in the US. The US 'voice' at Copenhagen and the new Obama administration's domestic actions have signaled some change in position from the climate change denial of the George W. Bush administration (climate skepticism is now espoused by most Republican political candidates), but the US is still not the leader it should be, nor, for all the efforts of the groups discussed here, is there much progress on legislation at the national level. US groups are aware that the effects of global climate change have global implications not only on the climate, but for agriculture, health, species preservation, water availability, poverty, and the global economic order. However, much of their effort has been aimed at motivating domestic action, and all too often, to raise awareness that global warming/climate change is (a) happening, (b) human-caused, (c) not desirable or beneficial, and (d) an issue that demands immediate legislative regulation, and not just voluntary actions. This is because there is well-funded, well-organized religious opposition to any governmental action, and indeed, to acceptance of the science of climate change. The hope is that if people won't listen to scientists, perhaps they will listen to religious figures of authority.

Other than general efforts to link religious faith with environmental action, several types of campaigns illustrate the type of organized religious activity aimed at combating global warming. First there has been almost twenty years of campaigns to put direct pressure on the government to sign or join other international efforts to cut the rate of CO_2 emissions, such as Copenhagen, through protests, lobbying, letter/petition campaigns, and election activities. Other campaigns (see Kearns 2007 for more detail than the discussion below) aim at federal, state, and local governmental action such as: increasing fuel and energy efficiency, promoting alternative energy sources, regulating CO_2 emissions by industry, working for a carbon tax or a cap-and-trade program, curtailing deforestation, and halting extractive industries in ecologically sensitive areas, such as oil drilling in Alaska, mountaintop coal removal in West Virginia and Kentucky, or oil shale and tar sands in Canada (by far the largest source of imported oil for the US) (US Energy 2009).

Given the difficulty and relative lack of success of the above campaigns aimed at governmental action, many groups have focused on voluntary action at the institutional and individual level, such as reducing energy consumption/increasing efficiency, green building and renovation practices, purchasing non-polluting energy or putting solar panels on houses of worship, promoting alternative transportation, or emphasizing local, organic foods and vegetarianism. The work of one coalition, the National Partnership for Religion and the Environment (NRPE, formed in 1992) demonstrates the range of activity, but examples of more local interfaith efforts can also be found, for instance in the work of GreenFaith, founded the same year.

An early effort in the mid-nineties to put pressure on the United States to sign the 1997 Kyoto Protocol was led by the National Council of Churches Eco-Justice Working Group (the NCC-EJWG, made up of mainline, but not evangelical, Protestants, Christian Ortho-dox, and Historic Black denominations) along with other members of the NRPE, the Evangelical Environmental Network (EEN), the US Catholic Conference of Bishops (USCCB), and the Coalition on Environment and Jewish Life (COEJL). Signatures on petitions were collected and delivered to Vice President Gore, and a video that argued for why it was a religious issue for all Christian and Jews to address was widely distributed. Each group compiled and distributed educational and worship materials for each tradition with a goal of reaching 100,000 congregations. The NRPE, through grant money, also helped fund state-level interfaith climate change campaigns to work toward transformation at the state and institutional level in over twenty states. These eventually became Interfaith Power and Light organizations (recognized at the ARC awards ceremony) and part of the Regener-ation Project, which in 2006 helped coordinate over 4,000 congregational viewings of former Vice President Gore's film, *An Inconvenient Truth*, or another film, *The Great Warming*.

In 2003, the NRPE constituent groups organized a visit to US automakers in Detroit by Jewish, Catholic, and Protestant leaders in a fleet of hybrid cars driven by Catholic nuns. The related action-alert e-mail on Faith and Fuel Economy asked: 'Have you heard the one about the rabbi, the priest, the pastor and the Toyota Prius? No, it's not a joke. And neither is global warming' (Global Ministries N.D.). The EEN's related 2003 'What Would Jesus Drive?' (WWJD) campaign, grounded in a New Testament 'love your neighbor as yourself' personal-ethics approach, demanded that one see the connection between an individual's auto emissions and air-quality issues as part of a larger evangelical focus on health (Goodstein 2006, 2007). It drew a great deal of national and international publicity (Berke-man 2002) and in many ways prepared the way for the 2006 US evangelical 'Call to Action' on climate change discussed below. This campaign is a good example of how to make something as global as climate change personal and real.

In the lead-up to Copenhagen, religious activity in the US and globally has obviously increased with too many examples to list. As recent polls in the US indicate, however, these efforts have not completely succeeded, with numbers of those who do not consider climate change to be anthropogenic increasing at the end of the decade after dipping in the middle (Pew 2009). With religious environmentalism growing, and with hundreds of religious NGOs active at Copenhagen (the NCC/US and the WCC globally encouraged candlelight vigils and the ringing of church bells on the middle weekend of the climate talks to signify support for action on climate), efforts to undermine and counter this work have also increased. Environ-mentalism has become a dividing issue among religious groups, indicating its importance. In order to understand the current roles of religion, I will now turn to the world of evangelicalism.

3 Green Evangelicals

Despite a long history of evangelical faith-based environmentalism (for instance, Schaeffer 1970, see Kearns 1997 for more), 'green' evangelicals still make news. This is in part because there is a widespread sentiment in the environmental community that religion, particularly Christianity, is part of the problem (in part due to variations/interpretations of the argument made by Lynn White in 1967) and in part because one doesn't have to look far to find negative Christian examples. Prominent anti-environmentalists, often associated with the 'wise use agenda' that was so successful in undermining environmental efforts under the Reagan, Bush, and Bush presidencies, are also often associated with conservative Christianity.[3] So environmentalists didn't expect such a thing as green Christians, and the media didn't expect evangelicals to espouse what was labeled a 'liberal' cause in the bifurcated world of US politics. Further, many key religious right figures such as Jerry Falwell, James Dobson, and Pat Robertson (until his 'eco-conversion' on climate change in 2006), who influence Christianity globally, have been vocal opponents of environmentalism and critical of fellow evangelicals who do embrace environmental concern. Such high-profile activity often prevented environmentalists from noticing positive Christian activity, as well as many Christians from becoming environmentalists. There are other factors that are also part of the conservative Christian hesitancy to respond to environmental concerns: the argument that the central focus of Christianity should be on salvation and saving souls, and not saving the creation; a related focus on the individual, the charge that religious environmentalists worship the creation and not the Creator, or are New Age; a hostility or distrust of any science because of creationism; an apocalyptic focus on eschatology, or the End Times, which seems to predict environmental degradation; a perception that environmentalism is hostile to capitalism, and the related accusation that environmentalists are socialists/communists bent on undermining the current global economic system. In its most extreme form, environmentalists are accused of advocating a one-world governmental regime and thus are associated with, or are, the Antichrist. These various factors give an insight into how complex the territory is for evangelical, or for that matter, many kinds of Christian, environmentalists, so that the term green evangelical can designate groups that actually oppose each other. I now differentiate between the most prominent green evangelicals, such as Richard Cizik, Joel Hunter, etc., who prefer to be called creation care advocates (and whom I once termed Christian stewards but no longer do) and the even more conservative Christian stewards, such as Calvin Beisner, who are often more linked to the Christian Right, and frequently oppose creation care evangelicals while still championing a constrained environmental stewardship.[4] These Christian stewards, while advocating the embrace of some environmental concerns (as opposed to outright denouncing them, as many on the Christian right do) are also ardent critics of any, Christian or otherwise, environmentalism that critiques and challenges the current economic order (and for that matter, neo-conservative politics), and are frequently climate change skeptics and deniers (or as one blogger, calling himself the 'Evangelical Ecologist' asserted, climate agnostics).

It was the 2006 statement 'Climate Change: An Evangelical Call to Action' (<http://www.christiansandclimate.org>) that brought the creation care evangelical voice to prominence with over 100 key evangelicals signing it, including megachurch pastors Joel Hunter, Bill Hybels, and Rick Warren, and the heads of key evangelical seminaries in the US (Goodstein 2006). This effort was initiated at an international gathering in Oxford, England, in 2002, and the introduction of Sir John Haughton of the Intergovernmental Panel on Climate Change to Richard Cizik, then vice president of governmental affairs for the National Association of Evangelicals (NAE). Houghton convinced Cizik that he 'couldn't shirk, shrug, rationalize or escape my Biblical responsibility to care for the environment' (Heltzel 2009: 154). Although the goal of a unanimous statement from the NAE failed and Cizik himself was forced to take his signature off the statement, due to pressure from high-profile religious right Christians (who ironically, were not members of the NAE), it was a major catalyst of activity.

Groups such as the Acton Institute for the Study of Religion and Liberty and the Cornwall Alliance, formerly the Interfaith Council on Environmental Stewardship (ICES) vociferously oppose these efforts of creation care evangelicals. Both are prominent examples of the well-funded, religiously based, free market, counter-mainstream environmentalism, conservative green movement (Bliese 2002), and have links to the organizations profiled in Dunlap and McCright's chapter in this volume. A central figure for these groups is E. Calvin Beisner, listed as the national spokesman for the Cornwall Alliance, and author of *Where Garden Meets Wilderness: Evangelical Entry into the Environmental Debate* (1997) and many other documents that criticize well-known green evangelicals such as the scientist Calvin DeWitt and the Evangelical Environmental Network. As the title indicates, there are competing voices that claim to be speaking for evangelicals, and contemporary Christian stewards take advantage of the broader world's ignorance of the differences both within conservative Christianity and within 'green' Christianity.

Two documents illustrate these tensions and the differences in frames: the 1994 'The Evangelical Declaration of Creation Care' (Berry 2000) and the 2000 'The Cornwall Declaration on Environmental Stewardship' (Acton 2007) written in response to the success of creation care evangelicalism, and sent to 37,000 religious leaders. The Cornwall Declaration argues that free-market forces can resolve environmental problems, and denounces the environmental movement for embracing faulty science and a gloom-and-doom approach. It reiterates the 'wise-use' movement's emphases (Helvarg 1992; Kearns 2005) on the continuing improvement of the environment through human technology, the investment of private property owners in caring for their properties, the abundance of resources put here for human use, and the critique of viewing more-than-human nature as an idyllic, harmonious state that must be preserved. In contrast, the Evangelical Declaration encourages Christians to become ecologically aware caretakers of creation and 'to resist the allure of wastefulness and overconsumption by making personal lifestyle choices that express humility, forbearance, self restraint and frugality' and choose 'godly, just, and sustainable choices.' Other differences revolve around the place and privileges of humans relative to nature, issues of biblical interpretation of key passages in Genesis, whether efforts toward saving the earth are replacing the central Christian emphasis on saving souls, and the role of God's sovereignty in solving environmental problems. On the latter, instead of climate change being a Pandora's box of catastrophes, with potentially huge economic costs as creation care advocates recognize, for Cornwall/Christian stewards, recognizing and responding to climate change becomes a threat to private property rights/liberty, capitalism, and a denial of belief in God's sovereignty. For if God is in charge, then global warming must be part of God's plan, and indeed is seen to fit various apocalyptic end time scenarios.

The tensions that are evident in these two documents make clear how religious framing can vary greatly, and how it masks other underlying frames. At stake in the discussion are what Nisbet calls the frames of 'Scientific Certainty/Uncertainty,' 'Pandora's Box of Catastrophes,' 'Social Progress,' and 'Economic Competitiveness.' The four central claims of the Evangelical Call to Action, a later document, easily illustrate these frames: (1) 'Human-induced climate change is real,' (2) 'The consequences of climate change will be significant, and will hit the poor the hardest,' (3) 'Christian moral convictions demand our response to the climate change problem,' and (4) 'The need to act now is urgent' (Christians and Climate N.D.).

4 SCIENTIFIC UNCERTAINTY

In that 'Call to Action,' creation care evangelicals had acknowledged that they 'have hesitated to speak on this issue until we could be more certain of the science of climate change, but the signatories now believe that the evidence demands action' (Christians and Climate N.D.). Similar admissions as part of Southern Baptist and Salvation Army statements demonstrate how pervasive this uncertainty framing is within the world of conservative Christianity, as do the surveys. Cornwall's recent 'An Evangelical Declaration on Global Warming' (N.D.) goes a bit farther than emphasizing uncertainty—it is certain that climate change isn't even an issue:

> We deny that Earth and its ecosystems are the fragile and unstable products of chance, and particularly that Earth's climate system is vulnerable to dangerous alteration because of minuscule changes in atmospheric chemistry. There is no convincing scientific evidence that human contribution to greenhouse gases is causing dangerous global warming. (Cornwall N.D.)

It then goes on to deny that 'carbon dioxide—essential to all plant growth—is a pollutant' (Cornwall N.D.). A 100+ page booklet that accompanies Acton's Effective Stewardship curriculum (distributed very widely for free) asserts: 'that the world is not experiencing overpopulation or destructive, manmade global warming or rampant species loss' (Acton 2007: 100). Another example is seen in an 'Open Letter' (Cornwall 2009) to the signers of the Evangelical Climate Initiative's 'Call to Action,' that argues 'against the extent, the significance, and perhaps the existence of the much touted scientific consensus on catastrophic human induced global warming.' This is the opposite of the positive role that religious authorities can play, if they use their authority to undermine any discussion of the science.

It is not surprising that scientific uncertainty is a dominant theme, for Nisbet (2009) reports that 'during the 1990s, based on focus groups and polling, top Republican consultant Frank Luntz helped shape the climate skeptic playbook, recommending in a strategy memo to lobbyists and Republican members of Congress that the issue be framed as scientifically uncertain, using opinions of contrarian scientists as evidence' (Environmental Working Group 2003). The media have abetted this effort, as they continue to fall prey to the common practice that in presenting two views on an issue, they thereby give equal weight to contrarian views on climate science, thus implying that there is an active, widespread scientific debate on climate change. The climate skeptics like to mention how many scientists challenge the consensus on global warming, who are often oil industry-backed researchers and conferences (Connor 2008), often without naming them as such, as their proof.[5] This is picked up by the press and reported, with little investigation into the credentials of the scientists, of which very few are respected, recognized climatologists.

This illustrates one of the contradictions or tensions present in the religious climate skeptic movement: despite fostering a general suspicion/distrust of science, in which climate change, like evolution, is presented as just a 'theory' (in this case used not in the scientific use of theory, but the popular usage to imply that it is hypothetical), references to science are still used to support the frame of uncertainty, indicating that science still has some master authority status. Yet the main effort is to undermine the authority of science, as seen in one set of conservative 'talking points' that constantly referred to the 'theology of global warming' or 'the religion of global warming' implying that like religion, one can choose what science to believe, or ignore if it is inconvenient. Perhaps this is evidence of the secularization of science, whose domain of influence and authority is diminished if it is implied that one can choose what science to believe (Fuller 2001). Climate skeptics point out what they see as failed predictions, overstated claims (in many scientific areas), and no longer held scientific truths (citing Galileo as a symbol for how hard it is to fight 'the established viewpoint'). Religious climate deniers capitalize on the sense of being a persecuted minority, who in both religion and science, know the real 'Truth' that is not recognized by the 'world.' Religious environmentalists, instead of focusing on the moral and faith imperative to respond to the climate crisis as it is appearing, end up debating the science, for which they are sometimes ill-equipped.

5 Economic Framing

In his memo, Luntz also suggested that a successful emotional appeal (he used the phrase an 'emotional home run') would be to emphasize the economic consequences of action, in order to make it seem an 'unfair burden' on Americans if other countries such as China and India did not sign international treaties and agreements. The recent international talks on climate change in Copenhagen demonstrate how successful this strategy has been, not only in the US, but also in other countries, such as China and India. Religious opponents to environmentalism almost always bring in an economic framing, whether it is that the economic costs will be too high and will hurt the poor, that responding will undermine our whole free-market economic system, or that environmentalism is a cover for socialism (Cornwall N.D.). Cornwall's 'Renewed Call to Prudence Truth, Prudence, and Protection of the Poor' states it the most bluntly: 'global warming alarmism' will

- destroy millions of jobs.
- cost trillions of dollars in lost economic production.
- slow, stop, or reverse economic growth.
- reduce the standard of living for all but the elite few who are well positioned to benefit from laws that unfairly advantage them at the expense of most businesses and all consumers.
- endanger liberty by putting vast new powers over private, social, and market life in the hands of national and international governments.
- condemn the world's poor to generations of continued misery characterized by rampant disease and premature death (Beisner 2006).

Alarmism indeed! The economic frame, as is evident above, is really the bottom line framing in the Acton/Cornwall materials. This had already been a successful strategy in the 'wise use agenda' playbook, with its invocation of the almost sacred rights of private property and protection from government regulation. Cornwall's 'Open Letter' (2009) also demonstrates this economic frame:

we believe it is far wiser to promote economic growth, partly through keeping energy inexpensive, than to fight against potential global warming and thus slow economic growth. And there is a side benefit, too: wealthier societies are better able and more willing to spend to protect and improve the natural environment than poorer societies. Our policy, therefore, is better not only for humanity but also for the rest of the planet.

While seeming similar, the Evangelical Call to Action switches the economic framing:

Poor nations and poor individuals have fewer resources available to cope with major challenges and threats. The consequences of global warming will therefore hit the poor the hardest, in part because those areas likely to be significantly affected first are in the poorest regions of the world. Millions of people could die in this century because of climate change, most of them our poorest global neighbors. (Christians and Climate N.D.)

6 JUSTICE AND ETHICS FRAMING

What is interesting is that in a Christian context, the debate over economic ideologies is masked under competing moral claims over who cares more about the poor and how to do so, so that the frame is seen to be about morality and justice, and not about economic philosophies. The Christian ethic 'to protect and care for the least of these as though each was Jesus Christ himself (Mt. 22: 34–40; Mt. 7: 12; Mt. 25: 31–46),' as the Call to Action states, is central for those for whom Christian morality is about justice, as is characteristic of most creation care evangelicals, more than personal morality. This also fits into the dominant justice framing of most religious activism on climate change. The NRPE video, 'God's Creation and Global Warming,' mentioned earlier, highlights the devastation already being faced by Pacific Island nations, as they confront the unjust reality of losing their land, and with it, becoming landless nations seeking asylum and refuge from other countries. Rabbi Warren Stone, representing North American Jewish organizations, articulated why scores of religious groups would be present at the failed 2009 climate change talks in Copenhagen: 'We are called by our religious traditions to serve as a bold voice for justice. Climate change will have a dramatic impact on hundreds of millions of the poorest people on our planet, especially those who live in coastal areas' (Bohan et al. 2009).

In looking back at 1.5 decades of prominent green evangelical activism, the battle over who cares about the poor became more central in the evangelical world as creation care evangelicals continued to make inroads in convincing Christians, and as justice became the shared language of religious environmentalism. This means it is even more difficult for the everyday lay person, on first glance, to discern key differences. The motto for Cornwall's more recent We Get It! Campaign promotes the linkage succinctly ('Caring for the environment and the poor—Biblically') and the campaign argues: 'Efforts to cut greenhouse gases hurt the poor. By making energy less affordable and accessible, mandatory emissions reductions would drive up the costs of consumer products, stifle economic growth, cost jobs, and impose especially harmful effects on the Earth's poorest people' (Wegetit.org N.D.). Arguments that seem to be about the poor are especially appealing, and when presented as the foremost reason to oppose treaties on climate change by Christian stewards who say they believe in caring for God's creation, many Christians listen. Of course, these are not the only dominant economic frames, as pitches of saved costs, and now of green jobs, more than moral claims, may succeed to convince people of energy efficiency and the merits of alternative technologies.

The greening of Evangelicalism continues to make the news, with key evangelicals and scientists testifying on both sides of the issue before the US Senate in November 2009 in order to urge, or oppose, action on climate legislation (Cox 2009). It has become a wedge issue that has divided evangelicals, and broken their once close allegiance with the agenda of the Republican Party. Whereas once, caring about the environment was not on their radar, Southern Baptists, Salvation Army, and Pentecostals have expressed environmental concern, often issuing statements outside of their official denominational agencies, as is the case with the Southern Baptists, or using the efforts of international branches, such as in the Salvation Army, to influence domestic practices and policies. Cornwall's response is a video series 'Resisting the Green Dragon' (2010) that 'exposes' environmentalism's 'lost for political power at the highest levels' and 'seducing our children' through education. As more conservative Christians recognize the connection between their faith traditions and environmental concern, telling the difference between groups becomes more crucial, and the frames of scientific uncertainty and threat to the economy that are pushed within conservative Christianity often become the framing that shapes conversations on religious environmentalism in many other places.

7 CONCLUSION

As the heated exchange between creation care and Christian steward advocates indicates, as well as the wide range of religious environmental activities discussed here, winning over people of faith to respond to climate change is critically important. As the activities of groups around the globe, like the ARC and the WCC, or the numerous US efforts, indicate, religions can make a difference in the response to climate change. By paying attention to how these appeals are framed, however, it should be evident that the correct solution is not just 'add religion and stir,' for there is plenty of seemingly 'green' religion that seeks to undermine efforts at treaties like Copenhagen. Although seemingly desirable, just putting something in a religious, moral, and ethical frame is insufficient, as both creation care evangelicals and Christian steward evangelicals appeal heavily to the religious ethic of concern for the poor, and all religious environmentalism invokes concerns over justice (some of it going beyond the realm of human justice to justice for all of creation) while disagreeing on what is just and unjust. Rather, it is how these groups invoke other dominant climate change frames, such as economic philosophies and the nature of certainty in science, and the actions that they take and promote, that is crucial in evaluating the role of religion in responding to climate change.

NOTES

1. Thomas Berry (2009), a key figure in what I term creation spirituality Christianity, proclaims the need for a planetary 'new story', found in understanding evolution, or the story of the universe. His work has influenced Palmer.

2. Claims such as this have led anti-environmentalists, such as on the website 'Environmentalism is Fascism' to assert, as part of their conspiracy theory, that 'Environmentalists covet Church wealth: donations, real estate, and stock/bond portfolios' (Environmentalism is Fascism N.D.).

3. I use the term conservative Christianity rather than just evangelicalism to recognize that it is a complex world made up of fundamentalists, Pentecostals, Christian right evangelicals, and many progressive evangelicals, who, while conservative on personal morality, are liberal on many social issues such as the environment, the unequal distribution of wealth, etc.

4. In this instance, Christian stewards are heirs to Gifford Pinchot, and not John Muir (as are most religious environmentalists), in arguing for the utilitarian conservation and wise use of natural resources primarily in relation to human uses. Like some leading conservation groups (MacDonald 2008), they are more favorable to corporate interests and funding, such as Exxon, than any of the other key Christian environmental groups. Thus the term stewardship has fallen out of favor among many religious environmentalists, as signifying a managerial approach.

5. For instance, they do so by mentioning the 30,000+ scientists who have signed the Oregon Petition (Petition Project 2008), which is filled with dubious signers, non-experts (only 35 percent have Ph.D.s) and very few peer-recognized climatology experts, or the Manhattan Declaration (from a climate change conference sponsored by groups who receive funding from oil companies), which has the same issues regarding credentials of signatories (Connor 2008).

REFERENCES

Acton Institute. 2007. The Cornwall Declaration on Environmental Stewardship. *Environmental Stewardship in the Judeo-Christian Tradition*, ed. MICHAEL BARKEY. Grand Rapids, CA: Acton Institute.

Alliance of Religions and Conservation (ARC). N.D.a. History. Available from: <http://www.arcworld.org/about.asp?pageID=2>. Accessed 18 January 2009.

—— N.D.b. Projects overview. Available from: <http://www.arcworld.org/projects_overview.asp>. Accessed 30 August 2010.

—— 2007. Sweden: FSC certification for church forests. Available from: <http://www.arcworld.org/projects.asp?projectID=169>. Accessed 14 January 2009.

BEISNER, E. C. 1997. *Where Garden Meets Wilderness: Evangelical Entry into the Environmental Debate.* Acton Institute for the Study of Religion and Liberty. Grand Rapids, MI: W. B. Eerdmans.

—— 2006. A call to truth, prudence, and the protection of the poor. 27 July 2006. Available from: <http://erlc.com/article/a-call-to-truth-prudence-and-protection/>. Accessed 26 April 2009.

BERKEMAN, O. 2002. 'What would Jesus drive' gas guzzling Americans are asked. *The Guardian*, 14 November.

BERRY, R. J. 2000. Evangelical declaration of creation care. Pp. 17–22 in R. J. Berry (ed.) *The Care of Creation: Focusing Concern and Action.* Downers Grove: InterVarsity Press.

BERRY, T. 2009. *The Sacred Universe: Earth, Spirituality, and Religion in the Twenty-first Century.* New York: Columbia University Press.

BLIESE, J. R. E. 2002. *The Greening of Conservative America*. Boulder, CO: Westview End Press.

BOHAN, C., COLVIN, R., and SPETALNICK, M. 2009. Obama heads to Copenhagen, sees progress with China. Reuters. Available from: <http://www.reuters.com/article/idUS-TRE5BG0MU20091218>. Accessed 17 January 2010.

Center for Research on Environmental Decisions. 2009. *The Psychology of Climate Change Communication: A Guide for Scientists, Journalists, Educators, Political Aides, and the Interested Public*. Available from: <http://cred.columbia.edu/guide/>. Accessed 5 December 2009.

Christians and Climate. N.D. *Climate Change: An Evangelical Call to Action*. Available from: <http://christiansandclimate.org/learn/call-to-action/>. Accessed 17 January 2010.

CONNOR, S. 2008. Tobacco and oil pay for climate conference. *The Independent*. 3 March. Available from: <http://www.independent.co.uk/environment/climate-change/tobacco-and-oil-pay-for-climate-conference-790474.html>. Accessed 17 January 2010.

Cornwall Alliance. N.D. <http://www.cornwallalliance.org/articles/read/an-evangelical-declaration-on-global-warming/>. Accessed 29 August 2010.

—— 2009. <http://www.cornwallalliance.org/docs/an-open-letter-to-the-signers-of-climate-change-an-evangelical-call-to-action-and-others-concerned-about-global-warming.pdf>. Accessed 2 June 2009.

COX, D. 2009. Scientists and evangelicals join forces on climate change. Available from: <http://www.publicreligion.org/blog/2009/11/19/scientists-evangelicals-climate-change/>. Accessed 8 January 2010.

DAVIS, J. E. 2002. *Stories of Change: Narratives of Social Movements*. New York: State University of New York Press.

Environmentalism is Fascism. N.D. *Environmentalism's Appropriation of Christianity*. Available from: <http://www.ecofascism.com/article16.html>. Accessed 14 January 2010.

Environmental Working Group. 2003. Luntz memo on the environment. March. Available from: <http://www.ewg.org/node/8684>. Accessed 1 December 2009.

Forum on Religion and Ecology. 2004. Available from: <http://fore.research.yale.edu/>. Accessed 15 January 2010.

FULLER, S. 2001. The re-enchantment of science: A fit end to the science wars?. Pp. 181–204 in K. Ashman and P. Barringer (ed.), *After the Science Wars*. New York: Routledge.

Global Ministries. N.D. Driven by Values: Protect God's Creation. Available from: <http://www.gbgm-umc.org/NCNYEnvironmentalJustice/transportation.htm>. Accessed 15 January 2010.

GOODSTEIN, L. 2006. Evangelical leaders join global warming initiative. *New York Times*, 8 February.

—— 2007. At home with Jim Ball: Living day to day by a gospel of green. *New York Times*, 8 March.

HALLMAN, D. 2005. Montreal UN Climate Change Conference. <http://www.oikoumene.org/resources/documents/wcc-programmes/justice-diakonia-and-responsibility-for-creation/climate-change-water/montreal-un-climate-change-conference.html>. Accessed 30 August 2010.

HELTZEL, P. 2009. *Jesus and Justice: Evangelicals, Race and American Politics*. New Haven: Yale University Press.

HELVARG, D. 1994. *The War against the Greens: The 'Wise-Use' Movement, the New Right, and Anti-Environmental Violence*. San Francisco: Sierra Club Books.

KEARNS, L. 1996. Saving the creation: Christian environmentalism in the United States. *Sociology of Religion* 57(1): 55–70.

—— 1997. Noah's Ark goes to Washington: A profile of evangelical environmentalism. *Social Compass* 44(3): 349–66.

—— 2004. The context of eco-theology. Pp. 466–84 in G. JONES (ed.), *The Blackwell Companion to Modern Theology*. Malden, MA: Blackwell Publishing.

—— 2005. Wise use movement. Pp. 1755–7 in B. TAYLOR (ed.), *Encylopedia of Religion and Nature*. New York: Thoemmes Continuum.

—— 2007. Cooking the truth: Faith, science, the market, and global warming. Pp. 97–124 in L. KEARNS and C. KELLER (ed.), *EcoSpirit: Religions and Philosophies for the Earth*. New York: Fordham Press.

MacDONALD, C. 2008. *Green, Inc.* Guildford, CT: Lyons Press.

NASH, R. 1989. *The Rights of Nature: A History of Environmental Ethics*. Madison: University of Wisconsin Press.

NISBET, M. 2009. Communicating climate change: Why frames matter for public engagement. *Environment* (March/April). Available from: <http://www.environmentmagazine.org/Archives/Back%20Issues/March-April%202009/Nisbet-full.html>. Accessed 14 January 2010.

PALMER, M. 2009. Religion still plays vital part in struggle for Earth's future. *Global Times*, 5 November. Available from: <http://opinion.globaltimes.cn/foreign-view/2009–11/482932.html>. Accessed 14 January 2010.

Petition Project. 2008. Global warming petition. Available from: <http://www.petitionproject.org/>. Accessed 30 August 2010.

Pew Research Center. 2009. Fewer Americans see solid evidence of global warming. 22 October. Available from: <http://pewresearch.org/pubs/1386/cap-and-trade-global-warming-opinion>. Accessed 18 January 2010.

SCHAEFFER, F. A. 1970. *Pollution and the Death of Man: The Christian View of Ecology*. Carol Stream, IL: Tyndale House Publishers.

SNOW, D., and BENFORD, R. 2000. Clarifying the relationship between framing and ideology. *Mobilization* 5(1): 55–60.

US Energy Information Administration. 2009. Crude oil and total petroleum imports top 15 countries. 30 December. Available from: <http://www.eia.doe.gov/pub/oil_gas/petroleum/data_publications/company_level_imports/current/import.html>. Accessed 16 January 2010.

WeGetIt.org. N.D. Declaration. Available from: <http://www.wegetit.org/declaration>. Accessed 4 June 2009.

WHITE, L. 1967. Ecologic roots of the environmental crisis. *Science* 155: 1203–7.

World Council of Churches (WCC). N.D. Public Campaign on Climate Change. Available from: <http://www.oikoumene.org/en/programmes/justice-diakonia-and-responsibility-for-creation/climate-change-and-water/public-campaign-on-climate-change.html>. Accessed 30 August 2010.

PART VIII

GOVERNMENT RESPONSES

CHAPTER 29

....

COMPARING STATE
RESPONSES

....

PETER CHRISTOFF AND ROBYN ECKERSLEY*

1 INTRODUCTION

....

DESPITE nearly two decades of international climate negotiations and near universal participation by states in the United Nations Framework Convention on Climate Change (UNFCCC) 1992, there has been no concerted or effective collective state response to the threat of global warming. Significant variations occur between the domestic climate policies of the world's 193 states and only a small handful of states have risen to the climate challenge despite the UNFCCC's requirement that developed countries must take the lead in combating climate change, and enact national policies and measures accordingly.[1] Indeed, many of the industrialized states that accepted quantified emissions reduction targets under the Kyoto Protocol are presently not on track to meet their modest first period commitments.[2]

This chapter provides a critical overview of research on comparative state responses to the challenge of climate change. We begin by considering some of the methodological challenges involved in assessing relative performance, focusing on the politics of measuring and judging, and then present key data that enable comparison between the twenty states that are collectively responsible for some 85 percent of total global emissions. Against this background, we provide a critical stocktaking of the existing literature on comparative state climate performance and conclude with some broad insights on what makes a climate leader or a climate laggard.

* We wish to thank Gwilym Croucher for research assistance.

2 MEASURING STATE PERFORMANCE

Environmental non-government organizations (ENGOs) and the mass media are particularly interested in publicizing 'league ladders' that rank countries in terms of their relative performance in response to climate change (e.g. Germanwatch/CAN Europe 2010a, 2010b; Ecofys 2009). Inevitably these rankings are determined by what is measured.

The ultimate measure of national performance, and the one most relevant to the UNFCCC's basic objective of minimizing the risk of dangerous climate change, is the level of, and changes in, a country's emissions. Yet data on changes in aggregate national emissions do not reveal the underlying causes of these changes. A decline in emissions or slowed emissions growth may result from policy decisions intended to produce those outcomes, or it might be the accidental consequence of economic downturn or collapse, as occurred among the Central and Eastern European 'economies in transition' in the post-Soviet period or is evident now in response to the Global Financial Crisis. Alternatively, a decline in national emissions might be the unintended outcome of international specialization and exchange. The relocation of energy-intensive manufacturing to the developing world has benefited developed countries insofar as a proportion of their emissions have been 'outsourced' while their affluent consumer lifestyles have been sustained or enhanced. It is now widely recognized, for instance, that a significant proportion of China's emissions is associated with the production of manufactured exports. Studies of flows of embedded carbon in international trade underline the extent to which trade-related emissions are registered at the point of production rather than attributed to those responsible for demand at the point of consumption (Wiedmann et al. 2008; Pan et al. 2009). Nevertheless attributing emissions to the state which hosts their production remains the pre-eminent means of accounting in the international climate regime. Any changes to this notion of responsibility for trade-related emissions would profoundly (re)shape assessments of national responses to climate change.

Emissions data also provide no indication of the relative difficulty or ease with which a nation achieves change relative to a particular baseline (Harrison and Sundstrom 2007). For instance, countries that reduced their fossil fuel dependency by improving energy efficiency before the chosen baseline year face a harder task in reducing emissions further than countries that made no such efforts and therefore often have more, easier, and cheaper abatement options. Indeed, it is impossible to devise a single metric that can provide an adequate basis for comparing relative performance for all states. Of course, the more parameters that are measured, the harder it is to make simple judgments about relative national performance. To interpret and assess comparatively how states perform—and to define their fair share of effort—also requires information about their relative circumstances such as national capacities and the limiting or enabling material, institutional, and cultural factors that shape and drive national commitments, policy output, and policy outcomes. Any comparative assessment of state performance necessarily makes judgments about effort relative to capacity. Indeed, a significant part of the debate over what constitutes an equitable response by individual developed and developing countries focused around the burden-sharing principle of 'common but differentiated responsibility' (CBDR) has dwelt on the differences in national wealth and capacities for action that have arisen as a result of historically different national exploitation of the planet's atmospheric commons. And while it is possible to develop a more comprehensive set of measures on emissions—including changes in aggregate emissions, per capita emissions, emission intensities, reductions from BAU (business as usual) and emissions in exports and imports (Burck et al. 2009), these data tell us little about policy commitment and policy output.

Given these complications, most attempts to 'score' performance have drawn upon a range of indicators that can be grouped into three clusters: indicators of measurable past performance; of current performance and identifiable trends; and of aspirational goals and commitments or 'policies for the future' (Table 29.1). Primary measures include aggregate and per capita emissions levels, emissions intensity, and rates and types of decarbonization

Table 29.1 Comparative indicators

Theme	Comparison	Indicator
Past Performance	Aggregate emissions over time	Trend in aggregate emissions from 1990
		% change in total emissions (with or without LULUCF) from 1990 baseline
	Per capita emissions over time	Per capita emissions from 1990
	Performance relative to Kyoto target (Annex I)	% change in emissions against Kyoto target
	Change in emissions intensity	Change in CO_2 emissions per unit GDP PPP[a]
Present Performance	International rank per GHG emissions	Volume of aggregate emissions
	Per capita emissions	CO_2 emissions from consumption of energy
	Sectoral performance (aggregate emissions)	Change in aggregate sectoral emissions (vol/% change against baseline for stationary energy sector; transport; agriculture; etc)
	Emissions intensity	CO_2 emissions per unit GDP PPP
	Decarbonization	Percentage of energy from renewable sources (excluding nuclear)
Policy Aspirations	Proclaimed targets	National aggregate emissions target National energy intensity target National renewable energy target International climate target
National factors and material capabilities	Development status	National GDP (gross and per capita)
	Energy configuration	% energy from domestic/imported fossil fuel sources
	Energy sources	Domestic availability of fossil fuels and other energy resources
	Population	Change in population since 1990 (or some other baseline)

[a] Purchasing power parity.

(for instance, uptake of renewable energy). Less common—partly because many comparisons are mainly made between highly industrialized Annex I countries with relatively slow rates of population growth—is the inclusion of additional measures of underlying factors affecting these primary indicators, such as changes (growth) in national GDP; the source, nature, and autonomy of energy supplies; population growth; and changes in per capita purchasing power. These measures are especially relevant to understanding changes in developing countries' profiles.

Some assessments and rankings have involved the aggregation of various data into a consolidated index. For instance, Germanwatch/CAN Europe (2010a, 2010b) surveys the 57 states responsible for more than 90 percent of the planet's emissions in its annual *Climate Change Performance Index*, and scores each state by combining indicator results into one index number via the following weightings: emissions trends by sector (50%), emissions levels (30%), and climate policy (20%). It justifies this approach by suggesting that 'with a weight of 70%, climate policy and emissions trend together . . . allows achievements in reducing emissions to be adequately reflected. As the absolute amount of greenhouse gases that a country emits can only be changed in long time periods, a stronger weighting of the current emissions would hardly allow movement within the ranking' (Germanwatch/CAN Europe 2010b: 5).

3 COMPARING STATE PERFORMANCE

Table 29.2 illustrates the performance of the twenty states that, together, are responsible for some 85 percent of total global emissions. When one examines a combination of quantifiable climate policy outcomes, measures of climate policy capacity, and climate policy commitments for these states, both predictable and unexpected results emerge.

3.1 Current Performance and Performance over Time

Germany and the United Kingdom appear clear leaders among Annex I states in terms of their reduction of aggregate and per capita emissions since 1990. However, this is only so when one discounts the performances of the Russian Federation and Ukraine and the inadvertent influence of economic downturn and restructuring on their emissions in the post-Soviet period. The data for the period 1990–2007 also fail to acknowledge the significant reductions by France and Japan in the 1970s and 1980s arising from their nuclear programs and energy efficiency measures. For instance, 43.5 percent of France's total energy now derives from nuclear reactors, compared with 7.7 percent for the United Kingdom, 11.0 percent for Germany, and 13.4 percent for Japan (IEA 2007). The development of this capacity in turn depended on the *dirigiste* nature of the French state and the absence of civil society opposition of the sort that halted similar development in Germany in the 1980s.

High rates of growth in emissions among Annex I countries are generally found among those relatively poor states which are pursuing strong economic growth. The exceptions here are Australia and Canada, which are among the wealthiest of the twenty developed and developing states under consideration, in terms of aggregate and per capita GDP, yet

Table 29.2 Twenty top emitters: ranking

PAST PERFORMANCE

Indicator	Argen	Austral	Brazil	Canada	China	France	Ger	India	Indon	Iran	Italy	Japan	Mexico	Russia	Spain	South Africa	South Korea	Ukraine	UK	USA	EU
EMISSIONS CHANGE % (1990–2007) GHG w LULUCF[a][b]	n/a	+82%	n/a	+46.7	n/a	−11.8%	−20.8%	n/a	n/a	n/a	+7.4%	+8.2%	n/a	−40.3%	+55.3	n/a	n/a	−54%	−17.8%	+15.8	−5.6
RANK (BEST TO WORST)	11	13	14	9	20	5	3	16	18	19	6	7	8	2	10		17	1	4	8	(5)[c]
PER CAPITA CHANGE % CO2 emissns (1990–2005)[d]	+14.9	+21.1	+35.3	+9.6	+109.8	+0.8	−17.6	+53.4	+89.8	+90.8	+10.3	+9.8	+8.4	−27.4	+48.3	−2.2	+72.8	−51.8	−9.5	+1	n/a
RANK (BEST TO WORST)	12	13	14	9	20	6	3	16	18	19	11	10	8	2	15	5	17	1	4	7	n/a
RE KYOTO TARGETS CHANGE % versus 2012 target, at 2007	n/a	+76	n/a	+52.7	n/a	−3.8	+0.2	n/a	n/a	n/a	+13.9	+14.2	n/a	−40.3	+45.3	n/a	n/a	−54%	−5.6%	+22.8	+2.4
RANK (BEST TO WORST)		11		10		4	5				6	7		2	9			1	3	8	(5)

PRESENT PERFORMANCE

Indicator	Argen	Austral	Brazil	Canada	China	France	Ger	India	Indon	Iran	Italy	Japan	Mexico	Russia	Spain	South Africa	South Korea	Ukraine	UK	USA	EU
CURRENT EMISSIONS 2007 GHG Mt (+LULUCF)[e]	196.7	825.9	1708.9	792.5	3421.8	463.4	934.0	1014.5	2857.4	347.7	482.9	1292.9	482.9	2005.8	414.3	351.1	467.0	392.6	638.5	6087.5	3792.5
RANK (HIGH TO LOW)	20	9	5	10	2*	15	8	7	3	19	=12	6	=12	4	16	18	14	17	11	1	(2)
PER CAPITA EMISSIONS (t) per capita CO2 emits[f] 2005 (t 1990)	3.77 (3.28)	18.71 (15.45)	1.88 (1.39)	17.30 (15.79)	4.28 (2.04)	6.55 (6.50)	10.05 (12.20)	1.12 (0.73)	1.67 (0.88)	6.47 (3.39)	8.14 (7.38)	9.77 (8.90)	3.98 (3.67)	10.95 (15.08)	8.45 (5.70)	7.19 (7.35)	9.83 (5.69)	6.43 (13.35)	8.94 (9.87)	19.87 (19.67)	7.56 (n/a)
RANK (HIGH TO LOW)	17	2	18	3	14	12	5	29	19	13	10	7	15	4	9	11	6	13	8	1	(12)

(continued)

Table 29.2 Continued

PRESENT PERFORMANCE

Indicator	Argen	Austral	Brazil	Canada	China	France	Ger	India	Indon	Iran	Italy	Japan	Mexico	Russia	Spain	South Africa	South Korea	Ukraine	UK	USA	EU
EMISSIONS INTENSITY CO_2/ GDP PPP (kg CO_2/ 2000 USD)[g]	0.28	0.59	0.22	0.55	0.61	0.21	0.34	0.33	0.45	0.84	0.28	0.34	0.37	0.99	0.46	0.67	0.32	0.95	0.29	0.50	n/a
RANK (WORST TO BEST)	17	6	19	7	5	20	12	14	10	3	17	12	11	1	9	4	15	2	16	8	
RENEWABLE (2007) % total energy use[h]	7.0%	5.6%	44.4%	16.2%	12.3%	7.0%	8.6%	29.1%	31.2%	1.3%	6.7%	3.4%	9.3%	3.3%	7.1%	10.4%	1.3%	1.2%	2.2%	5.0%	n/a
RANK (BEST TO WORST)	=10	13	1	4	5	=10	8	3	2	=18	12	15	7	16	9	6	=18	20	17	14	

ASPIRATIONS

Indicator	Argen	Austral	Brazil	Canada	China	France	Ger	India	Indon	Iran	Italy	Japan	Mexico	Russia	Spain	South Africa	South Korea	Ukraine	UK	USA	EU
EMISSION TARGET (2020)[i]	n/a	−5% below 2000	−36.1% /38.9% fr 2010	−17% below 2005	n/a	n/a	−40% below 1990	n/a	−26% below BAU	n/a	n/a	−25% below 1990	−30% below BAU	−20 to −25% (1990)	n/a	−34% below BAU	−30% below BAU	−20% below 1990	−34% below 1990	−17% below 2005	−20% below 1990
RANK (HIGH TO LOW)	n/a	10	6	8	n/a	n/a	1	n/a	7	n/a	n/a	3	n/a	4	n/a	n/a	n/a	5	2	8	n/a
ENERGY INTENSITY TARGET (for 2020)	n/a	n/a	n/a	n/a	−40% −45% fr BAU	n/a	n/a	−20%−25% fr BAU	n/a	n/a	n/a	n/a	n/a	n/a	n/a	n/a	n/a	n/a	n/a	n/a	n/a
RANK																					
RENEWABLES TARGET (%total energy x 2020)	n/a	20%	n/a	15% w nuclear	15%	23%	18%	n/a	n/a	n/a	17%	n/a	40%	4.5%	20%	n/a	n/a	n/a	15%[j]	none	20%
RANK (BEST TO WORST)		3		2		2	5		7		6		1	8	3				7		

NATIONAL CAPACITY

Indicator	Argen	Austral	Brazil	Canada	China	France	Ger	India	Indon	Iran	Italy	Japan	Mexico	Russia	Spain	South Africa	South Korea	Ukraine	UK	USA	EU
NATIONAL GDP (2007) ($US bill'n)[k]	584	851	2013	1281	8765	2108	2806	3862	962	828	1740	4159	1541	2109	1360	505	1364	302	2139	14256	14739
RANK	18	16	9	14	2	7	5	4	15	17	10	3	11	8	13	19	12	20	6	1	(1)
PER CAPITA GDP PPP $US (2007)[l]	14,560	38,910	10,513	38,025	6587	33,678	34,212	3176	4156	11,172	29,109	32,608	14,495	14,919	29,689	10,243	27,978	6650	34,619	46,381	29729
RANK	12	2	15	3	18	6	5	20	19	14	9	7	13	11	8	16	10	17	4	1	(8)

* More recent date for China and the USA indicate that China is now the largest aggregate emitter, followed by the USA.

[a] UNFCCC/SBI/2009/12

[b] As data for non-Annex-1 countries are only available (in some cases) for 1994 and 2000, these countries have been excluded from ranking as this would offer a misleading comparator.

[c] The European Union is also ranked as if it were a single entity among the 20 major emitters (with the ranking given in brackets).

[d] See Indicator 2 in PRESENT PERFORMANCE.

[e] 2007 data for Annex 1 states: UNFCCC/SBI/2009/12 2000 data for non-Annex 1 states: WRI, at: <http://earthtrends.wri.org/searchable_db/index.php?action=select_countries&theme=3&variable_ID=466>. Data here calculated by WRI in 2005. It is widely recognized that China has overtaken the USA in aggregate emissions since 2005.

[f] WRI: <http://earthtrends.wri.org/searchable_db/index.php?action=select_countries&theme=3&variable_ID=466>. Data presented here were calculated by WRI in June 2009.

[g] 2007 data for Annex 1 states: UNFCCC/SBI/2009/12 2000 data for non-Annex 1 states: WRI, at: <http://earthtrends.wri.org/searchable_db/index.php?action=select_countries&theme=3&variable_ID=466>. Data here calculated by WRI in 2005.

[h] Excluding nuclear power. IEA Statistics: Energy Indicators by country. At: <http://www.iea.org/stats/indicators.asp?COUNTRY_CODE>.

[i] Copenhagen Accord pledges, and national pledges, as at 1 February 2010. Annex 1 at <http://unfccc.int/home/items/5264.php> and non-Annex 1 at <http://unfccc.int/home/items/5265.php>.

[j] European data from EU Renewable energy policy. At: <http://www.euractiv.com/en/energy/eu-renewable-energy-policy/article-117536>.

[k] MF website. Washington, DC: International Monetary Fund. <http://www.imf.org/external/pubs/ft/weo/2010/01/weodata/index.aspx>. Retrieved 21 April 2010.

[l] MF website. Washington, DC: International Monetary Fund. <http://www.imf.org/external/pubs/ft/weo/2010/01/weodata/index.aspx>. Retrieved 21 April 2010.

are the poorest performers among both developed and developing countries in terms of their increase in per capita emissions, increase in aggregate emissions, and their performance against Kyoto targets. They are also, along with the United States, among the weakest of the developed countries in terms of the emissions intensity of their economies, while the Russian Federation and Ukraine together show striking evidence of the lingering influence of the inefficiencies of Soviet energy and economic planning. It is notable that Australia, Canada, and the United States are each federated states, with considerable and cheaply priced domestic fossil fuel resources and that the Australian and Canadian economies derive considerable income from primary commodity exports.

When performance is reviewed against economic capacity and development (best indicated by the measure 'per capita GDP') unsurprisingly the non-Annex I developing states show considerable increases in per capita emissions as they work to improve material conditions. However, changes in per capita emissions also need to be interpreted against data for population growth (not provided) and against changes in trade profile. As previously noted, China and India—the largest trade-oriented developing states in this group—have become increasingly integrated into the global economy over the past decade, resulting in growth in the proportion of their domestic emissions resulting from export-oriented production. (Reasons for the growth in per capita emissions in Iran are less clear.)

3.2 Aspirational Targets, Goals and Policy

Collectively, the current targets and goals of the twenty major emitting countries (along with other pledges contributed under the Copenhagen Accord) are insufficient to hold global average temperature to below 3.9 degrees (PIK 2010). However national commitments, which largely reflect trajectories established over the past two decades, vary considerably. Germany and the United Kingdom look set to continue to lead the Annex I bloc, providing the 'motor' for transformation within the European Union. Meanwhile Australia is projected to continue as the poorest performer. Canada has coordinated its emissions targets with those of its major economic partner, the United States, and both are likely to remain laggards. The Russian Federation and Ukraine are expecting a rebound in emissions based on future economic growth.

Many of the most populous and/or rapidly growing developing countries show growing levels of ambition in climate policy. In its Copenhagen pledge, Indonesia committed to reducing its aggregate emissions from BAU by 26 percent by 2020 while Brazil committed to reducing emissions by between 36 percent and 39 percent by 2020—both targets to be achieved by reducing logging of their tropical forests. Major emerging emitters such as China and India have committed to major improvements in emissions intensity by 2020 from a 2005 base (40–45% for China and 20–25% for India).

4 UNDERSTANDING STATE PERFORMANCE

Great powers, middle powers, small states, weak states, and failed states have different social identities, capacities, levels of development, interests, and preoccupations. These differences help to shape their 'emissions power' and hence bargaining power, their international legitimacy and consequently their ability to shape the international climate regime and their corresponding obligations and inclinations to pursue mitigation. Indeed, the key negotiating blocs that have emerged in the international climate negotiations provide a useful 'first brush' indication of their members' assessment of where their 'national interests' lie in the climate negotiations, how the burden-sharing norms of the climate regime should be interpreted, what kinds of policy commitments they consider 'fair' and acceptable, and how they might be expected to respond domestically. The leadership obligations of developed countries under the UNFCCC are derived from the principles of CBDR, which acknowledge that developed countries are largely responsible for the cumulative emissions in the atmosphere since the industrial revolution, that they have greater per capita emissions and greater capacity to pursue mitigation. In contrast, developing countries have significant unmet social and economic needs, relatively low per capita emissions, and many are especially vulnerable to the impacts of climate change yet lack the capacity to adapt. Developing countries have therefore been allowed to expand their share of global emissions relative to developed countries while undertaking 'nationally appropriate' mitigation actions relative to their national circumstances.

The developing country bloc has been a strong supporter of CBDR and has insisted that the global effort to reduce global aggregate emissions should not compromise its collective ability to meet its development needs. Within this bloc, those states that belong to the general category of 'least responsible, weakest capacity and most vulnerable', such as members of the Association of Small Island States (AOSIS) and of the African Group, have called on developed countries to perform most of the mitigation work and fund most of the assistance to developing countries in mitigation and adaptation.

Few of the developed states that belong to the general category of 'most responsible, strongest capacity and not especially vulnerable' have fully embraced their leadership obligations under the UNFCCC. Among the Annex I groupings, the EU has committed to the climate leadership obligations that flow from CDBR while the members of the loose coalition known as the Umbrella group (which comprises the US, Canada, Australia, New Zealand, Japan, Norway, Iceland, Russia, and Ukraine) have displayed varied, but typically much lower levels of commitment. Given that the climate regime requires developed states collectively to lead, then the crucial question arises as to why only some developed states have embraced this leadership responsibility. We turn to a domestic level of analysis to understand this variation. In what follows, we single out seven dimensions that shed light on individual state responses: general regime type; the character of domestic political institutions; national interests; national discourses; strategies of accumulation; and domestic actors (with a special focus on the role of veto coalitions).

4.1 Regime Type

All the developed states listed in Annex I of the UNFCCC are democracies, but the strong variation in climate performance within and across democratic and nondemocratic states makes it clear that this particular binary, based on whether a state holds free elections, serves as a poor predictor of performance and climate leadership. In any event, there are many different types of democracy, not all of which are equally adept at managing the complex challenge of climate change. Most environmentalists and green political theorists would expect stronger (i.e. more reflexive and deliberative) democracies to produce stronger climate policies (Eckersley 2004), but this claim cannot be put to the empirical test given the paucity of such democracies. However, among established democratic states, it appears that certain institutional forms of democracy have a greater propensity to foster better climate performance than others (we discuss these in the next section). The only clear conclusion that can be drawn from large-n studies comparing democratic and nondemocratic regimes is that 'good governance' and 'absence of corruption' are necessary but not sufficient conditions for robust environmental policies in general, and climate policies in particular.[3] There is no evidence to support the proposition that nondemocratic regimes are better equipped than democratic regimes to enact and implement robust climate policies.

4.2 Domestic Political Institutions

A number of broad insights gleaned from comparative work on different types of democracy are relevant to understanding variation in climate performance. A proportional electoral system or the presence of green parties in the legislature is conducive to, but not always necessary for, strong climate policy output, especially where there is vigorous advocacy by environmental NGOs in civil society, and climate issues and concerns have been absorbed by one or more of the major parties (Bomberg 1996, 1998; Weale et al. 2000: 246–50; Poloni-Staudinger 2008; Ward 2008: 401).

Presidential systems have been found to be less likely to provide public goods than parliamentary systems due to the greater number of checks and balances and the necessity for the trading of favors to secure the passage of legislation, both of which tend to advantage powerful minorities and the status quo (Dolsak 2001: 426; Ward 2008: 402). Leaders in parliamentary systems, where strong party discipline gives the ruling party control over the legislature, generally have more success in enacting legislation and ratifying treaties than leaders in presidential systems (Lantis 2009). However, leaders in federated parliamentary states may still face difficulties in implementing treaty commitments where provincial opposition is strong (Harrison 2007: 96–7).

Because peak organizations—the key political actors in corporatist systems—form long-term relationships within a small player group, corporatist systems typically display a greater capacity than do majoritarian parliamentary systems to work cooperatively towards long-term climate policies based on consensually agreed goals (Matthews 2001). By contrast, pluralist systems are more prone to short-term, self-aggrandizing behavior by interest groups because there are many more players, few repeat plays, and plenty of incentives to defect from cooperative activity.

For example, the US political system displays many of the institutional features that are not conducive to robust climate policy. Domestic climate advocates for a comprehensive energy tax or cap-and-trade bill have faced an uphill battle in the context of a presidential system and fragmented policy-making process that together create a strong bias in favor of the status quo. While the Constitution confers on the President a wide executive power to negotiate and sign international treaties, such treaties cannot become domestic law until they are ratified by a two-thirds majority, or 67 votes, of the Senate. Moreover, the procedural rules of Congress allow extensive filibustering, which can only be brought to an end to enable a vote by a cloture motion, which requires a three-fifths majority (or 60 votes). These supermajority requirements mean that the ratification of international climate treaties and the passage of national climate bills can be thwarted or considerably weakened by a well-organized minority. These problems are further exacerbated by long-standing, powerful ideological differences between Republicans and Democrats over the necessity of climate regulation (Dunlap and McCright 2008) and also over regulatory styles (incentives, voluntary programs, and technology funding vs. mandatory measures), which makes it difficult to build the requisite majority to enact energy taxes or cap-and-trade bills in Congress.

Finally, the evidence regarding the climate performance of federated states versus unitary states is generally mixed. On the one hand, experimentation with new climate initiatives at the state/provincial level can contribute to climate policy learning and create a 'bottom-up' momentum for national change. On the other hand, the concentration of negatively affected industries and communities in particular states or provinces can give rise to blocking coalitions in the federal legislature that can sometimes cross party lines (as in the US). According to Fisher (2006: 480), coal is extracted from twenty-six US states, while just over half of the US Senators (52 out of 100) come from states in which coal contributes to the state economy and employment. Democrats are generally more sympathetic to climate regulation than Republicans, but Democrats have showed a preparedness to cross party lines if they represent constituencies that are perceived to be negatively affected by climate change legislation (Fisher 2006: 485).

However, in other federal systems, much depends on the constitutional division of legislative powers. The Canadian provinces have regulatory authority over natural re-sources, including oil, coal, and gas and can therefore stymie efforts by the national government to implement treaty commitments (Harrison 2007: 97). The Australian states have legislative power over energy production and use, but the national government has overriding power to implement international treaties. Political outcomes therefore turn on which parties and which leaders are in power at the national and subnational levels, which can create different dynamics (sometimes mutually reinforcing, sometimes mutually op-posing) to develop or resist robust climate policies.

In quasi-federal regions like the EU, it has been argued that multiple levels of governance can provide mutually reinforcing opportunities for climate leadership (Schreurs and Tiber-ghien 2007) but this in itself does not explain why certain states (such as Germany and the United Kingdom) or bureaucrats within the European Commission choose to exploit these opportunities. It has been suggested that the Commission has used climate change to shore up its legitimacy by placing it at the center of the integration agenda in the face of high levels of climate concern among the European public, especially in the wake of the Stern Report and the IPCC's Fourth Assessment Report (Oberthür 2008). The EU is also a strong supporter of multilateralism, both internally and externally, and climate change provides a good opportunity for the EU to exercise its 'soft power,' with the US serving as the significant 'other' against which EU policies are compared (Vogler 2002: 20). Yet the EU emissions trading scheme (ETS) has not yet delivered significant cuts in emissions, many members are underperforming, and enlargement from EU-15 to EU-27 has introduced new strains in maintaining unity and enhancing policy coherence (such as between climate policy and energy market liberalization, trade, and agricultural policy and development assistance) (Oberthür 2008: 45). Environmental directives have been frequently violated by many member states, including so-called Northern leaders and Southern laggards (Börzal 2003).

4.3 'National Interests'

Rationalists predict that states facing high costs of mitigation (arising from high dependence on fossil fuels or high marginal costs of further energy efficiency improvements) can
be expected to be climate laggards, other things being equal. Conversely, states that are
highly vulnerable to climate change impacts and therefore face high domestic costs arising
from inaction, can be expected to be leaders (if they have strong capacity) or pushers, other
things being equal (e.g. Sprinz and Vaahtoranta 1994; Porter et al. 2000; Dolsak 2001). (The
term 'pushers' refer to states which accept the seriousness of the climate challenge but lack
the institutional capacity to take the lead and therefore push for strong climate action by
other states that are capable of taking effective action.).

Nevertheless, this explanation cannot fully account for the diversity in national climate
performance. First, relative vulnerability to climate change turns out to be a poor predictor of
leadership among developed states. Australia is highly vulnerable to the physical impacts of
global warming yet it has generally been a laggard. Second, while dependency on fossil fuels
in domestic energy use and/or the export sector certainly raises the bar for aspiring leaders
(particularly in the US, Canada, and Australia) it is not a barrier to leadership. Germany,
which has traditionally depended on coal for the generation of electricity, emerged as a
leader in the early stages of international climate negotiations and continues to play a major
positive role in shaping EU climate policy. It has adopted a 2020 emissions reduction target
of minus 40 percent and is the only Annex I country to commit to an unconditional target at
the top end of the IPCC's recommended target range of minus 25–40 percent. These
differences can only be understood by examining national discourses and policy framing.

4.4 National Discourses

One of the difficulties associated with rationalist approaches to climate policy is that they
take national interests as exogenous and 'given' for the purposes of cost/benefit calculations, and therefore fail to explore how national interests are constructed. Our review has
found that the meaning ascribed to 'national interests' in relation to climate change is
highly variable and turns on how the policy problem and the policy response are framed
and interpreted by domestic political actors, and how they are filtered through different
political cultures and institutions. Different national climate discourses can give rise to
quite different cost/benefit calculations (e.g. economic, political, and/or military) set over
varying time horizons. Indeed, one of the key insights emerging from our review of national
responses to climate change is that issue framing, particularly in the early phase of the
climate policy cycle, has played a major role in shaping the subsequent path of national
climate policy development (see e.g. Cass 2006).

At the same time, problem definition and policy framing are also constrained and shaped
by domestic political cultures and institutions. For example, climate policy delay is strongest in those jurisdictions in which climate science has been reduced to an ideological
marker between political adversaries (such as the US, Australia) than in those jurisdictions
where climate science received bipartisan respect and acceptance (such as Germany). In the
US and Australia, for example, despite the strong scientific consensus on the risks of
climate change represented by the IPCC's assessment reports, so-called climate skeptics
have been given considerable space and/or time in the media, and in US Congressional

hearings, creating doubt in the mind of a generally inattentive public over the veracity and robustness of climate change science (e.g. Boykoff and Boykoff 2004; McCright and Dunlap 2003). While this is not a reflection of the status of climate science in these countries, it does reflect the quality of national media reporting and of the strongly partisan and adversarial character of the American and Australian political systems. Instead of defending construct-ivism over rationalism, or the power of discourses over the power of interests, we seek to draw attention to the ways in which climate discourses, including the social construction of natio-nal interests in relation to the climate change challenge, are constrained and shaped by political institutions. National climate discourses are not 'free floating' and readily transpor-table; rather, they are anchored in, and, to a significant extent produced by, the particular configuration of actors, interests, institutions, and local circumstances in each jurisdiction.

4.5 Strategies of Accumulation

Of particular importance are the ways in which climate policies are linked with 'strategies of accumulation'. We noted earlier that the capacity and motivation of states to respond to climate change is shaped, in part, by their location in the global capitalist economy, their relative material capability, and their general status as a developed or developing country. These characteristics provide the broad parameters for understanding individual state responses to climate change, including whether they have the capacity to develop strategies of capital accumulation that aggressively promote energy efficiency and the development of low carbon technologies. Small trading nations with weak state capacity, low income levels, high levels of indebtedness, and a heavy dependence on commodity exports are not well placed to seize these opportunities. Likewise, developing countries that serve as the locus for heavy industry that has migrated from developed countries also face considerable challenges. Conversely, developed countries with strong state capacity, relatively high incomes, and a diverse, advanced economy with a strong research and development culture are, prima facie, well placed to pursue a strategy of ecological modernization understood here to mean constant technological innovation to improve the environmental efficiency of production.

   Nonetheless, even within the latter group there is a broad division between those states that have historically pursued an aggressive model of 'carboniferous capitalism', based on the exploitation of cheap fossil fuels and extensive land development (e.g. US, Canada, and Australia), and those that have pursued ecological modernization by enacting stringent climate regulation to force greater efficiency, environmental productivity, and environ-mental technological innovation as a new competitive strategy (e.g. Germany) (Dalby and Paterson 2008; Paterson 2009). These differences are broadly reflected in the division between the Umbrella group and the EU, but there are important exceptions (Japan in the former group, and the 'Southern laggards' and many new EU accession states in the latter). It is notable that most of the climate leaders are social democracies with a corporatist style of interest group intermediation, with the exception of the United King-dom. Moreover, global economic interdependence sets limits on the degree to which climate leaders can continue to ratchet up their carbon constraints and remain competitive without risking the migration of industry, and hence 'carbon leakage,' to jurisdictions with weaker climate policies.

4.6 Domestic Actors and the Role of Veto-Coalitions

While the presence of a strong national environment movement, and of green parties, is generally conducive to strong climate performance they are not always sufficient drivers, especially when they are faced with powerful oppositional players in a political system where public interest advocacy is overshadowed by well-organized sectional interests or domestic veto players. Domestic veto players refer to the negatively affected groups and constituencies (such as the fossil fuel and energy-intensive industries, unions working in such industries, and subnational regional communities dependent upon such industries) as well as institutional veto players that politically represent, or are otherwise sympathetic with, these interests and concerns and have the capacity to thwart, stall, or dilute national climate legislation. Veto players have also been shown to play a major role in shaping the political executive's international negotiations by limiting the set of possible international agreements that can be domestically acceptable (the 'win-set') (Putnam 1988; Tsebelis 2002; Lantis 2009).

The veto-player thesis can illuminate why some states are climate laggards but it cannot explain why some states emerge as climate leaders despite the existence of domestic opposition from negatively affected groups. To understand this variation it is necessary to explore why some negatively affected groups are able to exert a decisive influence on policy output while others are overruled. There is no single answer to this question. Much depends on the nature of the political institutions through which contending climate discourses are mobilized, mediated, and channeled into climate policy, the political ideology of the dominant party and political leaders, the relative strength of the domestic advocacy coalitions in favor of strong climate action, the status of public opinion, and the role of the media.

5 CONCLUSION: WHAT MAKES A LEADER OR A LAGGARD?

....

It should be clear from the foregoing analysis that the quest to find a single cause, or even a common set of drivers, to explain climate leaders or climate laggards is a near-futile exercise. This is underscored by the cases of the United Kingdom and Germany. The United Kingdom is a unitary state with winner-take-all voting system, negligible green party representation, and a pluralist system of interest-group intermediation and policy making. Germany is a federal state, with a proportional representation voting system, relatively strong green party representation, and a more corporatist style of interest-group intermediation and policy making. Despite these significant institutional differences, both states have emerged as climate leaders, albeit at different times. Germany's status as a leader arose at the beginning of the international climate negotiations whereas the UK emerged much later, in the mid-2000s. Weidner has shown that Germany's leadership rests on twenty years of positive policy path dependency, that survived despite changing governments and shifts in socio-economic conditions (Weidner and Mez 2008: 357). That is, starting out as a leader early in the climate negotiations has made it easier for Germany to remain a leader. But a delayed start did not stop the United Kingdom from moving into this space.

While there is no clear institutional 'identikit' of a likely climate leader, some general conditions for strong climate performance are discernible. These usually include an advanced economy, a strong civil society, a strong and respected tradition of scientific research, and a diverse media. With these background conditions (or even without, in the case of China), a committed political executive is in a better position to face down domestic opposition, dampen the fears of negatively affected groups and constituencies, and construct a climate response that highlights not only the environmental but also the economic advantages arising from aggressive carbon constraints, such as technological innovation, improved productivity from efficiency gains, new green jobs, and energy security. This is more likely to happen in corporatist states than pluralist ones. National discourses and policy framing, particularly in the early stages of the international climate negotiations, have also proved crucial in enabling or disabling climate leadership. Political elites in leading states have engaged in creative issue linkage between climate policy and energy, economic and security policy, and emphasized the benefits of early action while those in laggard states have focused on the costs of taking action. National climate discourses often bear no relationship to vulnerability to climate change—indeed vulnerability turns out to be a very poor predictor of strong climate performance. The degree of dependency on fossil fuel is a better predictor of 'negative framing' and hence poor performance but it is by no means decisive. Much depends on the geographic distribution of fossil fuel resources, and the political institutions through which national climate discourses are filtered. If US coal reserves were concentrated in and extracted from just one or two states, rather than twenty-six, then US climate policy, and the international climate negotiations, might have followed a different path.

Finally, while our primary focus has been on understanding relative state performance, it is clear that the collective performance of states is very far from what is required to achieve the ultimate objective of the UNFCCC. It is also clear that leadership by the political executive can make a real difference even in the face of considerable domestic challenges. However the lack of sufficient action to combat global warming among the world's 193 states tells a sobering general story about the institutional incapacity of states and the weak motivation of political elites to respond effectively to the complex challenge of climate change. Our observations about the contested nature of measuring and ranking climate performance helps to underscore why, despite routine rhetorical endorsement of the burden-sharing principles of the UNFCCC by most parties, there remains considerable disagreement over the meaning of equity, differentiated responsibility, and capabilities under Article 3(1). This has prevented the determination of a single, principled approach to the allocation of mitigation obligations to individual states and left it instead to individual countries to determine their own 'equitable share' of global effort—using sometimes genuine and more often disingenuous special pleading to justify or explain their weak performance and/or weak policy aspirations.

Notes

1. Article 3(1) and 4(2)(a) UNFCCC.
2. The phrase 'industrialized states' refers to the states listed in Annex I to the UNFCCC, which includes developed states and a number of states that are undergoing a process of transition to a market economy.

3. The empirical evidence on national environmental performance in general (Neumayer 2002; Esty and Porter 2005; cf. Li and Reuveny 2006; Ward 2008), and national climate performance in particular among democratic and nondemocratic states is ambiguous (Congleton 1992; Midlarsky 1998; Roberts et al. 2004; Gleditsch and Sverdrup 2003; cf. Battig and Bernauer 2009).

REFERENCES

BATTIG, M. B., and BERNAUER, T. 2009. National institutions and global public goods: Are democracies more cooperative in climate change policy? *International Organization* 63: 281–308.

BOMBERG, E. 1996. Greens in the European Parliament. *Environmental Politics* 5(2): 324–31.

—— 1998. *Green Parties and Politics in the European Union*. London: Routledge.

BÖRZEL, T. 2003. *Environmental Leaders and Laggards in Europe*. Aldershot: Ashgate.

BOYKOFF, M. T., and BOYKOFF, J. M. 2004. Balance as bias: Global warming and the US prestige press. *Global Environmental Change* 14: 125–36.

BURCK, J. BALS, C., and ACKERMANN, S. 2009. *The Climate Change Performance Index: Results 2009*. Germanwatch and CAN Europe. Available at <http://www.germanwatch.org/klima/ccpi09res.pdf>. Accessed 25 November 2009.

CASS, L. 2006. *The Failures of American and European Climate Policy: International Norms, Domestic Politics, and Unachievable Commitments*. Albany, NY: SUNY Press.

CONGLETON, R. D. 1992. Political institutions and pollution control. *Review of Economics and Statistics* 74(3): 412–21.

DALBY, S., and PATERSON, M. 2008. Over a barrel: Cultural political economy and oil imperialism. In F. Debrix and M. Lacy (eds.), *Insecure States: Geopolitical Anxiety, the War on Terror, and the Future of American Power*. London: Routledge.

DOLSAK, M. 2001. Mitigating global climate change: Why are some countries more committed than others? *Policy Studies Journal* 29(3): 414–36.

DUNLAP, R., and McCRIGHT, A. 2008. A widening gap: Republican and Democratic views on climate change. *Environment* 50(5): 26–35.

ECKERSLEY, R. 2004. *The Green State: Rethinking Democracy and Sovereignty*. Cambridge, MA: MIT Press.

Ecofys. 2009. G8 CLIMATE SCORECARD. At <http://www.worldwildlife.org/climate/Publications/WWFBinaryitem12911.pdf>

ESTY, D., and PORTER, M. E. 2005. National environmental performance: An empirical analysis of policy results and determinants. *Environment and Development Economics* 10: 391–434.

FISHER, D. 2006. Bringing the material back in: Understanding the US position on climate change. *Sociological Forum* 21(3): 467–94.

Germanwatch/CAN Europe. 2010a. *The Climate Change Performance Index 2010*. At <http://www.germanwatch.org/klima/ccpi.htm>.

—— 2010b. *The Climate Change Performance Index 2010—Background and Methodology*. http://www.germanwatch.org/klima/ccpi-meth.pdf>.

GLEDITSCH, N. P., and SVERDRUP, B. O. 2003. Democracy and the environment. Pp. 45–70 in E. A. Page and M. Redclift (eds.), *Human Security and the Environment: International Comparisons*. Cheltenham: Edward Elgar.

HARRISON, K. 2007. The road not taken: Climate change policy in Canada and the United States. *Global Environmental Politics* 7(4): 92–117.

—— and SUNDSTROM, L. M. 2007. The comparative politics of climate change. *Global Environmental Politics* 7(4): 1–18.

IEA (International Energy Agency). 2007. Selected 2007 indicators per country. At <http://www.iea.org/stats/indicators.asp?COUNTRY_CODE>.

LANTIS, J. S. 2009. *The Life and Death of International Treaties: Double-Edged Diplomacy and the Politics of Ratification in Comparative Perspective*. Oxford: Oxford University Press.

LI, Q., and REUVENY, R. 2006. Democracy and environmental degradation. *International Studies Quarterly* 50: 935–56.

McCRIGHT, A. M., and DUNLAP, R. E. 2003. Defeating Kyoto: The Conservative movement's impact on US climate change policy. *Social Problems* 50(3): 348–73.

MATTHEWS, M. M. 2001. Cleaning up their acts: Shifts of environment and energy policies in pluralist and corporatist states. *Policy Studies Journal* 29(3): 478–98.

MIDLARSKY, M. I. 1998. Democracy and the environment: An empirical assessment. *Journal of Peace Research* 35(3): 341–61.

NEUMAYER, E. 2002. Do democracies exhibit stronger international environmental commitment? *Journal of Peace Research* 39(2): 139–64.

OBERTHÜR, S. 2008. EU Leadership in international climate policy: Achievements and challenges. *The International Spectator* 43(3): 35–50.

PAN, J., PHILLIPS, J., and CHEN, Y. 2009. China's balance of emissions embodied in trade: Approaches to measurement and allocating international responsibility. Pp. 142–66 in D. Helm and C. Hepburn (eds.), *The Economics and Politics of Climate Change*. Oxford: Oxford University Press.

PATERSON, M. 2009. Post-hegemonic climate change? *British Journal of Politics and International Relations* 11(1): 140–58.

PIK (Potsdam Institute for Climate Impact Research). 2010. Ambitions of only two developed countries sufficiently stringent for 2 °C. Available at<http://www.pik-potsdam.de/news/in-short/ambition-of-only-two-developed-countries-sufficiently-stringent-for-2boc>. Accessed 10 March 2010.

POLONI-STAUDINGER, L. 2008. Are consensus democracies more environmentally effective? *Environmental Politics* 17(3): 410–30.

PORTER, G., BROWN, J., and CHASEK, P. S. 2000. *Global Environmental Politics*. Boulder, CO: Westview Press.

PUTNAM, R. 1988. Diplomacy and domestic politics: The logic of two-level-games. *International Organization*, 42(3): 427–61.

ROBERTS, T. J., PARKS, B. C., and VÁSQUEZ, A. A. 2004. Who ratifies environmental treaties and why? Institutionalism, structuralism and participation by 192 nations in 22 treaties. *Global Environmental Politics* 4(3): 22–64.

SCHREURS, M. A., and TIBERGHIEN, Y. 2007. Multi-level reinforcement: Explaining European Union leadership in climate change mitigation. *Global Environmental Politics* 7(4): 19–46.

SPRINZ, D. F., and VAAHTORANTA, T. 1994. The interest-based explanation of international environmental policy. *International Organization* 48(1): 77–105.

TSEBELIS, G. 2002. *Veto Players: How Political Institutions Work*. Princeton: Princeton University Press.

VOGLER, J. 2002. In the absence of the hegemon: EU actorness and the global climate change regime. Paper presented to the conference on *The European Union in International Affairs*, National Europe Centre, Australian National University, 30 July.

WARD, H. 2008. Liberal democracy and sustainability. *Environmental Politics* 17(3): 386–409.

WEALE, A., PRIDHAM, G., CINI, M., KONSTANDAKOPULOS, D., PORTER, M., and FLYNN, B. 2000. *Environmental Governance in Europe*. Cambridge: Cambridge University Press.

WEIDNER, H. 2005. Global equity versus public interest? The case of climate change policy in Germany. Discussion Paper SP IV 2005–102, WZB. Social Science Research Center, Berlin.

—— and MEZ, L. 2008. German climate change policy: A success story with some flaws. *The Journal of Environment and Development* 17: 356–78.

WIEDMANN, T., WOOD, R., LENZEN, M., MINX, J., GUAN, D., and BARRETT, J. 2008. *Development of an Embedded Carbon Emissions Indicator*. Report to the UK Department for Environment, Food and Rural Affairs by the Stockholm Environment Institute at the University of York and the Center for Integrated Sustainability Analysis at the University of Sydney, June. London: Department of Environment, Food and Rural Affairs.

CHAPTER 30

....

CLIMATE CHANGE POLITICS IN AN AUTHORITARIAN STATE: THE AMBIVALENT CASE OF CHINA

....

MIRANDA A. SCHREURS

1 INTRODUCTION

....

No single country will have a larger impact on the global community's ability to reduce its greenhouse gas emissions than China. China is among the most polluted countries on the planet. State-owned coal-fired power plants, petrochemical companies, and steel and cement factories spew huge amounts of pollution into the environment. China's rapidly growing automobile market has transformed the country from a land of bicycles to a land of super highways and traffic jams (Gallagher 2006). Technological inefficiencies mean that there is tremendous waste in the use of energy and other resources. Economic growth is privileged over environmental protection and corruption, especially at the local level, has inhibited the implementation of many environmental laws. China's greenhouse gas emissions are the highest in the world and those emissions are growing. This situation bolsters the position of those who argue that authoritarian states tend to do poorly at environmental protection (Feshbach and Friendly 1993; Shapiro 2001; Economy 2004).

This depiction of China is accurate, but it is also oversimplified. Although many Chinese cities fail to meet World Health Organization air quality standards and few Chinese would consider climate change a top policy priority, there have been significant changes in Chinese approaches to environmental pollution in general, and climate change more specifically since the early 1990s. The central government, a growing number of urban communities, and some provinces have strengthened substantially their environmental laws, programs, and institutional capacities. China has rapidly developed a wide range of environmental policies and programs that will contribute to a reduction in the growth of its greenhouse gas emissions (although not in total emissions). Environmental non-governmental organizations have been allowed to form and the government has forced the closure of thousands of the most energy-inefficient and highly polluting firms (Turner and Zhi 2006). These developments challenge theories that suggest that authoritarian states are largely unresponsive to environmental concerns.

This raises interesting theoretical questions as to whether China is developing a unique approach to climate change and other environmental matters. While on the one hand certain traditional elements of the authoritarian model continue to prevail in China, such as the restrictions placed on foreign environmentalists wishing to be active or to conduct research in Tibet or the lack of tolerance for protests against major dam projects or nuclear energy facilities, on the other China's authoritarian leaders appear to have accepted that the country's future is tied to improving its environmental performance. In some areas related to climate change, China is becoming a world leader, such as in the use of solar water heaters and the development of renewable energies. It has also introduced measures that would be hard to implement in most democratic systems, such as sudden top-down decisions to shut down on short notice hundreds of heavily polluting firms or to limit the use of automobiles and industrial activities during special events, such as during the Olympics in Beijing in 2008 or the World Expo in Shanghai in 2010. Mark Beeson (2010) argues that this kind of environmental authoritarianism may be necessary if China and other countries in Southeast Asia are to deal with the environmental challenges they face.

This chapter begins with an overview of China's energy mix and the country's growing contribution to global greenhouse gas emissions. It then considers how China's leaders have used China's status as the world's second largest economy to demand greater voice in the international climate negotiations, but also exploited its status as a transition economy with millions who still live at various levels of poverty to resist international demands that China take on binding climate mitigation goals. The chapter then discusses steps that have been taken in China to strengthen institutional capacity and plans, laws, and programs that have been introduced to combat climate change, develop renewable energies, and improve energy efficiency. It also considers the extent to which Chinese authorities are permitting critical voices, and specifically environmental groups, to form and be active. In a conclusion, the chapter considers whether China is developing a new form of environmental authoritarianism.

2 CHINA'S RISING GREENHOUSE GAS EMISSIONS

Fossil fuels are a dominant component of the energy mix in China. Coal accounted for approximately 70 percent, oil 20 percent, hydro 6 percent, natural gas 3 percent, nuclear 1 percent, and other renewables 0.06 percent of China's total energy consumption in 2006 (US Energy Information Administration 2009). Changing energy consumption patterns and improving energy efficiency will be critical to addressing climate change. China's per capita CO_2 emissions increased by about 46 percent between 2002 and 2005 (Perkins 2009).

With growth rates that hovered between 8 and 10 percent per year since approximately 1980, China's economy has surged and so has its dependence on energy. China is now the manufacturing base for much of the West. China's annual per capita carbon dioxide emissions from fossil fuels in 2006 (1.27 metric tons) are mid-way between those of India (0.37 metric tons) and Japan's (2.8 metric tons) although they are still well behind those of the US (close to 5.18 metric tons) (Boden et al. 2010).

3 DOUBLE FRAMES: CHINA AS DEVELOPING ECONOMY BUT ECONOMIC POWERHOUSE

China's mixed policy messages on climate change are also related to its status as a transition economy. In many ways, China is still very much a developing country with hundreds of millions living in various degrees of poverty. It is largely the urban coastal regions of China that are booming economically and are the biggest resource consumers. The inland and western regions of the country are where much of coal and other resources that are consumed by the richer regions of the country are mined. It is these areas that face some of the country's worst pollution problems. This inherent inequity—essentially an environmental justice problem—is known to the government. The government appears to be concerned about the numerous protests that have erupted in villages and towns where pollution problems are very serious (Sun and Zhou 2008; Moore 2009; L. Li 2009; Ma 2009). Programs like the Go West Campaign were conceived in part to bring more development prospects to less developed regions of the country. The Go West Campaign includes funding for roads, rail, airports, electricity grids, and coal-fired powerplants. It also is supposed to include environmental projects (Moxley 2010).

Within China, economic inequalities are large; first world/third world dichotomies can be found within cities, between urban and rural areas, and between the coastal regions and inland areas. Chinese leaders do not focus so much on the internal inequalities related to climate change as the international ones.

China's leaders have argued that it is Western countries that are responsible for global warming as they have emitted the vast bulk of the carbon dioxide emissions that have accumulated in the atmosphere over the past century. It is therefore, according to this reasoning, the responsibility of these developed countries to drastically cut back their greenhouse gas emissions and provide developing countries with technical and financial assistance to improve their energy efficiency and develop alternative energies. Central to this argument is that developing countries, including China, have the right to continue to develop and catch up to the economic level of the richer countries in Europe and North America. This means also that they should have a right to increase their per capita and total greenhouse gas emissions and not be bound by any binding emission targets (Heggelund et al. 2010).

At the same time, as the world's second largest economy, China is also very much a global economic heavyweight. Given China's relatively low per capita gross domestic product (GDP), per capita greenhouse gas emissions are about one-fifth of what they are in the United States and about half those in Europe. Yet, China overtook the United States as the world's largest greenhouse gas emitter in the mid-2000s (Netherlands Environmental Assessment Agency 2008). China accounts for over about one-fifth of global greenhouse gas emissions. China has the world's largest cement industry, second largest automobile market, a seemingly insatiable construction industry, and a large and growing middle class.

As a result of rapidly rising emission levels tied to this growing consumption and manufacturing, China is under considerable pressure to improve its sharply rising long-term greenhouse gas emission trajectory and to establish ambitious greenhouse gas mitigation targets.

4 Capacity Building for Climate Change, Energy, and the Environment

Many new environmental, energy efficiency, and renewable energy promotion laws and programs have been introduced in China since the country's leaders first began to pay more serious attention to climate change after the 1992 United Nations Conference on Environment and Development. There has been considerable institutional capacity building to address environmental, climate change, and energy concerns at both the national and the provincial levels.

The State Council's National Development and Reform Commission (NDRC), which has responsibility for overseeing and planning China's economic development, is also a central actor influencing China's climate change strategies. The NDRC's Department on Climate Change is responsible for formulating climate change strategies and policies and managing China's participation in the international climate negotiations and other international cooperation programs.

In 2007, a National Leading Committee on Climate Change comprising representatives of twenty ministries and government sectors to enhance the coordination of climate policies across policy sectors was established. Significantly, it is chaired by Prime Minister Wen Jiabao. The NDRC manages the National Leading Committee on Climate Change's daily operations as well.

Parallel structures are being set up at the provincial level on the order of the State Council. Interestingly, the provinces and national autonomous regions are setting up different types of leading groups. Some are more focused on climate change, others on energy saving, and at least one on energy saving and pollution reduction. Since 2008, governors at all levels have been required to report on their efforts to save energy and reduce pollutant discharge (Ministry of Environmental Protection 2007; Qi et al. 2007; Gang 2009).

In 2008, the State Environmental Protection Administration was replaced by an upgraded Ministry of Environmental Protection. This was a recommendation made by various international groups, including the Organization for Economic Cooperation and Development (2007) and the China Council on International Environment and Development, Environmental Governance Task Force (Xue et al. 2007). While still small in terms of personnel (about 300 in Beijing, 150 in regional offices, and another 2,000–2,500 in supporting agencies) and budget, the establishment of the ministry is nevertheless significant (He 2008). The Ministry of Environmental Protection is responsible for preparing and administering environmental laws and formulating environmental standards. It also prepares annual State of the Environment reports and participates in international environmental cooperation.

5 CHINA'S NATIONAL CLIMATE CHANGE STRATEGY

In June 2007, the Chinese government issued its first National Climate Change Program. The program argues that developed countries have primary historic responsibility for climate change and therefore must shoulder the bulk of the emission reductions necessary to prevent further global warming. Developing countries—and here China defines itself as one—have the right to further development and should not be expected to take actions that would hurt their economic growth. That said, the National Climate Change Program calls for the development of low carbon and renewable energy, promotes nationwide tree planting, and discusses family planning and China's one-child policy as an important aspect of a climate change policy. The program also points out the importance of further expanding climate change institutions and laws and research capacity. Of particular interest are the 2010 goals that were established by the program: a 20 percent reduction in energy consumption per unit GDP, a 10 percent share for renewables in the total energy mix (compared to the 3 percent level of 2003), and 20 percent forest coverage for the country.

After the central government issued the National Climate Change Action Plan in 2007 China's provinces and autonomous regions also began to issue their own plans. Xinjiang, an autonomous region experiencing desertification, for example, issued an action plan that promotes energy conservation and renewable energy. It has set a target to achieve a 20 percent improvement in energy intensity between 2005 and 2010 and a water use efficiency improvement target of 6 percent. Hebei Province, a major steel-producing region, set a goal of improving energy intensity by 8 percent by 2010 relative to 2005 levels and expanding forest cover to 26 percent. Energy conservation and improved energy intensity are key components of most of the plans (Qi et al. 2007).

China's national climate strategy focuses most intensively on means to reduce energy consumption per unit of GDP, develop new energy and environmental technologies, develop clean coal technologies, expand nuclear power, employ combined heat and power generation, sharply expand renewables, and expand forest coverage, among other measures (National Development and Reform Commission 2007). Another element taken up is the country's huge population. The one-child policy that was introduced in 1986 after the population hit the one-billion mark (compared with 550 million in 1950) has become an element of the national climate strategy.

As a developing country China is a non-Annex I party to the Kyoto Protocol, and therefore has no greenhouse gas emission target to meet under the agreement. China nevertheless has been able to benefit from the Clean Development Mechanism (CDM) of the Kyoto Protocol, which was established to permit industrialized countries to obtain emission reduction credits for greenhouse gas reduction projects taken in developing countries. China had hosted 737 CDM projects as of February 2010. China's CDM projects focused heavily on renewable energy, energy saving and efficiency improvements, and the capture and destruction of greenhouse gases with high global warming potentials (such as methane) (UNFCCC 2009).

6 RENEWABLE ENERGY AND ENERGY EFFICIENCY

··

As a planned economy, the Chinese government issues economic development plans every five years. Energy efficiency improvements and renewable energy expansion have become key energy policy goals. The Tenth Five-Year Plan (2001–5), for instance, made specific mention of the need to increase solar, wind, and geothermal energy. The plan also called for the development of large grid-connected wind, solar, thermal, and photovoltaic power. The Eleventh Five-Year Plan (2006–10) went one step further and established a target for reducing energy consumption per unit of GDP by 20 percent. Expectations are that the Twelfth Five-Year Plan (2011–15) will continue its focus on energy savings, renewable energy promotion, and sustainable development.

The 2005 Renewable Energy Law[1] and related enabling measures are based on a feed-in-tariff model, establishing fixed and government-guided subsidies for renewable energies. In the 2007 National Renewable Energy Development plan, Chinese authorities set a 15 percent renewable energy target as a share of the total energy mix for 2020.

The Renewable Energy Law requires grid operators to purchase renewable energy from registered producers. It also indicates the government's intentions to provide financial support for renewable energy development. A 0.25 yuan per kilowatt hour subsidy for renewable energy production was set by the law for the first fifteen years of a facility's operation, with subsidies decreasing by 2 percent per year relative to the price of the previous year for projects started after 2010.[2] A surcharge for renewable energy is to be paid by end users. The law mandates that technology must include 70 percent domestic Chinese content to receive subsidies, an incentive to the development of Chinese companies and engineers in this field. This provision was expected to help spur the development of China's renewable energy industry, eventually allowing China to become a world leader in related technologies. Renewable energy technology has in fact become a significant export market for China. In 2009, China became the world's largest manufacturer of wind turbines. It had already held that position for some years in relation to solar panels (Bradsher 2010).

An amendment to the law was passed in 2009 in the face of mounting problems with grid owners who were failing to connect renewable suppliers to the grid. The amendment requires grid operators to buy all renewable energy that has been produced. Operators failing to comply will be fined double the level of the loss experienced by the renewable energy producer (H. Li 2009).

6.1 Renewable Energies

Hydropower is by far the most significant (albeit at times sometimes rather controversial) renewable energy source in China. China is a leader in the field of small-scale hydropower generation and hosts the headquarters of the International Network on Small Hydro Power in Hangzhou. While small- and some medium-sized dams have been built in China without causing much societal unrest (at least visible to the outside world), in recent decades the Chinese government has planned several massive dam projects. For the Chinese government, dam development is viewed as very important to China's energy future. Hydroelectric power is touted as a clean energy source that will help the country to reduce growth in carbon dioxide emissions.[3]

Wind energy is the most rapidly expanding renewable source. China has been doubling its wind energy capacity on a yearly basis. In 2008, China was the world's fourth largest wind energy producer, with 12 GW of installed capacity. By the end of 2009, it had risen to the number 3 position worldwide after again doubling its installed capacity to 25 GW (Global Wind Energy Council 2010). The Chinese Renewable Energy Industry Association estimates that 1.12 million jobs were in the renewables sector (Bradsher 2010). China has for years been the world leader in the use of solar water-heating systems. More recently, it has been promoting photovoltaic energy technologies and biomass. Hydropower, however, remains the largest renewable energy source.

Renewable energy is only seen as a component of the solution, however. In addition, China is investing in nuclear energy and oil and gas exploration. The Chinese government is also investing significantly into carbon capture and storage.

6.2 Energy Efficiency

Energy efficiency improvements are a related key component of China's climate change strategies. In 2005, the Chinese government announced plans to reduce energy consumption per unit of GDP by 20 percent between 2005 and 2010. To help achieve this goal, a special program focusing on improving energy efficiency in China's top 1,000 energy-consuming industries was initiated as these industries accounted for approximately 47 percent of national industrial use in 2004. These include mainly enterprises in the steel, iron, utilities, petrochemicals, and construction materials industries. Enterprise-specific targets were negotiated with provincial and local governments based on each enterprise's specific conditions. Government authorities have an incentive to make sure that companies comply as their promotions can be tied to environmental and energy efficiency performance (Price et al. 2008).

Many other plans are in place as well. The Eleventh Five-Year Plan outlined numerous strategies for improving energy efficiency. These included the promotion of local combined heat and power generation, improving the efficiency of coal-fired boilers through, for example, the use of circulating fluidized bed or pulverized coal firing, energy conservation in buildings, energy efficient lighting, among other initiatives. The Chinese government has used climate and environmental goals to justify its decisions to shut down old and inefficient facilities. It has established incentive programs to encourage consumers to buy energy-efficient appliances.

7 China in the International Climate Negotiations on a Post-Kyoto Agreement

China had no obligations under the Kyoto Protocol. Developed countries, including in the European Union, Japan, and North America have made their positions clear that in any successor agreement, China and other transitioning economies must take on some commitments. Responding to these growing expectations and demands, at the United Nations General Assembly meeting in New York in September 2009 China staked out its emerging position.

Chinese President Hu Jintao stated that China would take steps to reduce the growth of its greenhouse gas emissions by 'a notable margin' by 2020 and increase the percentage of nuclear and renewable energies in the total energy mix to 15 percent by 2020 (MacFarquar 2009). In November 2009, the Chinese government went one step further announcing that it was prepared to reduce the country's greenhouse gas emissions per unit of GDP by 40 to 45 percent by 2020 (Graham-Harrison and Buckley 2009). This is the range China committed to when it signed the non-binding Copenhagen Accord in March 2010.

While the Chinese commitment was seen as unambitious and China's unwillingness to set a binding emissions target as obstructionist by the Western press and environmentalists, the Xinhua News Agency argued that the Western countries were to blame. The agency reported that during the Copenhagen negotiations there was debate about whether negotiations should continue within the United Nations Framework Convention on Climate Change (UNFCCC) and the Kyoto Protocol framework (i.e. a two-track approach) or only under the UNFCCC. They went on to write: 'While developing countries insisted on the two-track approach, developed countries were trying to throw away the Kyoto Protocol and replace it with a new single deal. Their real intention was to dodge their mandatory obligations under the Kyoto Protocol and force developing countries to do more, which ran counter to the principle of common but differentiated responsibilities' (Xinhua 2009). There is a clear difference in the framing of the climate change issue between China and many more developed countries, including European and North American ones.

8 Understanding the Driving Forces Behind China's Climate Change Programs

Over the course of the last decade, the Chinese government has become increasingly aware of the seriousness of its own environmental problems and the economic and human costs of pollution. Political attention to environmental problems has started to grow as the impact of pollution and natural resource degradation on the economy and quality of life has become increasingly large. Also important are the Communist Party's concerns with political stability and image.

8.1 Costs

For decades, pollution problems were largely ignored in China's quest for development (Shapiro 2001; Economy 2004). As a result, China suffers severely from essentially all forms of pollution. The poor air quality found in many urban and industrial areas is related to the burning of coal, oil, wood, and agricultural crops, all of which contribute to the release of greenhouse gases. Desertification, a consequence of over-grazing, drought, and excessive use of ground water, contributes to the country's problems with sand storms, especially in the spring months. Hillside deforestation has contributed to flooding and erosion problems. Climate change could contribute to more rapid desertification as water availability is reduced. The deforestation of vast parts of the country in the previous century has reduced China's natural carbon sinks.

Pollution has taken a heavy toll on China both physically and economically. The World Bank estimated that the total health and ecological costs of air and water pollution in 2003 equated to between 2.67 and 5.78 percent of GDP depending on the methodology employed (World Bank 2007).

8.2 Energy Security

A driving force behind China's climate policy initiatives that deserves special attention is the government's concern with the nation's growing energy dependence on imports. China, once energy self-sufficient, now must import coal, oil, and natural gas to meet domestic demand. This is pushing China to explore for energy contracts in other regions of the world, notably in the Middle East and Africa (Liangxing 2005). At the same time, Chinese leaders have pushed energy efficiency improvements and renewable energy development domestically. Essentially the development of all forms of energy is embraced and encouraged.

8.3 Political Stability and Party Image

Responding to this kind of information and growing pressures from within and without, Chinese leaders in their speeches now repeatedly highlight the importance of action. Prime Minister Wen Jiabao in an address to an annual national environmental Congress stated:

> We must be fully aware of the severity and complexity of our country's environmental situation and the importance and urgency of increasing environmental protection. Protecting the environment is to protect the homes we live in and the foundations for the development of the Chinese nation. We should not use up resources left by our forefathers without leaving any to our offspring. China should be on high alert to fight against worsening environmental pollution and ecological deterioration in some regions, and environmental protection should be given a higher priority in the drive for national modernization. (Address to the 6th National Conference on Environmental Protection, 17 April 2006)

Aware that pollution could derail Beijing's plans to host the Olympics, massive efforts were made to clean up the environment through a mix of long-lasting and temporary measures, including restricting diesel truck traffic, limiting automobile traffic through a license plate numbering system, shutting down heavily polluting factories, raising auto fuel emission standards, and halting construction. According to a study conducted by Cornell University (2009), carbon dioxide emissions in the city were 47 percent lower in 2008 than at the same time in 2007. Some cities have adopted some of the more permanent measures introduced in Beijing. Shanghai, Shenzhen, and Guangzhou followed Beijing's lead in adopting Euro 4 fuel standards (Earth Times 2008).

9 Environmental NGOs in an Authoritarian Regime: An Element of Pluralism?

There have been important changes in the government's relationship to civil society. Prior to 1994, there were no environmental non-governmental organizations in China because the government did not allow them. Wu Chenguang put it very directly: 'When Beijing launched her first bid for the Olympic Games in 1993 and was asked by officials from the International Olympic Committee (IOC), our delegation did not even know how to answer the question whether or not there were environmental NGOs in China.'[4] Now there are several thousand. These include both Chinese offices of international NGOs, government-backed groups, home grown environmental NGOs, student groups at universities, and groups formed by environmental journalists.

The first environmental NGO to register in China was the Academy for Green Culture (now called Friends of Nature). Global Village Beijing and Green Home were then set up in 1996. When environmental groups first emerged in China, they were primarily focused on environmental education and nature conservation (Yoon 2007). Now many deal with climate change as well.

World Wide Fund for Nature (WWF), China, for example, has launched a Low Carbon Cities Initiative.[5] The WWF's project was motivated by the critical role that China's cities will play in the environmental future of the country as well as the globe. The WWF notes that with the rapid growth of China's cities—as of 2005 there were 54 mega-cities (with populations of over one million) and 84 large cities—and the growing energy consumption of an increasingly wealthy country, pollution levels are severe and greenhouse gas emissions are rising. The Low Carbon City Initiative intends to work with various cities (starting with Shanghai and Baoding) to enhance energy efficiency in industry, construction, and transportation sectors and promote renewable energy. The WWF, China has also launched a campaign for public awareness raising called '20 ways to 20 percent'. The campaign has a network dimension to it. Fifty Chinese NGOs are working together to promote a 26 Degree Centigrade Air Conditioning Program, urging hotels, businesses, and residents not to set their air conditioners for below 26 degrees Centigrade in the summer; promoting energy efficiency labeling; green lighting; and office building projects.

Natural Resources Defense Council is working on winning support in China for green buildings and sustainable city planning with model projects and other initiatives. The efforts contributed to the Chinese government's decision to establish a green building standard in 2005. China's national goal is to achieve 50 percent energy-efficiency improvements in new buildings by 2020 and 65 percent in Beijing, Shanghai, Tianjin, and Chongqing (NRDC 2008; WRI 2009). Greenpeace China is working with academics to produce policy recommendations for the government. It is also working to win public and private support for a reduction in the country's dependence on coal.[6]

In the month leading up to the Copenhagen climate conference in December 2009, eight Chinese NGOs (Friends of Nature, Global Village of Beijing, Green Earth Volunteers, Institute of Public and Environmental Affairs, Greenpeace, Oxfam Hong Kong, World Wide Fund for Nature China, and Actionaid China), jointly issued a statement that called for developed countries to take responsibility and cut their emissions in the range of 40 percent of 1990 levels by 2020 and provide financial, technical, and capacity-building assistance to developing countries for mitigation and adaptation measures. It also called for developing countries to implement climate measures and adopt voluntary emission control targets. In relation to China, the NGOs called for Chinese leadership among developing countries in addressing climate change, consideration of social equity in the formulation and implementation of measures, and promotion of energy saving, renewable energies, a low-carbon society, and sustainable development.[7]

The China Youth Climate Action Network was formed in 2007. It brings together seven youth organizations who have agreed to work together to share and disseminate scientific information about climate change as well as reduce greenhouse gas emissions by 20 percent by 2012 at select higher education institutions.[8]

These examples suggest that environmental NGOs are increasingly accepted by the state and may even have become necessary for the government to achieve its internal goals and portray a degree of pluralism internationally. The existence of an NGO community bolsters China's leadership's efforts to portray itself as having a benevolent form of environmental authoritarianism. NGOs help the state meet international normative expectations, provide assistance in educating the Chinese population on environmental matters, and even provide some useful policy ideas. The ability of the NGOs to influence government policy, however, remains limited. Moreover, it must be remembered that there are still cases where the Chinese state clamps down on environmentalists who are too critical of state plans or the political leadership. Journalists and environmental activists are barred from entering some regions.

10 CONCLUSION: THE CLIMATE POLICY ACHIEVEMENTS AND CHALLENGES AND THE RISE OF ENVIRONMENTAL AUTHORITARIANISM

As a result of economic restructuring, energy-saving measures, research and development, and diffusion of energy-saving technologies, China's energy intensity decreased at an annual rate of 4.1 percent per year from 1990 to 2005. The share of coal in China's primary energy mix decreased from 76.2 percent in 1990 to 68.9 percent in 2005 and the share of renewables grew to 17 percent of the total electricity mix in 2006 (REN21 2008). These developments suggest that there has been some progress in meeting climate policy goals. On the other hand, the state remains strongly focused on economic development. Coal remains a dominant component of China's energy mix and can be expected to do so long into the future. The Chinese government continues to push the development of an automobile-centered transportation system and the infrastructure to support it. Western-style consumption is increasingly visible among China's rapidly growing middle class. These trends raise serious concerns about how China can eventually not just slow the growth in its emissions, but reduce them.

China appears to be developing a model of authoritarianism that is increasingly accepting of the need to address pollution problems and to act on climate change but only to the extent doing so does not interfere with economic growth objectives. At the same time that China is developing wind parks and promoting eco model cities (Schreurs 2010), the government is investing heavily in large-scale infrastructure development and fossil fuel and mineral exploration. Thus, while China may to some extent be adopting elements of the Singaporean (authoritarian) model of environmental protection,[9] government and economic leaders continue to see as core to the country's economic future the development of traditional manufacturing industries and the exploitation of fossil fuel energy sources.

China's new form of authoritarianism is a system in flux. The country is shifting rapidly from a socialist economic model to a capitalist one. The Communist Party still prohibits party competition but it has allowed for a degree of open debate about environmental matters. Still, the challenges to transforming the economy in a low-carbon direction remain daunting. Implementation challenges abound.

With continued rapid economic growth, the demand for energy can be expected to rise as domestic consumption expands. To combat rising greenhouse gas emissions, over the past decade China has introduced numerous new laws and programs. Successes have been achieved. Energy efficiency is improving, renewable energy production is growing, reforestation efforts have expanded the percentage of forested land. In the process of making these changes, China is becoming a global leader in clean energy technologies.

Chinese government leaders appear to be serious about their commitment to make environmental improvements albeit on their own terms, not those dictated by Europe, the United States, or Japan. In the long-run, the stability of Chinese society may well be tied to whether the government succeeds in forging a path towards sustainable development or continues down one of heavy pollution. China's choices will greatly influence global efforts to address climate change. If China's leadership uses a strong hand to guide the economy away from heavily polluting industries and production processes, is serious about enforcement, promotes nationwide environmental education campaigns that encourage the development of a less consumption-oriented societal model than has prevailed in the West, and remains responsive to the critiques of environmental groups and scientific experts, then it may well be able to develop a 'benevolent' form of environmental authoritarianism that could go far in mitigating China's increasingly large environmental footprint and addressing the climate change challenge. We must hope that this is indeed the direction that China's environmental authoritarianism takes.

NOTES

1. <http://www.martinot.info/China_RE_Law_Beijing_Review.pdf>.
2. <http://www.martinot.info/China_RE_Law_Guidelines_1_NonAuth.pdf>.
3. Certainly, the most controversial dam project has been the Three Gorges Dam. The idea of damming the Yangtze, one of the world's great rivers, was originally conceived by Sun Yatsen in 1919. Planning for the dam began in 1954 under the leadership of Mao Tsetung. Construction plans were approved by the National People's Congress in 1992 although one-third of the delegates either voted against it or abstained from voting, suggesting that there was considerable opposition to the project. The Three Gorges Dam is the world's largest dam. The dam was championed for a variety of reasons, but primarily for electricity production (the dam has over 18,000 MW installed capacity) and flood control. More recently, it is also presented as a major initiative that has been taken to contribute to greenhouse gas mitigation. Critics of the dam have condemned the environmental destruction, historical and cultural loss, and human resettlement the dam has required. Despite the opposition especially to large dams, the Chinese government plans to continue the expansion of its hydropower capacity. This is seen as critical to achieving energy security.

4. Wu Chenguang, translated by Wang Qian for China.org.cn, 13 July 2002.
5. <http://www.wwfchina.org/english/sub_loca.php?loca=1&sub=96>.
6. <http://www.greenpeace.org/china/en/campaigns/stop-climate-change/climate-change-what-we-do>
7. <http://www.eu-china.net/web/cms/upload/pdf/materialien/eu-china_2009_hintergrund_14.pdf>.
8. See H. Li 2009; <http://www.cycan.org/Category_33/index.aspx#1>.
9. Dr Stephan Ortmann, 'Environmental governance under authoritarian rule: Singapore and China,' paper presented to Deutsche Vereinigung fuer Politische Wissenschaft in Kiel Ad-hoc Group 'Vergleichende Diktatur- und Extremismusforschung,' 25 September 2009. <https://www.dvpw.de/fileadmin/docs/Kongress2009/Paperroom/2009VglDiktatur-pOrtmann.pdf>.

REFERENCES

BEESON, M. 2010. The coming of environmental authoritarianism. *Environmental Politics* 19(2): 276–94.

BODEN, T., MARLAND, G., and ANDRES, B. 2010. Carbon dioxide information analysis. Center Oak Ridge National Laboratory, doi 10.3334/CDIAC/00001.

BRADSHER, K. 2010. China leading global race to make clean energy. *New York Times*, 30 January.

Cornell University. 2009. Improved air quality during Beijing Olympics could inform pollution-curbing policies. *ScienceDaily*, 5 August. Retrieved from <http://www.sciencedaily.com /releases/2009/07/090724113548.htm>.

Earth Times. 2008. Champion of Beijing's Olympic air cleanup gets award. *Earth Times*, 12 November 2008.

ECONOMY, E. C. 2004. *The River Runs Black: The Environmental Challenge to China's Future.* Ithaca, NY: Cornell University Press.

FESHBACH, M., and FRIENDLY, A. Jr. 1993. *Ecocide in the USSR: Health and Nature under Siege.* New York: BasicBooks.

GALLAGHER, K. 2006. *China Shifts Gear: Automakers, Oil, Pollution and Development.* Cambridge: MIT Press.

GANG, C. 2009. *Politics of China's Environmental Protection: Problems and Prospects.* Singapore: World Scientific Publishing Co.

Global Wind Energy Council. 2010. Global wind power boom continues despite economic woes, 3 February. Retrieved from: <http://www.gwec.net/index.php?id=30&no_cache=1&tx_ttnews [tt_news]=247&tx_ttnews[backPid]=97&cHash=8a55b8eab5>.

GRAHAM-HARRISON, E., and BUCKLEY, C. 2009. China unveils carbon target for Copenhagen Deal, *Reuters*, 27 November.

HE, G. 2008. China's new Ministry of Environmental Protection begins to bark, but still lacks in bite. World Resources Institute. Submitted on 17 July. Retrieved from <http://earth-trends.wri.org/updates/node/321>.

HEGGELUND, G. Andresen, S., and Buan, I. F. 2010. Chinese climate policy: Domestic priorities, foreign policy, and emerging implementation. Pp. 229–59 in K. Harrison and

L. M. Sundstrom (eds.), *Global Commons, Domestic Decisions: The Comparative Politics of Climate Change*. Cambridge: MIT Press.

Lɪ, H. 2009. China amends law to boost renewable energy. *Xinhua News*, 26 December, <http://www.chinaview.cn>. Retrieved from <http://news.xinhuanet.com/english/2009–12/26/content_12706612.htm>.

Lɪ, L. 2009. China Youth Climate Action Network: Catalyzing student activism to create a low-carbon future. Pp. 114–16 in *China Environment Series*, Issue 10, Washington, DC: Woodrow Wilson International Center for Scholars.

Lɪᴀɴɢxɪɴɢ, J. 2005. Energy first: China and the Middle East. *Middle East Quarterly* 12(2): 3–10.

Mᴀ, T. 2009. Environmental mass incidents in rural China: Examining large-scale unrest in Donyang, Zhejiang. Pp. 33–49 in *China Environment Series*, Issue 10, Washington, DC: Woodrow Wilson International Center for Scholars.

MᴀcFᴀʀQᴜᴀʀ, N. 2009. US and China vow action on climate change but cite needs. *The New York Times* 23 September 2009. Available from <http://fromwww.nytimes.com>. Accessed 21 December 2009.

Ministry of Environmental Protection, People's Republic of China. 2007. China addresses challenge of climate change, 29 December. Retrieved from <http://english.mep.gov.cn/special_reports/Climate_change/domestic_actions/200801/t20080102_115790.htm>.

Mᴏᴏʀᴇ, M. 2009. China's middle-class rises up in environmental protest. *Telegraph*, 23 November. Retrieved from <http://www.telegraph.co.uk/news/worldnews/asia/china/6636631/Chinas-middle-class-rise-up-in-environmental-protest.html>.

Mᴏxʟᴇʏ, M. 2010. China renews 'go west' effort, *Asia Times* online, 23 July. Retrieved from <http://www.atimes.com/atimes/China_Business/LG23Cb01.html>.

National Development and Reform Commission (NRDC). 2008. Green buildings and sustainable cities. January. <http://china.nrdc.org/files/china_nrdc_org/greenbuildings.pdf>.

Netherlands Environmental Assessment Agency. 2008. Global CO_2 emissions: Increase continued in 2007. <http://www.pbl.nl/en/publications/2008/GlobalCO2emissionsthrough2007.html>. Accessed 16 December 2010.

Organization for Economic Cooperation and Development. 2007. *OECD Environmental Performance Reviews: China*. Paris: OECD.

Pᴇʀᴋɪɴs, S. 2009. Chinese carbon dioxide emissions eclipse efficiency gains. *Science News*, 6 March. Available from <http://www.sciencenews.org>. Accessed 21 December 2009.

Pʀɪcᴇ, L., Wᴀɴɢ, X., Yᴜɴɢ, J. 2008. *China's Top-1000 Energy-Consuming Enterprises Program: Reducing Energy Consumption of the 1000 Largest Enterprises in China*. Ernest Orlando Berkeley National Laboratory, Environmental Energy Technologies Division. June. LBNL-519E.

Qɪ, Y., Mᴀ, L., Zʜᴀɴɢ, H., and Lɪ, H. 2007. Translating a global issue into local priority: China's local government response to climate change. *Journal of Environment and Development*, 17(4): 379–400.

REN21. 2008. *Renewables 2007 Global Status Report*. Paris: REN21 Secretariat and Washington, DC: Worldwatch Institute.

Scʜʀᴇᴜʀs, M. A. 2010. Multi-level governance and global climate change in East Asia. *Asian Economic Policy Review*, 5: 88–105.

Sʜᴀᴘɪʀᴏ, J. 2001. *Mao's War against Nature: Politics and the Environment in Revolutionary China*. Ithaca, NY: Cornell University Press.

Sᴜɴ, Y., and Zʜᴏᴜ, D. 2008. Environmental campaigns. Pp. 144–62 in K. J. O'Brien (ed.), *Popular Protest in China*. Cambridge, MA: Harvard University Press.

TURNER, J., and ZHI, L. 2006. Building a green civil society in China. Pp. 152–70 in World Watch Institute, *World Watch 2006: Special Focus: China and India*. Washington, DC: World Watch Institute.

United Nations Framework Convention on Climate Change (UNFCCC). 2009. *CDM Statistics*. Retrieved from <http://cdm.unfccc.int/index.html>.

US Energy Information Agency. 2009. *Country Briefs: China*. July. Retrieved from <http://www.eia.doe.gov/cabs/China/Background.html>.

World Bank and State Environmental Protection Administration, PR China. 2007. *Cost of Pollution in China: Economic Estimates of Physical Damages*. Washington, DC: The World Bank.

World Resources Institute (WRI). 2009. Mitigation actions in China: Measurement, reporting, and verification. Working paper, retrieved from <http://pdf.wri.org/working_papers/china_mrv.pdf>.

Xinhua. 2009. Copenhagen accord marks new starting point for global fight against climate change. 25 December, <http://www.chinaview.cn>.

XUE, L., SIMONIS, U. E., and DUDEK, D. J. 2007. Environmental governance for China: Major recommendations of a task force. *Environmental Politics* 16(4): 669–76.

YOON, E., with LEE, S., and WU, F. 2007. The states and nongovernmental organizations in Northeast Asia's environmental security. Pp. 207–32 in I. Hyun and M. A. Schreurs (ed.), *The Environmental Dimension of Asian Security: Conflict and Cooperation over Energy, Resources, and Pollution* (Washington, DC: United States Institute of Peace Press).

....

CHAPTER 31

CITIES AND SUBNATIONAL GOVERNMENTS

....

HARRIET BULKELEY

1 INTRODUCTION

....

DESPITE the 'global' nature of the challenge, over the past two decades it has become increasingly apparent that cities and regions are critical places for addressing climate change. While estimates vary, and data is limited (Dodman 2009; Satterthwaite 2008a), the International Energy Agency (IEA 2009) suggests that cities may be responsible for over 70 per cent of global energy-related carbon dioxide emissions. The urban concentration of greenhouse gas (GHG) emissions is perhaps not surprising given the increasing proportion of the world's population who live and work in cities and the ways in which energy demand, buildings, waste and water services as well as industrial processes are centered in urban locations (Seto et al. 2010). Global and mega-cities may therefore represent a greater share of the world's GHG emissions than many nation-states. For example, London's GHG emissions are estimated at 44 Mt or 8 percent of the UK's total in 2006 (GLA 2007) and considered to be on a par with those of some European countries such as 'Greece or Portugal' (LCCA 2007: 1). Once regional levels of subnational government are considered, the contribution becomes potentially even more significant. For example, 'if compared directly with other nations, 17 [US] states and two [Canadian] provinces (Alberta and Ontario) would rank among the 50 largest governmental sources of greenhouse gases in the world' (Rabe 2007: 428). Furthermore, as the example of Hurricane Katrina in New Orleans in 2005 so vividly illustrated, cities create concentrations of infrastructure, economic activities, and communities that are potentially vulnerable to the impacts of climate change. Many of the world's major cities lie in delta areas which may experience more frequent and more severe flooding, while others are dependent on fragile water supplies. Increasing the resilience of cities to climate change is therefore also becoming a key issue for policy makers at urban and regional levels (World Bank 2010).

With growing recognition of the 'localized' causes and consequences of climate change, cities and subnational governments have been key actors and arenas in the development of policy responses. Individually and collectively, urban and regional governments have been amongst those leading the way in demonstrating the potential for reducing GHG emissions and are increasingly engaging with the issues of adaptation. This chapter examines the emergence of this phenomenon and the roles that cities and subnational governments have played in orchestrating the response to climate change over the past two decades. The second section considers how and why urban and regional governments came to be at the forefront of responses to climate change, and the political geographies of this movement. The third section examines the nature of urban and regional responses to climate change, and the fourth considers the tensions emerging between the rhetoric and reality of the possibilities of addressing climate change. In conclusion, the potential and implications for cities and subnational governments in responding to climate change are considered.

2 THE EMERGENCE OF CLIMATE CHANGE AS AN URBAN AND REGIONAL ISSUE

Urban and regional responses have taken different paths over the past two decades. In the main, municipal governments began to take action on climate change before their regional counterparts. Municipal action has also been characterized by the development of transnational networks, a phenomenon that has only recently started to develop amongst regional governments.

The history of municipal involvement with the issues of climate change can be traced back to the early 1990s and the start of international concern over the levels of GHGs in the atmosphere (Bulkeley 2010). Between 1991 and 1993 ICLEI, now known as ICLEI Local Governments for Sustainability, organized the *Urban CO₂ Reduction Project*,[1] funded by the US Environmental Protection Agency, the City of Toronto, and several private foundations, with the aim of developing city-level plans and tools for the reduction of GHG emissions (Bulkeley and Betsill 2003). By the mid-1990s, three transnational municipal networks—ICLEI Cities for Climate Protection, Climate Alliance, and energie-cites—had been formed. This first wave of municipal responses was characterized by issues of mitigation and the reduction of GHG emissions, and predominantly focused on issues of energy efficiency and the involvement of municipalities from North America and Europe (Kern and Bulkeley 2009). Nonetheless, by 1997 and the adoption of the Kyoto Protocol internationally, these networks could count between them several hundred municipal authorities and were beginning to recruit members from the global south.

Since the early 2000s, a second wave of municipal action on climate change can be identified, encompassing a broader range of transnational networks and a more geographically diverse range of cities, the growth of interest in climate adaptation alongside mitigation, and a more avowedly political approach to urban climate governance. Networks such as C40 Cities Climate Leadership Group, working with support of the Clinton Climate Initiative and targeting forty of the world's so-called 'global' cities, and the US Mayors Agreement (which now has over 900 members[2]) have sought to position cities as critical sites for addressing the issue of climate change, and in so doing advance claims for the strategic importance of urban governance. Through the emergence of these new networks and the steady growth of the three pioneering municipal associations—for example, the ICLEI CCP programme has expanded rapidly in Australia and New Zealand in the past decade, while energie-cites has developed associated networks in Eastern Europe— several thousand cities are now part of the movement to address climate change at the local level. The growing weight of this movement was evident at the 2007 Conference of the Parties to the United Nations Framework Convention on Climate Change, where representatives from municipalities formed a substantial constituency and the Bali World Mayors and Local Governments Climate Protection Agreement, drawing together for the first time the diverse networks discussed above, was signed.

The involvement of regional governments in the climate change issue has been prompted primarily by national political circumstances and the development of forms of cross-border cooperation. Most notably in the US, states, including New Jersey and Wisconsin, began to develop their own responses to the issue in the late 1980s and early 1990s (Rabe 2007). The increasingly hard-line position taken by the federal administration against action on climate change led during the 2000s to the development of a range of state-based energy and climate policies. Regarded as the 'most surprising development in recent years in American climate change policy' (Rabe 2007: 429) the strong response of the states is evident in actions such as 'statewide climate change action plans, mandating that electric utilities generate a specific minimum amount of power from renewable energy sources (so-called renewable portfolio standards) to be increased over time, and establishing public funds to support energy efficiency and/or renewable energy development' (Selin and VanDeveer 2009: 11). Cooperation amongst states, particularly in the Northeast where states have been involved in the development of both the 2001 New England Governors and Eastern Canadian Premiers (NEG-ECP) Climate Change Action Plan and the 2009 Regional GHG Initiative (RGGI), a regional cap and trade agreement, has been an important driver behind action, as well as the personal leadership of politicians, such as Arnold Schwarzenegger as Governor of California.

Similarly in Australia, as the federal government became increasingly unwilling to engage in international efforts to address climate change, states began to seize the initiative. Unlike the US, where differences in the political affiliations of state governments do not appear to have been a decisive factor, in Australia the predominantly Labor administrations incumbent in state governments in the early 2000s found climate change an area in which they could signal their opposition to the Liberal party federal government. For example, the 2002 *Victorian Greenhouse Strategy* set out a range of measures to encourage the development and use of renewable energy and reduce demand for energy, including the development of energy efficiency standards for buildings so that new developments were required to attain a 5* rating from 2005, the promotion of GreenPower energy, support for the Cities for Climate Protection program in regional and rural Australia, and the formation of regional partnerships between local governments to pool efforts and resources in addressing climate change (Bulkeley and Schroeder 2009). In Canada, however, a different picture emerges, with provincial governments proving to be less active on the issue of climate change than states in the US. For example, 'the five eastern Canadian provinces that signed onto a regional action plan with the six New England states in 2001 have taken less action to implement this plan compared to their U.S. counterparts' (Selin and VanDeveer 2009: 11). Nonetheless, in 2007 several provinces announced their intentions to pursue GHG emissions targets and by '2008, British Columbia had established itself as the provincial climate change leader with enactment of aggressive GHG reduction goals and a broad-based carbon taxation scheme' (Selin and VanDeveer 2009: 12; see also Rabe 2007; Selin and VanDeveer 2007). At the same time, opposition amongst provincial governments to federal action in Canada on the climate change issue remains. In particular, due at least in part to its fossil-fuel reserves 'Alberta has promoted an alternative, a homegrown initiative that emphasizes voluntary programs and a focus on carbon intensity as opposed to outright emission stabilization or reduction' (Rabe 2007: 442).

Despite these differences in the origins and nature of subnational level responses, evidence of a growing movement transnationally to address climate change at the regional level can be found. The Climate Group, a London-based nonprofit organization, launched a states and regions programme in 2005 and has since developed a series of commitments and areas for members to take action. Nonetheless, rather than being a driver of policy responses to climate change, as is the case with municipalities, transnational networking has provided a means through which regional governments have sought to increase the profile of their actions and to gain political leverage in national and international arenas.

The emergence of urban and regional responses to climate change is part of a broader trend that has seen, especially since the 1997 Kyoto Protocol, the emergence of various forms of climate governance being conducted by private actors (e.g. the Chicago Climate Exchange) and public-private partnerships (e.g. the Renewable Energy and Energy Efficiency Partnership) (Andonova et al. 2009; Bäckstrand 2008). These initiatives have been termed governance 'experiments,' signaling both the novelty of the actors and approaches to governing being deployed and the institutional vacuum within which they are emerging (Hoffman forthcoming; Bulkeley et al. forthcoming). In the broadest sense, processes of globalization can be seen to have led to the fragmentation of political authority (Rosenau 2000) while at the same time 'mega-multilateral' agreements such as that being forged internationally in response to climate change have become bogged down in seemingly endless negotiations leading many to question their utility in relation to the urgency of global (environmental) problems. While the effectiveness and legitimacy of the forms of climate governance emerging beyond the state is moot, these trends have created the political space for alternatives to flourish. These initiatives are driven by diverse motivations including the potential for profit (especially with regard to carbon markets), the urgency of the issue, the potential to expand political authority and claim additional resources, and various forms of ideological expression (Hoffman forthcoming; see also Bulkeley et al. forthcoming). The urban and regional responses discussed in this chapter are a result of complex mixtures of these drivers. For municipal networks that pioneered the idea of urban action on climate change, a combination of urgency and a belief that cities are critical sites for action could be identified as the primary motivations even while faith in international processes remained relatively intact. The decisions of individual cities and regions to take action, join networks, and campaign for national and international responses is frequently a matter of individual leadership, the potential to realize additional (financial) benefits, and of political ideology (Bulkeley and Betsill 2003; Bulkeley et al. 2009). Whatever the specific drivers, it is clear that over the past two decades urban and regional responses to climate change have gathered such momentum that they are now firmly part of the landscape within which policies will be forged and responses implemented.

3 GOVERNING THE CLIMATE IN
CITIES AND REGIONS

If climate change is now part of the agenda for many, though by no means the majority, of cities and regions, a set of questions then arise as to how this issue is being addressed and the consequent implications for the achievement of policy goals locally, regionally, nationally, and internationally. Municipal governments have a highly variable level of influence over GHG emissions through their roles in energy supply and management, transport, land-use planning, and waste management. In the case of regional subnational governments, these competencies are often more significant with states and provinces frequently being responsible for the development of energy, water, transport, and waste policies. Likewise, regional governments usually bear more responsibility for the protection of critical infrastructures than their urban counterparts, though cities may be the level at which emergency planning and land use zoning which can shape the potential for adapting to climate change take place. As discussed below, urban and regional responses to climate change are frequently constrained by their governance capacities in these critical areas. However, a feature of the development of policy responses by this group of actors has been the ways in which they have sought to move beyond existing competencies and develop policy innovations. For example, typical policy goals include the reduction of GHG emissions by between 10 and 20 percent below 1990 levels in the period 2010–20,[3] already well above the collective target of reducing emissions by 5 percent on 1990 levels by 2012 agreed in the Kyoto Protocol. There are examples of urban and regional authorities with much higher levels of ambition. For example, the Greater London Authority has a target 'to stabilise CO_2 emissions in 2025 at 60 per cent below 1990 levels, with steady progress towards this over the next 20 years' (GLA 2007: 19), while the 2006 California Global Warming Solutions Act mandates reductions in GHG emissions to 1990 levels by 2020 (equalling a 25–30 percent reduction) and an 80 percent reduction below 1990 levels by 2050 (Rabe 2007; Schroeder and Bulkeley 2008).

Differences in the approaches adopted in pursuit of such goals can be seen in terms of whether urban and regional governments seek to reduce GHG emissions and enhance the resilience of their own infrastructures and operations—through, for example, reducing the use of energy in municipal buildings or switching vehicles fleets to alternative sources of fuel—or whether they seek to act on behalf of the 'community' which they represent—with initiatives such as providing advice to residents on how to cut their energy use or enhancing flood defences. Traditionally, cities have focused on the reduction of municipal GHG emissions first and foremost, with initiatives at the community scale confined to a few urban areas. In the energy sector, this has been achieved through retrofits in commercial, domestic, and municipal buildings, by switching traffic lights to LEDs, improving street lighting, and purchasing green energy. In the transport sector, through increasing the number of hybrid cars in government fleets and by implementing transport planning policies which encourage alternatives to the private car. Land use planning has been used to promote the inclusion of renewable energy and energy efficiency measures in new buildings, and in some cases to mandate particular standards for domestic and commercial buildings. In the waste sector, authorities have increased programs for recycling and composting and have developed projects to capture methane at landfills (Bulkeley and Betsill 2003; Bulkeley and Kern 2006). However, despite the range of GHG emissions reductions activities that municipalities could engage with, research has found that 'attention remains fixed on issues of energy demand reduction' (Betsill and Bulkeley 2007: 450; see also Bulkeley and Kern 2006; Schreurs 2008), and primarily orientated around internal sources of municipal emissions.

However, the recent emergence of forms of public-private partnership and the mobilization of various types of private actors seeking to address climate change in the city—from community-based groups, nongovernmental organizations, and businesses—is leading to the generation of initiatives that are seeking to reduce GHG emissions beyond the confines of municipal governments. One such scheme is the Energy Efficiency Building Retrofit Program organised by the C40 network and the Clinton Climate Initiative which 'works with industry, financial, government and building partners to overcome market barriers and develop financially sound solutions that accelerate the growth of the global building efficiency market'[4] across some twenty so-called global cities. At the other end of the (ideological) spectrum local Carbon Reduction Action Groups are seeking to collectively ration their production of GHG emissions through setting personal or household targets and trading emissions locally in order to achieve these goals.[5] At the same time, several hundred low carbon energy, transport, housing, and planning schemes are being developed and implemented in cities around the world. A recent survey for UN-Habitat in conjunction with research being undertaken by the UK Economic and Social Research Council *Urban Transitions* project found over 400 such projects in 100 global and mega-cities.[6] Such initiatives, or experiments, demonstrate that the city is increasingly becoming a site through which public and private actors are seeking to mobilize responses to climate change. Despite this growth in activity at the urban level, and the growing involvement of a range of cities in the global south—particularly in rapidly industrializing countries (Bulkeley et al. 2009)—the focus remains on mitigation. Granberg and Elander (2007: 545) refer to this as a 'paradox, since the tangible effects on global warming of reducing GHG emissions in a single locality are microscopic, whereas measures of adaptation are crucial for preventing potential flooding and related natural catastrophes.' The reasons for this paradox are considered further below.

Regional governments, on the other hand, have usually sought to address issues of mitigation at a community scale from the outset. Critical here has been the legislative capacity of subnational governments. California, widely regarded as the subnational government in which the most ambitious GHG reduction measures have been taken, clearly illustrates the importance of this legislative capacity. In 2006, it passed the California Global Warming Solutions Act mandating reductions in GHG emissions to 1990 levels by 2020 (equalling a 25–30 percent reduction) and an 80 percent reduction below 1990 levels by 2050. The act includes a package of policies to be put in place by state agencies. In 2002 California became the first US state to regulate carbon dioxide emissions from motor vehicles by mandating the California Air Resources Board to develop and implement emission limits for vehicles The 2005 Renewable Energies Act requires that 20 percent of the electricity sold by investor-owned electric utilities in the state come from renewable sources by 2010 while the 2006 GHG Emissions Performance Standard requires the California Energy Commission and the California Public Utilities Commission to set GHG emissions standard for electricity used in California, regardless of whether it is generated in state or purchased from plants out of state (Rabe 2007; Schroeder and Bulkeley 2008).

This legislative capacity has also been critical in the development of cross-border agreements and institutions, including the Regional GHG Initiative (RGGI), a carbon dioxide emissions trading scheme developed among ten US states (Selin and VanDeveer 2007: 5). A focus on the development of carbon markets and market instruments can also be found in other subnational governments. In Australia, the New South Wales (NSW) government introduced the GHG Reduction Scheme, a cap and trade scheme with the 'stated policy intent of reducing GHG emissions created through NSW electricity consumption and encouraging activities that offset these emissions' (Passey et al. 2008: 2999), and subsequently states and territories developed the groundwork for establishing a national emissions trading scheme (Curran 2009). Equally, addressing climate change, and in particular the development of alternative sources of energy and energy technologies, is increasingly becoming part of mainstream economic development agendas at the regional level (Rabe 2004; Smith 2007).

4 RHETORIC AND REALITIES: THE POLITICS OF URBAN AND REGIONAL CLIMATE GOVERNANCE[7]

Despite the growing number of urban and regional governments taking part in governing the climate, and the range of schemes, measures, policies, and legal instruments that have been deployed to this end, research suggests that there is a gap between the rhetorical commitments towards climate change and the actions being undertaken. Systematic accounts of the reductions in GHG emissions achieved or the measures to adapt to climate change implemented are few and far between. Those estimates that do exist have been generated by those involved in the development of urban and regional responses. For example, a report on the progress of the 546 local governments in 27 countries who were members of the CCP program estimated that annual emissions has been reduced by 60 million tons of CO_2eq, which amounts to a 3 percent annual reduction among the participants and 0.6 percent globally (ICLEI 2006). Such estimates need to be treated with caution, both because of the uncertainties involved in the reporting mechanisms and the vested interests that cities, regions, and the organizations which are mobilizing their activities have in showing that their efforts are paying dividends. Despite these accounts, it is evident that much remains to be done in order to put urban and regional commitments into action. In seeking explanations for the gaps which are emerging between ambitious policy goals and political statements, the development and implementation of policy and the realization of emissions reductions or enhanced adaptive capacity, four key factors emerge: leadership; resources; multilevel governance; and political conflicts.

First, studies have frequently pointed to the roles of policy entrepreneurs and political champions in establishing climate change as an issue on municipal agendas and taking forward innovative action (Bulkeley and Betsill 2003; Bulkeley and Kern 2006; Qi et al.

2008; Rabe 2004, 2007; Schreurs 2008). However, these studies also suggest that individuals can only take climate change action so far and that 'a broader institutional capacity for climate protection is necessary' (Bulkeley and Kern 2006: 2253) to overcome the constraints encountered. Evidence also suggests that the opportunity for cities and regions to lead amongst their peer communities is important in driving climate change up the political agenda. Rabe (2007: 437) describes the competition amongst the US states to be at the forefront of responses to climate change as a 'race-to-the-top' as they seek to keep pace with the initiatives launched in California. 'Trigger' events, often tied to sporting initiatives or international conferences, can offer windows of opportunity within which such leadership can be displayed by providing both the motivation and opportunity to intervene in the physical landscape of the city (Bulkeley et al. 2009). Equally, adaptation measures often get adopted only in response to specific local or regional natural disasters, which may or may not be climate related. For example, in Mumbai, after the 2005 deluge flooding, the Greater Mumbai Disaster Management Plan was revised in 2007, strengthening the Municipal Corporation of Greater Mumbai's Disaster Management Committee and raising disaster preparedness of the city (Gupta 2007). However, in general, while political leaders have been able to create significant political capital on the issue of mitigation, this has not been the case in terms of addressing adaptation and may provide one explanation as to why this issue remains on the backburner for many cities and subnational governments.

A second issue which has been shown to critically shape urban and regional responses to climate change is that of resource. Holgate (2007) shows how in Cape Town the comparatively well-resourced municipality was able, with the help of additional resources from outside the local authority, to make significant advances in tackling the issues, while in Johannesburg one officer was responsible for addressing the range of environmental challenges facing the city and, at least partly as a result, the response to climate change was minimal. Satterthwaite (2008b) draws attention to the problem of a lack of municipal finance for providing basic infrastructures and the consequent implications for adaptation. While not as critical in life and death terms, many municipal governments in developed countries also lack the resources to address climate change, especially when it is not considered a political priority (Bulkeley and Kern 2006). In this context, the ability to secure funding from external sources—from national governments, the European Union, donor agencies, charitable foundations, and increasingly streams of carbon finance—has been shown to make a significant difference in the local capacity to address climate change (Betsill and Bulkeley 2007; Granberg and Elander 2007). Internal funding, in the form of novel financial mechanisms such as revolving energy funds (where financial savings from energy efficiency are reinvested in energy conservation or other climate change projects) or energy performance contracting (where external organizations invest in energy efficiency measures and profit from the financial savings made) (see e.g. Bulkeley and Kern 2006; Bulkeley and Schroeder 2008) can be invaluable in overcoming the 'inflexible budgetary structures' (Jollands 2008: 5) for which municipal authorities are usually renowned.

A third, and related, issue concerns the multilevel governance systems within which municipal and regional governments operate (Betsill and Bulkeley 2006). The powers and responsibilities of subnational levels of government vary considerably from country to country. In the main, local governments have limited powers and responsibilities with respect to environmental taxation, energy supply, and the supply of transport infrastructures (Jollands 2008; Lebel et al. 2007; Schreurs 2008: 353; Sugiyama and Takeuchi 2008:

425). State or provincial governments, as discussed above, often have considerably more authority in these areas. Nonetheless, the role of national government, and of relations between levels of government, can be critical. For example, the introduction in Japan of a mandate for local and regional governments to create climate change action plans was instrumental in the early lead taken on this issue (Sugiyama and Takeuchi 2008: 426), while in the US state and local government responses have been shaped by 'the State and Local Climate Change Program . . . established in 1990' which provided 'a series of grants and technical assistance that ultimately reached almost every state in the union' and 'the 1992 Energy Policy Act and 1990 Clean Air Act Amendments' that 'provided a number of incentives and opportunities for states to begin to develop expertise in renewable energy and emissions trading' (Rabe 2007: 435). Although 'neither bill was focused on climate per se . . . [they] significantly influenced the development of relevant policy capacity at the state level' (Rabe 2007: 435). Equally, the lack of leadership at the national level in the US and Australia, as discussed above, has created the political space within which climate policy has flourished, illustrating the ambivalent role that relations between different levels of government can have in creating governance capacity on this issue. Similar issues have also been found in relation to adapting to climate change. In a survey of ten Asian cities, Tanner et al. (2008: 18) identify the following constraints: (1) heavy top-down decision-making structures; (2) a lack of clarity between city-, state-, and national-level bodies, leading to inefficiencies and conflict; (3) poor coordination between departments and agencies; and (4) severe financial constraints, especially in Southern municipal governments.

The multilevel governance context within which cities and regions are responding to climate change goes beyond these vertical relations between tiers of government. New governance arrangements have also been identified as critical to the governing of climate change subnationally (Betsill and Bulkeley 2006). Transnational municipal networks—such as ICLEI CCP, Climate Alliance, energie-cities, C40—have been an important driver for municipal action on climate change mitigation since their inception in the early 1990s, providing resources, information, and political kudos to individuals seeking to gain support internally (Bulkeley and Betsill 2003; Collier 1997; Granberg and Elander 2007). There is some evidence to suggest that transnational municipal networks are most significant for 'leaders' on municipal climate action (Kern and Bulkeley 2009), and that they may be most important in the earliest stages of climate policy development as local policy actors seek ideas from cities and countries with similar politics globally (Schreurs 2008). Once examples of climate change strategy and action have developed nationally, some authors suggest that 'attention shifts to these cases' (Schreurs 2008: 353). On the other hand, the trend for involvement with transnational municipal networks has increased over the past decade, both within existing networks and through the formation of new networks such as the C40 Cities Climate Leadership Group and the Rockefeller Foundation Climate Change Initiative. These new networks represent a different approach, focusing not on accumulating an ever-larger membership and hence jurisdiction over an increasing proportion of global GHG emissions, but on the development of specific 'clubs' of cities which can gain privileged access to information, funding, and project implementation, in return for specific actions. As urban and regional governments increasingly seek to address climate change by working in 'partnership' with private and community sector organizations, the governance capacity challenges shift from issues concerning their direct competencies to those of creating financial incentives, persuading others to act and coordinating action

(Bulkeley and Kern 2006) as well as on the ability for local governments to create an 'enabling environment for local civil-society action' (Satterthwaite 2008b: 9).

Given the potentially significant issues and interests at stake, it is no surprise to find that issues of political consensus, compromise, and conflict have been identified as a fourth key factor shaping the gap between rhetorical commitments to act on climate change and on the ground realities. These tensions can be seen in the priority which climate change is accorded locally, an issue which is regarded as being particularly significant in the context of the global South due first to the limited resources and pressing agendas of meeting basic needs (Bai 2007; Jollands 2008; Romero-Lankao 2007) and second to broader questions of responsibilities and of development. As Lasco et al. (2007: 84) explain, 'for many developing countries GHG mitigation has a negative connotation because of the perception that this will deny them of their basic right to growth in human services and economic activities; the prospects of "reduced growth" or "no growth" are not feasible.' Such tensions are, however, also discernible in the politics of addressing climate change within cities in the north. In the US, for example, Zahran et al. 2008 observe that it is communities most likely to be affected by the impacts of climate change, and those with a 'liberal' political constituency in which climate change mitigation is likely to be prioritized. In their study of climate mitigation and transport policy in Cambridgeshire, Bulkeley and Betsill (2003) found that efforts to reduce the demand for travel and hence of GHG emissions locally had been confounded by the priority given to economic considerations within transport and land use planning and the stress on the need for increasing travel demand in the county.

Issues of political conflict also arise in relation to addressing climate change adaptation subnationally. In part, this reflects more systemic issues concerning the governance of vulnerable people, communities, cities, and regions, as those that face the greatest challenges are often those who are in the most vulnerable sites, often outside municipal jurisdictions and lacking competent and accountable local government (Satterthwaite 2008b: 15). In these cases it may be an absence of governance rather than any overt conflicts about how to address adaptation that is creating the biggest barriers to action on the ground, though as Huq et al. (2007: 14) have argued, the 'kinds of changes needed in urban planning and governance to "climate proof" cities are often supportive of development goals. But . . . they could also do the opposite—as plans and investments to cope with storms and sea-level rise forcibly clear the settlements that are currently on floodplains, or the informal settlements that are close to the coast.'

Given the potential ambiguous or overtly hostile responses to subnational climate change initiatives, a key factor in building capacity has been the ability of municipal and regional actors to reframe climate change as a 'local' problem and/or one that will have significant additional benefits (Betsill 2001). Bai (2007: 26) argues that there are plenty of 'local' hooks upon which responding to climate change might be hung within cities in the global south, including 'air pollution control, solid waste management, urban development and growth management, transportation and other infrastructure development, to name a few.' Other studies suggest that it is this process of reframing, 'localizing', or 'issue bundling' (Koehn 2008: 61) that has been effective in mobilizing local action on climate change in cities in the global south, and that this will remain an important aspect of building local capacity to act (Lasco et al. 2007; Romero-Lankao 2007). At the regional level, addressing climate change has often been linked to the development of a 'green' economy and the potential for technical innovation and new sources of employment that this might bring. At both urban and regional levels, the ability to link the development of alternative sources of energy and a reduction in energy use to issues of the security and costs of energy supply have also been significant in shaping policy responses. However, this focus on the 'win win' potential of addressing climate change can be seen as one of the key reasons why actions remain concentrated on issues of energy efficiency and focused on picking the 'low hanging fruit,' rather than engaging with more fundamental (and politically difficult) choices concerning how (and by whom) energy is provided, mobility demands satiated, and the production of wastes reduced. Furthermore, research suggests to date that there is an 'absence of issue framing that has linked adaptation to pressing urban social, economic and environmental issues with the result that adaptation has limited traction or support locally' (Bulkeley et al. 2009: 78). Relying on other issues as the Trojan horse through which to smuggle climate change past the city (or state) gates is therefore a potentially risky strategy.

5 CONCLUSIONS

Urban and regional governments have risen to prominence over the past two decades as critical actors in the governing of climate change. Although their influence over systems of energy use, transportation, and waste vary significantly, it is increasingly recognized that these actors have a significant role to play in mitigating climate change. More recently, as the realization has dawned that some degree of climate change is inevitable, these actors have also begun to mobilize on the issue of adaptation through their responsibilities for critical infrastructures and emergency response. At the same time cities in particular are becoming a critical site through which other actors—those in the private sector but also community-based groups and national governments—are seeking to experiment with responses to climate change or realize their own policy ambitions. Nonetheless, as this chapter has outlined, there are several factors—of leadership, resource, multilevel governance, and politics—that individually and collectively have worked to create significant gaps between the rhetorical commitment to addressing climate change subnationally and the reality on the ground. While 'localising' climate change as an issue that can be governed at the urban or regional scale has been an important strategy in surmounting some of these challenges, the issue remains marginal politically and subordinate to other social and economic goals. In this respect, the politics of climate change in the town hall or state house remain depressingly similar to those encountered in national governments and the international arena.

NOTES

1. Fourteen municipalities from North America and Europe participated in the *Urban CO₂ Reduction Project*: Ankara, Turkey; Bologna, Italy; Chula Vista, US; Copenhagen, Denmark; Dade County, US; Denver, US; Hanover, Germany; Helsinki, Finland; Minneapolis, US; Portland, US; Saarbrucken, Germany; Saint Paul, US; and Toronto, Canada.

2. For more information, see ⟨http://www.usmayors.org/climateprotection/list.asp⟩ (accessed April 2009).

3. A survey for the Tyndall Centre of 'non-state' actors commitments to reducing GHG emissions found that the majority of the 2,200 cities included in the analysis had these targets. A core group of 200 cities who had made commitments beyond their affiliation to climate networks had a typical target of reductions of 20 percent by 2010.

4. See <http://www.clintonfoundation.org/what-we-do/clinton-climate-initiative/our-approach/ cities/building-retrofit> (accessed November 2009).

5. See <http://www.carbonrationing.org.uk/> (accessed November 2009).

6. See <http://www.geography.dur.ac.uk/projects/urbantransitions> (accessed November 2009).

7. This section draws on Bulkeley et al. 2009.

REFERENCES

ALBER, G., and KERN, K. 2008. Governing climate change in cities: Modes of urban climate governance in multi-level systems. OECD International Conference, 'Competitive Cities and Climate Change', 2nd Annual Meeting of the OECD Roundtable Strategy for Urban Development, 9–10 October, Milan, Italy available from: <http://www.oecd.org/document/ 32/0,3343,en_21571361_41059646_41440096_1_1_1,00.html> (accessed January 2009).

ANDONOVA, L., BETSILL, M., and BULKELEY, H. 2009. Transnational climate governance. *Global Environmental Politics* 9(2).

BÄCKSTRAND, K. 2008. Accountability of networked climate governance: The rise of transnational climate partnerships. *Global Environmental Politics* 8(3): 74–102.

BAI, X. 2007. Integrating global environmental concerns into urban management: The scale and readiness arguments, *Journal of Industrial Ecology* 11(2): 15–29.

BETSILL, M. 2001. Mitigating climate change in US cities: Opportunities and obstacles. *Local Environment* 6(4): 393–406.

—— and BULKELEY, H. 2006. Cities and the multilevel governance of global climate change. *Global Governance* 12(2): 141–59

—— —— 2007. Guest editorial. Looking back and thinking ahead: a decade of cities and climate change research. *Local Environment* 12(5): 447–56.

BULKELEY, H. 2009. Planning and governance of climate change. In S. Davoudi, J. Crawford, and A. Mehmood (eds.), *Planning for Climate Change Strategies for Mitigation and Adaptation for Spatial Planners*. London: Earthscan.

—— 2010. Cities and the governing of climate change. *Annual Review Environment and Resources* 35. Forthcoming.

—— and BETSILL, M. 2003. *Cities and Climate Change: Urban Sustainability and Global Environmental Governance*. London: Routledge.

—— and KERN, K. 2006. Local government and the governing of climate change in Germany and the UK. *Urban Studies*, 43(12): 2237–59.

—— and SCHROEDER, H. 2008. Governing climate change post 2012: the role of global cities— London. Tyndall Working Paper 123, Tyndall Centre for Climate Change Research, <http:// www.tyndall.ac.uk/publications/working_papers/twp123.pdf> accessed 27 January 2009.

—— —— 2009. Governing climate change post 2012: the role of global cities—Melbourne. Tyndall Working Paper 138, September, Tyndall Centre for Climate Change Research

<http://www.tyndall.ac.uk/tyndall-publications/working-paper/2009/governing-climate-change-post-2012-role-global-cities-melbou> accessed November 2009.

—— —— JANDA, K., ZHAO, J. Armstrong, A., Chu S. Y., and Ghosh S. 2009. Cities and climate change: The role of institutions, governance and urban planning. Commissioned paper for the World Bank Urban Research Symposium Cities and Climate Change. Marseille, June. Available at <http://www.urs2009.net/papers.html> (accessed November 2009).

—— HOFFMAN, M., VANDEVEER, S., and MILLEDGE, T. Forthcoming. Transnational Governance Experiments: Evidence from the climate change arena. In F. Biermann and P. Pattberg (eds.), *Global Environmental Governance Reconsidered: New Actors, Mechanisms and Interlinkages.*

COLLIER, U. 1997. Local authorities and climate protection in the European Union: Putting subsidiarity into practice? *Local Environment* 2(1): 39–57.

CURRAN, G. 2009. Ecological modernisation and climate change in Australia. *Environmental Politics* 18(2): 201–17.

DHAKHAL, S. 2004. *Urban Energy Use and GHG Emissions in Asian Mega Cities: Policies for a Sustainable Future.* Kanagawa: IGES.

DODMAN, D. 2009. Blaming cities for climate change? An analysis of urban GHG emissions inventories. *Environment and Urbanization* 21(1): 185–201.

GLA. 2007. *Action Today to Protect Tomorrow: The Mayor's Climate Change Action Plan.* London: Greater London Authority, February.

GRANBERG, M., and ELANDER, I. 2007. Local governance and climate change: Reflections on the Swedish experience. *Local Environment* 12(5): 537–48.

GUPTA, K. 2007. Urban flood resilience planning and management and lessons for the future: a case study of Mumbai, India. *Urban Water Journal* 3(3): 183–94.

HOFFMANN, M. J. Forthcoming. *Climate Governance at the Crossroads: Experimenting with a Global Response.* New York: Oxford University Press.

HOLGATE, C. 2007. Factors and actors in climate change mitigation: A tale of two South African cities. *Local Environment* 12(5): 471–84.

HUQ, S., KOVATS, S., REID, H., and SATTERTHWAITE, D. 2007. Editorial: Reducing risks to cities from disasters and climate change. *Environment and Urbanization* 19(3): 3–15.

ICLEI. 2006. *ICLEI International Progress Report: Cities for Climate Protection.* Oackland: ICLEI.

IEA. 2008. *World Energy Outlook 2008.* Paris: International Energy Agency.

JOLLANDS, N. 2008. Cities and energy: a discussion paper. OECD International Conference, 'Competitive Cities and Climate Change', 2nd Annual Meeting of the OECD Roundtable Strategy for Urban Development, 9–10 October, Milan, Italy, Energy Efficiency and Environment Division, International Energy Agency <http://www.oecd.org/dataoecd/23/46/41440153.pdf> accessed 3 January 2009.

KERN, K., and BULKELEY, H. 2009. Cities, Europeanization and multi-level governance: Governing climate change through transnational municipal networks. *Journal of Common Market Studies* 47(2): 309–32.

KOEHN, P. H. 2008. Underneath Kyoto: Emerging subnational government initiatives and insipient issue-bundling opportunities in China and the United States. *Global Environmental Politics* 8(1): 53–77.

LASCO, R., LEBEL, L., SARI, A., MITRA, A. P., TRI, N. H., LING, O. G., and CONTRERAS, A. 2007. *Integrating Carbon Management into Development Strategies of Cities: Establishing a*

Network of Case Studies of Urbanisation in Asia Pacific. Final Report for the APN project 2004–07- CMY-Lasco.

LEBEL, L., HUAISAI, D., TOTRAKOOL, D., MANUTA, J., and GARDEN, P. 2007. A carbon's eye view of urbanization in Chiang Mai: Improving local air quality and global climate protection. Pp. 98–124 in R. Lasco et al. (eds.), *Integrating Carbon Management into Development Strategies of Cities: Establishing a Network of Case Studies of Urbanisation in Asia Pacific.* Final Report for the APN project 2004–07-CMY-Lasco.

London Climate Change Agency (LCCA). 2007. *Moving London towards a Sustainable Low-Carbon City: An Implementation Strategy.* London: London Climate Change Agency, June.

PASSEY, R., MacGILL, I., and OUTHRED, H. 2008. The governance challenge for implementing effective market-based climate policies: A case study of The New South Wales GHG Reduction Scheme. *Energy Policy* 36: 2999–3008.

QI, Y., MA, L., ZHANG, H., LI, H. 2008. Translating a global issue into local priority: China's local government response to climate change. *Journal of Environment and Development* 17 (4): 379–400.

RABE, B. 2004. *Statehouse and Greenhouse: The Evolving Politics of American Climate Change Policy.* Washington, DC: Brookings Institution Press.

—— 2007. Beyond Kyoto: Climate change policy in multilevel governance systems. *Governance* 20(3): 423–44.

ROMERO-LANKAO, P. 2007. How do local governments in Mexico City manage global warming? *Local Environment* 12(5): 519–35.

ROSENAU, J. 2000. Change, complexity and governance in a globalizing space. In J. Pierre (ed.), *Debating Governance.* Oxford: Oxford University Press.

SATTERTHWAITE, D. 2008a. Cities' contribution to global warming: Notes on the allocation of GHG emissions. *Environment and Urbanization* 20: 539–49.

—— 2008b. Climate change and urbanization: effects and implications for urban governance. United Nations expert group meeting on population distribution, urbanization, internal migration, and development, UN/POP/EGM-URB/2008/16.

SCHROEDER, H., and BULKELEY, H. 2008. Governing climate change post-2012: The role of global cities, case-study: Los Angeles. Tyndall Centre Working Paper 122, available at <http://www.tyndall.ac.uk/sites/default/files/wp122.pdf> (accessed November 2009).

SCHREURS, M. A. 2008. From the bottom up: Local and subnational climate change politics. *Journal of Environment and Development* 17(4): 343–55.

SELIN, H., and S. D. VANDEVEER. 2007. Political science and prediction: What's Next for U.S. climate change policy? *Review of Policy Research* 24(1): 1–27.

—— —— 2009. Climate leadership in Northeast North America. In H. SELIN and S. D. VANDEVEER (eds.), *Changing Climates in North American Politics: Institutions, Policy Making and Multilevel Governance.* Cambridge, MA: MIT Press.

SETO, K., SÁNCHEZ-RODRÍGUEZ, R., and FRAGKIAS, M. 2010. The new geography of contemporary urbanization and the environment. *Annual Review Environment and Resources* 35. Forthcoming.

SMITH, A. 2007. Emerging in between: The multi-level governance of renewable energy in the English regions. *Energy Policy* 35: 6266–80.

SUGIYAMA, N., and TAKEUCHI, T. 2008. Local policies for climate change in Japan. *Journal of Environment and Development* 17(4): 424–41.

TANNER, T., MITCHELL, T. POLACK, E., and GUENTHER, B. 2008. *Urban Governance for Adaptation: Assessing Climate Change Resilience in Ten Asian Cities.* Institute of Development

Studies. <http://www.ids.ac.uk/UserFiles/File/poverty_team/climate_change/IDS_Climate_
Resilient_Urban_Governance_RF_report_2008_2.pdf> (accessed January 2009).

World Bank. 2010. *World Development Report 2010: Development and Climate Change.*
Washington, DC: World Bank.

ZAHRAN, S., BRODY, S. D., VEDLITZ, A., et al. 2008. Vulnerability and capacity: Explaining
local commitment to climate change policy. *Environment and Planning C—Government
and Policy* 26(3): 544–62.

CHAPTER 32

ISSUES OF SCALE IN CLIMATE GOVERNANCE

DANIEL A. FARBER

1 INTRODUCTION

CLIMATE change is a global problem, but it stems from local emissions and its impacts are also felt differently in different locations. Thus, the problem spans geographic scales from the rural village or urban neighborhood to the planet. This poses a conundrum for policy makers. How should responsibility for addressing climate change be divided, coordinated, and enforced among governmental entities ranging from municipalities to international institutions? What factors influence the decision?

In the United States, there is a vigorous debate about the appropriate roles of the state and national government in reducing greenhouse gases and mitigating climate change (Arizona Law Review Editors 2009). Discussion is also beginning about how the tasks of adapting to climate change that can no longer be avoided should be allocated between levels of government. However, these issues transcend the US Institutional details and legal frameworks may differ in significant ways, but similar issues are presented in allocating authority between the German *Länder* and the German national government, or between France and the EU, or between the city of Los Angeles and California.

It is hard to argue with the conclusion that 'top-down, global, far reaching (and sometime homogenous) approaches to combating climate change must be blended with bottom-up, local, grass-roots schemes' (Sovacool and Brown 2009). The top-down approach provides economies of scale, limits leakage, and promotes harmonization. The bottom-up approach is adaptive to local conditions, which can vary for social, ecological, and technological reasons, while also enhancing public participation and promoting innovation. But difficult issues are posed when we ask how large-scale efforts and smaller-scale efforts should be coordinated.

This chapter considers some of the issues involved in determining which activities should be assigned to what geographic unit of government. Part II considers mitigation efforts, while Part III turns to adaptation. In both instances, the issue is between local tailoring of climate policy and a more uniform approach.

For convenience, throughout this chapter, I will refer to the geographically more extended unit of government as the 'nation,' and the smaller one as the 'state,' although these terms may not apply literally in all applications. In the event of a global climate agreement, for instance, national sovereigns might be the 'states' while the international authority might be the 'nation' in terms of this analysis. Within a single US state, the state government might be the 'nation' while a city with home role powers is the 'state.' The analysis applies whenever larger and smaller governance units have their own powers of regulation or their own fiscal authority. In particular, it applies to all federalist systems (for instance, the US, Canada, Australia, Germany), to the European Union, and to non-federalist systems in which cities or regions have some variety of regulatory or fiscal powers.

At the international level, the issues play out differently. The international community is not a government, and so it lacks the legal authority ability to impose regulations on unwilling nation-states or to create taxes in order to finance climate action. There is no such thing as centralized authority as a legal matter in the international system. At some points, however, there are analogies between issues in international laws and those arising within vertical governmental arrangements. We can still ask, as a matter of principle, whether issues should be addressed through international agreement or at the national level, but there is no recognized legal mechanism for enforcing an answer against unwilling states. However, where the major powers agree with each other, they may have de facto power (but not recognized legal authority) to impose their views on remaining states.

It should be noted that the division in this chapter between adaptation and mitigation is somewhat blurry at the edges. It is quite possible for the same action to reduce future climate change but also helps reduce harm from current climate change. An example is the use of increased insulation, which reduces energy demand (and hence reduces carbon emissions from electrical generators) while also helping to protect building inhabitants against unfavorable changes in climate. Ideally, mitigation and adaptation should be complementary. For convenience in analysis, however, it is helpful to deal with the two separately.

2 MITIGATION

Uniform national legislation has advantages. It equalizes burdens among the citizenry, gives firms a predictable environment and a level playing field, takes advantage of what is often greater technical expertise at the top level, and minimizes transaction costs.

Nevertheless, climate action is not limited to the top level. Some European member states such as Germany have been particularly aggressive and innovative in their efforts to decarbonize. In Canada, Manitoba has been a leader among the other provinces (Rabe 2009). For a variety of reasons that are still the subject of active research (DeShazo and Freeman 2007; Engell 2006; Esty 1996; Wiener 2007), US state governments also began to address climate mitigation even though the national government steadfastly opposed mitigation efforts until 2009 (Rabe 2009; Stein 2007; McKinstry 2004). By 2006, despite federal government hostility, every state had taken steps of some kind that address climate change. California has been the leader, with legislation aimed at reducing greenhouse emissions from automobiles and electrical generators, as well as an ambitious mandate to reduce emissions to 1990 levels by 2020. US cities are also active in addressing mitigation (Betsill 2001).

Coordinating local and national action can be a problem. Suppose that a national government adopts a regulation to reduce carbon emissions. Should it allow continued actions by states to reduce emissions? On the one hand, there are several reasons why the national government might welcome continued state activities. First, the national authorities might actually prefer a higher level of emission control, but be politically unable to adopt that higher level. If so, they may welcome state initiatives. Second, states may provide useful test beds for regulatory methods before they are adopted at the national level, and state regulation may develop useful information about abatement costs that could lead to a readjustment of national standards. Third, state regulation might force innovation with desirable national externalities.

On the other hand, state initiatives might disrupt the operation of the national scheme or create undesirable economic side effects. If so, national uniformity might be preferred. The question of how much scope to give state initiatives in light of national initiatives requires careful analysis.

The strength of the argument for a solely national approach depends in part on the national government's choice of regulatory instrument. If the instrument of choice is a carbon tax, the analysis is straightforward. If a state chooses to impose tough standards on local industry, those industries will exceed national standards. However, the operation of the tax in other regions will be unaffected. If there are ill-effects from the 'excessively' high standard, they will be felt by the local state in terms of reduced employment. The only negative external effects will be on tax revenues. Thus, the only reason for the national government to intervene would be fiscal, and this would be likely to arise only if the carbon tax was providing a large share of national revenue or if the proceeds had unwisely been spent or legally committed before the revenue level was known.

The situation with regard to a cap-and-trade scheme is more complex. A state might create incentives for a mandate for certain sources to achieve emission reductions. If sources are themselves covered by a national cap-and-trade system, they may want to sell the resulting allowances, or in an auction system, they will not bid for those allowances in the first place. Without some adjustment in the operation of the national system, the result will simply be to free up allowances for use elsewhere, resulting in no net reduction in emissions. Those allowances can be restricted, either by an adjustment in the federal cap or by a state or federal prohibition on trade in freed-up allowances—but this means fewer allowances in the system, a higher price on those allowances, and possibly lower liquidity in the market. Thus, the national government may be unfriendly to such state efforts. A similar problem exists if the state seeks to reduce emissions from sources that might otherwise be available as offsets.

This coordination problem must be addressed in designing the national legislation, or if the national government fails to do so, through judicial doctrines dealing with conflicting state and federal measures. For instance, in the US system, state efforts to restrict sale of freed-up allowances might be preempted by the federal cap-and-trade scheme (Glicksman and Levy 2008).[1] In the European Union, the issue might be the application of the Article

174(2) safeguard clause, which provides that measures harmonizing environmental protection should include a clause allowing member states to take provisional measures for noneconomic environmental reasons, subject to review by European institutions. Some measures, however, might fall under Article 133, which grants the EC exclusive power over external trade (Weatherill 2007; Faure and Johnson 2008).

Rather than using economic instruments such as a carbon tax or cap-and-trade system, a national government may choose to rely on direct regulation of sources—for instance, by banning new coal-fired plants or requiring a certain percentage of energy to come from renewable sources, or by mandating fuel efficiency standards for vehicles. In that situation, states that choose to impose even stricter mandates should be allowed to do so, since they will not interfere with achievement of national goals.

State efforts to mitigate climate change may also collide with another body of national law governing free trade. In the United States, this law has been created by the Supreme Court under what is called the 'dormant commerce clause' doctrine (Farber and Hudec 1994).[2] In the EU, Article 30 and 37 of the EEC Treaty guarantee the free movement of goods, and the European Court of Justice has enforced these provisions aggressively (Hunter and Muylle 1998). At the international level, the WTO has similar provisions (Articles III and XX) (Howse 2002; Charnovitz 1993). These restrictions are phrased differently, but in the end the key issue seems to be whether a regulation burdens firms outside the jurisdiction to an unreasonable extent given the purported environmental benefits.

The question of uniformity versus bottom-up local variation can arise in any system with multiple levels of governance. Even in a heavily decentralized system such as the international community, nations may find good reasons to agree on uniform centralized mechanisms to deal with climate issues, and when they do so, they may have good reasons to restrict national-level efforts of various kinds. But in the international system, these decisions can only be made by consensus and compliance is also an open question. In federalized systems, however, a voting mechanism may allow less-than-unanimous decision making, and mechanisms such as courts exist to obtain compliance. Thus, in those systems, the degree of decentralized action is in part a legal rather than purely policy/political issue.

3   ADAPTATION

States and local governments are in some ways the natural 'first responders' to climate change (Agrawal 2008). They may own or license critical infrastructure, provide health services, or control land use. Cities around the world have begun to rise to this challenge (as have regional and state governments). For instance, Chicago has issued a guide to adaptation for municipalities (City of Chicago 2008: 13). The guide considers a broad range of impacts including shoreline erosion, invasive species, health threats from heatwaves and increased ozone, damage to key infrastructure, and flood damage. Similarly, London has now launched a major adaptation planning effort (Mayor of London 2009). The national government, however, may step in to provide mandatory standards for adaptation efforts or to finance adaptation.

The key issues are (1) the extent to which the national government should impose mandates for adaptation on states or private actors, and (2) the extent to which the national government should fund adaptation efforts rather than leaving the funding to states. (Again, recall, that for the purposes of this chapter, the entities in question could be US federal government/state government, European Union versus member states, Australian government versus state government, German government versus *Länder*, state or regional government versus city.)

These issues have some analogy internationally. At the international level, adaptation standards cannot be imposed as express legal enactments, because no international institution has the legal authority to impose such regulations on nonconsenting nations. However, international institutions such as the World Bank do have some de facto ability to pressure nations to take desired actions. The finance issue does arise internationally, specifically with regard to whether the developed countries should collectively fund adaptation in less developed countries, particularly the least developed. Some financing arrangements are already in place, but debate continues on whether the past or future high emissions of developed countries require some form of compensation to developing countries.

3.1 Setting Adaptation Standards

To get some sense of the range of possibilities, we might consider the following as a sample of potential areas for national standards:

- Efforts to deal with the increased risk of floods, such as (a) limitations on infrastructure construction in flood plains or coastal areas, (b) mandates to relocate some existing infrastructure inland, or (c) minimum performance standards for levees, including those owned by state or local government or private parties.

- Building restrictions on areas that will be subject to wildfires in hotter, dryer conditions.

- Measures to deal with the threat of increased drought, including (a) water allocations between states during drought periods, (b) multistate or nationwide requirements for water conservation for irrigation or municipal water systems, or (c) bans on the production of crops such as rice or cotton in irrigated areas.

- Modification of biodiversity laws to preserve habitat for creatures when climate change eliminates their current living space.

In addition, climate impacts might be managed in part through current national law. Stricter control of water pollution may be needed to attain water quality standards in order to counter the effects of higher temperature or decreased flow. Similarly, higher temperatures could increase ozone levels, requiring stricter air pollution controls.

In considering how much of a role the national government should play, we can learn from the existing debate about the national role in environmental regulation. Why shouldn't environmental protection be left to the states? One reason is that environmental problems themselves may cut across jurisdictional lines. The argument for a coordinated solution in this situation is undeniable (Revesz 1996; Engel 2006). For similar reasons, adaptation efforts that cross state lines—for example, failure to conserve water in one state may decrease the amount available to users in another state or may impair other values of the water body such as biodiversity.

Such spillover effects are relevant to adaptation issues. For instance, in arid areas, failure by water users in one state to conserve may mean that less water is available for downstream users. The spread of invasive species due to climate change may also cross state lines, as do impacts on migratory native species or species whose ranges may be pushed northward due to climate change. As illustrated by Hurricane Katrina, major disasters may also displace large populations, imposing costs on communities outside their state of origin. Furthermore, infrastructure that is exposed to climate impacts such as highways, railroads, power lines, and pipelines, may suffer service interruptions that impact businesses and individuals well outside a state's borders.

Another type of spillover is economic rather than environmental. In a world of capital mobility, regulatory efforts may be stymied by capital flight. The result may be a race to the bottom, in which jurisdictions compete by progressively lowering their environmental standards until they hit rock bottom (Shapiro 1995; Hudec 1996). Only the intervention of a centralized authority can halt this destructive competition between jurisdictions. As Revesz explains, under this 'race to the bottom' model, local jurisdictions would face a prisoner's dilemma, so that national regulation can be seen as a mechanism by which states can improve the welfare of their citizens rather than an intrusion on the autonomy of states, as it is often portrayed (Revesz 1992: 1213; see also Engel 2006–7).

Considerable dispute exists among scholars regarding the 'race to the bottom' rationale, but it seems potentially applicable to climate adapation (Revesz 1992; Engel 2006–7). For instance, local jurisdictions that are eager to attract new development might allow building in flood plains or might encourage unwise development by providing subsidized insurance against storm or flood damage.

States may be particularly prone to under-regulation when the financial burden of resulting harms can be shifted to the national government, for instance, in the form of national disaster relief. National governments may feel compelled to provide assistance after national disasters, and if so, they may need to regulate before the fact to ensure that the prospect of national assistance does not lead to great risk taking.

Other national interventions may be ancillary to mitigation efforts—for example, requirements for water conservation in order to reduce water-related energy usage. In addition, some states may be lacking in the technical capacity to do their own adaptation planning effectively. No doubt much of the adaptation effort should be entirely within state control, but there is a definite argument for some national role in setting adaptation standards. Again, the international level is somewhat different, but for similar reasons, there may be an argument for using financial leverage to encourage nations to meet minimum adaptation standards, and in any event, for providing international expertise about adaptation requirements.

3.2 Financing Adaptation Efforts

Adaptation will not be cheap. A key emerging issue is how financial responsibility will be divided between the national government and the states. Should states finance all or most adaptation, or should the national government pick up most of the expense?

3.2.1 *State funding under the 'beneficiary pays' principle*

It is easy to construct an argument in favor of leaving the financial responsibility for climate adaptation with the states. Normally, people have to pay for goods and services if they want to consume them—at least, this is the theory of a market economy. When the private market is unable to produce certain goods, perhaps because of collective action problems, the government steps in. But the basic principle that the costs of producing goods should be borne by those who benefit from them remains appealing. On this theory, the individuals who benefit from adaptation should pay the cost. Usually, this will translate into state financing of adaptation efforts that benefit local citizens. Simply put, the argument is that people who choose to live in riskier areas cannot fairly demand that their fellow citizens pay to provide them protection from these risks (Rakowski 2000: 305). We can call this the 'beneficiary pays' principle.

The 'beneficiary pays' principle would seem to call for placing the responsibility for adaptation at the lowest possible governmental level, so that both costs and benefits would be concentrated on the same group. Thus, coastal measures might be financed by coastal states, or even better, by coastal counties within those states. Sometimes an adaptation project's beneficiaries will not correspond to any existing political entity. States might respond by creating a special purpose entity; it is easy to imagine Climate Change Adaptation Districts like today's local drainage or irrigation districts.

Alternatively, states might finance adaptation projects through special tax assessments on the beneficiaries, just as the owners of property may have to pay a special assessment to finance sidewalks or other improvements. For instance, if new varieties of wheat are needed because of climate change, wheat farmers might pay a special fee to help develop the new varieties. Or if a flood zone needs additional levees, landowners might pay a special tax.

'Beneficiary pays' is an appealing principle in terms of eliminating moral hazard and rent seeking. If project beneficiaries have to pay for projects, they are unlikely to want to overinvest beyond the project's benefits or to lobby the government for projects that will raise their taxes more than any corresponding benefit they receive. To the extent that we are concerned about overinvestment in adaptation, 'beneficiary pays' is the best solution.

On the other hand, 'beneficiary pays' does not advance other possible social goals. It provides no incentive for emitters to mitigate. It leaves the costs of climate change where it finds them, doing nothing to advance loss spreading. Furthermore, to the extent that we view emitters as culpable or unjustly enriched by their failure to mitigate, 'beneficiary pays' does not advance the concept of just deserts. Finally, because benefits and costs fall on the same individuals, 'beneficiary pays' also fails to serve any redistributive goal, and we may be concerned that poorer areas will lack the financial or institutional capacity to manage adaptation.

Whether these are serious shortcomings depends in part on whether these other social goals are viewed as important. It also depends on the availability of alternative methods to advance those goals. For example, given optimum mitigation requirements, complete insurance for all risks, and a fiscal system that achieves the desired income distribution, there is no need to use adaptation financing to help achieve those goals. Society may also think that just deserts is not a valid goal for social policy or that the circumstances of climate change are not such as to involve any principle of just deserts. Thus, evaluating the normative appeal of 'beneficiaries pay' may be complicated.

There are also practical issues to be considered. Determining the beneficiaries of a given project may be straightforward, thus limiting transaction costs. Yet this will not always be true. Adaptation projects may indirectly benefit other sectors of society. For example, a water storage project may primarily benefit users in the immediate area, but it may also offer a potential fallback supply to other users in unusual drought conditions. This is a particularly significant question with multi-jurisdictional waters. Or adaptation in one area may prevent local residents from moving elsewhere, which would have created the need for public services and infrastructure in those locations. Disputes over how benefits are allocated could become quite heated, with expert witnesses marshalling the evidence for attributing benefits in different ways.

'Beneficiary pays' clearly supports use of state financing when beneficiaries of an adaptation measure are geographically restricted, and when mitigation incentives, loss spreading, and just deserts are seen as unimportant or not relevant, or these other goals are addressed through other mechanisms. Thus, as a policy matter, much of the burden of adaptation should be on the state level. But this does not mean that the national government should be completely uninvolved.

3.2.2 *National funding for adaptation*

An alternative is for the cost of adaptation to fall nationally.[3] This approach rests in part on the premise that society as a whole should protect individuals from certain kinds of harm. This system achieves that maximum amount of loss spreading. It expresses the idea that climate change is a national problem, thus emphasizing national solidarity in the face of the threat. In conjunction with the operation of the tax system it may also help ensure the right distributional result for climate costs. (This factor looms particularly large in the international context because the disparities in wealth are so large.) To the extent that the national government is able to impose a carbon tax on emitters or charges them for the purchase of carbon allowances, using those proceeds to fund adaptation also fits well with the 'polluter pays' principle.

This approach could be implemented in several ways. The national government might simply take adaptation as its own responsibility and pay for projects directly from the Treasury. Alternatively, state and local governments might receive national grants to engage in adaptation, or private sector actors might receive tax credits or other subsidies.

Although it rates well in terms of loss spreading and may also further the 'polluter pays' principle under some circumstances, national funding is problematic along other dimensions. It maximizes the potential for moral hazard and rent seeking, since financial responsibility for adaptation is uncoupled from receipt of benefits. In the worst-case scenario, climate adaptation might become one of the biggest pork barrels in history.

As mentioned above, use of national funds may be particularly attractive when it can be used as a mechanism to shift adaptation costs onto carbon emitters. This can be justified on the basis of fairness and under the 'polluter pays' principle. Prior to the last quarter of the twentieth century, emitters may not have had strong grounds for believing that their conduct would cause serious harm. Nevertheless, the fact remains that they did cause harm (and are continuing to do so today); in the process, they have enjoyed lower costs than they would have incurred by using alternative technologies or by reducing output. Thus, there is a potentially plausible basis for reallocating some adaptation costs to emitters, which could be more easily done at the national level.

In short, national financing of adaptation is appealing when the need for adaptation is easily monitored (reducing the incentive to rent seek), when there is little risk that adaptation will cause undesirable reductions in self-protective action by beneficiaries, when doing so addresses wealth inequalities between states, and when the cost can be shifted to emitters in furthering the 'polluter pays' principle. As with setting adaptation standards, the strongest argument for national funding is probably spillover effects. When infrastructure projects affect multiple states, a national financing role is appropriate.

These arguments point toward a real but limited national role. Adaptation projects should not be nationally funded when there are no spillovers, no obstacles to state provision of the adaptation, no ability to shift costs to emitters, and no strong claim for national solidarity. But this leaves a substantial category of cases in which national funding is appropriate.

4 Conclusion

Because climate change is a global phenomenon, it is tempting to think that the ideal arrangement would be to address solely through international institutions, or failing that, at the highest level of government possible. On this view, action by lower levels of government is merely a stop-gap until higher levels are mobilized.

But this perspective overlooks the potential for smaller-scale units of government to contribute to addressing climate change even after larger-scale units are actively engaged. In terms of climate mitigation, smaller-scale units can provide a source of innovation in regulatory approaches and may also be more effective in addressing land-use issues relating to mitigation. If the higher-level jurisdiction relies primarily on a carbon tax, there is little potential for conflict between levels of government. The question becomes more complicated if the chosen instrument is a cap-and-trade system, but there is still considerable room for smaller-scale jurisdictions to play an active role. In terms of adaptation, smaller-scale jurisdictions are likely to play an even more important role. Although there may be countervailing considerations in some contexts, local authorities are presumptively the most appropriate decision makers and funding sources for adaptation.

Thus, we can expect to see the continuation of what Elinor Ostrom (2009) has called a polycentric approach to climate change. Coordinating the activities of different governmental units may not be easy, but there are significant potential benefits. Although care must be taken in integrating the activities, governmental units at all scales have significant roles to play.

Notes

1. See *Clean Air Markets Group v Pataki* 338 F 3d 82 (2d Cir 2003) (invalidating a New York law that interfered with the federal SO_2 trading scheme).
2. For examples of judicial opinions applying this doctrine in the environmental area, see *United Haulers Ass'n, Inc. v Oneida-Herkimer Solid Waste Management Authority* 550 US 330 (2007); *C & A Carbone, Inc. v Town of Clarkstown* NY 511 US 383 (1994).

3. For present purposes, it is irrelevant whether the government finances projects directly through taxes or by issuing bonds, which will later result in payments financed through taxes. It would make a difference, however, if imperfections in the bond market allowed the government to transfer some of the costs away from taxpayers to bondholders.

REFERENCES

Agrawal, A. 2008. The role of local institutions in adaptation to climate change. IFRI Working Paper #W08I-3.

Arizona Law Review Editors. 2009. Symposium on nationalism and climate change: The role of the states in a future national regime. *Arizona Law Review* 50: 1–938.

Betsill, M. M. 2001. Mitigating climate change in U.S. cities: Opportunities and obstacles. *Local Environment* 6(4): 393–406.

Charnovitz, S. 1993. The environment vs. trade rules: Defogging the debate. *Environmental Law* 23: 475.

City of Chicago. 2008. *Chicago Area Climate Change Quick Guide: Adapting to the Physical Impacts of Climate Change.*

DeShazo, J. R., and Freeman, J. 2007. Timing and form of federal regulation: The case of climate change. *University of Pennsylvania Law Review* 155: 1499.

Engel, K. 2006. State and local climate change initiatives: What is motivating state and local governments to address a global problem and what does this say about federalism and environmental law? *Urban Lawyer* 38: 1015.

——2006-7. Harnessing the benefits of dynamic federalism in environmental law. *Emery Law Journal* 56: 159.

Esty, D. C. 1996. Revitalizing environmental federalism. *Michigan Law Review* 95: 570.

Farber, D. A. 2009. Climate adaptation and federalism: Mapping the issues. *San Diego Journal of Climate and Energy Law* 1: 259–86.

——and Hudec, R. E. 1994. Free trade and the regulatory state: A GATT's-eye view of the dormant commerce clause. *Vanderbilt Law Review* 47: 1401.

Faure, M. G., and Johnston, J. S. 2008. *The Law and Economics of Environmental Federalism: Europe and the United States Compared.* Philadelphia: University of Pennsylvania Law School Institute for Law and Economics.

Glicksman, R. L., and Levy, R. E. 2008. A collective action perspective on ceiling preemption by federal environmental regulation: The case of global climate change. *Northwestern University Law Review* 102: 644.

Howse, R. 2002. The appellate body rulings in the shrimp/turtle case: A new legal baseline for the trade and environment debate. *Columbia Journal of Transnational Law.* 27: 491.

Hudec, R. E. 1996. Differences in national environmental standards: The level-playing-field dimension. *Minnesota Journal of Global Trade* 5: 1.

Hunter, R., and Muylle, K. 1998. European Community environmental law: Institutions, law making, and free trades. *Environmental Law Reporter* 38: 10477.

McKinstry, R. B. 2004. Laboratories for local solutions for global problems: State, local, and private leadership in developing strategies to mitigate the causes and effects of climate change. *Pennsylvania State Environmental Law Review* 12: 15.

Mayor of London. 2009. Leading to a greener London: An environmental programme for the capital. <http://www.london.gov.uk/mayor/publications/2008/08/climate-change-adapt-strat.jsp>.

OSTROM, E. 2009. *A Polycentric Approach for Coping with Climate Change*. Background Paper to the 2010 World Development Report. Washington, DC: World Bank.

RABE, B. G. 2009. Climate change policy and regulatory federalism: The divergent paths of Canadian provinces and American states in greenhouse gas reduction. Available at <http://www.allacademic.com//meta/p_mla_apa_research_citation/0/4/1/0/5/pages41058/p41058-1.php>.

RAKOWSKI, E. 2000. Can wealth taxes be justified? *Tax Law Review* 53: 263.

REVESZ, R. 1992. Rehabilitating interstate competition: Rethinking the 'race-to-the-bottom' rationale for national environmental regulation. *New York University Law Review* 67: 1210.

——1996. Nationalism and interstate environmental externalities. *University of Pennsylvania Law Review* 144: 2341.

SHAPIRO, D. L. 1995. *Nationalism: A Dialogue*. Chicago: Northwestern University Press.

SOVACOOL, B. K., and BROWN, M. A. 2009. Scaling the policy response to climate change. *Policy and Science* 27: 317 (doi:10.1016/j.polsoc.2009.01.003).

STEIN, E. S. 2007. Regional initiatives to reduce greenhouse gas emissions. P. 315 in M. B. Gerard (ed.), *Global Climate Change and U.S. Law*. Chicago: American Bar Association.

WEATHERILL, S. 2007. *Cases and Materials on EU Law*. 8th edn. New York: Oxford University Press.

WIENER, J. B. 2007. Think globally, act globally: The limits of local climate policies. *University of Pennsylvania Law Review* 155: 1961–79.

CHAPTER 33

....

DECARBONIZING THE WELFARE STATE

....

IAN GOUGH AND JAMES MEADOWCROFT[*]

OVER the course of the twentieth century the welfare state emerged as one of the most conspicuous features of the modern polity. Together with a market-mediated economy with concentrated private ownership of the principal productive assets, and political systems with multi-party elections and fairly extensive individual rights, the welfare state helps define the basic character of contemporary developed societies. The implications of human-induced climate change now pose significant challenges for each of these institutional pillars, raising profound questions about current economic practices, processes of political decision making, and welfare arrangements.

In this chapter we focus on linkages between climate change and the welfare state. Since welfare states are almost uniquely a feature of developed societies, we ignore all international aspects of climate change, unless these impinge directly or indirectly on the welfare states of the West. Unlike most other chapters in this Handbook, there is no systematic academic research, literature, or scholarly network on this particular topic, so we must gather material and build our arguments from what is available (but see Gough et al. 2008). In the absence of reliable comparative data, we have mainly used research findings on the UK.

The argument will proceed in three steps: first, a brief characterization of contemporary welfare states; second, a discussion of the challenges to the welfare state from climate change; and third, rethinking the welfare state in light of the decarbonization imperative.

1 CONTEMPORARY WELFARE ARRANGEMENTS

....

Social policy is often defined as the public management of social risks, usually idiosyncratic risks: individually unpredictable but collectively predictable, such as ill-health or

* Ian Gough acknowledges the support provided by the UK ESRC (grant reference number ES/H00520X/1). James Meadowcroft acknowledges the support of the Canada Research Chairs program. Thanks to Anna Coote for helpful comments on previous drafts.

unemployment. In order to meet those risks, welfare states transfer the allocation of goods and services from market determination to political determination. They substitute transfer payments and public services as 'social rights of citizenship' for income and services allocated by the market. Thus all rich OECD countries have extensive social security systems covering old age, disability, sickness, unemployment, and other contingencies, plus comprehensive public education systems. Most, excepting notably the US, also have universal healthcare entitlements and child allowances and other family programs (though the Obama healthcare bill of 2010 will bring the US closer to the mainstream). Today, in the long-standing OECD member states, average social expenditure, excluding education, accounts for around 23 percent of GDP. This expansion began in the first three decades after the Second World War, but it has continued since. Total social spending in the OECD has increased by five percentage points of GDP in the quarter-century since 1980. However, much of this was driven by big expansions in southern European countries (Spain, Portugal, and Greece) that were catching up after democratization in the 1970s; the rate of growth in the Anglophone countries was around three percentage points. This indicates substantial retrenchment in some countries over the past three decades. (This whole section draws on Castles et al. 2010.)

The three decades from the end of the Second World War to the mid-1970s was exceptionally favorable for welfare states for several reasons. Capital was relatively immobile in the initial period of the post-war settlement, so there was considerable room for redistribution and this was exploited by governments of all partisan complexions. The experience of war and depression paved the way for the emergence of a Keynesian consensus promoting high levels of employment, high tax and expenditures levels, and a dominant ideology favoring government management of demand and the business cycle in capitalist economies. Distributional conflicts were mitigated by a comparatively symmetric balance of power between the interest organizations of labor and capital and by relatively high rates of economic growth. Partisan competition as well as system competition in a world now divided by an Iron Curtain further fuelled welfare state expansion. Under these circumstances, social benefits were everywhere significantly raised, existing programs were extended to cover new groups of beneficiaries, and entirely new schemes were adopted. As a consequence, welfare state coverage as well as spending levels rose dramatically with important impacts on policy outcomes including a decrease in inequality and poverty, the limited 'decommodification' of labor, the guarantee of social rights, and improved macroeconomic performance.

Despite this massive expansion, however, the institutional differences laid down in the era of welfare state consolidation persisted or were transformed in path-dependent ways. Esping-Andersen in his classic *Three Worlds of Welfare Capitalism* identified three quite distinct 'welfare state regimes: (1) a social democratic or Nordic model manifesting high levels of decommodification, with cross-class solidarity resulting in a system of generous universal benefits and a strong state role; (2) a liberal or Anglophone model, with typically low levels of decommodification, more targeted welfare benefits, and a strong preference for private welfare spending; and (3) a conservative/continental model manifesting a moderate-high degree of decommodification, a narrower sphere of solidarity related to occupational status, and a commitment to subsidiarity and the preservation of traditional family structures typical of the countries of continental Europe (Esping-Andersen 1990). There has been considerable empirical support for this regime model since then. In analyzing the impact of climate change, we must therefore distinguish impacts common to all welfare states from those which differ according to type of welfare regime.

In the 1970s and early 1980s, the 'golden age' of welfare capitalism began to falter, and the 'silver age' began to dawn. The shift to a predominantly service economy and economic globalization entailed tighter constraints on public revenues, while societal modernization and changes in the economic structure produced mounting social needs, new risk patterns, and new priorities for social policy intervention, with education and social service provision on top of the list. This had to be managed by nation-states whose sovereignty, autonomy, and tax revenues were compromised by globalization. Nevertheless, the overall picture since the oil and stagflation crises of the 1970s has been one of resilience. This can be explained by recognizing the way that welfare states shape political interests, institutions, and forms of stratification, which then call forth political mobilizations that defend and extend these social programs. In general terms, welfare states are too important to users and voters to cut back drastically. This can also explain the strong persistence of distinct welfare state regimes within the OECD: different coalitions of professions, welfare beneficiaries, taxpayers, public sector workers, trades unions, and business interests can consolidate patterns of welfare provision along universal, selective, or corporatist lines.

The overall picture today then is one of slowly rising public expenditure on the welfare state, with cross-national variations fitting the traditional league table notion of a Nordic and continental vanguard of big spenders and an English-speaking rearguard of lower-spending countries. If anything the gap has widened since spending in the latter group increased the least, and from a low base. The share of social spending in total government expenditure also increased from 39 percent in 1980 to well over 52 percent in 2005—the welfare state has proved much more immune to expenditure retrenchment than other public policy areas such as education, defense, and economic affairs.

Yet, spending has not kept up with rising social needs, driven by demography, family change, and socio-economic shifts. By comparing indicators of social rights in 1995 with peak years, it is evident that retrenchment is pervasive. Governments have substantially scaled back pension promises in Italy, Sweden, and Germany, and sick pay and unemployment benefits in the Anglophone countries. Redistributive outcomes still vary widely: the Gini coefficient of inequality varies from 0.38 in the US to 0.23 in Denmark and Sweden, a difference of over 65 percent, and the poverty rate in the US is over three times higher. What is more the gap between countries has widened over the last three decades since the pioneering of neoliberal policies in the US and UK. But in all countries, welfare states effect some progressive redistribution of individual market incomes.

1.1 Coming Issues

Before discussing climate change we should briefly note two other critical challenges to contemporary welfare states: demographic change and the aftermath of the 2008 financial crisis. In all countries life expectancy continues to rise faster than predicted, resulting in a larger share of elderly in the population. Many European countries also exhibit fertility rates falling well below replacement rate and thus workforces shrinking in absolute size. *Ceteris paribus* the first trend engenders a growing demand for pensions and increasingly costly medical procedures, while the second reduces the size of the working population. Since public pension systems have been predominantly organized on the 'pay as you go' model, the next generation of workers will be required to support a much larger cohort of elderly. However *ceteris* is not necessarily *paribus*: social democratic and liberal regimes have higher birth rates than continental countries on average, and many countries have successfully experimented with adaptive changes, for example, raising the age of retirement in line with rising life expectancy and rewarding families with children.

These demographic pressures will be magnified by the continuing fallout from the 2008 financial crisis. Alongside the 'automatic stabilizers' (increased spending on unemployment and other social benefits plus reduced tax receipts), states have implemented large discretionary fiscal stimuli to prevent a major depression in the real economy and have spent unprecedented sums bailing out banks and other financial institutions. As a result of these three shifts, average government debt in the advanced G20 countries will increase by some 30 percentage points of GDP from 2008 to 2014, with higher rises in the UK and US. The implications for Western welfare states are somber. Unless taxes can be raised substantially, there will be intense pressure to cut resources across much of the welfare state. Nor will this pressure ease quickly; the British Institute for Fiscal Studies speaks of 'two parliaments of pain.' Thus a 'fiscal crisis of the welfare state,' much discussed in the 1970s, has returned as a central political issue (Gough 2010). The crisis may herald long-term stagnation in Western economies most exposed to the financial crisis including Britain, the US, Ireland, Greece, Spain, Portugal, and others. This is the potential scenario within which we must consider the impact of climate change.

2 THE CHALLENGE OF CLIMATE CHANGE

Climate change poses direct and indirect threats to public welfare in developed states. Although poorer countries in the South are especially vulnerable, developed countries will also be exposed to impacts from rising sea levels, extreme weather events, altered temperature and rainfall patterns, and the disruption of ecological systems (IPCC 2007a). This will generate risks to life, settlements, infrastructure, industrial and agricultural productivity (hydro power output, crop yields), and so on. *Direct* risks are expected to affect particularly (a) Australia and southern regions of Europe and the US, where heat and water stress will grow, and (b) coastal regions vulnerable to rising sea levels, such as the Netherlands. Over coming decades other parts of the developed world may experience more dramatic temperature changes—for example arctic areas of Canada and Europe. This will have significant impacts on local livelihoods and ways of life, but populations in these northern regions remain small.

Indirect risks include spillover from climate change impacts elsewhere: for example, the potential for distress migration from tropical regions. According to a report by Javier Solana and Benita Ferrero-Waldner, the EU should anticipate 'a flood of climate change migrants.' Broader concerns include the possibility that climate change may lead to: (1) international conflicts (particularly over water); (2) a breakdown of the global trade regime (if agreement on mitigation proves elusive and conflicts generated by climate-related 'border tax adjustments' get out of control); and/or (3) significantly higher food prices. In such cases economic losses could affect overall levels of social welfare in developed states. These spillovers point to an inherent tension within welfare states, that in delivering entitlements to citizens they discriminate against non-citizens and 'denizens' and can become 'fortress' welfare systems. Climate change, an ineluctably international phenomenon, will test the ability of national welfare states to internationalize and recognize collective responsibility for victims elsewhere in the world.

Focusing more particularly on the operation of contemporary welfare states, climate change presents three basic challenges to the existing institutional configuration: first, it introduces an expanded set of risks and distributional problems which will require active management by social institutions; second, it opens the possibility for conflict between climate-oriented measures on the one hand and traditional social policy goals on the other; and third, it may imply that the economic model that has underpinned the current welfare state is unsustainable. Let us look at each of these in turn.

2.1 Expanded Risks and Distributional Conflicts

Many of the risks associated with climate change are not new (societies have always had to cope with floods, droughts, violent storms, and so on), but their incidence, severity, and distribution will change, and welfare policies will have to be adjusted to cope. Moreover, the effects of climate change, of the measures taken to respond to a changing climate (adaptation), and of the policies put in place to slow further change (mitigation) could have profound distributional implications. Climate risks and burdens will press unevenly on different regions, economic sectors, communities, and individuals. The same may be true for adaptation and mitigation costs. So measures will be required to ensure an equitable sharing of risks and costs. For example, what support will be given to farmers in regions where agricultural production comes under stress? How will the burden of adjusting settlement patterns in flood plains or vulnerable coastal areas be distributed? How will workers be assisted in industrial sectors that are declining as a direct result of climate policy (for instance coal extraction)? How will regional concerns be balanced when local economies are differentially related to the causes and impacts of climate change? Should households with high carbon footprints unalterable in the short term be compensated for high carbon prices, and if so how? What will be the social entitlements of climate refugees?

2.2 Tension among Policy Objectives

Governments typically struggle to reconcile diverse policy goals and competing claims on the public purse. Climate change policies for adaptation and mitigation throw additional considerations into the mix, and there is ample opportunity for tensions with established social priorities. Traditionally, welfare policies have trumped environmental policies, because direct human impacts from social ills such as poverty and disease usually bite harder and/or more rapidly than do indirect effects of environmental deterioration, and because welfare systems have nurtured interest coalitions in their support. But as worries about climate change become more pronounced they will increasingly preoccupy decision makers. Climate policy measures (but also a failure to enact such measures) could undermine established social objectives. For example, carbon taxes will press more heavily on the poor, who spend a greater proportion of their income on energy. To date environmental protection absorbs a tiny proportion of state budgets (less than 1 percent according to OECD figures), and one influential economic analysis suggests that a relatively robust mitigation response could be organized for no more than 1–2 per cent of GDP (Stern Review 2007). This is small compared to the overall scale of transfers involved in the welfare system, but to the extent that it consumes new social resources it will cause difficulties for further expanding entitlements. The real worry is that a serious mitigation response will be delayed for one or more decades, and that (1) the *subsequent severity* of the climate impacts, (2) *the scale of the necessary adaptation* activities, and (3) *the cost of the delayed crash mitigation program* that would ultimately be introduced, will result in much more serious economic losses. At that point the urgency of addressing climate change might more significantly weaken state capacity to promote traditional welfare policy objectives.

2.3 Viability of the Current Economic Model

Contemporary welfare states are predicated on an expansionary economic model, which assumes steadily rising material living standards, a gradually increasing population, and continuous economic growth. This provides jobs and business opportunities, generates tax revenues which finance welfare programs, and provides opportunities that discourage radical demands for wealth redistribution. But—at least up until this point—it also produces a growing environmental footprint of which greenhouse gas emissions are one manifestation. 'Decoupling' economic activity from environmental pressures, so that societies pollute less even as they grow more prosperous, provides one way out of this dilemma. By changing 'the quality of growth,' so that it does not harm the environment, development could become sustainable (WCED 1987). But so far evidence for such decoupling is weak: it has been achieved for some problems, in some countries, over fairly limited periods of time. In principle it should be possible to increase resource efficiencies, introduce innovative technologies, and reduce pollutant releases so that imposed environmental burdens fall dramatically.

But to realize *absolute* reductions in environmental pressures while *population* increases, and *material consumption* per capita also rises, would demand high and continuous performance improvements year after year. And given the magnitude of the absolute greenhouse gas emissions reductions required in coming decades to limit climate change to two degrees of temperature rise, achieving such decoupling represents an epochal challenge. Serious efforts to 'decouple' carbon emissions from economic activity have not yet been made, so it is too early for a conclusive assessment. But if room is to be made for peoples in developing countries to raise their living standards (requiring higher resource use and waste generation), some argue that developed countries will have to turn their backs on the expansionary economic model that has so far provided the economic foundation for the welfare state. This does not mean that 'development' will cease; industrialized societies will still be able to increase well-being: the moral, social, cultural, and material position of their citizens. But this cannot be predicated on continuously expanding appropriation from limited natural endowments (arable land, forests, fish stocks, water, the absorptive capacity of the atmosphere, and so on).

3 RETHINKING WELFARE STATES: DECOUPLE AND DECARBONIZE

Addressing climate change requires a transformation of production/consumption practices that produce greenhouse gas emissions; but it will also require a rethink of social welfare institutions built up over the last century. The ultimate consequences for Western welfare states are not clear. But following on from the discussion of economic models above, we can identify two broad, and very different, scenarios:

(a) using technological innovation to decouple economic growth from carbon emissions and at the same time decarbonize and reorient the welfare state;

(b) move from a growth to a steady-state economy and radically transform the meaning of welfare and the institutions for achieving it.

We consider each in turn: scenario (a) in this section and scenario (b) in section 4.

Even within the more benign scenario (a) there remain severe implications for Western welfare states. Proposals usually entail the idea of 'policy integration'—drawing together environmental, economic, and social decision making (Lenschow 2002; Nilsson and Eckerberg 2007; WCED 1987). Achieving it in practice is critical to addressing the challenges cited above. Over time, welfare institutions will need to be adjusted to address climate risks; and climate policy must be structured to take account of equity. Potentially important issues and avenues for welfare state reform include the following.

3.1 Green Taxes Plus Adjustments to Social Security Systems

Green taxes have been much discussed but little implemented. General carbon taxes exist in Sweden and Denmark, and more specific taxes, notably on transport fuels, are high in several European countries, such as the UK and Germany. However, the overall yield is small as a share of GDP and has fallen in the 2000s. A recent UK fiscal commission studied the effect of raising green taxes to 20 percent of total tax revenues by 2020, to be offset by lower employer social security contributions, plus 10 percent spent on retrofitting houses and eco-innovation. The modeling suggested that this alone could achieve the UK's commitment to reduce GHGs by 34 percent by 2020 (over 1990 levels). Macroeconomic effects would be minimal, except that employment would actually rise substantially due to lower employment costs (Green Fiscal 2009).

However, the effects of the carbon taxation would be regressive, even with the tax offsets and the boost to employment, and policies to ensure fairness would be an essential corollary to ensure public support. Although lower-income households spend less on energy on average, it accounts for a higher share of their income. And in the UK 30 percent of the poorest quintile of households actually use more energy than the national average, mainly because they live in very fuel-inefficient houses or are rural/suburban residents more reliant on cars. Thus carbon taxation requires complementary social policies, both to invest in low-emission housing, transport, and communities, and to protect those with low incomes but high carbon consumption (Hills 2009).

3.2 Develop Eco-social Investment

By promoting eco-social investment governments can push down greenhouse gas emissions while simultaneously addressing social issues. *Housing*, a neglected part of welfare states, is an obvious area here. The IPCC Forth Assessment Report shows that baseline carbon emissions could be reduced in the residential sector by 29 percent at little cost—the highest scope for reductions in any sector (IPCC 2007b). Countries with very inefficient houses, such as the UK, could achieve a win-win outcome by improving quality and reducing emissions. Building standards are much more stringent in, for example, Norway, Sweden, and Germany: houses meeting their building codes use around one-quarter of the energy of houses meeting the required standards in England and Wales (Monbiot 2006). However, since new building constitutes a tiny fraction of the housing stock, such improvement requires retrofitting millions of properties to a high standard. This could be supported through grants and tax relief, but existing research suggests it requires coordinated local government and community action to achieve the severe carbon reduction targets. The recent UK Climate Change Committee called for street-by-street retrofitting, in essence a new form of social investment policy (Committee on Climate Change 2009).

This would mark a shift towards an eco-social investment state. In some respects traditional welfare states have been reprioritizing social investment over social protection in the last two decades, but this would mark a step-change. Other areas for such investment include the development of public transportation and the transformation of urban forms. In each case social policy goals (the improvement of the living conditions of citizens, disadvantaged groups, the elderly, and so on) can be combined with climate mitigation and adaptation efforts through public investment strategies. This strategy lies at the heart of recent 'Green New Deal' proposals which envisage a transformational program to reduce the use of fossil fuels and in the process tackle the decline in demand caused by the credit crunch (NEF 2008).

3.3 Decarbonize Existing Social Services

At the same time existing social services will need to decarbonize rapidly. The welfare state itself has a substantial carbon footprint. For example, the British National Health Service carbon footprint in 2004 was 18.6 $mtCO_2$, some 25 percent of English public sector emissions, and it is rising fast (SDC 2008). Transforming energy and transport systems, developing green public procurement, and altering modes of service delivery could make substantial inroads here. There is huge scope to reduce the emissions footprint of government service delivery.

3.4 Change Consumer Behavior

Welfare states all affect some domains of consumer behavior, either explicitly (alcohol, drugs, parental care, job search, etc.) or implicitly. Social policy affords valuable lessons here, for example in the successful reduction of smoking. Most countries have used all three basic means available to governments to shift behavior: education and persuasion; taxation, subsidies, and other monetary incentives; and regulation (including prohibition). But there is critical experience in social policy of their limits. Incentives that appeal solely to self-interest may fail when they degrade intrinsic motivations such as altruism and solidarity. Others recognize the limits of top-down approaches and stress the need to engage people and communities in changing behavior. Thus social policy can provide valuable lessons and templates in bringing about the much more epochal changes required to mitigate climate change. Yet, anti-smoking policies took thirty years to achieve their present impact, and even now about 30 percent of adults continue to smoke.

The three areas of individual consumption that most directly effect carbon emissions are housing, transport, and food: housing primarily relates to space heating/cooling, water heating, and household appliances; transport relates to automobile usage and air travel; and food to meat consumption and 'food miles.' In each case there is a complex relation between the potential for collective and individual action: changing consumer attitudes can result in different consumption choices that can have a substantial impact on aggregate emissions; but shifts in regulatory policy (for example building codes, product energy consumption standards, automobile emission standards, and so on) are also important.

3.5 Utilize Synergies

On the bright side, there are considerable potential synergies between climate and, for example, health policies. One UK study concludes that a shift in transport from driving to walking and cycling could bring about significant reductions in heart disease/stroke (10–20 percent), breast cancer (12–13 percent), dementia (8 percent), and depression (5 percent). Similarly, a 30 percent reduction in livestock production and consumption would reduce heart disease by 15 percent (excluding effects on all other obesity-related diseases) (*Lancet*). These improvements would, *ceteris paribus*, reduce demands on health services and save money. In 2009 overweight and obesity cost the NHS £4.8 billion. If the incidence of obesity in all social classes had been the same as for the most affluent social class 1, the cost would have been £2.2 billion, a reduction of 54 per cent. By 2025 the estimated total cost to the NHS of £8.9 billion would be reduced by £4.8 billion or 46 per cent if the effects of class inequalities were eliminated. This amounts to *c*.10 percent of the current NHS budget (McPherson et al. 2009).

Managing the five issues discussed above will tax the administrative capacities of democratic political systems. It may well be that different welfare states will prove more or less capable of arriving at effective and equitable solutions, returning us to the distinction between different welfare regimes noted earlier on. There is growing evidence within the developed world that welfare regimes map on to environmental regimes. Dryzek concludes: 'social democratic welfare states and what Hall and Soskice call coordinated market economies . . . are better placed to handle the intersection of social policy and climate change than the more liberal market economies with more rudimentary welfare states' (in Gough et al. 2008). This follows because their institutions and political culture enable an interventionist state to act to promote the public good, often using the discourse of 'ecological modernization.' This is supported in a recent cross-national analysis of environmental governance regimes which identifies six 'thick eco-states' combining high levels of government involvement with high scores for civic involvement: Denmark, Norway, Sweden, Finland, Germany, and Austria (Duit 2008). The first four are social democratic welfare states and the latter two are paradigm coordinated market economies. Theory and history suggest that different types of welfare state and welfare regime will vary in their abilities to transform into 'eco-states' (Meadowcroft 2005).

4 RETHINKING WELFARE STATES: ZERO GROWTH AND RADICAL TRANSFORMATION

Others profoundly doubt that decoupling of the sweep and speed necessary can be achieved, and especially question whether we can move to a sustainable low carbon world whilst still maintaining growth *in the rich countries* (Jackson 2009). The case rests on arithmetic and ethics. To stabilize climate change on relatively optimistic assumptions may require global carbon emissions of below 4 billion tons per annum by 2050. To achieve this with continued global population growth (0.7 percent a year) and income growth (1.4 percent a year) would require a *twenty-fold* improvement on the current global average carbon intensity (grams of carbon dioxide per dollar of GDP). But even if this were achieved, it would allow for no greater catch-up by the developing world. The world in 2050 would be one of similarly egregious inequalities and suffering to the present; indeed absolute inequalities would be greater. And it would be a world of continuing cumulative income growth in the affluent West, with average incomes more than doubling again. To achieve a world where the entire population enjoyed an income comparable with EU citizens today, the world economy would need to grow six times between now and 2050, implying a technical shift of still higher orders of magnitude if climatic disaster is to be avoided. Jackson concludes: 'There is as yet no credible, socially just, ecologically sustainable scenario of continually growing incomes for a world of nine billion people' (Jackson 2009: 86).

If then the 'growth state' on which the welfare state was built is unsustainable in the West, the welfare state would have to transform. This would raise the following issues among others.

4.1 Redistributing Carbon

As well as green taxes and regulation there would need to be a more explicit distribution and redistribution of carbon. One way to do this could be through some form of Personal Carbon Allowances and Trading (PCAT) (Environmental Audit Committee 2008). There is a wide variety of such proposals, but all entail a cap on a country's total GHG emissions (decreasing year by year) and a division of this amount into equal annual allowances for each adult resident (often with a lower allowance for each child). In effect a dual accounting standard and currency is developed—energy has both a money price and a carbon 'price.' Those who use less than the average could sell their surplus and gain, while higher users would pay a market price for their excess. Advocates claim many benefits: a PCAT scheme covering domestic energy, road fuel, and air travel would be on average quite progressive; it would make real the carbon rationing required and could bring about behavioral change more directly and quickly. It could be implemented using personal carbon cards and smart metering, though the administrative difficulties should not be underestimated. In effect it would constitute a carbon form of the Basic Income idea.

Though PCAT would be inherently progressive, it raises similar issues of fairness to carbon taxation, concerning those living in inefficient or underutilized housing, or dependent on car travel, or with special needs. Too may exceptions to the standard allowance could undermine the scheme, but too few would result in 'rough justice,' which could undermine public support. For these and other reasons the UK government is now winding down its support for testing the idea.

4.2 Redistributing Work and Time

Employment policy has always been at the core of the welfare state. The post-Second World War assumption was that adult men would work full time and adult married women would undertake full-time unpaid housework perhaps with intermittent part-time labor. Since the

1960s, women have entered the paid labor force in growing numbers and policies have (slowly and variably) adapted to encourage this. In the 1990s there was a further policy shift notably in the Anglophone states to force or encourage benefit recipients to enter paid wage labor. In all this the official recognition of housework and care work has been absent or sporadic until recently.

However, it is clear that moving towards a steady-state economy must entail a significant cut in the share of time spent in paid work. This is so for several reasons, including: to break the habit of working to earn to consume, to distribute working time more evenly across the population, to reduce the ill-being associated with unemployment, and to enable a better balance between paid work and the variety of unpaid activities, such as childcare, personal care, engagement in local activities, etc. (This goes well beyond the typical economists' trade-off between work and 'leisure'.) In the simulations of the Canadian economy under-taken by Victor (2008), a reduced working week emerges as a crucial necessary condition for a high-quality, no-growth economy. However, it is unquestionable that this policy shift too would raise serious distributional problems, including the risk of increasing poverty among the low paid and trade union opposition to its impact on earnings in all income brackets. The welfare state could play a role in radically redistributing work and time opportunities among individuals, but redistribution of incomes and wealth would also be necessary.

4.3 Redistributing Income and Wealth

Welfare states have always been compatible with substantial inequalities of wealth and income; but the more comprehensive social democratic welfare states are also those where economic inequalities are more restrained. To the extent that a low carbon economy slows traditional economic growth it may spark calls for more redistributive policies. Why might this be so? In the first place, resources to deal with climate change adaptation and mitigation will have to come from somewhere, and the argument can be made that the affluent can afford to contribute more. Second, if everyone is being asked to watch their carbon footprint, then the luxury consumption of the rich may fall under the spotlight. Third, since the conspicuous consumption of the affluent is about positional goods and helps drive fashion, it would be disproportionately important to curb excesses. Fourth, there is evidence that large income inequalities erode the social solidarity required for an active public policy oriented to deal with common problems such as climate change. The traditional redistributive case for welfare states is enhanced in a future of radical climate change mitigation.

The upshot is that in a steady-state economy, a radically different welfare system would need to *integrate the redistribution* of carbon, work/time, and income/wealth. At present these are mainly studied, and policies developed, within separate silos, but that would need to change. More generally, this scenario would also require a new indicator system to monitor final well-being and sustainability, as distinct from throughput measures such as GDP (Stiglitz et al. 2009). There is now substantial evidence that excessive economic growth beyond some point (that has been exceeded in most OECD countries) can harm both objective well-being and subjective well-being as well as environmental sustainability (Kasser 2002). The idea and measurement of well-being would be progressively dissociated from that of income and commodity consumption.

4.4 Rethinking Population Policy

To the extent that welfare states in developed countries have engaged with population size and growth rates, the concern has mainly been to reward larger families (in countries where birth rates have fallen well below replacement levels), to address the problem of ageing, and to manage immigration. A steady-state economy would ultimately be predicated on stable population levels. And since (other things being equal) more people imply more greenhouse gas emissions, climate change raises anew the question of appropriate population size. Jonathan Porritt, the recent Chair of the UK Sustainable Development Commission, for example, has warned that Britain must drastically reduce its population if it is to build a sustainable society. And the Optimum Population Trust advocates a goal of halving the UK's present size to 30 million people. Yet immigration is a sensitive political issue in most developed states and issues related to reproductive rights and family planning provoke serious controversy. How this is to be handled during the protracted period where population growth in developing countries remains high, and the economic inequalities between North and South remain pronounced, is unclear. Yet it is hard to imagine that this issue can be bracketed indefinitely, and sooner or later arguments about appropriate family size and population levels and growth rates will come to the fore.

5 CONCLUSION

Climate change will raise extra demands for 'traditional' social policy measures, add new demands to manage harmful consumption, generate additional fiscal requirements for environmental policies and expenditures, and pose novel distributional dilemmas for welfare states. This transformed landscape will impose major adaptations on existing welfare states, even if decoupling is successful and continuing green growth can supply additional revenues to fund these new policy demands. In this case it is likely that universal redistributive welfare states and coordinated economic systems will be better able to adapt to a welfare-eco state model.

But if proponents of steady-state economics are right and continued economic growth in the rich world is incompatible with sustainability, then all forms of existing welfare state would need to radically transform. In which case we might see what would amount to a second decommodification of capitalism. The first decommodification, so memorably described by Polanyi, ultimately created welfare states—citizenship entitlements to common need satisfiers and social benefits mainly provided by public services paid for by taxes and social contributions. However, though entitlements were decommodified, the services were produced in a commodified form. This second stage would entail a move towards decommodified *production*—reducing working hours and commodity purchases, developing 'coproduction' (comprising civic and household economies), and fostering preventive social behavior (NEF 2009).

However, can either scenario evolve in resilient, path-dependent, inertial institutions such as established welfare states? There are at present few signs of the collective agency such a radical shift will require. The current conjuncture of economic crisis and dangerous climate change presents us with an unprecedented problem of system (dis)integration but without a coherent social movement to advance what appears to be the only sustainable solution. This leaves elite self-interest as the main stimulus for reform—a not inconsiderable resource. But the lessons from the history of welfare states suggest that radical reforms are most successful and durable when elite self-interest is combined with mobilization and pressure from below. This chapter has not considered the politics of climate change and the structure of interests and mobilizations around this issue, a critical gap which other chapters address.

REFERENCES

CASTLES, F. G., LEIBFRIED, S., LEWIS, J., OBINGER, H., and PIERSON, C. (eds.) 2010. *The Oxford Handbook of the Welfare State.* Oxford: Oxford University Press.

Committee on Climate Change. 2009. *Meeting Carbon Budgets: The Need for a Step Change.* London: HMSO.

DUIT, A. 2008. *The Ecological State: Cross-national Patterns of Environmental Governance Regimes.* Berlin: Ecologic.

Environmental Audit Committee. 2008. *Personal Carbon Trading.* London: House of Commons.

ESPING-ANDERSEN, G. 1990. *The Three Worlds of Welfare Capitalism.* Cambridge: Polity Press.

GOUGH, I. 2010. Economic crisis, climate change and the future of welfare states. *Twenty-First Century Society* 5: 51–64.

—— MEADOWCROFT, J., DRYZEK, J., GERHARDS, J., LENGFELD, H., MARKANDYA, A., and ORTIZ, R. 2008. Climate change and social policy: A symposium. *Journal of European Social Policy* 18: 325–44.

Green Fiscal Commission. 2009. *The Case for Green Fiscal Reform.* London: Green Fiscal Commission.

HILLS, J. 2009. Future pressures: Intergenerational links, wealth, demography and sustainability. In J. Hills, T. Sefton, and K. Stewart (eds.), *Towards a More Equal Society? Poverty, Inequality and Policy since 1997.* Bristol: Policy Press.

Intergovernmental Panel on Climate Change. 2007a. *Impacts, Adaptation and Vulnerability: Contribution of Working Group II.* Cambridge: Cambridge University Press.

—— 2007b. *Mitigation of Climate Change: Contribution of Working Group III.* Cambridge: Cambridge University Press.

JACKSON, T. 2009. *Prosperity without Growth: Economics for a Finite Planet.* London: Earthscan.

KASSER, T. 2002. *The Value of Materialism: A Psychological Enquiry.* Cambridge, MA: MIT Press.

Lancet. 2001. Health effects of climate change in the UK: An expert review for comment. *Lancet.*

LENSCHOW, A. 2002. *Environmental Policy Integration.* London: Earthscan.

McPHERSON, K., BROWN, M., COOTE, A., MARSH, T., and LOBSTEIN, T. 2009. Social class and overweight: Modelling the anticipated effects of social inequality on health service treatment costs until 2025. Unpublished paper commissioned by the UK SDC and NEF for the Marmot Review Task group 5.

MEADOWCROFT, J. 2005. From welfare state to ecostate? Pp. 3–23 in J. Barry and R. Eckersley (eds.), *The State and the Global Ecological Crisis.* Cambridge, MA: MIT Press.

MONBIOT, G. 2006. *Heat: How We Can Stop the Planet Burning.* London: Penguin.

NEF. 2008. *A Green New Deal.* London: NEF.

—— 2009. *Green Well Fair: Three Economies for Social Justice.* London: NEF.

NILSSON, M., and ECKERBERG, K. 2007. *Environmental Policy Integration in Practice: Shaping Institutions for Learning.* London: Earthscan.

SDC USDC. 2008. *NHS England Carbon Emissions: Carbon Footprinting Report.* London.

Stern Review. 2007. *The Economics of Climate Change.* Cambridge: Cambridge University Press.

STIGLITZ, J. E., SEN, A., and FITOUSSI, J. 2009. *Report by the Commission on the Measurement of Economic Performance and Social Progress.* Paris.

VICTOR, P. 2008. *Managing Without Growth: Slower by Design, not Disaster.* Cheltenham: Edward Elgar.

WCED. 1987. *Our Common Future.* Oxford: World Commission on Environment and Development, Oxford University Press.

CHAPTER 34

DISCOURSES OF THE GLOBAL SOUTH

SIVAN KARTHA*

1 INTRODUCTION

THE global South is not a monolith. And, as such, no single discourse can faithfully reflect its varied perspectives and interests. 'The South,' for the purposes of the UNFCCC, has come to mean a diverse assemblage of 150 states, cast together into the so-called 'non-Annex I' group of Parties.[1] At its core, accounting for 95 percent of its population,[2] is the 'Group of 77.' Created in 1964, the G-77 has since expanded and evolved into a caucus bloc for developing country engagement across the multilateral system. Its membership ranges from the Least Developed Countries (LDCs) to the major emerging economies including China and the Asian 'Tigers.' It encompasses the Alliance of Small Island States (AOSIS), forest-rich countries such as Brazil, the Congo, and Indonesia, and the oil-rich OPEC nations. It spans the political spectrum from democracy to dictatorship.

Eclipsing this diversity, however, is the unifying fact that it is home to the overwhelming majority of the world's poor, and is the locus of the world's most profound development challenges. Even while the South witnesses an acceleration of economic growth in some regions, its average individual income is still only one-sixth that of the average Annex I citizen. More to the point, the South contains virtually the entire global population of people living in extreme poverty, and practically its entire undernourished population. It is home to every country with life expectancy below 65 years, and to several countries with life expectancy below 50 years. It is home to every country with under-5 mortality rate exceeding 2.5 percent, and several with rates *ten times* higher.

It is against this backdrop that the South, in spite of its diversity, has hung together as a coherent force within the UNFCCC. One cannot say that the positions and tactics of non-Annex I countries have been rigorously consistent, nor deny that vigorous debate sometimes rages within the G-77. The South does not, and cannot, advance a single coherent and

* The author (who takes full responsibility for any errors of fact or judgment) would like to thank Tom Athanasiou, Paul Baer, Eric Kemp-Benedict, and Matthew Stilwell for intellectual input, and the Rockefeller Brothers Fund and the Swedish International Development Cooperation Agency (Sida), who provided financial support (but who do not necessarily subscribe to the views presented here).

consistently articulated discourse. Nevertheless, certain arguments and persistent themes do resonate deeply within the many Southern discourses, and thus arise in various forms in the rhetoric and strategies of the South. The objective of this chapter is to outline those fundamental commonalities of the Southern discourses.

2 THE RIGHT TO DEVELOPMENT

Firmly imbedded in the Southern climate discourses is the *right to development*. If it can be said that the many discourses share a core tenet, this would be it.

In both the North and the South, it is understood that climate change impacts, if left unmitigated, are a challenge to fulfillment of the right to development. What is more keenly felt in the South, however, is the threat that climate change *action* poses to the right to development. A simple thought experiment illustrates the nature of that threat. Imagining that a global consensus emerged to keeping warming below 2 °C, Figure 34.1 shows (black line) a scientifically realistic assessment of the sort of emission pathway that would be required. The efforts implied by this 2 °C pathway are heroic indeed; global emissions are kept within a total budget of 1,000 billion tons of carbon dioxide ($GtCO_2$) for the first half of the twenty-first century, achieved by forcing annual emissions to peak before 2015 and decline to 85 percent below 1990 levels by 2050. This path is enormously ambitious, and could only be realized under circumstances of a true, global, emergency mobilization. Yet, even this scale of effort would not mean we were 'safe.' Society would still suffer major climate impacts and risk potentially catastrophic impacts. Not only would we be virtually certain to exceed 1.5 °C, but would face a 15–30 percent chance[3] of exceeding 2 °C. Thus, this is what the Intergovernmental Panel on Climate Change would refer to as a pathway that was 'likely,' but not 'very likely' to keep warming below 2 °C (IPCC 2007).

If we next imagine that Annex I countries undertook bold efforts, starting immediately, and were able to virtually eliminate their emissions by 2050, Figure 34.1 shows (dashed line) the path they would follow and the portion of the small available global emissions budget they would consume. It shows Northern emissions declining at more than 5 percent annually, falling 40 percent below 1990 levels by 2020, and 90 percent by 2050. The emission path shown here is far more aggressive than the current pledges from Annex I, and in fact matches the demands made by many of the non-Annex I countries.

Having stipulated a global trajectory and an Annex I trajectory, simple subtraction reveals the carbon budget (shown with dotted line) that would remain to support the South's development. Despite the apparent stringency of the Annex I trajectory, the atmospheric space remaining for developing countries would be alarmingly small. Developing country emissions would have to peak only a few years later than those in the North—still before 2020—and then decline by more than 5 percent annually through 2050. And this would have to take place while most of the South's citizens were still struggling out of poverty and desperately seeking a meaningful improvement in their living standards.

It is precisely this last point—one that is not fully appreciated in the North—that animates the Southern discourses. For the only proven routes to development—to water and food security, improved healthcare and education, secure livelihoods—involve expanding access to energy services, and, consequently, a seemingly inevitable increase in fossil fuel use and thus carbon emissions. Indeed, in the absence of environmental constraints, emissions in the South would rise much more rapidly than the North's. This would be the route by which the South's

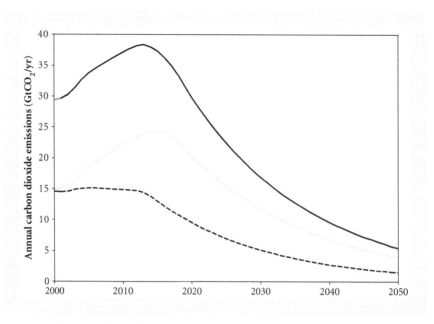

FIGURE 34.1 **The South's Dilemma.** The black line shows a global emission pathway consistent with a 2 °C goal, in which global CO_2 emissions peak by 2015 and fall to 80 percent below 1990 levels in 2050. The dashed line shows an ambitious mitigation future for the North, in which emissions decline to 90 percent below 1990 levels by 2050. The dotted line shows, by straightforward subtraction, the emissions space that would remain for the South.

citizens finally gained access to basic energy services, built the infrastructure that they have so long needed, and, eventually, moved toward some sort of parity with the citizens of the North. As a recent report from the United Nations Development Program and the World Health Organization (UNDP/WHO 2009) underscores once again, access to energy services is fundamental to the fulfillment of Millennium Development Goals.

None of this is to suggest that the Southern discourse around the term 'development' is not fraught. The South, like the North, is dominated by proponents (including most states, and the elites with whom they are generally aligned) of the view that development is more or less equivalent to macroeconomic growth. They would be quite content if the South were to follow a development path that mirrored the North's. But alternative voices, often from indigenous and other grassroots movements, can of course also be heard, raising issues of distributive justice and critiquing the fixation on GDP growth to the exclusion of alternative dimensions of human welfare and empowerment. Some go further, asserting that just and sustainable development is inconsistent with capitalism (People's Agreement 2010). But even across these widely varying conceptions of development, it is difficult to identify a vision in which lives improve significantly, especially for the impoverished majorities, that does not entail the dramatic expansion of access to energy services.

And so, the Southern discourses underscore the very real fear that the imperatives of climate stabilization will deprive the South of access to the cheap fossil energy sources that made development possible for the North. Both China and India have long counted on their vast coal reserves to fuel their long-awaited growth, but as they ponder a future with climate policies stringent enough to spur a rapid, low-carbon energy transition, they have analyses such as those of the International Energy Agency to consider. Reporting on his organization's analysis of carbon-pricing policies that would be serious enough to spur an energy transition, Executive Director Nobuo Tanaka warned: 'this is the price of carbon [necessary] to make this historic change possible, but, it means a very high price of energy for consumers. So, we are saying, . . . [the] cheap energy age is simply over. And we have to accept that. And we have to live with these higher prices' (Tanaka 2010).

Unfortunately, higher prices, for the poor majority in developing countries, may well mean the difference between access and no access. Which is exactly why, in the absence of a proven alternative route to development, it is extremely difficult for the South to imagine an equitable future in which its emissions decline as precipitously as Figure 34.1 suggests they must. The South is deeply and justifiably concerned that an inequitable climate regime will force a choice between development and climate protection. It is for this reason that developing countries remain unambiguous in their insistence that, as important as it is to deal with climate change, a solution cannot come at the expense of their right to development.

That poverty—rather than climate change—is foremost in the minds of Southern negotiators is of course unsurprising. The development crisis has shown itself to be not merely a challenge but an intractable crisis, badly in need of greater resources and political attention. To make matters worse, the impacts of climate change are now directly affecting the world's poor, not as some abstract future threat, but as a tangible force undermining food security, water security, and livelihoods. With even the minimal Millennium Development Goals being treated as second-order priorities, and little demonstrated interest in meeting them on the part of the North, the South has little reason to assume that the North would not willingly allow the exigencies of the climate crisis to eclipse the poverty crisis.

Thankfully, the conflict between climate protection and the right to development is not irreconcilable. After all, clean energy alternatives exist—but the point is that they still exist only in potential, as 'alternatives' that have not been seriously pursued. The North has not led the world in developing them, and indeed continues to pursue measures that inhibit their development and that further entrench the conventional options (through, for example, subsidies to fossil fuel exploitation). As things stand, these alternative paths are not yet real, certainly not for the poor.

With respect to the negotiations and the politics surrounding them, the key point is that the right to sustainable development is not merely ethically justifiable. It is also, fundamentally, a non-negotiable foundation of greenhouse-age geopolitical realism. Unless Southern negotiators are offered a global climate deal that explicitly preserves such a right, they may quite justifiably conclude that their countries have more to lose than to gain from any truly earnest engagement with a global climate regime that, after all, must significantly curtail access to the energy sources and technologies that historically enabled those in the industrialized world to realize their right to development.

This, precisely, is the problem that lies at the core of much of the Southern discourse. A solution must be offered before any true global climate mobilization can begin.

3 EQUALITY

A second persistent element of Southern discourses is, not surprisingly, *equality*. It has been framed in various ways, perhaps none more influential than the seminal piece by Anil Agarwal and Sunita Narain (1991), *Global Warming in an Unequal World*, which introduced many in the climate community to the argument for equal per capita emission rights. The global climate system is, after all, a public commons, as is the atmosphere into which our GHG emissions flow. As such, the privilege of using the finite atmospheric commons, they argued, must be shared equally among all people.

One can measure the atmospheric commons in terms of its total capacity to accept our carbon dioxide emissions, starting from the dawn of the industrial age (say 1850, when fossil fuel burning began in earnest) and ending in, say, the middle of the twenty-first century (by which time the fossil era must be essentially ended). Based on a path that maintains a reasonable chance of holding warming below 2 °C (as in Figure 34.1), the total available global emissions budget, over this entire period, amounts to somewhat less than 2,000 gigatons of fossil fuel carbon dioxide ($GtCO_2$). When Agarwal and Narain proposed that emissions rights should be allocated henceforth on a per capita basis, only one-third of the atmospheric commons ($\sim$650 $GtCO_2$) had already been appropriated.[4] (In Figure 34.2, see the black area.) As there was still two-thirds remaining, the notion of equally sharing access to the remaining space could reasonably be advocated as a fair enough way to share the atmospheric commons.

Over the intervening years, the situation has markedly changed. The depletion of the atmospheric commons has not slowed, as Agarwal and Narain had optimistically proposed, but rather has accelerated. Whereas it took more than 130 years to consume the first one-third of the atmospheric commons, the next one-third has taken barely twenty years. If these past two decades had been spent weaning our societies from fossil fuels, it might not be a great problem that so little of the atmospheric commons remains. But, in fact, we have not made substantive progress toward decarbonization; we remain as dependent on fossil fuels as twenty years ago. Moreover, the urgency of the climate problem has in the meantime become only more firmly supported by the scientific evidence. It was until recently thought by many that climate protection could be achieved by stabilizing temperatures at 2 °C and atmospheric carbon dioxide concentrations in the 450–500 ppm range (e.g. Stern 2007). Now, the need to keep warming below 1.5 °C and concentrations below 350 ppm is increasingly cited by scientists (Hansen et al. 2008; Pachauri 2009), UNFCCC Parties (AOSIS/LDCs 2009), and civil society (350.org, World Council of Churches 2009). The remaining atmospheric commons is shrinking not only as we consume it, but also as we learn that our earlier estimates of its size were overly optimistic.

For these reasons, many in the South are now arguing that Agarwal and Narain's notion of equality is no longer fair enough.[5] In its place has arisen the notion that equality means an equal sharing of the *entire* atmospheric commons, both the remaining portion (as Agarwal and Narain proposed) as well as the portion already consumed. Such an approach, of course, draws attention to the past and ongoing overconsumption of the industrialized nations. On a per capita basis, the North has consumed atmospheric space at a rate ten times greater that the South, accruing a large and still growing *carbon debt*.

Figure 34.3 illustrates the extent of this carbon debt.[6] The black area shows what the North would have emitted if it had kept within its equal per capita share of global emissions throughout the 200-year period shown. What the North has actually emitted has been, of course, much greater than its per capita share, and is shown here by the black area *plus* the striped area. Conversely, the actual emissions of the South (the white area) have been much less than its per capita share (the white area plus the striped area). The striped area thus shows us how much of atmospheric commons the North has 'borrowed' from the South. It is significant—the

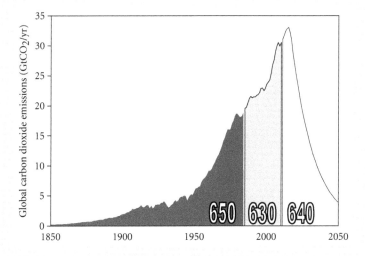

FIGURE 34.2 **The depletion of the global atmospheric commons.** If global warming is to be held below 2 °C with reasonable likelihood, society will need to limit total emissions of carbon dioxide from fossil fuels below approximately 2,000 gigatons. Although fossil fuels had been in use for significantly more than a century by the time of Agarwal and Narain's analysis, only one-third of the available budget had yet been expended (black area), so their proposed 'equal per capita entitlements' to the remaining two-thirds could reasonably be advocated as an equitable approach to sharing the atmospheric commons. In the short intervening time, another one-third of the budget has been quickly expended (grey area), leaving only one-third remaining (white area). On a per capita basis, the North has been consuming atmospheric space at a rate ten times greater that the South. And with the atmospheric commons being steadily and inequitably depleted, an equal per capita allocation of the diminishing remainder has become an increasingly inadequate proxy for an equitable sharing of the atmospheric commons.

*over*consumption of the North is nearly equal to the entire *consumption* of the South. And this is true even though there is only one consumer in the North for every five in the South.

Climate diplomats and civil society in the South is increasingly drawing attention to the historical responsibility of the North, and its resulting carbon debt. This is not to propose that the North should now reverse course, putting in place massive geoengineering schemes to extract all its excess carbon dioxide back from the atmosphere. Nor is it simply to demand reparations for a historical injustice, which would only further entrench North–South antagonisms. The intent is, rather, to underscore the fact that the North has gained its wealth by way of depleting a common resource that is therefore no longer available for others who need it. It is to provide a further justification for, and perhaps a means of quantifying,[7] the North's obligation to provide the technological and financial resources that the South needs in order to survive and develop within the limited atmospheric space that the North has left it. The North has greatly overexploited a shared resource, but the salient point is that it has thereby gained the financial and technological where-withal to enable—in the North and the South—the necessary global energy transition.

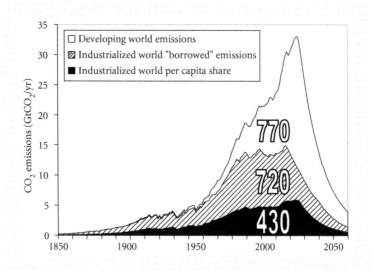

FIGURE 34.3 **Sharing the finite global commons.** Emissions from the North exceed (the striped area, 720 GtCO2) its per capita share (the black area, 430 GtCO$_2$) of the global budget for fossil fuel CO$_2$ emissions available if warming is to be held below 2 °C. Conversely, actual emissions from the South (the white area, 770 GtCO$_2$) fall far below its per capita share of global emissions (the white plus the striped area, 1490 GtCO2). The striped area thus shows us how much of atmospheric commons the North has 'borrowed' from the South.

4 THE NORTH TAKES THE LEAD

From the perspective of the South, a politically viable and equitable global climate regime means even more than an agreement that would in theory safeguard a right to development and preserve equal access to the atmospheric commons. Any agreement that would legally curtail its emissions still appears to the South to be too great a risk. Nor is this reluctance hard to understand. To this point, industrial development has been almost entirely driven by fossil fuels, and why, without the North's demonstrated willingness to help chart out, and indeed pave an alternative course, should the countries of the South sign away their rights to follow along this proven pathway?

The North, it must be said, has not demonstrated such a willingness thus far. Quite the contrary, given Annex I's neglect of its Rio promise to return emissions to 1990 levels by 2000 (notwithstanding its formal compliance, unwittingly delivered by virtue of the Soviet economic collapse), and given the past decade of half-hearted efforts to meet Kyoto commitments (and, in the case of the United States, of entirely shunning them). Understandably, the South's distrust of binding mitigation commitments is directly linked to the North's inattention to its own emission constraints.

Equally importantly, the North has also neglected its UNFCCC and Kyoto promises to provide technological and financial support for mitigation and adaptation in the South. Here, at the risk of appearing pedantic, it is useful to summarize the legal commitments formally made by developed countries under the UNFCCC. Specifically, the developed countries agreed[8] that they shall 'provide such financial resources, including for the transfer of technology, needed by the developing country Parties to meet the agreed full incremental costs of implementing measures' (UNFCCC, Art. 4.3) necessary to fulfill their own obligations to 'Formulate, implement, publish and regularly update . . . measures to mitigate climate change . . . and measures to facilitate adequate adaptation to climate change' (UNFCCC, Art. 4.1(b)).

Additionally, the developed countries agreed to provide developing countries the agreed full incremental costs of implementing measures to 'Promote and cooperate in the development, application and diffusion, including transfer, of technologies, practices and processes that control, reduce or prevent anthropogenic emissions of greenhouse gases . . . in all relevant sectors, including the energy, transport, industry, agriculture, forestry and waste management sectors' (UNFCCC, Art. 4.1(c)).

The UNFCCC underscores that the provision of necessary funding

> shall take into account the need for adequacy and predictability in the flow of funds and the importance of appropriate burden sharing among the developed country Parties. (UNFCCC, Art. 4.3)

The UNFCCC emphasizes that developing country implementation of its own commitments is contingent on the provision of developed country funding:

> The extent to which developing country Parties will effectively implement their commitments under the Convention will depend on the effective implementation by developed country Parties of their commitments under the Convention related to financial resources and transfer of technology and will take fully into account that economic and social development and poverty eradication are the first and overriding priorities of the developing country Parties. (UNFCCC, Art. 4.7)

The developed countries restated in the Kyoto Protocol (Article 11.2(b)) these same commitments to provide new and additional funding to cover the full incremental costs of mitigation, adaptation, and technology transfer in developing countries.

However, over the last eighteen years, the amount of support that developed countries have actually delivered to developing countries clearly shows that they have not fulfilled their UNFCCC and Kyoto Protocol commitments for providing financial resources. As the World Bank put it, 'Current climate-dedicated financial flows to developing countries, though growing, cover only a fraction (less than 5%) of the estimated amounts that developing countries would need over several decades.'[9]

With Annex I Parties having fallen short on fulfilling both their mitigation and finance commitments, the perspective across much of the South is that the North has failed to 'take the lead in combating climate change and the adverse affects thereof,' as UNFCCC Article 3.1 legally obliges them. In effect, the actions of the North over the past eighteen years are observed in the South as strong evidence that urgent climate action is prohibitively expensive, scientifically unjustified, and politically unviable, rather than the precisely the opposite.

5 STRATEGY

Given this state of affairs, progress in the climate regime is unlikely without the North, finally and definitively, fulfilling their obligation to 'take the lead' on climate. The Southern strategy, consequently, is to compel the North to undertake aggressive mitigation at home, and to make a good faith effort toward delivering new and additional resources to finance climate actions in the South during the coming period. From the perspective of the South, only this is consistent with the UNFCCC, the Kyoto Protocol, and the Bali Road Map. Countries of the North that insist on making their own efforts contingent on the efforts of the South are advancing a disruptive and risky strategy. Whereas by taking the lead, the North can still establish the foundation for a fair and ambitious global climate regime in which the South can feel confident their participation will not undermine their development priorities.

To be sure, the strategy of the South is aggressively challenged by the North, nowhere more evidently so than in Copenhagen. Indeed, the UNFCCC negotiations have been contentious precisely because of this North/South antagonism, much more so than because of any particular differences of opinion within each bloc. (Such as, for example, the relatively mild disagreements among Annex I nations regarding the centrality of market mechanisms, or even the disparate levels of ambitiousness of their individual mitigation efforts or financial contributions. Or, among non-Annex I nations, disagreements about how funding for adaptation should be allocated, and the relative validity of the OPEC-favored notion of compensation for the impacts of 'response measures.')

As countries of the North are quick to point out, the South is, after all, the source of virtually all recent growth in emissions, and is already responsible for nearly two-thirds of annual global GHG emissions. The implication is indisputable: no climate regime can be effective if it does not bring about considerable mitigation activity in the South. In the words of Northern government actors (expressed with various levels of bluntness), this is taken to mean that the South cannot be exempted from 'specific scheduled commitments to limit or reduce greenhouse gas emissions,' 'meaningful participation,' 'a symmetrical legally binding treaty,' 'appropriate actions,' or, most generously, 'contribut[ing] adequately according to their responsibilities and respective capabilities'.[10] To divine the actual meaning of these oblique phrases, one need only examine the actions of the Annex I countries.

The most recent and illuminating actions follow from COP-15 in Copenhagen in December 2009, which led Annex I countries to formally put forward quantitative mitigation pledges. In aggregate, these pledges amount to a reduction of between 12 and 18 percent below 1990 levels,[11] a level of effort remarkable only in its inadequacy. This is especially clear given the range of accounting loopholes that would allow Annex I countries to comply with their nominal pledges with significantly less mitigation action than the pledges would seem to imply, and possibly no real action at all.[12] The only way these pledges can be interpreted as consistent with the position of these same Annex I governments, that warming should be held below 2 °C, is if a far larger overall mitigation effort is expected from the South.

As support for the South in its efforts, the governments of the North have coupled their mitigation pledges with a 'commitment . . . to provide . . . resources . . . approaching USD 30 billion for the period 2010–2012 . . .' (UNFCCC 2010). This sum, spread across both mitigation and adaptation needs, is not large, especially considering that the stated objective is to hold warming below 2 °C, which, after all, would require a global mobilization that would force emissions to peak by roughly 2015 (as shown in Figure 34.1). And considering further that this sum is the outcome not only of Copenhagen, but of the eighteen years of negotiations since Rio, the $30 billion seems barely significant, especially with no guarantee that it will truly be 'new and additional' finance, rather than existing aid streams relabeled or diverted (Stadelmann et al. 2010). Indeed, Tuvalu's lead negotiator bitterly dismissed it as 'thirty pieces of silver to sell our future.'

It is in the face of these inadequate steps by the North that the South faces its most significant point of internal tension. At the risk of oversimplifying, this tension arises between two abiding concerns within the developing countries. The first of these is the desire to preserve 'environmental integrity', and it is best characterized by the countries of AOSIS and the LDCs. Not only do these vulnerable countries keenly feel the threat of climate change, but fear climate change as a threat to their very existence. Consequently, these countries have called for a climate regime that would hold warming well below 1.5 °C. Seeing environmental integrity as a matter of survival, they are understandably dismayed and frustrated that non-Annex I has not put forward this demand as a unified position.

The lack of a consensus around a stringent climate goal is because of the second abiding concern—the desire to secure 'developmental justice.' It is most clearly embodied by countries that are facing intense pressure by the North to take on mitigation commitments. In particular, the so-called BASIC countries—Brazil, South Africa, India, and China— which together account for roughly one-half of non-Annex I greenhouse gas emissions.

The source of the tension between 'environmental integrity' and 'developmental justice' is evident. A stringent climate goal such as 1.5 °C—or even 2 °C for that matter—implies an extremely tight global carbon budget. Simple arithmetic is all it takes to determine the budget left for the South once the North has laid claim to the portion implied by its mitigation commitments. (Recall Figure 34.1.) But, the North has not demonstrated a willingness to undertake mitigation at a level that could in any sense be considered consistent with a stringent global budget, nor has it demonstrated a willingness to provide financial and technological support to enable the South to live within the scant remaining budget. Nor, in the broader negotiations, has the North shown an interest in engaging the South on the topic of principles for equitable burden sharing. Yvo de Boer, Executive Secretary of the UNFCCC for the five years leading up to Copenhagen, observed that for the South to commit to a stringent global budget before the North had put forward commensurate mitigation targets and finance plans 'was like jumping out of a plane and being assured that you are going to get a parachute on the way down' (de Boer 2009). Under such circumstances, it is understandable that countries under intense pressure to curtail emissions might allow developmental justice to trump environmental integrity.

Some have interpreted this as a fissure in the Southern bloc (e.g. Gupta 2010). One can also see it as an expected tension within a bloc of diverse countries whose interests are otherwise largely aligned, and that continue to pursue a common overall strategy. From the Rio Earth Summit in 1992 where the UNFCCC was adopted, up to the ongoing negotiations in preparation for the 16th Conference of Parties in Cancun, the South has remained united in its primary objective: to compel the industrialized countries to bear principal responsibility for addressing the climate crisis. In the current negotiations, this approach implies a close adherence to the principles and provisions of the UNFCCC, including the protective firewall that the Annex I/non-Annex I divide provides. With the Bali Road Map, Parties agreed to a negotiating track for launching a second phase of the Kyoto Protocol with deeper mitigation commitments for the Annex I countries, along with a parallel track for negotiations focused on financial and technological support for climate-related actions in developing countries. After Copenhagen, which saw a determined attempt by the North to eliminate the firewall and discard the Kyoto Protocol system of mitigation commitments, the non-Annex I countries reasserted their insistence on the two-track negotiating process.

This does not excuse the South from earnestly engaging, and despite the rhetoric, this point is well understood. Indeed, a solution to the climate crisis is only scientifically meaningful if it involves the South deeply. The South will need to be developing and implementing adaptation strategies, and transitioning to carbon-free development paths. But the point is it cannot be expected to bear the costs, and that, for the meanwhile, its actions will be voluntary rather than binding. Legally binding, principle-based Southern commitments might ultimately be necessary, as is acknowledged by many of the burden-sharing proposals emerging in the South, but the time for them has not yet come. This next period will be one in which the developing countries, though they must act, aggressively and in many ways, will do so under agreements that are softer and more implicit than many in the North might wish. Nor should this be seen as unfair and unreasonable. The South, though it insists on latitude unavailable to the industrialized countries, is not (as many believe) obstinately persisting in an outdated and legalistic interpretation of the UNFCCC and the Kyoto Protocol, in the hopes of indefinite free-riding. Rather, its wariness is fully understandable, confronted as it is by both a climate crisis and a development crisis, and skeptical that both poverty and carbon-based growth can be simultaneously left behind. With the North for the past eighteen years showing a comparable wariness, despite its much less compelling justification, the South is holding firm in its demand that the North must now decisively take the lead.

Notes

1. It comprises, in other words, all Parties to the UNFCCC except the forty 'Annex I' countries: Australia, Belarus, Canada, Croatia, Iceland, Japan, Liechtenstein, Monaco, New Zealand, Norway, Russian Federation, Switzerland, Turkey, Ukraine, United States of America, and the member states of the European Union (excluding Cyprus).
2. Twenty-three countries account for the remaining 5 percent of non-Annex I population: twelve former Soviet republics and eastern European states, six (of thirty-nine) AOSIS countries, two OECD countries (Mexico and South Korea), two European countries (Cyprus and San Marino), and Israel.
3. For details, see Baer and Mastrandrea (2006) and Meinshausen et al. (2009).
4. Figure 34.2 is compiled from data from the Carbon Dioxide Information Analysis Center (CDIAC 2009) of the US Department of Energy, which compiles for all nations emissions of CO_2 from all fossil fuel combustion, as well as cement production and natural gas flaring, which together comprise the majority of greenhouse gas emissions. If CO_2 from land-use change and non-CO_2 gases are included, the budget is correspondingly larger.
5. Even less fair are the weaker variants such as 'Contraction and Convergence' and the Indian Prime Minister's proposal, both of which offer the developing world less than a per capita share of the remaining space. Contraction and Convergence combines grandfathering with per capita emission rights, with a gradual transition from the former to the latter over a specified number of decades (e.g. Global Commons Institute 2000). The Indian Prime Minister expressed his proposal as follows: 'At the last G-8 Summit at Heiligendamm I made a commitment on behalf of India on carbon emissions. India is prepared to commit that our per capita carbon emissions will never exceed the average per capita emissions of developed industrial countries' (Singh 2008).

6. Figure 34.3 shows actual emissions up to the present, plus future emissions assuming the North and South each follows its path as in Figure 34.1, in order to hold warming below 2 °C. The per capita shares shown in Figure 34.3 are based on the North and South share of global population in each year, which varies over the 200-year span shown.
7. Several analysts have used an equal per capita access to the full atmospheric space as a basic for quantifying obligations under a global climate regime: Bode (2003), Pan et al. (2008), Kanitkar et al. (2010).
8. These commitments apply specifically to the Annex II countries, a subset of Annex I that includes the wealthier (OECD) industrialized countries. See Mace (2003) and (2005) and UNFCCC (2006) for a comprehensive treatment of adaptation funding commitments in particular.
9. World Bank (2009). Porter et al. (2008) conclude that 'Despite the fact that the GEF was designated as the financial mechanism for the climate and biodiversity conventions, the funding provided by donor countries was never at the level required to produce significant progress in reversing the threats to climate stability and biodiversity conservation.' See also Yu (2009) and Oxfam (2009).
10. These are statements of the US Senate's Byrd-Hegel resolution (S. Res 98), the Clinton Administration (AGI 1998), the Obama Administration (Vidal 2010), the European Commission (2009), and the European Council (2009), respectively.
11. Several analyses (Stern and Taylor 2010; WRI 2010; del Elzen et al. 2010) have examined the Annex I countries' pledges and get consistent figures in this range, which expresses the span between the lower unilateral pledges and high pledges that are contingent on other countries' actions.
12. The case clearly presented by Rogelj et al. (2010). See also Point Carbon (2009), den Elzen et al. (2010), and Terry (2010).

REFERENCES

AGI (American Geological Institute). 1998. *Global Climate Change Update: Government Affairs Program 29 December 1998.* (<http://www.agiweb.org/legis105/climate.html>).

AOSIS. 2009. Aggregate Annex-1 reductions for 2020. In *Compilation and Analysis presented by the Alliance of Small Island States to the Ad Hoc Working Group on Further Commitments for Annex I Parties under the Kyoto Protocol.* 11 June.

—— /LDCs. 2009. Small islands and least developed countries join forces on climate change. Press release, <www.independentdiplomat.org/documents/joinforces>.

BAER, P., and MASTRANDREA, M. 2006. *High Stakes.* London: IPPR.

BODE, S. 2003. Equal emissions per capita over time: A proposal to combine responsibility and equity of rights. Discussion Paper 253. Hamburgisches Welt-Wirtschafts-Archiv (Hamburg Institute of International Economics), ISSN 1616–4814.

CDIAC (Carbon Dioxide Information Analysis Center). 2009. *National CO_2 Emissions from Fossil-Fuel Burning, Cement Manufacture, and Gas Flaring: 1751–2006.* Contrib. T. A. Boden, G. Marland, and R. J. Andres. Carbon Dioxide Information Analysis Center of the United States Department of Energy. <http://cdiac.ornl.gov/>.

DE BOER, Y. 2009. Quoted in: G8 makes scant progress to Copenhagen climate pact. Reuters article by Alister Doyle, 9 July.

DEN ELZEN, M. G. J., HOF, A. F., MENDOZA BELTRAN, M. A., ROELFSEMA, M., VAN RUIJVEN, B. J., VAN VLIET, J., VAN VUUREN, D. P., HÖHNE, N., and MOLTMANN, S. 2010. *Evaluation of the Copenhagen Accord: Chances and Risks for the 2 °C Climate Goal.* Netherlands Environmental Assessment Agency (PBL), May.

European Commission. 2009. Communication from the Commission to the European Parliament, the Council, the European Economic and Social Committee and the Committee of the Regions: Towards a comprehensive climate change agreement in Copenhagen. COM (2009) 39 final, 28 January, Brussels.

European Council. 2009. Council Conclusions on the further development of the EU position on a comprehensive post-2012 climate agreement (Contribution to the Spring European Council). 2928th Environment Council Meeting, 2 March, Brussels.

Global Commons Institute. 2000. GCI briefing: Contraction and convergence. Available at <www.gci.org.uk/briefings/ICE.pdf> (originally published as A. Meyer, *Engineering Sustainability* 157(4) (2000): 189–92).

GUPTA, J. 2010. Fissure over numbers as some countries block climate study. *Thaindian News* 10 June.

HANSEN, J., SATO, M., KHARECHA, P., BEERLING, D., BERNER, R., MASSON-DELMOTTE, V., PAGANI, M., RAYMO, M., ROYERM, D. L., and ZACHOS, J. C. 2008. Target atmospheric CO_2: Where should humanity aim? *Open Atmospheric Science Journal* 2: 217–31. <http://www.columbia.edu/~jeh1/2008/TargetCO2_20080407.pdf>.

IIASA (International Institute for Applied Systems Analysis). 2009. Analysis of the proposals for GHG reductions in 2020 made by UNFCCC Annex I parties, analysis and report by Fabian Wagner and Markus Amann. Laxenberg: IIASA.

IPCC. 2007. *Climate Change 2007: The Physical Science Basis. Contribution of Working Group I to the Fourth Assessment Report of the Intergovernmental Panel on Climate Change*, ed. S. SOLOMON, D. QIN, M. MANNING, Z. CHEN, M. MARQUIS, K. B. AVERYT, M. TIGNOR and H. L. MILLER. Cambridge: Cambridge University Press. See in particular Box TS.1, p. 23.

KANITKAR, T., JAYARAMAN, T., D'SOUZA, M., SANWAL, M., PURKAYASTHA, P., and TALWAR, R. 2010. Meeting equity in a finite carbon world. Background Paper for the Conference on Global carbon budgets and equity in climate change 28–9 June. Mumbai: Tata Institute of Social Sciences.

MACE, M. J. 2003. Adaptation under the UN framework convention on climate change: The legal framework. Presented at Justice in adaptation to climate change, an international seminar organized by the Tyndall Centre, FIELD, IIED, and CSERGE at the Zuckerman Institute for Connective Environmental Research University of East Anglia, 7–9 Sept.

—— 2005. Funding for adaptation to climate change: UNFCCC and GEF Developments since COP-7. *RECIEL* 14(3).

MEINSHAUSEN, M., MEINSHAUSEN, N., HARE, W., RAPER, S. C. B., FRIELER, K., KNUTTI, R., FRAME, D. J., and ALLEN, M. R. 2009. Greenhouse-gas emission targets for limiting global warming to 2 °C. *Nature* 458: 1158–63 (<http://www.nature.com/nature/journal/v458/n7242/full/nature08017.html>).

Oxfam. 2009. Beyond aid: Ensuring adaptation to climate change works for the poor. Oxfam Briefing Paper 132. 16 Sept.

PACHAURI, R. 2009. Interview with Rajendra Pachauri, Chairman of the IPCC, quoted in *Agence France-Presse* article by Marlowe Hood, Top UN climate scientist backs ambitious CO_2 cuts, 25 Aug.

PAN, J., CHEN, Y., WANG W., and LI, C. 2008. Global emissions under carbon budget constraint on an individual basis for an equitable and sustainable post-2012 international climate regime. Working Paper, Research Centre for Sustainable Development, Chinese Academy of Social Sciences, Beijing.

People's Agreement. 2010. Statement from the World People's Conference on Climate Change and the Rights of Mother Earth. 22 April, Cochabamba.

Point Carbon 2010. Assigned amount unit: Seller/buyer analysis and impact on post-2012 climate regime. A report by Point Carbon for CAN Europe. 26 Oct.

PORTER, G., BIRD, N., KAUR, N., and PESKETT, L. 2008. New finance for climate change and the environment. Washington, DC: Heinrich Böll Stiftung and WWH Macroeconomics Office.

ROGELJ, J., NABEL, J., CHEN, C., HARE, W., MARKMANN, K., MEINSHAUSEN, M., SCHAEFFER, M., MACEY, K., and HÖHNE, N. 2010. Copenhagen Accord pledges are paltry. *Nature* 464 (22 April).

SINGH, M. 2008. Text of Dr Manmohan Singh's address at the Delhi Sustainable Development Summit. Available at <http://www.domain-b.com/economy/general/20080207_dr-manmohan.html>.

STADELMANN, M., ROBERTS, J. T., and HUQ, S. 2010. Baseline for trust: Defining 'new and additional' climate funding. IIED Briefing. London: International Institute for Environment and Development. <www.iied.org/pubs/display.php?o=17080IIED>.

STERN, N. 2007. *The Economics of Climate Change: The Stern Review.* Cambridge: HM Treasury and Cambridge University Press.

—— and TAYLOR, C. 2010. What do the appendices to the Copenhagen Accord tell us about global greenhouse gas emissions and the prospects for avoiding a rise in global average temperature of more than 2 °C? Policy paper, Centre for Climate Change Economics and Policy Grantham Research Institute on Climate Change and the Environment in collaboration with the United Nations Environment Programme (UNEP).

TANAKA, N. (International Energy Agency Executive Director) 2010. Interview at CERA Week, Houston, 11 Mar. <http://www.cleanskies.com/videos/tanaka-the-impact-north-american-shale-gas>.

TERRY, S. 2010. *Integrity Gap: Copenhagen Pledges and Loopholes.* Sustainability Council of New Zealand.

UNFCCC. 2006. Overview of existing programmes and policies to assist adaptation activities (including an overview of existing decisions relating to assistance for adaptation). Background paper prepared for the UNFCCC Workshop on the Adaptation Fund, Edmonton, Alberta, 3–5 May.

—— 2010. *The Copenhagen Accord.* FCCC/CP/2009/11/Add.1 30 March. Decision 2/CP.15.

—— Secretariat. 2009. Compilation of information relating to possible quantified emission limitation and reduction objectives as submitted by Parties. 12 June.

UNDP and WHO (United Nations Development Program and World Health Organization). 2009. *The Energy Access Situation in Developing Countries.* New York: UNDP and WHO.

UN-OHRLLS. 2009. Small islands and least developed countries join forces on climate change. Press release <http://www.unohrlls.org/en/newsroom/archives/?type=2&article_id=17>.

VIDAL, J. 2010. Confidential document reveals Obama's hardline US climate talk strategy. *The Guardian,* 12 Apr.

World Bank. 2009. Monitoring and reporting on financial flows related to climate change. Discussion Paper. 29, December. <http://beta.worldbank.org/climatechange/node/5290>.

World Council or Churches (WCC). 2009. Statement on eco-justice and ecological debt, adopted 2 September. <http://www.oikoumene.org/resources/documents/central-committee/geneva-2009/reports-and-documents/report-on-public-issues/statement-on-eco-justice-and-ecological-debt.html>.

WRI (World Resources Institute). 2010. Comparability of Annex I emission reduction pledges. Kelly Levin Rob Bradley WRI working paper. Washington, DC: World Resources Institute. Available online at <http://www.wri.org>.

Yu, V. P. 2009. Have Annex I parties met their commitments under the UNFCCC and its Kyoto Protocol? South Centre Research Paper #25, October. Geneva: South Centre.

PART IX

....

POLICY INSTRUMENTS

....

....

CHAPTER 35

ECONOMIC POLICY INSTRUMENTS FOR REDUCING GREENHOUSE GAS EMISSIONS

....

DAVID HARRISON, ANDREW FOSS,
PER KLEVNAS, AND DANIEL RADOV*

1 INTRODUCTION

....

ECONOMIC instruments have played a major role in policies and proposals to address climate change in many countries and regions, with the European Union Emissions Trading Scheme (EU ETS) the most prominent example. The EU ETS is a cap-and-trade program in which an overall cap is set for carbon dioxide emissions and covered sources must submit emission allowances, which can be traded. As with the other major economic policy instrument—a carbon tax—a cap-and-trade program 'puts a price on carbon' and relies upon market responses (along with monitoring and enforcement provisions) to achieve reductions in greenhouse gases (GHGs).[1] In contrast to other policy approaches—including subsidies or mandates for specific technologies (e.g. renewables and energy efficiency)—putting a price on carbon has the economic advantage of reducing the overall cost of meeting a target level of GHG emissions because it provides incentives for firms and households to undertake the cheapest options across a wide range of potential abatement measures. Considerable experience with economic instruments—notably the US Acid Rain Trading Program, which was estimated to achieve more than a 50 percent cost

* This chapter draws on several prior evaluations of economic policy instruments, including Harrison (2002), Ellerman et al. (2003), and Harrison et al. (2008). The chapter covers policy developments through August 2010. The authors are grateful to the editors of this volume for very helpful comments on a draft of this chapter although responsibility for the chapter and for any errors or omissions it may contain is that of the authors.

savings relative to a less flexible approach—indicates that this economic advantage can be achieved in practice.

This chapter considers the use of economic instruments to address climate change, including lessons from previous experience as well as a list of the key design elements.[2] It focuses on the cap-and-trade approach and complementary credit-based programs because they have been most prominent in existing policies and proposals. We begin with an overview of the conceptual similarities and differences between cap-and-trade programs and carbon taxes. The second section summarizes experiences with emissions trading and taxes that provide lessons on how the programs work in practice. The third section describes key policy issues that arise in designing a GHG cap-and-program, many of which apply to carbon taxes as well. The final section provides brief concluding remarks.

2 OVERVIEW OF CAP-AND-TRADE AND CARBON TAX

As noted, cap-and-trade programs and carbon taxes both work by putting a price on emissions. They have much in common, notably the creation of incentives to minimize the costs of reducing emissions. The basic distinctions between the two approaches relate to how prices are set and how overall emissions are determined.[3]

- *GHG cap-and-trade.* In a cap-and-trade program, the government sets a cap on total emissions by issuing a fixed quantity of emission allowances (i.e. rights to emit a ton of GHGs) and requiring covered sources to submit allowances equal to their emissions. Allowances (which may be distributed freely or sold by the government) can be bought and sold, creating a market that places a price on emissions and a value on emissions reductions. This approach fixes the total emissions, but leaves the price of allowances uncertain and up to the market.

- *Carbon tax.* A carbon tax works by requiring covered sources to pay a fixed amount per unit of emissions. As with cap-and-trade programs, each unit of emissions has a cost. Unlike cap-and-trade programs, however, it is the price that is fixed, and the total quantity of emissions that is uncertain.

In theory, the two policies could lead to equivalent results and welfare effects: an 'optimal' carbon tax could be set to reduce emissions to the 'optimal' level, or an 'optimal' emissions cap could be set so that allowances have an 'optimal' price.[4] In practice, uncertainty about abatement measures and their costs means that the approaches may well lead to different outcomes. A fundamental trade-off is that a carbon tax would not guarantee achievement of an emissions target, although the tax would provide greater certainty on compliance costs. In terms of theoretical welfare gains, which approach is superior depends upon both the nature of the expected costs and benefits and on the uncertainties in those costs and benefits (Weitzman 1974).

Many differences between cap-and-trade and carbon tax programs have been discussed in the literature, although most of the apparent differences tend to diminish when one considers real-world implementation and how one or the other approach might be modified to deal with its 'limitations.' With regard to lack of cost certainty, for example, a cap-and-trade program can be modified to include a 'safety valve' (i.e. an allowance price ceiling) or a 'price collar' (i.e. a combination of a price ceiling and a price floor), reducing the potential 'cost certainty' advantages of the carbon tax approach (and also possibly reducing the potential 'emissions certainty' advantages of the cap-and-trade approach). Moreover, the possibility of revising carbon taxes over time (to achieve environmental goals or for other reasons) detracts from the apparent advantages of price predictability. Similarly, although commentators often point to the advantages of using carbon tax revenues to replace relatively inefficient taxes, revenues from auctioned allowances under a cap-and-trade program could provide the same revenues. In the same vein, allocations of allowances in a cap-and-trade program can be used to compensate potential losers, thereby providing greater likelihood that some policy will be adopted, but tax exemptions in theory could be used to achieve similar results.

In practice, preference for one type of policy over another may be driven less by economic efficiency or theoretical advantages than by other factors, such as which is more politically feasible and which is more likely to be well designed (Furman et al. 2007; Stavins 2008). The cap-and-trade approach has been more prominent in terms of actual policies and legislative proposals. Given the greater practical experience with cap-and-trade, we focus on this policy here, highlighting parallels to taxes where appropriate.

3 Experience with GHG Emissions Trading and Taxes

The first initiatives to put a price on the emission of GHGs were taxes introduced by various European countries in the 1990s. Early examples included taxes on CO_2 emissions in Finland, Sweden, and Norway (Sterner and Köhlin 2003), with later examples including the Climate Change Levy in the United Kingdom. A lesson from these initiatives was the difficulty of unilateral action by individual countries given the global nature of the issue and the competitive effects of acting alone. Although some of the tax rates were very high, virtually all of the countries included exemptions that applied to selected industrial sectors to protect them from potential competitive disadvantage. The 1990s also saw numerous attempts to introduce an EU-wide carbon tax that were ultimately unsuccessful. Instead, efforts for EU climate change policy coalesced around the use of EU-wide emissions trading, resulting in the 2003 decision to adopt the EU ETS, which we discuss below.

The European experience has been repeated elsewhere, with a marked preference for cap-and-trade over taxes. The closest examples of using carbon taxes as an alternative to cap-and-trade are in Canada (Quebec and British Columbia). However, there are no major examples of national governments opting for carbon taxes as an *alternative* to cap-and-trade. Instead, recent proposals have focused on using taxes as a complement to emissions trading. For example, Ireland introduced a carbon tax in 2010, designed to cover sources not already capped under the EU ETS, and a similar (but abortive) proposal was recently made in France. Recent proposals in the United Kingdom have a different focus, aiming to use a carbon tax to put a 'floor' under the effective price of carbon faced by emitters, aimed primarily at those covered by the EU ETS.

Most policy effort over the last decade thus has been devoted to initiatives to introduce cap-and-trade programs for GHGs. The most prominent of these efforts include the provisions under the Kyoto Protocol, the EU ETS, and various regional programs in the United States.

The design of these cap-and-trade programs for GHGs has drawn on past experience with emissions trading for other emissions. Cap-and-trade programs (and other emissions trading programs) have been used in the United States since the 1970s to improve local and regional air quality (Harrison 2002). The major programs (developed in the 1990s) are the Acid Rain Trading Program to control emissions of sulfur dioxide (SO_2), and the Regional Clean Air Incentives Market (RECLAIM) and Northeast Nitrogen Oxides (NO_x) Budget Trading Program to control NO_x emissions.

The experience with these programs offers both a demonstration that the theoretical advantages of cap-and-trade can be realized in practice and valuable lessons for program design. All three programs have helped reduce emissions substantially, and at lower cost than could have been achieved by equally stringent policies without the flexibility of trading. For example, the Acid Rain program achieved reductions in SO_2 emissions of 40 percent (US EPA 2009), with cost savings estimated to be more than 50 percent relative to a hypothetical equally stringent policy without trading (Ellerman et al. 2000). All programs showed that trading was feasible, with significant transfers of allowances between participants and over time.

GHGs are in some ways even better suited to the use of the cap-and-trade approach than are other emissions, as many commentators have pointed out (e.g. Ellerman et al. 2003). GHGs remain in the atmosphere for decades or centuries, and they mix uniformly throughout the world. Programs thus can be national or even international in scope, and can include significant flexibility in their chronological trajectories while still delivering a target reduction in the stock of GHGs over time. Both features increase the potential cost savings of emissions trading relative to command-and-control policies.

The negotiation of the Kyoto Protocol—signed in 1997 and coming into force in 2005—provided significant impetus for the use of emissions trading. Countries can meet part of their emission reduction requirements through three flexibility mechanisms:

1. *Emissions trading*—either through the transfer of emission units between governments or through the pooling of individual commitments to form a 'bubble';

2. *Clean Development Mechanism (CDM)*—participation in emissions reduction projects in developing countries without reduction requirements; and

3. *Joint Implementation (JI)*—participation in emissions reduction projects in developed countries with reduction requirements.

Emissions trading under the Kyoto Protocol provides direct scope for the use of cap-and-trade principles at the level of national governments and was the backdrop for the adoption of an EU cap-and-trade program. The CDM and JI are different in nature. Rather than emissions caps, these are examples of 'offsets'—i.e. emissions credits created through projects to reduce emissions from sources that are not subject to a cap on emissions. The use of offset emissions trading forms an important complement to cap-and-trade emissions trading, as we discuss in more detail below.

The EU ETS was adopted in 2003 as a cost-effective mechanism to help the EU meet its reduction requirement under the Kyoto Protocol, and it began operation in January 2005.[5] It has demonstrated the feasibility of using cap-and-trade to regulate a large and diverse set of emissions sources and constitutes by far the largest actual program to put a price on GHG emissions. The scheme includes some 11,500 installations that collectively emit about 2 billion tons of CO_2 per year, or about 45 percent of total EU CO_2 emissions. The covered sectors include power generation, virtually all industrial sectors with significant emissions, and numerous smaller combustion activities. EU Directives adopted in 2008 and 2009 will expand the scope of the EU ETS to include aviation from 2012 and additional industrial activities and GHGs from 2013.

Emissions allowances have been issued and allocated by national governments ('member states'), but the process will be harmonized on an EU level from 2013. In the period to 2012, virtually all allowances have been awarded free of charge. Allocation typically has been based on historical emissions ('grandfathering') or industry-specific benchmarks (i.e. emissions per unit of capacity, input, or output). From 2013, a much larger share of allowances will be auctioned, starting with the power sector and with the aim of gradually transitioning away from free allocation for all sectors. Member states typically have reserved a portion of total allowances for new installations. Allowances also may enter the program through the CDM and JI mechanisms of the Kyoto Protocol, a provision that has led to extensive use of CDM and (to a lesser extent) JI projects.

The EU ETS drew on lessons from previous emissions trading programs, and in turn offers both a model and valuable lessons (some of them cautionary) for programs in other countries and jurisdictions. Reforms to the EU ETS reflect some of this 'learning by doing', with Phase III (2013–20) including a single and longer-term EU cap trajectory and harmonized allocation rules implying much greater use of auctioning. These trends, as well as provisions to address concerns about impacts on industrial competitiveness and provisions to allow continued but restricted use of offsets, have been mirrored elsewhere, as we discuss below.

The EU ETS has been followed by initiatives in several other countries and jurisdictions. In the United States, various states and regions have developed or are developing cap-and-trade programs. The Regional Greenhouse Gas Initiative (RGGI) began in 2009 and covers CO_2 emissions from power plants in ten northeastern states. Prices in RGGI have been low (in the range of $2–3 per ton), reflecting the relatively lax caps. The California Global Warming Solutions Act of 2006 calls for reducing California's GHG emissions in 2020 to their level in 1990; this could involve creating a GHG cap-and-trade program in California (Market Advisory Committee 2007). Other regional programs include initiatives by governors in the West and Midwest, both of which might include Canadian provinces (Pew Center 2009a).

Bills creating national emissions trading programs have been proposed in the US Congress for many years, including the Waxman–Markey bill that was passed by the House of Representatives in June 2009. In contrast to the phased and more narrowly focused approach in the EU ETS, most of the proposals (including the Waxman–Markey bill) set caps long into the future and would cover a large share of total US emissions by including 'upstream' fuel supply as well as 'downstream' emissions at their source. The complex allocation provisions in many of these bills are designed to achieve many objectives, including mitigating price impacts on consumers, maintaining industrial competitiveness, promoting development of clean energy technologies, and reducing federal budget deficits. Auctioning percentages start low in these US bills but increase over time.

As of August 2010, the US Congress has not adopted a federal GHG cap-and-trade program, and the immediate prospects for a comprehensive US GHG trading program appear dim (Hulse and Herszenhorn 2010). Some supporters of the cap-and-trade approach have developed more limited proposals such as programs that would only cover emissions from the power sector, and various commentators have questioned the entire approach (Broder 2010). Whether opposition to cap-and-trade relates to the approach itself (relative to equally stringent alternatives) or to federal comprehensive climate change legislation per se is difficult to sort out. Current economic concerns in the United States make it difficult to pass legislation that virtually everyone agrees would be wide ranging.

Emissions trading programs to address climate change also have been proposed and adopted in developed countries outside Europe and the United States.[6] In 2008 New Zealand formally adopted an Emissions Trading Scheme, initially covering the forestry sector. This expanded in July 2010 to include major industrial sectors and the power sector. The scheme is a hybrid approach, with a fixed price for (unlimited) allowances until the end of 2012, after which a cap is planned (New Zealand Ministry for the Environment 2010). The Australian federal government has proposed a national cap-and-trade program, but Parliament has rejected the proposal. South Korea also has announced its intention to introduce a cap-and-trade program, and the Japanese government has indicated an interest in the use of emissions trading, although there appear to be considerable differences of opinion among government agencies.[7]

There also has been significant development of credit-based emissions trading, often referred to as 'offsets.' Instead of capping emissions, this type of emissions trading works by defining a baseline level of emissions and issuing credits for reductions below this level. One major motivation for offsets is that they can achieve reductions from sources in countries or jurisdictions that are unlikely to agree to overall emissions caps for many years. These additional abatement measures can reduce the overall cost of reducing global emissions.

The potential gains from spreading emissions reductions globally could be very large (e.g. Lazarowicz 2009; Edmonds et al. 1999). Studies of US proposals show large cost savings from the use of offsets (e.g. US EPA 2010). In addition, proponents argue that international offsets can offer opportunities for technology transfer and additional investment in poor countries, and also can create institutional 'buy-in' for eventual global agreement to reduce emissions.

The most significant credit-based trading for GHGs are the CDM and JI provisions of the Kyoto Protocol.[8] Both are under United Nations auspices, and have seen extensive activity; for example, there are more than 100 approved CDM 'methodologies' covering a wide range of emissions reductions projects, including non-CO_2 GHGs. As of August 2010, some 14,600 CDM projects had been registered (40 percent of these in China), corresponding to potential emissions reductions of 6.5 billion metric tons of CO_2 equivalents. Many are still awaiting processing or approval, and the number of credits actually issued is around 425 million metric tons of CO_2e (UNFCCC 2010). The volume of potential JI credits thus far is much smaller, with projects corresponding to around 90 million tons of CO_2e in the pipeline.

The main source of demand for CDM and JI credits (referred to as Certified Emissions Reductions or 'CERs,' and Emissions Reduction Units or 'ERUs,' respectively) stems from their validity for compliance with Kyoto commitments. The EU ETS has provisions for the use of CERs and ERUs. In addition to funding CDM and JI projects on their own, governments and private companies have bought stakes in carbon funds that purchase credits for their members (World Bank 2010).

Other credit-based programs include the New South Wales scheme for electricity retailers, in operation since 2003, and credits sold through the Chicago Climate Exchange (which is primarily a voluntary program). Proposals for a US offset program as part of a federal cap-and-trade program would create very large volumes of credits from projects in farming and forestry.

The use of credit-based trading has generated substantial controversy. The main contentious issue has been whether credits represent genuine emissions reductions—i.e. reductions additional to what would have occurred anyway (in a counterfactual scenario without the credits). Critics of credit-based trading have argued that the procedures and institutions used to assess whether or not reductions are real have not been robust enough, thus jeopardizing environmental integrity. For these and other reasons, existing cap-and-trade programs and proposals often include restrictions on the use of credits, as we discuss below.

The debate about the future global role of credit-based trading is far from settled, and in the absence of an international agreement after the expiration of the Kyoto Protocol in 2012, the future of CDM also is uncertain. It seems likely that some form of credit-based arrangements will continue to be used as a complement to cap-and-trade programs or carbon taxes.

4 Key Policy Issues for a GHG Cap-and-Trade Program

Many issues arise in developing and using economic instruments to address climate change. We organize the issues into two major areas: (1) basic program design elements; and (2) elements related to concerns about international competitiveness.

4.1 Basic Program Design Elements

The following are major program design elements for a GHG cap-and-trade program, most of which would apply to carbon taxes as well.

4.1.1 Coverage of sources

Defining either a GHG trading program or a carbon tax requires determining which emissions are to be covered and also the point(s) in the production chain at which emissions are covered. Both the potential environmental benefits and cost savings typically increase as the number of covered sectors increases. The EU ETS is an example of a 'downstream' program, regulating at the point of direct emission. This puts some practical limitations on the feasible coverage, as small sources, such as motor vehicles and households, would incur undue administrative costs if included. Alternatively, emissions taxes or allowance liabilities can be imposed 'upstream' at the point of production or first sale of fossil fuels, thereby including emissions from many small sources by regulating only a small number of fuel suppliers. Recent US GHG trading proposals have involved a combination of downstream implementation for large emitters (e.g. electricity generators) and upstream implementation for other sectors with a large number of emitters (e.g. motor vehicles). As a result, the proposed coverage is significantly more complete than under the EU ETS.

4.1.2 Cap or price trajectory

The overall cap on emissions and its time path determine the environmental goals of the cap-and-trade program. It also is the single most important influence on the trajectory of the allowance price (and thus overall societal costs). Conversely, the trajectory of a carbon tax would determine the trajectory of emissions over time. To be consistent with a given economy-wide emissions target, the cap or tax trajectory should consider potential reductions from non-covered emissions sources. It is also important to take into account the implications of other climate policy targets and instruments—such as renewable energy mandates, energy efficiency policies, or appliance and technology standards—that influence and are influenced by emissions trading and tax programs (Harrison et al. 2005).

4.1.3 Banking and borrowing

Banking refers to the ability to use allowances issued in one year to cover emissions in future years. It thus provides an option to trade 'across time,' shifting abatement effort from periods with higher costs to ones with lower costs. Experience from prior programs indicates that allowing banking increases cost savings from emissions trading (see e.g. Ellerman et al. 2003). Experience also has shown that banking can provide a buffer against short-term price volatility and uncertainty due to changes in demand conditions. Both the EU ETS (from Phase II (2008–2012) onward) and virtually all US GHG proposals allow banking. The reverse of banking is borrowing, which allows firms to use allowances they will receive in the future to cover current emissions, with borrowed allowances repaid. Borrowing also can reduce overall costs (especially if the cost of emissions reductions is expected to fall over time), but has been controversial. The principal concerns are that it might delay emission reductions and, as with any loan, borrowers may default. Some borrowing is possible within the EU ETS because each year's allowances are issued two months before companies must surrender allowances to comply with the previous year, and any allowances issued in a given phase are valid for compliance within that phase. Many US congressional cap-and-trade proposals also implicitly allow borrowing through similar arrangements.

4.1.4 Use of offsets/credits

As noted above, offsets can reduce the cost of meeting emissions targets, and provisions for their use have been included in the Kyoto Protocol, the EU ETS, RGGI, the New Zealand ETS, and most US federal cap-and-trade proposals. Nonetheless, concerns about whether emissions reductions are genuine have led to various restrictions on the use of credits, including restrictions on allowed quantities, types, and the conditions under which their use is permitted. The Kyoto Protocol and EU ETS require that the use of credits be 'supplemental' to other emissions reductions, which in the EU ETS has been interpreted in practice to mean that no more than half of a country's emissions reductions can be met through credits. Many US federal cap-and-trade proposals also have included quantitative limits on the use of credits (Paterson in this volume, EPRI 2007, EPRI 2010, and Pew Center 2009b). The EU ETS also has qualitative restrictions on the type of credits, notably the exclusion of credits from land use and forestry. RGGI allows offsets only if allowance prices reach a given level.

4.1.5 Other cost containment measures

The inclusion of offsets, and to some extent also banking, are often discussed in terms of 'cost containment,' that is, methods to reduce the overall cost of reducing emissions. A 'safety valve' would protect against 'excessive' GHG emission allowance prices (see Pizer 1997 and Jacoby and Ellerman 2004). Under a safety valve, the government would agree to sell an unlimited number of allowances at a pre-specified price, effectively capping the allowance price. As noted, this gives emissions trading some of the features of a carbon tax, with more certainty about the future level of prices, but less certainty about the quantity of emissions reductions. Recent US proposals have extended the safety valve approach to include a 'price collar' (a combination of price floor and price ceiling). Proposals have also included provisions that the additional allowances created must be offset by future reductions in allowances issued (thus constituting a version of scheme-wide borrowing). One potential drawback is that these measures may make it more difficult to create an international emissions trading market through the linking of individual cap-and-trade schemes. This is because a safety valve in one program would effectively apply to any linked program (Harrison et al. 2006), which the authorities in the 'linking' jurisdiction may not support. As noted, there is discussion in Europe about using taxes as an alternative method to ensure that allowance prices do not fall too low.

4.1.6 Allocation of allowances or tax revenue

Both cap-and-trade programs and carbon taxes create assets or revenues of significant value—either the value of emissions allowances or the revenues from taxes. In emissions trading programs and proposals the rules for the initial allocation of allowances have been among the most contentious issues. There is an extensive literature on different allocation options (see e.g. Harrison and Radov 2002). One important issue is the mix between free allocation and government auctioning. Existing emissions trading programs generally have allocated allowances largely for free. The arguments for free allocation have included the 'grandfathering' of past rights to emit, compensation for adverse impacts, or pragmatic considerations of political support (Harrison et al. 2007a). Arguments for auctioning include the 'polluter pays' principle and alternative distributional priorities. As noted, the EU ETS establishes much greater use of auctioning in Phase III, with continued free allocation only to sectors deemed to create risk of 'carbon leakage' (discussed below). However, the use of either auctioning or a carbon tax does not bypass distributional issues, since the government must decide how the revenues will be spent. Economists have noted the possibility of a 'double dividend,' whereby an emissions tax (or allowance auction) could lead to environmental gains as well as reduction of distortionary taxes such as corporate income taxes (Bovenberg and de Mooij 1994). As many commentators have pointed out, however, it is not clear that new revenues would actually result in the reduction of other taxes.

4.1.7 Allocation method for free allowances

Designers of emission trading programs also must decide how to distribute free allowances. Free allocation in the EU ETS has generally been based on historical emissions ('grand-fathering'), with some use of 'benchmarking' (i.e. linking allocations to emissions per unit of capacity, input, or output), especially in the power sector (Harrison et al. 2007a). Benchmarking is to be the dominant method in future years, in part to avoid reliance on increasingly outdated emissions data. Simple theory suggests that different approaches do not affect incentives for abatement, provided they are based on historical information, and thus the choice is 'just' one of distribution. When complications are introduced that violate these simple conditions, however, the incentive-equivalence of the major allocation methods no longer applies. Notably, 'updated' allocation methodologies—where the quantity of allowances received depends on participant decisions—can produce different incentive effects, as we discuss below.

In existing US programs and the EU ETS, free allocations generally have been limited to covered sources. There is no reason why allowances cannot be allocated to others, such as downstream producers and consumers who face higher prices. Recent US proposals include substantial allocations to entities that are not directly covered by the program but that would be affected by price increases (e.g. to electricity local distribution companies for consumer benefit, or to states).

4.2 International Competitiveness and 'Carbon Leakage'

Regulation of GHG emissions varies significantly internationally, ranging from no restrictions to the relatively high prices on emissions in the EU ETS. Differences in carbon prices are likely to persist for many years. For international product markets, some producers thus face costs not faced by their competitors. This raises two related concerns. First, emissions reductions in regulated regions may be counteracted by increases elsewhere—commonly referred to as emissions 'leakage.' Second, companies in countries imposing more stringent GHG policies will be at a competitive disadvantage, possibly resulting in losses of jobs and investment. Various proposals have been developed to address these concerns.

4.2.1 Program scope

Leakage and competitiveness issues are especially pronounced when small regions limit their emissions. As noted, the early carbon taxes exempted most energy-intensive industries for this reason. The same concern applies to emissions trading programs. Estimates suggest, for example, that a cap imposed only in the RGGI states would have a leakage rate of about 40 percent, i.e. for every 10 tons of CO_2 emissions reduced in the state, emissions would increase elsewhere by 4 tons (Harrison et al. 2007b). The overall leakage rate is likely to be significantly smaller for a US national program, although it may still be high in individual sectors. An attractive way to reduce leakage and competitiveness concerns thus is to broaden coverage. Recent discussions of this topic have focused on 'sectoral' approaches, aiming to ensure that the constraints and costs faced by producers within a given sector are similar across jurisdiction. However, broadening coverage may be much easier said than done.

4.2.2 Linking programs

The effective scope also can be extended through linking programs. The potential gains from linkage are clear—since gains from trade depend upon differences in costs, linking trading programs with different sources (and thus different abatement costs) promises to amplify the overall cost savings from trading, while also reducing leakage concerns by equalizing costs across regions. Moreover, linkage can promote efficient emission reductions within and between companies with operations in multiple countries. Linking can take a variety of forms, including allowing participants to trade in all markets, providing specific exchanges for inter-program transactions, or limiting cross-program exchanges to governments. So far, linking of cap-and-trade programs has been limited—for example, Norway has linked its program to the EU ETS. Proposals to link different state, regional, national, or international programs must deal with design features that reduce the compatibility of different programs. As noted above, one important design issue concerns the presence of a safety valve. Other design differences (for example, in monitoring and verification procedures) might also lead to difficulties (Harrison et al. 2006). Some harmonization of program design is likely to be required for linking to be desirable or politically feasible (Kruger et al. 2007). In practice, much of the concern about leakage and competitiveness relates to differences between regions that introduce carbon prices, and ones that do not. Given this, linking is unlikely to be a major option to address the issue.

4.2.3 Allocation methods that reduce price effects

The de facto approach in the EU has been to address potential leakage and adverse competitiveness impacts through free allocation of allowances designed to counter and reduce the potential adverse impacts. As noted, free allocation in future phases of the EU ETS will be limited to sectors deemed to be at risk of carbon leakage. There is some debate about the efficacy of this approach. Simple theory suggests that free allocation based on historical information would not be effective at reducing leakage, as the opportunity cost of allowances would be passed through to product prices. However, actual evidence of cost pass-through by EU industry outside the power sector is scant, and energy-intensive industries have argued that free allocation is crucial to enable exposed sectors to stay in business until the competitive distortion created by unequal international regulation of emissions is removed or reduced.

Another approach has been to use 'updating' methods of allowance allocation, which can lead to emissions costs not being reflected in prices. For example, if producers know that their future allocations will fall if they reduce output, they will have incentives to avoid pass-through of allowance costs to prices. Without leakage, this would just be an inefficiency that would increase the overall cost of the program. However, if output reductions are offset by increased imports, and thus increased emissions in countries without emissions regulations, the disincentive to reduce production provided by updated allocations may be less of a problem

(Harrison et al. 2007a). The EU ETS has stayed clear of updated allocations to existing emitters, though it does provide new entrant allocations and confiscates allowances upon closure (thus providing a disincentive to close down, which may reduce incentives to relocate production outside the EU). Several US proposals have included more direct updating provisions that would link allocations directly to production and employment levels.

4.2.4 *Border tax adjustments*

A final approach to addressing leakage and competitiveness issues is to subject imports to border taxes that reflect the GHG emissions associated with production of the imported goods. This approach is relatively straightforward if applied to the carbon content of imported fuels, but accurately applying taxes becomes much more complicated when it is necessary to calculate the GHG emissions associated with imports' *production* processes. In addition to administrative hurdles, adjustments to border taxes also may face legal and political hurdles. Current WTO rules allow for some tariffs to be imposed in lieu of domestic taxes, but it is unclear either that this would extend to emissions trading, or that the type of adjustments that have been proposed could be introduced without resulting in significant trade disputes. Despite these potential obstacles, various US proposals have included provisions for border taxes on goods produced in countries that do not introduce their own emissions controls.

5 CONCLUSION

Economic policy instruments have been prominent in programs and proposals around the world to address climate change. Many countries now have experience with economic instruments, particularly cap-and-trade programs and environmental taxes. These approaches seem particularly appropriate for dealing with GHGs because the gases mix uniformly throughout the globe—so that the location of emissions does not matter—and it is difficult to design cost-effective command-and-control policies for emissions (especially CO_2) that arise in practically all form of energy-using economic activity.

Despite extensive experience, designing an effective and efficient GHG cap-and-trade program or carbon tax is challenging in large part because so much of the economy is affected and thus the stakes are high. Although economic policy approaches can be designed to be less costly and intrusive than command-and-control alternatives, a major GHG cap-and-trade program or a carbon tax could impose substantial costs on businesses and ultimately on consumers. Moreover, the distributional effects of the means used to allocate allowances initially (or account for any carbon tax exemptions or revenue 'recycling') could be substantial.

All firms—particularly those that are large energy producers or consumers—need to evaluate how they would be affected by a GHG program and how their choices on investment and other decisions should be modified (Harrison and Johndrow 2009). Indeed, taking maximum advantage of the flexibility provided by economic instruments would enable firms to improve their profitability—relative to less flexible regulatory approaches—

and at the same time allow the programs to achieve environmental goals at the least overall cost to society.

Notes

1. Credit-based programs—which allow trades in emission reductions (relative to a projected baseline) rather than in emissions allowances—are another type of economic policy instrument for reducing GHG emissions.
2. See Jordan et al. in this volume for analyses of why policy makers choose economic instruments and/or other policy approaches to deal with climate change. Paterson focuses on the processes that led to the development of the Kyoto Protocol and the EU ETS.
3. For more detailed discussion of the two approaches, see e.g. Nordhaus (2007), Metcalf (2007), and Stavins (2007).
4. An 'optimal' result would maximize the sum of consumer and producer surplus and thus take into account both the value and the cost of reducing emissions.
5. Jordan et al. in this volume discuss the political process by which the EU ETS was developed as well as some of the political considerations involved in choices of other climate policy instruments.
6. There are some proposals to develop intensity-based trading programs for developing countries—such as India and China—that would cap emissions intensity relative to GDP rather than overall emissions, in order to reduce effects on economic growth (Frankel 2007). China was recently reported to be considering a voluntary emissions trading program (Jing 2010).
7. For a summary of developments in Japan, see Toda (2010) and Institute of Energy Economics, Japan (2010).
8. For discussions of issues related to CDM and JI programs, see e.g. EPRI (2007), EPRI (2010), and World Bank (2010).

References

Bovenberg, A. L., and de Mooij, R. A. 1994. Environmental levies and distortionary taxation. *American Economic Review* 84(4): 1085–9.

Broder, J. M. 2010. 'Cap and trade' loses its standing as energy policy of choice. *New York Times* 25 March.

Edmonds, J., Scott, M. J., Roop, J. M., and MacCracken, C. N. 1999. *International Emissions Trading and Global Climate Change: Impacts on the Cost of Greenhouse Gas Mitigation.* Prepared for the Pew Center on Global Climate Change. Washington, DC: Battelle.

Electric Power Research Institute (EPRI). 2007. *A Comprehensive Overview of Project-Based Mechanisms to Offset Greenhouse Gas Emissions.* EPRI Product 1014085. Palo Alto, CA: EPRI.

—— 2010. *Emissions Offsets: The Key Role of Greenhouse Gas Emissions Offsets in a US Greenhouse Gas Cap-and-Trade Program.* EPRI Product 1019910. Palo Alto, CA: EPRI.

ELLERMAN, A. D., JOSKOW, P. L., SCHMALENSEE, R., MONTERO, J.-P., and BAILEY, E. M. 2000. *Markets for Clean Air: The US Acid Rain Program.* Cambridge: Cambridge University Press.

—— —— and HARRISON, D. 2003. *Emissions Trading in the US: Experience, Lessons, and Considerations for Greenhouse Gases.* Prepared for the Pew Center on Global Climate Change. Arlington, VA: Pew.

FRANKEL, J. 2007. Formulas for quantitative emission targets. Chapter 2 in J. Aldy and R. Stavins (eds.), *Architectures for Agreement: Addressing Global Climate Change in the Post Kyoto World.* New York: Cambridge University Press.

FURMAN, J., BORDOFF, J., DESHPANDE, M., and NOEL, P. 2007. *An Economic Strategy to Address Climate Change and Promote Energy Security.* Hamilton Project Strategy Paper. Washington, DC: The Brookings Institution, Oct.

HARRISON, D. 2002. Tradable permit programs for air quality and climate change. In T. Tietenberg and H. Folmer (eds.), *International Yearbook of Environmental and Resource Economics,* vol. vi. London: Edward Elgar.

—— and JOHNDROW, J. 2009. What every company should do to prepare for a mandatory US greenhouse gas cap-and-trade program. *NERA Climate Policy Economics Insights,* 1.

—— and RADOV, D. 2002. *Evaluation of Alternative Initial Allocation Mechanisms in a European Union Greenhouse Gas Emissions Allowance Trading Scheme.* Prepared for the European Commission Directorate-General Environment. Cambridge, MA: NERA Economic Consulting.

—— KLEVNAS, P., and RADOV, D. 2005. *Interactions of Greenhouse Gas Emission Allowance Trading with Green and White Certificate Schemes.* Prepared for the European Commission Directorate-General Environment. Boston: NERA Economic Consulting.

—— —— —— and FOSS, A. 2006. *Interactions of Cost-Containment Measures and Linking of Greenhouse Gas Emissions Cap-and-Trade Programs.* Prepared for the Electric Power Research Institute (Product #1013315). Palo Alto, CA: EPRI.

—— —— —— —— 2007a. *Complexities of Allocation Choices in a Greenhouse Gas Emissions Trading Program.* Prepared for the International Emissions Trading Association. Boston: NERA Economic Consulting.

—— RESCHKE, P., NAGLER, D., FOSS, A., and KLEVNAS, P. 2007b. *Effects of the Regional Greenhouse Gas Initiative on Regional Electricity Markets.* Boston: NERA Economic Consulting.

—— KLEVNAS, P., NICHOLS, A. L., and RADOV, D. 2008. Using emissions trading to combat climate change: Programs and key issues. *Environmental Law Reporter News & Analysis* 38: 10367–84.

HULSE, C., and HERSZENHORN, D. M. 2010. Democrats call off climate bill effort. *New York Times.* 22 July.

Institute of Energy Economics, Japan. 2010. Emission trading in Japan is controversial. In *Japan Energy Brief No. 8.* Tokyo: IEE. July.

JACOBY, H., and ELLERMAN, A. D. 2004. The safety valve and climate policy. *Energy Policy* 32 (4): 481–91.

JING, L. 2010. Carbon trading in pipeline. *China Daily* 22 July.

KRUGER, J., OATES, W. E., and PIZER, W. A. 2007. *Decentralization in the EU Emissions Trading Scheme and Lessons for Global Policy.* Resources for the Future Discussion Paper 07–02. Washington, DC: RFF.

LAZAROWICZ, M. 2009. *Global Carbon Trading: A Framework for Reducing Emissions.* London: The UK Stationery Office.

Market Advisory Committee to the California Air Resources Board 2007. *Recommendations for Designing a Greenhouse Gas Cap-and-Trade System for California.* Sacramento, CA: CARB.

METCALF, G. E. 2007. *A Proposal for a US Carbon Tax Swap: An Equitable Tax Reform to Address Global Climate Change.* Hamilton Project Discussion Paper 2007–12. Washington, DC: Brookings Institution, Oct.

New Zealand Ministry for the Environment. 2010. Questions and answers about amendments to the New Zealand Emissions Trading Scheme. ⟨http://www.climatechange.govt.nz/emissions-trading-scheme/building/policy-and-legislation/faq-amendment-act.html⟩.

NORDHAUS, W. D. 2007. To tax or not to tax: Alternative approaches to slowing global warming. *Review of Environmental Economics and Policy* 1(1): 26–44.

Pew Center on Global Climate Change. 2009a. *Climate Change 101: State Action.* Arlington, VA: Pew Center.

—— 2009b. *Greenhouse Gas Offsets in a Domestic Cap-and-Trade Program.* Arlington, VA: Pew Center.

PIZER, W. 1997. *Optimal Choice of Policy Instrument and Stringency under Uncertainty: The Case of Climate Change.* Washington, DC: Resources for the Future.

STAVINS, R. N. 2007. *A US Cap-and-Trade System to Address Global Climate Change.* Hamilton Project Discussion Paper 2007–13. Washington, DC: Brookings Institution, October.

—— 2008. Cap-and-trade or a carbon tax? *Environmental Forum* 25(1): 16.

STERNER, T., and KÖHLIN, G. 2003. Environmental taxes in Europe. *Public Finance and Management* 3(1): 117–42.

TODA, E. 2010. *Recent Development in Cap and Trade in Japan.* Tokyo: Japanese Ministry of the Environment. June.

United Nations Framework Convention on Climate Change (UNFCCC). 2010. Number of CERS requested and issued. 30 August. ⟨cdm.unfccc.int/Statistics/Issuance/CERsRequestedIssuedBarChart.html⟩.

US Environmental Protection Agency (EPA). 2009. *Acid Rain and Related Programs: 2008 Highlights.* Washington, DC: EPA.

—— 2010. *EPA Analysis of the American Power Act in the 111th Congress.* Washington, DC: EPA.

WEITZMAN, M. 1974. Prices vs. quantities. *Review of Economic Studies* 41(4): 477–91.

World Bank. 2010. *State and Trends of the Carbon Market 2010.* Washington, DC: World Bank.

CHAPTER 36

POLICY INSTRUMENTS IN
PRACTICE

ANDREW JORDAN, DAVID BENSON,
RÜDIGER WURZEL, AND ANTHONY ZITO*

1 INTRODUCTION

CLIMATE change constitutes a governance challenge of breathtaking and possibly unprece-dented complexity (see Steffen, this volume). The difficulties associated with governing (or steering society with respect to) such an inherently 'wicked' problem come into particularly sharp focus when governments and organizations with quasi-governmental competences such as the European Union (EU) select, calibrate, and deploy policy instruments to implement their political promises. In many ways, the application of policy instruments, tools, and techniques (different terms which are often used interchangeably in the existing literature) constitutes the very essence of governing (Hood 2007: 142–3), or what used to be termed policy design (Bobrow and Dryzek 1987).

This chapter argues that by adopting a policy instruments perspective, analysts can illuminate crucially important aspects of governing that are not addressed by other perspectives. For example, a perspective that focuses only on political commitment, whether expressed in general terms or legal targets, may fail to appreciate that commitment is easily expressed (and retracted) as and when circumstances dictate. Policy instruments, by contrast, arguably take more time and require more political energy to select and deploy. Consequently they are often held to be better 'signifiers' of political commitment than policy objectives or targets (Lascoumes and Le Galès 2007: 9). Similarly an exclusive focus on discourse tends to emphasize the various ways in which instruments are described, justified, and explained, but downplays the downstream politics emerging around their practical application or, as is often the case, partial implementation. A policy instrument

* Andrew Jordan acknowledges the support of the European Commission (FP6 project ADAM) and the Leverhulme Trust. Rüdiger Wurzel thanks the British Academy (grant number: SG46048) for financial assistance. All remaining errors and omissions are solely the responsibility of the authors.

perspective also illuminates the reasons informing and underlying particular governance choices—e.g. whether to govern hierarchically by regulating or in more networked ways (ibid.). Whereas talk and rhetoric can be cheap, policy instruments demand more time and resources—at least if the aim is to govern rather than simply pacify critics. For these reasons, policy instrument analysts believe that they provide a more concrete (but not entirely unproblematic) measure of policy change (Hall 1993).

For those subscribing to a policy instruments perspective, the differences between the way in which instruments function in theory and 'in practice' is also hugely important and thus deserving of more detailed analysis. In spite of this (and with some notable exceptions (Andersen 1994; Andersen and Sprenger 2000)), the existing literature tends to focus either on one or the other. Economists often dominate the more theoretical literature. Their primary interest lies in the inherent characteristics (such as cost-effectiveness) of particular instruments (see Harrison et al., this volume); they tend to ignore or actively downplay contextual factors such as institutional constraints or the messy politics commonly dividing theory from reality. Often their arguments are motivated by the normative urge to prescribe certain actions rather than simply describe and/or explain instrument deployment 'in practice' (Howlett and Ramesh 1993: 7). Consequently, the politics of instrument adoption and implementation tend to be decried as an unwelcome departure from the black and white world of theory.

The theoretical literature is not of course entirely dominated by economists; public policy scholars have developed numerous classificatory schemes (for example, Peters and Nispen 1998a; Linder and Peters 1989: 39) identifying the theoretical advantages and disadvantages of particular instruments. Again, the messy world of everyday politics tends to be kept at arms length.

By contrast, political scientists dominate the more empirical literature on how policy instruments are used 'in practice.' Following the lead of Hood (1983: 9) and Majone (1976), they tend to view instrument choice as an inherently political act which involves many different actors operating within certain institutional constraints. Crucially, this literature emphasizes the tendency for instruments to be selected according to criteria other than economic efficiency (Bressers and Huitema 1999). For those adopting such a perspective, rather different questions emerge, including who shapes instrument choices and how, who wins and loses from their deployment, and what factors lead to their (non)delivery (Lascoumes and Le Galès 2007: 4; Kassim and Le Galès 2010). The price paid for empirical completeness is, however, a lack of clarity in relation to key issues (such as causality or the basis of the criteria used to make evaluative judgments).

How have these two sub-streams in the literature approached the question of instrument selection in the area of climate policy? The theoretical literature on climate policy instruments is vast and still expanding with many notable contributions from environmental economists. Consequently, the menu of potentially usable climate policy instruments is rather broad. To carbon taxes and emissions trading schemes have been added even more innovative ideas such as personal carbon quotas and labels that indicate the carbon intensity of particular goods and services. Therefore should one infer from this vibrant discussion that climate policy making is an incredibly active and dynamic area of instrument innovation and application? Or have the messy politics of governing somehow intruded, leaving their imprint on what is (or is not) used 'in practice'? This chapter weighs the evidence by looking at the differences between the evolving theory of climate policy instruments and accounts of their use 'in practice.' Throughout, our assumption is that the two are not the same. Moreover, the differences between the two are hugely important, whether one is interested in climate change policy, the politics of policy instruments, or simply governing in the more general sense.

The rest of our argument proceeds as follows. The next section surveys the theoretical literature on instruments, identifying and unpacking key terms. Section 3 sets out two ways to theorize and thus explain observed patterns of instrument selection 'in practice,' the first based on actor behavior and the second on the mediating effect of institutions. Section 4 investigates instrument choices in one important governing context, namely the EU, which comprises 27 Member States and hosts the world's largest single market. For more than a decade, it has pledged to use more non-regulatory instruments (CEC 2001). It is also a world leader in developing ambitious climate policies (Jordan et al. 2010) in relation both to mitigation and adaptation. Therefore, if there is one context in which one would expect to discover dynamic and clear-cut climate policy instrument practices, it is surely the EU. This assumption is investigated from both actor-centred and institutional perspectives. Given space constraints, the analysis mainly focuses on instrument selection and adoption practices rather than performance (i.e. in the language of policy instruments, their 'effectiveness'—but see Haug et al. 2010 and Huitema et al. 2011). It also focuses on the instruments associated with mitigating climate change rather than adapting to its unfolding impacts (but for preliminary reflections, see Rayner and Jordan 2010). The final section draws together our main arguments and identifies future challenges in this rapidly developing area of climate change research and policy.

2 POLICY INSTRUMENTS

Policy makers worldwide have introduced a diversity of instruments in response to the threat of climate change. In Europe, for example, climate policy instruments have been developed by both Member States and the EU, although the latter is fast becoming *the* dominant policy setter and instrument adopter (Jordan et al. 2010). To analyze and understand the constantly evolving patterns of instrument selection and use in general but specifically in the EU, one must first understand how policy instruments are defined and theorized.

2.1 Definitions and Categorizations

The policy instruments literature is awash with definitions and typologies, none of which enjoys unanimous support (Howlett and Ramesh 1995: 81). Put very simply, policy instruments are the 'myriad techniques at the disposal of governments to implement their policy objectives' (Howlett 1991: 2). Other analysts have tried to flesh out this broad definition by differentiating between particular subtypes of instrument, each of which embodies slightly different approaches to governing. Bemelmans-Videc et al. (1998: 50–2) for example distinguish between: regulation or '*sticks*' (i.e. highly constraining of societal choices); market based instruments or '*carrots*' (i.e. moderately choice constraining); and informational devices or '*sermons*' (i.e. which instead try to persuade societal actors to take certain choices). To this may be added voluntary agreements, through which state and societal actors pledge to tackle problems jointly in formalized but essentially non-hierarchical ways. Finally, some scholars believe there is a fifth category which focuses on instruments of a more 'organizational' nature i.e. establishing new ministries or agencies (Hood 1983; Howlett 1991: 81), but space constraints prohibit any further exploration of it in this chapter. The next subsection unpacks these categories, using EU environmental policy as a running example. Then Section 3 focuses on their use specifically within the area of climate change mitigation.

2.2 Different Instrument Types in Theory

Regulatory instruments constitute a relatively prescriptive form of governing, through which targets are established by government and then implemented by other public and private actors (Howlett and Ramesh 1995: 87). Failure to meet them usually triggers some form of punitive action. Environmental policy makers in the EU have relied heavily on regulation in the past, due to the EU's basic inability to engage in more distributional and redistributive forms of governing. Majone's (1994) characterization of the EU as a 'regulatory state' certainly seems apt judging by the total number of environmental laws it has adopted since the late 1960s—some 1,000 separate items (Haigh 2009).[1]

Market-based instruments 'affect [the] estimates of costs of alternative actions open to economic agents' (OECD 1994: 17). Eco-taxes and emissions trading schemes, long advocated by economists on cost-efficiency grounds, are seen as being the most relevant in the environmental field. In a very general sense, economic theories have already informed everyday practice: since the 1970s, eco-tax usage in the OECD has grown significantly, but particularly in parts of northern Europe (Jordan et al. 2005; Wurzel et al. 2012). However, EU decision making rules require that proposals for new taxes must secure the unanimous support of every single Member State, so they have not—as explained below—yet been systematically applied at EU level. As we shall see, this institutional rule creates a very high threshold for policy change; one which continues to frustrate policy innovation at EU level.

Informational instruments seek to provide information to social actors with the aim of changing their behavior (Howlett and Ramesh 1995: 91). In the environmental field, eco-label schemes and eco-management and audit schemes have been extensively applied in countries such as Austria and Germany, which are characterized by relatively high levels of green consumerism. By contrast, their use in less affluent parts of Europe has been rather more limited. In 1992, the EU introduced its own eco-label scheme, but this has continually suffered from a low public profile. In part, this reflects the strong desire of producers and retailers for either Member State-led schemes (that they know and have already invested in) or their own 'in house' schemes (which of course they control). At present, the latter greatly outnumber the former.

Finally, in theory *voluntary instruments* are agreed between particular target groups (e.g. industry) and public authorities (OECD 1999: 4). They can either be unilateral (i.e. self-declarations of good intent by industry) or formally negotiated between state and non-state actors and enshrined in law. In practice though, the boundary between the latter and 'ideal type' regulation is rather more blurred than is commonly supposed. For example, in the Netherlands voluntary agreements often take the form of formal contracts whose achievement can be legally enforced (Mol et al. 2000). Furthermore, the use of voluntary agreements 'in practice' is also rather mixed, with everyday practice concentrated more heavily in Germany and the Netherlands than in Southern and Eastern Europe (CEC 2002).

Attempts by the EU's executive, the European Commission, to harmonize their use by adopting EU-wide agreements have mostly floundered. Industry has generally been unwilling to sign them (preferring either no involvement or the predictability of regulation) and the EU's parliament, the European Parliament, has tended to be suspicious about the lack of external scrutiny. Finally, there are other, more theoretical problems associated with voluntary agreements, such as free-riding, against which legal enforcement in the EU's supranational court cannot easily be sought or achieved. In all these respects, instrument use 'in practice' departs strongly from what one reads in textbooks.

3 UNDERSTANDING THE SELECTION OF POLICY INSTRUMENTS 'IN PRACTICE'

3.1 Patterns of Deployment

By now it should be clear that policy instrument theory and practice are two totally different things. Take non-regulatory instruments for example. Although all EU Member States deploy them, the level of commitment (and hence usage) varies considerably within and between Member States. Broadly speaking, commitment to them has been far higher in the Northern than in the Southern and Eastern parts of Europe. Second, actors tend to hold sharply inconsistent preferences. Thus some Member States have been continually enthusiastic about some instrument types while remaining hostile towards others. For example, the UK pioneered the use of one type of market-based instrument at EU level (emissions trading) but has consistently opposed another (eco-taxation). Third, the number and diversity of instruments has steadily increased over time although this trend has been more marked at the national than at EU level. So while the EU remains politically committed to deploying a greater number of 'new' instruments, 'in practice' it has adopted very few voluntary agreements, struggled to develop a popular eco-label scheme, and failed outright to adopt eco-taxes (Jordan et al. 2005). Fourth, the practical use of instruments does not neatly correspond to the ideal types outlined in Section 2. In fact there can be as much difference within the main subtypes as there is between them. Take regulation for example. The older 'command-and-control' forms are significantly different to the more flexible or framework forms which the EU tends to prefer today (Jordan et al. 2005). In many respects, these newer forms have a lot in common with negotiated agreements. Finally, similar types of instrument may be employed in highly dissimilar ways by governors in contiguous jurisdictions. For example voluntary agreements in the Netherlands tend to be legally enforceable, whereas in Austria and Germany they are akin to self-commitments (Jordan et al. 2003a). In a word, the overall pattern is not just messy but in constant flux. For Dryzek (1997: 117), this is to be expected:

> proposals for . . . instruments can never enter in the clean and straightforward fashion of the economic textbooks. Instead, their entry and so their design is heavily dependent on the configuration of political forces and the prevailing political economic context.

The next section begins the task of uncovering and explaining this messiness.

3.2 Investigating Policy Instrument Choices

In an attempt to explain this dynamic and highly differentiated pattern of use and non-use, some analysts have tried to move away from examining the theoretical characteristics of particular instruments in their neat, 'textbook' form, towards more contextual approaches that investigate how they are used and thus become embedded in broader policies and institutional settings (Peters and Nispen 1998b). Broadly speaking, the instruments 'in practice' literature either adopts an actor-centered approach or is more institutional in its framing assumptions (for reviews, see Eliadis et al. (2007a) and Linder and Peters (1989)). The former starts from the assumption that policy systems are populated with multiple actors, including (but not limited to) governors (i.e. governments) and various target groups. The selection and calibration of all instruments (not just regulation) is assumed to be strongly affected by bargaining between different actor coalitions (Bressers and O'Toole 1998).

By contrast the latter approach argues that instrument choices in complex governing contexts such as the EU are institutionally bounded, where institutions comprise legal norms, decision-making procedures, and accepted administrative practices (Bulmer 1994). Crucially, when making choices about instruments, actors will devote some of their energies to shaping the prevailing institutions in order, in part, to affect the next cycle of instrument design choices (Majone 1976). The politics of policy instruments in other words, are not wholly to do with policy instruments; many other considerations typically intervene such as symbolic politics (firmly penalizing polluters by regulating them for example) or ideology (note the normative appeal of market forces in the EU's single market, for example). Moreover, once adopted and institutionalized, policy instruments themselves will often structure subsequent choices (Lascoumes and Le Galès 2007: 9). Regulation, for example, certainly casts a long shadow over more experimental instruments such as personal carbon trading, as well as better-established instruments such as voluntary agreements (see below). The next section employs these two broad approaches to interpret and decode instrument selection patterns found within the climate policies of the EU.

4 The Instruments of EU Climate Policy: An Interpretation

Climate policy is an area in which the EU has attempted to exert international leadership. Yet, EU climate policy only really gained momentum after 2000 (Jordan et al. 2010) and then mostly in relation to targets and policy goals rather than instruments. The EU has adopted several eye-catching emission reduction targets in the last twenty years, but curiously (and for reasons that are explored more fully below) its ability to select from the available menu of instruments remains more constrained than that of Member States. Table 36.1 summarizes the main types of policy instrument found at EU level.

Table 36.1 EU climate change policy: major instruments 1992–2009

Regulatory instruments	• 1992 Monitoring CO_2 emissions • 2001 Electricity from renewable energy • 2003 Energy performance of buildings • 2003 Biofuels • 2004 Promotion of combined heat and power • 2009 Climate change and energy package of instruments (covering: CO_2 emissions; carbon capture and storage; renewable energy; revision of emissions trading; 'Effort sharing' agreement)
Market-based instruments	• 2004 Upper and lower limit for national fuel taxes • 2003 Emissions trading (functioning from 2005)
Informational instruments	• 1992 Energy labeling • 1992 Eco-label • 1993 Eco-management and Audit Scheme (EMAS)
Voluntary instruments	• 1999/2000 Car emissions (supplanted by 2009 EU Regulation)

Sources: Haigh (1996); Krämer (2006), Jordan et al. (2010), and own research.

4.1 Market-Based Instruments

The Commission—and particularly its environmental department, Directorate-General (DG) Environment—has continually advocated the adoption of eco-taxes at EU level. During the final negotiations of the UN Framework Convention on Climate Change in 1992, the then Environment Commissioner, Carlo Ripa di Meana, tried to secure political agreement from Member States on a highly ambitious proposal for a common, EU-wide carbon dioxide (CO_2)/energy tax. However, he was thwarted by a powerful coalition of industrialists who feared it would impose a huge competitive disadvantage on them in world markets. *The Economist* magazine reported that it was *the* most heavily lobbied proposal that the Commission had ever produced (quoted in Paterson (1996: 88)). This coalition was firmly supported by those who were deeply and almost ideologically opposed to giving the EU the power to intervene in national fiscal matters. Put simply, the need for unanimity on all fiscal matters in the EU allowed the most sceptical states (the UK, principally on sovereignty grounds; Spain on economic ones), aided and abetted by the fossil fuel industry, to veto the Commission's proposal, even though there were many companies specializing in renewable energy and energy efficiency technologies that stood to benefit from higher carbon prices. Thereafter, a group of like-minded actors (namely states) organized informal meetings between 1994 and 1998 to keep the initiative alive, believing that the adoption of an EU-wide scheme would be inherently better than the alternative—a messy patchwork of different national eco-taxes. However, political support gradually drained away and the EU was eventually only able to adopt a directive (in 2003), which established a framework for harmonizing taxes on energy products including electricity. But even this was embellished with so many derogations and transition periods that it could hardly be described as an instrument of 'command and control.' Either way, this fifteen-year episode undermines the view that policy instruments are a slightly mundane, almost administrative, aspect of contemporary political life. Far from it. Inter-actor dynamics were very important, as were constraints of a more institutional nature.

The development of the EU's emissions trading scheme (ETS) was breathtakingly speedy by comparison. The Commission first issued a proposal in 2001 which was adopted by Member States just two years later in 2003. The EU's emissions trading scheme, which became operational in 2005, is the world's first transnational scheme (Jordan et al. 2010). There were three main reasons why it was adopted so quickly. First, several actors (the UK, the Netherlands, Denmark, and Sweden) acted as pioneers. Having already adopted (or planned to establish) their own national trading systems, they saw EU-level action as a means to avoid competitive disadvantages. Second, institutional factors also had a big effect: not being a tax, the EU ETS could be adopted by qualified majority voting rules. Third, the instrument itself was designed in such a way as to buy off political opponents (Ellerman et al. 2007). For example, the Commission had originally proposed that all emissions allowances should be auctioned (i.e. the approach advocated in economics textbooks), but one highly influential group of actors—the Member States—insisted on their right to distribute most of the permits (or allocations) for free in the first few phases.

The first phase (2005–7) was described as being one of 'learning by doing,' but even so it solicited criticism due to the perceived leniency afforded by some governments to their industries. But during the second phase (2008–12), the Commission secured support for a tougher stance. The European Parliament has continually insisted on full auctioning, but Member States opposed this. In summary, the (relatively decentralized) design of the ETS reflects as much political considerations (i.e. who should govern future EU and Member State climate change and energy policies) as those of a more technical or economic nature. Moreover, the ETS does not mean that the instrument of emissions trading has somehow won the day, and that the EU will henceforth establish schemes to trade in other climate-related entities. During the development of the EU's 2009 'climate-energy package,' which was designed to prepare the EU for the post-Kyoto period, there was strong opposition from many Member States to a Commission proposal to facilitate the trading of certificates in the supply of renewable energy (Jordan et al. 2010). The politics of instrument design and adoption tends it seems to defy such simple predictions.

4.2 Informational Devices

The most salient informational instruments adopted at EU level are the 1992 EU eco-label and the 1993 Eco-management and Audit Scheme (EMAS). These seek to enhance the visibility of firms and products that meet minimum environmental criteria. Neither focuses specifically on climate policy although they do partially relate to the reduction of green-house gases. However, both instruments have struggled against better-established national standards (in the case of eco-labelling) and the International Standard Organization's much less demanding ISO 1400 standards. The EU's energy label has arguably enjoyed more success. It rates the energy consumption of traded products such as lightbulbs and large electrical appliances (e.g. refrigerators and washing machines). However, attempts to make it more stringent have become mired in a deep controversy, splitting Member States and EU institutions (ENDS Europe 2009).

4.3 Voluntary Instruments

The most high-profile voluntary agreement is that between the association representing European automobile manufacturers (ACEA) and the Commission. Established in 1999, it aims to reduce CO_2 emissions from passenger cars, which are one of the fastest growing sources of greenhouse gas emissions in the EU. The Commission initially proposed to regulate, but ACEA fought hard for what it perceived to be an administratively less burdensome instrument—a voluntary agreement. It won that particularly battle and a voluntary agreement was adopted, quickly followed by similar agreements involving Japanese and Korean vehicle manufacturers.

At first, the producers lived up to their agreements and achieved sufficient reductions, but gradually progress faltered and in 2004 the Commission warned that unless things improved it would have to regulate. They did not and it duly proposed a new regulation, which was eventually adopted in late 2008. In fact things have not simply come full circle because the new regulation is significantly *more* hierarchical than the one originally proposed by the Commission in the late 1990s. For example, it enables the Commission to levy fines on car manufacturers (i.e. *not* states) that exceed their targets.

4.4 Regulation

Table 36.1 graphically illustrates the EU's continuing reliance on regulation. Despite some politically high-profile (but isolated) examples of instrument change (namely the ETS, see above), the most common instrument of EU climate policy (at least in terms of the number of adopted measures) is still regulation. Many of these regulations address fairly mundane issues such as the energy performance rating of traded products such as lightbulbs, the free trade in which is an integral part of the EU's single market. Indeed, as the EU's desire for international climate leadership has grown (expressed, that is, in terms of the adoption of ambitious reduction targets) its climate policy instruments have if anything become *more* not less regulatory (i.e. binding). For example, its 2009 climate-energy package contained no less than six separate pieces of legislation. These will accelerate the EU's impact on national climate change policies and, to a lesser degree, energy policies (Jordan et al. 2010).

It is also important to note that two of the EU's most significant market-based instruments were originally established via regulations. For example, the 2003 and 2009 Directives established and then revised the ETS. The EU ETS, as a hybrid of hierarchical and market-based modes of governing, amply demonstrates the gap between the neat manner in which instruments appear in textbooks and the rather more messy way in which they are enacted and function 'in practice.' Similarly, one of the first items of EU climate policy (a 1993 Decision establishing a mechanism to monitor emissions) has allowed the Commission to pool data provided by particular Member States. This practice has engendered a culture of 'naming and shaming' which is actually more strongly associated with network-based modes of governing (Schout et al. 2010) such as the open method of coordination, than classic 'top-down' forms of regulation.

5 SUMMARY AND CONCLUSIONS

The post-war period has witnessed a rapid proliferation of policy instruments; a trend which in many ways is constitutive of the wider development and expansion of the modern state. In a very early account of this trend, Dahl and Lindblom (1953: 8) referred to it as 'perhaps the greatest political revolution of our times.' They saw instruments—the exact term they used was 'politico-economic techniques' (ibid. 6)—as a means to govern social and economic problems 'without the intense cleavages and ideological debates that might otherwise occur' (Schneider and Ingram 1990: 511). Were they writing today, one wonders what they would make of the EU, which manages to engage in a great deal of governing which is relatively technocratic and relatively non-ideological in manner. The single market and the single currency, for example, were established peacefully and without recourse to coercion or significant political contestation at the national level. Perhaps this is because the EU is not a state in a classical sense, i.e. it does not engage in the messy politics of taxation, has relatively little money to spend, and is very much limited in its choice of policy instruments.

Be that as it may, recent experience has shown—whether in relation to climate change or other problems—that instrument design and selection in *all* governance systems including the EU is anything but politically and ideologically uncontested. This chapter has revealed that selecting, adopting, and implementing policy instruments 'in practice' can be every bit as political as the framing of problems, and the establishment and implementation of targets, thus adding credence to the claim that a policy instruments approach reveals things about governing that other perspectives leave in shadow (Eliadis et al. 2007b: 4).

But what precisely does a policy instruments perspective offer beyond a recognition that instrumentation is both important and inherently political? As yet, there is still no parsimonious theory of policy instrument selection or even an accepted typology of the main instrument types (Peters and Nispen 1998b). In fact, instruments continue to play a somewhat 'peripheral' role in public policy analysis (Lascoumes and Le Galès 2007: 1). They do not, for example, appear in the indexes of comprehensive reviews of the field such as Parsons (1995), Peters and Pierre (2006), or Sabatier (2007). Climate change scholars have started to show more interest in things like emissions trading, but there is still no comprehensive account of climate policy instruments 'in practice' that students can turn to. In the main, the literature focuses on one instrument at a time. But things are changing. In recent years scholarly interest has burgeoned (Salamon 2002; Eliadis et al. 2007a), particularly in the EU (Jordan et al. 2003b; Zito et al. 2003; Kassim and Le Galès 2010), for no other reason than that it is the main locus of climate policy development in Europe. The literature on climate policy instruments in Europe in particular has mushroomed in recent years (Jordan et al. 2010; Wurzel et al. 2012).

By adopting a combination of actor-centered and institution-centered approaches, this chapter has tried to show what each reveals in relation to the pattern of climate policy instrument use in one important jurisdiction—the EU—which has adopted a leading position in the global climate debate. It has confirmed that even in this politically highly committed jurisdiction, the patterns of use are messy and in constant flux; policy instrument theory and practice are and show every sign of remaining very different things. What one is likely to encounter is less the smooth adoption of ideal types and more a 'case of muddling through in which [choices are] shaped by the characteristics of the instruments, the nature of the problem at hand, past experiences ... of dealing with the same or similar problems, the subjective preference of the decision makers and the likely reaction to the choice by affected social groups' (Howlett and Ramesh 1993: 13). In short, a complicated skein of different factors. To this long list, one could also add the defining features of the prevailing governing context, which in the case of the EU includes a widely distributed set of steering capacities, a deep commitment to trade liberalization, and ongoing interaction effects, both vertically between instruments at different levels of governance (e.g. international, EU, and national), and horizontally across policy areas (e.g. between transport, energy, or the environment) (Zito et al. 2003).

The rather messy pattern of policy instruments that one encounters in the EU certainly presents an analytical challenge, to which scholars have still not fully responded. However, it also presents a series of more immediate political challenges to those seeking to build popular support for the kinds of policy changes demanded by climate scientists and environmental pressure groups. And it also complicates the task of legitimating supranational policy interventions at the national level. Lawyers are not alone in sensing that the 'complexity and diversity' of EU policy makes 'it ... hard for citizens to understand what [EU climate policy] really means, and whether [the overall instrument mix] is equitable and effective' (Deketelaere and Peeters 2006: 19). Even in a seemingly propitious context such as the EU which has been very supportive of higher standards, this complexity does not bode well for future policy success (Jordan et al. 2010).

To conclude, textbook accounts of policy instruments offer a very poor guide to how they appear and function 'in practice.' Having adopted a 'lead by example' approach in relation to climate, one might have imagined that the EU would have gone a lot further in closing the gap between theory and practice. It has certainly innovated in the way its selects and adopts some climate policy instruments, but overall patterns of use remain inherently bounded (Halpern 2010; Jordan et al. 2003b; Jordan et al. 2005). To be sure, the EU has successfully imported instruments first used outside Europe (emission trading was originally pioneered in the USA) and built on preexisting instrument choices made at the Member State level, but overall it remains a *regulatory state*—hinting at limitations on policy instrument selection and use not fully accounted for in actor-centered theoretical approaches.

In large part, this is a function of the diversity of actor preferences. In the EU, the EU institutions continually seek to extend their powers whereas Member States remain anxious to retain control, a struggle which is continually played out in relation to the selection and adoption of policy instruments. Instruments, in a sense, are indeed 'signifiers' of much deeper political conflicts (Wurzel et al. 2012). But this chapter has also demonstrated the inherent complexities associated with governing such a 'wicked' matter as climate change at a supranational level. A combination of actor preferences and institutional factors has meant that the EU now uses few voluntary agreements (with rather mixed success), struggles to use labeling schemes, and still cannot agree ambitious eco-taxes. Innovation is only really discernible with respect to emissions trading, but even then the EU ETS exhibits a curious hybrid form that one would not find in an economics textbook. In this case, one actor (the Commission) was able to behave entrepreneurially because of help received from other actors (principally Member States and influential business groups) and a permissive set of institutional conditions (not least the availability of majority voting). Time will tell whether similarly benign political conditions appear in the future, facilitating further innovations in policy instrumentation.

For those who are keen to push for much stronger climate policies, a focus on policy instruments as they are used 'in practice' generates a rather mixed picture. The good news is that it is capable of offering a more politically realistic picture of the opportunities for and obstacles to governing than that found in the older and more theoretical accounts of policy instrumentation. The bad news is that it still tends towards fairly descriptive accounts of particular instruments and struggles to furnish unequivocal guidance on precisely how to design and deploy them to achieve radical policy change. This is probably not the kind of message that scholars of governance will want to hear, or for that matter those pushing for early and strong action on climate change.

Note

1. Note that Haigh's (2009) manual also lists relatively minor amendments and/or revisions of existing legislation.

References

ANDERSEN, M. S. 1994. *Governance by Green Taxes*. Manchester: Manchester University Press.

—— and SPRENGER, R.-U. (eds.) 2000. *Market-Based Instruments for Environmental Management: Politics and Institutions*. Cheltenham: Edward Elgar.

BEMELMANS-VIDEC, M.-L., et al. 1998. *Carrots, Sticks and Sermons*. New York: Transaction Publishers.

BOBROW, D., and DRYZEK, J. 1987. *Policy Analysis by Design*. Pittsburgh: University of Pittsburgh Press.

BRESSERS, H., and HUITEMA, D. 1999. Economic instruments for environmental protection. *International Political Science Review* 20(2): 175–96.

—— and O'TOOLE, L. 1998. The selection of policy instruments. *Journal of Public Policy* 18(3): 213–29.

BULMER, S. 1994. The governance of the European Union. *Journal of Public Policy* 13(4): 351–80.

CEC 2001. *European Governance: A White Paper*. COM (2001) 428 final. Brussels: Commission of the European Communities.

—— 2002. *Environmental Agreements at Community Level*. COM (2002) 412 final. Brussels: Commission of the European Communities.

DAHL, R., and LINDBLOM, C. 1953. *Politics, Economics and Welfare*. New York: Harper & Brothers.

DEKETELAERE, K., and PEETERS, M. 2006. Key challenges of EU climate change policy: Competences, measures and compliance. In M. Peeters and K. Deketelaere (eds.), *EU Climate Change Policy*. Cheltenham: Edward Elgar.

DRYZEK, J. 1997. *The Politics of the Earth*. Oxford: Oxford University Press.

ELIADIS, P., HILL, M., and HOWLETT, M. (eds.) 2007a. *Designing Government: From Instruments to Governance*. Montreal: McGill Queens University Press.

—————— 2007b. Introduction. In P. ELIADIS, M. HILL, and M. HOWLETT (eds.), *Designing Government: From Instruments to Governance*. Montreal: McGill Queens University Press.

ELLERMAN, D., BUCHNER, B., and CARRORO, C. 2007. *Allocation in the European Emissions Trading Scheme*. Cambridge: Cambridge University Press.

ENDS Europe. 2009. States give little ground in energy labelling talks. ENDS Europe, 11 November. <http://www.endseurope.com>.

HAIGH, N. 1996. Climate change policies and politics in the European Community. In T. O'Riordan and J. Jäger (eds.), *Politics of Climate Change*. London: Routledge.

—— 2009. *Manual of Environmental Policy*. Leeds: Maney Publishers.

HALL, P. 1993. Policy paradigms, social learning and the state. *Comparative Politics* 25(3): 275–96.

HALPERN, C. 2010. Governing despite its instruments? Instrumentation in EU environmental policy. *West European Politics* 33(1): 59–70.

HAUG, C., RAYNER, T., JORDAN, A., et al. 2010. Navigating the dilemmas of climate policy in Europe: Findings from a meta-policy analysis of policy evaluations. *Climatic Change*.

HOOD, C. 1983. *The Tools of Government*. Basingstoke: Macmillan.

—— 2007. Intellectual obsolescence and intellectual makeovers: Reflections on the tools of government after two decades. *Governance* 20(1): 127–44.

HOWLETT, M. 1991. Policy instruments, policy styles and policy implementation. *Policy Studies Journal* 19(2): 1–21.

—— and RAMESH, M. 1993. Patterns of policy instrument choice. *Policy Studies Review* 12(1–2): 3–24.

—— —— 1995. *Studying Public Policy*. Oxford: Oxford University Press.

HUITEMA, D. JORDAN, A. J., MASSEY, E., RAYNER, T., et al. 2011. The Evaluation of Climate Policy: theory and emerging practice in Europe. *Policy Sciences* (In press).

JORDAN, A., WURZEL, R. K. W., and ZITO, A. R. (eds.) 2003a. *New Instruments of Environmental Governance*. London: Frank Cass.

—— —— —— and BRUECKNER, L. 2003b. European governance and the transfer of 'new' environmental policy instruments. *Public Administration* 81(3): 555–74.

—— —— —— 2005. The rise of 'new' policy instruments in comparative perspective: Has governance eclipsed government? *Political Studies* 53(3): 477–96.

—— et al. (eds.) 2010. *Climate Policy in the European Union*. Cambridge: Cambridge University Press.

KASSIM, H., and LE GALÈS, P. 2010. Exploring governance in a multi-level polity: A policy instruments approach. *West European Politics* 33(1): 1–22.

KRÄMER, L. 2006. Some reflections on the EU mix of instruments on climate change. Pp. 279–96 in M. Peeters and K. Deketelaere (eds.), *EU Climate Change Policy*. Cheltenham: Edward Elgar.

LASCOUMES, P., and LE GALÈS, P. 2007. Introduction: Understanding public policy through its instruments. *Governance* 20(1): 1–22.

LINDER, S., and PETERS, B. G. 1989. Instruments of government. *Journal of Public Policy* 9(1): 35–58.

MAJONE, G. 1976. Choice among policy instruments for pollution control. *Policy Analysis* 2: 589–613.

—— 1994. The rise of the regulatory state. *West European Politics* 17(3): 77–101.

Mol, A., Lauber, V., and Liefferink, D. 2000. *The Voluntary Approach to Environmental Policy*. Oxford: Oxford University Press.

OECD 1994. *Managing the Environment: The Role of Economic Instruments*. Paris: OECD.

—— 1999. *Economic Instruments for Pollution Control and Natural Resources Management in OECD Countries*. Paris: OECD.

Parsons, W. 1995. *Public Policy*. Cheltenham: Edward Elgar.

Paterson, M. 1996. *Global Warming and Global Politics*. London: Routledge.

Peters, B. G., and Nispen, F. (eds.) 1998a. *Public Policy Instruments*. Cheltenham: Edward Elgar.

—— —— 1998b. Epilogue. In B. G. Peters and F. Nispen (eds.), *Public Policy Instruments*. Cheltenham: Edward Elgar.

—— and Pierre, J. (eds.) 2006. *Handbook of Public Policy*. London: Sage.

Rayner, T., and Jordan, A. J. 2010. Adapting to a changing climate: An emerging European Union policy? In A. J. Jordan et al. (eds.), *Climate Policy in the European Union*. Cambridge: Cambridge University Press.

Sabatier, P. 2007. *Theories of the Policy Process*. 2nd edn., Boulder, CO: Westview.

Salamon, L. (ed.) 2002. *The Tools of Government*. Oxford: Oxford University Press.

Schneider, A., and Ingram, H. 1990. Behavioural assumptions of policy tools. *Journal of Politics* 52(2): 510–29.

Schout, A., Jordan, A. J., and Twena, M. 2010. From old to new governance in the European Union: Explaining a diagnostic deficit. *West European Politics* 33(1): 154–70.

Wurzel, R. K., Zito, A. R., and Jordan, A. J. 2012. *Environmental Governance in Europe: A Comparative Analysis of 'New' Policy Instruments*. Cheltenham: Edward Elgar.

Zito, A. R., et al. 2003. Introduction to the symposium on 'new' policy instruments in the European Union. *Public Administration* 81(3): 509–12.

....

CHAPTER 37

CARBON TRADING:
A CRITIQUE

....

CLIVE L. SPASH

1 INTRODUCTION

....

CARBON emissions trading schemes (ETS) have become a dominant approach to Greenhouse Gas (GHG) reduction as a national policy and global governance issue, being advocated by orthodox economists and powerful vested interest groups. ETS schemes, as established in Europe and proposed in the USA and Australia, involve creating multi-billion dollar markets and a new monetary instrument offering a source of profits for the financial and banking sectors. Nevertheless, various levels of government are also implementing numerous other policy changes (e.g. energy efficiency standards, renewable energy incentives, improved electricity grid policies, vehicle fuel efficiency, public transport subsidies) to encourage GHG reduction. Such alternatives, especially direct regulation, contrast with the recommendation of those economists and politicians who believe that everything should follow from simply 'getting the price of carbon right.' An establishment discourse has then become trapped in advocating a single regulatory approach, backed by mainstream economists, despite the rich array of potential options.

This approach has arisen from highly simplified economic textbooks by authors who advocate theories where efficiency is the primary societal goal to the exclusion of all else. In practice public policy involves multiple goals and conflicting values. Pricing is not a simple or straight forward solution to anything. Markets are structured institutions set-up by social rules of governance and public policy. Pollution emission control requires that regulators understand the science and its limits, control powerful vested interest groups, and address the psychological and ethical motives for public action (see Spash 2010: section 5). This chapter covers ETS as a regulatory instrument choice where design and implementation are integrally entwined with issues of power.

There are two broad sets of concerns over applying economic pollution control theory to mitigating GHGs. First, a simple pollutant model assuming known impacts between a limited number of known source point polluters proves inadequate for capturing the essential characteristics of the problem. Second, a lack of realism in terms of market structure, and a total absence of anything in the economic model relating to power in society, mean analysts implicitly adopt the existing political economy without awareness as to the consequences for public policy. The first is a more technical economics debate covered in section 2 and the second relates to political economy and is covered in section 3. I then turn to the idea that ETS can be redesigned to make it better and eventually all problems with the approach can be solved. This is criticized in section 4 using the examples of GHG accounting problems and permit allocation. Design interacts with international politics and one result has been the creation of the 'offsets' market, which is the topic of section 5. I then draw some conclusion in the final section.

2 PROBLEMS IN APPLYING ORTHODOX ECONOMICS TO GHG CONTROL

GHGs are all pervasive in the economic structure of the modern economy and, due to embodied fossil-fuel energy, relate to all products and processes. The simple economic model assumes changing the price of a pollutant will have a limited impact which relates to a specific isolated product from a single sector. Pollutants are regarded as minor aberrations in an otherwise perfectly functioning economic system. Thus, control of a pollutant has limited knock-on effects and can be analyzed in a partial equilibrium framework. Due to the all-pervasive character of GHGs, changing their price affects all the prices in the economy and is highly unpredictable in consequence. This prevents any simple claims of economic efficiency in regulatory tool selection and policy design. Consider a few issues this raises.

There is path dependency. ETS schemes are implemented in partial ways on some sectors, and selected GHGs, while excluding others. Which gas is controlled from which source and in what order will then influence future costs, but the implications are unknown even at a single point in time, let alone for entire cost functions over time and across sectors. For example, if electricity-generating sources are regulated by ETS first the outcome will differ compared to regulating transport sectors first. Different price structures will result and resources will be allocated in different ways over time. Economic analysis proves lacking. Static equilibrium models fail to capture such dynamics while the role of human behavior in creating social indeterminacy affecting control costs is ignored completely.

Then there is interdependence and endogenous determination of prices and costs. The costs of controlling pervasive pollution are a function of any price placed on the pollutants. For example, carbon pricing changes relative energy source prices and energy versus labor cost, both of which affect the cost of different types of products and processes including methods of emissions abatement. So under an ETS there is a fundamental, but unpredictable, interdependence between permit price and control costs.

Then there is inseparability between allocative equity and economic efficiency. Abatement costs are meant to be technically determined and independent of ETS allocations (Rose et al. 1998). However, as Vira (2002) has argued, the initial allocation of permits can influence incentives to search for low-cost abatement options and so the pollution control function, e.g. due to affecting the size of the market and so technological innovation and change. Endowment effects also predispose people to stick with their initial allocation rather than actively trading (Thaler 1980). This means equity and efficiency are linked, so violating another basic assumption of mainstream analysis.

Mainstream economics focuses exclusively on efficiency analysis and recommends ETS on the basis that it can reduce a known set of technically determined abatement control costs. Pollution control costs are typically assumed to be straightforward market prices, like going shopping to buy, say, a filter to purify your water. They are regarded as easy to calculate compared with the benefits of control, which relate to damages involving issues such as loss of life and incommensurable values. Yet, control costs themselves can involve all the same aspects that typically affect benefit assessment. First, determining nomenclature as cost or benefit is a matter of the adopted status quo from which pollutants are to be adjusted (Spash 1997; Spash 2002: 172–3). For example, assuming emissions preexist and must be reduced is different from assuming no emissions and an activity (e.g. new plant) will add them afresh. In the former case avoided damages are a benefit and lost firm output a cost, while in the latter case damages are a cost and firm output a benefit. Second, within control cost calculations contentious value categories may arise. For example, a statistically recognizable number of people may die in the process of implementing control strategies (e.g. in the construction or production industries) with some control methods far riskier than others. Third, where negative costs arise this is just another approach to assessing (secondary) benefits. For example, planting trees to absorb carbon may have positive impacts on wildlife and biodiversity, provide recreational opportunities, and protect watersheds. Under economic efficiency criteria, all such beneficial outcomes should be taken into account as reducing the social costs of any project aimed at GHG reduction. So, claims of social efficiency, by regulators or polluters, require calculating the GHG marginal abatement cost function which is far from straightforward, known, or even knowable.

3 MARKET POWER AND MARKET STRUCTURE

Economic assumptions are also challenged by the reality of market structure. In basic economic theory firms are price takers with no market power. In practice most markets involve mixed structures, often with considerable concentrations of power amongst some large corporations and multinationals, e.g. the energy and mining sectors. Rather than being price takers and setting prices according to marginal production costs, the powerful corporation is able to engage in such practices as mark-up pricing, price discrimination, and monopsony. The potential for price manipulation and variation due to market structure means the standard assumptions of marginal costs rising under an ETS, and price signaling working to indicate social costs of pollution, no longer hold. An ETS then will fail to send the expected signal that GHGs are getting scarce and need to be used less; the price incentive just does not operate and the outcomes cannot be predicted.

Market power also means more than just the ability to manipulate prices, collude, or use mark-ups. Galbraith (2007 [1967]) explained the modern industrial economy as consisting of two sectors: one in which producers are small, lack power, and are subject to competition, and the other in which producers are large, have considerable power, and are run by professional managers (the technostructure). The problem which Galbraith then exposed was the close relationship the technostructure develops with politicians and regulators. This is particularly relevant to climate change because the energy and transportation sectors are dominated by large national and international corporations that are able to access considerable resources and lobby politicians to achieve institutional arrangements suited to their own ends. Thus we find US business cited as spending $US100 million to fight the Kyoto Protocol (Grubb et al. 1999: 112), high control costs estimates funded by US electric power generators (Chapman and Khanna 2000; Spash 2002: 160),[1] and climate skeptics organized and funded by polluting corporations (Lohmann 2006b: 41). The Australian Government (2008) proposed an ETS that would reduce petrol taxes to protect road transport against price rises, and give free permits to trade exposed large point sources, such as aluminium smelters. The scheme, covering 1,000 firms, also showed Galbraithian characteristics in proposing large polluters be 'compensated' with free permits while the smaller more numerous competitive fringe faced buying theirs at auction.

Powerful vested interest groups may then be encouraged to support ETS for good self-interested reasons. Polluting industries see the potential for massive financial rewards in return for their participation. For example, under the EU ETS, Europe's largest emitter, the German power company RWE, is estimated to have received a windfall of $US6.4 billion in the first three years of the system (Kantner 2008), and made €1.8 billion in one year by charging customers for permits it received for free (Lohmann 2006a: 91). The Australian ETS proposed 'compensation' to polluters on the basis of emissions intensity which would deliver substantial windfall gains to the worst polluters (brown coal-fired power stations). In the first five years of the proposed scheme the electricity-generating industry has been estimated to be in line for over 130 million free permits worth $AUS3.9 billion in nominal terms (Macquarie Capital Group Ltd. 2009: 8). This rose to $AUS7.2 billion through negotiation during 2009.

A clear aim has been to prevent impairing industrial competitiveness by using ETS design (European Commission 2001: 5). Specific industrial sectors or industries have argued for protection, from price effects due to carbon pricing, and particularly exporters. However, the type of large wealth transfers that have been occurring under the EU ETS now look like illegal hidden subsidies (Grubb and Neuhoff 2006). Where cost price rises occur in one country, due to carbon pricing, but not in another due to 'compensation,' a case can be made for an unfair advantage of the latter over the former. The relative difference might be avoided if all competing sectors were facing similar GHG control measures in all countries. However, if European experience is typical, excessive free permit allocation in some countries and to specific sectors is likely to create competitive distortion between different countries. Governments seem to fear the political consequences of 'underallocation' to specific sectors more than those of collective 'overallocation' (Grubb et al. 2005: 130).

Banking and finance is another powerful sector aiming to profit from ETS and related sequestration projects. Financial speculators and bankers see permits as financial instruments which provide money-making opportunities. Professional financial intermediaries, advisors, and investment banks have identified a new advisory role and potential for commission and brokerage fees. Already many have established climate change and emissions-trading specialist groups. A clear attraction of ETS schemes for countries with established international finance and banking sectors is to establish superiority in the new financial markets trading carbon. GHG control has been termed a 'pro-growth strategy' (Stern 2006: iii) which offers great opportunities for banks and the financial sector in funding pollution reduction (i.e. defensive expenditures) and partaking in carbon trading (Stern 2006: 270). The fact is that, while necessary due to past mistakes or systems failures, defensive expenditures add nothing to, or even detract from, human welfare because they are countering a societal problem, and should be distinguished as such in GDP measures (Spash 2007: 711). Being forced to make expenditures to protect yourself against increasing threats of harm (e.g. pollution, crime, violence, war, floods, and fires) is not a sign of societal progress, improved welfare or raised living standards. In addition, the transaction costs inherent in an ETS appear to be viewed by some (such as Stern et al.) as a source of economic growth, rather than a deadweight loss.

4   SOLVING PROBLEMS BY CHANGING ETS DESIGN

....

While many recognize problems with ETS there is a general belief that these can be designed away. Hence, the EU ETS Phase I (2005–8) has been regarded by advocates as a trial or test run showing the faults which Phase II should correct. Yet, the redesign of carbon ETS cannot simply remove basic properties of physical, economic, and social systems. The problems involved are illustrated here with reference to GHG accounting and permit allocation.

4.1 Greenhouse Gas Accounting

Achieving national emissions reduction under any regulatory approach requires knowing the responsibility of different sources for emissions and being able to monitor or otherwise estimate their compliance. Identifying and regulating key contributors would be the aim to achieve effective control. The difference under an ETS is the attempt to make GHGs themselves a valuable item of exchange which then implies having a comprehensive accounting system to achieve the claimed efficient outcome. This assumes a level of certainty about sources and sinks which is unattainable. In practice the carbon budget is surrounded by unknowns, ignorance, and social indeterminacy (Spash 2002). Rather than accepting such strong uncertainty, and developing social and institutional mechanisms whereby it might be addressed, the pretence remains that perfect knowledge can be obtained by more research and idealized carbon accounting can be achieved.

In reality baseline GHG levels, necessary for ETS source permissions and sink credits, have become matters for political negotiation. International caps are highly contested. For example, under Phase II of the EU ETS some countries have taken legal action against the EC.[2] During the attempts to get Kyoto ratified in 2001 Russia and Japan refused to sign until they received additional carbon credits for their domestic forests (Lohmann 2006b: 53). They succeeded and so effectively increased their carbon sinks on paper. Famously, under Kyoto, Russia and the Ukraine were awarded excess carbon allowances due to baseline projections calculated for the economic structure of the former USSR. These permissions became known as 'hot air' due to being meaningless in terms of actual carbon. That is, rather than giving permits relating to an existing pollution source they related to historically existing sources no longer operational. No actual existing pollution is then reduced by the sale of the permits. In the late 1990s, environmental NGOs were particularly concerned about the ability of polluters to buy Russian hot air and so avoid controlling their own emissions (Wettestad 2005: 10). Purchasing these cheap hot air credits was seen by the US as a means of avoiding control and used as such in the Kyoto Protocol bargaining negotiations (Lohmann 2006b: 52). Avoiding real emissions reductions by using such purchases remains an issue in light of the expected failure of countries to meet Kyoto targets (Grubb et al. 2005: 131).

4.2 Permit Allocation

Allocating permits is equivalent to attributing polluters a property right. Much is sometimes made of the temporary aspect of permits (e.g. only being valid for a year) and that they are not a transfer of property rights. However, once permits systems are established, and permits have been allocated, a government has created property rights for pollution which the courts may well protect. Subsequent attempts to reduce the numbers of permits (i.e. tighten the cap) could then require the government to buy back permission initially given away for free. Countries subject to a carbon cap and wishing to establish an ETS must therefore decide how to distribute permits knowing the potential for a shift in property rights and the need to tighten the cap over time.

Permits could be auctioned with the revenues going to the public purse, which would allow reduction of discretionary taxes (e.g. on labor and savings) or funding of infrastructure change for GHG reduction. However, the political preference has been for giving away permits to existing polluters. Typically this involves reference to historical emissions—termed grandfathering. Allocation under grandfathering tends to use a 'business as usual' baseline. In economic textbooks actual emissions would be the reference, assuming perfect knowledge. In practice data are unavailable, costly to obtain, and uncertain. Phase I of the EU ETS saw member states rely on companies' self-reported emissions estimates. These estimates are susceptible to self-interested framing and manipulation or simple overoptimism. The 'business as usual' baseline requires forecasting economic growth and other factors influencing output and so becomes influenced by numerous assumptions reflecting the primary concerns of those producing the estimates. Government growth promotion and protection of industrial exports leads to high baseline estimates. This overallocates permits risking a collapse of the trading system and little or no abatement. A prime example is Phase I of the EU ETS, which within a year had run into problems, having created too many permits, freely given to major polluters. This drove the trade price from €30 to €12 from April to May 2006 and eventually it reached a low of €0.1 (Skjærseth and Wettestad 2008: 276, 280). European industry 'played a major role in weakening Phase I allocations to a point that may undermine the credibility of emissions trading as an effective instrument' (Grubb et al. 2005: 135).

Large price fluctuations also point to the potential instability of an ETS. A frequently stated aim of ETS is to provide certainty to industry, yet there is no reason why carbon prices should be any less volatile than for other commodities and good reasons why it might be more so. Carbon price volatility is subject to the 'vagaries of near-term economic and emissions growth trends and related variables such as weather and gas-coal price relationships' (Grubb et al. 2005: 135). It can be exacerbated by speculators using the market purely to gain trading profits (e.g. selling high and buying low), which has nothing to do with pollution control. Firms may seek to reduce their exposure to price volatility through the use of forward contracts and hedging; however, such deals can usually only be made a few years in advance (due to counterparty risk). There is also the significant matter of uncertainty surrounding changes in government rules and regulation which are liable to be greater in a regulatory ETS than in most other markets.

Emissions trading in itself cannot therefore provide polluting firms with certainty about future carbon prices (despite the confident predictions of economic modelers). Its attraction is more likely to relate to the potential windfall gains of free permits. Indeed over-allocation, market power, profiteering, and speculation can actually increase investment in polluting technologies. For example, Lohman (2006a: 91) cites the case of Czech electricity giant CEZ being allocated a third of the country's allowances, selling them in 2005 when the price was high, being able to buy them back after the price collapsed, and then using the trading profit to invest in coal energy production.

5 EMISSIONS OFFSETS OR HOW TO AVOID CONTROLLING YOUR EMISSIONS

The concept of 'offsets' was created under the Kyoto Protocol to refer to emissions reductions not covered by the cap in an ETS. A standard ETS requires a seller to have controlled their source emissions to be able to sell a permit. Offsets are based upon projects which are disassociated from the polluting source and either reduce GHG emissions elsewhere or increase the capacity of a sink to absorb GHG pollution beyond 'business as usual'. Offsets are also now widely traded outside the Kyoto-compliance market, including by individuals and firms voluntarily aiming to offset their GHG emissions. Kyoto offsets were intended to provide industrialized countries with greater flexibility in meeting their caps whilst supporting sustainable development, and are also referred to as 'flexibility mechanisms'. Under Kyoto most offset projects occur in unregulated industrially developing countries under the Clean Development Mechanism (CDM) and are called certified emission reduction (CER) credits.

There is a major potential for Kyoto offsets to take over the functioning of any ETS. In June 2009 there were 1,600 registered CDM projects creating 300 million CERs, and by 2012 around 2,900 million CERs are expected.[3] A CER is equal to one metric tonne of CO_2-equivalent. Offsets are therefore a growth industry for supplying GHG credits for sale to polluters on the open market. European countries already know they will fail to meet their Kyoto targets and so a number of governments plan to buy offset credits totaling around 550 million tonnes of CO_2 and have budgeted some €2.9 billion for these purchases. In addition, businesses in the EU ETS are expected to purchase 1.4 billion tonnes of CO_2 offsets from 2008 to 2012 (European Commission 2008: 24).

Despite the 'emission reduction' *nom de plume* these offsets do not require a polluting source to reduce emissions, but instead allow them to increase emissions and then aim to offset them elsewhere. They could just as sensibly be called certified 'emission increase' units. While net global emissions reductions should occur for source offset, where sink offsets are involved the total scale of systemic GHG cycling will be expanded (e.g. via more sources justified by more sinks). Such a process seriously risks further enhancing the Greenhouse Effect. Offsets also suffer from a range of other problems.

The Kyoto Protocol specifies offsets be supplementary to domestic action. This 'supplementarity principle' is referred to in Article 6(1)(d), Article 12(3)(b), and Article 17, but all three Articles leave the exact meaning vague. The Marrakesh Accords,[4] which elaborate on the rules for offsets, state that 'use of the mechanisms shall be supplemental to domestic action and that domestic action shall thus constitute a significant element of the effort made by each party included in Annex I to meet its quantified emission limitation and reduction commitments.' Exactly what constitutes 'a significant element' is open to interpretation. The extent to which international offset mechanisms are permitted clearly has a great impact on whether domestic emissions are actually reduced.

The potential for exporting emissions control also raises serious concerns over the credibility of offsets. Foremost is the issue of determining what would have been undertaken in any case, i.e. baseline scenarios. This means credits for sink offsets should relate to additional GHG absorption beyond that which would have occurred in any case, e.g. excluding all forests planted anyway for other reasons such as conservation or commercial forestry. Once again vested interests may profit by manipulating projections. For example, stating two new coal-fired plants would be built when only one was ever intended and then claiming the emissions from the fictional plant has been reduced. The inclusion of offsets within ETS is argued to lower overall compliance costs, but seems to actually risk compromising the integrity of any emissions reductions.

Verification, enforcement, and monitoring become major concerns because of the potential abuse of 'business as usual' projections, the neglect of social and environmental impacts, and disregard for local communities and their interests in some of the poorest areas of the world. For example, inappropriate exotic species and monocultures used for forestry have resulted in communities trying to maintain trees which die due to poor soil conditions, are lost due to fires and drought, and which displace more sustainable practices and take away people's livelihoods (Lohmann 2006c). This is more damning because of the aforementioned claim of offsets to support sustainable development.

The purely physical carbon-accounting aspect of equivalence between source and sink is technically fraught with problems. 'A tonne of carbon in wood is not going to be "sequestered" from the atmosphere as safely, or as long, as a tonne of carbon in an unmined underground coal deposit' (Lohmann 2006a: 155). For example, the amount of added uptake in trees and soils is highly variable on the basis of local environmental conditions, skills of foresters, management practices, and enforcement of regulations. Forestry can also cause disturbance, erode soils, and release carbon. Human intervention, pest infestation, fires, climatic change, and so on all affect forestry, and then it has a natural rotation cycle in which carbon is released. Offsets assume physical equivalence for diverse points in a GHG's cycle where serious non-equivalence prevails. The inability of sinks to compensate economically for increased sources adds further complexity. That is the source-related harms must equate to the sink-created goods. This means assessing the damages and social problems created by offset schemes, e.g. peasant dislocation and resistance, privatization of common lands, acidification of soils, local environmental degradation, and health impacts.

One response to the problems with CDMs has been the 'Gold Standard' offset, initiated by the World Wide Fund for Nature to evaluate additionality and address concerns over neglected environmental and social impacts. Projects gaining Gold Standard certification can claim a price premium, but if carried out effectively will cost considerably more. The problem then is that in a competitive market bad drives out good as long as cheap credits can be obtained and treated as equivalent to those of a higher standard. The Australian government (2008: 11–12) has, for example, stated that it 'does not consider it necessary to accept only those CERs that meet additional criteria, such as the Gold Standard . . . neither does it consider that it should assess the broader environmental and social impacts of CERs.' Justifications offered are that this would increase costs, there are too few of such certified projects, and Kyoto standards are good enough. That CDM projects may be positively harmful both socially and environmentally is apparently compensated by obtaining a plentiful supply of cheap permits.

6 CONCLUSIONS

Much attention has been focused upon the efficient means of controlling GHGs for minimal reductions, rather than effective means for meeting a set of targets necessary to minimize human enhancement of the Greenhouse Effect. The targets under the Kyoto Protocol have been framed as part of an economic discourse where priority is given to creating gains from trade, extending the role of markets, and protecting the profits of potentially vulnerable polluters. In this debate economic efficiency has been used as an argument favoring the trading of pollution permits. The rhetoric of textbook theory has then been adopted as the grounds for creating new multi-billion dollar international carbon markets. The divorce between the assumptions of economic theory and complex reality has been neglected.

What should be clear is that regulatory instruments (whether taxes, permits, or direct regulation) are not neutral either politically or ideologically. They play to specific groups within society. The grounds upon which the ETS approach is then promoted, as a gain for public welfare, diverges from the reality of who actually advocates the scheme due to the potential for private gain or gains by their social group or organization. This becomes even clearer once actual ETS design is considered.

All the problems outlined here mean claims of efficiency gains for any regulatory instrument are far from clear or determinate. Transferring textbook predictions will lead to exaggerated and unrealistic expectations, and ignore complex interactions. Claims for an ETS being the most efficient policy instrument cannot then be substantiated. This means mainstream economists' main argument for policy choice is inoperable. Economists pursue efficiency as a narrow, professionally defined, technical matter, which then becomes the dominant form of discourse, negating other concerns. The assumption is that other goals can be dealt with as totally separate matters, in an unspecified political process, without impacting on the economics. As most people recognize, efficiency is but one goal and its pursuit a societal choice. Other goals may be adopted as more easily substantiated and more important (e.g. precaution, effectiveness, equity), and be achieved more easily using a variety of policy measures (e.g. direct regulation, taxes, subsidies, nationalization, setting behavioral norms, design of institutions) which treat them as primary rather than secondary concerns. What is most needed today, to tackle human-induced climate change, is an effective public policy approach which redirects the economy away from energy and material intensive production and avoids placing all the emphasis on market pricing for hypothetical efficiency gains.

NOTES

1. Control costs have been inflated which means in a cost-benefit framing of the problem action looks unfavorable compared to low control costs for the same benefits.
2. In September 2009, the European Court of First Instance found in favor of Poland and Estonia, who had challenged the EC over their EU ETS caps for Phase II. Six other countries have launched appeals against national allocation plan decisions: Hungary, Czech Republic, Bulgaria, Romania, Latvia, and Lithuania. See <http://www.carbon-financeonline.com/index.cfm?section=lead&action=view&id=12416>. Accessed 9 December 2009.
3. <http://cdm.unfccc.int/Statistics/index.html>. Accessed 16 June 2009.
4. The Marrakesh Accords are the aggregate Decisions (2/CP.7 through to 24/CP.7) of the Conference of the Parties to the UNFCCC set down in its seventh session, held at Marrakesh, Morocco, from 29 October to 10 November 2001. Those decisions were adopted in Montreal in November 2005.

REFERENCES

Australian Government. 2008. *Carbon Reduction Scheme: Australia's Low Pollution Future.* Vol. i. Canberra: Commonwealth of Australia.

CHAPMAN, D., and KHANNA, N. 2000. Crying no wolf: Why economists don't worry about climate change and should. *Climatic Change* 47(3): 225–32.

European Commission. 2001. *Proposal for a Directive of the European Parliament and of the Council Establishing a Framework for Greenhouse Gas Emissions Trading within the European Community and Amending Council Directive 96/61/EC.* COM (2001) 581. Brussels: European Commission.

—— 2008. *EU Action against Climate Change: EU Emissions Trading, European Commission.* <http://ec.europa.eu/environment/climat/pdf/brochures/ets_en.pdf>. 6 February.

GALBRAITH, J. K. 2007 [1967]. *The New Industrial Estate.* Princeton: Princeton University Press.

GRUBB, M., and NEUHOFF, K. 2006. Allocation and competitiveness in the EU emissions trading scheme: Policy overview. *Climate Policy* 6(1): 7–30.

—— VROLIJK, C., and BRACK, D. 1999. *The Kyoto Protocol: A Guide and Assessment.* London: Earthscan and Royal Institute of International Affairs.

—— AZAR, C., and PERSSON, U. M. 2005. A commentary. *Climate Policy* 5(1): 127–36.

KANTNER, J. 2008. Clean carbon copy not enough for US. *Australian Financial Review* 12 December.

LOHMANN, L. 2006a. Lessons unlearned. *Development Dialogue* 48(September): 71–218.

—— 2006b. 'Made in the USA': A short history of carbon trading. *Development Dialogue* 48(September): 31–70.

—— 2006c. Offsets: The fossil economy's new era of conflict. *Development Dialogue* 48(September): 219–328.

Macquarie Capital Group Ltd. 2009. *Australia's Carbon Pollution Reduction Scheme: Looking Behind the Black and the White.* Sydney, Renewable Energy and Climate Change, Macquarie Capital Group Ltd.

ROSE, A., STEVENS, B., EDMONDS, J., and WIS, M. 1998. International equity and differentiation in global warming policy: An application to tradable emissions permits. *Environmental and Resource Economics* 12: 25–51.

SKJÆRSETH, J. B., and WETTESTAD, W. 2008. Implementing EU emissions trading: Success or failure? *International Environmental Agreements* 8: 275–90.

SPASH, C. L. 1997. Reconciling different approaches to environmental management. *International Journal of Environment and Pollution* 7(4): 497–511.

—— 2002. *Greenhouse Economics: Value and Ethics.* London: Routledge.

—— 2007. The economics of climate change impacts à la Stern: Novel and nuanced or rhetorically restricted? *Ecological Economics* 63(4): 706–13.

—— 2010. The brave new world of carbon trading. *New Political Economy* 15(2): 169–95.

STERN, N. 2006. *Stern Review on the Economics of Climate Change.* London, UK Government Economic Service. <http://www.sternreview.org.uk>.

THALER, R. 1980. Toward a positive theory of consumer choice. *Journal of Economic Behavior & Organization* 1(1): 39–60.

VIRA, B. 2002. Trading with the enemy? Examining North–South perspectives in the climate change debate. Pp. 164–80 in D. W. Bromley and J. Paavola (eds.), *Economics, Ethics, and the Environmental Policy.* Oxford: Blackwell Publishing.

WETTESTAD, J. 2005. The making of the 2003 EU Emissions Trading Directive: An ultra-quick process due to entrepreneurial proficiency? *Global Environmental Politics* 5(1): 1–23.

CHAPTER 38

REDESIGNING ENERGY SYSTEMS

MARK DIESENDORF

1 INTRODUCTION

ENERGY use is by far the largest source of global greenhouse gas emissions (IPCC 2007: fig. TS.1b). Apart from some methane emissions from coal-mines, petroleum wells, and gas pipelines, energy-related emissions are almost entirely in the form of carbon dioxide (CO_2).

Leading climate scientists recommend a target CO_2 concentration in the atmosphere of 350 parts per million (ppm) (Hansen et al. 2008), which is considerably less than the actual CO_2 concentration in 2010 of 390 ppm. To enable such a global target to be achieved in the face of the current growth rate of 2 ppm per year, the vast majority of countries will have to transform their energy systems to ones with almost zero emissions. If we accept the growing recognition that carbon capture and storage (CCS) is unlikely to be commercially available before 2025–30 (see section 2), then a precautionary approach demands a complete redesign of systems for using and producing useful energy. This transformation must reduce substantially the quantities of energy used by developed countries; greatly improve the efficiencies of energy use in all countries; rapidly reduce energy generation from fossil fuels; and rapidly increase energy production from renewable energy and other low-carbon sources. The transformation must address all three categories of energy use: electricity (including mechanical energy), transportation, and heating/cooling.

To understand what has to be done, we must reject the outdated and damaging framework of projecting future energy demand from present demand and then projecting future energy supply to fill 'the gap.' The notion that demand must grow endlessly is a substantial part of the problem. Stabilizing energy demand, initially in the developed countries and in the longer term in all countries, is essential for both greenhouse mitigation and energy security.

At one conceptual level, greenhouse gas emissions have three drivers: consumption per person (sometimes called 'affluence'), population, and technology choice. For simplicity, let's consider CO_2 emissions C resulting from energy generation E by a population P. In this case consumption per person is simply energy use per person (E/P) and technology choice can be measured by CO_2 emissions per unit of energy use (C/E). Then we can disaggregate CO_2 emissions into the three driving factors as follows:

$$C = P \times (E/P) \times (C/E) \tag{1}$$

By cancelling the Es and Ps, we obtain $C = C$, showing that the relationship (1) is identically true. Each factor on the right-hand side of (1) can be measured separately. To address each factor requires different kinds of policies and so this decomposition of emissions is meaningful and useful. Clearly, to reduce emissions substantially, we must address each factor.

The former head of economics of the World Bank, Lord Nicholas Stern, has been reported as saying that the rich countries must aim to achieve steady-state economies by 2030 (Watts 2009). Economic growth, even with highly efficient industries, inevitably involves using more materials, energy, and land. Population growth is of particular concern in countries with very high per capita emissions, such as the USA and Australia. On a finite planet endless population growth is impossible (Ehrlich 1968).

Because time is of the essence in tackling emissions, this chapter focuses on transforming the technologies and associated industries of the energy system. In theory technological change can be implemented more rapidly than nonviolent changes to population and the economic system. Effective technological transformation must address the hardware, the associated software that determines how the hardware is used, and the 'orgware,' that is, the social and economic institutions and organizations that support or impede the dissemination and use of that technology.

This chapter shows that we already have most of the hardware parts of the technologies needed for a zero emission energy system, although not all of them are commercially available as yet. Disseminating these technologies in the absence of appropriate 'orgware' and in the face of existing practices and vested interests is still a formidable challenge. Doing so will inevitably encourage new industries and disadvantage some older greenhouse-polluting industries, leading to changes in the economic structures of nations and the world. Even if a zero emissions energy system can be achieved, emissions may continue to grow from nonenergy industrial sources, land-use changes, and wastes. Zero emissions energy technologies will buy us time to develop an economy with zero growth in populations and the use of energy, materials, and land—a steady-state economy (Daly 1977, 2008; Jackson 2009).

Because energy systems in industrialized economies are responsible for the vast majority of emissions, this chapter focuses on developed countries and the industrialized component of developing economies.

Climate scientists show that the principal measure needed for rapidly cutting CO_2 emissions from the energy sector is to phase out emissions from coal-fired power stations by 2030 (Hansen et al. 2008). This is in the context that oil production is already close to its peak, that gas is much less polluting than coal or oil, and that gas too will reach its production peak within several decades. So, as long as conventional oil and gas are not replaced with oil from shale and tar sands (in the absence of CCS), reducing coal emissions is the key. In the absence of CCS, first coal and oil, then later gas, the least polluting of the fossil fuels, must be replaced with a portfolio of low-carbon technologies. If CCS becomes available during this process, well and good. But, to delay the redesign in the hope that it will, could be disastrous.

The principal contributions to cutting energy-related greenhouse gas emissions over the next decade can only come from technologies that are either commercially available or at the pre-commercial stage now. To this end, section 2 summarizes the maturities of various low-carbon technologies for stationary energy and transportation. Section 3 puts the viable technologies together into scenarios, while refuting the fallacy that a modern energy supply system cannot be based entirely upon renewable energy. Section 4 summarizes the strategies and policies needed to facilitate the essential redesign of the energy system.

2 MATURITY OF ENERGY TECHNOLOGIES

2.1 Stages of Maturity

Classification of the status of technologies is a complex matter, because it may involve technological, scale, economic, social, cultural, and institutional factors. Any two-dimensional display is likely to be an oversimplification. Boundaries between different stages are blurred. With these caveats, Table 38.1 attempts to define the different stages of maturity of technologies. The present definitions are chosen to be independent of the economics, since economics depends upon indefinite future government policies, such as the future time trends in carbon prices.

Despite the classification into stages, it should be emphasized that the development of technologies does not always follow a linear process from research and development (R&D) to commercially mature. At the demonstration and pre-commercial stages, there may be several different designs competing for the same task, several generations of a single design, and dead-ends and loops in the process. R&D may be needed at all the other stages, including commercial, since the market doesn't always reward the best designs.

Table 38.2 classifies a number of low-carbon energy technologies into these stages of maturity. Again, some of the boundaries are blurred and some of the classifications may be controversial. For instance, although a few parts of the coal+CCS system are commercially available in special cases (e.g. compression of CO_2 and its transmission by pipeline), the table treats the system as a whole as still at the R&D stage.

Table 38.2 doesn't indicate the different potential rates of growth of different technologies. For instance, with policy assistance, global wind power has grown at about 20 per cent per year for the past two decades, while wind power in China has grown at 100 per cent per year since 2005 (GWEC website). Solar photovoltaic (PV) power has grown at greater than 30 percent p.a on the global scale through the 2000s. On the other hand, nuclear power has grown slowly over the past thirty years, despite large subsidies. Its global generating capacity stabilized in the mid-2000s.

Table 38.3 gives a rough indication of the size of the additional potential for different technologies on a global scale for 2030 and 2050. The 2030 potential reflects both the resource potential and the speed at which the technology could be implemented, while the 2050 potential reflects mainly the available resource. Only technologies currently at the demonstration stage or beyond are considered in Table 38.3. The descriptions 'large,'

Table 38.1 Descriptions of stages of technological maturity

Stage of maturity	Description
R & D	Undergoing laboratory or small-scale field study. Main focus on proof of concept. Design is not directed towards future production processes.
Demonstration	Field deployment of a single unit or very few units of the technology. Unit is large, although rarely full-size. Designed with mass production in mind, but not all main features are fully understood at this scale.
Pre-commercial, early diffusion	Field deployment of several large-scale systems having designs whose main features are well understood; however, some detailed design features and production methods still being optimized in preparation for mass production.
Commercially mature	Mass production of well-tested systems at an accurately predictable price, aiming for a large market. The need for a carbon price does not invalidate this classification.

Sources: Modified from UNFCCC (2008) and Foxon et al. (2005).

'medium,' and 'low' are relative to the size of the end-use sector to which the technology belongs: either electricity, heat, or transport. For instance, the use of landfill gas is commercial, but the resource is so tiny that it is omitted from the table, while solar thermal electricity is still pre-commercial, but the resource is huge.

The table cannot reflect the large variations in resource size by country or geographic region. For instance, wind potential is generally high at high latitudes and low at low latitudes. Insolation (solar energy input) is in theory highest at low latitudes, but in practice many tropical regions are high in humidity and diffuse cloud and so are not highly suitable for solar concentrators—inland areas at mid-latitudes are generally better. Countries with small land areas have limited potential for solar, biomass, and onshore wind (MacKay 2009). Therefore, long-range transmission of energy, especially diffuse renewable sources, must play an important role.

Some additional comments on some of the technologies and measures listed in Table 38.2 follow.

2.2 Demand Reduction

The first step in designing a zero emissions energy system is to reduce energy demand. The fastest and least expensive measures to do this are efficient energy use (called 'energy efficiency' for brevity), and energy conservation. Energy efficiency is using less energy to provide the same energy services, while energy conservation is choosing to have fewer energy services.

Studies by McKinsey and Company (2009) and ClimateWorks (2010) identify a wide range of economic savings from demand reduction measures that could offset a large proportion of the increased costs of clean energy supply measures. Although many energy efficiency technologies and measures are highly cost-effective, with their higher capital costs offset by lower annual energy costs, their widespread dissemination is limited by

Table 38.2 Stages of maturity of different low-carbon energy technologies

Stage	Technologies
Commercial	Energy conservation; many energy efficiency technologies and designs for buildings, appliances, stationary equipment, and motor vehicles; passive solar house design; urban consolidation around public transport routes and nodes; solar hot water; onshore wind power; landfill gas; biomass cofiring with coal; biomass combustion for generation & cogeneration; solar PVs based on silicon crystal and amorphous silicon; generation 2 nuclear power; hybrid vehicles; cogeneration fuelled by gas; conventional geothermal power; geothermal heat pumps for heating large buildings; tidal power with dam; pumped hydro storage; compressed air storage; bicycle; heavy rail and light rail; lithium-ion and nickel-metal hydride batteries for electric vehicles.
Pre-commercial, early diffusion	Offshore wind power; solar thermal electricity; molten salt thermal storage; generation 3 nuclear power; trigeneration fueled by gas or biomass; thin film PVs; plug-in hybrid vehicles; modern electric vehicles; several advanced batteries.
Demonstration	2nd generation biofuels from lignocellulose; cogeneration & trigeneration based on concentrated solar thermal; CCS at gas field; wave and ocean current power; thermal energy storage based on graphite blocks and ammonia dissociation; hot rock geothermal power (just entering this stage); fast reactor, excluding integral fast reactor.
R&D	Novel PVs; hydrogen production by clean pathways; fuel cells for hydrogen and methanol; liquid fuels from algae; coal burning+CCS; integral fast reactor; nuclear fusion.

Sources: Adapted by the author from Foxon et al. (2005).

Table 38.3 Additional global potential relative to sector, for selected low-carbon commercial, pre-commercial, and demonstration energy technologies, 2030 and 2050

Sector	Technology or measure	Potential 2030	Potential 2050	Comment
Electricity	Energy conservation plus energy efficiency	Large	Large	Especially in buildings, electric motors, and appliances.
	Wind (both onshore and offshore)	Medium	Large	Needs upgraded transmission systems, more peak-load plant or storage to balance fluctuations, and improved offshore wind turbines.
	Solar PV	Small	Medium	Needs further cost reductions and increased grid-connected storage.
	Solar thermal with thermal storage	Medium	Large	On the brink of rapid growth.
	Biomass	Small	Medium	Limited by need to avoid competition with food production.
	Geothermal (conventional)	Small	Small	Limited to a few regions with high potential, but small on global scale.
	Geothermal (hot rock)	Small	Medium	Large potential in several regions, but only just entering demonstration stage.
	Tidal (conventional)	Small	Small	ditto
	Tidal current	Small	Small	ditto
	Ocean thermal	Small	Small	ditto
	Hydro	Small	Small	Ditto, subject to environmental constraints.
	Wave	Small	Medium	More experience needed to identify best of the wide variety of existing designs.
	Nuclear fission (generations 2 & 3)	Small	Small	Slow to construct and limited by CO_2 emissions in long term.
	Nuclear fission (generation 4)	Small	Medium	Slow to develop, complex, dangerous, and very expensive; low CO_2 emissions.
Heat	Energy conservation & efficiency; passive solar housing	Large	Large	Especially in buildings.
	Solar & heat pump water heating	Small	Small	Although small relative to total heat, could provide majority of hot water.

Sector	Technology or measure	Potential 2030	Potential 2050	Comment
	Solar space heating/ cooling and industrial heat	Small	Medium	Hybrid heating & cooling systems look promising.
	Combined heat, power and cooling	Small	Medium	Cogeneration is commercially available and trigeneration is pre-commercial for industry and commercial buildings using gas and biomass as a fuel. Concentrated solar versions are at demonstration & pre-commercial stages.
	Geothermal space heating	Small	Medium	Application to large commercial buildings and industries requiring low-temperature heat.
Transport	Demand reduction, cycling, and walking	Medium	Large	Subject to improving urban design over decades.
	Urban rail (heavy and light) and buses, integrated with urban consolidation	Small	Medium	As proportion of global population living in cities increases, public transport becomes increasingly important.
	Biofuels	Small	Small	Limited land availability; lowest environmental impacts from residues.
	Electric	Medium	Large	Rate of deployment depends on improving batteries. Effectiveness entails charging by renewable energy.

Key: small = 5–15% of sector; medium = 16–25%; large = 26–50%
Note: Other relevant dates: peak oil 2010–20; peak gas and peak coal 2025–50.

Sources: IPCC (2007); Diesendorf (2007a); the author.

several market failures, such as split incentives between landlord and tenant and by the irrational demand for very short repayment periods compared with those for additional energy supply.

A common objection to cost-effective energy efficiency is that it is undermined by the so-called 'rebound effect.' In this concept, when money saved by means of energy efficiency is spent, it increases energy use. In practice, the rebound is generally small. Since expenditure on energy is only a small fraction of GDP, only a small fraction of every dollar saved by energy efficiency will be spent on additional energy use. This fraction can be made even smaller by policies to encourage emission reductions, such as regulations and standards for buildings and appliances. It can be eliminated altogether by structuring greenhouse response measures for consumers into packages of energy efficiency and renewable energy that have no net cost savings: for example, the savings from energy efficiency could be hypothecated into paying for the increased costs of renewable energy (Diesendorf 2009: 44). The more difficult broader issue, the dissipation of the benefits of a wide range of production efficiency improvements by economic growth, can only be addressed by moving towards a steady-state economy.

2.3 Solar Heat

The simplest form of solar heating is the passive solar design of buildings, which can also be considered to be a form of energy efficiency.

Active solar space-heating systems that provide both winter space heating and summer air conditioning are at the demonstration and pre-commercial stages, as are solar cogeneration plants (combined heat and power) and trigeneration plants (power, heat, and cooling). Solar thermal systems to provide industrial heat at low temperatures (less than about 250 °C) are under development. To provide higher-temperature industrial heat requires a high degree of concentration of sunlight. For this, combined solar-gas systems are feasible in appropriate climatic locations as gas prices increase. Liquid and gaseous biofuels, produced sustainably, could gradually replace gas as back-up for solar heat.

2.4 Cogeneration and Trigeneration with a Liquid or Gaseous Fuel

Cogeneration by burning gas or biomass is commercially available for industrial use and large commercial buildings; for small buildings, it is pre-commercial. Trigeneration is becoming commercial.

2.5 Solar Thermal Electricity

Solar thermal electricity, otherwise known as concentrated solar thermal power, works well, as witnessed by the 354 megawatt (MW) Luz power station that has operated for over twenty years in the Mojave Desert, California. Since 2006, stimulated by government policies in Spain and a few states of the USA, there has been a renaissance of this technology with several new designs being built. Some of the Spanish power stations, both operating and under construction, have low-cost thermal storage for 7.5 hours or more in tanks of molten salt, enabling them to meet summer peak demands with a high degree of reliability. In principle, in regions with high availability of marginal land, such as in North America, North Africa, and Australia, storage could be extended to 24 hours or more, giving this technology the ability to supply base-load power.

2.6 Wind Power

Wind power is the least expensive of the non-hydro renewable electricity technologies. Onshore wind power is a mature industry, with 159,000 MW of global installed capacity and tens of thousands of jobs at the end of 2009 (GWEC website). In each of 2008 and 2009, wind power provided the greatest growth in the European Union's generating capacity of all electricity generation technologies. Offshore wind power in shallow waters is a pre-commercial technology with high growth potential in northern Europe and North America.

2.7 Geothermal Power and Heating

Conventional geothermal power, where electricity is generated from steam that comes to the surface in volcanic regions, is limited to just a few locations in the world, such as Iceland, New Zealand, California, Indonesia, and the Philippines. However, a new form of geothermal power, known as hot rock or engineered or enhanced geothermal, has huge potential in large parts of North America, Europe, and Australia. To tap heat at 180 °C upwards, drilling to depths of 3–5 km is generally required. Currently there are very small (megawatt-rated) pilot plants generating electricity in France, Germany, and the USA and numerous drilling operations are under way elsewhere. It will be several years before demonstration plants of rated power about 50 MW are operating.

Geothermal heat for the space heating of buildings, is based on heat pumps, does not need a high temperature difference, and so involves only shallow digging (tens of meters or less). It is already commercially available on a limited production scale for large buildings.

2.8 Nuclear Power

Conventional nuclear power has been declining through the 2000s in terms of its percentage contribution to global electricity generation, falling to 13.7 percent by the end of 2009. Over the 2000s its capital costs have been escalating rapidly (Schneider et al. 2009; Diesendorf 2010). Because no nuclear power stations have been built in the USA in the past thirty years and only two (called generation 3) are under construction in Western countries (Finland and France), there is little basis for determining current costs. Since neither of the above generation 3 reactors is going to be built within budget, they cannot be classified as commercial. Construction in Finland of the 1,600 MW Areva reactor commenced in 2005 and is already three years behind schedule and at least 1.7 billion euros over budget, yielding a total capital cost (including the French export incentive) so far of 5.5 billion euros (Hollinger 2008; World Nuclear News 2009).

Contrary to common belief, every step in the nuclear fuel cycle, apart from reactor operation, emits CO_2. At present, when high-grade uranium ore (at least 0.1 percent uranium oxide) is mined and milled to produce nuclear fuel, total emissions from the nuclear fuel cycle are quite small, about 60 g CO_2/kWh (Lenzen 2008). However, reserves of high-grade uranium ore are limited, with estimates ranging from 40 to 80 years at present usage rates. Once low-grade ore (0.01 percent or less of uranium oxide) is mined and milled using fossil fuel (diesel), emissions from the nuclear fuel cycle jump to 130 g/kWh (Lenzen 2008). However, if waste mountains at uranium mines are covered, this estimate could double. Thus it appears that generation 2 and 3 nuclear power stations may become too CO_2 intensive to be a significant part of the long-term solution to climate change mitigation.

Developing generation 4 nuclear reactors (e.g. fast breeders with reprocessing of spent fuel) would in theory solve the long-term CO_2 problem in exchange for a further jump in risks and costs, but breeders and other fast neutron reactors are only at the demonstration stage and reprocessing is at best pre-commercial, considering that only two non-military plants (in France and Japan) are currently in continuous operation in the world.

All generations of nuclear power are slow to build, and so new nuclear power cannot be a significant part of the short-term (to 2020) mitigation strategy and is unlikely to become a large part of 2030 energy mixes.

2.9 Storage of Renewable Energy

Wind and solar PV, and to a lesser degree wave and tidal current power, are fluctuating sources that do not have any cheap or built-in forms of storage. When small amounts are integrated into an electricity grid, existing peak-load power stations, hydro, and gas turbines can handle the renewable energy fluctuations in the same way that they already handle fluctuations in demand and supply. For larger penetrations into the grid, the simplest and cheapest form of 'storage' is often additional peak-load plant, that is only operated occasionally, when needed.

It is expensive to store electricity on a large scale. The cheapest forms of electrical storage are indirect: compressed air stored in underground reservoirs and pumped hydro, which are both commercial. For direct storage, flow batteries, such as vanadium redox batteries, are undergoing demonstration. The advent of electric cars whose batteries are charged from a 'smart' grid that allows a two-way flow of electricity could potentially provide quite large amounts of storage that would enable wind and PV to play much greater roles than in the more cautious renewable electricity scenarios.

Most renewable sources of heat already have built-in storage in the form of biomass, hot water, molten salt, graphite blocks, etc.

3 SCENARIOS FOR REDESIGNING THE ENERGY SYSTEM

3.1 Technological Scenarios

Now we have a picture of a wide range of different low-carbon technologies at different stages of maturity at the present time and different resource potentials for 2030 and 2050. From these and reasonable assumptions about future development of technologies that are not yet commercially available, scenarios can be constructed for redesigning energy systems from 2010 to 2050. In the early years, before 2020, the main contributions would come from the least-cost technologies: energy efficiency and conservation, solar hot water, gas, wind power, and bioenergy (both heat and electricity) from organic residues of existing crops and plantation forests. Between 2025 and 2030, solar thermal power and solar PV could catch up with the early starters. Beyond 2030, it is possible that solar thermal with thermal storage will be the dominant source of electricity globally, closely followed by wind and biomass, with significant contributions from PV, marine power (especially wave), and hot rock geothermal.

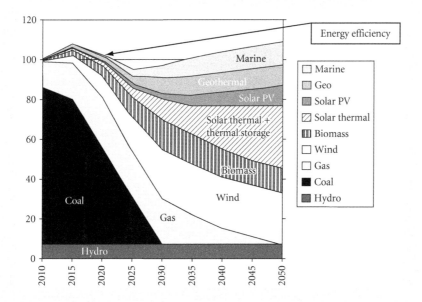

FIGURE **38.1** Scenario of transition to 100 percent renewable electricity for a coal-dependent country, Australia, 2010–2050.

Sources: The author's extension of scenarios from Saddler et al. (2007) and Diesendorf (2007b).

Notes: The graph measures annual electrical energy generation relative to the 2010 level, which is chosen to be 100 units. Graph should be read in conjunction with Table 38.3.

Figure 38.1 shows an electricity generation scenario for a country, Australia, which is highly dependent upon coal for electricity generation. Rigorous demand reduction from energy efficiency and conservation is responsible for the initial leveling off of demand growth. The increase in demand after 2025 reflects the growth in electric vehicles. Subsequently, demand is stabilized by leveling off economic and population growth. If such a scenario, with the phase-out of coal by 2030, is feasible for Australia, it should be feasible (with different mixes of renewable energy sources) for most other countries.

The above scenario is presented in the context of uncertainty about the future viability of coal with CCS and conventional nuclear power as greenhouse solutions in the foreseeable future, and the likelihood that a peak in global oil production in 2010–20 will be followed by peak gas and peak coal in 2025–50 (Heinberg 2009). Therefore, the principal interest is in scenarios that achieve close to 100 per cent renewable energy by 2050, with a transitional contribution from gas. On regional and national scales, such scenarios have been developed for Europe (Lehmann and Drees 1998; EREC 2010), northern Europe (Sørensen 2008), Britain (Centre of Alternative Technology 2010), the USA (Makhijani 2007), Denmark (Lund and Mathiesen 2009), Japan (Energy Rich Japan 2003), Germany (Klaus et al. 2010) and New Zealand (Mason et al. 2010). On the global scale there is a 100 per cent renewable energy scenario by Sørensen and Meibom (2000) and 80 per cent scenarios by the Energy Watch Group (Peter and Lehmann 2008) and Greenpeace (Teske et al. 2010).

These scenarios are 'conservative', in the sense that they are based on applying the Precautionary Principle to climate mitigation. To continue with business as usual would put human civilization and many nonhuman ecosystems at risk. Several of the renewable energy scenarios, including the present author's, meet Hansen et al.'s (2008) requirement for a phase-out of coal emissions by 2030. They show that 80–100 percent renewable energy is now technically possible, however it may still be a few years before good estimates of the economics of the solar components of these systems are available.

But is renewable energy too unreliable and too diffuse to provide a complete energy system, as some critics claim? Can renewable energy provide base-load (that is, 24-hour) power?

3.2 Reliability of Renewable Energy

Energy efficiency and solar hot water can substitute for base-load power stations. Bio-electricity and geothermal power are base-load. Solar thermal with thermal storage can play the role of peak-, intermediate- or base-load power stations, depending upon how much collector area and how much storage are installed relative to generator capacity.

Fluctuating renewable energy sources can also make large contributions. Existing electricity generation systems are designed to handle fluctuations in demand and unexpected breakdowns of large power stations. With fast-response peak-load plant and hot reserve base-load plant, generation systems can handle small contributions (up to 5–15 percent of annual energy generation) from fluctuating wind and solar PV without any additional back-up. Larger penetrations of fluctuating sources may require some additional peak-load plant to maintain the reliability of the generating system. Even in the extreme case where all the wind power in the grid is installed at a single site, the back-up peak-load capacity (rated power) required to maintain reliability is only about half the wind power capacity (Martin and Diesendorf 1982). In the usual case, where there are many geographically distributed wind farms in different wind regimes, the fluctuations in wind power are much smaller and so the back-up capacity becomes a small fraction of the wind capacity (Grubb 1988a, 1988b, Boyle 2007). Thus large penetrations of wind power into the grid can replace base-load power stations with the same annual average electricity generation, provided some additional peak-load plant is installed. For wind energy penetrations of less than 20 percent, the peak-load back-up is not used frequently and has low capital cost, and so can be considered to be reliability insurance with a low premium (Diesendorf 2007a: 117–21).

Alternatively, connection to a neighboring grid can provide all the back-up power and the sink for excess wind power. An example is the balancing of Danish wind power with Norwegian hydro-power.

Because of the high cost of electrical storage, solar PV modules without storage initially provide daytime (peak- and intermediate-load) power. When PV collectors are oriented towards the west, the peak in electricity generation can coincide with the peak in summer demand. Although this causes a loss in annual energy generation, it can produce a large gain in economic value of PV when there is time-of-day electricity pricing.

Thus an electricity generating system with large contributions from renewable sources will have less base-load plant as a percentage of total capacity and more peak-load. The remaining base-load fossil-fuelled power stations can be replaced by a mix of different renewable energy sources together with demand reduction.

3.3 Is Renewable Energy too Diffuse?

MacKay (2009) shows by calculation that the UK's current demand for energy could not be replaced by renewable energy on the available land and shallow offshore sites in the UK. Either the demand would have to be reduced or some renewable energy would have to be imported from continental Europe or North Africa. This is the consequence of the well-known fact that renewable energy sources, like fossil fuels, are not uniformly distributed across the planet (for example, the UK has low levels of sunlight) and that some countries (again the UK) have insufficient land area to capture sufficient renewable energy.

However, other countries and regions, such as the USA, North Africa, and Australia, have more than enough marginal land and renewable energy potential to provide for their own demands and to export large quantities of renewable energy, either by transmission line or in the form of hydrogen or methanol.

4 STRATEGIES AND POLICIES

This section first argues that a carbon price, in the form of either a carbon tax or a cap-and-trade emissions trading scheme, is not sufficient to redesign the energy system, and that several different types of government policies are needed to facilitate the development and commercialization of a wide range of technologies, not only the cheapest substitutes for fossil fuels.

4.1 Pricing is not Sufficient

The fallacy that pricing is sufficient is based on the incorrect assumption that the energy sector is a perfectly competitive market. In practice, efficient energy use is characterized by several market failures, such as split incentives between landlord and tenant; incomplete consumer information; lack of consumer understanding on how to trade off a higher capital cost against a lower operating cost; and the custom of demanding much higher rates of return from energy efficiency than energy supply (Hirst and Brown 1990; Greene and Pears 2003). In the absence of regulations and standards for buildings, appliances, and energy-using equipment, many efficient energy use measures that are highly cost-effective are not being widely implemented.

The market also fails to provide the necessary infrastructure for research institutes, transmission lines, railways, and gas pipelines. Government funding is needed for both the hardware and software, which includes good design and planning.

Furthermore, the market, by its nature, cannot provide long-term planning. Some renewable energy technologies with large potential are still at the pre-commercial stage—such as offshore wind, solar thermal electricity with thermal storage, and innovative types of solar PV cells. To mitigate one of the most dangerous challenges to human civilization on a timescale extending from the present to 2050 and beyond, we cannot afford to risk limiting our response to commercially available technologies that are cheapest at the margin. Contrary to neoliberal economic dogma, we must pick a portfolio of potential winners and build up their markets now.

A low initial carbon price of (say) US$25 per tonne of CO_2 will allow natural gas in some countries to compete with conventional coal for base-load electricity. However, such a low carbon price would not enable renewable energy to compete with coal or gas. Without complementary measures to a slowly increasing carbon price, renewable energy could be held back for decades until gas becomes scarce and its price very high. It is vital that policy measures for greenhouse gas mitigation promote the development *now* of a range of zero emission technologies with high potential and low degrees of physical and financial risk, so that their industries and contributions can make growing contributions. This argument does not justify very large support for highly risky technologies, such as CCS, generation 4 nuclear fission, and nuclear fusion.

Solar electricity in particular has huge potential for the longer term. Since it will benefit little from an emissions trading scheme for the foreseeable future, because the price gap is too large, it needs separate policy measures, such as feed-in tariffs or bands within a Renewable Portfolio Standard, to build its market until its costs are much lower (Buckman and Diesendorf 2010). It also needs R&D funding to stimulate second and third generations of solar technologies.

Finally, the possibility must be considered that the form of carbon pricing most favored by industry and government, cap-and-trade emissions trading, may fail to be effective in reducing emissions. Emissions trading schemes look good in economic theory, but the versions being introduced in Europe, North America, and Australia have been influenced by the big greenhouse polluters and so bear little resemblance to the ideal. For instance, the proposed Australian scheme is designed to pay the polluters instead of making the polluters pay. Therefore, we need 'policy insurance,' in the form of specific complementary policies to carbon pricing that will foster specific technologies of high potential and low risk.

Government policies must cover all stages and aspects of the process for innovation (Foxon et al. 2005; Stern 2006: 349). They must foster a stable and consistent framework to assist new technologies to achieve their potential. As well as expanding markets, they

> should be aimed at improving risk/reward ratios for demonstration and pre-commercial technologies. This would enhance positive expectations, stimulate learning effects leading to cost reductions, and increase the likelihood of successful commercialisation. (Foxon et al. 2005)

4.2 Summary of Policies

In the light of the above discussion, there is a wide range of basic policies that, taken together, would achieve an effective greenhouse mitigation strategy for the energy sector:

- An international target for atmospheric CO_2 concentration of 350 ppm or lower.
- An international agreement to set the nations on the pathway known as 'Contraction and Convergence,' with the goal of achieving the same average per capita greenhouse gas emissions by all countries within several decades (Global Commons Institute website).
- Science-based national greenhouse gas emissions targets, both short term and long term, to set the initial direction of change and the final goal.
- Targets, both short term and long term, for renewable electricity, renewable heat, and efficient energy use.

- A significant carbon price that increases over time. In view of the ineffectiveness of existing and proposed emissions trading schemes, Hansen (2008) recommends a national carbon tax from which the revenue is rebated in equal amounts to all adults in the nation, with half-shares to children. Border tax adjustments could be made for energy-intensive trade-exposed industries (Saddler et al. 2006).

- Until the carbon price has reached a sufficiently high level, a ban on all construction, refurbishment, and expansion of conventional coal-fired power stations.

- A ban on oil produced without CCS from coal, shale, and tar sands.

- Termination of subsidies to the producers and users of fossil fuels (de Moor 2001).

- Additional policies to expand markets for renewable energy until the carbon price is high enough to take over. These policies include gross feed-in tariffs, renewable portfolio standards with tradable certificates, tax concessions, and loan guarantees.

- Regulations and standards for efficient energy use in all buildings and energy-using appliances and equipment.

- A socially just transition for low-income earners and workers in greenhouse-intensive industries. Measures include funding for retraining, relocation, and pensions for displaced workers.

- Increased education and training to provide the skilled workforce for the transition.

- National industry development policies and programs to foster local manufacture of renewable energy systems appropriate to the region concerned.

- Essential infrastructure—such as railways, gas pipelines, and transmission lines—to enable the transition to an energy efficient, renewable energy future.

- Grants for research and innovation, especially for efficient energy use, renewable energy, biosequestration of CO_2, cleaner industrial processes, new transport technologies, social change processes, and a steady-state economic system.

Table 38.4 Policies for various stages of maturity of sustainable energy technologies

Stage	Policies
R&D	Research grants; tax deductions; support for patents
Demonstration	Capital grants; tax deductions; support for patents
Pre-commercial	Feed-in tariffs; renewable energy certificates; new infrastructure (e.g. transmission lines; railways; cycleways); institutional changes; government purchases; education and training of workers & professionals; ban on additional conventional coal power; removal of subsidies for fossil fuels; regulations and standards for energy efficiency; standards for renewable energy technologies; loan guarantees
Commercial	Increasing carbon price; feed-in tariffs and renewable energy certificates to be phased out only after carbon price is sufficient; new infrastructure; continuing institutional changes; government purchases; education & training

Source: the author, drawing upon Foxon et al. (2005)

- National legislation to develop the appropriate institutions/organizations to set and monitor quality standards and shape the markets for the transition.

- Leveling of population growth, especially in countries that have very high per capita greenhouse gas emissions, such as the USA and Australia.

- Development and implementation of a steady-state economic system (Daly 1977, 2008; Jackson 2009), initially in the developed countries.

Table 38.4 suggests how some of these policies can be allocated to different stages of technological maturity. The policy list gives a program to benefit society and the environment over the next two generations. If these policies are implemented widely and rapidly, there is still a chance of avoiding the worst potential impacts of global warming, as well as improving energy security, creating many new jobs, and reducing air and water pollution and land degradation.

5 CONCLUSION

Based on technologies that are currently commercial or semi-commercial, it is now feasible to consider a transition spanning several decades to national (and possibly global) energy systems based predominantly on the efficient use of renewable sources of energy. Wind and solar technologies in particular are capable of very rapid growth. If technologies currently at the demonstration stage are added to the mix, then 100 percent renewable energy may be possible before 2050. In view of the uncertainties that coal power with carbon capture and storage and generation 4 nuclear power could make significant contributions before 2030, a precautionary approach is to aim for achieving a very large contribution from renewable energy by that date. To drive this transition, a portfolio of different types of policies and programs is needed. In the face of market failures, a carbon price alone is not sufficient.

REFERENCES

BOYLE, G. 2007 (ed.). *Renewable Electricity and the Grid: The Challenge of Variability.* London: Earthscan.

BUCKMAN, G., and DIESENDORF, M. 2010. Design limitations in Australian renewable energy policies. *Energy Policy* 38: 3365–76; Addendum doi:10.1016/j.enpol.2010.06.016.

Centre of Alternative Technology. 2010. *Zero Carbon Britain 2030: A New Energy Strategy.* The second report of the Zero Carbon Britain project, <http://www.zerocarbonbritain.org>.

ClimateWorks Australia. 2010. *Low Carbon Growth Plan for Australia.* Melbourne: Climate-Works Australia.

DALY, H. E. 1977. *Steady-State Economics.* San Francisco: W. H. Freeman & Co.

—— 2008. *A Steady-State Economy.* London: Sustainable Development Commission, 24 April.

DE MOOR, A. 2001. Towards a grand deal on subsidies and climate change. *Natural Resources Forum* 25(2): 167–76.

DIESENDORF, M. 2007a. *Greenhouse Solutions with Sustainable Energy.* Sydney: UNSW Press.

—— 2007b. *Paths to a Low-Carbon Future: Reducing Australia's Greenhouse Gas Emissions by 30 Per Cent by 2020.* Greenpeace Australia Pacific, September.

—— 2009. *Climate Action: A Campaign Manual for Greenhouse Solutions.* Sydney: UNSW Press.

—— 2010. *Comparing the economics of nuclear and renewable sources of electricity.* Solar 2010 Conference. Australian Solar Energy Society, Canberra, 1–3 Dec.

EHRLICH, P. R. 1968. *The Population Bomb.* New York: Ballantine Books.

Energy Rich Japan. 2003. *Energy Rich Japan,* <http://www.energyrichjapan.info/en/download.html>.

EREC 2010. *Rethinking 2050: A 100% Renewable Energy Vision for the European Union.* European Renewable Energy Council, <http://www.erec.org>.

FOXON, T. J., GROSS, R., CHASE, A., et al. 2005. UK innovation systems for new and renewable energy technologies: Drivers, barriers and systems failures. *Energy Policy,* 33: 2123–37.

Global Commons Institute 2009. <http://www.gci.org.uk/contconv/cc.html>. Accessed 9 December.

GREENE, D., and PEARS, A. 2003. *Policy Options for Energy Efficiency in Australia.* Australian CRC for Renewable Energy.

GRUBB, M. J. 1988a. The potential for wind energy in Britain. *Energy Policy* 16: 594–607.

—— 1988b. The economic value of wind energy at high power system penetrations: An analysis of models, sensitivities and assumptions. *Wind Engineering* 12: 1–26.

GWEC (Global Wind Energy Council). 2009. <http://www.gwec.net>. Accessed 9 December.

HANSEN, J. 2008. Dear Michelle and Barack. 29 December. <http://www.columbia.edu/~jeh1/mailings>.

—— 2009. Never-give-up fighting spirit: Lessons from a grandchild. *The Observer,* 29 November, <http://www.columbia.edu/~jeh1/mailings/2009/20091130_FightingSpirit.pdf>.

—— SATO, M., KHARECHA, P., et al. 2008. Target atmospheric CO2: Where should humanity aim? *Open Atmospheric Science Journal* 2: 217–31.

HEINBERG, R. 2009. *Blackout: Coal, Climate and the Last Energy Crisis.* Gabriola Island, Canada: New Society Publishers.

HIRST, E., and BROWN, M. 1990. Closing the efficiency gap: Barriers to the efficient use of energy. *Resources, Conservation & Recycling* 3: 267–81.

HOLLINGER, P. 2008. Areva in talks with TVO over EPR delays. *Financial Times* FT.com 16–10–2008.

IPCC 2007. *Contribution of Working Group III to the Fourth Assessment Report of the Intergovernmental Panel on Climate Change.* Technical Summary. <http://www.ipcc.ch>.

JACKSON, T. 2009. *Prosperity without Growth: Economics for a Finite Planet.* London: Earthscan.

KLAUS, T., et al. 2010. *Energieziel 2050: 100% Strom aus erneuerbaren Quellen.* Dessau-Rosslau: Umweltbundesamt.

LEHMAN, H., and DREES, B. 1998. Scenario for a sustainable future energy system. Pp. 37–126 in LTI-Research Group (ed.), *The Long-Term Integration of Renewable Energy Sources into the European Energy System.* Heidelberg: Physica-Verlag.

LENZEN, M. 2008. Life cycle energy and greenhouse gas emissions of nuclear energy: A review. *Energy Conversion and Management* 49: 2178–99.

LOVINS, A. B., and SHEIKH, I. 2008. *The Nuclear Illusion.* November, <http://www.rmi.org/sitepages/pid257.php>.

LUND, H., and MATHIESEN, B. V. 2009. Energy system analysis of 100% renewable energy systems: The case of Denmark in years 2030 and 2050. *Energy* 34(5): 524–31.

LUNDVALL, B.-A. (ed.) 1992. *National Systems of Innovation: Towards a Theory of Innovation and Interactive Learning*. London: Pinter.

MACKAY, D. J. C. 2009. *Sustainable Energy: Without the Hot Air*. Cambridge: UIT.

McKinsey & Company. 2009. *Pathways to a Low-Carbon Economy*. Version 2 of the global greenhouse gas abatement cost curve, <http://www.mckinsey.com/clientservice/ccsi/pathways_low_carbon_economy.asp>.

MAKHIJANI, A. 2007. *Carbon-Free and Nuclear-Free: A Roadmap for U.S. Energy Policy*. Muskegon, MI: RDR Books; Takoma Park, MD: IEER Press.

MARTIN, B., and DIESENDORF, M. 1982. Optimal thermal mix in electricity grids containing wind power. *Electrical Power & Energy Systems* 4: 155–61.

MASON, I. G., PAGE, S. C., and WILLIAMSON, A. G. 2010. A 100% renewable electricity generation system for New Zealand utilising hydro, wind, geothermal and biomass resources. *Energy Policy*, doi: 10.1016/j.enpol.2010.03.022.

PETER, S., and LEHMANN, H. 2008. *Renewable Energy Outlook 2030: Energy Watch Group Global Renewable Energy Scenarios*. Bonn: Energy Watch Group, <http://www.energy-watchgroup.org?Studien.4+M5d637be38d.0.html>.

SADDLER, H., DIESENDORF, M., and DENNISS, R. 2004. *A Clean Energy Future for Australia*. Sydney: Clean Energy Future Group, <http://wwf.org.au/publications/clean_energy_future_report.pdf>.

—— MULLER, F., and CUEVAS, C. 2006. *Competitiveness and Carbon Pricing: Border Adjustments for Greenhouse Policies*. Discussion paper 86. Canberra: Australia Institute, April.

SCHNEIDER, M., et al. 2009. *World Nuclear Industry Status Report 2009: With Particular Emphasis on Economic Issues*. Commissioned by German Federal Ministry of Environment, Nature Conservation and Reactor Safety, August.

SØRENSEN, B. 2008. A renewable energy and hydrogen scenario for northern Europe. *International Journal of Energy Research* 32(5): 471–500.

—— and MEIBOM, P. 2000. A global renewable energy scenario. *Int. J. of Global Energy Issues* 13(1/2/3): doi: 10.1504/IJGEI.2000.000869.

STERN, N. 2006. *Stern Review: The Economics of Climate Change*. <http://www.webcitation.org/5nCeyEYJr>.

TESKE, S., et al. 2010. *Energy [R]evolution: A Sustainable World Energy Outlook*. 3rd edn. European Renewable Energy Council and Greenpeace International, June, <http://www.greenpeace.org/international/en/publications>.

UNFCCC. 2008. *Identifying, Analysing and Assessing Existing and Potential New Financing Resources and Relevant Vehicles to Support the Development, Deployment, Diffusion and Transfer of Environmentally Sound Technologies*. Interim report by the chair of the Expert Group on Technology Transfer. FCCC/SB/2008/INF.7, <http://unfccc.int/resource/docs/2008/sb/eng/inf07.pdf>.

WATTS, J. 2009. Rich nations will have to forget about growth to stop climate change. guardian.co.uk, 29 September, <http://www.guardian.co.uk/environment/2009/sep/11/stern-economic-growth-emissions/print>.

World Nuclear News. 2009. Oikiluoto 3 losses to reach 1.7 billion. <http://www.world-nuclear-news.org/newsarticle.aspx?id=24732>.

PART X

PRODUCERS AND CONSUMERS

CHAPTER 39

....

CORPORATE RESPONSES

....

SIMONE PULVER

THE fascination with corporate responses to climate change is motivated by both the lure of the transformative power of corporations and the caution required by their generally poor records of environmental performance. Corporations have been at the heart of reshaping how society interacts with the environment over the past century. Moreover, the consequences of this transformative power for environmental quality are well established. For much of the past century, corporations have used resources and disposed of waste with little concern for environmental impacts. However, the past twenty-five years have seen a significant shift in perceptions about the appropriate role for corporations in environmental protection—from targets of regulation and activism to partners in clean technology innovation (Schmidheiny 1992). The debate over these two corporate identities—corporations as environmental polluters and corporations as green leaders—is contentious and ongoing, as evidenced by British Petroleum's recent fall from grace. The corporation, feted as a leader on corporate climate action in the late 1990s, in 2010 became infamous for the oil spill disaster in the Gulf of Mexico. While viewing corporations as either polluters or innovators may seem simplifying, it is of little use in analyzing the variation in corporate responses to climate change. The goal of this review is to move away from a simplistic, dualistic framing of corporate action by presenting an overview that highlights the heterogeneity of corporate responses. Corporations vary in their awareness of climate change, their climate practices and policies, in the motivations guiding their climate strategies, and in the meanings corporate executives and others ascribe to their actions.

The review follows a three-part structure. The first section identifies three areas of corporate activity that can be used to compare and rank corporate responses to climate change: business practices, political action, and corporate governance. Each highlights a distinct dimension of corporations' climate responses and presents a unique set of indicators by which corporate responses can be assessed and compared. The second section analyzes drivers of corporate climate action, which include the physical, regulatory, reputational, and liability risks of climate change and the associated technological, regulatory, and reputational opportunities. Such risks and opportunities, and firm perceptions thereof, vary by sector, by national political and social context, and by firm-specific factors. In the final section, I theorize the likely contribution of corporations to the global project of climate protection, based on the empirical findings synthesized in the prior sections.

1 COMPARING CORPORATE RESPONSES: FRAMEWORKS AND METRICS

Initially, comparative assessments of corporate responses to climate change used an inductive approach. Scholars tallied the elements of a corporation's climate strategy and then broadly categorized its approach to the climate issue. For example, the climate strategies of multinational oil corporations differed in their recognition of the climate problem, acceptance of the scientific findings of the Intergovernmental Panel on Climate Change (IPCC), views on the Kyoto Protocol, investment in non-fossil energy technologies, internal greenhouse gas emissions reductions goals, memberships in business lobby groups, and relationships with environmental NGOs. These differences were used to classify oil company climate strategies as 'proactive,' 'wait and see,' and 'reactive' (Skjaerseth and Skodvin 2001; van den Hove et al. 2002), 'collaborative' versus 'adversarial' (Pulver 2007b), and 'leaders,' 'followers,' and 'laggards' (Dunn 2002). Skjaerseth and Skodvin (2001) call this a 'soft indicators' approach.

More deductive approaches distinguish between three elements of corporate climate activity: business practices, political activities, and corporate governance. Changes in business practices are of greatest interest to policy makers and publics because of their direct impact on climate outcomes. Are corporations reducing the greenhouse gas emissions from their own facilities and operations? Are new products climate friendly? Are corporate investments moving towards decarbonizing the economy? Equal attention is garnered by corporate political activities. Do individual corporations support the regulation of carbon dioxide and other greenhouse gases in various governance arenas? Are lobbying and media efforts focused on undermining or bolstering the emerging scientific consensus on climate change? Finally, corporate business practices and political activities often depend on a corporation's internal governance framework. Shifts in business practices and political activities are possible only if senior management prioritizes climate change as a core issue of corporate strategy.

1.1 Business Practices

The most common business practice associated with climate change is greenhouse gas emissions accounting. The Carbon Disclosure Project (CPD) documented that in 2009 almost 70 percent of the 500 largest corporations on the FTSE Global Equity Index series reported on their greenhouse gas emissions (PriceWaterhouseCoopers 2009). However, such reports are of variable quality. Only 50 percent of CPD corporations have their emissions data verified. A 2009 Trucost comparison of greenhouse gas emissions in the S&P 500, the 500 largest US companies, found that '66% of companies did not publish adequate data on direct emissions from operations' to allow for their inclusion in the Trucost study (2009: 5). Despite this, it is worth noting that the majority of companies in carbon-intensive sectors do disclose precise emissions data.

For many corporations, greenhouse gas emissions accounting is a precursor to investments in emissions reductions, a second climate-related business practice (Capoor and Ambrosi 2009). To motivate and benefit from such investments, some corporations experiment with internal emissions trading systems (Victor and House 2006) or trade in voluntary carbon markets (Hoffman 2005). Others are mandated to participate in emissions markets such as the European Union Emissions Trading Scheme (EU ETS) (Brewer 2005). Engels et al. (2008) identified two EU ETS carbon market strategies in their survey of firms in the United Kingdom, Germany, Denmark, and the Netherlands; those aimed merely at complying with nationally mandated targets versus those aimed at trading on the carbon market as a source of revenue. In developing countries, the Clean Development Mechanism (CDM) under the Kyoto Protocol has catalyzed corporate carbon emissions reduction investments across sectors ranging from sugar to cement (Lutken and Michaelowa 2008; Pulver et al. 2010). Firms in China, India, and Brazil have been most active in making CDM investments.

A third and most transformative category of business practice is the development of new climate-friendly products and/or markets. For example, renewable energy technologies are a key growth market within the United States, Europe, and internationally (IEA 2009). Chinese and Indian companies are among the fastest-growing wind turbine manufacturers in the world (Lewis 2007). Likewise, carbon accounting, consulting, and trading has emerged as a new product/market opportunity (Kolk et al. 2008; Knox-Hayes 2010). However, the most innovative carbon neutral technologies, such as carbon neutral cement or electric vehicles, are still in the research, development, and demonstration phase (Cain Miller 2010). Moreover, for established corporations, such new products and markets represent a small fraction of their overall portfolios.

Kolk and Pinkse (2005) developed a typology of corporate responses to climate change that integrates emissions accounting, emissions reductions, and new product/market development. Their template highlights both corporations' *strategic intent* and *scope of action*. They define strategic intent as the degree to which corporations are using climate change to create new business opportunities. Are they innovating their internal processes, developing new products, and pioneering new markets? Or are they responding to the climate challenge via compensation strategies, simply by the external acquisition of emissions credits? Scope of action differentiates between an internal focus, a supply chain focus, and corporate action that extends beyond the supply chain. Kolk and Pinske used their template to classify the climate actions of 136 global corporations. Six clusters of corporate responses to climate change emerged. 'Cautious planners' and 'emergent planners,' focused on internal compensation responses, constituted the largest category. Nearly 70 percent of corporations in Kolk and Pinkse's sample were either cautious or emergent planners. The remaining 30 percent were split between 'emissions traders,' 'internal explorers,' 'vertical explorers,' and 'horizontal explorers.' Only 5 percent of corporations were among the most innovative 'horizontal explorers.'

In sum, very few corporations are proactive in changing their climate-related business practices. The rhetoric regarding climate-related business opportunities outpaces actual changes (Jones and Levy 2007). A 2007 global survey by McKinsey & Company of over 2,000 corporate executives supports this insight. The survey found that while climate change was considered of great importance by an average of 60 percent of firms, only half of those consistently included climate considerations in corporate strategy (Enkvist and Vanthournout 2007).

1.2 Political Activities

A second approach relevant to analyzing corporate responses to climate change focuses on the ways in which corporations attempt to shape the organizational, informational, legal, and political contexts in which they operate. Levy and Egan (1998) identify on three channels of corporate influence: Structural power, instrumental power, and discursive power. Structural power refers to the central role of corporations in maintaining economic growth. Instrumental power refers to the ability of corporate elites to affect political outcomes through financial support, lobbying activities, and social and business connections. Finally, discursive power refers to the capacity of corporations to frame issues and influence public opinion through advertising and other reputation-building activities.

Suchman (2009) offers a useful framework for categorizing different types of corporate political engagement. His matrix categorizes firms on two dimensions—policy stance (for or against) and level of engagement (high or low)—and identifies four types of firms: proponents (for, high engagement), acceptants (for, low engagement), opponents (against, high engagement), and reluctants (against, low engagement). Corporate political activities can be further distinguished as informational, financial, or stakeholder; as issue-by-issue or relational; and as individual or collective (Kolk and Pinkse 2007).

Certain key corporations have clearly emerged as proponents of and opponents to action on climate change. On specific policy issues, they can be distinguished by their lobbying stances (Davies and Sawin 2002). A more general approach categorizes corporations based on their memberships in various business and industry associations. Efforts to organize a business voice on the climate issue date back to the late 1980s, with the formation of several new business and industry NGOs (BINGOs) and the constitution of climate groups within established BINGOs. The most well-known include the Global Climate Coalition (GCC), the International Chamber of Commerce (ICC), and the World Business Council on Sustainable Development (WBCSD) (Newell 2000). BINGOs can be arrayed on a scale charting their degree of opposition to/support for national and international efforts to regulate greenhouse gas emissions. In her analysis of corporate lobbying on carbon capture and storage under the CDM, Vormedal (2008) identifies ten business NGOs and arranges them on a scale from grey to green. Pulver's (2002) analysis uses BINGO membership to chart the shifting climate strategies of multinational oil corporations. As companies reconsidered their climate strategies, they defected from certain BINGOs and sought other BINGO memberships that better reflected their new or updated approach to the climate issue. In particular, corporate membership in the US-based GCC was a key indicator of an opponent stance. Over the course of the international climate negotiations, leading corporations defected from the GCC, causing it to disband in 2001. Withdrawal from the GCC indicated a shift in a corporation's position on climate change.

In addition to lobbying directly and indirectly through business associations, corporations have used their discursive power to frame debates about climate change. Several companies have made climate change a core issue in advertising campaigns (Kolk and Pinkse 2007). Others have sought NGO partners to lend credibility to their corporate climate change initiatives (Newell 2000). Others still attempt to shape climate debates by funding climate science, both mainstream and skeptical. Mainstream initiatives include a British Petroleum-funded research program at Princeton University and ExxonMobil's contributions to the Global Climate and Energy Project at Stanford University (Ellison 2007).

However, greater attention has been given to the links between opponent corporations and funding for the group of scientists who challenge the IPCC consensus on climate change (Leggett 1999). McCright and Dunlap (2003) trace the affiliations between conservative US think tanks, US climate change skeptics, and fossil-fuel industry funding, pointing to the conservative counter-movement as a key cause of US inaction on climate change.

While activities of proponent and opponent corporations receive most scholarly and public attention, the majority of corporations are passive political actors, either acceptants or reluctants, with low levels of engagement. Paralleling their research on corporate business practices, Kolk and Pinkse (2007) analyzed the political activities of over 200 corporations from among the FTSE Global Equity Index Series 500. They found that only 54 percent reported on political activities. Among those, the dominant political strategy was to provide information to policy makers.

1.3 Corporate Governance

Changing corporate business practices and engaging in political activities implies management attention to climate change. A third framework for comparing corporate responses focuses on corporate climate governance and assesses the extent to which climate concerns are integrated into corporate management and decision-making structures. Here again, there is wide variation in corporate practice. Some corporations identify climate change as a key issue in corporate strategy and assign responsibility for the climate issue to the CEO and top management (Pulver 2007b). Others constitute trans-departmental climate teams to access expertise from across the organization (Zisa 2007). Others still see climate change as one of several environmental issues and delegate responsibility to lower-level environment, health, and safety managers.

According to Ceres, a US-based environmental investor network, best practices related to corporate climate governance include: constituting a board of directors' committee with oversight of corporate climate change programs; periodic board of directors' review of such programs; a clear articulation by the chief executive of a company's views on climate change; executive compensation linked to meeting corporate climate policy goals; the pursuit of opportunities to reduce emissions, minimize exposure to physical and regulatory risks, and maximize low-carbon initiatives; public disclosure of corporate climate liabilities and corporate climate response strategies; externally verified carbon emissions data; and setting emissions reduction targets (Cogan 2006: 3).

As with the business practices and political activities frameworks for comparing corporate responses, empirical analyses of corporate climate governance reveal generally low levels of engagement and significant divergence between high and low performers (Cogan 2006, 2008; Cogan et al. 2008). In a ranking of 100 US corporations from the ten most carbon-intensive sectors by their corporate climate governance performance, individual company scores ranged from 3 to 90 out of 100 possible points. British Petroleum and DuPont scored 90 and 85 points, respectively, both having implemented many of the best practices. Williams Natural Gas Company and United Airlines were tied for the lowest score of 3 points. On average, top performing sectors included the chemical industry, electric power, and the auto industry. Sectors that had made least progress in terms of corporate governance included the airline industry, food industry, coal industry, and oil and gas industry. With the exception of the chemical industry, all sectors surveyed averaged below 50 out of 100 potential points (Cogan 2006). In terms of particular sub-elements of corporate governance, emissions accounting and disclosure were most widely implemented. Conversely, the average performance on strategy planning, including decisions related to identification of business opportunities, managing climate risks and liabilities, and greenhouse gas emissions target setting for facilities and products was weakest (Cogan 2006).

2 EXPLAINING VARIATION IN CORPORATE RESPONSES TO CLIMATE CHANGE

Comparing corporate responses to climate change using the business practices, political activities, and corporate governance frameworks highlights the variation in corporate behavior. Such variation is to be expected given the scope of the climate change challenge. Yet it also presents an empirical puzzle. What distinguishes firms that see climate change as a business opportunity from those that consider it an additional compliance burden? Why do some corporations support action to mitigate climate change while others oppose efforts at greenhouse gas regulation? Why do corporations in the same sector have widely divergent views of climate-related risks and opportunities?

All corporate responses to climate change are due to real and/or perceived changes in business environments, related to the direct and indirect effects of climate change. Direct effects impinge on firm operations due to physical changes in the climate. Indirect effects originate from policy and social initiatives to mitigate greenhouse gas emissions and/or to adapt to climate change. Climate change thus presents corporations with a series of operating risks and related opportunities, which may be physical, technological, regulatory, or reputational (Cogan 2006; KPMG 2008; Hoffman 2005). These risk and opportunities, and corporate perceptions thereof, vary by sector, national political and social context, and by firm-level characteristics. There are competing perspectives on the links between external drivers and firm behavior. Rational actor models depict corporate executives as articulating climate strategies based on definitive information regarding risks and opportunities. Variation in firm behavior is a product of variation in external pressures, variation in firm operational characteristics, or the intersection of the two. Socially embedded models emphasize the social construction of knowledge, particularly under conditions of uncertainty. Perfect information about climate risks and opportunities is absent, and thus the social networks in which firm decision makers are embedded shape their perceptions of climate risks and opportunities (Pulver 2007b).

2.1 Sector Differences

Business sectors vary in their exposure to direct and indirect climate risks. For example, the insurance industry is most concerned with direct climate change impacts. Insurance companies were early pioneers in considering the physical risks of climate change. Damages from severe weather events have resulted in rising disaster losses and the need to 'advance loss-prevention solutions' (Mills 2005: 1040). In contrast, the greater concern of most oil companies has been the potential effects of mitigation policies on oil demand, on emissions from oil facilities, and on stock prices (Austin and Sauer 2002). For some sectors, such as the tourism industry, direct and indirect effects are of equal concern. An increase in extreme weather events or changes in snowpack due to warming constitute direct effects. Changes in travel patterns due to increases in fuel prices are indirect effects that may significantly alter the sector's financial performance (Bicknell and McManus 2006).

Differences in sector exposure to climate risks and in sector availability of business opportunities provide a first-cut explanation of variation in corporate responses to climate change. Emissions data offer a first metric for assessing corporate exposure to climate change. They highlight which sectors contribute most to climate change. Among the S&P 500, electric utilities are the biggest emitters, accounting for 59 percent of total direct greenhouse gas emissions, followed by oil and gas, accounting for 20 percent of emissions. Similar patterns hold when supply chain emissions are added to direct emissions. The five largest emitters among the S&P 500 are the oil companies ExxonMobil, Chevron, and ConocoPhillips, and the electric utilities, American Electric Power Company and the Southern Company. Emissions intensity data (emissions per unit of revenue) provide a further metric for comparing corporations, enabling performance ranking of corporations operating within the same sector. For example, Allegheny Energy underperforms com-pared to other utilities. It is the most carbon-intensive company in the S&P 500, although it only ranks 26th in terms of absolute emissions (Trucost 2009). In theory greenhouse gas emissions and intensity data can also be used to assess corporate progress in emissions reductions over time. However, such time series data are not yet available in aggregated, standardized formats (PriceWaterhouseCoopers 2009; Trucost 2009).

A meta-analysis aggregating fifty studies on corporate exposure to climate risks provides a summary overview of risk across sectors (KPMG 2008). Regulatory rather than physical risks were of greatest concern to most firms. Sectors identified as facing high regulatory risks include automotive, construction/materials, mining/metals, oil/gas, transportation, and utilities, mirroring sectors with high levels of direct greenhouse gas emissions and/or carbon intensity. In terms of physical risks, the most highly impacted sectors were agriculture, forestry, healthcare, pharmaceuticals, insurance, and tourism. Some firms are documenting sector-specific climate impacts and identifying coping strategies. The increas-ing salience of adaptation to climate change is broadening corporate climate awareness (Berkhout et al. 2006). However, corporations tend to systemically underestimate physical risks, limiting their analyses to damages due to weather-related events. The KPMG authors suggested that the physical risks of climate change are more extensive. Climate change may disrupt the availability of labor, force relocation of operations, and affect commodity prices. Finally, risks due to reputation and litigation were ranked below regulatory and physical risks. There was a perception of low risk to reputation across almost all sectors mentioned. Litigation risk was considered a concern only in the United States.

Corporations in high-risk sectors are more likely to address climate change issues, either as proponents or opponents. Moreover, high-risk sectors also have the potential to be high-reward sectors. A Lehman Brothers report on carbon challenges and opportunities projected that the automobile, utility, integrated oil and gas, and chemicals sectors would be best able to benefit from technological innovation spurred by climate change (Llewellyn 2007). Other opportunities, such as operational improvement, accessing new capital, improving risk management, elevating corporate reputation, identifying new market opportunities, and enhancing human resources management are less sector specific (Hoffman 2005).

Variation by sector in corporate responses to climate change is to be expected. More puzzling is variation within sectors. Multiple sectors in the climate debates have been characterized by climate policy splits between leading corporations (Falkner 2009). Explanations of such intra-sector variation focus on national political and social contexts and on firm-specific factors.

2.2 National Political and Social Contexts

National contexts shape corporate responses to climate change through a range of channels. In their research on climate policy divergences in the oil industry, Skjaerseth and Skodvin (2001) identify government supply of climate policy, societal demand for action on climate change, and the nature of institutions linking supply and demand as three key elements of national context.

Government supply of climate policy is constituted by the particular climate and energy policies implemented by national governments (IEA 2002). Societal demand for action on climate change is driven by levels of social concern (Dunlap 1998), public perceptions of climate science, and societal norms about corporate behavior. Elite and grass-roots social movements for and against action on climate change are central to social responses (Dunlap and McCright, this volume; Fisher 2008). Linking institutions include national regulatory cultures and other domestic political institutions such as legislative structures and party systems (Vogel 1986). For example, national economies can be characterized by the degree to which they rely on state versus private actors to spur innovation and provide public goods, such as greenhouse gas mitigation (Dobbin 2004). The UK is characterized as a liberal market economy, which fosters market-based solutions and inter-firm competition. In contrast, in Germany, which is described as a coordinated-market economy, negotiated agreements between firms and government shape firm operating environments (Hall and Soskice 2001). As a result of these differences, British firms were more likely to embrace emissions trading than German firms (Engels et al. 2008). More generally, Engels et al.'s (2008) analysis finds that cross-national differences in the market orientation of the economy, national policy styles, and the emergence of specialized carbon-accounting service industries account for variation in firm participation in the EU ETS across the United Kingdom, Denmark, the Netherlands, and Germany. Differences in social and political contexts are also central to explaining the generally adversarial climate strategies of US corporations in comparison to their European counterparts (Levy and Kolk 2005; Pulver 2007b). DeSombre argues that the US senate is particularly subject to the influence of domestic industry versus other constituencies, accounting in part for the US refusal to ratify the Kyoto Protocol (DeSombre 2005).

2.3 Firm-Level Strategic, Operational, and Organizational Characteristics

A third source of variation in corporate responses to climate change operates at the firm level. Specific factors internal to a corporation affect managers' perceptions of climate-related risks and opportunities. Internal factors can be categorized as strategic, operational, and organizational.

Strategic factors include a corporation's financial situation, market position, and core competencies (Kolk and Levy 2001; Hoffman 2002). Firms with financial resources can dedicate investment to new climate initiatives, which corporations in weaker financial positions cannot afford (Schneider et al. 2009). In terms of market position, environmental issues may create opportunities for challenger firms to erode the market share of incumbent firms (Schurman 2004). Core competencies also influence corporate climate strategies. Pulver (2007a) demonstrates that an alignment between preexisting core competencies and the expected benefits of adopting an emissions reduction target led Mexico's national oil company to develop a climate strategy that set it apart for other developing country oil companies.

At the operational level, technical characteristics such as geographic location of operations, firm size, and stage in technology replacement cycle shape corporate responses to climate change. Focusing specifically on the oil industry, Rowlands (2000) identifies the carbon intensity of a corporation's fossil fuel portfolio and the regional distribution of operations as potential operational drivers of the divergence between Exxon and British Petroleum's climate strategies, although his data supports neither hypothesis. Austin and Sauer (2002) demonstrate that smaller oil companies are more vulnerable to international climate regulation than their multinational counterparts. Likewise, Schneider et al. (2009) find that firm size is a predictor of CDM participation in the Indian pulp and paper industry.

Finally, corporate culture and organizational histories influence corporate strategy. Cogan's (2006) survey identifies CEO support as central to proactive corporate climate governance. Others point to internal learning and decision-making processes (Lowe and Harris 1998; Berkhout et al. 2006). Firms' access to climate expertise is also a key factor driving variation (Begg et al. 2005; Sullivan 2008).

3 CORPORATIONS: BOTH OBSTACLE AND SOLUTION TO THE CLIMATE CHALLENGE?

Three broad questions motivate scholarship on corporate responses to climate change: How have corporations responded to climate change? What drives corporate action and what explains patterns and divergences in corporate climate policies and practices? And finally, what is the contribution of corporations to the global project of environmental protection? My assessment of corporate climate practices and drivers of action demonstrates that variation is the cross-cutting theme in the answers to the first two questions. Corporations vary in the extent to which they implement climate-related business practices, engage in political activities, and embed climate concerns into management structures. This variation is driven by actual and perceived differences in climate risks and opportunities across sectors, across national political and social contexts, and across firms.

What does the variation in corporate responses imply for the third question? Can the transformative power of corporations be harnessed to address the climate challenge? In a *New York Times* editorial in anticipation of the 2009 international climate negotiations in Copenhagen, Jared Diamond posed the question 'Will big business save the earth?'

Diamond argued that while some corporations are destructive, 'others are among the world's strongest positive forces for environmental sustainability' (Diamond 2009). Subsequent letters to the editor challenged the basis and content of Diamond's arguments and his positive assessments of Wal-Mart, Coca-Cola, and Chevron's environmental practices. The interchange was representative of the widely divergent perspectives on the implications of corporate environmentalism and specifically corporate responses to climate change.

This review provides the empirical terrain on which to analyze claims and counterclaims regarding the potential of corporations to green the global economy. Viewing corporations as either polluters or as 'saviors,' borrowing Diamond's phrasing, offers little insight on this question. Corporations figure in the climate arena as greenhouse gas emitters, as technology and market pioneers, and as political actors. Each provides an avenue for corporate participation in addressing the climate challenge. However, the degree and type of participation depend on the political, legal, and social contexts in which corporations operate. Overall, the data suggest at best an ambiguous role for corporations in propelling decarbonization of the global economy.

First, empirical findings clearly demonstrate that most corporations, though aware of climate change, have not altered their business practices to minimize climate impacts or to adapt to climate change. Rhetoric regarding climate-related business opportunities continues to outpace action. Among the corporations that have embraced action on climate change, climate-related activities remain a small fraction of the overall scope of business activities. These limitations are multiplied when one considers that most research on corporate responses has focused on large enterprises, headquartered in industrialized countries, i.e. the firms most likely to act on climate change. Small to medium-sized enterprises have been mostly ignored, and little is known about the patterns of corporate responses to climate change in developing countries. Looking ahead, the technology and infrastructure investments of the latter will be crucial to determining global greenhouse gas emissions trajectories.

Second, research also clearly demonstrates that corporate action is driven primarily by the risks and opportunities created by regulation. The history of corporate responses shows that corporate behavior follows governmental supply of climate regulation rather than the reverse. When climate change emerged as a domestic and international policy issue in the late 1980s, energy and greenhouse gas intensive industries were the first to engage with the climate issue. Their initial corporate response, led primarily by US corporations, was opposition to greenhouse gas regulation and to the findings of the IPCC. However, regulatory and societal action in the European and international arenas caused this united front to fracture, and some companies came to view action on climate change as a business opportunity. Others continue to pursue blocking strategies, to great effect in the United States and potentially internationally.

Looking ahead, continued variation in corporate responses is likely. Levy and Rothenberg (2002) predicted that the initial heterogeneity in firm climate strategies would decline as homogenizing pressures from the international arena came to dominate firm policy and organizational fields. However, with the collapse of the 2009 international climate change negotiations in Copenhagen and the rise of multiple climate governance projects around the globe, the regulatory, informational, normative changes in corporate business environments will continue to be divergent and conflicting. Moreover, as governance efforts spread across the globe to include emerging economies, the mediating factors related to national political and social contexts and firm characteristics will increase the heterogeneity of corporate voices in the climate debates. Only a globally coordinated, long-term regulatory push will continue to motivate corporate action, although societal demand also has a role to play in creating political and economic environments that reward proactive climate initiatives and punish political and corporate laggards.

REFERENCES

AUSTIN, D., and SAUER, A. 2002. *Changing Oil: Emerging Environmental Risks and Share-holder Value in the Oil and Gas Industry.* Washington, DC: World Resources Institute.

BEGG, K., VAN DER WOERD, F., and LEVY, D. (eds.) 2005. *The Business of Climate Change: Corporate Responses to Kyoto.* Sheffield: Greenleaf Publishing.

BERKHOUT, F., HERTIN, J., and GANN, D. 2006. Learning to adapt: Organizational adaptation to climate change impacts. *Climatic Change* 78: 135–56.

BICKNELL, S., and McMANUS, P. 2006. The canary in the coalmine: Australian ski resorts and their response to climate change. *Geographical Research* 44(4): 386–400.

BREWER, T. 2005. Business perspectives on the EU emissions trading scheme. *Climate Policy* 5(1): 137–44.

CAIN MILLER, C. 2010. Mixing in some carbon. *New York Times* 21 March.

CAPOOR, K., and AMBROSI, P. 2009. *State and Trends of the Carbon Market 2009.* Washington, DC: The World Bank.

COGAN, D. G. 2006. *Corporate Governance and Climate Change: Making the Connection.* Boston: Ceres.

—— 2008. *Corporate Governance and Climate Change: The Banking Sector.* Boston: Ceres.

—— GOOD, M., KANTOR, G., and McATEER, E. 2008. *Corporate Governance and Climate Change: Consumer and Technology Companies.* Boston: Ceres.

DAVIES, K., and SAWIN, J. 2002. *Denial and Deception: A Chronicle of ExxonMobil's Efforts to Corrupt the Debate on Global Warming.* Washington, DC: Greenpeace.

DESOMBRE, E. 2005. Understanding United States unilateralism: Domestic sources of U.S. international environmental policy. In R. S. Axelrod, D. L. Downie, and N. J. Vig (eds.), *The Global Environment: Institutions, Law and Policy.* Washington, DC: Congressional Quarterly Press.

DIAMOND, J. 2009. Will big business save the earth? *New York Times*, 6 December.

DOBBIN, F. 2004. The sociological view of the economy. In F. Dobbin (ed.), *The New Economic Sociology: A Reader.* Princeton: Princeton University Press.

DUNLAP, R. 1998. Lay perceptions of global risk: Public views of global warming in cross-national context. *International Sociology* 13(4): 473–98.

DUNN, S. 2002. Down to business on climate change. *Greener Management International* 39: 27–41.

ELLISON, K. 2007. Of foxes and henhouses. *Frontiers in Ecology and the Environment* 5(10): 568.

ENGELS, A., KNOLL, L., and HUTH, M. 2008. Preparing for the 'real' market: National patterns of institutional learning and company behaviour in the European Emissions Trading Scheme (EU ETS). *European Environment* 18: 276–97.

ENKVIST, P.-A., and VANTHOURNOUT, H. 2007. *How Companies Think about Climate Change.* McKinsey & Company.

FALKNER, R. 2009. *Business Power and Conflict in International Environmental Politics.* New York: Palgrave Macmillan.

FISHER, D. 2008. Who are climate change activists in America? *Environmental Law Reporter* 38(December): 10864–7.

HALL, P. A., and SOSKICE, D. 2001. *Varieties of Capitalism: The Institutional Foundations of Comparative Advantage.* Oxford: Oxford University Press.

HOFFMAN, A. J. 2002. Examining the rhetoric: The strategic implications of climate change policy. *Corporate Environmental Strategy* 9(4): 329–37.

—— 2005. Climate change strategy: The business logic behind voluntary greenhouse gas reductions. *California Management Review* 47(3): 21–46.

IEA. 2002. *Dealing with Climate Change: Policies and Measures in IEA Member Countries.* Paris: International Energy Agency.

—— 2009. *Energy Technology Roadmaps: Charting a Low-Carbon Energy Revolution.* Paris: International Energy Agency.

JONES, C. A., and LEVY, D. 2007. North American business strategies towards climate change. *European Management Journal* 25(6): 428–40.

KNOX-HAYES, J. 2010. The developing carbon financial service industry: Expertise, adaptation and complementarity in London and New York. *Journal of Economic Geography* 9(6): 749–77.

KOLK, A., and LEVY, D. 2001. Winds of change: Corporate strategy, climate change, and oil multinationals. *European Management Journal* 19(5): 501–9.

—— and PINKSE, J. 2005. Business responses to climate change: Identifying emergent strategies. *California Management Review* 47(3): 6–20.

—— —— 2007. Multinationals' political activities on climate change. *Business & Society* 46(2): 201–28.

—— LEVY, D., and PINKSE, J. 2008. Corporate responses in an emerging climate regime: The institutionalization and commensuration of carbon disclosure. *European Accounting Review* 17(4): 719–45.

KPMG 2008. *Climate Changes your Business: KPMG's Review of the Business Risks and Economic Impacts at Sector Level.* Amsterdam: KPMG International.

LEGGETT, J. 1999. *The Carbon War: Global Warming and the End of the Oil Era.* London: Penguin Books.

LEVY, D. L., and EGAN, D. 1998. Capital contests: National and transnational channels of corporate influence on the climate negotiations. *Politics & Society* 26(3): 337.

—— and KOLK, A. 2005. Multinational responses to climate change in the automotive and oil industries. In K. Begg, F. van der Woerd, and D. Levy (eds.), *The Business of Climate Change: Corporate Responses to Kyoto.* Sheffield: Greenleaf Publishing.

—— and ROTHENBERG, S. 2002. Heterogeneity and change in environmental strategy: technological and political responses to climate change in the global automobile industry. In A. J. Hoffman and M. Ventresca (eds.), *Organizations, Policy, and the Natural Environment: Institutional and Strategic Perspectives.* Stanford, CA: Stanford University Press.

LEWIS, J. 2007. Technology acquisition and innovation in the developing world: Wind turbine development in China and India. *Studies in Comparative International Development* 42(3/4): 233–55.

LLEWELLYN, J. 2007. *The Business of Climate Change: Challenges and Opportunities.* New York: Lehman Brothers.

LOWE, E. A., and HARRIS, R. J. 1998. Taking climate change seriously: British Petroleum's business strategy. *Corporate Environmental Strategy* 5(2): 23–31.

Lutken, S. E., and Michaelowa, A. 2008. *Corporate Strategies and the Clean Development Mechanism*. Cheltenham: Edward Elgar Publishing Ltd.

McCright, A. M., and Dunlap, R. E. 2003. Defeating Kyoto: The conservative movement's impact on US climate change policy. *Social Problems* 50(3): 348–73.

Mills, E. 2005. Insurance in a climate of change. *Science* 309: 1040–3.

Newell, P. 2000. *Climate for Change: Non-state Actors and the Global Politics of the Greenhouse*. Cambridge: Cambridge University Press.

PriceWaterhouseCoopers 2009. *Carbon Disclosure Project 2009: Global 500 Report*. London: Carbon Disclosure Project.

Pulver, S. 2002. Organizing business: Industry NGOs in the climate debates. *Greener Management International* 39(Autumn): 55–67.

—— 2007a. Importing environmentalism: Explaining Petroleos Mexicanos' proactive climate policy. *Studies in Comparative International Development* 42(3/4): 233–55.

—— 2007b. Making sense of corporate environmentalism: An environmental contestation approach to understanding the causes and consequences of the climate change policy split in the oil industry. *Organization & Environment* 20(1): 44–83.

—— Hultman, N., and Guimaraes, L. 2010. Carbon market participation by sugar mills in Brazil. *Climate and Development* 2(2), in press.

Rowlands, I. H. 2000. Beauty and the beast? BP and Exxon's positions on global climate change. *Environment and Planning C: Government and Policy* 18: 339–54.

Schmidheiny, S. 1992. *Changing Course: A Global Business Perspective on Development and the Environment*. Cambridge, MA: MIT Press.

Schneider, M., Hoffmann, V. H., and Gurjar, B. R. 2009. Corporate responses to the CDM: The Indian pulp and paper industry. *Climate Policy* 9(3): 255–72.

Schurman, R. 2004. Fighting 'Frankenfoods': Industry opportunity structures and the efficacy of the anti-biotech movement in western Europe. *Social Problems* 51(2): 243–68.

Skjaerseth, J. B., and Skodvin, T. 2001. Climate change and the oil industry: Common problem, different strategies. *Global Environmental Politics* 4(1): 18–42.

Suchman, M. 2009. Taming the market for medical information: 'Sharing is (s)caring' on the digital divide. In *Annual Meeting of the Law and Society Association*. Las Vegas, NV.

Sullivan, R. (ed.) 2008. *Corporate Responses to Climate Change: Achieving Emissions Reductions through Regulation, Self-Regulation, and Economic Incentives*. Sheffield: Greenleaf Publishing.

Trucost. 2009. *Carbon Risks and Opportunities in the S&P 500*. Boston: Investor Responsibility Research Center Institute (IRRCi).

van den Hove, S., Le Menestrel, M., and de Bettignies, H.-C. 2002. The oil industry and climate change: Strategies and ethical dilemmas. *Climate Policy* 2(1).

Victor, D., and House, J. 2006. BP's emissions trading system. *Energy Policy* 34: 2100–12.

Vogel, D. 1986. *National Styles of Regulation: Environmental Policy in Great Britain and the United States*. Ithaca, NY: Cornell University Press.

Vormedal, I. 2008. The influence of business and industry NGOs in the negotiation of the Kyoto mechanisms: The case of carbon capture and storage in the CDM. *Global Environmental Politics* 8(4): 36–65.

Zisa, S. 2007. *WalMart and Sustainability: Closing the Eco-Efficiency Gap*. Environmental Studies, Brown University, Providence, RI.

CHAPTER 40

IS GREEN CONSUMPTION PART OF THE SOLUTION?

ANDREW SZASZ

1 INTRODUCTION: CONCERN ABOUT CLIMATE CHANGE AND A NEW KIND OF 'GREEN CONSUMING'

CLIMATE scientists had sought for years to convince the American public that climate change was real and that, left unchecked, it would have dire, if not catastrophic, consequences for life as we know it. Their efforts were largely unsuccessful (Brechin 2003). Then, in 2007, wide distribution of Al Gore's film, *An Inconvenient Truth*, and publication of the Intergovernmental Panel on Climate Change's fourth, and most dire, set of reports occasioned a surge of media coverage (Brulle 2009). New polls confirmed a sea change in public opinion. Suddenly most Americans were agreeing that 'climate change' or 'global warming' is a serious problem.

Let's also observe that by this time many Americans had learned to express environmental concern by changing their consumption choices: to shun furs, if they believed in animal rights; to buy things made of recycled materials, if they wished to help protect natural resources; to buy organic foods, filter their water, purchase 'natural' or 'non-toxic' goods, if they were concerned about ingesting toxic substances (Szasz 2007). Given that, it was perhaps inevitable that concern about climate change would prompt a new form of green consuming, buying commodities that promise to have a *lower carbon footprint* than some other, comparable commodities. (From here on, I will use the abbreviation 'LCF consuming' or 'LCF' when referring to 'lower carbon footprint.')

LCF alternatives, such as the compact fluorescent lightbulb, began to appear on the market. Hybrid cars are probably the most iconic example of the new LCF commodity as auto-makers rushed to offer consumers a growing variety of ever more attractive hybrid models.

Climate activism through consumption was endorsed by the Sierra Club, the Natural Resources Defense Council, and other environmental movement organizations. Their websites, online environmental news sites such as GRIST, and many green blogs offered advice about how one could shop differently to reduce one's carbon footprint. Progressive consumers could find similar advice in magazines in the print media, in magazines like Mother Jones or books with titles like *50 Simple Things You Can Do to Save the Earth*, and *Climate Change: Simple Things You Can Do to Make a Difference*.

At first glance, mobilizing citizens as consumers, getting American consumers to do something about climate change every time they go shopping, seems an attractive political strategy. It is a truism that ours is a consumer society. The average American may not be particularly engaged politically (Putnam 2000), but she or he does shop every day and, typically, invests quite a bit of psychic energy doing so. Harnessing that to political ends? It seems possible; promising.

Individual consumption accounts for a substantial fraction of the nation's carbon emissions, on the order of around 40 percent (Padgett et al. 2008). If millions of consumers made the LCF choice, there would be significant reductions in aggregate CO_2 emissions. In addition, it would send a powerful signal, encouraging corporations to bring more LCF products to market.

Finally, making the LCF choice might be a politicizing experience. It might be a first step toward a greater commitment to doing more about the climate issue. Few Americans see themselves, first and foremost, as activists, as engaged political actors. On the other hand, everyone shops. An initial awareness/concern may encourage one to buy LCF goods. That consciously ethical decision, choosing the alternative that is kinder to the climate, especially if that decision is made not just once but repeatedly, may stimulate a further deepening of caring, a growing sense of responsibility or sense of stewardship. It may trigger a process through which the person organizes a new identity for herself or himself, an identity as someone who *really* cares about the environment.

In spite of all the attention, LCF consuming is still a minor phenomenon. Hybrid cars account for less than 3 percent of auto sales in the US (Green Car Congress 2008). Or consider electricity from solar power: California leads the rest of the nation, by far, in solar power, with 50,000 solar installations, generating more than 500 megawatts of electricity (Barringer 2009). An impressive amount, but that amounted to less than one half of one percent of the total power generated in California in 2009 (California Energy Commission 2010).

Low levels of market penetration translate into only very modest reductions in CO_2 emissions. But more may be possible. Would it be worth making the effort to try to increase levels of LCF consuming? In other words, should changing consumer behavior be an important goal for climate activism? This is a question of political strategy, made urgent by climate scientists' overwhelming consensus that time is growing short and that we must take effective action soon.

In this chapter I enumerate the conditions that would have to exist if LCF consuming is ever to become a significant factor in meeting the climate change challenge. I assess whether or not these conditions currently exist. When they do not, I explore what would have to be different to create those favorable conditions.

Before I continue, I wish to note two limitations of this discussion.

I wish to emphasize, first, that I am interested in individual, not institutional LCF consumption. Quite a lot has been happening at the level of institutions. Many universities have initiated campus sustainability programs that, among other goals, aim to reduce those campuses' carbon footprints. Cities, counties, states, and regional organizations of states have declared their intention to go carbon neutral. The proliferation of such institutional-level efforts is very encouraging. However, this sort of initiative is not what comes to mind first when 'green consumption' is the topic. Usually, we mean by that expression the individual consumer, entering a market, faced with choices, and that individualized type of consumption is what I wish to examine here.

Second, some of the conditions I discuss might apply only to the United States. In some Western European nations, for example, political culture may be such that green consum-ing and political activism are more seamlessly linked than they are in the US. It should be understood, then, that the analysis presented in this chapter is meant to describe conditions in the US and may not be applicable to the climate politics of other nations.

2 How Can Lower Carbon Footprint Consumption Become a Mass Phenomenon?

2.1 Condition 1: Consumers Believe Climate Change is Real and is Urgent

If LCF consuming is to become significant, a substantial fraction of Americans have to believe that climate change is *real*, that it is caused by human activity, not by natural variation in the Earth's climate (in which case human intervention is irrelevant), and that it is *serious enough* to warrant immediate changes in one's behavior. Polling data shows that most Americans think the problem is real, but the percent that think so has been falling. Worse, these polls show that to most Americans the issue doesn't seem particularly urgent.

As I noted in the Introduction, as recently as the early 2000s, Americans were poorly informed about climate change and seemed to care little about it (Brechin 2003). Polls conducted only a few years later show, however, that public opinion was evolving quite dramatically. By 2007 and 2008, large majorities were agreeing that climate change is real and that left unchecked, climate change would have serious consequences (Dunlap 2008; Ray 2009; World Public Opinion.org 2009).

Such a shift in public opinion might lead to optimism about the prospects for increased carbon footprint consuming. Unfortunately, more recent polls show opinion moving in the opposite direction, and that raises doubts about the depth and seriousness of the public's understanding of and caring about the threat.

In polls conducted in 2009, the percentage of Americans who agree that there is 'solid evidence' for climate change fell from 71 to 57 percent (Pew 2009b). The percentage agreeing that climate change is caused by human activity fell from 47 to 36 percent (Pew 2009b). What percentage believe global warming has already begun? In 2007, 61 percent; in 2009, 53 percent (Saad 2009). Worry a 'great deal' or a 'fair amount' about it? 66 to 60 percent (ibid.). Do you believe it is going to be a serious threat in your lifetime? 40 to 38 percent (ibid.). Does media exaggerate the problem? In 2007, 33 percent said yes; in 2009,

41 percent (ibid.). Should it be one of the government's top priorities? In 2008, 35 percent said yes; in 2009, 30 percent (Pew 2009a).

The recession has undoubtedly been the principal cause of this decline, as concern about pocketbook issues, jobs, and falling home prices trump all other concerns (Newport 2009; Pew 2009a; Saad 2010). A big drop in media coverage in the second half of 2007 and in 2008 (Brulle 2009) may have contributed, as well, since it has been shown that public opinion waxes and wanes as a function of media coverage (Iyengar and Kinder 1989).

Perhaps even more important, even if many Americans agree that the climate issue is real and that the threat is serious, that doesn't mean that they consider the issue *urgent*. When asked about issue importance or salience, respondents rank climate change far lower than most other issues that pollsters ask about; in fact, concern about climate—and concern about 'environment,' in general—ranks *at or close to the bottom* on people's lists of what they worry about, well below concerns about the economy, jobs, terrorism, social security, Medicare, healthcare, education, budget deficits, and crime (Pew 2009a).

How ready are Americans not just to acknowledge climate change but to do something about it? Consider the findings of the Yale Project on Climate Change's study, 'Global Warming's Six Americas 2009.' The study found that only 18 percent of the public is 'alarmed' enough to be 'already making changes in their own lives and support an aggressive national response' (Yale 2009: 3). Another 33 percent are 'concerned . . . convinced that global warming is a serious problem, but . . . distinctly less involved in the issue—and less likely to be taking personal action—than the Alarmed' (ibid.). (The remaining 49 percent are 'cautious,' 'disengaged,' 'doubtful,' or 'dismissive' (ibid.)).

Early in 2010, climate 'skeptics' and the conservative 'blogosphere' seized upon a handful of e-mails hacked from a server used by climate scientists at the University of East Anglia and upon two errors in one of the IPCC's 2007 reports to ratchet up their claim that climate change is unproven, that climate scientists are guilty of bad science and guilty of hyping their findings. New polls show that 'Climategate,' as this episode has come to be called, has had some traction; Americans' belief that climate change is really occurring have continued to decline (Newport 2010).

These reversals in polling trends, the speed with which the percentage of the population agreeing that the issue is real, serious, etc. has fallen, plus the evidence that the issue has never achieved significant levels of salience, suggest that this first condition for large numbers of consumers embracing LCF consuming has not yet been met. Climate scientists are beginning to realize that doing the science is not enough, that they (and the political activists who advocate immediate action on climate) will have to find more effective, more persuasive ways to communicate their findings—and their sense of urgency—to the general public.

2.2 Condition 2: Affordable, Attractive Lower Carbon Footprint Alternatives Exist

Even if Condition 1 is met at some point in the near future and significant numbers of citizens become convinced that buying LCF products is something immediate and tangible that they can do to act on their concern, LCF alternatives would have to be (a) available; (b) competitive in terms of price; and (c) competitive in terms of quality.

2.2.1 *Availability*

Hybrid cars, energy-efficient appliances, energy-efficient electronics, the compact fluores-cent bulb, rooftop solar panels, making one's existing home more energy efficient with better insulation, double-paned windows; these are the kinds of consumer items that come most readily to mind when one thinks about what consumers can do now to lower their carbon footprint. Auto-makers have been especially innovative recently, bringing to market an ever expanding variety of hybrid models, moving quickly to develop a practical all-electric car.

It is instructive to consider how such items have come to exist. In some cases, it is simply the private sector offering a new commodity where it perceives new demand. Often, though, government intervention has played a key role, sometimes via subsidies, sometimes via regulations that force change. Billions of dollars in federal subsidies have spurred technological innovation (Block 2008; Ling 2009). Standards, such as fleet mileage stand-ards for cars, require manufacturers to achieve greater aggregate fuel efficiency. At the extreme, the federal government can outright ban products that consume too much energy, as in the forthcoming ban on production of incandescent light bulbs (Sheppard 2009; Mufson 2009), thereby simply giving consumers no choice other than to buy the LCF alternative.

When we imagine consumers choosing the green alternative, we gravitate to thinking of the consumer electronics store, the appliance store, the auto dealership. Some vitally important forms of consumption, though, transcend that scene. Instead of looking at product X versus product Y, the consumer is presented with certain 'facts on the ground.' Consider the spacial organization of American cities and suburbs. The Census reports that more than half of all Americans now live in suburbs. Increasingly, 'community' in the US means suburb, edge city, exurb, the 'post-suburban' social geography exemplified by Orange County, CA (Kling et al. 1991). Yes, one chooses to buy this or that house, but it was the housing industry that determined, years ago, where all those houses were built, how they were built. Heating, air conditioning, and lighting the typical suburban home takes prodigious amounts of energy. The spatial organization of American communities—the post-suburban sprawl that separates the functional nodes of residential, shopping, work, and entertainment spaces, then connects those nodes with freeways; the profound absence of a viable public transit alternative to driving—also constrains consumer choice. Yes, one can choose a fuel-efficient car or a hybrid instead of an SUV or monster pickup truck, but if there is no mass transit one cannot choose to not drive. *Social geography*, how communities are spacially organized, the type of housing available to consumers in the market for a home, the availability or absence of public transportation, constitute external constraints on individual consumer choice.

Here, even more than in the case of store-bought goods, intervention at the level of social policy is what is required. Retrofitting existing homes does reduce energy consumption, but most homes are unlikely to be retrofitted unless retrofitting is either mandated by policy (as in the case, say, of low-flow toilets) or is subsidized (see next section on 'price'). 'Retro-fitting' suburban/exurban sprawl with a mass transit system, such as light rail, obviously requires intervention at a level other than the individual yearning for an alternative to single-occupancy commuting.

In the long run, the availability of ever more substantial LCF alternatives depends upon social policy that, over years, could construct a different social geography, perhaps one that embodies the principles of what is known as 'smart growth,' along with building codes that require new homes to be built with energy efficiency in mind (Environmental News Service 2009; Galbraith 2009; Granade et al. 2009; Mazria 2009), investment in mass transit alternatives to the private automobile.

I conclude from this discussion that LCF goods are available today—and their availability will continue to increase—not just because, perhaps not even primarily because, individual consumers' preferences have aggregated into a powerful market signal. Some combination of planning, regulation, and incentives has been required, and will continue to be required if such goods are to be increasingly available to consumers.

2.2.2 Price

Some consumers are so committed to environmental causes that, for them, the politics trumps factors that typically determine buying choices. Such consumers are willing to pay more for the green alternative product, even if that product is, in some ways, inferior to its conventional counterpart (Griskevicius et al. 2010). But such ideology-driven consumer behavior will always be restricted, I believe, to a rather small demographic. To the average American consumer, price (and quality) continues to matter even if s/he brings environmental concerns to her/his consuming decisions. Price is certainly a major issue given current economic conditions. Income inequalities in the US have been growing since about 1973 (Saez 2009). Most Americans' real wages have either stagnated or have actually fallen (Piketty and Saez 2003). Add to that the fact that the Great Recession that started late in 2007 has thrown millions of Americans out of work. Most consumers simply cannot allow themselves to make the LCF choice unless that choice is competitively priced.

Even if the LCF alternative saves money over the useful life of the product, consumers may not be able to afford it if the purchase price is too high. It takes years of saving at the pump to make up for the higher cost of a hybrid car. Weatherizing or paying a contractor to install solar panels on the roof will save on utility bills in the long run, but require a substantial initial investment. A rooftop solar system can cost $20,000 or more.

Government intervention, in the form of incentives and subsidies, may succeed where the market fails to offer LCF goods at competitive prices. Substantial federal tax credits have made hybrid vehicles more affordable. Federal and state governments offer subsidies to help homeowners retrofit their homes with rooftop solar arrays. Solar's growth in California, from 500 installations ten years ago to 50,000 units installed today, 33 percent growth in just one year, from 2007 to 2008, was made possible by generous state, and in some places additional local, subsidies (Barringer 2009). Electric utilities offer incentives that make it more attractive for their customers to purchase newer, more energy-efficient appliances. Following on the success of the 'cash for clunkers' program, the federal government launched a 'cash for appliances' program that will, similarly, encourage consumers to replace ageing appliances with newer, more energy-efficient models (Glover 2009) and a 'cash for caulkers' program that would help homeowners weatherize their homes, saving energy (and creating jobs) (Leonhardt 2009; Roberts 2009).

Policy could influence buyers' choices another way, by raising the price of conventional goods. A tax on products' carbon emissions throughout those products' life cycles (emissions generated in extracting the necessary raw materials; emissions generated in production; emissions generated during a product's useful lifespan; and emissions generated when that item is disposed or recycled) would likely raise their prices, making the LCF alternative more attractive. Given the current political climate in the US, it seems unlikely that a similar carbon tax could be enacted any time soon.

2.2.3 Quality

Beyond price, if an LCF product is to be competitive, it has to have other, non-monetary qualities roughly comparable to the qualities of its conventional counterpart. It has to meet consumers' expectations of that sort of product; it must perform as expected, be sturdy, reliable, attractive, easy to use. If such qualities are lacking, if the product is significantly inferior compared to its conventional counterpart, that's a strong disincentive to choose it.

Honda's first generation Insight hybrid had outstanding gas mileage, but it was completely impractical, cramped. It was exceedingly odd looking, to boot. Consumers shunned it. Toyota's first generation Prius was a bit too small, even for a subcompact, and it had little curb appeal. Prius sales improved dramatically after the car got bigger, closer to what a buyer of a subcompact expects, and was given a distinctive, appealing, aerodynamic, somewhat futuristic look.

The compact fluorescent bulb, the first widely available alternative to the Edison lightbulb, has issues—manufacturing quality controls seem poor, leading to a high rate of initial failure; it doesn't give off as much light; it doesn't last as long as promised; because it contains mercury, a hazardous substance, it can't just be thrown in the trash when the bulb dies—which make it unlikely to succeed. Better alternatives, such as LED technology, are in development, however, and promise to displace compact fluorescents as the preferred alternative to the incandescent bulb.

These examples suggest to me that once a market for an LCF good is established manufacturers will respond to buyers' expectations and will, over time, bring to market better versions than what they originally offered, versions with qualities that make them more attractive to buyers. So this issue is probably not as important as, say, price.

2.3 Condition 3: Motivation, Trust

2.3.1 Motivation

Consuming is not just a straightforward satisfaction of need. Other motives, deep ones, are in play. Cultural processes in modern society have infused acts of consumption with other meanings. A person can articulate an identity, for themselves, through the things they buy. As Herbert Marcuse put it, in *One Dimensional Man*, 'people recognize themselves in their commodities; they find their soul in their automobiles, hi-fi set, split-level home, kitchen equipment' (Marcuse 1964: 9). One's consumption choices, properly displayed, are an important way to lay claim to a particular social identity (Veblen 1912; Bourdieu 1984).

People are powerfully motivated by the wish to have a positive self-image (to one's self) and a socially-honored identity (for others). Believing that climate change is real and is urgent (Condition 1) may predispose a consumer toward buying LCF products. The motivation to do so is heightened if acquiring and subsequently using/displaying the LCF product enhances self-esteem and/or enhances one's social reputation, if consuming LCF goods is seen as something that accrues social approval, confers a good name, and enhances the consumer's social status.

Marketers of LCF products appeal to such motives with words, images, packaging, advertising that say, 'This will be not only be good for you; it will be good for the Earth,' and that imply, 'It will confirm your identity, to you and to others, as someone who *cares*.'

Appealing as that message is, green motivations may be in conflict with other motives, as powerful and perhaps more powerful—deeply entrenched, culturally sanctioned identity aspirations, the wish to be seen as someone with wealth, status, power, the wish to be seen as attractive, desirable, virile, cool. Such yearnings and aspirations are operationalized via consumption choices that are anything but climate friendly: fast, powerful cars; big homes that consume copious amounts of energy for heating and lighting; lots of meat on the table; expensive vacations to far-away places.

My point, here, is that the motive to shop green must go up against some stiff competition from other motives, quite powerful ones, that drive consumer choice, especially when we know, from opinion polling, that most Americans aren't all that certain that climate change is real, and even many of those who are convinced it is real do not think it is all that urgent.

Other motives could come into play. Energy security—our dependence on foreign oil; the wars and other entanglements that dependence requires—may motivate consumers to make the LCF choice. Such messages are certainly out there. They don't seem to have affected consumer choice yet, and such messages may lead not to more mindful consuming but to support for exploiting ecologically damaging extraction of domestic coal and oil supplies.

2.3.2 *Trust*

If consumers are to make the green choice, it's not enough that alternative goods are available, affordable, and of good quality. The consumer has to trust that these alternative goods do, indeed, have the eco-friendly qualities they claim to have.

Do consumers trust manufacturers' claims? A recent study, conducted by BBMG, a New York branding and marketing firm, finds that mostly they do not (Environmental Leader 2009). That distrust seems warranted. One recent report states that 98 percent of green claims in advertising/marketing are, to some degree, misleading (TerraChoice 2009). Other articles assert, likewise, that greenwashing is rampant (GRIST 2008; Cornwall 2008), and that the Federal Trade Commission does not have the manpower to enforce regulations against false advertising (Watson 2009).

Uncertainty about green claims undermines trust, leaving consumers hesitant to spend money on purportedly greener goods. How can trust be established?

Environmental organizations are making substantial efforts both to warn consumers about greenwashing and also to endorse products that they judge to be worth trusting. Advice from such organizations is likely to be trusted by consumers who are already at least somewhat committed to green politics.

Consumers seem to trust government certification that a product meets certain standards. UDSA organic certification is a good recent example. In the case of climate change, the federal government could require manufacturers to disclose to consumers the total carbon footprint of a product over its whole life cycle: how much CO_2 was emitted manufacturing the product and shipping it to market; how much CO_2 would be emitted as the product is used; how much will be emitted subsequently, when the product is thrown away or recycled.

Such data would help consumers make better informed decisions, but mandating the collection and disclosure of such information may not be politically feasible. Governmental efforts to mandate similar disclosures have generated considerable conflict. Manufacturers fight for more permissive standards when the government decides to certify products. Witness the protracted and intense conflicts over what would be required of products that wished to display federal certification that those products are 'organic' (Kindy and Layton 2009). Manufacturers can simply try to kill proposals that would have them disclose information that seems to them potentially damaging. Monsanto, the manufacturer of rBGH, fought to prevent certain milk producers from putting on their containers the fact that their milk comes from cows not injected by bovine growth hormone (Dupuis 2000). That's just one battle in a bigger war over food labeling, a war that has brought us various states' 'food libel' or 'agricultural disparagement' laws. One can expect to see similar conflicts over what kinds of climate-relevant information the government can mandate for products. In 2009, the housing industry successfully fought off a proposal that when an older building goes on the market the seller should disclose to the potential buyer that building's expected future energy consumption (Rahim 2009).

Third-party certification, independent of both manufacturers and activists, promises greater objectivity, hence trustworthiness. That third party can be a federal or state regulator (EPA's Energy Star program), or it can be a nongovernmental certification program (LEED standards for green home construction; Green Seal). Polling data suggests that consumers do have more trust in third-party certification. However, I have also seen the argument that the proliferation of third-party certification programs has caused some consumer confusion (DeBare 2008).

2.4 Condition 4: LCF Consumption as an Organizing and Recruiting Strategy

Finally, if making the LCF choice, again and again, deepened a consumer's awareness and fostered a deeper political commitment, LCF shopping would also have important psychological/political impacts.

For someone who is already an environmentalist, LCF consumption is an expression of, not a path to, political commitment. LCF consumption is one way to act on beliefs they already hold dear. Consider, though, the potentially much larger fraction of the public who may have known or cared little about climate change but who suddenly become aware and concerned. Is it possible that an initial, perhaps initially quite shallow, environmental 'concern' leads, first, to acts of LCF shopping, which then serve to further sensitize the consumer, increases her or his environmental awareness, makes that awareness a more continuous and more central feature of their thought process, their values, their identity? If so, LCF shopping might be a way to organize millions into a movement to meet the climate challenge.

The Yale study referred to earlier found that concern about climate change, even when it does exist, leads more often to nothing than to some kind of personal decision to act. Let's assume, though, for the moment, that dawning awareness of and concern about climate change does lead to some act of LCF consuming. Would that lead to a deepening of awareness, concern, engagement, and the intent to do more? Or would it lead, instead, nowhere, politically speaking?

Gert Spaargaren, one of Europe's pre-eminent scholars of green consumption, refers in his papers to the 'citizen-consumer' (Spaargaren and van Vliet 2000; Spaargaren 2003). I take the phrase to mean that one can *be* a citizen, one can discharge some of the obligations of being a citizen, by *being* a consumer. To be a political actor, a citizen, one doesn't necessarily have to engage in what are ordinarily considered to be political behaviors, joining an organization, attending meetings, demonstrating, writing letters, contributing to causes, voting. No, one can be a political actor, in a meaningful sense, simply by shopping correctly. Citizen and consumer can be a seamless unity/identity. That may well be true in Western Europe, given its current political culture. I don't think that one can just assume that such a Subject exists in the context of American society, American political culture. In a society that is comparatively weak in civic engagement, low in measures of political participation (Putnam 2000), a society which valorizes shopping to such an extreme, sees shopping as an excellent way to spend leisure time and as a way to feel good about one's self, there may, in fact, be little relationship between 'citizen' and 'consumer.'

A first sweep through the literature on consumer psychology suggests that, to date, researchers have focused mostly on the question, 'why do consumers make one choice and not the other?' and have paid little attention to the quite different question, 'once the consumer has made a decision and bought one thing and not another, what is the political *aftermath* of that choice?' (See e.g. de Groot and Steg 2009; Heffner et al. 2005; Kahn 2007; Rego et al. 2007).

In the absence of much actual research, one can imagine two very different sequences or scenarios. In one, taking the time to be mindful when making choices, spending the extra amounts that have to be spent to make the green choice, deepens awareness, strengthens commitment. One might argue that the well-known mechanism of 'cognitive dissonance' (Festinger 1957) comes into play: if one spends the time and money, one experiences psychic pressure to want to believe that the extra effort was worth it. Beyond such purely individual-level mechanisms, one can consider group-level motivations. Green consuming may increase one's identification with others whose values one respects. One may start to see one's self (if only in the imagination, at first) as a fellow traveler of a *movement* that is trying to do great things, and, in this way, green consuming can be a first step toward actual, active participation in that movement.

One can, however, imagine another sequence in which a person begins to believe that climate change is, indeed, a serious problem. She or he would like to do something. But what? That's not obvious. The problem is both huge and abstract. The solutions described by scientists and respected opinion makers seem impossible even to imagine, much less to inspire confidence that I, one person, can realistically do anything to really affect the outcome. At the same time, there are so many more immediate things to worry about . . .

But the newly concerned citizen is told that they *can* do something. There are, in fact, many things one can do, *Simple Things*, as several book titles say. One can help by buying some LCF goods. The newly concerned, not knowing what they can do as 'citizen,' can shift to the register of 'consumer' and buy some LCF goods. They feel good, feel virtuous, feel that they have 'done something' about the climate problem. The felt need to do more diminishes. They have done what they can, for now. They have done enough.

In this sequence or scenario, LCF consumption leads not to increased engagement or commitment; it leads, instead, to satisfaction of the felt need to act, and that satisfaction leads to decreased attention and concern, freeing the person to think about, worry about all those other, more immediately, personally pressing matters. Working in 'decision theory' tradition developed by Baruch Fischhoff, Amos Tversky and Daniel Kahneman, Elke Weber, of Columbia University's Center for Research on Environmental Decisions, identifies two psychological mechanisms that would tend to produce this kind of result, 'finite pool of worry' and 'single-action bias.' Journalist Jon Gertner describes the latter as: when 'prompted by a distressing emotional signal [i.e. we are suddenly made worried about climate change], we buy a more efficient furnace or insulate our attic . . . —a single action that effectively diminishes global warming as a motivating factor' (Gertner 2009).

3 Promising, but Conditions for Success Are Far from Assured

Is individual consumption change part of the solution? It could be, if tens of millions of citizens 'voted with their pocketbooks' by buying LCF alternatives. Such consumer action could prove even more powerful if it led the consumer to engage in other forms of climate activism.

As I have argued throughout this chapter, though, several conditions would have to be present if consumer behavior were to be a major factor in climate change politics. Today, few of those conditions are in place. Let's summarize what would have to happen, and who would have to take what kind of action to make those things happen:

A substantial fraction of the public has to understand that climate change is happening and it is urgent enough not just to acknowledge it but to do something personally about it. Scientists' and climate activists' efforts to date seem not to have worked particularly well. If the public is to be convinced, new ways to communicate the facts and the urgency of the situation must be found.

More alternatives to conventional goods must become available and all such alternatives must be comparable to their conventional counterparts in both price and in quality. Developing, producing, and marketing a lower carbon alternative to this or that individual consumer item has proven to be fairly straightforward. Government subsidies can accelerate the pace of technological innovation that may be required to develop competitive lower carbon alternative goods. Subsidies and tax incentives will continue to be needed in order to make alternative goods affordable and to help the consumer with high up-front costs (as in the case of retrofitting homes with rooftop solar).

To make really substantial gains in the longer term, the existing social geography of suburban sprawl will have to be replaced by a different organization of social space, one that will allow millions to live, work, and play, to move through the ordinary day, every day, in ways that require a lower expenditure of energy than such activities require today. When we contemplate changes at that scale, we are still talking about changing consumption, but in ways that will require interventions, land-use planning, home construction standards, investment in transportation infrastructure, that are very far from our usual image of intervention at the level of the individual consumer.

Assuming Conditions 1 and 2 are met (many citizens believe climate change is real and urgent; alternative goods are available, of good quality, affordable), one has still to address issues of trust and motivation.

Improving trust is the easy part, at least in the abstract, although these measures may be difficult to implement in practice: stronger green certification programs; more comprehensive labeling standards; better regulation of green claims.

Motivation is a greater challenge. What will make making the green choice enhance self-esteem and confer social honor/status? Advertisers already do this when they encourage consumers to buy their green products. Something bigger than that is needed, though, something like the building of a movement for a sustainable society, only one facet of which would be to valorize and encourage green consuming, LCF consuming. Here again we are contemplating intervention at a level far beyond the level of the individual consumer, or, more precisely, contemplating an intervention or a strategy in which much more global change needs to be fostered before intervention at the level of the individual consumer becomes meaningful.

Organizing the carbon footprint consumer: As discussed above, it is far from certain that even if the consumer is encouraged to make the LCF choice that that act will deepen their commitment. It may well lead, instead, to a decrease in concern and to disengagement. Challenging as it is, it is not enough to get the consumer to shop green. Politically motivated consuming will open consumers to deeper engagement only if a movement teaches them that shopping green is good but that it is not enough.

The title of this chapter asks, is consumption part of the answer to the challenge of climate change? There is no simple answer to that question. Consumption could be part of the answer, given the conditions I have described. Most of those conditions don't exist today. Creating those conditions looks daunting. Given that time grows short, targeting the individual consumer, trying to change her or his consumption choices, probably should not be a priority. Some of the more systemic interventions enumerated here, strengthening federal regulations, subsidizing technological innovation, finding better ways to build homes, reducing dependency on the automobile, reorganizing the built environment, are more promising *and* are necessary if we are to see substantial change at the level of individual consumption.

REFERENCES

BARRINGER, F. 2009. With push toward renewable energy, California sets pace for solar power. *New York Times* 16 July.

BLOCK, F. 2008. Swimming against the current: The rise of a hidden developmental state in the United States. *Politics & Society*, 36(2): 169–206.

BOURDIEU, P. 1984. *Distinction: A Social Critique of the Judgement of Taste.* Cambridge, MA: Harvard University Press.

BRECHIN, S. R. 2003. Comparative public opinion and knowledge on global climatic change and the Kyoto Protocol: The U.S. versus the world? *International Journal of Sociology and Social Policy,* 23(10): 106–34.

BRULLE, R. 2009. Nightly news coverage of global warming (NBC, CBS, & ABC). Data graph courtesy of Robert Brulle.

California Energy Commission 2010. *Total Electricity System Power.* <http://energyalmanac.ca.gov/electricity/total_system_pwer.html>. Accessed 3 June.

CORNWALL, W. 2008. Companies try to cash in on green trend, but should consumers buy it? *Seattle Times* 28 June.

DeBARE, I. 2008. Green product seals are gray area. *San Francisco Chronicle* 19 April.

de GROOT, J. I. M., and STEG, L. 2009. Mean or green: Which values can promote stable pro-environmental behavior? *Conservation Letters* 2: 61–6.

DUNLAP, R. E. 2008. Climate-change views: Republican-Democratic gaps expand. <http://www.gallup.com/poll/107569/ClimateChange-Views RepublicanDemocratic-Gaps-Expand.aspx>. Posted 29 May.

DUPUIS, E. M. 2000. Not in my body: rBGH and the rise of organic milk. *Agriculture and Human Values* 17: 285–95.

Environmental Leader. 2009. Study: Consumers lack trust in green claims. <http://www.environmentalleader.com/2009/04/03/study-consumers-lack-trust-in-green-claims/>. Posted 3 April.

Environmental News Service. 2009. *Report: Energy Efficiency Could Halve U.S. Greenhouse Gases by 2050.* <http://www.ens-newswire.com/ens/jul2009/2009-07-30-095.asp>. Posted 30 July 30.

FESTINGER, L. 1957. *A Theory of Cognitive Dissonance.* Evanston, IL: Row Peterson.

GALBRAITH, K. 2009. Efficiency drive could cut energy use 23% by 2020, study finds. *New York Times* 29 July.

GERTNER, J. 2009. Why isn't the brain green? *New York Times Sunday Magazine.* 19 April.

GLOVER, M. 2009. Energy-efficient appliance program may begin by December. *Sacramento Bee* 5 September.

GRANADE, H. C., CREYTS, J., DERKACH, A., FARESE, P., NYQUIST, S., and OSTROWSKI, K. 2009. Unlocking energy efficiency in the U.S. Economy. McKinsey & Company, <http://www.mckinsey.com/clientservice/electricpowernaturalgas/us_energy_efficiency/>, July.

Green Car Congress. 2008. US sales of hybrids in October down 10%: Monthly share up to 2.6%. <http://www.greencarcongress.com/2008/11/us-sales-of-hyb.html>. Posted 5 November.

GRISKEVICIUS, V., TYBUR, J. M., and VAN DEN BERGH, B. 2010. Going green to be seen: Status, reputation, and conspicuous conservations. *Journal of Personality and Social Psychology* 98(3): 392–404.

GRIST. 2008. Is it really green? Advice for navigating the wild world of products with eco-claims. *DailyGrist: Top Environmental News from around the World.* <http://www.grist.org/article/is-it-really-green/>, 28 January.

HEFFNER, R. R., KURANI, K., and TURRENTINE, T. 2005. *Effects of Vehicle Image in Gasoline-Hybrid Electric Vehicles.* Davis, CA: Institute of Transportation Studies, University of California.

IYENGAR, S., and KINDER, D. R. 1989. *News That Matters: Television and American Opinion.* Chicago: University of Chicago Press.

KAHN, M. E. 2007. Do greens drive Hummers or hybrids? Environmental ideology as a determinant of consumer choice. *Journal of Environmental Economics and Management* 54(2): 129–45.

KINDY, K., and LAYTON, L. 2009. Purity of federal 'organic' label is questioned. *Washington Post* 3 July.

KLING, R., OLIN, S., and POSTER, M. (eds.) 1991. *Postsuburban California: The Transformation of Postwar Orange County since World War II*. Berkeley and Los Angeles: University of California Press.

LEONHARDT, D. 2009. A stimulus that could save money. *New York Times* 17 November.

LING, K. 2009. DOE makes $30B available to jumpstart renewable energy, 'smart grid' projects. *New York Times* 30 July.

MARCUSE, H. 1964. *One Dimensional Man: Studies in the Ideology of Advanced Industrial Society*. Boston: Beacon Press.

MAZRIA, E. 2009. Oh, those sexy building codes: More powerful than 100 nuclear plants. *DailyGrist: Top Environmental News from around the World*, <http://www.grist.org>. 23 July.

MUFSON, S. 2009. New lighting standards announced: Changes could save up to $4 billion each year, Energy Department says. *Washington Post* 30 June.

NEWPORT, F. 2009. Americans: Economy takes precedence over environment. <http://www.gallup.com/poll/116962/Americans-Economy-Takes-Precedence-Environment.aspx>. Posted 19 March.

——2010. Americans' global warming concerns continue to drop. <http://www.gallup.com/poll/126560/Americans-Global-Warming-Concerns-Continue-Drop.aspx>. Posted 11 March.

PADGETT, J. P., STEINMANN, A. C., CLARKE, J. H., and VANDENBERGH, M. P. 2008. A comparison of carbon calculators. *Environmental Impact Assessment Review* 28: 106–15.

Pew Research Center for the People & the Press 2009a. Economy, jobs trump all other policy priorities in 2009: Environment, immigration, health care slip down the list. <http://people-press.org/report/485/economy-top-policy-priority>. Posted 22 January.

——2009b. Fewer Americans See Solid Evidence of Global Warming. <http://pewresearch.org/pubs/1386/cap-and-trade-global-warming-opinion>. Posted 22 October.

PIKETTY, T., and SAEZ, E. 2003. Income inequality in the United States, 1913–1998. *Quarterly Journal of Economics* 118(1): 1–39.

PUTNAM, R. D. 2000. *Bowling Alone: The Collapse and Revival of American Community*. New York: Simon & Schuster.

RAHIM, S. 2009. The real estate industry quietly removes a label showing energy use. *New York Times* 4 August.

RAY, J. 2009. In major economies, many see threat from climate change. <http://www.gallup.com/poll/121526/Major-Economic-Threat-Climate-Change.aspx>. Posted 8 July.

REGO, J., STEMPEL, J., and MINTZ, R. 2007. *The Prius Effect: Learning from Toyota*. White Paper. Beverly Hills, CA: Brand Neutral.

ROBERTS, D. 2009. Jobs we can believe in: Merkley wants Senate jobs bill to help finance building efficiency retrofits. *DailyGrist: Top Environmental News from around the World*. <http://www.grist.org>. 21 November.

SAAD, L. 2009. Increased number think global warming is 'exaggerated.' <http://www.gallup.com/poll/116590/Increased-Number-Think-Global-Warming-Exaggerated.aspx>. Posted 11 March.

SAAD, L. 2010. *Americans Firm in Prioritizing Economy Over Environment.* <http://www.gallup.com/poll/126788/Americans-Firm-Prioritizing-Economy-Environment.aspx>. Posted 18 March.

SAEZ, E. 2009. *Striking it Richer: The Evolution of Top Incomes in the United States* (updated with 2007 estimates). <http://elsa.berkeley.edu/~saez/saez-UStopincomes-2007.pdf>.

SHEPPARD, K. 2009. Obama announces new efficiency initiatives as part of big clean energy push. *DailyGrist: Top Environmental News from around the World.* <http://www.grist.org>, 29 June.

SPAARGAREN, G. 2003. Sustainable consumption: A theoretical and environmental policy perspective. *Society and Natural Resources* 16: 687–701.

——and VAN VLIET, B. 2000. Lifestyles, consumption and the environment: The ecological modernisation of domestic consumption. *Environmental Politics*, 9(1): 50–77.

SZASZ, A. 2007. *Shopping our Way to Safety: How We Changed from Protecting the Environment to Protecting Ourselves.* Minneapolis: University of Minnesota Press.

TerraChoice Environmental Marketing. 2009. Greenwashing affects 98% of products including toys, baby products and cosmetics. Press release, 15 April.

VEBLEN, T. 1912. *The Theory of the Leisure Class.* New York: Macmillan.

WATSON, T. 2009. Green claims by marketers go unchecked. *USA Today* 22 June.

World Public Opinion.org. 2009. *Publics Want More Government Action on Climate Change: Global Poll.* <http://www.worldpublicopinion.org>, 29 July.

Yale Project on Climate Change 2009. *Global Warming's Six Americas 2009: An Audience Segmentation Analysis.* <http://environment.yale.edu/climate/>.

PART XI

GLOBAL GOVERNANCE

CHAPTER 41

SELLING CARBON: FROM
INTERNATIONAL CLIMATE
REGIME TO GLOBAL CARBON
MARKET

MATTHEW PATERSON

THIS chapter discusses the principal dynamics in the international climate change negotiations since they started in 1991. It does not provide a detailed history of the negotiations; these can be found elsewhere (Paterson 1996; Bodansky 1993; Depledge 2000, 2005). Instead, it focuses on a set of ongoing tensions and features of the negotiations, which help us to understand most of the specific processes and outcomes from the UNFCCC in 1992 to the latest round of negotiations in Copenhagen in 2009.

The chapter first explores four principal dynamics, namely the relationship between scientific developments and the negotiations, North–South politics, the particularities of the US political situation, and the power of global business. After this discussion, the chapter explores how the international governance of climate change has become oriented around the construction of a series of markets where what is traded is carbon emission rights or credits. It shows how this marketization of climate change in the interstate regime is an outcome to a large extent of the interaction of the main dynamics already outlined.

1 SCIENCE AND POLITICS

Scientific developments have been important in shaping international climate negotiations (see also the chapters by Weart, by von Storch, and by Dunlap and McCright in this volume). First, scientific developments, as organized through the Intergovernmental Panel on Climate Change (IPCC), have shaped the 'limits of the possible' in climate change negotiations. The decision in late 1990 to start negotiations was in part prompted by the first IPCC report issued earlier that year. The IPCC Second Assessment Report, which claimed to identify a 'discernible human impact' (WGI 1995: 11) on the climate, was widely seen also to constrain the discourse of states, such as Saudi Arabia or Australia, that wanted to use remaining uncertainties as a justification for inaction (Kassler and Paterson 1997; Depledge 2008).

Second, scientific developments also provided opportunities for various actors to pursue strategies they favored, and deploy scientific claims instrumentally. Early on, states like the US, Saudi Arabia, and the Soviet Union/Russia used remaining scientific uncertainties to justify inaction. Such strategies still exist—Saudi Arabia in particular, alongside many conservatives in the US, using the 'Climategate' episode, with stolen e-mails from the University of East Anglia and apparent attempts to exclude certain scientists from IPCC processes and manipulate climatic data, to further the claim that the science remains too uncertain as a basis for policy action (Pearce 2010).

Third, climate change science has provided significant framing effects on the negotiations. For example, the market-based discourse has depended on an initial framing about the universality and ubiquity of greenhouse gases (GHG) in the atmosphere. This has enabled a narrative that 'the climate doesn't care where the emissions come from' facilitating in turn projects like emissions trading or carbon offsets. Similarly, the establishment of Global Warming Potentials by the IPCC, which contain significant elements of judgment beyond a simple scientific 'equivalence,'[1] have become the basis both for the general ability of states to focus on one or other GHG in their abatement strategies, but also specifically trading between these gases once a market had been established.

2 North–South Politics

A second prevalent dynamic in international climate negotiations has been that of North–South politics (see also Kartha, this volume). In early negotiations, alongside the question of how industrialized countries were going to limit and reduce their emissions, this was the most prominent, recurring, and fractious issue in the negotiations (Paterson 1996; Rowlands 1995; Gupta 2001). Underlying this issue in the negotiations is the fact that while historic and current emissions are both heavily skewed—with industrialized countries representing around 25 percent of the world's population emitting about 75 percent of current global emissions (and more like 90 percent if historic emissions are counted), the future trajectory of the distribution of emissions is one where, even without conscious policies to reduce them, emissions in industrialized countries are relatively stable, while emissions in developing countries are growing rapidly.

As a consequence, developing country framing of climate change as an issue has been couched either in terms of retributive notions of justice, focusing variously on the historic responsibility of industrialized countries for causing climate change, or in terms of egalitarian accounts of justice, focused on the equal rights of all to emit (resulting in proposals that emissions should converge at common per capita levels). Conversely, in industrialized countries, the framing has been either in pragmatic terms emphasizing the commonality of interests in responding to climate change, or in normative accounts emphasizing that there should be a rough equality of the burden each state shoulders in addressing its emissions (Shue 1992; Vanderheiden 2008; Okereke 2008; Page 2006; see also Baer, this volume).

The term 'common but differentiated responsibilities' arose as a fix for these competing frames, and is core to the objective of the UNFCCC (UN 1992: article 3.1). It sets up a hard distinction between industrialized countries (Annex I countries, in FCCC parlance) and developing countries (non-Annex I countries). The former have obligations to reduce their emissions under both the Convention and the Kyoto Protocol, while the latter do not. But this framing failed to end the conflict. First, domestic actors in the US in particular never accepted the basic argument that developing countries should have no obligations to reduce their emissions, and this was central to the failure to find an agreement at Kyoto that could be ratified in the US. Second, as responses focused on allowing countries 'flexibility' in meeting their commitments (see below), tensions were focused on whether the projects that ensued implied that developing countries themselves had obligations to limit their emissions. Third, the rapid rise in the emissions of a number of large developing countries, notably China and India, started to mean that industrialized countries, and some of the more vulnerable developing countries started to support de facto the US position that at least some developing countries needed to take on commitments to limit their emissions. Those countries started to adapt to this growing pressure, but at the same time to see other opportunities and spin-off benefits of acting to reduce their emissions. Nevertheless, countries like China and India are still insistent that the justification for them acting to limit emissions is by no means equivalent to the obligations industrialized countries have to do so, and the remaining North–South conflict here, crystallized in the US–China conflict, was a key cause of the collapse of negotiations in Copenhagen in December 2009.

3 US Exceptionalism?

A third dynamic has to do with the ongoing particularities of the US. At the base of the problem here is that the US, until 2008, has been the largest single emitter, accounting for around 25 percent of global emissions. So a long-term multilateral agreement without the US is bound to be ineffective, and the short-term competitiveness consequences for many other states of US non-participation risks causing negotiations to fall apart. At the same time, the US is still in effect the only global superpower, able to exert veto power and to act more positively to get other states to bend to its will. This US state power is evident throughout the international climate negotiations. Before 1992, the US was the key hold-out amongst industrialized countries in rejecting a legally binding target to stabilize industrialized country emissions at 1990 levels by 2000. In the Kyoto negotiations, the US used its power in a more positive way, persuading other states to accept a series of institutional mechanisms, known collectively as 'flexibility mechanisms' (emissions trading, joint implementation, and the Clean Development Mechanism) as a condition of the US signing the Protocol (Grubb et al. 1999: ch. 3). During the period when the details of Kyoto were being negotiated, from 1997 to 2002, the rejection of Kyoto by the incoming Bush Jr administration in March 2001 had a decisive effect on the negotiations. Here, it had the opposite effect than that desired by Bush. Condoleeza Rice hubristically declared that 'Kyoto is dead' (quoted in Ochs and Sprinz 2008: 151), but the other participating states, led by the EU, decided to aggressively pursue ratification to bring it into effect. Finally, since the election of Obama and the switch in US strategy back to positive engagement with the climate regime, the negotiations have nevertheless been dominated by the complexity of the problems facing any US administration in getting a climate treaty ratified, as witnessed in the Copenhagen negotiations in 2009 and beyond (Paterson 2009a; Andresen 1991).

The preponderance of the US would be less of a problem but for two particularly awkward features of US domestic politics. First, the US constitution creates significant hurdles to the ratification of treaties. Ratification requires a two-thirds majority in the Senate, and also requires that there is a plan in place to implement the measures required to meet the treaty obligations. Second, the US political system is particularly open to direct lobbying of individual congresspeople (rather than through government ministries or political parties as in many other states), making it relatively easy for blocking coalitions to form to prevent the adoption of implementing measures and thus treaty ratification (Dryzek et al. 2003). This problem has plagued US politics and as a consequence the international negotiations. Key moments here have been the adoption of the Byrd–Hagel resolution in July 1997, where the Senate voted 95 : 0 to reject any treaty that either harmed US economic interests and/or failed to have 'meaningful participation' by developing countries, and the run-up to the Copenhagen negotiations, where Obama's negotiators were heavily constrained by the requirements for ratification in the US, with arguably disastrous consequences for the negotiations as a whole.

4 THE POWER OF BUSINESS

Fourth, the negotiations have been shaped by the power of, and shifting coalitions of, big business. At the outset of negotiations, two organizations formed—the Global Climate Coalition and the Climate Council—to lobby governments on behalf of business interests. The GCC in particular claimed to be 'the voice of business' on climate change, and largely succeeded in this claim until the late 1990s. In practice, the members of these organizations were almost all US-based multinationals, and all from sectors either involved in fossil-fuel production and distribution, or who were heavy energy users (e.g. electricity, steel, automobiles). Its goal was fairly simple, to slow down negotiations and attempt to block action. It did so both at the international negotiations themselves, and within particular countries, especially the strategically crucial US. In the narrow context of the negotiations, it built alliances not only with the US but also with key oil-exporting states like Saudi Arabia and Kuwait. It went as far as supplying negotiating text to those delegations at various points in the negotiations (Gelbspan 1997; Newell 2000: ch. 5; Lacy 2005: ch. 4).

But the ability of the GCC to present itself as the sole voice of business fell apart in the second half of the 1990s (Newell and Paterson 2010: ch. 3). Since then, we have had a much more complex landscape of business interests in the negotiations. There are the 'sunrise' industries of renewables and energy efficiency, which have clear vested interests in emissions reductions policies. There are the insurers, worried about climate change impacts. There are also splits within the fossil-fuel firms, whose departure from the GCC in the late 1990s caused it to collapse in the early 2000s. For example, some car manufacturers (led by Ford in leaving the GCC) increasingly saw their interests in promoting alternative fuels, or at least in hedging their bets to take advantage of climate change regulations. The oil companies Shell and BP led the way in that sector in advocating a more 'constructive' engagement with climate negotiations (Skjærseth and Skodvin 2001). These shifts in business strategy helped to create space in the negotiations to resolve ongoing tensions in the Kyoto Protocol, and in particular to enable the states who had ratified the Protocol to build the political support for implementing the measures necessary to meet its obligations.

5 MARKET MECHANISMS IN CLIMATE NEGOTIATIONS

These four dynamics have all, in differing ways, contributed to a major trend in the international climate regime, that towards the organization of climate governance through the creation of markets in rights to emit GHGs (for details of the policies involved, see Harrison et al., this volume). Behind these specific features of climate negotiations is the broader shift to neoliberalism which favors, for ideological reasons, markets as solutions to environmental problems (Dryzek 2005; Bernstein 2001; see also Hajer and Versteeg, this volume). But at the start of international climate negotiations in early 1991, it was far from obvious that the global response would be constructed around what are now called carbon markets. Marketized climate governance has however its origins in two central issues put on the table by negotiators and commentators at the start of negotiations, which constitute the central discursive frames shaping climate change responses around market creation.

The first discourse was that of flexibility. Many industrialized countries, led by Norway and the US, argued that they would need to have flexibility in how they met their commitments to reduce emissions. The principal product of this demand for flexibility was what would become known as 'joint implementation.' In the UNFCCC, this became embodied in the phrase that Annex I countries could meet their emissions reductions goals 'individually or jointly' (UN 1992: article 4.2(b)). The demand for flexibility also meant the UNFCCC focused on a batch of GHGs rather than just CO_2, and to address both emissions and sinks, both crucial to the later construction of carbon markets.

There was much resistance in the run-up to COP-3 in Kyoto in 1997 to the idea of 'joint implementation,' especially from developing countries worried it was a backdoor way of imposing commitments on them. Nevertheless, US pressure succeeded in making sure a mechanism for joint implementation was included in the Kyoto Protocol. The US leapt on a proposal by Brazil for a mechanism that would penalize industrialized countries that failed to meet their targets by requiring them to pay into a compensating mechanism that would invest in 'clean development' in developing countries. The US turned this stick into a carrot; instead of penalizing countries that failed to meet their targets, it would enable investment in the South as a means of meeting those targets. This became the basis for the Clean Development Mechanism (CDM) in the Kyoto Protocol (UNFCCC 1998: article 12), dubbed at the time as 'Kyoto's surprise' (Werksman 1998).[2]

Alongside a discourse of flexibility was one of efficiency. For neoclassical economists, proposed policy solutions should be efficient—that is, they should equalize the marginal costs of abatement for all actors, and thus minimize the overall costs of meeting specified emissions reductions goals. Michael Grubb was the first to apply this logic in the climate change case (Grubb 1989; UNCTAD 1992). In 1996, the US proposed emissions trading

(ET) formally, and over the next year in the run-up to Kyoto, negotiated progressively harder to make sure it was included. Other states gradually backed down and accepted the US's proposals as in effect the condition of US signature of the Kyoto Protocol, and ET became accepted as a part of that agreement (UNFCCC 1998: article 6).

The Kyoto architecture did some key things that have become the infrastructural basis for the carbon markets that have followed.[3] First, in order to make the emissions reductions obligations tradable, it created a unit of account, known as the Assigned Amount Unit and equivalent to a tonne of carbon dioxide equivalent, or tCO_2e.[4] So a target to reduce emissions by say 6 percent (Canada's obligation) becomes an allocation of 94 percent of Canada's 1990 levels, and then a specific amount of tCO_2e, enabling them to be unbundled and traded. The credits created by Joint Implementation (Emissions Reduction Units) or the CDM (Certified Emissions Reductions) are then made equivalent with the Assigned Amount Unit so they can be traded for each other.

Second, the Kyoto system created a means of tracking emissions trades in order to regulate the trade in these various units. This involves a transaction log, where all trades are to be registered, and a registry, where ownership of all allowances is to be recorded. These systems have been borrowed and developed in other carbon markets.

Third, the Kyoto system created a complex arrangement for approving the creation of its principal credits. With an Assigned Amount Unit, it is a relatively straightforward task of measuring emissions in 1990 and in the commitment period—2008–12—and comparing the two. But creating a Certified Emissions Reduction unit is much more complex, involving not only direct measurement of emissions but making claims about a set of counterfactuals. In the CDM, Annex B countries invest in individual projects in the developing world.[5] The claim is that the project reduces emissions in that country compared to what they would have been if the investment had not gone ahead. This involves making baseline assessments of projected emissions without the project, and comparing actual emissions with the project to this estimated baseline. Even for simple choices like replacing coal with wind power, this is not obvious, and for much more uncertain projects like in forestry, this is an extremely difficult claim to sustain. The CDM system has thus spent considerable effort developing systems for verifying the claims made by project developers.

The complicated methodological questions here are the basis for considerable controversy over the CDM. Critics claim that these methodological problems are insurmountable and that all carbon offset systems including the CDM are plagued by 'carbon fraud' (Bachram 2004; Lohmann 2006; Smith 2007; Böhm and Dabhi 2009). Conversely, those involved in developing projects complain routinely that the CDM system is overly bureaucratic and creates major delays for project developers that threaten the viability of the market (Dornau 2008; Michaelowa and Jotzo 2005).[6]

The organization of the Kyoto Protocol around these market mechanisms can be understood as the interaction between the four principal dynamics outlined earlier. In the background, the science-politics interface has been important in creating the infrastructural basis for carbon markets. It established the basic claim underlying the market logic, that from a climate point of view it doesn't matter where geographically the emissions are reduced. Beyond this, the IPCC created the Global Warming Potential (GWP) as a way to compare the climate impacts of different GHGs. The GWP initially could be used by states to create flexibility in how they meet emissions reductions goals, but it is also now the basis for trading between different gases in carbon markets. The emphasis of both scientific developments and the international regime on measurement of emissions and sinks, and the development of this measurement and reporting capacity, has similarly been crucial in enabling carbon markets, which need extremely precise measurement of emissions, to emerge.

The role of the other three core dynamics in producing the marketization of climate governance is more immediately visible. The North–South dynamic has been important in creating the elaborate market mechanism in the CDM. As industrialized states have sought to reduce and externalize the costs of emissions reductions, developing countries have first refused to take on commitments to reduce their own emissions, and then been increasingly enthusiastic about using the CDM to attract investment. The particular power of the US, combined with its domestic political dynamics—the power of business and the resistance of US business to emissions reductions, as well as the particular dominance of market ideology in the US—produced the strong push for market mechanisms in the run-up to the Kyoto negotiations, and the sense amongst other countries that they had to accede to US demands. Finally, the power of business helped to intensify the search for relatively inexpensive emissions reductions, to accommodate business sectors hostile to those reductions, and increasingly, the marketization was driven by the realization of the various market opportunities for a whole range of actors, from finance, to project development, to new energy technologies, created by the emergence of carbon markets themselves.

6 THE FLOURISHING OF CARBON MARKETS

These dynamics favoring marketized climate governance were well established in the international negotiations by the late 1990s. Once set up, carbon markets have rapidly proliferated, and have until recently been in a phase of very rapid expansion. As they have done so, the business-politics dynamic has become the most important one in driving forward their development.

Many jurisdictions have developed ET schemes. And following the lead of the EU, most such ET schemes have mechanisms that enable firms regulated to purchase offsets to count against their obligations to reduce their own emissions. As a result, ET based on an efficiency logic and carbon offset markets based on a flexibility discourse have become closely linked.

First, and still the largest carbon market in operation, was the system in the EU (Skjærseth and Wettestad 2008; Ellerman et al. 2010; see also Jordan et al., this volume). The rapid developments in the EU were largely a surprise since in the Kyoto negotiations the EU had opposed the use of emissions trading. A number of things combined to cause a shift in the EU's position, notably a change in personnel in the key directorate, and the realization that ET only needed a qualified majority to get through rather than the unanimity required for a carbon tax. The EU Commission developed the proposal for an EU ETS between 1998 and 2003, and the first phase of the system came into operation in 2005.

The EU ETS regulates around 11,000 installations amounting to around 45 percent of EU emissions. These installations are required to hold allowances equivalent to their emissions.

An overall target is set for all the installations covered, and then the distribution of this target has been negotiated among EU member states. To date, the member states then allocate allowances to firms within their jurisdiction, although in Phase III of the scheme (2013–20) the Commission will directly allocate the allowances to firms. Firms may then meet their obligations by reducing their own emissions, by buying extra allowances from other firms that have been able to reduce their emissions further than their prescribed limit, releasing allowances into the market. And in 2004, the EU issued a Linking Directive, which enabled firms to purchase credits issued by the Kyoto system, and to hold them against their own emissions reductions obligations.

The EU ETS, and the demand in the CDM it drives, accounted in 2008 for over 80 percent of global carbon market trades (Capoor and Ambrosi 2009: 33). But since then, many other jurisdictions have developed ET systems. Some are now in operation, while others are expected to come into effect in the coming years. Three different groups can be identified.

First are those developed at subnational levels in Australia, Canada, and the US. In the US at least, this process was part of a broader move by states to take action in the absence of federal leadership (Rabe 2004; Selin and VanDeveer 2009). Those already in operation are the systems in New South Wales in Australia, whose system came into effect in 2003; the Regional Greenhouse Gas Initiative (RGGI), a group of ten states in the northeastern US that issued its first allowances in September 2008; and a carbon offset system in place in Alberta, Canada. Other jurisdictions in the US and Canada are planning to institute systems at subfederal levels.

Second is the more recent momentum for emissions trading at national levels. ET schemes have been going through legislation recently in Canada, the US, Australia, New Zealand, and Japan. While not all of these processes are sure of success, there is a good chance that practically the whole of the OECD will be covered by an ET system in the near future. Indeed, where these policies are still contested (as in the US and Australia) the contestation is less about the use of markets in the design of climate policy, but more a question of a reinvigorated opposition to climate change action per se.

The third site of ET development is the Chicago Climate Exchange (CCX). This is a nonstate initiative, whose members are private companies. Most were initially in the US with a few in Canada, but more recently firms have joined from other industrialized countries and also now in China and India. To join, the firms sign up to an emissions reductions commitment. Firms join for a mixture of reasons, either in relation to gaining green PR advantages, or as learning exercises in preparation for the emergence of regulatory markets. At the moment this means their emissions being 6 percent below 'baseline' (average emissions between 1998 and 2001) by 2010.[7] As in other systems, the firms can then do so by reducing their own emissions, trading amongst themselves, or buying in offsets.

7 EXPLAINING THE TAKE-OFF OF CARBON MARKET POLICIES

ET systems are usually discussed in terms of their pros and cons—their efficiency, effectiveness in reducing emissions, and so on (Skjærseth and Wettestad 2008; Michaelowa and

Jotzo 2005; Aldy and Stavins 2007). Rarely is the question addressed of why they have managed to take off so widely. While other factors can be identified, I will focus here on the dynamic relationship between policy makers and business interests in favoring ET.[8]

The gradual realization of this dynamic can be seen in the way that the growth of policy makers' interest in ET and the growth of business activity around carbon markets co-evolved. The key period of take-off in this dynamic was 1997–2000, where many actors shifted considerably from skepticism, if not hostility, towards carbon markets, to positions of cautious acceptance and increasing enthusiasm. Various processes can be outlined. First was the establishment of a number of start-up companies engaged in carbon trading. Second, a number of major banks established carbon-trading offices. Others bought up some of the early start-up companies. Third was the establishment of a variety of organizations supporting the carbon market. Key here was the lobby organization the International Emissions Trading Association (IETA) and the news organization Point Carbon. Fourth was the organization of regular, ongoing sites of debate about carbon markets. Currently, the principal one is the Carbon Expo conference organized annually by IETA. Fifth was the ongoing development of governmental and intergovernmental activity, notably the dramatic switch in approach in the EU from 1998 onwards. Also important is the entrepreneurship of various members of the UNFCCC Secretariat in developing the Kyoto mechanisms into a workable scheme (Depledge 2005: ch. 6).

During this period therefore a significant part of finance capital, especially that based in London, started to realize the potential for creating cycles of investment and accumulation of capital around carbon markets, and build business strategies around them. At the same time, they started to become part of a coalition of forces with an interest in continued development of carbon markets and by extension climate policy, both at national and international levels. As a consequence, the balance of political forces promoting and opposing climate policy shifted. The capacity of the fossil-fuel coalition represented by the Global Climate Coalition could no longer claim effectively to represent the interests of all business, and policy makers began to see how they might forge a coalition of forces that could support more ambitious emissions reduction goals. As the implementation of ET schemes became more certain, more firms developed business strategies around them, giving government officials more confidence about support for these policies.

8 CONCLUSIONS

This chapter has tried to demonstrate that a dominant element in political responses to climate change has been the creation of a series of carbon markets. It has also tried to show that this marketization of governance (Newell 2008) has occurred principally because of a structural feature of carbon markets as responses—that they create concentrated, immediate benefits for powerful actors—and the political dynamic which has resulted from this structural feature. Three things are worth elaborating by way of conclusion.

One is to re-emphasize that while carbon markets have expanded rapidly and continue to do so, this should not be understood as a simple story of 'roll-out neoliberalism' (Peck and Tickell 2002; Heynen et al. 2007). Rather, they have been contested from the outset and have been shaped in part through the character of those contests. Two specific examples illustrate this. First, the CDM largely excludes forests in part because of concerted lobbying by environmental NGOs during the period that the detailed rules for the CDM were being agreed (1998–2002). There are now pressures to reverse this. Most industrialized countries, and a number of developing countries, now seem intent on including forestry, now reframed in particularly problematic ways as 'Reduced Emissions from Deforestation and Forest Degradation in Developing Countries', or REDD, in the CDM or some similar international offsets mechanism. But the ongoing contestation of the role of forests in carbon markets will be an enduring feature of this politics. Second, the development of nonstate certification systems for the carbon offset markets has also shaped how those markets work (Paterson 2009b; Newell and Paterson 2010: ch. 7). At the Copenhagen COP in December 2010, a substantial part of the object of the many protests organized there was to resist carbon markets as a policy mechanism. Its effect on the negotiations themselves was probably minimal, but if such protests are maintained, its effect on the type of carbon markets that develop, and how they are governed, is likely to be important.

The second point is about the hiatus we currently find ourselves in, given the shift in strategy of both the US and more recently China. Up to COP-15 in Copenhagen, most observers assumed a model of carbon markets that would continue as a multilateral bargain about the 'grand design:' a single unit of account, the tCO_2e, to which all individual carbon markets would be designed; and a multilateral mechanism like the CDM which could be a means of integrating markets worldwide.

But the pressure for a 'new architecture' for the international climate regime (Aldy and Stavins 2007) coming from the US in the early 2000s, with their denouement in Copenhagen, now make that outcome very considerably in doubt. It is now just as possible that we will have a patchwork of carbon markets, where each jurisdiction sets its own rules and where linkage between carbon markets is highly doubtful. In side-events at Copenhagen, for example, there were many debates that focused on precisely what you need to be able to link two markets to each other. Do you need a single unit of account? Do you need emissions reductions targets and thus aggregate carbon prices to be broadly similar? Do you need consistent registry and transaction log systems? Do you need similar rules about how much access to offsets firms have? And so on. In emerging carbon markets that lack a clear multilateral architecture, these questions will dominate, and most of the interventions seemed to imply that linkage would be very limited in this situation.

The third concluding point to make is that as markets have become central to broad social responses to climate change, it is less and less useful to characterize climate politics as solely about states. Many of the novel forms of global climate governance discussed by Biermann (this volume) and others (Jagers and Stripple 2003; Andonova et al. 2009; Okereke et al. 2009; Hoffmann forthcoming) have been driven by the marketization of governance. This has occurred as private actors have sought to develop private governance mechanisms to govern carbon markets, for example through the development of certification schemes, registries, or exchanges (Lövbrand and Stripple 2010; Hoffmann forthcoming), or have sought to develop parallel governance of related practices such as investment practices, as in the Carbon Disclosure Project (Newell and Paterson 2010: ch. 4). So even if the events of Copenhagen do create a challenge for climate change governance, this does not undermine the overall trend; that international climate politics is organized around the construction of a series of markets in emissions rights.

NOTES

1. This is principally to do with the time period that different GHGs persist in the atmosphere, and thus judgments about which timeframe matters.
2. Kyoto also contains another similar mechanism, known as Joint Implementation. This enables investment by an industrialized country (confusingly known as Annex B countries in Kyoto, but the same group as Annex I countries under the UNFCCC)—in another Annex B country, in order to claim credit against its own emissions reductions obligations. The CDM allows such investment by Annex B countries in non-annex B countries.
3. The fullest description of these technical details can be found in Yamin and Depledge (2004).
4. The Kyoto Protocol regulates a basket of six greenhouse gases—carbon dioxide, methane, nitrous oxide, hydrofluorocarbons, perfluorocarbons, and sulfur hexafluoride. The system allocates an equivalence of each to the principal GHG, CO_2, in order that they can be made commensurable. CFCs and HCFCs are excluded as these are covered under the Montreal Protocol on Substances that Deplete the Ozone Layer.
5. For accounts of the CDM, see for example Streck (2004) or Lövbrand et al. (2009).
6. For an analysis focusing on this contested legitimacy of such markets, see Paterson (2010).
7. See the description of the commitment at CCX's website at <http://www.chicagoclimatex.com/content.jsf?id=72>, accessed 11 January 2010.
8. These political advantages can also be understood in terms of the distribution of costs and benefits. Given that climate change is a quintessential public goods problem on a global scale, the benefits of climate change abatement are diffuse, experienced across the global population. However those benefits are both distributed over time and largely invisible to the beneficiaries. Moreover, the chain of causality from emissions reductions to climate impact benefits is too long to be able to discern any clear relationship between the two. By contrast, reducing emissions places immediate costs that are focused on particular actors. In this context, carbon markets are useful politically as they create concentrated and immediate benefits for a range of actors involved in the markets that result. This fits well with a typology of public policy problems first developed by James Wilson (1974: 331–7), although he sees the combination of 'distributed costs' and 'concentrated benefits' as generating a politics of resentment of 'special interests,' a charge which is starting to be made about the financial interests that benefit from carbon markets but which has not yet dominated the overall process.

REFERENCES

ALDY, J., and STAVINS, R. 2007. *Architectures for Agreement: Addressing Global Climate Change in the Post-Kyoto World*. Cambridge: Cambridge University Press.
ANDONOVA, L., BETSILL, M., and BULKELEY, H. 2009. Transnational climate change governance. *Global Environmental Politics* 9: 52–73.
ANDRESEN, S. 1991. US greenhouse policy: Reactionary or realistic? *International Challenges* 11: 17–24.
BACHRAM, H. 2004. Climate fraud and carbon colonialism: The new trade in greenhouse gases. *Capitalism, Nature, Socialism* 4.

BERNSTEIN, S. 2001. *The Compromise of Liberal Environmentalism.* New York: Columbia University Press.

BODANSKY, D. 1993. The United Nations Framework Convention on Climate Change: A commentary. *Yale Journal of International Law* 18: 451–558.

BÖHM, S., and DABHI, S. (eds.) 2009. *Upsetting the Offset: The Political Economy of Carbon Markets.* London: Mayfly Books.

CAPOOR, K., and AMBROSI, P. 2009. *State and Trends of the Carbon Market 2009.* Washington, DC: World Bank.

DEPLEDGE, J. 2000. *Tracing the Origins of the Kyoto Protocol: An Article-by-Article Textual History.* Technical Paper, prepared for the UNFCCC Secretariat . . . Document FCCC/TP/2000/2. Available at: <http://unfccc.int/resource/docs/tp/tp0200.htm>, accessed 29 June 2009.

—— 2005. *The Organization of Global Negotiations: Constructing the Climate Regime.* London: Earthscan.

—— 2008. Striving for no: Saudi Arabia in the climate change regime. *Global Environmental Politics* 8: 9–35.

DORNAU, R. 2008. Defending the integrity of the CDM. Pp. 77–81 in K. Carnahan (ed.), *Greenhouse Gas Market 2008: Piecing Together a Comprehensive International Agreement for a Truly Global Carbon Market.* Geneva: International Emissions Trading Association.

DRYZEK, J. 2005. *The Politics of the Earth: Environmental Discourses.* 2nd edn. Oxford: Oxford University Press.

—— DOWNES, D., HUNOLD, C., SCHLOSBERG, D., and HERNES, H.-K. 2003. *Green States and Social Movements: Environmentalism in the United States, United Kingdom, Germany, and Norway.* Oxford: Oxford University Press.

ELLERMAN, A. D., CONVERY, F. J., and DE PERTHUIS, C. 2010. *Pricing Carbon: The European Emissions Trading Scheme.* Cambridge: Cambridge University Press.

GELBSPAN, R. 1997. *The Heat is On: The High Stakes over Earth's Threatened Climate.* Reading, MA: Addison-Wesley.

GRUBB, M. 1989. *The Greenhouse Effect: Negotiating Targets.* London: Royal Institute of International Affairs.

—— BRACK, D., and VROLIJK, C. 1999. *The Kyoto Protocol: A Guide and Assessment.* London: Earthscan.

GUPTA, J. 2001. *Our Simmering Planet: What to Do about Global Warming?* London: Zed Books.

HEYNEN, N., McCARTHY, J., PRUDHAM, S., and ROBBINS, P. 2007. Introduction: False promises. Pp. 1–21 in N. Heynen, J. McCarthy, S. Prudham, and P. Robbins (eds.), *Neoliberal Environments: False Promises and Unnatural Consequences.* London: Routledge.

HOFFMANN, M. J. forthcoming. *Into the Void: Experimenting with Climate Governance after Kyoto.* New York: Oxford University Press.

JAGERS, S., and STRIPPLE, J. 2003. Climate governance beyond the state: Contributions from the insurance industry. *Global Governance* 9: 385–99.

KASSLER, P., and PATERSON, M. 1997. *Energy Exporters and Climate Change Politics.* London: Royal Institute of International Affairs.

LACY, M. 2005. *Security and Climate Change: International Relations and the Limits of Realism.* London: Routledge.

LOHMANN, L. 2006. Carbon trading: A critical conversation on climate change, privatization and power. *Development Dialogue.*

Lövbrand, E., and Stripple, J. 2010. Carbon market governance beyond the public–private divide. In F. Biermann, P. Pattberg, and F. Zelli (eds.), *Global Climate Governance Post 2012: Architectures, Agency and Adaptation*. Cambridge: Cambridge University Press.

—— Rindefjall, T., and Nordqvist, J. 2009. Closing the legitimacy gap in global environmental governance? Lessons from the emerging CDM market. *Global Environmental Politics* 9: 74–100.

Michaelowa, A., and Jotzo, F. 2005. Transaction costs, institutional rigidities and the size of the clean development mechanism. *Energy Policy* 33: 511–23.

Newell, P. 2000. *Climate for Change: Non-State Actors and the Global Politics of the Greenhouse*. Cambridge: Cambridge University Press.

—— 2008. The marketization of global environmental governance: Manifestations and implications. Pp. 77–95 in J. Park, K. Conca, and M. Finger (eds.), *The Crisis of Global Environmental Governance*. London: Routledge.

—— and Paterson, M. 2010. *Climate Capitalism: Global Warming and the Transformation of the Global Economy*. Cambridge: Cambridge University Press.

Ochs, A., and Sprinz, D. 2008. Europa riding the hegemon? Transatlantic climate policy relations. In D. B. Bobrow and W. Keller (eds.), *Hegemony Constraint: Evasion, Modification, and Resistance to American Foreign Policy*. Pittsburgh: University of Pittsburgh Press.

Okereke, C. 2008. *Global Justice and Neoliberal Environmental Governance*. London: Routledge.

—— Bulkeley, H., and Schroeder, H. 2009. Conceptualizing climate change governance: beyond the international regime. *Global Environmental Politics* 9: 56–76.

Page, E. 2006. *Climate Change, Justice, and Future Generations*. Cheltenham: Edward Elgar.

Paterson, M. 1996. *Global Warming and Global Politics*. London: Routledge.

—— 2009a. Post-hegemonic climate politics? *British Journal of Politics and International Relations* 11: 140–58.

—— 2009b. Resistance makes carbon markets. Pp. 244–54 in S. Böhm and S. Dabhi (eds.), *Upsetting the Offset: The Political Economy of Carbon Markets*. London: Mayfly Books.

—— 2010. Legitimation and accumulation in climate change governance. *New Political Economy* 15.

Pearce, F. 2010. 'Climategate' was PR disaster that could bring healthy reform of peer review. *The Guardian*, 9 February, online edition, at <http://www.guardian.co.uk/environment/2010/feb/09/climate-emails-pr-disaster-peer-review>, accessed 29 March 2010.

Peck, J., and Tickell, A. 2002. Neoliberalizing space. *Antipode* 34: 380–404.

Rabe, B. 2004. *Statehouse and Greenhouse*. Washington, DC: Brookings Institute Press.

Rowlands, I. 1995. *The Politics of Global Atmospheric Change*. Manchester: Manchester University Press.

Selin, H., and VanDeveer, S. D. (eds.) 2009. *Changing Climates in North American Politics Institutions, Policymaking, and Multilevel Governance*. Cambridge, MA: MIT Press.

Shue, H. 1992. The unavoidability of justice. In A. Hurrell and B. Kingsbury (eds.), *The International Politics of the Environment*. Oxford: Oxford University Press.

Skjærseth, J. B., and Skodvin, T. 2001. Climate change and the oil industry: Common problems, different strategies. *Global Environmental Politics* 1: 43–64.

—— and Wettestad, J. 2008. *EU Emissions Trading: Initiation, Decision-Making and Implementation*. Aldershot: Ashgate.

Smith, K. 2007. *The Carbon Neutral Myth: Offset Indulgences for your Climate Sins*. Amsterdam: Carbon Trade Watch.

STRECK, C. 2004. New partnerships in global environmental policy: The Clean Development Mechanism. *Journal of Environment and Development* 13: 295–322.

UN 1992. *Framework Convention on Climate Change*. New York: United Nations.

UNCTAD (ed.) 1992. *Combating Global Warming: Study on a Global System of Tradable Carbon Emission Entitlements*. New York: United Nations Conference on Trade and Development.

UNFCCC. 1998. *Kyoto Protocol to the United Nations Framework Convention on Climate Change*. United Nations. Available from <http://unfccc.int/resource/docs/convkp/kpeng. pdf>.

VANDERHEIDEN, S. 2008. *Atmospheric Justice: A Political Theory of Climate Change*. New York: Oxford University Press.

WERKSMAN, J. 1998. The clean development mechanism: Unwrapping the 'Kyoto surprise.' *Review of European Community and International Environmental Law* 72: 147–58.

WGI, IPCC 1995. *Climate Change 1995: The Science of Climate Change: Contribution of Working Group 1 to the Second Assessment Report of the Intergovernmental Panel of Climate Change*. Cambridge: Cambridge University Press.

WILSON, J. 1974. *Political Organizations*. New York: Basic Books.

YAMIN, F., and DEPLEDGE, J. 2004. *The International Climate Change Regime: A Guide to Rules, Institutions and Procedures*. Cambridge: Cambridge University Press.

CHAPTER 42

IMPROVING THE PERFORMANCE OF THE CLIMATE REGIME: INSIGHTS FROM REGIME ANALYSIS

ORAN R. YOUNG

1 INTRODUCTION

WHAT can we learn from studying the performance of multilateral environmental agreements (MEAs) and other environmental regimes that will help us either to strengthen the existing climate regime or, alternatively, to terminate this arrangement and replace it with a more effective regime? No two problems and no two regimes are alike; we must be careful about applying 'insights' gleaned from the study of regimes in general to the specific case of climate change. Still, we are accumulating a significant body of knowledge about the determinants of effectiveness in MEAs and other regimes. Because the climate regime in its current form does not have the capacity to solve the problem of climate change and because the problem itself is becoming more acute with the passage of time, there is a sense of urgency about the need to put in place a more effective governance system to address this problem. So long as we proceed with care and add a note of caution to any recommendations we advance regarding this challenge, then, it makes sense to approach the climate regime as a member of the universe of environmental regimes and to apply findings derived from analyses of this universe to the case of climate.

This article addresses this issue in three steps. The second section examines the effectiveness of the climate regime itself. It applies insights from regime analysis to explain the weak performance of this regime in its current form, discusses briefly what it would take to create a more effective regime to tackle the problem of climate change, and asks whether recent developments in the climate regime are heading in the right direction. The third section turns to interactions between the climate regime and other relevant regimes or what regime analysis knows as the problem of interplay. It focuses on matters of architecture and explores the question of whether the existing climate regime is structured in such a way as to promote synergy rather than conflict in its interactions with other regimes. The final section then turns to the implications of the arguments articulated in the preceding sections for climate policy. The main message is that we should be thinking about restructuring the current regime rather than tinkering with this arrangement at the margins.

2 HOW CAN WE IMPROVE THE PERFORMANCE OF THE CLIMATE REGIME?

Much of regime analysis directs attention to what has become known as the problem of fit (Galaz et al. 2008). Rather than endeavoring to devise some sort of generic indicator that allows us to rank environmental problems from benign to malign or from tractable to wicked, the key observation here is that we need to focus on devising regimes that are well matched to the characteristics of the problems they are designed to solve (Ostrom 2007; Young 2008). Efforts to solve comparatively easy problems can go astray and produce disappointing results if policy makers create inappropriate regimes to deal with them. It is possible to make significant progress in dealing with seemingly hard problems (e.g. transboundary air pollution) when appropriate institutional arrangements are put in place. Nevertheless, regime analysis has yielded some general conclusions about require-ments for success in solving international environmental problems (Young 1999; Mitchell 2008; Underdal 2008). What can these conclusions tell us about the weaknesses of the climate regime in its current form and the changes that would be needed to improve its performance during the next phase?

The core of the climate regime is embedded in the provisions of the United Nations Framework Convention on Climate Change (UNFCCC) and the Kyoto Protocol (KP) along with several related arrangements, such as the Global Environment Facility, estab-lished to help with the implementation of these provisions. The Framework Convention, which was adopted in 1992 and entered into force in 1994, articulates the goal of the regime as the 'stabilization of greenhouse gas concentrations in the atmosphere at a level that would prevent dangerous anthropogenic interference with the climate system' (Art. 2).[1] It draws a distinction between Annex I or developed countries and non-Annex I or develop-ing countries, calls on Annex I countries to adopt measures to reduce greenhouse gas (GHG) emissions to 1990 levels by 2000, establishes several subsidiary bodies (e.g. the Subsidiary Body for Scientific and Technological Advice or SBSTA), and requests the parties to make available national inventories of GHG emissions on a regular basis. The KP, which was adopted in 1997 and entered into force in 2005, commits Annex I countries to reduce GHG emissions by an overall average of 5.2 percent by the end of the first commitment period running from 2008 through 2012.[2] Several flexibility mechan-isms—emissions trading, joint implementation, and the Clean Development Mechanism or CDM—have been developed under the KP to assist Annex I countries in fulfilling their emissions reduction obligations and, in the case of the CDM, to encourage developing countries to move toward more climate-friendly forms of economic development.

The first general insight from regime analysis that seems important in this connection is that it helps to articulate common goals in forms that are easy to grasp and straightforward to use in assessing regime performance. Examples include phasing out the production and consumption of particular ozone-depleting substances (ODSs) by specified dates, maintaining Antarctica in a completely demilitarized state, or making the Rhine River safe for salmon. The goal of the climate regime, however, is hard to understand in any concrete sense and even harder to use in judging whether we are making progress toward solving the problem of climate change. Many participants have made efforts to translate the goal of avoiding dangerous anthropogenic interference with the climate system into terms that are measurable. Popular formulas include avoiding a doubling of preindustrial concentrations of carbon dioxide in the Earth's atmosphere or keeping increases in surface temperatures below 2 °C. But these are arbitrary formulas. We do not know whether fulfilling these objectives would suffice to prevent serious disturbances of the climate system. Growing numbers of scientists now take the view that substantially lower concentrations are desirable (e.g. limiting CO_2 in the atmosphere to 350 ppm), and the idea of limiting temperature increases to 1.5 °C became a powerful rallying cry at the Copenhagen climate meeting in December 2009 (Hansen 2009). But such adjustments do little to solve the basic problem. The climate regime provides no straightforward way to calculate whether or not we are making progress in addressing this problem. The absence of a measurable objective invites confusion and opens up opportunities for manipulative activities on the part of those who do not support the climate regime and desire to block progress in this realm, though they may find it politically incorrect to say so openly.

Regime analysis also suggests that real success is virtually impossible in the absence of the development of principles of fairness that are satisfactory to all the major parties or groupings of parties to a regime. Much of the debate about the climate regime has been cast in terms of calculations of benefits and costs. Are the costs attributable to climate change likely to exceed the costs of taking action to solve this problem (Stern 2009)? Which policy instruments are likely to prove most efficient in the sense that they minimize the costs to society of reducing GHG emissions (Aldy and Stavins 2007)? But whereas calculations of the relevant costs and benefits quickly become esoteric, issues of fairness and equity arise in direct, blunt, and inescapable terms. The problem here is that the climate regime is not based on principles of fairness that are broadly acceptable to all the major players. The non-Annex I countries (largely the G-77 plus China) assert that the OECD countries caused the problem, that these countries must take decisive steps toward solving it in a manner that will not thwart the economic objectives of the developing world, and that adequate funding mechanisms addressing adaptation as well as mitigation will be required to ensure meaningful participation on the part of victims of climate change. The United States and some other Annex I countries, by contrast, maintain that the terms of the KP are unfair because they provide developing countries with inappropriate trade advantages and because the GHG emissions of leading developing countries are growing rapidly and will soon account for the majority of global emissions.[3] The lesson of regime analysis in this realm is straightforward. With regard to a problem like climate change, where more or less universal participation is needed to achieve a real solution, a widely shared sense that the basic deal embedded in the regime is fair is essential to success. All the technical sophistication in the world cannot make up for the absence of a shared sense of fairness. Conversely, all sorts of useful experiments with technically sophisticated mechanisms become feasible in a setting in which the major players are committed to making a good faith effort to solve the problem.

Simple solutions may suffice to solve simple problems. The commitment to demilitarize Antarctica under the terms of the 1959 Antarctic Treaty does not require complex institutional arrangements. But complex problems like climate change give rise to a different story. Regime analysis suggests that it is hard to solve complex problems featuring a number of tightly connected components on a piecemeal basis. In the case of climate change, this would mean developing a package of provisions addressing, at a minimum, mitigation, adaptation, and the financial and technological mechanisms needed both to bring non-Annex I countries into the fold with regard to measures to reduce GHG emissions and to address the challenge of adaptation for those experiencing the impacts of climate change through no fault of their own (e.g. small island developing states). There is ample room for debate regarding the pros and cons of different ways to assemble such a package. There is a world of difference, for example, between strategies emphasizing major financial transfers from developed to developing countries as envisioned in the December 2009 Copenhagen Accord and those based on ideas like distributing emissions rights or permits globally on a per capita basis. Yet the basic point is clear. The regime articulated in the UNFCCC/KP is lopsided. It places great emphasis on mitigation but has relatively little to say about adaptation. Even with regard to mitigation, the provisions of the UNFCCC and the KP are notable for a tendency to address the issue of emissions reductions at the margin, focusing on modest reductions from current levels on the part of major emitters rather than setting overall emissions levels and confronting the question of how to allocate emissions permits on a global scale.

It will come as no surprise that regime analysis also makes it clear that implementation is crucial. Of course, much of the responsibility for implementing the terms of international agreements like the UNFCCC and the KP rests with the individual parties. But several international concerns are also important in this realm. One has to do with what has become known as the issue of monitoring, reporting, and verification (MRV) in current discussions regarding strengthening the climate regime. In a regime that requires ongoing actions to fulfill obligations (e.g. meeting emissions reduction targets), there is no substitute for devising tracking systems that generate a sense of confidence regarding actual performance on the part of all parties to the agreements. Similar concerns arise with regard to the problem of 'leakage' or the displacement of production facilities from one country to another to avoid emissions restrictions and to the administration of arrangements like the CDM to minimize incentives for parties to manipulate the system to claim credit for actions that they would have taken even in the absence of the CDM or that they are initiating solely to obtain CDM credits. The current climate regime leaves a lot to be desired in these terms. Admittedly, implementation is a difficult challenge with respect to matters of public policy in all settings. But regime analysis makes it clear that implementation measures well designed to fit the characteristics of the problem at hand constitute an important determinant of success regarding arrangements like the climate regime.

Finally, regime analysis points to the need for an effective climate regime to be both flexible and nimble. In addressing problems involving complex and dynamic systems, it is essential to recognize the significance of changes that are nonlinear, often abrupt, and frequently irreversible. The climate problem is almost certainly the most dramatic exemplar of large-scale environmental concerns raising issues of this sort. Given the character of the Earth's climate system, we cannot hope to address the problem effectively without creating a regime that is easy to adjust and capable of responding quickly to major changes in the nature of the problem or the character of our understanding of it. Even at the national level, it is extremely difficult to adjust systems for managing human-environment relations in a prompt and timely manner. The negotiations required under the terms of the existing climate regime even to make modest adjustments of targets and timetables or to fine tune the flexibility mechanisms invariably give rise to hard bargaining that is time consuming at best and that can easily lead the parties to settle for incoherent or inappropriate institutional compromises which are then subject to the vagaries of ratification processes. Are there potential solutions for this problem in the case of the climate regime? Following the lead of the ozone regime in allowing for some adjustments without triggering a requirement for ratification on the part of the parties is one interesting possibility. Encouraging member states to experiment with a variety of policy instruments designed to bring about emissions reductions in a cost-effective manner is another. Launching a series of pilot projects involving different approaches to funding adaptation in areas impacted by climate change is still another. There is a huge need for innovation in this realm. A prerequisite for success is recognizing that problems involving complex and dynamic systems pose special challenges and that finding ways to respond to shifting circumstances that circumvent time-consuming and convoluted processes is essential in building effective regimes to address problems of this sort.

This analysis makes it clear that the regime established under the provisions of the UNFCCC and the KP will require major restructuring in order to prove effective in solving the problem of climate change. There are a number of ways to characterize the critical issues in this regard. But I would argue that we should be thinking in terms of: (i) fleshing out the goal of the regime in such a way as to make it easy to track performance; (ii) articulating principles of fairness that are appealing to all major groups of players; (iii) developing systems of monitoring, reporting, and verification that inspire confidence on all sides; (iv) establishing credible funding mechanisms; and (v) experimenting with procedures that meet the need for flexibility and nimbleness.

What would make the goal of the climate regime straightforward, appealing, and easy to track? As I have said, there is no objective answer to this question. But one appealing solution to this problem would be to set a globally acceptable level of GHG emissions (e.g. calculated in terms of gigatons of CO_2 equivalent per year) and then to measure progress in terms of movement toward or away from the agreed-upon level.[4] At present, of course, the adoption of this procedure would reveal that the world is moving fairly steadily away from the common goal with respect to the problem of climate change rather than toward solving the problem (Ramanathan and Feng 2008; Le Quere et al. 2009). An alternative might be to reframe the goal of the climate regime in terms of the idea of decarbonizing industrial societies. We would then need to construct a measure that would allow us to track progress toward reducing the carbon-intensity of economic activities throughout the world, including both Annex I and non-Annex I countries. But whatever the approach ultimately adopted, the test of progress will be the ease and confidence with which we can track performance in terms of goal attainment.

A much more contentious matter that also requires agreement at the international level centers on responding effectively to the challenge of fairness. The Annex I countries are correct in arguing that all major players must accept obligations to reduce emissions in order to make a dent in the problem of climate change. No climate regime can succeed without finding ways to persuade the major non-Annex I countries (e.g. China, India, Brazil, Indonesia) to accept the need to cap and then to reduce emissions sharply over time. But it is equally clear that we cannot make headway in dealing with the problem until the Annex I countries respond in a serious manner to the legitimate concerns of the developing countries, who point out that the OECD countries relied heavily on burning hydrocarbons to fuel their economic development, that the emissions of these countries still account for most of the rise in GHG levels in the Earth's atmosphere relative to pre-industrial levels, and that developing countries are likely to suffer from the effects of climate change first and most severely. There are numerous ways to address this problem. But the crux of the matter is apparent (Stern 2009). The Annex I countries must find ways to facilitate the efforts of the developing world to transition toward a green economy and to cushion the impacts of climate change on non-Annex I parties in return for meaningful steps on the part of these countries to reduce emissions by moving promptly toward low-carbon development pathways.

To be successful, any climate regime must be able to count on the member states themselves to take the lead in dealing with matters of implementation, enforcement, and adaptation. Yet it will not do simply to delegate responsibility for these matters entirely to the members. Although it is unlikely that an international climate regime will be in a position to impose serious penalties on those who fail to meet their obligations, international arrangements with access to state-of-the art technologies (e.g. remote sensing) must play an active role with regard to monitoring, reporting, and verification. As we learned some time ago in developing arms control agreements, invasive forms of MRV that arouse sovereignty sensitivity within powerful member states will almost certainly be non-starters. But technologies already available should make it possible to address most issues of MRV in a manner that can satisfy concerns about compliance without being unduly provocative in terms of sensitivity of this sort.

No climate regime can succeed without addressing the problem of funding. It is widely understood that a successful regime must engage the major non-Annex I countries and that these countries simply will not accept the necessary obligations without the establishment of funding mechanisms capable of alleviating the financial burden associated with making the transition to low-carbon development and with undertaking necessary adaptation measures. The current arrangement featuring reliance on the Global Environment Facility is not working well. Although it is important to avoid facile comparisons, the Multilateral Fund (MLF) created to deal with a somewhat analogous problem in the case of strato-spheric ozone depletion is worth looking at seriously in this connection.[5] A comparable arrangement in the case of climate change would require funding in the tens to hundreds of billions in contrast to the 2–3 billion US dollars needed in the case of the ozone regime. But experience with the MLF will be an important source of lessons when we decide to take the climate problem seriously.

Meeting the need for flexibility is a tall order. Desirable as they are in terms of high-lighting norms of good governance and reducing democracy deficits, recent trends toward maximizing participation and negotiating under a fairly strict consensus rule actually complicate this challenge. The difficulties of seeking to achieve consensus among 192 delegations operating under intense public scrutiny can easily become insurmountable. The way forward in this regard is not obvious. But it is notable that only twenty-four parties adopted the Montreal Protocol at the outset in 1987 and that many large developing countries, including China and India, were not among the original parties to what has now become a regime with universal membership (Benedick 1998). This suggests the importance of finding ways to move forward on the basis of leadership provided by a coalition of critical players or what has become known in regime analysis as a k-group (Schelling 1978). In the case of climate change, some 12–15 countries dominate the situation; a concerted effort on their part could make a decisive difference in efforts to strengthen the climate regime. As in the case of the Montreal Protocol, persistent efforts to universalize a restructured climate regime over time would be essential. But there is much to be said for adopting an approach allowing a k-group to move forward the process of restructuring the climate regime in a timely manner.

What can we say about the outcome of the 15th Conference of the Parties (COP) to the UNFCCC/5th Meeting of the Parties (MOP) to the KP, meeting in Copenhagen in December 2009, in terms of this analysis of the effectiveness of the climate regime? Does the Copenhagen Accord represent real progress in addressing the issues touched on in this section (Copenhagen Accord 2009)? The facts that the formal efforts to complete the work of the Bali Action Plan (Bali Action Plan 2007) failed, that the accord is only a political document which carries no formal commitment on the part of those who crafted it, and that this document was not formally adopted by the COP/MOP are clearly disappointing.[6] Beyond this, the accord presents a mixed picture. It does not address the problem of fleshing out the goal of the climate regime, and it offers no more than a hint of progress regarding matters of MRV. On the other hand, the accord does take several steps toward acknowledging the concerns of the non-Annex I countries, and it demonstrates an awareness of the need to address the issue of funding in a serious manner. Perhaps the most critical feature of the Copenhagen Accord is the message it sends regarding the need for political leadership regarding climate change. The substantive terms of the accord are far from satisfactory, and the processes involved in its development provoked understandable complaints from many quarters. Nonetheless, the emergence of the accord does reflect both an awareness of the need for leadership regarding efforts to solve the climate problem and an initial effort to meet this need. The next stage must feature major improvements in both the substance and the style of this leadership, rather than a retreat in the face of the vocal protests emanating from many quarters in the aftermath of the Copenhagen meeting (Young 1991).

3 CAN WE MANAGE INTERPLAY TO ACHIEVE SYNERGY WITH OTHER REGIMES?

There is a marked tendency in regime analysis to examine regimes one at a time, focusing attention on the details of the regime at hand and largely ignoring the interplay between individual regimes and others with which they interact in a more or less continuous manner. Although the sources of this tendency are easy to understand, especially at the international level, the rapid growth during the postwar era in the number of governance systems addressing a wide range of issues makes the practice of analyzing regimes one at a time problematic. Sometimes the linkages become so extensive and so tight that it makes sense to speak of the emergence of institutional complexes, such as the Antarctic Treaty

System or the governance system for plant-genetic resources (Raustiala and Victor 2004). But even when this is not the case, it is increasingly important to consider the interplay between or among regimes in efforts to handle specific problems effectively and, more generally, to address what has become known as the problem of institutional architecture (Underdal and Young 2004; Gehring and Oberthür 2008; Biermann et al. 2009; Oberthür and Stokke forthcoming).

From the outset, the climate regime has interacted with a variety of other institutional arrangements in ways that are critical to solving the climate problem. Some of these links involve interactions with other environmental regimes (e.g. the ozone regime). Others arise from interactions with economic regimes (e.g. the global trade regime) and with scientific arrangements (e.g. the Intergovernmental Panel on Climate Change). The result is a complex mosaic of differentiable institutional arrangements that play an important role in regulating emissions of GHGs and in determining the effectiveness of efforts to solve the problem of climate change. Some of these interactions are synergistic; others pose actual or potential problems.

A good place to start in this case is with interactions between the climate regime and other MEAs. Most ozone-depleting substances (e.g. CFCs, halons) are also GHGs, so success in the operation of the regime for the protection of stratospheric ozone contributes to the effort to slow the growth of concentrations of GHGs in the Earth's atmosphere (Kaniaru 2007). A striking observation in this regard is that reductions in emissions of GHGs brought about through the operation of the ozone regime exceed by a sizeable factor all reductions achieved or even mandated under the provisions of the UNFCCC/KP (Velders et al. 2007; Molina et al. 2009). If anything, this gap is growing. Since forests are carbon sinks, somewhat similar observations are in order regarding international efforts to protect forests, such as the development of the International Tropical Timber Agreement and the effort to follow up on the forest principles articulated at the 1992 UN Conference on Environment and Development. Forests actually withdraw carbon from the atmosphere by sequestering it in individual trees and underground biomass, a significant fact if we decide to limit concentrations of carbon dioxide in the atmosphere to 450 ppm, much less the more ambitious target of 350 ppm.

An interesting challenge in this realm focuses on the role of short-lived climate forcers, like black carbon or carbon soot (Ramanathan and Carmichael 2008; Tollefson 2009). Because black carbon resides in the atmosphere only for days or weeks (in contrast to CO_2), cutting back sharply on emissions of black carbon could contribute significantly to addressing the problem of climate change in short order, buying time to devise effective measures to reduce emissions of GHGs (Wallack and Ramanathan 2009). But how can we pursue this goal (Molina et al. 2009)? Black carbon is neither an ordinary greenhouse gas nor an ozone-depleting substance. Regulating the sources of black carbon under either the UNFCCC/KP or the Montreal Protocol would be difficult. But there are other mechanisms that might serve well in addressing this source of climate change. The UN Development Program, for example, could play an effective role in efforts to replace the burning of fuelwood in rural areas, a major source of health problems as well as a large producer of black carbon. There is room to regulate the diesel electric engines widely used in maritime commerce under the terms of the 1973/1978 International Convention for the Prevention of Pollution from Ships (MARPOL) and other legally binding agreements administered by the International Maritime Organization.

While other regimes can bolster the climate regime, we need to turn this relationship around as well and note that a failure to create an effective climate regime is likely to prove harmful consequences with regard to the effectiveness of other environmental regimes. The effects of rapid climate change are already threatening key species (e.g. polar bears) in the high latitudes, thereby undercutting efforts to protect biological diversity under the auspices of the Convention on Biological Diversity. Climate change is expected to make many of the Earth's drylands (e.g. the Sahel in Africa, the American Southwest) drier, complicating the challenges facing those struggling to achieve the goals of the UN Convention to Combat Desertification. A similar story arises in connection with efforts to administer regimes dealing with large river systems. As climate change leads to the melting of the snow pack and glaciers located at the headwaters of major rivers (e.g. the Ganges, the Mekong, the Yangtze), these vital sources of freshwater will experience greater seasonal fluctuations at best and may be reduced to a trickle for sizeable segments of the year. If left unchecked, climate change may generate collateral effects that produce serious challenges for those responsible for the operation of a wide range of other environmental regimes. Conversely, success in addressing the problem of climate change will generate substantial benefits in other domains as well.

Interactions between climate change and governance systems operating beyond the realm of environmental issues are also important. The most prominent case involves interactions between the climate regime and the global trade regime administered by the World Trade Organization (Hufbauer et al. 2009). The critical issues here center on actions like the use of trade sanctions as a means of securing compliance with the provisions of the climate regime and the imposition of climate-related trade restrictions (e.g. border tariffs on energy-intensive goods) in order to exert pressure on countries to take the problem of climate change seriously. Policy makers may also resort to similar measures that generate tensions with the provisions of regional trade agreements, such as the North American Free Trade Agreement (NAFTA). Synergy may be hard to achieve in this realm; those committed to the defense of free trade have a tendency to react skeptically to any restrictions on the movement of goods and services across international borders. But it is critical to find ways to avoid clashes between the climate regime and trade regimes that could prove disruptive on all sides.

Interactions between the climate regime and arrangements focused on the conduct of science and the application of scientific results to matters of public policy are both important and good candidates for generating synergy. The work of the Intergovernmental Panel on Climate Change (IPCC), sponsored by the World Meteorological Organization and the UN Environment Program, is an obvious case in point. The IPCC pre-dates the UNFCCC. It is an autonomous body that is not a part of the climate regime in any formal sense (Schneider 2009). But its periodic assessments of the state of knowledge regarding the behavior of the Earth's climate system have emerged as critical inputs to policy making under the auspices of the climate regime. Somewhat similar remarks are in order regarding the role of the Earth System Science Partnership (ESSP), a joint venture of the four large global environmental change research programs designed to tackle questions that no one research community can address successfully by itself (e.g. why do some societies succeed in adapting to climate changes, while others collapse?) (Diamond 2005). A notable development in this realm is the emergence of a practice featuring regular ESSP briefings at meetings of the UNFCCC's SBSTA. Scientific progress may lead to recommendations (e.g. the need to cut back concentrations of CO_2 to 350 ppm) that are difficult for the existing climate regime to embrace, much less to implement (Hansen 2009). Nonetheless, the interactions described here are largely synergistic in nature.

Nothing in this discussion of interplay between the climate regime and other institutional arrangements constitutes an excuse for slacking off on efforts to improve the effectiveness of the climate regime itself. Yet it is clear that we must think more systematically about matters of what regime analysts have come to describe as institutional architecture as we move forward in efforts to avoid dangerous anthropogenic interference with the climate system (Biermann et al. 2009). Just as other regimes affect the climate regime, the (non)development of the climate regime affects efforts to address a range of other problems. There is a growing need to develop a consultative process among those responsible for the climate regime and their counterparts in other environmental regimes as well as major regimes operating in other issues areas (e.g. the global trade regime). This process will be time consuming and may involve at least temporary roadblocks in some areas, developments that will make it all the harder to maintain the nimbleness of the climate regime. But efforts to push forward in addressing climate change without accounting for institutional interplay are bound to prove costly in the long run.

4 WHAT ARE THE POLICY IMPLICATIONS OF THIS ANALYSIS?

Any serious effort to avoid dangerous anthropogenic interference with the climate system will require major changes in the lifestyles and living arrangements of large numbers of ordinary people. We can imagine cuts of 30–40 percent of current GHG emissions resulting from a combination of initiatives featuring conservation, increased energy efficiency, and reliance on alternative sources of energy along with an effective use of forest sinks that do not require fundamental changes in existing economic and social systems. But achieving the 80+ percent reductions in current emissions that now seem necessary to protect the climate system is another story. There is no way to reach this goal without encouraging far-reaching shifts in the behavior of consumers in developed countries and persuading consumers in developing countries (e.g. China, India) to set their sights on lifestyles that differ markedly from those characteristic of the United States and other advanced industrial countries today.

Still, there is no reason to throw up our hands and simply accept the onset of climate change as a fact of life. What is needed is a climate regime that is well matched to the problem at hand. It will be hard to fulfill this requirement through a strategy based on incremental reforms or, in other words, adjustments to the current regime at the margin. This means that there is a need to think about restructuring the climate regime at the constitutive level rather than simply tinkering with its operating rules. This would take us beyond debates about precise targets and timetables and about the fine points of the operation of flexibility mechanisms like the CDM. An effective alternative must rest squarely on a bargain between the Annex I countries and the non-Annex I countries in which the former make a firm commitment to reduce current emissions of GHGs drastically and promptly and the latter make an equally firm commitment not to increase emissions.[7] The precise means of fulfilling these commitments must be left to the discretion of individual countries. But the Annex I countries would need to use some combination of taxes and subsidies to curb the consumption of fossil fuels and increase the use of alternative fuels, and the less developed countries would need access to an international mechanism capable of providing substantial funding on concessional or highly affordable terms to carry out well-designed initiatives in this realm.

Because the Earth's climate system is subject to changes that are nonlinear, abrupt, irreversible, and quite likely nasty from the perspective of human welfare, a restructured climate regime must be nimble in the sense that it can adjust to new information about the problem of climate change promptly and efficiently. This is a tall order even at the domestic level, much less at the international level. In some ways, current developments that feature decentralizing authority regarding environmental issues in the interests of providing mean-ingful opportunities for stakeholder engagement in policy making run counter to the need to make timely adjustments to global arrangements like the climate regime. The resultant tension between the democratization of policy making and the need to act promptly and decisively in dealing with problems like climate change could turn out to be the most fundamental challenge we face in efforts to address the problem of climate change.

How can we make progress in meeting this challenge? Drawing on our experience with the control of ozone-depleting substances, there is a strong case for authorizing the COP/ MOP of the climate regime to make a variety of adjustments in the provisions of the regime without triggering a requirement for ratification by the parties. It also seems critical to encourage a proliferation of experiments with approaches to low-carbon development at the subnational level that can generate insights and give rise to innovative ideas that can be recast for adoption at the international level. There are no simple solutions in this realm. Yet successful social practices are often remarkably supple, shifting their character to meet changing circumstances in the absence of any formal process of decision making. Under the circumstances, it is a mistake to concentrate only on devising formal procedures designed to enhance flexibility at the expense of cultivating the development of social practices that can adjust to changing conditions through informal processes (Young 2002).

It is essential as well to avoid thinking about the climate regime as a stand-alone arrangement to be developed and evaluated without reference to interactions with a variety of other regimes. This may turn out to be good news in cases where other regimes can play a role in controlling emissions of GHGs. The current effort to suppress the production and consumption of HFCs through initiatives undertaken under the auspices of the Montreal Protocol is a striking example (Molina et al. 2009). In other cases, such as the interplay between the climate regime and the trade regime, this feature of the problem of climate change will prove more challenging. But the basic message is clear. We must think in terms of institutional architecture rather than focusing narrowly on the restructuring and strengthening of the climate regime itself. Creating an institutional complex in which management arrangements for a variety of problems are joined together to enhance synergy and minimize conflict is a daunting challenge. But this is what is required in a world in which human actions are major drivers of biophysical changes and in which the major drivers interact with one another in a variety of ways.

The problem of climate change will not go away. Many analysts believe that we are already committed to a trajectory that will lead to increases in surface temperatures of more than 2 °C (Hansen 2009). Unlike ozone depletion, where the challenge is to phase out completely the production and consumption of a range of man-made chemicals, dealing with climate change involves a balancing act in which we need to keep concentrations of GHGs in the atmosphere within a limited band that is compatible with the maintenance of the benign and relatively stable climate that has been a major feature of the Earth's history during the last 10,000 years. Although a variety of biophysical forces play important roles in this realm, anthropogenic drivers have now become critical forces affecting the maintenance of this balance. To avoid dangerous anthropogenic interference with the climate system, we will need to learn how to control human actions leading to emissions of GHGs. We can tackle this problem effectively. But to do so will require the restructuring of the existing climate regime to improve its match with prevailing biophysical and socio-economic conditions and to direct attention to the issue of institutional architecture or, in other words, the linkages among a number of distinct regimes that are relevant to solving the problem of climate change.

Notes

1. The article goes on to say that 'Such a level should be achieved within a time frame sufficient to allow ecosystems to adapt naturally to climate change, to ensure that food production is not threatened and to enable economic development to proceed in a sustainable manner.'
2. The commitments of individual parties vary substantially; some are even allowed to increase emissions. The United States, the leading emitter until 2007–8, has not become a party to the KP.
3. China overtook the United States during 2007–8 as the leading emitter of GHGs. Indonesia is now the third largest emitter.
4. Presumably, the approved level would decline over time in order to arrive at an overall goal set by the parties.
5. The MLF is an internationally administered fund created under the terms of the Montreal Protocol to provide supplemental funding needed to allow developing countries to proceed with their economic development plans using alternatives to ODSs. It is managed by an Executive Committee that has equal representation from the developed and developing parties to the Montreal Protocol.
6. In the final hours of the meeting, the COP/MOP decided to 'take note' of the Copenhagen Accord rather than to adopt it as a product of the meeting.
7. China is a special case because it must undertake to reduce current levels of GHG emissions, even though it is a non-Annex I country (CRS 2008).

References

ALDY, J. E., and STAVINS, R. N. (eds.) 2007. *Architectures for Agreement: Addressing Global Climate Change in the Post-Kyoto World*. Cambridge: Cambridge University Press.

Bali Action Plan 2007. <http://unfccc.int/files/meetings/cop_13/application/pdf/cp_bali_action.pdf>.

BENEDICK, R. E. 1998. *Ozone Diplomacy: New Directions in Safeguarding the Planet.* Enlarged edn. Cambridge, MA: Harvard University Press.

BIERMANN, F., et al. 2009. *Earth System Governance: People, Places, and the Planet.* IHDP Report No. 20. Bonn: IHDP.

Congressional Research Service (CRS). 2008. China's greenhouse gas emissions and mitigation policies. CRS Report for Congress, RL34659, 10 September.

Copenhagen Accord. 2009. <http://unfccc.int/resource/docs/2009/cop15/eng/107.pdf>.

DIAMOND, J. 2005. *Collapse: How Societies Choose to Fail or Succeed.* New York: Viking.

GALAZ, V., et al. 2008. The problem of fit among biophysical systems, environmental and resource regimes, and broader governance systems: Insights and emerging challenges. Pp. 147–86 in O. R. Young, L. A. King, and H. Schroeder (eds.), *Institutions and Environmental Change.* Cambridge, MA: MIT Press.

GEHRING, T., and OBERTHÜR, S. 2008. Interplay: Exploring institutional interaction. Pp. 187–223 in O. R. Young, L. A. King, and H. Schroeder (eds.), *Institutions and Environmental Change.* Cambridge, MA: MIT Press.

HANSEN, J. 2009. *Storms of my Grandchildren: The Truth about the Coming Catastrophe and our Last Chance to Save Humanity.* New York: Bloomsbury.

HUFBAUER, G. C., CHARNOVITZ, S., and KIM, J. 2009. *Global Warming and the World Trading System.* Washington, DC: Peterson Institute for International Economics.

KANIARU, D. (ed.) 2007. *The Montreal Protocol: Celebrating 20 Years of Environmental Progress.* London: Cameron May.

——et al. 2007. Strengthening the Montreal Protocol: Insurance against abrupt climate change. INECE Working Paper.

LE QUERE, C., et al. 2009. Trends in the sources and sinks of carbon dioxide. *Nature Geoscience* 2 (December), <http://www.nature.com/naturegeoscience>.

MITCHELL, R. B. 2008. Evaluating the performance of environmental institutions: What to evaluate and how to evaluate it? Pp. 79–114 in O. R. Young, L. A. King, and H. Schroeder (eds.), *Institutions and Environmental Change.* Cambridge, MA: MIT Press.

MOLINA, M., ZAELKE, D., MADHAVA SARMA, K., ANDERSON, S. O., RAMANATHAN, V., and KANIARU, D. 2009. Reducing abrupt climate change risk using the Montreal Protocol and other regulatory actions to complement cuts in CO_2 emissions. *Proceedings of the National Academy of Sciences* <http://www.pnas.org/content/106/49/20616.full>.

OBERTHÜR, S., and STOKKE, O. S. (eds.) Forthcoming. *Institutional Interaction and Global Environmental Change.* Cambridge, MA: MIT Press.

OSTROM, E. 2007. A diagnostic approach for going beyond panaceas. *Proceedings of the National Academy of Sciences* 104: 15181–7.

RAMANATHAN, V., and CARMICHAEL, G. 2008. Global and regional climate changes due to black carbon. *Nature Geoscience* 1: 221–7.

——and FENG, Y. 2008. On avoiding dangerous anthropogenic interference with the climate system: Formidable challenges ahead. *Proceedings of the National Academy of Sciences* 105: 14245–50.

RAUSTIALA, K., and VICTOR, D. G. 2004. The regime complex for plant genetic resources. *International Organization* 55: 277–309.

SCHELLING, T. C. 1978. *Micromotives and Macrobehavior.* New York: W. W. Norton.

SCHNEIDER, S. H. 2009. *Science as a Contact Sport: Inside the Battle to Save Earth's Climate.* Washington, DC: National Geographic.

STERN, N. 2009. *The Global Deal: Climate Change and the Creation of a New Era of Progress and Prosperity.* New York: Public Affairs.

TOLLEFSON, J. 2009. Climate's smoky spectre. *Nature* 460 (2 July): 29–32.

UNDERDAL, A. 2008. Determining the causal significance of institutions: Accomplishments and challenges. Pp. 49–78 in O. R. Young, L. A. King, and H. Schroeder (eds.), *Institutions and Environmental Change.* Cambridge, MA: MIT Press.

——and YOUNG, O. R. (eds.) 2004. *Regime Consequences: Methodological Challenges and Research Strategies.* Dordrecht: Kluwer Academic Publishers.

VELDERS, G. J. M., et al. 2007. The importance of the Montreal Protocol in protecting climate. *Proceedings of the National Academy of Sciences* 104: 4814–19.

WALLACK, J. S., and RAMANATHAN, V. 2009. The other climate changers: Why black carbon and ozone also matter. *Foreign Affairs* 88: 105–13.

YOUNG, O. R. 1991. Political leadership and regime formation: On the development of institutions in international society. *International Organization* 45: 281–308.

——1999. Hitting the mark: Why are some international environmental agreements more successful than others? *Environment* 41 (October): 20–9.

——2002. *The Institutional Dimensions of Environmental Change: Fit, Interplay, and Scale.* Cambridge, MA: MIT Press.

——2008. Building regimes for socioecological systems: Institutional diagnostics. Pp. 114–44 in O. R. YOUNG, L. A. KING, and H. SCHROEDER (eds.), *Institutions and Environmental Change.* Cambridge, MA: MIT Press.

CHAPTER 43

....

RECONCEPTUALIZING
GLOBAL GOVERNANCE

....

PAUL G. HARRIS

1 INTRODUCTION

....

GOVERNANCE can be conceived of as a process whereby actors cooperate to achieve common objectives (cf. Barnett and Sikkink 2008: 78) or 'the achievement of collective goals, and the collective processes of rule through which order and goals are sought' (Rosenau 2000: 175).[1] Conceptualizing governance involves thinking about both these processes and the goals to which they are directed. Climate governance has been characterized by a historically unprecedented amount of negotiation among governments, leading to a collection of international agreements, notably the 1992 Framework Convention on Climate Change and the 1997 Kyoto Protocol, as well as related agreements regarding implementation measures and ongoing efforts to achieve the Framework Convention's core objective of preventing dangerous changes to the Earth's climate system. In addition to interstate cooperation on climate change, there have been measures and actions at other levels of governance. These include voluntary national policies and programs to limit greenhouse gas (GHG) emissions not required by the Kyoto Protocol (e.g. measures by developing-country governments to improve energy efficiency; see Christoff and Eckersley in this volume); regulations by substate actors to limit pollution that contributes to climate change (e.g. national regions and municipalities; see Bulkeley in this volume); changes to business practices that use alternative, less-polluting fuels (see Pulver, this volume); initiatives by non-governmental actors to encourage less atmospheric pollution and to respond to the consequences of climate change (see Lipschutz and McKendry, this volume), and changes in individual behaviors that do likewise (see Szasz, this volume).

These actions are having significant impacts on GHG pollution. However, they have been grossly inadequate. Despite a quarter-century of negotiations among governments, emissions of carbon dioxide and other GHGs continue to increase, and current concentrations of carbon dioxide in the atmosphere mean that many effects of climate change—including impacts that will lead to widespread hardship among billions of people, suffering for hundreds of millions of them, and almost certainly the deaths of millions more—are inevitable. Thus, while there has been much in the way of climate governance, that governance has been largely ineffective relative to the scale of ecological change, destruction, suffering, and death that climate change portends. Without a radical increase in activity to combat the causes of climate change, in a short space of time (perhaps a decade or two), and much more aggressive action to deal with the inevitable impacts, too many people are doomed to hardship or worse, as are other species and ecosystems.

Few people knowledgeable of the impacts of climate change and efforts by government to address them would argue that relevant institutions and regimes do not require revitalization, unless of course they reject existing institutions outright. If we accept that it is unrealistic to discard extant institutions completely, or simply that there is not enough time for that given the pace of climate change, a fundamental question is which concepts will undergird this revitalization effort, should it happen. This raises the vital question of how the process of climate governance might be made more effective. Routinely the answer is to create new interstate institutions, such as a Global or World Environmental Organization (Biermann and Baur 2005; Esty 2009). But a follow-on question is who or what the revitalization process is for.[2] In an attempt to help answer this question, this chapter summarizes and critiques fundamental concepts underlying climate governance today before proposing a new way to conceptualize global governance in this context.[3]

The failures of a climate change regime premised on prevailing interstate doctrine suggest that an alternative concept is needed to underlie the climate change regime. If that concept is not to replace the doctrine of inter*national* environmental justice, it must at least improve upon it. A cosmopolitan conception of who is most important in world affairs is more suited to the realities of climate change, modernity, and globalization than are extant forms of justice premised on the morality of states. Cosmopolitan or global justice brings people more explicitly into the climate change regime while also helping to create conditions for much more aggressive international action. In short, what is needed is a reconceptualization of climate governance away from the state and toward cosmopolitan values and interests.

2 CONCEPTS UNDERLYING GLOBAL GOVERNANCE

For centuries the world has been guided by, and governments have sought to reinforce, Westphalian norms of state recognition, sovereignty, and non-intervention. According to these norms, sovereign states are the legitimate expressions of human organization, and it is to states that people ought to turn for governance and for solutions to major challenges. These norms have so far largely guided discourse, thinking, and responses to transboundary environmental problems: environmental diplomacy, regimes, and treaties have been based on the responsibilities, obligations, and capabilities of states to limit pollution and to cooperate to cope with the effects of environmental harm and resource exploitation. The inter*national* norms have been so powerful as to result in what is, effectively, a doctrine of international (that is, inter*state*) environmental justice, manifested in the principle of common-but-differentiated responsibility among states. This inter*national* doctrine has guided and permeated a number of international environmental agreements and regimes, such as the treaties to combat stratospheric ozone depletion and manage biological diversity.[4]

Like other international environmental agreements, those that comprise the climate change regime have been premised on Westphalian norms of state sovereignty and states' rights. The norms, discourse, and thinking associated with this state-centric doctrine have resulted in diplomatic delay, minimal action (relative to the scale of the problem), and a blame-game between rich and poor countries that has prevailed even as global GHG emissions have exploded in recent years. Domestic and international responses to climate change have been preoccupied with protecting perceived national interests. This focus on the rights and interests of states has been written into the climate change agreements, including the Framework Convention, the Kyoto Protocol, and subsequent agreements and diplomatic negotiations on implementing the protocol and devising its successor. Although some major industrialized countries, notably in Europe, have started to restrict and even reduce their emissions of GHGs, these responses pale in comparison to the major cuts (at least 80 percent) demanded by scientists. Just as profoundly, many large developing countries are experiencing huge emissions increases as their economies grow, bolstered by increased exports and by the material consumption of millions of their citizens joining the global middle class. These growing emissions will dwarf planned cuts by developed states. In short, most of the world's expanding wealthy classes continue to aggressively consume and pollute, usually with few if any legal restrictions, regardless of the growing impact on the planet.

Despite provisions in the climate change agreements for inter*national* justice designed to garner participation by more states, particularly in the developing world, these agreements have failed to bring about substantial changes in behavior relative to the scale of the problem.[5] This is a typical tragedy of the commons, albeit on a truly monumental scale. Fundamental to this tragedy of the atmospheric commons is the 'you go first' mindset among most states, which has resulted in only modest GHG cuts in Europe and almost no cuts at all in most of the developed world, not to mention continuing increases in the developing countries and globally (because increases in developing countries far surpass cuts in developed countries, meaning global increases well into the future). The conference of the parties to the climate change conventions held in Copenhagen in December 2009 reveals the fundamental problem. As Baer (in this volume) points out, the negotiations among diplomats at Copenhagen were 'dominated by efforts of the developed countries to increase the emissions reductions commitments of the developing countries, and the resistance of the developing countries to those efforts.' The problem that faces the climate change regime is how to get around this even while working from the otherwise commendable core concept of common but differentiated responsibility of states for climate change.

3 Reconceptualizing Climate Governance

One potential antidote to the tragedy of the commons that characterizes today's responses to climate change can be found in cosmopolitan ethics and global conceptions of justice that focus on *people* as well as states. Cosmopolitans recognize the rights, obligations, and duties of capable individuals regardless of nationality. Inter*national* justice considers national borders to be both the practical and ethical foundation for justice. But cosmopolitan justice, while recognizing that national borders have practical importance, views them as an inadequate basis for deciding what is just—including in the context of climate change. In contrast to the Westphalian norms that have guided and indeed defined the international system for centuries—a set of ethics premised on protecting the interests of the state—cosmopolitans envision an alternative way of ordering the world. Cosmopolitans want to 'disclose the ethical, cultural, and legal basis of political order in a world where political communities and states matter, but not only and exclusively. In circumstances where the trajectories of each and every country are tightly entwined, the partiality, one-sidedness and limitedness of "reasons of state" need to be recognized' (Held 2005: 10).

Scholars of cosmopolitanism often disagree on the concept's fundamental tenets.[6] Nevertheless, Thomas Pogge sums up three common core elements of cosmopolitanism:

> First, *individualism*: the ultimate units of concern are *human beings*, or *persons*—rather than, say, family lines, tribes, ethnic, cultural, or religious communities, nations, or states. The latter may be units of concern only indirectly, in virtue of their individual members or citizens. Second, *universality*: the status of ultimate unit of concern attaches to *every* living human being *equally*—not merely to some sub-set, such as men, aristocrats, Aryans, whites, or Muslims. Third, *generality*: this special status has global force. Persons are ultimate units of concern for *everyone*—not only for their compatriots, fellow religionists, or such like. (Pogge 2008: 175)

For cosmopolitans, 'the world is one domain in which there are some universal values and global responsibilities' (Dower 2007: 28).

Lorraine Elliott argues that environmental harms crossing borders 'extend the bounds of those with whom we are connected, against whom we might claim rights and to whom we owe obligations within the moral community' (Elliott 2006: 350). She describes this as a 'cosmopolitan morality of distance,' which effectively creates 'a cosmopolitan community of duties as well as rights' (ibid.). Elliott believes that this obtains for two reasons:

> the lives of 'others-beyond-borders' are shaped without their participation and consent [and] environmental harm deterritorialises (or at least transnationalises) the cosmopolitan community. In environmental terms, the bio-physical complexities of the planetary ecosystems inscribe it as a global commons of a public good, constituting humanity as an ecological community of fate. (Elliott 2006: 351).

Consequently, the cosmopolitan standpoint provides a better 'theoretical and ethical road map for dealing with global environmental injustice' than does inter*national* doctrine (ibid. 363). Robin Attfield goes further, arguing that only cosmopolitanism can do 'justice to the objective importance of all agents heeding ethical reasons, insofar as they have scope for choice and control over their actions, and working towards a just and sustainable world society' (Attfield 1999: 205). He believes that criticisms of state responses to environmental problems will inevitably be based on cosmopolitanism because 'the selective ethics of nation states are liable to prioritize some territories, environments, and ecosystems over others. If this meant nothing but leaving the other environments alone, this might not be too pernicious. [But] it often means not leaving alone the others but polluting or degrading them' (ibid. 41). Derek Heater also critiques what he calls the 'traditional linear model of the individual having a political relationship with the world at large only via his state' because, at least if we are concerned about 'the integrity of all planetary life, the institution of the state is relegated to relative insignificance—if not, indeed, viewed as a harmful device' (Heater 1996: 180).

Andrew Dobson (2006) makes a case for cosmopolitan obligation arising from the causal impacts of globalization in its many manifestations, including global environmental change. What is especially important about his argument is that he goes beyond cosmopolitan morality and sentiment. Dobson identifies what he calls 'thick cosmopolitanism,' which he argues will motivate people (and other actors) to not only accept that every person is part of a common humanity but will also motivate them to act accordingly:

> Recognising the similarity in others of a common humanity might be enough to undergird the principles of cosmopolitanism, to get us to 'be' cosmopolitans (principles), but it doesn't seem to be enough to motivate us to 'be' cosmopolitan (political action). (Dobson 2006: 169).

Common humanity is one basis for cosmopolitanism, but it does not create the 'thick' ties between people that arise from causal responsibility. However, as Dobson points out, changes to the climate 'thicken the ties that bind us to "strangers," to bring these strangers "nearer" without having to rely on empathetically constructing them as surrogate neighbours' (Dobson 2006: 175–6). He invokes Linklater's suggestion that if we are 'causally responsible for harming others and their physical environment' we are far more likely to act as cosmopolitans should (Linklater 2006: 3). Relationships of causal responsibility 'trigger stronger senses of obligation than higher-level ethical appeals can do' (Dobson 2006: 182).

One reason that climate change is a matter of cosmopolitan justice, which fully considers the rights and duties of individuals—is that millions of people geographically and temporally distant from the sources of global warming suffer from its consequences. Steve Vanderheiden points out that, 'insofar as a justice community develops around issues on which peoples are interdependent and so must find defensible means of allocating scarce goods, global climate change presents a case in which the various arguments against cosmopolitan justice cease to apply. All depend on a stable climate for their well-being, all are potentially affected by the actions or policies of others, and none can fully opt out of the cooperative scheme, even if they eschew its necessary limits on action. Climate change mitigation therefore becomes an issue of cosmopolitan justice by its very nature as an essential public good . . .' (Vanderheiden 2008: 104).

Another reason is that people presently causing future global warming no longer live almost exclusively within the states that have historically caused it. Until quite recently we could talk in both moral and practical terms about climate change as a problem caused by the world's developed countries and their citizens. They were by far the primary sources of GHG pollution and thus (if we focus on national causality) the logical bearers of responsibility to end that pollution, to make amends for it, and to aid those who will suffer from it. The climate change regime, insofar as it acknowledged this responsibility, is premised on this notion of developed states—and, indirectly, their people and commercial entities— having primary responsibility for addressing this problem. However, climate change is no longer solely or even predominantly caused by the relatively affluent people of the world living in the affluent states of the world (cf. Botzen et al. 2008). Increasingly, pollution of the atmosphere comes from affluent people in developing countries who are joining the long-polluting classes of the developed countries (see Chakravarty et al. 2009).

Given the developing countries' large populations, this change does not alter the *international* moral calculus very much; after all, their national per capita emissions remain well below those of the developed states. What is different is the growing number of 'new consumers' (see Myers and Kent 2004) in the developing world, many of them extremely affluent persons, who are living lifestyles similar to those of most people in the developed countries. Indeed many of these new consumers are living more like the wealthiest classes in the industrialized world. They already number in the many hundreds of millions. For example, in China a few hundred million people, and in India many tens of millions, have higher standards of living—and higher levels of pollution—than people in the developed countries (see Myers and Kent 2004; Wheary 2009/10). They produce GHGs through voluntary consumption at a pace and scale never experienced in human history. While some societies in the West are starting to make changes to limit their GHG emissions, the new consumers in developing countries are going in the opposite direction. The consequences for the atmospheric commons will be severe.

The new consumers have been largely omitted from climate change agreements, diplomacy, and most discourse about climate change. At present, they face few if any legal obligations to mitigate the harm they do to the global environment, and they have so far escaped moral scrutiny. If solutions to climate change are to be found, this will have to change, in part because 'old consumers' in developed countries are watching these new consumers consume in ways that the former are being told they must stop for the sake of the Earth's atmosphere. This makes it extremely difficult politically for developed-country governments to implement regulations for major GHG cuts at home. As long as the new consumers are able to hide behind their own states' overall poverty, practical and politically viable solutions to climate change will remain elusive globally. Thus, the rise of the new consumers makes cosmopolitan justice more vital than ever. It offers an opportunity for the governments of developing countries to, in effect, join in GHG limitations without having to take on new ethical or legal burdens. They can do this by demanding that affluent polluters within their borders behave as affluent polluters everywhere should behave. Implementing cosmopolitan justice here means that obligations to act on climate change, and to aid people harmed by it, apply to all affluent polluters regardless of where they live.

4 A COSMOPOLITAN COROLLARY TO CLIMATE GOVERNANCE

While a wholesale reconceptualization of climate governance may be desirable, more realistically we can say that cosmopolitanism points to a corollary (or supplement) to the international governance of climate change which acknowledges the responsibilities and duties of developed states while also explicitly acknowledging and acting upon the responsibilities of all affluent people, regardless of nationality. This cosmopolitan corollary offers an alternative to the status quo climate change regime, premised as it is on the rights and duties of states while ignoring the rights and duties of too many people. Such a corollary is more principled, more practical, and indeed more politically viable than the current doctrine and norms of *international* justice applied to climate change. It suggests a direction for converting a global or cosmopolitan ethic into a new set of understandings that can constrain and coordinate the actions of states that comprise the climate change regime. An objective here is to start outlining how this alternative to inter*national* doctrine might be implemented, although there will be other ways to actualize cosmopolitanism in this context.

Putting cosmopolitanism into practice could mean a number of possible forms of governance and institutions. It could mean a world government, but it need not do so. As Hayden points out, 'cosmopolitanism is not inherently opposed to the state *per se*.... Rather cosmopolitanism is generally concerned to develop varied modes of governance—from the local to the global—with the goal of facilitating the rights and interests of individuals *qua* human beings. Indeed, states may be one mode of governance well suited to this end...' (Hayden 2005: 21). While it might be ideal for the world to be governed by truly cosmopolitan institutions, they are unlikely anytime soon. Axel Gosseries believes that adopting the assumptions of cosmopolitanism need not stop us from 'using states as our point of reference [because states can be] most able to represent the individuals that constitute them and because they are currently the most relevant units in the context of global attempts of curbing [greenhouse gas] emissions' (Gosseries 2007: 280). Future institutions could be premised on the understanding that the 'central argument of contemporary cosmopolitan political thought is that the demands of justice must be decoupled, at least to some degree, from the territorial bounds of the states' (Maltais 2008: 594). Simon Caney (2007) advocates a kind of 'revised statism' in which states do more to promote and implement global justice.

We are left with a continuing role for states but with institutions informed by cosmopolitanism's advocacy of the equal worth of all persons and the need for protecting their basic rights. Consequently, following the cosmopolitan belief that 'every human being has a global stature as the ultimate unit of moral concern' (Pogge 2002: 169), but without rejecting the state system generally and existing climate change institutions in particular, a cosmopolitan corollary to inter*national* doctrine would seek to establish a much more prominent place for human beings—their rights, responsibilities, and duties—in the evolving climate change regime. Diplomacy surrounding climate change would be premised on the rights and duties of human beings. Persons would be at (or at least very near) the center of climate change discourse, negotiations, and policies, and they would be viewed as the primary *ends* of diplomacy and government policy. Thus, while governments would inevitably retain a central role, as even most cosmopolitans recognize (because states are unlikely to surrender their role), a cosmopolitan corollary would have governments playing the role of facilitating the implementation of cosmopolitan obligation.

5 INSTITUTIONALIZING A COSMOPOLITAN COROLLARY

A cosmopolitan corollary to the doctrine of inter*national* environmental justice would start with recognition of global justice and the rights of all persons and the duties of capable persons, and includes conscious efforts to actualize global justice in agreements and the national and multinational institutions of states. The corollary would include two fundamental changes to the manner in which climate change is governed: (1) a change in the official discourse so that it acknowledges and affirms the rights and duties of all people in the context of climate change, and (2) the explicit incorporation of human rights and responsibilities into international agreements on climate change. The corollary would be layered with extant national and international responses to climate change. It is, in essence, cosmopolitanism grafted onto extant Westphalian norms and the doctrines of international relations that have so far guided climate governance. In this respect, cosmopolitanism is a kind of bridge across the divide between the nation-state system and the imperative of climate protection. This process of making *both* persons (and their rights and responsibilities) and states (and their rights and responsibilities) objectives of the climate change regime is the central feature of a cosmopolitan corollary. By more explicitly encompassing both states and persons, it builds on and helps to correct the existing state-centered regime. A cosmopolitan corollary could have the effect of helping states to free themselves from the myopia of states' rights and obligations by focusing on how practical progress on climate change can be made through recognizing and trying to actualize the rights and obligations of affluent, capable persons *everywhere*.

How might a cosmopolitan reconceptualization aid in strengthening the global climate regime? It might be ideally the case that the regime should be reformed from the bottom up, with states being guided by cosmopolitan ethics rather than self-interest. But such a bottom-up transformation, even if it is desirable, would come much too late; robust action to lower GHG emissions, and to respond to the inevitable consequences of climate change, especially for the world's poor, is needed in the short term. Thus a corollary to current concepts underlying the regime is more realistic than wholesale change. With this in mind, and given problems described above with the climate change regime that arise from its underlying Westphalian norms, what would this corollary look like in institutional terms? We can answer this by returning to Pogge's identification of the fundamental features of cosmopolitanism: individualism, universality, and generality (Pogge 2008: 175).

5.1 Individualism

Cosmopolitans are ultimately concerned with human beings, while states are an incidental concern because human beings live within them. As it happens, those who are ultimately affected by climate change are also human beings (although some small-island states do face an existential threat to their territorial integrity as a consequence of sea-level rise). What might this mean for climate-related global institutions? Those institutions ought to be ultimately concerned with human beings: their rights, responsibilities, needs, and capabilities. While it would be natural for states to continue to give priority to people within their own borders, even here there would be benefits. For example, many newly industrialized developing countries (e.g. China and Brazil) that do not believe they are facing immediate major threats from climate change are focusing on the responsibility of developed countries for past pollution of the atmosphere. This translates into developing-country opposition to major global cuts in GHGs that might one day require them to make cuts alongside wealthy states. The merits of this argument from the perspective of purely interstate justice are almost unassailable, but on a practical level they could amount to harming their own citizens, and of course people in other political jurisdictions, for the sake of upholding interstate norms and expectations.

One example of how individualism can be brought into the climate regime is the proposal from Baer et al. (2008) for Greenhouse Development Rights, which allocate climate-related obligations on the basis of a combined indicator of per capita responsibility and capacity (equated to income), exempting both income and emissions below a designated threshold so that people's basic needs can be met (see Baer in this volume). Similarly, Chakravarty et al. (2009) propose a scheme whereby the emissions of the world's 1 billion highest-emitting individuals are the basis for allocating responsibility for action on climate change. These proposals would, in principle, require high-emitting individuals everywhere to reduce their pollution, while people polluting at low levels would be allowed to increase their emissions to fulfill their needs. These frameworks are then translated into rights and obligations for states based on how many low- and high-emitting people live within their respective borders, with obligations and resources being allocated to states accordingly. The focus of these plans on *individuals*, and their rights, needs, and responsibilities, is the direction that cosmopolitanism takes the climate change regime. Having said this, a flaw in these particular plans may be that they convert individual rights and obligations into national rights and obligations, which could in turn be abused by states. Should this happen, a climate change regime based on these or similar plans would require mechanisms for possibly bypassing states so that individuals in need within state borders could be aided directly, whether by international organizations or by deputized non-governmental organizations (Harris and Symons 2010).

5.2 Universality

From a cosmopolitan perspective, every human being is of ultimate concern, and this applies equally to every person regardless of their nationality, ethnicity, or other characteristic. This concern is crucial for reshaping climate-related global institutions because it requires that individuals be treated the same no matter where they live. Consequently, high-polluting individuals in, say, Brussels would be treated the same by climate institutions as would a high-polluting individual in, say, Beijing. At present an affluent, high-polluting person in the latter city has no legal obligation, neither via the Chinese state nor individually, to limit, let alone reduce, those emissions, despite the future consequences. The former high-polluter is required to limit her or his emissions, at least indirectly, as a consequence of the European Union's (and Belgium's) commitments under the Kyoto Protocol and no doubt under successor agreements to come. A modified institution based on global conceptions of climate justice instead of interstate norms of justice would take this into consideration. This would mean that future institutions, while not necessarily requiring China to reduce its GHG emissions *as a state* would nevertheless place obligations on *affluent and capable Chinese people* to take on the same obligations as affluent and capable people in developed countries. This might have significant political advantages at the level of interstate negotiations because governments in the West could not blame rich Chinese for future pollution any more than they could blame their own citizens.

5.3 Generality

From a cosmopolitan perspective, the special status of individuals as the ultimate unit of concern applies to and for everyone; persons are of ultimate concern for everyone regardless of whether those persons live in the same state, follow the same religion, or are members of the same group. This reinforces the requirement that climate change institutions focus above all on the well-being of individuals, regardless of nationality. But it also places demands on individuals to be global citizens—or, more accurately in the context of climate change, good environmental (or atmospheric) citizens—and to act accordingly. To be sure, this is largely something for individuals, and one might argue that many people, particularly young people, are already moving in this direction (Attfield 1999: 191–206) (e.g. young people are more concerned about climate change and willing to act accordingly than are their parents (Pew Research Center 2009)). But a role for global institutions still obtains: one of their objectives should be to encourage global citizenship—and environmental citizenship—by supporting education programs that aggressively raise awareness of the causes and consequences of climate change. Where education is already deficient, international assistance would have to be enhanced so that children have a basis upon which to build environmental awareness and action.

This is only a small sample of the kind of changes to climate-related institutions that could be implemented to actualize both cosmopolitan aims and a healthier atmosphere.

6 Utility of a Cosmopolitan Reconceptualization

A cosmopolitan corollary to extant global climate governance would be principled, practical, and politically viable. It would be principled for a variety of the reasons (e.g. causality, capability, vital interests, harm, and so forth). Perhaps most simply, a cosmopolitan corollary is more principled and just because it attaches duties to those who cause global warming and advances the rights of those who suffer the most from it, regardless of their nationality.[7] A cosmopolitan corollary stops ignoring humans, their rights, and their duties. It recognizes and promotes the rights of people, especially the least well off, while incorporating the duties of all capable people, notably affluent ones all over the world, into efforts to address climate change. To be sure, for a cosmopolitan corollary to work, the principle of it will have to be taken seriously. This is not so much because principle matters but because, if diplomats see cosmopolitanism as just another instrument for promoting state interests, they and their governments will fall back into the same tragic behavior. By aiming the climate change regime at promoting cosmopolitan principles, meaning putting people at the center of global governance, diplomats can direct more attention to the causes and consequences of global warming and less to how traditional state-centric policies present problems for their states' perceived interests and long-held positions in the international relations of climate change.

A cosmopolitan corollary to the doctrine of inter*national* environmental justice is also practical. It reflects climate change realities rather than assuming that the problem, and all of the solutions to it, must or even can comport with the Westphalian assumptions of state sovereignty, rights, autonomy, and independence. Unlike current doctrine, it focuses on the actual source of much of the world's GHG pollution—individuals. It directly addresses the increasingly important role played by millions of newly affluent people in the developing world while still fully encompassing most people in the industrialized world and while recognizing that many poorer people in the latter are relatively minor contributors to the problem or not really capable of taking on obligations related to climate change. To be sure, the duty of most people in developed countries to act is undiminished, but we must stop letting all of the new consumers, and even rich elites, in developing countries pollute as their counterparts in the West have done for generations. Even if we accept that cosmopolitanism is idealistic in other spheres of human activity, in the case of climate change it is utterly practical and necessary.

A cosmopolitan corollary is also politically viable and likely to be politically essential if the world is to salvage the climate change regime and move it toward much more robust outcomes, especially in terms of mitigation but also in terms of adaptation. Because the corollary is not intended to replace the doctrine of inter*national* justice underlying the climate change regime, but rather to supplement it and build upon it, the corollary is unlikely to face the kind of opposition from people and governments that would be experienced by wholesale change. What is more, if the world's new consumers are brought into the climate change regime, as the corollary would require, people in rich countries will see affluent people in poor countries responding. This will make it far easier for governments of the developed countries to sell the climate change regime to their citizens. At the same time, governments of developing countries can agree to limit the greenhouse pollution of their affluent citizens without undermining long-standing demands for inter*national* justice. They need not do what they insist that they will not do—take on mandatory *national* commitments to cut emissions of GHGs. They can tell their citizens that they have won the argument in this respect. But the result on the ground is that large numbers of affluent people in developing countries will start to limit their emissions even as majorities of people there (the poor) are not required to do so and instead are aided to improve their living standards. In other words, pollution among some groups *within* developing states will decline even though pollution *of* those states is not required to do so before the affluent countries effectively take robust action.

7 CONCLUSION

Global climate governance must ultimately be about limiting harm to the Earth's atmosphere from GHG emissions and dealing with the consequences of emissions that cannot be avoided. From a cosmopolitan perspective, the reason for achieving these objectives is to ensure human well-being, or at least to limit human suffering. Equally from this perspective, it is human beings themselves who must bear much of the responsibility for dealing with the problem. The climate change regime, and the institutions of which it is comprised, exist to enable human well-being and human action. Conceiving of climate governance from this perspective, and indeed the idea of a cosmopolitan corollary to extant conceptions of climate governance, are not politically idealistic, least of all utopian. An advantage of conceptualizing climate governance in this way is its political palatability. It gives states and diplomats the political cover they need by allowing them to stick with their long-standing principles. States take on few new obligations when agreeing to a cosmopolitan corollary. Diplomats from developing states can go home and say, in all truthfulness, that they have not compromised on their demands for inter*national* justice, and even that people of the developed countries have, through their governments, agreed to take on new commitments to cut GHG emissions. Diplomats from developed states can go home and claim, accurately, that people in the developing countries have finally agreed, via their own governments, to take on new commitments to cut their GHG pollution. This gives an important political concession to the developing countries, and it gives the developed countries what they want and that which is required—involvement of the developing world in emissions cuts. The psychological impact of this new paradigm on populations in Australia, Canada, the United States, and other developed countries where people (and governments) have been waiting for developing countries to commit to GHG cuts could be very powerful. It gives governments the political insulation they need to finally do what they know, as states, they ought to have been doing for some time.

Climate change cries out for a cosmopolitan response. It is a global problem with global causes and consequences. As Held puts it, 'cosmopolitanism constitutes the political basis and political philosophy of living in a global age' (Held 2005: 27). The idea that states can continue to control what happens within their borders is no longer valid because 'some of the most fundamental forces and processes that determine the nature of life chances within and across political communities are now beyond the reach of individual nation-states' (Held 2000: 399). Climate change is one contemporary phenomenon that creates 'over-lapping communities of fate' requiring new cosmopolitan institutions (Vanderheiden 2008: 89). Generally speaking, however, the most that individuals must do at present is pay taxes and comply with minimal regulations imposed by some developed-country governments. Where cosmopolitan justice is especially important is in locating obligation—to stop harming the environment on which others depend and to take steps to aid those who suffer from harm to the environment—on the shoulders not only of governments but also of capable individuals. As Attfield points out, 'the global nature of many environmental problems calls for a global, cosmopolitan ethic, and for its recognition on the part of agents who thereby accept the role of *global citizens* and membership of an embryonic global community' (Attfield 2003: 182). Cosmopolitan justice, and the associated obligations, should therefore at minimum supplement the traditional inter*national* view of climate governance, although it should not dilute the common but differentiated responsibilities of states. This has significant implications for reshaping the climate change regime—in line with cosmopolitan and environmentally friendly objectives—without ignoring the reality of continuing state dominance of related institutions.

Notes

1. This can be distinguished from *government*, which typically refers to the formal institutions and organs through which states implement their rule. (For an elaboration of the distinctions between government and governance, see Rosenau 1992.)
2. Esty (2009: 428–9) lists a number of principles that should underpin the global environmental governance system, but these are generally focused on states and their interests (see also Saran 2009).
3. Parts of this chapter recount and build on ideas in Harris (2010 and 2011).
4. Like all doctrines, this one is often practiced in the breach (see Harris 2010).

5. For a summary of international justice and climate change, see the chapter by Baer in this volume. While Baer addresses cosmopolitan climate justice, he defines cosmopolitanism very broadly, essentially to encompass most obligations across borders. Baer does not address questions of individual responsibility, which are fundamental to what follows here.
6. For a examinations of the various permutations of cosmopolitan thought, see Brock and Brighouse (2005) and Dower (2007).
7. To be sure, individual behaviors that contribute to global warming are influenced by economic systems, infrastructures, and other forces, but most people are capable of overcoming some of these constraints, and indeed the newly affluent in the developing world are, like those before them and still in the developed world, willingly and enthusiastically embracing high-emitting lifestyles.

REFERENCES

ATTFIELD, R. 1999. *The Ethics of the Global Environment*. Edinburgh: Edinburgh University Press.
—— 2003. *Environmental Ethics*. Cambridge: Polity Press.
—— 2005. Environmental values, nationalism, global citizenship and the common heritage of humanity. Pp. 75–104 in J. PAAVOLA and I. LOWE (eds.), *Environmental Values in a Globalizing World*. London: Routledge.
BAER, P., ATHANASIOU, T., KARTHA, S., and KEMP-BENEDICT, E. 2008. *The Greenhouse Development Rights Framework: The Right to Development in a Climate and Aid Constrained World*. 2nd edn., Berlin: Heinrich Böll Foundation, ChristEcoEcquity, and the Stockholm Environment Institute.
BARNETT, M., and SIKKINK, K. 2008. From international relations to global society. Pp. 62–83 in C. REUS-SMIT and D. SNIDAL (eds.), *The Oxford Handbook of International Relations*. Oxford: Oxford University Press.
BIERMANN, F., and BAUER, S. 2005. *A World Environment Organization: Solution or Threat for Effective International Environmental Governance?* Aldershot: Ashgate.
BOTZEN, W. J. W., GOWDY, J. M., and VAN DEN BERGH, J. C. J. M. 2008. Cumulative CO_2 emissions: Shifting international responsibilities for climate debt. *Climate Policy* 8: 569–76.
BROCK, G., and BRIGHOUSE, H. (eds.) 2005. *The Political Philosophy of Cosmopolitanism*. Cambridge: Cambridge University Press.
CANEY, S. 2007. Cosmopolitanism, democracy and distributive justice. Pp. 29–64 in D. Weinstock (ed.), *Global Justice, Global Institutions*. Calgary: University of Calgary Press.
CHAKRAVARTY, S., CHIKKATUR, A., DE CONINCK, H., PACALA, S., SOCOLOW, R., and TAVONI, M. 2009. Sharing global CO_2 emissions reductions among one billion high emitters. *Proceedings of the National Academy of Sciences* early edition: 1–5.
DOBSON, A. 2006. Thick cosmopolitanism. *Political Studies* 54: 165–84.
DOWER, N. 2007. *World Ethics: The New Agenda*. 2nd edn., Edinburgh: Edinburgh University Press.
ELLIOTT, L. 2006. Cosmopolitan environmental harm conventions. *Global Society* 20(3): 345–63.
ESTY, D. C. 2009. Revitalizing global environmental governance for climate change. *Global Governance* 15: 427–34.

GOSSERIES, A. 2007. Cosmopolitan luck egalitarianism and the greenhouse effect. In Weinstock 2007: 279–310.

HARRIS, P. G. 2010. *World Ethics and Climate Change*. Edinburgh: Edinburgh University Press.

—— 2011. Cosmopolitan diplomacy and the climate change regime. In P. G. HARRIS (ed.), *Ethics and Global Environmental Policy: Cosmopolitan Conceptions of Climate Change*. Cheltenham: Edward Elgar Publishing.

—— and SYMONS, J. 2010. Justice in adaptation to climate change: Cosmopolitan implications for institutions. *Environmental Politics* 19(4).

HAYDEN, P. 2005. *Cosmopolitan Global Politics*. Aldershot: Ashgate.

HEATER, D. 1996. *World Citizenship and Government*. London: Macmillan.

HELD, D. 2000. Regulating globalization? The reinvention of politics. *International Sociology* 15: 394–408.

—— 2005. Principles of cosmopolitan order. Pp. 10–27 in G. BROCK and H. BRIGHOUSE (eds.), *The Political Philosophy of Cosmopolitanism*. Cambridge: Cambridge University Press.

LINKLATER, A. 2006. Cosmopolitanism. Pp. 109–30 in A. Dobson and R. Eckersley (eds.), *Political Theory and the Ecological Challenge*. Cambridge: Cambridge University Press.

MALTAIS, A. 2008. Global warming and the cosmopolitan political conception of justice. *Environmental Politics* 17(4): 592–609.

MYERS, N., and KENT, J. 2004. *The New Consumers*. London: Island Press.

Pew Research Center for the People and the Press. 2009. Fewer Americans see solid evidence of global warming. News release.

POGGE, T. 2002. *World Poverty and Human Rights*. Cambridge: Polity.

—— 2008. *World Poverty and Human Rights*. 2nd edn., Cambridge: Polity.

ROSENAU, J. (ed.) 1992. *Governance without Government: Order and Change in World Politics*. Cambridge: Cambridge University Press.

—— 2000. Change, complexity, and governance in a globalizing space. Pp. 169–200 in J. PIERRE (ed.), *Debating Governance: Authority, Steering, and Democracy*. Oxford: Oxford University Press.

SARAN, S. 2009. Global governance and climate change. *Global Governance* 15: 457–60.

VANDERHEIDEN, S. 2008. *Atmospheric Justice*. Oxford: Oxford University Press.

WHEARY, J. 2009/10. The global middle class is here: Now what? *World Policy Journal* 26(4): 75–83.

CHAPTER 44

THE ROLE OF INTERNATIONAL LAW IN GLOBAL GOVERNANCE

WALTER F. BABER AND ROBERT V. BARTLETT

1 Introduction

THE journey from the 1992 Framework Convention on Climate Change (UNFCCC) to its 1997 Kyoto Protocol to the 2009 non-binding Copenhagen Accord was an unsettling one for members of the international law community. Disagreement exists over how to characterize the Copenhagen Accord. Some regard it as an unmitigated disaster, a fig leaf for failure that might actually be worse than nothing at all. Others argue that, although the Accord fails to establish the binding carbon emissions targets that were hoped for, it at least creates a new multi-polar order that engages the developed nations whose participation is vital to future progress (see Bäckstrand, this volume). What seems clear, however, is that years of preparation by professional staff and months of arduous diplomatic negotiations over a draft agreement (Betsill 2011) were thrown overboard in negotiating the Accord. Politics prevailed and international law seems, once again, to have failed to live up to expectations.

As with so many other things, however, frustrated expectations occur in a context and that context needs to be understood if the frustration is to be dealt with effectively. To this end, we pose here three questions that provide the structure for the rest of this chapter: (1) Of what does international law consist and (perhaps more important) what does it not contain? (2) What can international law reasonably be expected to do about global climate change? (3) Could the reach of international law regarding global climate change be increased?

2 The Contours of International Environmental Law

2.1 The Elements of International Law

At the most general level, international law is a collection of agreements that represent the will and consent of nation-states with respect to the rules that govern their relationships. It might best be understood as the law of a global polity that consists of only about 200 individuals, each of whom is corporate in character. One finds the law of this polity in a number of places.

Treaties are the nearest thing that the international community has to legislation and they are the increasingly predominant form of international law (Weiss et al. 2007). Unlike domestic legislation, however, treaties must pass through two distinct legislative procedures. The international negotiation that produces the text of the 'legislation' is followed by a ratification process in which each of the signatories employs domestic legal institutions to commit itself formally to the agreement that it has already negotiated. For example, the UNFCCC was negotiated prior to the 1992 United Nations Conference on Environment and Development, where it was signed by 154 countries. But it did not enter into force until after it had been ratified by the fiftieth country (the minimum specified in the treaty itself) in December 1993. It is as if the law is actually adopted by a promise of obedience from each of the polity's members, acting in what Kelsen characterized as their international legal personalities (Kelsen 2006 [1949]).

In this way, treaties share more in common with contracts than with domestic laws that are the expression of will of a single legal personality. They codify an agreement to behave in a certain way, but admit no necessary obligation to do so. In fact, treaties may be something even less than a formal contract. They may be most nearly similar to a letter of intent, creating a record of the agreement in more careful language than everyday speech, clarifying the terms of cooperation, providing a basis for mutually beneficial interaction, but anticipating no form of external enforcement (Goldsmith and Posner 2005). In the environmental arena, one can chart a progression in the use of treaties from simple bilateral accords, like those between the United States and Canada respecting their boundary waters, to multilateral agreements designed to confront genuinely global challenges like protection of marine life or prevention of atmospheric ozone depletion.

International law also can be found in international custom. Of minimal direct relevance to climate change so far, customary law consists of those sets of norms derived from the actual practices of states undertaken in the belief that those practices are required of them by law (Weiss et al. 2007). Customary international law is found, therefore, at the intersection of habitual behavior and mutual obligation. Customary law is recognized (at least in principle) as obligatory by nations and legal scholars alike, in spite of the fact that it has never been memorialized in writing (Young 1994). Customary international law is, however, more difficult to invoke than is treaty law. One must be able to both articulate the terms of the putative law and then show that the law is accepted by states as law (Hunter et al. 2007).

A number of observations about this process are helpful to an understanding of its potential and its limitations. First, the invocation of customary law is an empirical process rather than a normative one. It attempts to describe existing rules that govern the interaction of nations rather than to propose new rules. Second, invoking customary law requires a showing that state compliance with the rule is both extensive and virtually uniform and that it particularly includes the state(s) affected by the case at hand. It must be clear that state conduct inconsistent with the putative rule has generally been treated as a breach of international obligations. Third, the pattern of compliance must be attributable to a sense of legal obligation (*opinio juris sive necessitatus*). It is insufficient that states follow the putative rule either out of a sense of moral obligation or in pursuit of political expediency. Finally, even if these requirements are satisfied, states may insulate themselves from the application of customary law if they can demonstrate that they have been persistent objectors. If a state can show a consistent record of conduct reflecting an unwillingness to be bound by the rule or to recognize it as law, it can exclude itself from any obligations under the rule (Bodansky 1995).

Examples of customary law in the environmental area might include the principle that a state should not use its territory in a way that causes environmental harm outside that territory. The precautionary principle has been suggested as another candidate for customary law status, despite the fact this putative rule is subject to significant contestation (Peel 2011). Both principles have been invoked with respect to climate change, but there is no evidence that either has yet guided actual behavior. The difficulty in proffering examples of customary law thought to be at work in the environmental area reflects the limitations implicit in that form of law. It is difficult, for example, to determine whether a reluctance to cause harm outside of one's territory is grounded in a sense of legal obligation or merely a desire to maintain positive relations among neighboring states. The precautionary principle may eventually fail as customary law because its application to concrete circumstances raises concerns of over-regulation that undermine its universality as a rule. Customary international law is thus plagued by both an inability to characterize accurately the behavioral regularities that are its foundation and a fundamental ambiguity about how to apply its rules to contested cases (Goldsmith and Posner 2005). The real influence of customary law principles may be 'to frame the debate rather than to govern conduct' (Bodansky: 2010a: 203).

The decisions of international tribunals interpret and apply treaties and customary law and as such they influence the evolution of international law. But the potential of prior decisions actually to guide the development of international law is largely unrealized and will remain so until international jurisprudence makes them relevant by recognizing the doctrine of *stare decisis* that decisions are precedents binding on future decisions and precedents should not be ignored or overturned unless there are strong reasons to do so (Bhala 1999, 2001) and until the number of decided cases rises to the level that makes possible the drawing of generalizations about what they teach.

Additional sources of international law are worthy of mention, although they continue to be of lesser importance than treaties and customary law. General principles of law are behavioral rules that are widely recognized in nations with formally organized legal systems. By definition, these principles are characteristic of all (or nearly all) domestic legal systems and, for that reason, enjoy a presumptive validity at the international level. An example, perhaps, is the 'polluter pays' principle, which is encountered widely in domestic law and has been so well received among nations and international organizations that it has become virtually synonymous with environmental dispute resolution (Larson 2005).

International administrative law is that law which governs the authorization and operation of intergovernmental organizations. Largely procedural in character, this form of law could be expected to be relevant to environmental protection only indirectly and only to the extent that it structures the actions of organizations with environmental responsibilities (Weiss et al. 2007). For example, much of the monitoring, reporting, and other technical work of the UNFCCC Secretariat is done pursuant to rules that are not contained in formal treaties. And under the provisions of the Vienna Convention for the Protection of the Atmosphere, the Montreal Protocol, and its amendments, some high global-warming potential gases—notably HCFCs—are regulated in part through adjustments made to annex lists of chemicals and control levels rather than through formal amendment of the treaties themselves.

Finally, soft law occupies an ambiguous status. It can manifest itself in aspirational or hortatory language contained in an otherwise binding document or in apparently obligatory language in a document which is not itself binding (Weiss et al. 2007). Perhaps the most significant example of soft law in relation to international climate change is the 2009 Copenhagen Accord, which was adopted by the UNFCCC Conference of the Parties in the face of inability to agree on terms of a substantive protocol to succeed the Kyoto Protocol. It is worth remembering, however, that soft law can stand in one of several relationships to hard law. It can serve as a next-best alternative to hard law where no hard law has been developed and, occasionally, even encourage the development of hard law. It can also be a complement to hard law that can facilitate dialogic and experimentalist transnational and domestic processes transformative of norms, understandings, and perceptions of state interests. But the presence of distributive conflicts among states, in particular among powerful states, coupled with the coexistence of hard- and soft-law regimes within a regime complex, can undermine the smooth and complementary interaction of hard and soft law to such an extent that the two become adversaries (Shaffer and Pollack 2010). Whether this oppositional relationship is positive or negative depends, of course, on one's estimation of the legitimacy and effectiveness of the hard law involved. It can be either a creative tension that gives the regime a necessary reflexive capability or a source of internal contradiction that promotes gridlock. It will be some time before it becomes clear, for example, whether the impact of the 2009 Copenhagen Accord on the development of hard international climate change law will be significantly positive or negative or both.

2.2 What International Environmental Law Is Not

Having discussed in broad outline what international law is, we now turn our attention to what international law is not. One way of orienting ourselves is to ask, what elements of domestic legal systems are not to be found in the international legal system? When approached in this way, three observations occur immediately.

First, as a system of governance, international law lacks the capacity to grow. It is incapable of extending its reach and developing new solutions to problems as they arise in the policy environment. Any new competencies that the system ever enjoys are introduced from outside. This stands in marked contrast to domestic systems of common law, which can develop new legal doctrine and new behavioral norms by working their way through a large number of concrete cases, breaking legal ground and experimenting with new rules as they go (Baber and Bartlett 2009). International law is not prevented from doing this by the lack of institutional infrastructure. International tribunals like the International Court of Justice (ICJ) exist and deal with cases, often guided (as Article 38 of the ICJ Statute provides) by the writings of eminent legal scholars and the decisions of judges in national courts, though none of those sources are international law in themselves. The United Nations' International Law Commission (ILC) is fully capable of restating rulings of the ICJ in the same way that, for example, the American Law Institute restates the common law in a wide variety of fields for the guidance of legal practitioners in the United States. It is true that the ICJ and ICL do not enjoy the same huge 'database' of court decisions that exist in common law countries like the United States. But even that limitation (which, like the inexperience of youth, would correct itself over time) is not the heart of the problem.

The critical element that is missing from international law that prevents it from learning and growing is the absence of the doctrine of *stare decisis* (Hunter et al. 2007). Decisions of international tribunals, even those of the ICJ, are binding only upon the parties to the dispute at hand. Although it is true that prior decisions are used by later courts as guides in finding the content of the law, the fact that international adjudications are not binding even on the courts that render them encourages exception mongering by parties to future disputes and discourages progressive development of the law through the careful analysis of precedent that characterizes common law legal systems. Some commentators view this as a largely mythical problem, or a self-inflicted wound, born of the fact that international jurists hew to 'non-binding' precedent while maintaining the fiction that their prior decisions are not actually binding (Bhala 1999). Be that as it may, the more pertinent observation from our current perspective is that a de jure doctrine of *stare decisis* in international adjudication would help assure the coherence in international environmental law that both states and international organizations ought to expect if they are serious about building a well-ordered multilateral system of environmental protection (Bhala 2001).

Second, as a form of political praxis, international law lacks democratic legitimacy. In the leading forms of democratic government, democratic legitimacy is grounded in an elected legislature. The key feature of both parliamentary systems and presidential systems is that citizens choose representatives to legislate on their behalf. Whether one affirms a democracy that is representative or directly participatory (Barber 2003), the essence of the matter is that citizens are empowered to live under laws of their own choosing. 'To govern oneself, to obey laws that one has chosen for oneself, to be self-determining...' is the objective of democratic practice (Dahl 1989: 89). Contrast this emphasis on democratic legislation with the observation that at the international level, it is the legislative function that is the least well developed. International 'legislators' lag behind their judicial and executive counterparts for a variety of reasons, 'ranging from constituents' suspicion of 'junkets' to their own inability to stay in office long enough to develop enduring relationships with their foreign counterparts' (Slaughter 2004: 127). The Campaign for a United Nations Parliamentary Assembly notwithstanding, it is not clear how one would go about restructuring international institutions to create elected legislators with more meaningful roles, even assuming that that such institution building would be desirable and the difficulties arising from domestic political circumstances could be overcome. A more fundamental question is precisely what legislators should be doing within the confines of an international legal system, conceived of as part of a larger system of 'global governance in a disaggregated world order' (130)?

Third, as a form of environmental problem solving, international law lacks sufficient implementation mechanisms. Within the context of international law, the issue of implementation is sometimes reduced to nothing more than a question of coercive enforcement and promptly dismissed as impossible in an international polity that remains fundamentally anarchic. This is unfortunate for at least two reasons. One is that coercion and cooperation are not discrete categories, even at a conceptual level. In many respects, the two are functionally identical, consisting of 'acts, threatened acts, or offers of action' directed at one state by another state with the intent of inducing a change in the latter state's behavior (Goldsmith and Posner 2005). Conflating implementation and coercive enforcement is also to be avoided because it obscures the fact that implementation of international agreements is a varied collection of specific actions that international actors (both states and IGOs) take to make international agreements operative in national legal systems (Faure and Lefevere 2011). Ignoring this obstructs our efforts to imagine and perfect implementation mechanisms that do not depend on conventional adjudication, which has not been and is not likely to become an effective means of inducing compliance with international environmental obligations (Knox 2001). Finally, the inadequacy of implementation mechanisms in international environmental law leads to an inability of the system to adapt to changing circumstances. If every element required to implement the mandates of international law must be negotiated in detail by the principals, that system will be inflexible and slow to react to new challenges. This will be a particularly serious flaw in the area of climate, which is subject to nonlinear and sometimes abrupt changes (see Young, this volume).

In summary, international environmental law exists within a system of governance that features institutions answering to the general description of courts, legislatures, and executive agencies. But those institutions each lack at least one essential element of their domestic counterparts. International courts lack the ability to bind themselves (and potential future litigants) through their decisions and to accumulate binding decisions into a coherent superstructure of legal norms that can be studied and, perhaps, codified. They must wait to be supplied with legal rules by 'legislation' to an extent that courts in common law countries need not. International 'legislatures' lack the democratic legitimacy provided by the electoral processes that have come to be regarded as the *sine qua non* of democracy at the domestic level. The paths of public accountability between citizen and the government official who represent them in international negotiations are so attenuated that the results are immediately suspect from the democratic perspective. And, finally, international executive agencies are hamstrung by a lack of the authority to craft appropriate implementation mechanisms. Without this form of authority, which has made the modern regulatory state possible, it is hard to imagine how systems of global environmental regulation could ever be developed or maintained.

3 WHAT INTERNATIONAL LAW CAN (AND PROBABLY CANNOT) DO ABOUT CLIMATE CHANGE

3.1 The Record to Date

At the outset, it will be helpful to review what international law has accomplished with respect to climate change. Generalizations about potential future accomplishments based upon that analysis are more likely to prove reliable than more speculative predictions. So, let us begin at the beginning.

The United Nations Framework Convention on Climate Change (UNFCCC) was signed in 1992. It set out a broad structure for climate governance, including both general objectives and principles and an institutional structure for international cooperation in confronting the challenge of climate change. The UNFCCC, however, set no binding targets or timetables for the reduction of greenhouse gas (GHG) emissions and empowered the international organizations that it established to do little more than prepare for future negotiations on the subject (Bodansky 1993). In part as a result of its own modesty, the Convention achieved rapid acceptance by nearly every country in the world. It is worth noting that the UNFCCC is only one example of a capacity that international law possesses that most systems of domestic law lack. As a framework convention, it constitutes a commitment on the part of its signatories to continue negotiating as members of an ongoing 'community of fate' with respect to climate change (see Harris, this volume). The Framework was, so to speak, an agreement not to abandon the search for agreement. This was, of course, a vital first step in adding climate to the agenda of global governance.

The Kyoto Protocol, adopted by the third Conference of the Parties to the UNFCCC (COP-3) in 1997, entered into force in 2005 without the participation of the United States. Unlike the UNFCCC, the Protocol contained a concrete commitment on the part of the European Union and other developed nations to cut their GHG emissions, to be achieved during an initial commitment period expiring in 2012. That the EU and other industrialized nations agreed to proceed with the Kyoto commitments in the absence of US participation can be seen as an encouraging indication of the level of commitment to climate governance that has been achieved as a result of the UNFCCC process. Moreover, the vast majority of countries are in compliance with their reporting commitments under that framework.

The 2009 Copenhagen Accord is the result of the 15th Conference of the Parties of the UNFCCC. COP-15 was to have focused on what steps would follow the first commitment period under the Protocol. The expectation was that a new set of targets, extending the reach of Kyoto to include the United States and fast-growing economies of the developing world, would be forthcoming. The decision by more than 100 heads of state to attend only served to heighten expectations of a major agreement. But a lack of negotiating progress in the months leading up to the meeting foretold a different outcome—a purely political non-binding agreement negotiated by a group of roughly twenty-eight nations and explicitly objected to by a handful of others (Hunter 2010) but ultimately agreed to by over 135. The Copenhagen Accord sets no emissions targets. It merely establishes a goal of holding long-term climate change to an increase of two degrees Celsius. It also contains a system of 'pledge and review' for mitigation actions by individual countries as well as a commitment by developed nations to provide substantial new financial resources to aid developing nations in meeting the challenge of climate change (Bodansky 2010b).

3.2 The Actually Possible and the Merely Plausible

What one believes to be possible going forward from Copenhagen depends to a considerable degree upon what one concludes actually happened in Copenhagen. Dubash describes the Accord as perhaps the worst possible outcome, a thin layer of success concealing a deeply flawed outcome that serves only to perpetuate entrenched differences (Dubash 2009). Even less pessimistic observers concede that the Accord must be viewed as a major setback for anyone seeking a hard, binding agreement containing specific performance requirements (Hunter 2010). If this perspective is the basis for our judgment, Copenhagen was a disaster and international environmental law stands convicted of impotence in the face of climate change. It is, however, possible to sustain a different perspective.

The Copenhagen Accord can be viewed as having introduced new and potentially positive elements to the ongoing UNFCCC process. It can be regarded as an embryonic statement that is evidence of a real struggle to build a climate consensus that includes all nation-states (Burleson 2010). The choice to accept a non-binding statement reflected disagreement not over whether there should be a binding agreement but, rather, on what requirements it should contain and to whom they should apply. Although this is, of course, no small matter, the fact that the argument is now joined and that both developed and developing nations have accepted that they will have climate change responsibilities opens a door that Kyoto had left closed (Hunter 2010). The Copenhagen Accord constitutes an agreement to participate in a bottom-up process in which all nations will list their climate activities internationally and submit those actions to international scrutiny. It also articulates, for the first time, a quantified long-term goal of holding global warming below 2 degrees Celsius. Moreover, a number of potentially positive process developments can be discerned. There was an apparent shift among the BASIC countries (Brazil, South Africa, India, and China) in moving away from a long-standing hardline position against accepting any responsibility for mitigation that signals a lowering of the wall between the developed and developing nations. That these emerging economies have agreed to the 'internationalization' of their national climate change policies (although without agreeing to forgo any degree of sovereignty) suggests that the monolithic divisions that plagued Kyoto might eventually be overcome. A more fluid constellation of forces holds out at least some hope for future negotiations (Bodansky 2010b).

This more positive appraisal of Copenhagen makes it at least plausible that international law will be able to find an improved balance between equity and efficiency considerations in supporting the climate change activities of developing countries, expanding and strengthening market-based mechanisms like the cap-and-trade system, and finally achieving an ecologically sustainable climate regime (Burleson 2010). If that future regime is achieved, it will be because we have learned that the narrow legal question of whether international commitments are binding is not really very important. We will have discovered that the appropriate criteria for evaluating international environmental law is not whether a binding climate agreement was adopted but whether action was taken that actually reduced the risk of significant climate disruption (Hunter 2010). So it would be wise to give some consideration to what it is unlikely that international law will be able to do with respect to climate change.

3.3 The Implausible

Very few things are genuinely impossible (given sufficient time for experimentation and observation). But it seems clear that certain desirable climate-relevant outcomes are implausible if we are limited to the current processes of international law. We focus here on three such implausible outcomes, three things of which international law would seem to be incapable. They are current limitations in the practice of international environmental law that are both significant obstacles to progress on climate change and within our ability to address. These limitations correspond to the general shortcomings of international law described in section 2.2 above.

First, international environmental law as a practice is generally unable to derive substantive norms from its own concrete experiences. For example, in the context of climate change the fundamental norm of common but differentiated responsibility did not emerge organically from the practice of resolving environmental disputes. A comparison to common law practice may be useful to show the importance of this point. The doctrine of contributory negligence was used to deny damages to a plaintiff whose own lack of ordinary care had contributed to his loss. This doctrine, which originally developed to avoid rewarding people for their own carelessness, eventually produced such troubling results in so many cases that it was replaced by the doctrine of comparative negligence. This new approach did not deny recovery to a careless plaintiff. Rather, it reduced any damages by that proportion which represented the plaintiff's contribution to his own injury. The doctrine of common but differentiated responsibility, it could be argued, operated under the Kyoto Protocol as a mirror image of contributory negligence. It excused from responsibility for the problem of climate change those parties whose contribution to it were slight. It allowed what was different in the climate change problem to obscure what was common to the detriment of responsibility as a whole. But because it was developed as an abstract principle, rather than within the context of a continuing process of problem solving, it lacked any compelling provenance and was not susceptible of refinement through cumulative experience.

Second, whether one regards Copenhagen as a step forward or a retreat (from either a legal or an ecological perspective), it is clear that it was a failure from the perspective of democratic legitimacy. Despite the enormous turnout of civil society representatives, the Accord was developed and adopted behind closed doors with little or no opportunity for those represented by the negotiators to be heard when it really counted (Bodansky 2010b). This, however, is not a feature unique to COP-15. As discussed above, the democratic deficit in international negotiations is systemic and pervasive. Indeed, given the evidence that Kyoto enjoyed the support of most Americans at the time that it was negotiated (Nisbet and Myers 2007), a failure of democratic representation would seem to have created the background conditions that made the COP-15 negotiations so intractable in the first place. If a democratic deficit of this sort is allowed to continue, the results could be particularly serious in the area of climate change because addressing that problem implicates the most basic processes of modern industrialized economies. Reliable public support for whatever preventative measures are developed will be vital, but ultimately unavailable if the policy process has been anti-democratic.

Finally, the Copenhagen Accord might be regarded as an important step forward because it recognizes a limitation of international environmental law that Kyoto ignored. The Accord establishes a bottom-up process of climate policy development in which signatories list their national actions internationally and subject them to some as yet under-defined process of global scrutiny (Bodansky 2010b). Although it is difficult to know precisely how this process will be developed in the coming years, it represents a realization that implementing any climate change regime will be unavoidably dependent on the member states to play the leading role in matters of implementation (see Young, this volume). Again, this dependence is neither unique to the area of climate change nor is it a fatal flaw in the Copenhagen approach. It is, however, an implicit recognition that the implementation capacities that are required to make any climate regime effective exist (or must be developed) at the level of the state and even below that level. Financial and technical assistance can be organized internationally. Effective action to prevent environmental degradation probably can never be.

4 WHAT WE MIGHT CHANGE

The development of fundamental environmental norms, the democratic choice among competing policy prescriptions, and the organization of implementation plans are probably beyond the reach of international environmental law as presently conceived. It is not impossible, however, to imagine means by which these limitations could be addressed, if not completely overcome.

As we have observed, international law (unlike domestic common law systems) is incapable of generating legal norms internally by adjudicating a large number of cases under the doctrine of *stare decisis*. Generally speaking, the influence of precedent in international law has been limited by the myth that *stare decisis* does not apply in that system. But even if that doctrine were adopted, development of international law would proceed at a glacial pace compared to that of the common law simply because so many fewer cases arise when there are only about 200 legal personalities in the jurisdiction (even if they are unusually litigious when compared to the nearly 7 billion natural personalities they represent). One solution that has been proposed is for international governmental organizations to initiate a process of adjudicating hypothetical disputes, each of which would be crafted so as to present clear choices between contending legal norms. Citizen juries, on a worldwide basis, would be asked to resolve such disputes consistent with their collective sense of justice. Legal scholars would then analyze and summarize the verdicts in the same way that lines of case law are subjected to 'restatement' (Baber and Bartlett 2009). Such restatements could then provide the precedential foundation for international tribunals to resolve real world disputes over environmental issues in ways that would lead to the progressive development of international environmental law. If international environmental law could teach itself to learn in this way, the results might be quite valuable. In the United States and elsewhere, a majority of environmental cases are filed under common law theories. Moreover, common law provides the conceptual foundation for most statutes and regulations. Legislators and bureaucrats alike rely on common law to fill in gaps in public law and to guide both courts and agencies in their interpretation of statutes and rules (Plater et al. 2004).

Restatements of this sort might also serve as the raw material for a process of codification (through conventional international negotiations) that could produce results with a stronger democratic pedigree than is currently typical. As an alternative, democratic input for the process of regime negotiation could be provided by adapting the process of deliberative polling to the needs of international diplomacy. Deliberative polling brings together a stratified random sample of citizens in the sponsoring jurisdiction to participate in both small-group discussions and larger plenary sessions that allow them to assimilate substantial and balanced information about a pending issue of public policy, exchange views on competing policy alternatives, and come to considered judgments (as opposed to unreflective opinions) about what the group considers to be the most reasonable course of action (Ackerman and Fishkin 2003). In effect, average citizens are permitted to play the role of legislator for a limited period with respect to a single policy problem involving a limited number of alternative solutions. While the results of such exercises are not binding on actual legislators, eventual policies that deviate markedly from patterns of decision revealed by deliberative polls are immediately suspect and may be avoided by elected and appointed officials for precisely that reason (Gastil and Levine 2005).

Finally, the development of implementation plans in environmental policy has long been viewed as an ideal opportunity for democratic decentralization and deliberative innovation. If the logic of the Copenhagen Accord prevails, first national and then subnational plans for climate action will be encouraged. Stakeholder partnerships would provide a useful model of how such processes of implementation planning could be developed. A revealing example is the watershed partnership. The management of watershed areas presents serious challenges that are both ecological and political. The areas under management are usually large and environmentally complex. This scale and complexity poses problems that are legal and political as well as scientific. Watersheds often cross jurisdictional boundaries. Their resources are exploited by a variety of stakeholder groups in ways that are sometimes complementary and sometimes competing. A watershed partnership responds to this complexity and conflict by convening a planning group composed of any and all interested parties. It has no authority to issue binding regulations. But these groups often contain corporations or large landholders who are encouraged to moderate their behavior by civil society groups who then refrain from pursuing oppositional strategies in response to which governmental officials ratify these voluntary agreements in their management of watershed resources (Baber and Bartlett 2005). Stakeholder partnerships are widely used in other areas of environmental regulation and clearly provide a template for the creation of climate action plans at the national and subnational level and the restrained role of international governmental organizations contemplated by the Copenhagen Accord.

5 A Conclusion—Of Sorts

Even though the Copenhagen Accord left many with the sense that the UNFCCC process is fractured beyond repair, there is at least one irrefutable reason for trying to make the most of it. There really is no available alternative. The ad hoc processes that COP-15 set in motion will undoubtedly continue, driven by small groups of international specialists working to reassemble a coherent climate policy out of the broken shards of Kyoto. As Bodansky has observed, however, 'if world leaders were unable to make further progress through direct negotiations, under an intense international spotlight, there is little reason to expect midlevel negotiators to be able to achieve a stronger outcome anytime soon' (Bodansky 2010b: 240). Given this rather dark appraisal of our shared prospects, we close with a simple question. If world leaders have failed and functionaries faltered, why not let the world's citizens have a go at the problem? Success would be far from certain. Even the most democratic polities make mistakes. It may turn out that the poet Piet Hein was correct when he said that we are all 'global citizens with tribal souls' (Barnaby 1988: 192). On the other hand, there are significant commonalities among the chthonic legal traditions that typify the world's many tribal societies (Glenn 2010). To the hopeful mind, these repeating themes and recurring traditions are enough to suggest that we may be able to discover our obligations to one another and embrace what we have in common if we distance ourselves sufficiently from all that differentiates us.

References

ACKERMAN, B., and FISHKIN, J. 2003. Deliberation Day. In J. Fishkin and P. Laslett (eds.), *Debating Deliberative Democracy*. Malden, MA: Blackwell.

BABER, W. F., and BARTLETT, R. V. 2005. *Deliberative Environmental Politics: Democracy and Ecological Rationality*. Cambridge, MA: MIT Press.

—— —— 2009. *Global Democracy and Sustainable Jurisprudence: Deliberative Environmental Law*. Cambridge, MA: MIT Press.

BARBER, B. R. 2003. *Strong Democracy: Participatory Politics for a New Age*. Berkeley and Los Angeles: University of California Press.

BARNABY, F. (ed.) 1988. *Gaia Peace Atlas*. London: Pan Books.

BETSILL, M. M. 2011. International climate change policy: Toward the multilevel governance of global warming. In R. Axelrod, S. D. Vandeveer, and D. L. Downie (eds.), *The Global Environment: Institutions, Law, and Policy*. Washington, DC: CQ Press.

BHALA, R. 1999. The myth about stare decisis and international trade law (part one of a trilogy). *American University International Law Review* 14: 845–955.

—— 2001. The power of the past: Towards de jure stare decisis in WTO adjudication (part three of a trilogy). *George Washington International Law Review* 33: 873–978.

BODANSKY, D. 1993. The U.N. framework on climate change: A commentary. *Yale Journal of International Law* 18: 451–558.

—— 1995. Customary (and not so customary) international environmental law. *Indiana Journal of Global Legal Studies* 3: 105–20.

—— 2010a. *The Art and Craft of International Environmental Law*. Cambridge, MA: Harvard University Press.

—— 2010b. The Copenhagen climate change conference: A postmortem. *American Journal of International Law* 104: 230–40.

BURLESON, E. 2010. Climate change consensus: Emerging international law. *William & Mary Environmental Law and Policy Review* 34: 543–87.

DAHL, R. A. 1989. *Democracy and its Critics*. New Haven: Yale University Press.

DUBASH, N. 2009. Copenhagen: A climate of distrust. *Economics and Politics Weekly*.

FAURE, M. G., and LEFEVERE, J. 2011. Compliance with global environmental policy. In R. S. Axelrod, S. D. Vandeveer, and D. L. Downie (eds.), *The Global Environment: Institutions, Law, and Policy*. 3rd edn., Washington, DC: CQ Press.

GASTIL, J., and LEVINE, P. (eds.) 2005. *The Deliberative Democracy Handbook: Strategies for Effective Civic Engagement in the Twenty-First Century*. San Francisco: Jossey-Bass.

GLENN, H. P. 2010. *Legal Traditions of the World: Sustainable Diversity in Law*. New York: Oxford University Press.

GOLDSMITH, J. L., and POSNER, E. A. 2005. *The Limits of International Law*. New York: Oxford University Press.

HUNTER, D. 2010. Implications of the Copenhagen Accord for global climate governance. *Sustainable Development Law & Policy* 10: 4–15.

—— SALZMAN, J., and ZAELKE, D. 2007. *International Environmental Law and Policy*. New York: Foundation Press.

KELSEN, H. 2006 [1949]. *General Theory of Law and State*. New Brunswick, NJ: Transaction Publishers.

KNOX, J. H. 2001. A new approach to compliance with international environmental law: The submissions procedure of the NAFTA Environmental Commission. *Ecology Law Quarterly* 28: 1–122.

LARSON, E. T. 2005. Why environmental liability regimes in the United States, the European Community, and Japan have grown synonymous with the polluter pays principle. *Vanderbilt Journal of Transnational Law* 38: 541–75.

NISBET, M. C., and MYERS, T. 2007. The polls—trends: Twenty years of public opinion about global warming. *Public Opinion Quarterly* 71: 444–70.

PEEL, J. 2011. Environmental protection in the twenty-first century: The role of international law. In R. S. Axelrod, S. D. Vandeveer, and D. L. Downie (eds.), *The Global Environment: Institutions, Law, and Policy*. 3rd edn., Washington, DC: CQ Press.

PLATER, Z. J. B., ABRAMS, R. H., GOLDFARB, W., HEINZERLING, L., WIRTH, D., and GRAHAM, R. L. 2004. *Environmental Law and Policy: Nature, Law, and Society*. New York: Aspen Publishers.

SHAFFER, G. C., and POLLACK, M. A. 2010. Hard vs. soft law; Alternatives, complements, and antagonists in international governance. *Minnesota Law Review* 94: 706–99.

SLAUGHTER, A.-M. 2004. *A New World Order*. Princeton: Princeton University Press.

WEISS, E. B., MCCAFFREY, S. C., MARGRAW, D. B., and TARLOCK, A. D. 2007. *International Environmental Law and Policy*. New York: Aspen Law & Business.

YOUNG, O. R. 1994. *International Governance: Protecting the Environment in a Stateless Society*. Ithaca, NY: Cornell University Press.

PART XII

RECONSTRUCTION

CHAPTER 45

THE DEMOCRATIC LEGITIMACY OF GLOBAL GOVERNANCE AFTER COPENHAGEN

KARIN BÄCKSTRAND

1 INTRODUCTION

THE UN climate summit or the 15th conference of parties (COP-15) in Copenhagen in Denmark 7–18 December 2009 marked a turning in point in global climate politics. The Swedish Presidency of the European Union dubbed the meeting as a 'disaster' while President Barack Obama described it as a considerable success (Egenhofer and Gieorgiev 2009). The summit can be conceived of as a watershed in climate politics that marked the rise of a new multi-polar climate order geared toward the US–China simultaneously with a decline of UN climate multilateralism. Expectations of a new global legally binding climate treaty with universal participation and binding carbon emission targets for the industrialized countries were not met by the Copenhagen Accord,[1] which is a nonbinding two-and-half-page political declaration 'taken note' of by fifty-five parties rather than adopted in consensus (United Nations 2009). The Copenhagen summit has also been interpreted as the return to the geopolitics of climate change as the future direction for global climate diplomacy emerged from bargaining between major nation-states (Grubb 2010). Even if grand drama and collapse has been the stuff of international climate diplomacy from Kyoto to Hague, COP-15 has been framed as a legitimacy crisis for international climate diplomacy. The multilateral negotiations between 193 states were replaced by a G-2 bargaining between China and the US during the last day of frantic efforts to secure a political agreement. The framing of the failure in Copenhagen is inextricably linked to failure to establish a universal legally binding agreement modeled on the Kyoto Protocol containing negotiated targets and timetables.

This chapter links the legitimacy crisis for international climate politics after Copenhagen to scholarly debates on the prospects for global democracy. It examines how international climate governance after Copenhagen taps into ideals of global democracy, by bringing normative theories of global democracy to the empirical practice of contemporary global climate change governance. However, a critical question is whether global democracy is the solution to the pressing problems of climate change and a remedy to the weak outcomes in Copenhagen. While cosmopolitan scholars argue that global democracy is necessary to adequately respond to the climate change threat, skeptics question if it is possible or even desirable to democratize transnational climate change governance (Cerny 2009; Dahl 1999). The answer, of course, depends ultimately on the standards of democracy you apply. This chapter starts from the premise that 'experimenting with what democracy can mean is an essential part of democracy itself' (Dryzek 2000: 135). Three alternative accounts for democratizing global governance are advanced to assess the democratic credentials of the global climate order: democratic intergovernmentalism, transnational deliberative democracy, and global stakeholder democracy. They represent three different paths to global democratization beyond the cosmopolitan visions of world government, the realist dismissal of global democracy and critical perspectives' rejection of international democracy as inevitably hegemonic (Bäckstrand 2006: 494). However, they also overlap in rethinking notions of global democracy, which are compatible with the non-electoral, horizontal, and non-hierarchical features of world politics and move beyond state-centric models of democracy. While the efforts to democratize transnational climate governance may seem insurmountable, we should bear in mind that climate change also poses challenges to national democracies. In liberal democracies with short-term electoral cycles and interest politics, climate change as a 'wicked problem'—defined by long-term time frames, cross-border impacts and scientific uncertainties—poses numerous challenges for democratic decision making. No liberal democratic state has yet met these challenges effectively (Held and Hervey 2009; Holden 2002).

The perceived legitimacy crisis of international climate policy precipitated by COP-15 taps into debates on the 'democratic deficits' of international organizations and global governance arrangements, and how to make these more accountable, transparent, and inclusive. In the wake of the anti-globalization protest since the late 1990s, the language of democracy has gained currency in debates on the reform of global institutions and efforts to counter the democratic deficit in European Union (Moravcsik 2004). Democratic values, such as participation, representation, deliberation, inclusion, and accountability are part and parcel of the mainstream rhetoric of multilateral institutions. Simultaneously, there has been a virtual explosion of policy-oriented literature on a post-Kyoto climate treaty as the deadline for finalizing an agreement was approached (and missed) in Copenhagen in December 2009. While the study of global democracy and international climate policy are separate academic fields, they are united by efforts to reconcile normative ideals with political realities, such as aspirations for transnational democracy or a fair and effective post-2012 global climate treaty in a world divided by sovereign states.

The second section advances three paths for the democratization of global governance. The third section briefly summarizes the outcome of COP15—the Copenhagen Accord—and examines how the three accounts of global democracy interpret democratic credentials of the outcome in Copenhagen. The fourth section analyzes how theories of global democracy position themselves in the contemporary debates on the future direction of international climate policy, where the merits and dangers of a polycentric, complex and pluralist global climate order are highlighted.

2 BEYOND REALISM AND COSMOPOLITANISM: THREE PATHS TO DEMOCRATIZING GLOBAL GOVERNANCE

....

Proponents of cosmopolitan democracy argue that structures and processes of global governance must be transformed to recognize individuals rather than sovereign states as the legitimate actors and moral agents. Cosmopolitan democracy favors a world order based on a global constitution and rule of the law at a regional and global level (Held 1995). Democratizing global governance entails a global and divided authority system based on confederalism, the enactment of a cosmopolitan law and extension of the mandate of international courts (Held 1995: 22). This encompasses cross-national referenda, global assembly, or a world parliament based on a reformed UN General Assembly. The move from a territorial principle to a more encompassing principle of 'affectedness' is to be realized by establishing new, permanent territorial organizations and mechanisms at the regional and global level.

The cosmopolitan model has been criticized for its naive utopian ambitions, its neglect on issues such as the role of coercive force in international relations, the political economy of global capitalism, and for underplaying the role of more sporadically deployed and hybrid mechanisms of governance. Dahl (1999) is the most outspoken critic of global democracy, which he views as neither desirable nor feasible. Communitarians argue that transnational democracy is not possible because of the lack of coherent global constituency or *demos* in the politically, socially, and culturally divided community of 193 states. Realists claim that the dynamics of *realpolitik*, international anarchy, and sovereign power are structural features preventing the emergence of global democracy. Critical perspectives are deeply skeptical to the ideas of global democracy that ignore the ways in which power inequalities permeate international institutions and mask power relationships (McGrew 2002: 152–4). Global democracy is a shorthand for neoliberal market economy, liberal democracy, and Western hegemony that operates through the language of 'good governance' and civil society participation (O'Brien et al. 2002).

Between the outright skeptics of global democracy and optimistic proponents of cosmopolitan democracy three pragmatic paths to global democracy can be identified—democratic intergovernmentalism, transnational deliberative democracy, and global stakeholder democracy. They overlap in emphasizing the need for transparency, deliberation, participation, and public accountability of global governance, but differ on the most viable path to democratizing global governance. Democratic intergovernmentalism stresses the accountability of international organizations. Global stakeholder democracy emphasizes institutional representation of affected stakeholders, while transnational deliberative democracy highlights the importance of transnational public spheres containing a multiplicity of discourses and discursive contestation. They are pragmatic in the sense that they recognize the continued importance of state sovereignty in global governance. It should be stressed that these three models or ideal types of democratization of global governance are not mutually exclusive, but rather represent different lenses that allow us to see what is going on. They are also complementary in their quest to assess the democratic legitimacy of new emerging complex, fragmented, and hybrid global governance arrangements.

2.1 Democratic Intergovernmentalism

Democratic intergovernmentalism or liberal-institutionalism (McGrew 2002: 158) represents an effort to reform international institutions towards adopting democratic values. In the absence of global democracy, transnational legitimacy can be enhanced by strengthening democratic values, such as accountability, transparency, inclusion, and participation (Buchanan and Keohane 2006). Democratic intergovernmentalism accepts that sovereignty remains an entrenched principle of the international system. States are recognized as the only actors with rights, obligations, and the capacity to form and be bound by international treaties. Legitimacy in the global arena is derived from intergovernmental negotiations among sovereign states that have the ultimate decision-making authority.

Consequently, democratic intergovernmentalism is associated with executive multilateralism, which entails that states coordinate their interests internationally and implement them domestically (Zürn 2004). However, with the rise of transnational activism, the acceptance of executive multilateralism as a decision-making mechanism is increasingly problematic (Zürn 2004: 277). Legitimacy deficits arise from the weak accountability of political elites and negotiators, who increasingly have to justify decisions both to national and transnational publics. Consequently, a shift can be traced from executive multilateralism to societally backed multilateralism, where the latter entails full multi-media coverage and access to non-state actors (Zürn 2004: 286).

To counter these problems, democratic intergovernmentalism aims to strengthen democratic values such as accountability, transparency in existing institutions, and intergovernmental decision processes (Keohane and Nye 2003). This represents largely a proceduralist view of democracy, where accountability of international institutions is framed as an issue of access and openness to external stakeholders outside the community of legitimate decision makers (states). Increased non-state actor participation and civil society consultation in the WTO and the World Bank, as well as at global multilateral summits, typifies this approach.

Accountability, through shared global publics (NGO contestation), professional norms, and transnational networks (epistemic communities) and markets (rating agencies and corporate standards) can be enhanced (Keohane and Nye 2003). Civil society participation is encouraged since it adds accountability, openness, and expertise. Transnational civil society complements intergovernmental bargaining. However, the assumption is that sovereign state-centric practices are prevailing and cannot be replaced by non-state actors. In sum, global legitimacy is advanced by institutional reform to extend transparency and accountability, rather than by challenging the primacy of state-centric decision making.

2.2 Transnational Deliberative Democracy

Deliberative democracy is frequently conceived of as the most viable model for democracy beyond the state (Risse 2004). It is conducive to international society as it downplays the role of boundaries. It represents democracy without electoral representation and demos: deliberative accounts of democracy do not rely on a predefined community. As Nanz and

Steffek (2004: 318) argue: 'A deliberative understanding of democratic collective decision making is particularly suited for global governance, where there is a lack of competitive elections'. Deliberative democracy is compatible with the structures of world politics, which are characterized by their non-hierarchical, non-electoral, and non-territorial features. The democratization of global governance goes through strengthening the discursive quality of transnational public spheres rather than institution building towards a world polity. The more discursive contestation, the better. Legitimacy stems from free, inclusive, and unconstrained public deliberation between equal participants, revolving around the notion of ideal procedure for deliberation and decision making. Deliberative accounts of democracy have been taken up by scholars in environmental politics as a route to shape both more legitimate and effective policies (Bäckstrand et al. 2010).

International regimes can be conceived of as public spheres to promote deliberation on norms such as transparency and participation (Payne and Samhat 2004). The transnational public sphere and global civil society play a central role in deliberative accounts of global democracy (Dryzek 2006). Civil society participation within regimes will stimulate the use of arguments and public scrutiny of policy choices (Nanz and Steffek 2004: 321). A vital discursive contestation in civil society can counter global technocracy of scientific elites and diplomats found in interstate bargaining and international bureaucracies. Two kinds of deliberative processes can be identified: Deliberation within institutions, regimes, and international negotiations and deliberation in an independent transnational public sphere outside the venues of authoritative decision making. Accordingly, in the international climate regime, civil society can exercise discursive influence on formal negotiations and collective decision making as well as promoting a transnational public sphere independent from sovereign authority. In elaborating a deliberative system that can be applied to local, national, and global governance levels Dryzek (2009: 5) distinguishes between the public space exhibiting a diversity of viewpoints and the empowered space constituting sites of authoritative decision making. Civil society can act as a transmission belt to empowered spaces (of international organizations and formal negotiations) from public transnational spheres of civil society activism and protest.

2.3 Global Stakeholder Democracy

Global stakeholder democracy represents the third route to democratizing global governance. The stakeholder democracy model overlaps with deliberative accounts of democracy by stressing the importance of communicative action and dialogue in a transnational public sphere. However, questions of power, voice, representation, and accountability of non-state actors are central to stakeholder democracy (Bäckstrand 2006; MacDonald 2008). Global stakeholder democracy can be conceived as a subset of pluralism in which democratic legitimacy in enhanced through the participation and representation of non-state actors in transnational governance arrangements (Dingwerth 2007: 22). The institutionalized representation of affected stakeholders in decision making is an important complement to interstate aggregate decision making (Bäckstrand 2006). Multi-stakeholder multilateralism denotes the employment of hybrid mechanisms, such as stakeholder consultations and public-private partnerships associated with the 2002 World Summit on Sustainable Development in Johannesburg (WSSD) (Maartens 2007).

Global stakeholder democracy represents a 'bottom-up' theory of democratization of the global order, where the primary agents are critical social movements and 'communities of fate' (McGrew 2002: 158). It reinforces ideas that a global civil society can act as a democratizing force. However, the democratic credentials of NGOs and global civil society have been questioned due to problems of self-selection and lack of accountability (Scholte 2002). Variants of stakeholder democracy aim to represent stakeholder interests from market (including transnational corporations), government, and civil society spheres in decision making. Group-based 'meso' deliberation at the interface between state and society has been advanced for result-based environmental governance (Meadowcroft 2004: 188). In contrast to transnational deliberative democracy, global stakeholder democracy promotes interactive decision-making between affected societal interests and government representatives.

Since there are no procedures for overcoming disagreement in deliberative processes, in most global governance arrangements institutionalized representation of stakeholders would not provide a 'stand-alone framework for global public decision-making and negotiation' (MacDonald 2008: 161). Most likely, it will complement a representative sovereignty-based framework of bargaining between territorial states (Bäckstrand 2006: 475). Global stakeholder democracy recognizes the limitations of democratic intergovernmentalism to facilitate deliberative processes between stakeholder communities and decision makers (MacDonald 2008: 152). But in contrast to ideals of transnational deliberative democracy to leave deliberation to the sphere of civil society, it is problematic to expect inclusive public deliberation, given the weak and uneven structure of global civil society across different regions in the world. Consequently, institutional representation of multi-stakeholder interests is important as 'a voice but not a vote' (MacDonald 2008: 162).

The following sections will illustrate how we can we make sense of the democratic credentials of the global climate order by examining the Copenhagen summit through the three perspectives on global democracy.

3 A DEMOCRATIC GLOBAL CLIMATE GOVERNANCE ORDER AFTER COPENHAGEN?

....

How do liberal intergovernmentalism, transnational democracy, and global multi-stakeholder democracy interpret the Copenhagen outcome and the different types of multilateralism in play in terms of the prospects for democratic governance beyond the nation-state? Before analyzing the Copenhagen summit through these three democracy lenses, a brief account of the Copenhagen Accord is given.

3.1 The Copenhagen Accord

The original goal of the conference in Copenhagen was to finalize a new international climate agreement serving as a successor to the Kyoto Protocol, whose first commitment period comes to an end in 2012. After the UN climate summits on Bali in Indonesia 2007, two years of intense negotiations built unrealistic expectations of a new global climate agreement with binding targets and timetables to be signed in Copenhagen. COP-15 failed to produce a treaty with binding mechanisms to keep the global mean temperature below 2 degrees Celsius, a temperature goal based on the assessment of the Intergovernmental Panel on Climate Change (IPCC). The Accord, which was negotiated in the last twenty-four hours in an all-night session with a small group of heads of states and government, has an unclear legal status as it was not adopted in consensus but only notified by fifty-five parties in the final plenary. Six countries—none of them major emitters—objected to the Accord.[2] The Copenhagen Accord represents an aspirational blueprint rather than a legal agreement. The two-and-a-half-page document acknowledges the 2-degree temperature target and states that deep cuts in global emission are required to stabilize greenhouse gas emissions. Some progress was made in financing, deforestation, and adaptation, such as a REDD mechanism (reduced emissions from deforestation and forest degradation), the Copenhagen Green Climate Fund, climate finance of US$30 billion for the two years 2010–12 and non-committal pledges of US$100 billion a year until 2020 (Stavins and Stowe 2010).

The Accord does not contain any binding timetables or figures for quantitative emission reductions and lacks monitoring and verification mechanisms to enforce compliance. Instead, it consists of two appendices for industrialized (Annex I) and developing countries (non-Annex I) that invite Parties to submit voluntary pledges by 31 January 2010. Industrialized countries were invited to submit voluntary action plans for quantified emission targets reductions while developing countries were to submit 'nationally appropriate mitigation actions.' By June 2010, 138 states (including the 27-member EU) that represent more than 85 percent of the global greenhouse gas emissions had joined the Accord. However, according to calculations from several NGOs and independent research institutes the accumulated submissions of voluntary targets will not be sufficient to reach the 2-degree target (Rogelj et al. 2010; WRI 2010). Judging from the submitted action plans, a temperature increase of 3 degrees Celsius or more is expected, which will pave the way for significant risks for human society and ecosystems.

3.1 Executive Multilateralism and Intergovernmentalism

Despite an unprecedented participation of non-state actors, from the perspective of democratic intergovernmentalism, the Copenhagen summit confirmed the primacy of interstate bargaining between sovereign states and the dim prospects for a global democratic polity as envisioned by cosmopolitans. COP-15 consolidated the return to power politics and multi-polar climate order with a grand bargain between emerging economies and the US. In this respect, the liberal reform agenda of democratic intergovernmentalism to increase stakeholder consultations and strengthen public accountability was limited in Copenhagen. The bargaining process toward the Copenhagen Accord is captured by the notion of executive multilateralism. Multilateral principles of transparency, inclusion, and participation were compromised to reach an agreement. COP-15 has been criticized on procedural grounds for being undemocratic and non-transparent. In order not to return empty-handed, a 'coalition-of-the-willing' of twenty-five heads of state and government leaders—branded Friends of the Chair—drafted the Copenhagen accord in the last two days of the conference and thereby sidelined the official UN negotiations (Müller 2010). The Danish Chair of COP-15, Prime Minister Lars Løcke Rasmussen, received criticism for the ad hoc selection of states to be included in the 'Friends of the Chair' group.

The US negotiations with China and the BASIC (Brazil, South Africa, India, China) countries helped to seal the deal. The EU, the self-proclaimed global leader on climate change, was largely marginalized from the process (Haug and Berkhout 2010). The draft negotiation text of 200 pages was reduced to a political declaration written by state leaders under heavy time constraints (Müller 2010: i). This is a rather unorthodox practice in the history of environmental diplomacy. Usually heads of state arrive at the end to resolve the few remaining bracketed text segments and sign the agreement. Partly as a response to mistrust and the lack of transparency, the Copenhagen summit has been described as the most leaky conference in the history of environmental diplomacy (Vidal and Watts 2009). The repeated leaks of draft negotiation texts emanated from suspicion from developing countries, the Friends of the Chair process, and confusion over what texts to negotiate. The Copenhagen Accord's pledge and review process marked the turn to a bottom-up and decentralized fragmented climate policy architecture in which voluntary pledges from major emitters replaced multilaterally negotiated targets (Stavins and Stowe 2010).

In the new climate order, 'minilateralism' emerges as a governance mode, represented by technology agreements such as the Asia-Pacific Partnership for Clean Development and Climate (APP), the Carbon Sequestration Leadership Forum (CSLF), and the Methane to Markets (Kellows 2006). These represent voluntary agreements between governments involving cooperation for clean technology, renewable energy, clean coal, and carbon sequestration and can be conceptualized as transgovernmental networks. A majority of the technology partnerships are initiated or led by the US and can be framed both as alternatives or complements to the international climate regime (McGee and Taplin 2006). Mostly, they are 'multifunctional' agreements, simultaneously addressing issues such as energy security, technology transfer, clean coal, economic development, and pollution abatement. A key element of these networks is the cooperation between 'like-minded' governments. Outside the UNFCCC process, climate policy is negotiated in G-8, G-20, and Major Economies Forum (MEF). Fragmentation and small n agreement (coalition of willing, carbon clubs) can increase effectiveness, but also lead to a 'chilling effect' and 'race-to-the bottom'.

COP-15 did not live up to the ideals of democratic intergovernmentalism to make interstate decision making more transparent and accountable to external stakeholders. Instead it can be conceived of as a prime example of executive multilateralism or mini-lateralism. As will be discussed in the next section, while the summit broke the world record for the largest number of registrants of non-state actors, these were largely excluded from the negotiations in the decisive second week. The Copenhagen Accord was negotiated in an opaque Friends of the Chair process, limiting accountability, transparency, and participation by stakeholders.

3.2 Deliberative Multilateralism

Deliberative democrats would make a more optimistic assessment of the summit beyond the intergovernmental bargaining between sovereign states. Climate diplomacy epitomizes emergent forms of deliberative or participatory multilateralism, which promote public deliberation. The yearly UN climate summits since Kyoto 1997 can be conceived of as an emergent 'global agora' (Stone 2008), 'transnational deliberative system' (Dryzek 2009), 'global public domain' (Ruggie 2004) or 'transnational neopluralism' (Cerny 2009). Compared to other policy areas, environmental diplomacy has generous accreditation procedures for non-state actors. Most formal negotiation sessions at COP-15 were open to non-state observers and web cast. The plural sites for debate and contestation in official negotiations, organized side events, and alternative civil society forums can be conceived of as a transnational public sphere where social movements and civil society actors engage in a dialogue attempting to influence policy makers. Civil society acted both as insiders and outsiders. COP-15 entailed an unprecedented attendance of non-state actors. There were three times more registered parties and observers in Copenhagen compared to the previous UN climate summits. The total number of registrants for the formal negotiation venue—Bella Center—was more than 30,000. Of these around 20,000 were NGO observers and roughly 8,000 parties and the remainder media representatives and journalists (Fisher 2010: 12).

Apart from the official negotiation session, where NGO observers could make statements, an important forum for civil society debate was the side events held in the conference venue. Among the more than 200 formal side events, NGOs have, on average, organized more than 50 percent. There are several functions of side events, such as information dissemination and capacity building and introducing items for negotiations with the aim to influence decision making. Secondly, parallel with the official negotiations was the alternative People's climate summit—Klimaforum09—located a few kilometers away from the Bella Center (Klimaforum09 2010).[3] Thirty-one Danish NGOs and sixty-three international grassroots organizations organized the forum for twelve days. It aspired to represent authentic voices from the South. According to their organization's self-evaluation, more than 50,000 people visited Klimaforum09 to participate in more than 300 organized events, such as debates, seminars, concerts, and films. The attendance of NGO observers increased during the final days of COP-15 when NGO access to the formal negotiation sessions was severely limited due to security reasons. Thirdly, demonstrations were another transnational space for articulating critical and alternative discourses on climate change. The major demonstration in central Copenhagen to Bella Center gathered between 60,000 and 100,000 participants (Fisher 2010: 14). The demonstration was part of an internationally coordinated Global Day of Action around climate change. Moreover, Copenhagen saw the mobilization of the global climate justice movement, which can be seen as an offspring from the anti-globalization movement in Seattle in 1999. Civil society activist groups such as Climate Justice Now called for non-violent protest and direct action. While not successful, they planned to storm the Bella Center and turn the negotiations into a People's Assembly (Fisher 2010: 15).

Despite the multitude of civil society activists, COP-15 has been described as a backlash against civil society, even reinforcing the disenfranchisement of social movements and stakeholders (Fisher 2010). There were two reasons for this. First, the poor conference planning, organization, and logistics from the UNFCCC secretariat and the Danish host contributed to the limitations in access for non-state observers. The Bella Center could accommodate 15,000 participants, yet the secretariat allowed over 30, 000 participants to register. In the second week, access was severely restricted. The undersized conference venue caused hours of lines both for parties and observers. In the last two days only 90 non-state actors out of the totally 8,000 registrants were allowed into the Bella Center. Second, direct action and protests by the climate justice movement outside the conference venue led to restrictions by civil society actors inside the conference venue, to ensure security of the parties. 'Ironically, the more civil society actors try to participate—and the diversity of the perspectives represented by the civil society actors involved—the less access they are likely to have' (Fisher 2010: 17).

COP-15 in many respects marked the peak of deliberative multilateralism. It conforms to ideals of transnational discursive democracy with multiple and overlapping public spheres of discursive contestation, from Bella Center to the alternative People's summit. It brought the largest civil society community ever to a UN climate summit, with a diversity of perspectives from insiders to outsiders. However, civil society influence on agenda setting and decision making was limited.

3.3 Multi-stakeholder Multilateralism

To what extent does the Copenhagen summit conform to ideals of global stakeholder democracy? Compared to negotiations on trade, security, and human rights, the climate change negotiations are more inclusive and open to access. As discussed above, COP-15 reflected the notion of deliberative multilateralism with an unprecedented participation of civil society actors representing a diversity of discourses from vocal outsiders to professionalized insiders. A range of stakeholders participated, such as business associations, trade unions, research institutes, municipalities and cities, and transnational companies. COP-15 represented multi-stakeholder multilateralism *par excellence*. The Copenhagen summit featured the largest presence of business community and carbon market actors. Just a few minutes' walk from Bella Center, the International Emissions Trading Association (IETA) arranged its largest business-led side-event program ever at a UN climate summit, covering topics such as post-2012 climate policy, the Clean Development Mechanism (CDM), carbon market governance, and the role of forestry as mitigation option (IETA 2009). But to what extent were affected stakeholder interests from NGOs and industry institutionalized at COP-15? What was the degree of interaction between negotiators, civil society, and business?

First, the UNFCCC negotiations have less elaborate mechanisms for stakeholder representation than other international environmental negotiations. There has been experiments with global stakeholder democracy in multilateral processes on sustainable development, which has institutionalized representation of UN 'major groups', such as NGOs, business, trade unions, farmers, youth, women, scientific communities, municipalities, and indigenous communities (Bäckstrand 2006). In contrast, climate negotiations have no formalized procedure to represent affected stakeholders or promote interaction and deliberation between government, business, and civil society. While the organized constituencies for scientists, business, and NGOs are recognized as focal points by the UNFCCC secretariat, exemplified by RINGO (Research and Independent Non-Governmental Organizations), BINGOs (Business and Industry NGOs), and ENGOs (Environmental NGOs), there are no formalized links to the negotiations. While some national delegations at COP-15 included NGOs, scientists, and business representatives as delegates, stakeholder input was limited to statements in the negotiation.

There was a wide array of activities by non-state actors in the alternative civil society summit, the carbon market forum, and the official side-event forum, where both observers and parties could apply for organizing a side event. However, while negotiators attended some of these parallel events, and NGOs attended the negotiations, there were no formalized mechanism for 'meso' deliberation between negotiators and stakeholders. While statements by non-state actors could be presented in negotiations, the restricted access to enter the Bella Center during second week's ministerial segment limited stakeholder input radically. Furthermore, the Friends of the Chair process and the last twenty-four hours' bargaining of the text of the Copenhagen Accord between political leaders did not include any stakeholder communities but were restricted to parties.

In sum, the model of global stakeholder democracy has not gained currency in the climate change negotiations. The alternative civil society summit, the business summit, and the formal negotiations represent rather distinctive spheres. Fisher (2010: 16) argues that the Copenhagen summit may even mark the end of stakeholder inclusion in climate governance.

4 The Democratic Legitimacy of a New Global Climate Governance Order after Copenhagen?

How do contemporary global climate governance processes reinforced by the Copenhagen summit resonate with normative ideals of global democracy? What is the potential for democratic legitimacy of climate governance given the increasing fragmentation of global governance and the transformation of multilateralism, which relies increasingly on collaboration between private and public actors reflected by notions of 'complex multilateralism,' 'public-private multilateralism,' and 'market multilateralism'?

The judgment of the Copenhagen outcome has been harsh both from activists and scholars: 'The conference was a failure whose magnitude exceeded our worst fears and the resulting Copenhagen Accord was a desperate attempt to mask that failure' (Dimitrov 2010: 18). The Copenhagen summit has been interpreted as the decline of multilateral climate diplomacy and the consolidation of a fragmented global climate order, where the UN negotiations were hostage to pending climate legislation in the US Senate. The verdict of COP-15 has varied strongly between the US, the EU, civil society, and developing countries. The desired solution for civil society and the EU was an international comprehensive climate regime with universal participation containing binding long-term timetables and targets for the industrialized countries. Consequently, the dominant framing of a legitimate outcome was a legally binding accord and universal accord negotiated among the 193 parties. Among negotiators and civil society, the legitimacy of climate diplomacy rests on both securing an open, transparent, and inclusive process and an effective outcome in terms of a single comprehensive treaty curbing and stabilizing carbon emissions to prevent adverse outcomes for society and natural ecosystems. COP-15 has been criticized both for procedural shortcomings and weakness in outcome. While the Copenhagen Accord has been interpreted as a setback for the EU's model of liberal multilateralism, it can be conceived as a success for the US preferences for domestic flexibility and a voluntary pledge and review system:

> [The Copenhagen Accord] carries promise, because it recognises core geopolitical realities and works with rather against them . . . If successfully elaborated over the next few months,

the Accord can provide anchoring agreement but will need in addition, a diversity of bilateral and regional agreements and—yes—'coalition of the willing'. (Giddens 2010)

It is frequently argued that COP-15 turned into an undemocratic process where multilateral norms were sidelined and replaced by a bargaining process between major powers. The debate after COP-15 is replete with competing claims of what constitutes legitimate global climate governance and how to shape a fair and effective climate order after Copenhagen. A pessimistic account is that UN-centered climate diplomacy is weakened at the expense of a new climate pact between US and emerging economies in the BASIC alliance.

The contested debate among climate activists and in the climate policy community on the failure in Copenhagen are mirrored in debates in the climate policy research community on competing ideals of an international climate policy architecture for a post-Kyoto world: one universal/centralized, the other fragmented/decentralized (Aldy and Stavins 2007). In the former, which is associated with positions of the EU, civil society, and small developing countries, legitimacy rests on a single universal legally binding treaty negotiated under UN auspices and modeled on the Kyoto Protocol. In contrast, in the fragmented/decentralized climate architecture, which has been advanced by the US and China, climate policies would emerge from below in terms of 'domestic portfolios' and a voluntary pledge and review system. The Copenhagen Accord institutionalized a fragmented decentralized climate governance architecture at odds with the aspiration to establish a universal climate regime.

The debates about legitimacy of the new global climate order post-Copenhagen can be transferred to the scholarly debate on the prospects for a global democracy in an era with increasingly complex, multi-layered, and polycentric governance arrangements. The cosmopolitan interpretation of COP-15 differs sharply from the complementary perspectives of liberal intergovernmentalism, transnational deliberative democracy, and the global multi-stakeholder democracy. A cosmopolitan perspective links the 'disaster' in Copenhagen to the failure to negotiate a universal legally binding successor agreement to the Kyoto Protocol. While such a comprehensive climate change regime would not amount to world polity as it lacks supranational elements where states have ceded sovereignty to global institutions, the aspiration to build a single climate regime from above resonates with cosmopolitan efforts to enact universal law and international institutions. The three accounts of global democracy overlap in their view that the failure of UN climate diplomacy does not mean the failure of broader processes of transnational climate governance. Liberal intergovernmentalists would argue that efforts to establish a comprehensive integrated UN climate regime are likely to be unsuccessful. The Copenhagen summit demonstrated that this strategy is politically unfeasible. The sheer complexity and problem diversity of the global climate problem requires institutional diversity (Keohane and Victor 2010: 10). A loosely linked 'regime complex' has been advanced as a more viable approach compared to attempts to set up a single comprehensive international regime modeled on the Kyoto Protocol. A regime complex allowing for flexibility, innovation, and institutional differentiation would be a more realistic alternative to a single international regime in responding to the problem of diversity of climate change. The UNFCCC could serve as an umbrella for facilitating negotiations, without aspiring to be a central locus for the building of a comprehensive universal regime (Keohane and Victor 2010). From the perspectives on transnational deliberative democracy and global stakeholder democracy, the COP-15 represented the peak of participatory multilateralism with the largest civil society mobilization and business participation hitherto, in the negotiation venue and in the adjacent People's Summit and business forum. The slogan 'system change not climate change' by the civil society forum represented an alternative discourse on climate change.

The three accounts to democratization of the global order all move beyond the study of international regimes in evaluating the democratic legitimacy of climate governance. The multifaceted and polycentric nature of climate change requires new approaches to assess the legitimacy of networked governance. Contemporary climate governance features a range of new mechanisms, fora, and actors, which are characterized by the rise of public, private, and hybrid networks engaged in governance function such as rule making, information sharing, and capacity building (Andonova et al. 2009; Biermann, this volume). Beyond the UNFCCC, climate policies emerge at municipal and subnational levels entailing corporate actors, civil society actors, public-private partnerships, and the global and regional carbon markets (Paterson, this volume). Transnational networks of global climate governance can be conceived as deliberative systems (Dryzek 2009). Consequently, a key empirical question for models of global democracy is to whether processes of networked climate governance can live up to standards of democratic legitimacy.

5 CONCLUSION

The Copenhagen Accord has been framed as a total failure of multilateral climate diplomacy as well as a new opening for international climate politics, as the major carbon emitters in the South and North are bought under a common framework. The democratic legitimacy of the Copenhagen Accord is fundamentally contested, and depends on what criterion for democracy is applied. Between the realists' rejection of global democracy and cosmopolitans' proposals of world government, liberal intergovernmentalism, transnational deliberative democracy, and global stakeholder democracy are three viable routes to a democratization of global governance. They have in common that they reinterpret democratic legitimacy and turn attention to emerging networked and fragmented climate governance processes. Democratic intergovernmentalism focuses on accountability of international organizations as a key element of legitimacy, transnational discursive democracy stresses the importance of a vibrant transnational public sphere of discursive contestation, while global stakeholder democracy emphasizes the importance of institutionalized representation of affected stakeholders from sectors of civil society and the market.

These three ideals of transnational democracy have served as lenses for interpreting the outcomes of the COP-15 as well as the implications for transnational climate governance at large. The Copenhagen Accord marked a new multi-polar global climate order where multilateral principles were marginalized and replaced by a non-transparent bargaining process between coalitions of willing states. It paved the way for decentralized climate governance architecture building on voluntary pledges rather than mandatory emissions cuts and timetables. While democratic intergovernmentalism recognizes the primacy of sovereignty, Copenhagen was a setback for efforts to increase the accountability and transparency of climate diplomacy. From the perspective of transnational deliberative democracy, COP-15 gathered the largest civil society mobilization hitherto at a UN summit both in the official negotiations and in the alternative People's climate forum. The diversity of discourses, from professionalized NGOs on the 'inside' to the protesting global climate justice movement on the 'outside' amounts to a transnational public sphere. Finally, the Copenhagen summit epitomized multi-stakeholder multilateralism, in terms of the unprecedented participation of stakeholders from the industry, NGO, and government sectors. However, COP-15 lacked formal mechanisms to represent stakeholder interests in the agenda setting and negotiations, as well as participatory policy innovations to promote institutional interaction between stakeholders and decision makers. Global climate governance after Copenhagen does not live up to models of global democracy as envisioned by cosmopolitans. Skeptics of transnational democracy will continue to argue that it does not make sense to speak about democracy in the context of global governance of climate change. However, even if contemporary global climate governance does not fulfill normative ideals of 'democracy as we know it' (Cerny 2009: 769), it represents an ongoing experiment in global democratic practice, where democratic values of transparency, deliberation, accountability, participation voice, and inclusion, are integral in efforts to shape a fair and effective post-Kyoto global agreement.

NOTES

1. For the text of the Accord, see <http://unfccc.int/resource/docs/2009cop15/eng/107.pdf>.
2. Bolivia, Cuba, Nicaragua, Sudan, Tuvalu, and Venezuela.
3. The main funder of the Klimaforum09 was the Danish Ministry of Foreign Affairs contributing with around 8 million Danish Crowns.

REFERENCES

ALDY, J. E., and STAVINS, R. N. 2007. *Architectures for Agreement. Addressing Global Climate Change in the Post-Kyoto World.* Cambridge: Cambridge University Press.

ANDONOVA, L., BETSILL M., and BULKELEY, H. 2009. Transnational climate governance. *Global Environmental Politics* 9: 52–73.

BÄCKSTRAND, K. 2006. Democratising global environmental governance: Stakeholder democracy after the world summit on sustainable development. *European Journal of International Relations* 12: 467–98.

——KHAN, J., KRONSELL, A., and LÖVBRAND, E. (eds.) 2010. *Environmental Politics and Deliberative Democracy: Examining the Promise of New Modes of Governance.* Cheltenham: Edward Elgar.

BIERMANN, F., PATTBERG, P., VAN ASSELT, H., and ZELLI, F. 2009. The fragmentation of global governance architectures: A framework for analysis. *Global Environmental Politics* 9: 14–40.

BUCHANAN, A., and KEOHANE, R. O. 2006. The legitimacy of global governance institutions. *Ethics and International Affairs* 20: 405–37.

CERNY, P. G. 2009. Some pitfalls of democratisation in a globalising world: Thoughts from the 2008 Millennium Conference. *Millennium* 37: 767–90.

DAHL, R. A. 1999. Can international organizations be democratic? Pp. 19–36 in I. Shapiro and C. Hacker-Gordon (eds.), *Democracy Edges*. Cambridge: Cambridge University Press.

DIMITROV, S. 2010. Inside Copenhagen: The state of climate governance. *Global Environmental Politics* 10: 18–24.

DINGWERTH, K. 2007. *The New Transnationalism: Transnational Governance and Democratic Legitimacy*. Basingstoke: Palgrave Macmillan.

DRYZEK, J. S. 2000. *Deliberative Democracy and Beyond. Liberals, Critics, Contestations*. Oxford: Oxford University Press.

——2006. *Deliberative Global Politics*. Cambridge: Polity Press.

——2009. Democracy and earth system governance. Paper prepared for presentation at the Amsterdam Conference in Human Dimensions of Global Environmental Change 'Earth System Governance: People, Places and the Planet', 2–4 December 2009.

EGENHOFER, C., and GEORGIEV, A. 2009. The Copenhagen Accord. A first stab at deciphering the implications for the EU. *CEPS Commentary* 25 December. Brussels: Centre for European Policy Studies.

FISHER, D. R. 2010. COP-15: How the merging of movements left civil society out in the cold. *Global Environmental Politics* 10: 11–17.

GIDDENS, A. 2010. Big players, a positive accord. *Policy Network* January. Accessed 8 July 2010 at <http://www.policy-network.net/uploadedFiles/Articles/Big%20players,%20a%20positive%20Accord.pdf>.

GRUBB, M. 2010. Copenhagen. Back to the future? *Climate Policy* 10: 127–30.

HAUG, C., and BERKHOUT, F. 2010. Learning the hard way? European climate policy after Copenhagen. *Environment* 52: 20–7.

HELD, D. 1995. *Democracy and the Global Order: From the Modern State to Cosmopolitan Governance*. Stanford, CA: Stanford University Press.

——and HERVEY, A. F. 2009. Democracy, climate change and global governance: Democratic agency and the policy menu ahead. Policy Network Paper. Accessed at <http://www.policy-network.net/publications/3406/Democracy-climate-change-and-global-governance>.

HOLDEN, B. 2002. *Democracy and Global Warming*. New York: Continuum.

IETA (International Emission Trading Association). 2009. Schedule of side events at COP15 and CMP5. IETA.

KELLOWS, A. 2006. A new process for negotiating multilateral agreements? The Asia–Pacific climate partnership beyond Kyoto. *Australian Journal of International Affairs* 60: 287–303.

KEOHANE, R. O., and NYE, J. 2003. Redefining accountability for global governance. Pp. 386–411 in M. Kahler and D. Lake (eds.), *Governance in the Global Economy. Political Authority in Transition*. Princeton: Princeton University Press.

——and VICTOR, D. G. 2010. The regime complex for climate change: The Harvard Project on International Climate Agreements. Discussion paper 10–33. Cambridge, MA: Harvard Kennedy School.

Klimaforum09 2010. *Peoples' Climate Summit: Evaluation Report 09*. Copenhagen: Foreningen Civilsamfundets Klimaforum.

MAARTENS, J. 2007. *Multistakeholder Partnerships: Future Models of Multilateralism?* Occasional Papers 29. Berlin: Friedrich-Ebert Stiftung.

MACDONALD, T. 2008. *Global Stakeholder Democracy. Power and Representation beyond Liberal States*. Oxford: Oxford University Press.

McGee, J., and Taplin, R. 2006. The Asia–Pacific partnership on clean development and climate: A competitor or complement to the Kyoto Protocol? *Global Change, Peace and Security* 18: 173–92.

McGrew, A. 2002. Democratising global institutions: Possibilities, limits and normative foundations. Pp. 149–70 in J. Anderson (ed.), *Transnational Democracy: Political Spaces and Border Crossings*. London: Routledge.

Meadowcroft, J. 2004. Deliberative democracy. Pp. 183–218 in R. F. Durant, D. J. Fiorini, and R. O'Leary (eds.), *Environmental Governance Reconsidered*. Cambridge, MA: The MIT Press.

Moravscik, A. 2004. Is there a 'democratic deficit' in world politics? A framework for analysis. *Government and Opposition* 29: 336–63.

Müller, B. 2010. Copenhagen 2009: Failure of final wake-up call for our leaders? Oxford Institute for Energy Studies, February.

Nanz, P., and Steffek, J. 2004. Global governance, participation and the public sphere. *Government and Opposition* 39: 314–34.

O'Brien, R., Goetz, A. M., Scholte, J. R., Williams, M. 2002. *Contesting Governance. Multilateral Institutions and Global Social Movements*. Cambridge: Cambridge University Press.

Payne, R. A., and Samhat, N. H. 2004. *Democratizing Global Politics: Discourse Norms, International Regimes, and Political Community*. New York: State University of New York Press.

Risse, T. 2004. Global governance and communicative action. *Government and Opposition*, 39: 288–313.

Rogelj, J., et al. 2010. Copenhagen Accord pledges are paltry. *Nature* 464: 1126–8.

Ruggie, J. G. 2004. Reconstituting the global public domain: Issues, actors, and practices. *European Journal of International Relations* 10: 499–531.

Scholte, J. A. 2002. Civil society and democracy in global governance. *Global Governance* 8: 281–304.

Stavins, R. N., and Stowe, R. C. 2010. What hath Copenhagen wrought? A preliminary assessment. *Environment* 52: 8–14.

Stone, D. 2008. Global public policy, transnational policy communities and their networks. *Policy Studies Journal* 36: 19–38.

United Nations. 2009. The Copenhagen Accord. Decision -/CP. 15. Advanced edited version.

Vidal, J., and Watts, J. 2009. Copenhagen: The last-ditch drama that saved the deal from collapse. *The Guardian* 20 December.

WRI (World Resource Institute). 2010. *Summary of Climate Finance Pledges Put Forward by Developed Countries*. Washington, DC: WRI.

Zürn, M. 2004. Global governance and legitimacy problems. *Government and Opposition* 29: 260–87.

CHAPTER 46

....

NEW ACTORS AND MECHANISMS OF GLOBAL GOVERNANCE

....

FRANK BIERMANN

1 INTRODUCTION

....

THE development of global governance to coordinate national responses to climate change has become a key challenge for politicians and political scientists alike. Unsurprisingly, the December 2009 conference of the parties to the United Nations Framework Convention on Climate Change ('climate convention') evolved into one of the largest diplomatic gatherings in history. The current negotiation of a long-term governance system on climate change has been compared to the creation of the United Nations Organization in 1945. It is one of the most demanding global institution-building efforts ever.

This tremendous challenge has given rise to a number of governance innovations in world politics. These were largely driven by the particular problem structure in climate governance, which is undoubtedly one of the most 'wicked' problems of world politics at present (Biermann 2007; see also Steffen, this volume). Climate governance is marked, first, by high uncertainties both analytically (regarding its scientific basis) and normatively (regarding the political and ethical principles that are applicable). Second, climate governance is characterized by high degrees of functional, spatial, and temporal interdependence that require comprehensive coordination and integration of governance responses. Thirdly, climate governance is characterized by high degrees of stakes—not the least for governmental actors. The impacts of climate change may be severe for many nations, threatening economic systems or food production and maybe even requiring relocation of affected communities. Conversely, also the need to mitigate will pose high burdens on some nations, in particular those with high emissions compared to the size of their populations. These uncertainties, interdependencies, and high costs of both regulation (mitigation costs) and non-regulation (costs of climate change impacts) place high burdens on negotiations, which makes the development of a global climate regime tedious and fragile.

Yet this situation has also given rise to new approaches in global governance that have made climate governance one of the most innovative and experimental areas of world politics today.

Three broad clusters of innovation are outlined in more detail in section 2 below. First, uncertainties and complexities of the overall governance system have given rise to a stronger role of actors beyond the nation-state. Second, this new emergence of multiple-actor governance, along with spatial and functional interdependencies, has stimulated the emergence of new mechanisms of global climate governance, namely transnational regimes, transnational public policy networks (often in the form of public-private partnerships), and transnational market-based governance mechanisms (such as emissions trading). Thirdly, the overall complexity of the governance arena, along with the stakes involved and resulting negotiation stalemates have led to an overall fragmentation of the policy system with multiple spheres of authority that require new types of interplay management. In section 3, I will further review these three trends in climate governance followed by a sketch of policy reform measures that could help drive the development of climate governance forward.

2 New Modes of Climate Governance

2.1 New Actor Constellations

First, global climate governance is no longer confined to nation-states but is characterized by increasing participation of actors that have so far been largely active at the subnational level. This transnational multi-actor governance includes private actors such as networks of experts, environmentalists, and multinational corporations, but also new agencies set up by governments, including intergovernmental bureaucracies. Novel is not simply the increase in numbers, but the ability of non-state actors to take part in steering the political system. Agency—understood as the power of individual and collective actors to change the course of events or the outcome of processes—is increasingly located in sites beyond the central governments of nation-states (Biermann et al. 2009a: 37–43).

There are three elements to this new development. First, the number of actors and the degree of their participation in global climate governance has increased substantially. Second, the variety of types of organizations increased. Next to governments, intergovernmental bureaucracies, non-governmental organizations, and business actors, novel forms of organizations have emerged, such as private rule-making organizations for example in forest management (e.g. the Forest Stewardship Council) or industry certification (e.g. the Global Reporting Initiative). Third, established organizations have adapted new roles and responsibilities. For example, intergovernmental bureaucracies became more autonomous from their principals, that is, the governments that created them. In addition, non-governmental organizations have taken a different role by more often directly engaging in agenda setting, policy formulation, and the establishment of rules and regulations.

From all non-nation-state actors that at present influence global climate governance, advocacy groups have been analyzed early and extensively (Lipschutz and McKendry, this volume). Research has shown that activist groups provide research and policy advice, monitor the commitments of states, inform governments and the public about the actions of their own diplomats and those of negotiation partners, and give diplomats at international meetings direct feedback (Betsill and Corell 2001). Climate governance is marked also by the emergence of networks of subnational public actors, notably networks of cities, in particular the largest forty cities that work together in committing to, and implementing, climate-mitigation targets (see here Bulkeley and Betsill 2003; also Bulkeley, this volume).

Highly relevant in climate governance—and to some extent unique—is the emergence of transnational networks of scientists. Such scientific advice for political decision-making is not new in world politics; negotiations on fishing quotas for example have long been assisted by the International Council for the Exploration of the Sea. These early examples, however, have been significantly increased in both number and impact, and many scientific networks have now assumed a role in providing complex technical information that is indispensable for policy making on issues marked by analytic and normative uncertainty (for example, Haas 2000; Mitchell et al. 2006).

While this new role of experts in world politics is evident in many policy areas, it is particularly prevalent in the field of global climate policy. The initial high uncertainties about causes, timing, and consequences of climate change had stalled negotiations in the late 1980s and 1990s, leading to the increasing institutionalization of scientific assessment and advice through the Intergovernmental Panel on Climate Change, which now comprises several thousands of scientists, who jointly assess the state of knowledge and condense it to succinct policy advice. The Panel is a hybrid organization that combines private actors—largely scientists on an individual basis, but also scientific organizations—with governmental selection, oversight, and funding. The final policy recommendations that flow from the scientific assessment are reviewed by experts appointed by governments, turning the 'summary for policy makers' into a 'summary of policy makers.' Because of the intrinsic analytical and normative uncertainties in climate governance, assessment bodies such as the IPCC became autonomous political actors, eventually diminishing the political influence of so-called 'climate skeptics' and thus affecting the bargaining influence of countries that first denied the global warming problem.

Yet this increasing political role of the IPCC resulted in its own further institutionalization and politicization. Complaints by under-represented developing countries that showed little trust in the first IPCC reports that were largely compiled by OECD-based scientists resulted in complex quota systems that almost resembled intergovernmental, UN types of institutional patterns (Biermann 2002). In addition, civil-society organizations were better integrated in IPCC deliberations, which also broadened the scope of literature that was to be assessed in the IPCC reports.

The IPCC is now a heavily institutionalized network that relies on scientific input and the cooperation of scientists and scientific organizations, but that at the same time builds on a complex set of norms that evolved from the public role that the panel has acquired and that seek to balance the diversity of interests and knowledge claims from a variety of governments and non-state communities. While other international policy arenas know similar assessment mechanisms (such as fishing, whaling, or food production), it is fair to assume that the new 'governance by scientific assessment' is unique in climate governance in terms of the relevance and degree of transnational institutionalization that the IPCC underwent in the last two decades.

The complexity of global climate governance also increased the role of intergovernmental bureaucracies and their civil servants. Intergovernmental bureaucracies provide in modern global governance important functions in the synthesis and dissemination of knowledge and the shaping of global policy discourses. They also influence negotiations by informing governments about actions and commitments by other actors, by reporting on the overall problem assessment, and by providing compromise solutions that may eventually influence negotiations (Biermann and Siebenhüner 2009).

In climate governance, these roles are largely performed by the secretariat to the climate convention, which also serves the 1997 Kyoto Protocol. This secretariat is largely independent from the overall UN system, and it has evolved into one of the largest intergovernmental bureaucracies in the environmental field (Busch 2009). While the climate secretariat has initially shown sizeable restraint in developing and propagating its own policy visions, this restraint decreased under its new executive director, Yvo de Boer, who assumes a more visible political role in negotiations. Important sources of influence and political power for the secretariat stem from the overall complexity of the negotiation system and the underlying problems that require in particular smaller countries to rely on information and advice from intergovernmental bureaucracies to the extent that these bureaucracies manage to maintain the trust upon which their formal and informal influence relies. While a strong role of intergovernmental bureaucracies is not new to world politics but evident in many areas, it is apparent that the analytical and normative uncertainties in climate governance, along with strong interdependencies that generate an overall complex political arena, have given rise to a particularly prominent role of technocratic policy actors such as the climate secretariat.

2.2 New Transnational Mechanisms of Governance

Climate governance is also marked by the emergence of new mechanisms of global governance in addition to the intergovernmental regime and negotiation system (Stripple and Pattberg 2010; Pattberg 2010).

Over the last decades, many non-state actors became formally part of global norm-setting and norm-implementing institutions and mechanisms, which denotes a shift from intergovernmental regimes to public-private and increasingly private-private cooperation and global policy making (Cutler et al. 1999; Higgot et al. 1999; Hall and Biersteker 2002; Pattberg 2005, 2007). Public-private cooperation has received more impetus with the 2002 Johannesburg World Summit on Sustainable Development and its focus on partnerships of governments, non-governmental organizations, and the private sector—the so-called Partnerships for Sustainable Development. More than 330 such partnerships were registered with the United Nations around or after the Johannesburg summit (Andonova and Levy 2003; Bäckstrand 2006; Glasbergen et al. 2007), and many address climate-related issues.

In the climate arena, it is in particular global networks of (major) cities that have provided most impetus to negotiations, providing new standards for subnational entities even where the national government is unsupportive of global regulation. In addition, sectoral solutions have been explored in a variety of networks, ranging from agreements on policy measures for particular pollutants (e.g. the Methane to Markets partnership), activities (e.g. the Global Gas Flaring Reduction Partnership), or industries (e.g. the Renewable Energy and Energy Efficiency Partnership or the Solar Energy Society (ISES)) up to novel networks that emerge around new issues such as carbon sequestration or geo-engineering.

Climate governance is also the most important policy domain in which both governments and non-governmental actors experiment with markets as new governance mechanisms (see Paterson, this volume). While markets have been used as environmental policy mechanism in a number of national contexts (notably in the United States) and have been experimented with in the 1987 Montreal Protocol on Substances that Deplete the Ozone Layer (which allowed for limited joint implementation of reduction commitments among industrialized countries), the 1997 Kyoto Protocol was the first intergovernmental agreement to provide for large-scale development of market mechanisms. These range from joint implementation among industrialized countries to the project-based North–South joint implementation through the Clean Development Mechanism and to a future, not yet fully specified global mechanism for emissions trading. In addition to this intergovernmental mechanism, a number of regional markets have been developed, including the EU emissions trading scheme and regional markets in North America, Australia, and New Zealand. In addition to these essentially public or publicly controlled market mechanisms, a private trading system has evolved around various offsetting schemes that rely on private commitments of private or semi-public actors largely in the industrialized countries (Stripple and Lövbrand 2010).

The environmental effectiveness of these new mechanisms of global climate governance remains subject to academic and policy debate, and a general assessment of the effectiveness of the entire system of non-state regulation is probably impossible in the first place given the variation in mechanisms, commitments, and types of implementation. Overall, there are strong indications that the effectiveness of many of the more than 300 public-private partnerships registered with the United Nations is low; several of such registered networks appear to be non-operational (Biermann et al. 2007). Emissions trading schemes are (except for the EU system) too recent to be evaluated. The networks of cities might have some effects, even though the causal link between transnational networks and local policies is difficult to make, and many changes in city management will eventually have been motivated not by global networks but by local concerns.

However, the overall effects of these novel mechanisms are likely to be larger than their direct impact on the reduction of emissions. Maybe even more important are the discursive effects that help raise awareness in many countries and that may shape public debates and decision making. Forest certification schemes, for example, do not only inform consumers about production methods, but also generally raise awareness about the need to make choices, thus contributing to changing lifestyles. Also, in many countries that have at present rather weak governmental regulations on climate governance, these new mechanisms beyond the realm of central government policy making may provide a (non-binding) regulatory environment that helps stimulate innovation and action. For example, the non-binding self-commitment of a university or a local community to achieve carbon neutrality at a certain date might at present merely lead to small measures and succeed merely through purchasing credits in private offsetting schemes (the added benefit of which might be debatable). Yet in the longer run, such self-commitments might set in motion changes in local perception and behavior, institutionalize innovative search processes, and thus in the end provide for emissions reductions that would not have occurred through governmental programs.

2.3 New Multiplicities of Spheres of Authority

Finally, global climate governance is characterized by an increasing segmentation of different layers and clusters of rule making and rule implementing, fragmented both vertically between supranational, international, national, and subnational layers of authority (multi-level governance) and horizontally between different parallel rule-making systems maintained by different groups of actors (multi-polar governance).

First, the increasing global institutionalization of climate politics does not occur, and is indeed not conceivable, without continuing policy making at national and subnational levels. Global standards need to be implemented and put into practice locally, and global norm setting requires local decision making and implementation. This results in the coexistence of policy making at the subnational, national, regional, and global levels, with the potential of both conflicts and synergies between different levels of regulatory activity. Likewise, the increasing global institutionalization of climate politics does not occur in a uniform manner that covers all parts of the international community to the same extent. Instead we observe the emergence of a multitude of institutions and networks that include different actors and may over time develop into divergent regulatory regimes in global climate governance (Biermann et al. 2009b).

Students of global environmental governance have highlighted the significant challenges that divergent policy approaches within such a horizontally and vertically segmented policy arena pose. First, lack of uniform policies may jeopardize the success of the policies adopted by individual groups of countries or at different levels of decision making. For example, global emissions trading will be less effective if major nations are not party to the mechanism, which has given rise to an intense debate on border tax adjustments to counter such leakage effects (Biermann and Brohm 2005). Likewise, a segmentation of governance architectures may complicate positive linkages with other policies, whereas a universal and coordinated architecture could allow systematic and stable agreements between institutional frameworks (Biermann et al. 2009b; see also on interlinkages Young, this volume). Since a segmented architecture decreases entry costs for participants, it is also conceivable that business actors use regulatory diversity to choose among different levels of obligation, thereby starting a race-to-the-bottom within and across industry sectors. Power differentials are also crucial, since fragmented governance systems give powerful states the flexibility to opt for a mechanism that best serves their interests, and to create new agreements if the old ones do not fit their interest anymore (Hafner 2004; Benvenisti and Downs 2007).

On the other hand, segmented climate governance may also have advantages. Distinct institutions allow for the testing of innovative policy instruments in some nations or at some levels of decision making, with subsequent diffusion to other regions or levels (Vogel 1995; Busch and Jörgens 2005). Regulatory diversity might increase innovation. Fragmentation could enhance innovation at the level of the firm or public agency and increase innovation in the entire system of climate governance. Important here is the notion of diffusion of innovation, including innovations of policies, technologies, procedures, and ideas. One example of this line of thought is Stewart and Wiener (2003), who proposed that the United States should stay outside the Kyoto Protocol and seek instead to establish a new framework with China and, possibly, other key developing countries. In their view, this would address the world's two largest greenhouse gas emitters and allow for experimentation of alternative regulatory frameworks.

Despite this contestation in the literature, fragmentation of global governance architectures appears on balance to bring more harm than positive effects and can generally be seen as a burden on the overall performance of the system, increasing the economic (e.g. leakage effects) and political costs (e.g. regulatory chill and race-to-the-bottom) of comprehensive climate policies at the national level (in more detail Biermann et al. 2009b). This raises the policy question of how to minimize extreme cases of fragmentation and how to address its negative effects. This policy question is particularly important for climate governance. To increase synergies within UN climate governance, it seems for instance crucial to better integrate processes under the climate convention and the Kyoto protocol. Regarding fragmentation between UN climate governance and climate arrangements outside this umbrella, it is imperative to open these institutions to additional members. For example, the Asia-Pacific Partnership could be broadened to include least developed countries and small-island developing states, and to ensure through formal declarations or clauses better integration with overall UN processes. The UN climate regime also needs to be better coordinated with non-environmental institutions in order to minimize conflictive fragmentation, most importantly with regard to the WTO. This is recognized by decision makers, and in 2007 trade ministers met for the first time during a conference of the parties to the climate convention. Other useful mechanisms are joint tasks forces and committees to study for example the trade-related aspects of various national climate policies.

3 CONSEQUENCES

The emergence of new modes of governance in climate policy poses a number of important questions for both politics and political science.

First, to the extent that new actors and new modes of governance prove to be effective steering mechanisms, questions of power and of shifting power constellations away from the central governments arise. By a large measure, the emergent influence of experts and bureaucrats can be seen as a technocratization of global (climate) politics that further reduces the role of democratic oversight and, in particular, the effective role of parliaments.

Second, and related to this point, the new role of actors beyond the nation-state and of new modes of governance such as transnational regimes, partnerships, and markets raises questions about the legitimization of (global) decision making. Some transnational regimes—notably the Forest Stewardship Council—have developed elaborated systems of interest representation in different chambers that include North–South quotas and different caucuses, yet this type of representation could also be criticized, for example for over-representing Northern constituencies. As for non-rule-setting public-private partnerships, the analysis of the Amsterdam-based Global Sustainability Partnerships Database (which includes more than 300 partnerships, not all from the climate domain) has shown that many partnerships tend to be dominated by larger countries, intergovernmental bureaucracies, and non-governmental organizations and give little space to the voice and the interests of marginalized groups (Biermann et al. 2007).

Thirdly, the institutionalization of climate governance is still largely focused on the regulation of mitigation, which stands at the center of most new mechanisms in climate governance. However, recent scientific findings suggest that global warming can no longer be halted with a sufficient degree of certainty, making global adaptation governance a key requirement for institutional development. While many adaptation efforts are well under way at local and national levels, the intergovernmental governance arena lacks strong and specific mechanisms to deal with possibly drastic climate changes. Many intergovernmental regimes and organizations are in place that could develop timely response programs to drastic climate change, notably the international organizations in the field of food, agriculture, development, and possibly even the UN Security Council. Yet it remains doubtful to what extent existing governance mechanisms are able to react effectively should global crises because of global warming accelerate. For example, the needs of maybe millions of future climate refugees are at present hardly addressed in existing governance mechanisms (Biermann and Boas 2010).

Finally yet importantly, from a theoretical point of view it remains debatable how the new modes of governance in the climate realm—which are essentially governance beyond the central governments as key actors—relate to the remaining role of the state. It might well be that networks, markets, and partnerships that are populated and pushed forward by non-state actors are a direct response to the complexities of the climate problem, which states can no longer handle without strong non-state involvement. Yet at the same time, it is also possible that the current experimentation with 'governance beyond the state' is not related to an incapability of the modern state, but merely to temporary inaction and negotiation stalemates in the intergovernmental system, notably the conflicts around ratification and implementation of the Kyoto Protocol. It may well be possible that once intergovernmental consensus on the key parameters of a strong global regime emerges, also parallel networks, institutions, and parameters that have evolved in recent years 'beyond the state' might lose their influence and be surpassed by stronger public regulation again.

However, the 2009 conference of the parties to the climate convention in Copenhagen might have, for the time being, rather increased the search for novel mechanisms of governance beyond the traditional intergovernmental process. Stalemates in intergovernmental negotiations, combined with little political progress in many key countries, has reduced in many quarters the optimism that a strong global agreement with quantitative, demanding targets will be in place when the commitment period of the Kyoto Protocol ends. The events in Copenhagen, perceived by many as a failed summit, have spurred a variety of reactions, including a renewed research focus on areas as diverse as climate change adaptation, geoengineering, and large-scale social transformations.

In parallel, Copenhagen gave fresh impetus to those research programs and political projects that focus on the critique of the 'UN system' and try to explore novel ways of global governance that go beyond the current core system of multilateral diplomacy, legally binding intergovernmental agreements, and regular mega-sized political and diplomatic summits. Many of these mechanisms have been outlined in this chapter, and it is likely that their relevance will increase because of the Copenhagen event, both as a research object and as political strategy.

4   Conclusion: Revitalizing Global Climate Governance

In conclusion, the current system of largely fragmented, multi-layered climate governance is in need of reform to provide for a strong global regulatory environment. I propose in this final section a number of policy reforms, all of which have been developed elsewhere in more detail.

First, intergovernmental bureaucracies have shown their increasing role in climate govern-ance, as well as in many other areas of international environmental policy, and they have proven vital in particular as regards smaller countries that often rely on the information, analysis, and advice provided by intergovernmental bureaucracies. International organiza-tions have also proven to be important negotiation arenas, in particular in areas with a long-standing institutionalization of norm setting (such as labor or health). Here it seems important to strengthen the role of the climate secretariat and to increase its autonomy and independence. In parallel, increased policy integration and the mainstreaming of climate concerns in other environmental issue domains could be strengthened by the upgrading of the UN Environment Program to a world environment organization or 'United Nations Environment Organization,' as it is now supported by more than fifty nations. This organiza-tion would be an independent specialized agency within the UN system with independent, contribution-based funding, possibly operational tasks with field offices and project teams around the world, and a mandate to initiate intergovernmental agreements (comparable to the International Labor Organization) and possibly to oversee the numerous multilateral environmental agreements and their mostly independent secretariats (see Biermann 2005 in more detail).

Second, transnational governance systems outside the traditional intergovernmental process might lack legitimacy with certain constituencies. In particular, they might over-represent Northern interests and disempower actors from the South. Such problems could be resolved by a parallel strategy of institutionalizing non-state regulation and collabora-tion to the effect that Southern interests become better represented, and at the same time of opening up the intergovernmental system to the institutionalized and balanced involve-ment on non-state actors. One way forward could be the institutionalization of chamber(s) of civil society in intergovernmental decision making as has been proposed by the Com-mission on Global Governance (1995). Interesting recent experiences with the delimitation of constituencies of civil society in transnational governance exist in the realm of private rule setting, for example the Forest Stewardship Council and the Marine Stewardship Council, where stakeholders such as business or environmentalist groups have been organized in separate voting chambers. The current nine 'major groups' (business, indige-nous people, women, etc.) of civil society within the United Nations is a similar develop-ment in the organization of caucuses of civil society at the transnational level. Such experiences are at this stage certainly no model that can be easily adopted in intergovern-mental decision making, and the resistance by countries—notably those that rely on autocratic steering modes—is predictably strong. However, the parallel developments of increasing reliance on global decision-making—not the least in the climate arena—and decreasing legitimacy of such types of rule setting create a need to further explore options for better civil-society representation also at the level of global rule making.

Thirdly, it will be important to counter the increasing fragmentation of decision making by stronger efforts in collaboration and integration of rule-making systems. In particular, this calls for the strengthening of the United Nations Framework Convention on Climate Change as the central focus of normative development and intergovernmental consensus.

Last but not least, it is crucial to intensify research and rule making on global adaptation. Protecting the most vulnerable members of humankind to possible drastic climate change requires strong reforms, ranging from the development of *sui generis* legal titles for climate refugees to the set-up of new funding mechanisms and programs to help developing countries become 'climate proof.'

Climate governance is unprecedented in its problem structure that combines new types and degrees of uncertainty, interdependence, and impacts. This makes the negotiation and development of new institutions and modes of governance conflictive and tedious, but creates at the same time room for new ideas and innovations in governance. The effectiveness of the overall global effort of halting global warming and leading human societies to a low-emissions trajectory is still uncertain. Yet despite these uncertainties, it is well possible that many of the innovations developed in the climate arena will eventually spill over also to other policy domains.

REFERENCES

ANDONOVA, L. B., and LEVY, M. A. 2003. Franchising global governance: Making sense of the Johannesburg type II partnerships. Pp. 19–32 in O. Schram Stokke and Ø. B. Thommessen (eds.), *Yearbook of International Cooperation on Environment and Development*. London: Earthscan.

BÄCKSTRAND, K. 2006. Multi-stakeholder partnerships for sustainable development: Rethinking legitimacy, accountability and effectiveness. *European Environment* 16(5): 290–306.

BENVENISTI, E., and DOWNS, G. W. 2007. The empire's new clothes: Political economy and the fragmentation of international law. *Stanford Law Review* 60: 595–632.

BETSILL, M. M., and CORELL, E. 2001. NGO influence in international environmental negotiations: A framework for analysis. *Global Environmental Politics* 1: 65–85.

BIERMANN, F. 2002. Institutions for scientific advice: Global environmental assessments and their influence in developing countries. *Global Governance* 8(2): 195–219.

—— 2005. The rationale for a world environment organization. Pp. 117–72 in *A World Environment Organization: Solution or Threat for Effective International Environmental Governance?* Aldershot: Ashgate.

—— 2007. 'Earth system governance' as a crosscutting theme of global change research. *Global Environmental Change: Human and Policy Dimensions* 17: 3–4, 326–37.

—— and BOAS, I. 2010. Preparing for a warmer world. Towards a global governance system to project climate refugees. *Global Environmental Politics* 10(1): 60–88.

—— and BROHM, R. 2005. Implementing the Kyoto Protocol without the United States: The strategic role of energy tax adjustments at the border. *Climate Policy* 4: 289–302.

—— and SIEBENHÜNER, B. (eds.) 2009. *Managers of Global Change: The Influence of International Environmental Bureaucracies*. Cambridge, MA: The MIT Press.

——CHAN, S., MERT, A., and PATTBERG, P. 2007. Multi-stakeholder partnerships for sustainable development: Does the promise hold? In Glasbergen et al. 2007: 239–60.

—— BETSILL, M. M., GUPTA, J., KANIE, N., LEBEL, L., LIVERMAN, D., SCHROEDER, H., and SIEBENHÜNER, B., with contributions from Conca, C., da Costa Ferreira, L., Desai, B., Tay, S., and Zondervan, Z. 2009a. *Earth System Governance: People, Places, and the Planet. Science and Implementation Plan of the Earth System Governance Project*. Earth System Governance Project 1, IHDP Report 20. Bonn: IHDP: The Earth System Governance Project.

—— PATTBERG, P., ASSELT, H. VAN, and ZELLI, F. 2009b. The fragmentation of global governance architectures: A framework for analysis. *Global Environmental Politics* 9(4): 11–40.

—— PATTBERG, P., and ZELLI, F. 2010. *Global Climate Governance beyond 2012: Architecture, Agency, and Adaptation.* Cambridge: Cambridge University Press.

BULKELEY, H., and BETSILL, M. M. 2003. *Cities and Climate Change: Urban Sustainability and Global Environmental Governance.* Oxford: Routledge.

BUSCH, P.-O. 2009. The climate secretariat: Making a living in a straitjacket. In Biermann and Siebenhüner 2009: 245–64.

—— and JÖRGENS, H. 2005. International patterns of environmental policy change and convergence. *European Environment* 15: 80–101.

Commission on Global Governance 1995. *Our Global Neighbourhood: The Report of the Commission on Global Governance.* Oxford: Oxford University Press.

CUTLER, A. C., HAUFLER, V., and PORTER, T. (eds.) 1999. *Private Authority and International Affairs.* Albany, NY: State University of New York Press.

GLASBERGEN, P., BIERMANN, F., and MOL, A. P. J. (eds.) 2007. *Partnerships for Sustainable Development: Reflections on Theory and Practice.* Cheltenham: Edward Elgar.

HAAS, P. M. 2000. International institutions and social learning in the management of global environmental risks. *Policy Studies Journal* 28: 558–75.

HAFNER, G. 2004. Pros and cons ensuing from fragmentation of international law. *Michigan Journal of International Law* 25: 849–63.

HALL, R. B., and BIERSTEKER, T. J. (eds.) 2002. *The Emergence of Private Authority in Global Governance.* Cambridge: Cambridge University Press.

HIGGOT, R. A., UNDERHILL, G. D., and BIELER, A. (eds.) 1999. *Non-State Actors and Authority in the Global System.* London: Routledge.

MITCHELL, R. B., CLARK, W. C., CASH, D. W., and DICKSON, N. M. (eds.) 2006. *Global Environmental Assessments: Information and Influence.* Cambridge, MA: MIT Press.

PATTBERG, P. 2005. The institutionalization of private governance: How business and non-profits agree on transnational rules. *Governance: An International Journal of Policy, Administration, and Institutions* 18: 589–610.

—— 2007. *Private Institutions and Global Governance: The New Politics of Environmental Sustainability.* Cheltenham: Edward Elgar.

—— 2010. The role and relevance of networked climate governance. In Biermann et al. 2010: 146–64.

STEWART, R. B., and WIENER, J. B. 2003. *Reconstructing Climate Policy: Beyond Kyoto.* Washington, DC: The AEI Press.

STRIPPLE, J., and LÖVBRAND, E. 2010. Carbon market governance beyond the public-private divide. In Biermann et al. 2010: 165–82.

—— and PATTBERG, P. 2010. Agency in global climate governance: Setting the stage. In Biermann et al. 2010: 137–45.

VOGEL, D. 1995. *Trading Up: Consumer and Environmental Regulation in a Global Economy.* Cambridge, MA: Harvard University Press.

CHAPTER 47

....

RESILIENCE

....

W. NEIL ADGER, KATRINA BROWN,
AND JAMES WATERS

1 INTRODUCTION

....

THE focus of study of this book is change. Resilience is the ability of a system to absorb change while retaining essential function; to have the ability for self-organization; and to have the capacity to adapt and learn. Resilience can apply to people, places, and ecosystems. Hence resilience ideas provide a framework for understanding change: planned and unplanned, desirable and undesirable.

Individuals, communities, and states are now faced with a range of challenges associated with the impacts of climate change and the policy imperatives to decarbonize economic and other activities. But at the same time societies face the challenges of economic globalization and financial crises, political upheaval, and the interactions between these phenomena. These challenges go beyond the experience of many institutions and governance structures and may fundamentally reframe notions of well-being and human security. Furthermore the scientific evidence on the scale of the potential impacts of climate change beyond 2 °C during the twenty-first century suggests the need to consider radical transformations. Resilience ideas may be useful in framing and defining the nature of these changes.

This chapter outlines the key elements of resilience ideas—or a resilience theory—as they have emerged from ecology and other disciplines. Those elements are often described as radical blueprints for change—making both development and environmental management more flexible and more focused on process rather than outcome. Resilience has been taken up as a battle cry for advocacy groups and even government agencies in the arena of climate policy. However as we discuss here, resilience as commonly applied in climate change policy is not used to promote radical change. In fact, the policy discourses associated with resilience often run counter to the scientific understandings and ideas at the heart of resilience theory. Here we outline the core elements of resilience thinking and evaluate the salience and provide a critique of the use of social-ecological resilience as a normative goal for the climate change challenge.

2 WHAT IS RESILIENCE?

2.1 A Brief History and Core Principles

Resilience is the ability of a system to deal with, and respond to, a spectrum of shocks and perturbations whilst retaining the same structure and function (Walker et al. 2004; Nelson et al. 2007). Box 47.1 shows definitions of some key terms as they emerged in ecology and related sciences. These debates emerged long before their application to the climate change challenge. The need to consider resilience emerged in part from pioneering analysis of ecological systems (Holling 1973). In realizing that populations of predators and prey do not exist in a singular pattern of dynamics and relationships, but rather may exist in a number of potential configurations, or multiple 'stable states,' the value of considering persistence within current states became clear. Holling's analysis used a diagram of interconnected cycles to illustrate the dynamics of ecosystem functions which he described as adaptive cycles (see Box 47.1), and suggested that 'adaptive management' was necessary to enhance resilience in these systems (see Gunderson and Holling 2002: 32–5 for a good explanation and the Resilience Alliance website at ⟨http://www.resalliance.org⟩). Further study showed how overexploitation and simplification reduces overall ability to cope with perturbations and change (Berkes and Folke 1998; Gunderson and Holling 2002). Evidence for multiple regimes has since built up across varied ecosystems, as have empirical examples of the shift between one state and another (see Folke 2006).

Crucially, resilience theory has since been applied to the interface of environment and society. In doing so, the unit of analysis has generally become the 'social-ecological system' (Gallopin 2006)—the interactions between human action, environmental system consequences, and feedbacks between the two (see Box 47.1). Ultimately human society has affected and been affected by the climate system through altering bio-geochemical cycles and adapting to the consequences of that change. But social-ecological interactions between climate and society operate at various temporal, spatial, and political scales. Gunderson and Holling (2002) suggested that such processes operate through adaptive cycles of change that describe how systems undergo phases of shifting connectedness and availability of 'capital;' this may be 'locked up,' followed by periods of release and finally reorganization.

These insights on social-ecological systems, much associated with the global science network of the Resilience Alliance, suggest new modes of management and governance, and a move 'away from policies that aspire to control change in systems assumed to be stable, towards managing capacity of social-ecological systems to cope with, adapt to and shape change' (Folke 2006: 254). Further research has sought to examine the interactions that occur within social-ecological systems and the links between other analytical frameworks used in climate change analysis such as vulnerability (Smit and Wandel 2006). The resilience approach emphasizes the need to manage for change and to see change as an intrinsic part of any system, social or otherwise. However resilience encompasses more than just the ability to absorb shock or disturbance. It is also about the opportunity that arises from disturbance and the ability to evolve in that period of change. Therefore the resilience of a system may be usefully broken down into a set of three key components (following Carpenter et al. 2001): the capacity to buffer against change; the ability to self-organize; and the ability to build capacity to learn and adapt. In order to understand each of

Box 47.1 Key resilience concepts defined

Resilience

Resilience is the capacity of a system to absorb disturbance and reorganize while undergoing change so as to still retain essentially the same function, structure, identity, and feedbacks.

Social–ecological system

Social-ecological systems are complex, integrated systems in which humans are part of nature.

Adaptation

Adaptation is the decision-making process and the set of actions undertaken to maintain present systems and to realize the opportunities for future change.

Adaptive capacity

The capacity necessary to enable adaptation, including social and physical elements, and the ability to mobilize these elements. In the context of climate change such capacity can be specific to particular weather-related risks and indirect impacts of climate, or can be a generic capacity to adapt to multiple or unknown stresses and perturbations.

Adaptive cycle

The adaptive cycle is a metaphor used to describe four commonly occurring phases of change in complex systems. The four phases are: exploitation, conservation, creative destruction, and renewal.

Transformability

The capacity to create a fundamentally new system when ecological, economic, or social conditions make the existing system untenable.

Source: Adapted from Resilience Alliance glossary <http://.resalliance.org/608.php>, Intergovernmental Panel on Climate Change glossary at <http://.ipcc.ch> and Nelson et al. (2007).

these, one must consider both the current system state, as well as the processes that allow it to maintain and manage change.

Resilience theory contributes some core concepts to help describe these components and navigate such change. Systems are subject to nonlinear dynamics and may exist in a number of potential 'states,' which involve fundamentally different structures and processes that maintain their stability. Thresholds demarcate transitions between such states (Folke 2006). Thirdly, influences at multiple scales will directly affect the resilience of a system. For instance governance and ecosystem services are interlinked at multiple scales. The scale of management must be taken into account when trying to maintain ecosystem services for human well-being (Carpenter et al. 2009).

Taking these ideas of dynamic, multi-scale social-ecological systems together, we suggest that the analysis of climate change requires understanding of both slow and fast drivers of change at multiple scales, as outlined in Figure 47.1 (derived from Carpenter et al. 2009). Resilience theory distinguishes between fast and slow variables in systems and how people and places react to multiple stresses and drivers of change, as illustrated in Figure 47.1. This has particular implications for considering societal

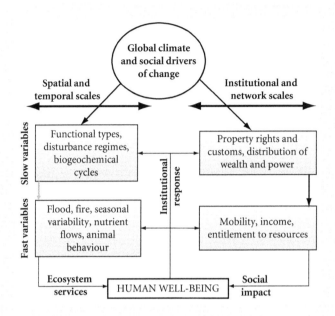

FIGURE 47.1 Interaction between slow and fast variables in social and ecological dimensions of systems perturbed by climate change

Source: Adapted from Carpenter et al. (2009).

responses to climate change and the impacts into the future. Weather-related disasters act on short timeframes and often have significant consequences, including triggering societal demand for new roles and actions for government to protect its citizens and in regulating risk. However, there may be trade-offs between short-term and long-term responses to climate change. Again a resilience analysis may help to evaluate these trade-offs (see Nelson et al. 2007).

Uncertainty lies at the heart of the challenges in devising responses to climate change. Resilience theory deals explicitly with issues of uncertainty by highlighting how systems move to alternative stability domains, triggered by small initial changes. The prevalence of uncertainty in climate futures suggests that attempting to control the natural world and attempting to generate perfect foresight will never be achievable (Dessai et al. 2009). Thus adaptation decisions should, it is argued, be taken such that they are robust to present circumstances and value systems and not ruled out by potential future states. Resilience theory highlights how stability is not the same as resilience—indeed there may be trade-offs between managing for stability and managing for adaptability and change. These trade-offs may have significant moral dimensions: for example choosing adaptation rather than prevention has potentially negative consequences for the vulnerable (Eakin et al. 2009).

Adaptive capacity is an important concept in resilience theory—it concerns the necessary resources for systems to be able to adapt and learn (see Box 47.1). Climate change will bring significant impacts and change to social-ecological systems all over the world, some potentially rapid and transforming. To minimize the injustice of this imposed impact on human well-being, the resilience literature suggests a critical analysis of both the drivers of change and the capacities that each system has to deal with change.

In the following sections we describe each of three key elements of resilience and their implications for understanding society and climate change: the importance of fast and slow variables, adaptive capacity, and a consideration of potential trajectories beyond the current stable state. Before examining how these insights might be applied and influence policy, we review specific contributions from wider research disciplines.

2.2 Slow and Fast Drivers of Change

The likelihood of a system remaining in a certain state is governed by both external perturbations and internal changes to the system. These two types of variable operate on a very broad range of timescales (Gunderson and Holling 2002). Slow variables are key to understanding thresholds in systems. In general it is relatively straightforward to identify slow variables in ecological systems, such as land use, nutrient shocks, soil properties, freshwater dynamics, and biomass of long-lived organisms. The impacts of climate change in social ecological systems are manifest through changes in the provision of ecosystem services and yet the specific services that most affect the resilience of ecosystems, namely regulating services, are given much less attention than provisioning services. In coastal areas, for example, changes to sea temperatures affect fish species distribution and abundance, directly affecting the viability of coastal fisheries and communities reliant on them. Yet the loss of regulating services of coastal protection by natural habitats such as dune systems and mangroves is amplified by coastal erosion or storm surges that may have greater impact on coastal communities in the short run.

Social structures and institutions necessary for adaptation to climate change are also often slow to change. Furthermore, rapidly implemented policy interventions may become entrenched, thereby becoming slow variables. For example, as people adjust to the presence of levees in low-lying areas for their protection, they become locked into a system that necessitates the existence of these structures. Such adjustment can involve a change in mental models of risk such as loss of trust in governance structures, or expectation of bailout to hazards. Other social drivers that operate on relatively slow timescales include education, cultural norms, urbanization, and isolation from nature. Taking into account these slowly changing variables may sharpen awareness of upcoming transitions, and help to manage system dynamics that build resilience.

2.3 Adaptive Capacity

Adaptive capacity refers to the ability of a system to evolve to external changes and thereby expand the range of variability with which it can cope (Nelson et al. 2007). This process occurs through a series of 'adaptive responses,' which will be underpinned by physical and social preconditions, or a set of capitals and the ability to mobilize them. Adaptive capacity may apply at various levels from the individual to the community, and may be generic, or specific to particular climate changes.

The social science of adaptive capacity is uncertain. What constitutes adaptive capacity at one level may be detrimental to adaptive capacity at another. The uncertainty is fundamentally due to the contested nature of development, progress, and well-being. Yohe and Tol (2002) suggest eight determinants of adaptive capacity for climate change including the range of available technological options for adaptation, the availability of resources and their distribution, the structure of critical institutions, the stocks of human and social capital, access to risk-spreading mechanisms, the ability of decision makers to manage risks and information and the public's perceived attribution of the source of the stress and the significance of exposure to its local manifestations. They conclude, however, that 'many of these variables cannot be quantified, and many of the component functions can only be qualitatively described' (2002: 27).

More importantly, the direction of causation is contested (Adger and Vincent 2005). For the world's poorest countries, economic integration into the world economy and economic growth is often portrayed as the route to reduce vulnerability to climate change and overcome so-called 'destiny' factors of geography. Thus, wealth and economic growth are positive elements of adaptive capacity. Yet, structuralist accounts of dependency and aid portray economic integration as contributing to exposure to forces of economic volatility that amplify risks to land use, to marginalized populations. Such a route leads to double exposure of vulnerable populations to economic globalization and climate change (Leichenko and O'Brien 2008). Hence economic growth often comes about at the expense of resilience and adaptive capacity. But it is clear that economies, regions, and communities that have generic and specific adaptive capacity have the ability to buffer change that, in the case of climate change, may radically alter and transform the availability and stability of resources.

2.4 Transitions and Transformations

The impacts of climate change are likely to make some areas economically unattractive or physically uninhabitable. Similarly, some current energy-intensive technologies and consumption patterns are not consistent with policy aims to decarbonize economies. In these cases, simply adapting to changing conditions within the current system is not possible and systems are forced to change in more significant ways. Resilience posits that there are multiple stable states in most systems. The transition between those states is often not gradual, but rather nonlinear across a certain threshold. In the transition from one stable state to another, the nature and extent of internal reinforcing processes shifts to a new set of feedbacks. Such shifts may be irreversible, or only possible to reverse over long periods of time (Scheffer and Carpenter 2003). Transitions may be triggered as much by internal (slow) variables as external perturbations. Attempts to find ways of predicting transitions, or early-warning systems, have come up with generic indicators across a range of systems (Scheffer et al. 2009). However, while it may be possible to 'pull a system back from the brink,' most current indicators cannot detect serious shifts much in advance, at least not such that policy makers are likely to roll out recovery schemes (Biggs et al. 2009).

When current conditions or pathways of a system become undesirable or untenable, resilience may no longer be a desirable attribute. Instead there may be a need to 'transform' a system to a completely new state. Dealing with climate change may require such transformations, for example, to decarbonize economies or reduce exposure to major impacts associated with sea-level rise or water availability as a result of climate change. Smith and Stirling (2010) discusses the common ground but also important difference between literatures on socio-technical transitions and social ecological resilience. Transitions in technologies and use of energy tend to progress through replication of a technology; through scaling-up of technologies; or through diffusion such that new technologies become the mainstream. Some of these transitions are steered or managed (such as the drive for renewable energy), while others, such as the adoption of mobile telephone technology in Africa, are much less predictable.

Smith and Stirling show how normative assumptions, underlying problem framing, and acknowledgement of politics and power, are played out in transitions management approaches compared to governing of transformations. In particular they highlight the potential tensions between a focus on resilience of structures and resilience of functions. In other words, what is to be made more resilience importantly determines whether a resilience approach makes transformation more or less likely. Much of the current literature focuses on making current systems more resilient and not on transforming the systems.

Transformations, such as large-scale adoption of renewable energy, or planning to move settlements in the face of changing hazards, may be planned and navigated. Indeed major transformational changes in urbanization and population over the next half-century will necessitate radical change in land use even in the absence of climate change. Migration, for example, will be an involuntary and forced transformation for some. The government of the Maldives is contemplating such a transformation. Almost 80 percent of the 1,200 islands of the Maldives are no more than 1 meter above sea level and, without the large sea defenses that protect the capital Malé, the islands may be uninhabitable if sea levels rise by more than 1 meter. Hence the Maldives President Mohamed Nasheed talks of a 'survival deal' on global climate change and the potential for the 400,000 residents to move en masse to a 'new home' in Asia or Australia in land purchased through a sovereign wealth fund (New York Times 2009). Many questions still exist in this arena, such as the trade-offs and commonalities between adaptive and transformative capacity, and the trade-offs between protecting the vulnerable in the short term versus building a longer-term resilience of societies and ecosystems (Eakin et al. 2009).

2.5 Challenging Resilience Theory

Resilience theory provides an underpinning for contested concepts such as human security and vulnerability of people and places to the impacts of climate change. Thus resilience analysis demonstrates the resources that constitute adaptive capacity of communities to cope with different types of shocks or disturbances (e.g. Eriksen et al. 2005; Marshall 2010). Second, resilience insights allow cross-scale and dynamic examination of security that applies to individuals, communities, states, and global systems such as energy, food, and water that may well suggest radical policy outcomes for avoidance of maladaptation, traps, and thresholds (Peterson 2009; Barnett and O'Neill 2010). Normative analysis of resilience shows that there are likely to be winners and losers from implementing resilience. So-called poverty traps and lock-in to fossil-fuel technologies are both resilient properties within the relevant systems. Finally in some instances we may have to accept that it is not possible to maintain the status quo; on these occasions resilience analysis provides tools to consider future scenarios and build our capacity to adapt and transform.

Resilience thinking challenges many standard elements of social science. Critiques of resilience theory, for example, suggest that issues of power and of significant social relations are underplayed (Leach 2008). Indeed Hornborg (2009) and others have suggested that policy prescriptions on resilience fail to recognize asymmetry in power between actors and are thus relatively weak and benign towards the status quo. These critiques resonate with structuralist accounts of how states 'see' problems and make others invisible (Scott 1998). Walker et al. (2009), for example, argue for a shake-up of global institutions and greater coordination between them to help construct and maintain a global-scale social contract. But others suggest that resilience requires moving away from globalized governance structures towards greater local autonomy and diversity (Leichenko and O'Brien 2008). While the principles of a resilience theory for social-ecological systems are in place, it would appear, therefore, that resilience as a goal for policy is the most contested area. Hence we now explore how resilience is applied across the arenas of the climate change challenge.

3 APPLICATIONS, POLICIES, AND DISCOURSES

The concept of resilience is increasingly infusing climate policy as well as scientific debates. Resilience has, for example, growing purchase in the security and international relations domain: Cascio (2009) suggested resilience imperatives as highlighting the inability of states to encompass profound uncertainty and nonlinearities and surprise in social systems (also Evans et al. 2010). Resilience ideas are evident in global change and policy statements such as the Millennium Ecosystem Assessment; in the Human Development Reports from UNDP; the Swedish government sponsored Commission of Climate Change and Development; and initiatives such as the World Bank's Program for Climate Resilience. Non-governmental organizations including Christian Aid and Oxfam, and think-tanks such as Institute for Public Policy Research and World Resources Institute have used the term to frame their policy documents and analysis.

The adoption of resilience rhetoric in policy is matched neither by action, nor by engagement with the analytics of resilience as an emergent property of a system. Generally we can observe that resilience is presented as a uniformly beneficial and desirable trait—of systems, including social ecological systems, the economy, and communities. But resilience is used, applied, and interpreted in a number of different ways in the prescriptive arenas. Table 47.1 shows selected examples of these policy prescriptions in the climate change arena highlighting the different meanings.

Each policy statement in Table 47.1 interprets resilience rather differently—in terms of its framing and focus; the main actors, agents, and protagonists; and how threats and processes of change are characterized. This is turn shapes the types of interventions deemed necessary to enhance resilience. Thus the World Resources Institute in its 2008 Report *Roots of Resilience* highlights resilience as a property of the rural poor and how they are able to respond to environmental and social challenges including climate change, loss of traditional livelihoods, political marginalization, and breakdown of customary village institutions. By contrast both Christian Aid statements and UNISDR in their resilient cities checklist describe resilience very much as a property associated with responding to disasters, within a context where disasters are increasing in frequency and impact—increased vulnerability to natural hazards. In the area of national security, IPPR (2009) views increasing vulnerability due to complex, densely networked, and heavily technology-reliant society where there is little margin for error because of globalization and

Table 47.1 Dissecting policy prescriptions on resilience

Focus	How resilience is defined	Policy prescription
1. Resilient communities	Resilience is the ability to withstand the impact of shocks and crises. It is determined by people's assets and their ability to access services provided by external infrastructure and institutions.	The Building Disaster Resilient Communities project supports local partner organizations . . . to strengthen communities' abilities to manage and recover from crises and to prepare for and reduce the risk of future disasters. (p. 8)
2. Resilient people	Resilience is the ability of a joint social and ecological system—such as a farm—to withstand shocks, coupled with the capacity to learn from them and evolve in response to changing conditions. Building resilience involves creating strength, flexibility, and adaptability.	Makes the case for investing in building up the resilience of vulnerable farming communities as a critical stepping stone to addressing the global challenges of food security; climate change adaptation; and climate change mitigation. (p. 7)
3. Resilient nation	Resilience is the opposite of vulnerability and is a character of critical infrastructure; enterprise; and communities.	A commitment to building national resilience, especially in our infrastructure, by measures including educating and increasing the self-reliance of communities, is an integral part of security policy. (summary p. 4)
4. Climate resilience	Resilience policy (not specifically defined) aims to 'integrate climate risk and resilience into core development planning, while complementing other on-going activities'.	Seeks to provide incentives for scaled-up action and transformational change in integrating consideration of climate resilience in national development planning consistent with poverty reduction and sustainable development goals. (section B paragraph 4)
5. Resilient cities	Building on the Hyogo Framework of Action 2005–15, resilience is the capacity to adapt by 'resisting or changing in order to reach and maintain an acceptable level of functioning and structure'. It is focused on potential exposure to hazards, and adaptation is via social organization and learning.	The checklist prescribes putting in place disaster risk reduction measures: appropriate organization and coordination to reduce risk, budget, data on hazards and vulnerabilities and risk assessments, and critical infrastructure. Also includes education programs, building regulations, early warning systems, and protection of ecosystems that provide natural hazard protection.

1. Christian Aid, *Overexposed: Building Disaster-Resilient Communities in a Changing Climate* (no date).
2. Oxfam, *People-Centred Resilience* (2009).
3. Commission on National Security in the 21st Century, *Shared Responsibility* (2008).
4. World Bank, *Pilot Program for Climate Resilience under the Strategic Climate Fund* (2008).
5. UNISDR, *Ten-Point Checklist For Local Governments: Ten Essentials for Making Cities Resilient* (2009).

Source: Brown (2010)

privatization. All of these policy prescriptions however view climate change as a significant threat to resilience, whether acting in synergy or in addition to other important drivers of change.

The divergent policies in Table 47.1 highlight two very important questions that should be applied in resilience thinking: 'resilience of what, and to what?' These policies differ on whether they are aimed at rural or urban peoples, on natural disasters or agriculture. For clearly identifying paths to attain policy objectives, both questions must be answered early on. In considering resilience in cities for example, one would examine broader perturbations and impacts, and sources of adaptive capacity, if the focus was not purely on natural hazards.

3.1 Resilient Cities

Cities are likely to be at the forefront of transformation over the incoming century and therefore it is worth expanding on this example of resilience policy. Cities face massive changes in population dynamics, as 1.3 million people a week are moving to cities to add to the already predominant global urban population (Grove 2009). Urban people face risks from climate change impacts including changing incidence of storms, rising sea levels, and eroding coasts, in addition to more indirect impacts such as changing incidence of infectious disease or altered linkages to distant sources of resource production. To consider this diversity of impacts, more generically resilient urban forms have moved beyond previous foci on robust infrastructure and emergency planning. Ecosystem-mediated impacts should be taken into account, especially for coastal cities. For example, following Hurricane Katrina in New Orleans in 2005, unforeseen consequences were a factor of wetland loss and erosion of coastline by development, combined with sea-level rise (Kates et al. 2006; Ernstson et al. 2010). To take this one step further, interactions between 'engineered' resilience in physical structures and building natural buffering capacities may involve a transformation of relationship between urban dwellers and their physical and natural environment.

In Table 47.1, the resilient cities policy from UNISDR's 'Making Cities Resilient' campaign focuses on disaster risk reduction, including integrating protection of ecosystems and natural buffers. However it notes that most efforts to build a 'disaster resilient city' will also contribute to resilience to climate change. From this example, it is clear that care is needed to ensure what so-called 'resilient' policies, plans, and institutions are resilient *to*, and importantly which influences or impacts are yet to be considered. Moreover, there is little mention of attributes often associated with resilience such as flexibility, adaptability, and redundancy. Obviously it is critically important to protect people against natural disasters; the point is that the recommended actions seem focused around defending the status quo approaches to doing so. How city development plans may be adaptive to future uncertainty requires further work.

Finally, plans for resilient cities build on great opportunity in these areas: cities offer prospects for significant decarbonization, and as home to vast populations they have the potential to be hubs of innovation, social networks, and labor power, thereby building great adaptive capacity.

3.2 Climate Resilient Development

Table 47.1 also illustrates how ideas about 'climate resilient development' have been widely promoted within debates on international development under climate change. Resilience is seen as the main means by which vulnerability can be reduced and ongoing development efforts and investments defended and promoted in the face of climate change. Thus the World Bank summarizes 'climate resilient development' (for Africa) according to four pillars:

1. Making adaptation a core component of development, with a particular focus on sustainable water resources, land, and forest, integrated coastal zone management, increased agricultural productivity, health problems, and conflict and migration issues;

2. Focusing on knowledge and capacity development by improving weather forecasting, water resources monitoring, land-use information, improving disaster preparedness, investing in appropriate technology development, and strengthening capacity for planning and coordination, participation and consultation;

3. Benefiting from mitigation opportunities through access to carbon finance against land-use changes and avoided deforestation, promoting clean energy sources (e.g. hydropower) and energy efficiency, and adopting cost-effective clean coal energy generation and reduced gas flaring;

4. Scaling up financing for investment in low carbon growth and reducing sensitivity to climate change impacts.

The discourse implicit in these 'pillars' interprets resilience as part of a strategy which mainstreams climate change adaptation and mitigation into development efforts. It sees climate change as a challenge to current development but also as providing opportunities— especially for attracting investments (via carbon markets and market-based mitigation especially in the case of Africa in land use, forest management, and renewable energy). It suggests that adaptation is 'fundamentally about sound, resilient development.' Resilience in this context is bestowed by mainstreaming climate change; by making adaptation and risk management core development elements; by fostering knowledge and capacity (particularly in terms of access to information and forecasting); and by scaling up financing. There are explicit links made to disaster preparedness and disaster risk reduction; to providing layers of insurance protection, and safety nets where appropriate; and to building 'climate smart' systems. The approach therefore mixes a number of different aspects of resilience thinking, including multiple and cross-scale dynamics; the emphasis on shocks and disturbances to the system; but also aspects of engineering-like resilience.

Importantly climate resilient development is not presented as anything fundamentally different to current development—it emphasizes that current plans need to be 'climate proofed,' in other words the potential future impacts of climate change and associated risks must be accommodated. But the premise of continued growth and the benefits of this strategy are not questioned. Climate change, in the 'climate resilient' development discourse, reinforces the need to do development 'better,' more effectively and with emphasis on shifting vulnerabilities and how they may reconfigure the distribution of costs and benefits within society.

4 RECONCILING RESILIENCE SCIENCE
AND PRACTICE

....

There is no one agreed definition of resilience in either climate science or climate policy realms. We suggest here that the science of social-ecological systems is useful in understanding how society can respond to climate change. Resilience is currently widely promoted in policy communities as a positive characteristic of social and economic, ecological, and political systems which can support climate change adaptation. Most of the normative prescriptions currently made suggest resilience is a characteristic that can be added or enhanced—can be somehow enriched without making fundamental changes in how systems operate.

Communities facing a changing climate in the incoming decades will have to deal with slow changing variables and creeping change; with changing risks associated with extreme events; and with unforeseen and possibly rapid onset regime shifts in climate and resource availability. There are three major likely climate change drivers that will, for example, directly affect settlements and locations. These are changes in sea level; flood and other extreme weather-related risks; and declining water availability through for example desiccation and glacial retreat and snow pack loss.

The scale of change and interconnectedness of impacts, and increasing sensitivity of both natural and social systems to relatively small impacts (Smith et al. 2009) may be such that the window of opportunity for adaptation is smaller than previously imagined. There has been, in effect, virtually no global greenhouse gas emissions reduction in this century. With no prospect of emissions peaking before 2020, there will have to be a major turnaround in policy, planning, and behavior to avoid atmospheric concentrations that pose a significant risk of global mean warming of 2 °C or beyond (Meinshausen et al. 2009). With higher levels of cumulative emissions there is a risk of mean warming of 4 °C or more above pre-industrial levels with serious implications in terms of impacts (Parry et al. 2009; New et al. 2011). Further, present concentrations have already locked the world in to commitments to slow long-term sea-level rise and large reductions in dry season rainfall in several regions (Solomon et al. 2009).

An emerging literature on climate change therefore addresses the need for more fundamental change; seeing climate change as a symptom of a wider set of problems rather than the main driver of change. Herman Daly expresses the problem thus: 'Climate change, important as it is, is nevertheless a symptom of a deeper malady, namely our fixation on unlimited growth of the economy as a solution to nearly all problems' (Working Group on Climate Change and Development 2009: 3). Other analysts identify the fundamental problem as one of global inequality (Roberts and Parks 2007), neoliberalization and commoditization (McMichael 2009), or as a fundamental divergence in values over progress and sustainability (Hulme 2009).

Clearly the orthodox development paradigm seeks to maximize growth and productivity and handles climate change only as an unforeseen externality. This paradigm has shaped national and sectoral development planning around the globe and taken managerialist, technocratic, and market-based approaches to dealing with environmental change and particularly climate change (Brooks et al. 2009). Brooks and colleagues (2009) argue that a new paradigm is necessary to really deal with climate change as a development issue.

We contend that resilience, in its normative goal of promoting the ability to change, represents an alternative paradigm to growth. Policies that promote resilience should see change and variability as intrinsic system characteristics and see equity and redistribution and lowering vulnerability as priorities. The policy implications of a more resilience-based approach have implications for agriculture and rural development, economic management, institutional design, urbanization, and for how well-being is understood and promoted. Thus a resilient world needs a more fundamental set of transformations and transitions than currently promoted in international and national policy arenas, and an alternative economic pathway for a post-carbon world.

REFERENCES

ADGER, W. N., and VINCENT, K. 2005. Uncertainty in adaptive capacity. *CR Geoscience* 337: 399–410.

BARNETT, J., and O'NEILL, S. 2010. Maladaptation. *Global Environmental Change* 20: 211–13.

BERKES, F., and FOLKE, C. (eds.) 1998. *Linking Social and Ecological Systems: Management Practices and Social Mechanisms for Building Resilience.* Cambridge: Cambridge University Press.

BIGGS, R., CARPENTER, S. R., and BROCK, W. A. 2009. Turning back from the brink: Detecting an impending regime shift in time to avert it. *Proceedings of the National Academy of Sciences* 106: 826–31.

BROOKS, N., GRIST, N., and BROWN, K. 2009. Development futures in the context of climate change: Challenging the present and learning from the past. *Development Policy Review* 27 (6): 741–65.

CARPENTER, S., WALKER, B., ANDERIES, J. M., and ABEL, N. 2001. From metaphor to measurement: Resilience of what to what? *Ecosystems* 4: 765–81.

——MOONEY, H. A., AGARD, J., CAPISTRANO, D., DEFRIES, R. S., DÍAZ, S., DIETZ, T., DURAIAPPAH, A. K., OTENG-YEBOAH, A., PEREIRA, H. M., PERRINGS, C., REID, W. V., SARUKHAN, J., SCHOLES, R. J., and WHYTE, A. 2009. Science for managing ecosystem services: Beyond the Millennium Ecosystem Assessment. *Proceedings of the National Academy of Sciences* 105: 1305–12.

CASCIO, J. 2009. The next big thing: Resilience. *Foreign Policy* April, accessed at <http://. foreignpolicy.com/2009/04/15>.

DESSAI, S., HULME, M., LEMPERT, R., and PIELKE JR., R. 2009. Climate prediction: A limit to adaptation? Pp. 64–78 in W. N. Adger, I. Lorenzoni, and K. L. O'Brien (eds.), *Adapting to Climate Change: Thresholds, Values, Governance.* Cambridge: Cambridge University Press.

EAKIN, H., TOMPKINS, E. L., NELSON, D. R., and ANDERIES, J. M. 2009. Hidden costs and disparate uncertainties: Trade-offs in approaches to climate policy. Pp. 212–26 in W. N. Adger, I. Lorenzoni, and K. L. O'Brien (eds.), *Adapting to Climate Change: Thresholds, Values, Governance.* Cambridge: Cambridge University Press.

ERIKSEN, S., BROWN, K., and KELLY, P. M. 2005. The dynamics of vulnerability: Locating coping strategies in Kenya and Tanzania. *Geographical Journal* 171: 287–305.

ERNSTSON, H., VAN DER LEEUW, S. E., REDMAN, C. L., MEFFERT, D. J., DAVIS, G., ALFSEN, C., and ELMQVIST, T. 2010. Urban transitions: On urban resilience and human-dominated ecosystems. *Ambio* 39: 531–45.

EVANS, A., JONES, B., and STEVEN, D. 2010. *Confronting the Long Crisis of Globalization: Risk, Resilience and International Order*. Washington, DC: Managing Global Insecurity Program, Brookings Institute.

FOLKE, C. 2006. Resilience: The emergence of a perspective for social-ecological systems analysis. *Global Environmental Change* 16: 253–67.

GALLOPÍN, G. C. 2006. Linkages between vulnerability, resilience, and adaptive capacity. *Global Environmental Change* 16: 293–303.

GROVE, J. M. 2009. Cities: Managing densely settled social-ecological systems. Pp. 281–94 in F. S. Chapin, G. Kofinas, and C. Folke (eds.), *Principles of Ecosystem Stewardship*. New York: Springer.

GUNDERSON, L. H., and HOLLING, C. S. (eds.) 2002. *Panarchy: Understanding Transformations in Human and Natural Systems*. Washington, DC: Island Press.

HOLLING, C. S. 1973. Resilience and stability of ecological systems. *Annual Review of Ecological Systems* 4: 1–24.

HORNBORG, A. 2009. Zero-sum world: Challenges in conceptualizing environmental load displacement and ecologically unequal exchange in the world-system. *International Journal of Comparative Sociology* 50: 237–62.

HULME, M. 2009. *Why we Disagree about Climate Change*. Cambridge: Cambridge University Press.

IPPR Commission on National Security in the 21st Century. 2009. *Shared Destinies: Security in a Globalised World*. London: Institute for Public Policy Research.

KATES, R. W., et al. 2006. Reconstruction of New Orleans after Hurricane Katrina: A research perspective. *Proceedings of the National Academy of Sciences* 103: 14653–60.

LEACH, M. (ed.) 2008. *Re-framing Resilience: A Symposium Report*. Brighton: ESRC STEPS Centre, University of Sussex.

LEICHENKO, R. M., and O'BRIEN, K. L. 2008. *Environmental Change and Globalization: Double Exposures*. Oxford: Oxford University Press.

MCMICHAEL, P. 2009. Contemporary contradictions of the Global Development Project: Geopolitics, global ecology and the 'development climate'. *Third World Quarterly* 30(1): 247–62.

MARSHALL, N. A. 2010. Understanding social resilience to climate variability in primary enterprises and industries. *Global Environmental Change* 20: 36–43.

MEINSHAUSEN, M., et al. 2009. Greenhouse-gas emission targets for limiting global warming to 2 °C. *Nature* 458: 1158–62.

NELSON, D. R., ADGER, N., and BROWN, K., 2007. Adaptation to environmental change: contributions of a resilience framework. *Annual Review of Environment and Resources* 32: 395–419.

NEW, M., LIVERMAN, D., SCHRODER, H., and ANDERSON, K. 2011. Four degrees and beyond: The potential for a global temperature increase of four degrees and its implications. *Philosophical Transactions of the Royal Society A* 369: 6–19.

New York Times 2009. Wanted: A new home for my country. *New York Times* 8 May. Accessed <http://.nytimes.com/2009/05/10/magazine/10MALDIVES-t.html>.

PARRY, M., LOWE, J., and HANSON, C. 2009. Overshoot, adapt and recover. *Nature* 458: 1102–3.

PELLING, M. 2010. *Adaptation to Climate Change: From Resilience to Transformation*. London: Routledge.

PETERSON, G. 2009. Ecological limits of adaptation to climate change. Pp. 25–41 in W. N. Adger, I. Lorenzoni, and K. O'Brien (eds.), *Adapting to Climate Change: Thresholds, Values, Governance*. Cambridge: Cambridge University Press.

ROBERTS, J. T., and PARKS, B. C. 2007. *A Climate of Injustice: Global Inequality, North-South Politics and Climate Policy*. Cambridge, MA: MIT Press.

SCHEFFER, M., and CARPENTER, S. R. 2003. Catastrophic regime shifts in ecosystems: Linking theory to observation. *Trends in Ecology and Evolution* 18: 648–56.

——et al. 2009. Early-warning signals for critical transitions. *Nature* 461(7260): 53–9.

SCOTT, J. C. 1998. *Seeing like a State: How Certain Schemes to Improve the Human Condition have Failed*. New Haven: Yale University Press.

SMIT, B., and WANDEL, J. 2006. Adaptation, adaptive capacity and vulnerability. *Global Environmental Change* 16: 282–92.

SMITH, A., and STIRLING, A. 2010. The politics of social-ecological resilience and sustainable socio-technical transitions. *Ecology and Society* 15(1): 11. <http://.ecologyandsociety.org/vol15/iss1/art11/>.

SMITH, J. B., et al. 2009. Assessing dangerous climate change through an update of the Intergovernmental Panel on Climate Change (IPCC) reasons for concern. *Proceedings of the National Academy of Sciences* 106: 4133–7.

SOLOMON, S., et al. 2009. Irreversible climate change due to carbon dioxide emissions. *Proceedings of the National Academy of Sciences* 106: 1704–9.

WALKER, B., HOLLING, C. S., CARPENTER, S. R., and KINZIG, A. 2004. Resilience, adaptability and transformability in social-ecological systems. *Ecology and Society* 9(2): 5.

——BARRETT, S., POLASKY, S., et al. 2009. Looming global-scale failures and missing institutions. *Science* 325: 1345–6.

Working Group on Climate Change and Development 2009. *Other Worlds are Possible: Human Progress in an Age of Climate Change*. London: New Economics Foundation.

YOHE, G., and TOL, R. S. J. 2002. Indicators for social and economic coping capacity: Moving toward a working definition of adaptive capacity. *Global Environmental Change* 12: 25–40.

NAME INDEX

Subject Index

Milton Keynes UK
Ingram Content Group UK Ltd.
UKHW030026040923
428007UK00002B/3